NEUROVASCULAR IMMUNOLOGY

Vasoactive Neurotransmitters and Modulators in Cellular Immunity and Memory

Thomas C. Moore, M.D., Ph.D. (Cantab.)

Professor of Surgery
University of California – Los Angeles
and
Chief of Pediatric Surgery and
Director of Surgical Research in Molecular Immunology
Harbor – UCLA Medical Center
Torrance, California

CRC Press
Boca Raton Ann Arbor London Tokyo

Cover — Photomicrograph of the memory-critical center of the mammalian (rat) brain presenting the whole hippocampus and showing the familiar ram's horn configuration (Cornu Ammonis; CA1, CA2, CA3) as outlined autoradiographically by immunoreactivity to the vasoactive neurotransmitter and "memory" molecule serotonin. See also Figure 4 in Chapter 1. (Adapted from Oleskevich, S. and Descarries, L., *Neuroscience,* 34, 19, 1990. With permission.)

Library of Congress Cataloging-in-Publication Data

Catalog record is available from the Library of Congress.

Direct all inquiries to CRC Press, Inc., 2000 Corporate Blvd., N.W., Boca Raton, Florida, 33431.

International Standard Book Number 0-8493-6894-4

Printed in the United States of America 1 2 3 4 5 6 7 8 9 0

Printed on acid-free paper

Dedicated to Beth

THE AUTHOR

Thomas C. Moore, M.D., Ph.D. (Cantab.) is Professor of Surgery in the UCLA School of Medicine and Chief of Pediatric Surgery and Director of Surgical Research in Molecular Immunology at the Harbor–UCLA Medical Center in Torrance, California. From 1968 to 1978, he also was Chief of Clinical Kidney Transplantation Surgery at the Harbor–UCLA Medical Center.

Professor Moore received his undergraduate education at Dartmouth College. He was elected to the Phi Beta Kappa scholastic honor society during his junior year at Dartmouth and was awarded his B.A. *summa cum laude* in 1942, with distinction in his major subject (English Classics). He received his M.D. in 1945 from the Harvard Medical School, and in 1982 he received a Ph.D. in Molecular Immunology from the University of Cambridge (Trinity College) in England.

Professor Moore is a broadly trained and broadly based General Surgeon. He did his surgical training at the Peter Bent Brigham Hospital in Boston (Harvard), the Childrens Hospital in Boston (Harvard), and the Indiana University Medical Center in Indianapolis. He has served on surgical faculties at the professorial level at Indiana University, the University of Kentucky, and the Medical College of Virginia prior to coming to UCLA in 1968 as a full-time Full Professor of Surgery. Two years (1966–68) were spent in clinical and experimental transplantation surgery with the late Professor David M. Hume at the Medical College of Virginia, and 1 year (1977–78, on sabbatical from UCLA) was spent at the Transplantation Unit of Professor Sir Roy Calne at the University of Cambridge in England.

Professor Moore is board certified in General Surgery, Thoracic Surgery, and Pediatric Surgery, and he has specialty credentials in Vascular Surgery (member of the Society for Vascular Surgery since 1957) and Transplantation Surgery (member of the founding and organizing Committee of Four for the American Society of Transplant Surgeons and first Chairman of its Committee for Scientific Studies). Professor Moore served for 4 years in the National Institutes of Health Surgery Study Sections with responsibilities in organ preservation and transplantation immunobiology. He was previously an officer (Treasurer, 1956–59) of the Society of University Surgeons and has been a member of the American Surgical Association since 1960. His society memberships, in addition to the above, include the American Association for Thoracic Surgery, the American Pediatric Surgical Association, the British Society for Immunology, the American Association of Immunologists, and the Society for Neuroscience. Of Professor Moore's 312 scientific publications, 195 have been devoted to subjects relating to this book — the vascular system, the immune response system, and vasoactive neurotransmitter and modulator molecules. These are listed in Appendix B.

ACKNOWLEDGMENTS

To Leslie L. Iversen, Salvador Moncada, and Sami I. Said for their friendship, collaboration, advice, and assistance which, over many years, have enriched the lives and efforts of the author and his wife, Beth, to whom this book is dedicated.

To the late Professor U. S. von Euler for his friendship and many outstanding scientific contributions to multiple areas covered in this volume, including the discovery and study of substance P, the prostaglandins, and noradrenaline.

KAROLINSKA INSTITUTET

FYSIOLOGISKA
INSTITUTIONEN I
Tel. 33 99 58
Prof. U. S. v. Euler, M.D.

S-104 01 STOCKHOLM
Solnavägen 1
SWEDEN

October 7, 1982

Professor Thomas C. Moore
Department of Surgery
Harbor/UCCLA Medical Center
1000 West Carson Street
TORRANCE, CA 90509 U.S.A.

Dear Tom:

Very many thanks for the enjoyable luncheon and afternoon on the last day of my stay in Los Angeles with scenic views and interesting discussions. I trust the summer pavillion in your garden has already come into use and that weather allows you to enjoy it. Strangely enough the good weather continues here, a kind of extended Indian summer.

With best wishes, also to your charming wife.

Sincerely,

(U.S. von Euler)

The author wishes to thank the Joseph Drown Foundation for their generous and vital financial aid in support of the preparation and completion of this book. He also wishes to thank Sue Rizzo, his long-time secretary, for her timely and flawless typing of the text, tables, legends, and references from his nearly illegible script. He also wishes to thank Deanna Hockett of Multigraphica medical illustration services for all drawings and illustrations and to thank Emma Alegre and her outstanding associates at the Parlow Library of the Harbor–UCLA Medical Center for their superb, cheerful, and invaluable assistance in the transport and photocopying of large volumes of scientific articles and related material over many months.

TABLE OF CONTENTS

Chapter 4
Histamine

Chapter 5
Serotonin (5-HT)

Chapter 10
Substance P

Chapter 11
Neurotensin

Chapter 1

INTRODUCTION TO THE NEUROVASCULAR IMMUNE RESPONSE AXIS

I. INTRODUCTION

The neuroimmune axis is being recognized increasingly as a major factor in the immune response. The two major divisions of this axis are neurovascular immunology and neuroendocrine immunology.

Neurovascular immunology is concerned with potent, short-lived vasoactive neurotransmitter and neuromodulator molecules that first were identified and studied as mediators of inflammation and immediate hypersensitivity. These soluble mediators are simple molecules: amino acids, decarboxylated amino acids (amines), short sequences of amino acids (peptides), and short-carbon-chain lipids. The amino acids include aspartate, glutamate, glycine, and GABA (γ-aminobutyric acid). The amines include histamine, 5-hydroxytryptamine (serotonin), and dopamine and other catecholamines. The short-chain peptides include met-enkephalin, CCK-8 (cholecystokinin-8), bradykinin, substance P, neurotensin, bombesin, VIP (vasoactive intestinal peptide), and CGRP (calcitonin gene-related peptide). The short-chain lipids include the prostaglandins, thromboxane, the leukotrienes, the lipoxins, and other lipoxygenase-pathway metabolites.

These small molecules have a long evolutionary history of involvement in biological defense. They provide the pain-, inflammation-, and shock-producing effects of the stings of stinging nettles of the plant kingdom, and of the stings and venoms of multiple animals, from primitive invertebrates (coelenterates) to amphibians and reptiles. The vasoactive neurotransmitter stinging substances and venoms of these animals include those of jellyfish and the Portuguese Man-of-War; of the octopus; of wasps, bees, spiders, and many other stinging insects; of venomous toads and frogs; and of rattlesnakes, adders, and vipers.

In addition to stinging defenses, these vasoactive neurotransmitters have been identified in the sensory cells and neurites of some of the most primitive invertebrates. Serotonin, substance P, and histamine are major neurotransmitter molecules in many invertebrates and early vertebrates, as well as being major components of their stings and venoms, along with bradykinin, CCK-8, and others.

These dual actions of serotonin and substance P, in both stinging defense and neurotransmission in evolutionary antiquity, are well illustrated, as an example, by recent histochemical studies of the cnidarian coelenterates *Renilla koellikeri* and *Hydra*. These cnidarian coelenterates are the most primitive existing animal organisms endowed with a nervous system and generally are considered to be the contemporary representatives of the earliest organisms known to possess a nervous system. Their small neurons are bipolar or multipolar and are linked by chemical synapses or gap junctions. They are observed to form one or many network-like integrating units that are characterized as "nerve nets". The unmyelinated projections of these neural cells are unrecognizable either as axons or dendrites and are called "neurites". Multiple varicosities have been identified along these neurites.

Prior to these *Renilla* and *Hydra* studies, serotonin and substance-P neurons and pathways, present in mammals and man, had been tracked down the phylogenetic tree to flatworms (platyhelminthes). The evolutionary connections of these small vasoactive neurotransmitter and neuromodulator molecules are most impressive and surely of contemporary functional significance in more advanced animals.

In vertebrates, mammals, and man, the defense-oriented actions of these vasoactive neurotransmitter and modulator molecules have been incorporated into the internal milieu organization of biological defense, with particularly important involvement in the immune response defense system and in the increasingly complex molecular mechanisms of memory- and learning-related responses.

In man and mammals these primitive defense and communication molecules play major roles in the immune response system. The anatomic diffuseness and functional diversity of the lymphoid immune response system render it especially well suited for local and systemic defense needs. The efficient and organized operation of this system requires well-coordinated local, as well as central, controls for synchronous and purposeful actions. These actions necessarily include modulation and site-limitation controls as well as turn-on and turn-off systems and backups. The mobilization and functionally directed transport of important cells and substances of the immune defense response system to reaction centers and to and into target areas are essential parts of the proper working of this system *in vivo*.

Two major communications and transport systems exist in mammals, primates, and man: the nervous system and the blood and lymph vascular system. They are closely interrelated anatomically and in function. These two systems, in various degrees of complexity and sophistication, appeared long before the evolutionary development of vertebrates and mammals and their new and unique lymphoid immune response system. This new lymphoid immune response system would appear to have hijacked and insinuated itself into the blood-lymph and nervous systems anatomically and to have adopted for its own use the shared effector and modulator molecules of these two systems — the vasoactive neurotransmitter and modulator molecules cited above.

These molecules are first-line molecules in both communications and transport and make important contributions to target area and lymphoid tissue microenvironments. They influence blood and lymph flow and vascular permeability. Their actions on blood vessels are at the postcapillary venule level anatomically.[1-3] This area also is of major importance in the immune response system as the area of transport and transit of the especially important recirculating small lymphocytes in the circulating arterial blood supply to lymph nodes, to the interior substance of these lymph nodes, and, thence, into the lymphatics of the lymph circulatory system and back into the blood vascular system at lymphatic-venous communications.[4-6]

There are receptors for them on both neurons and immune response cells. Their stimulatory/excitatory actions in both immune response and nervous systems are exerted in transmembrane signaling by the phosphatidyl inositol (PI) and the cyclic GMP (guanosine monophosphate) signaling mechanisms, while their inhibitory/modulatory actions are by cyclic AMP (adenosine monophosphate) and adenosine transmembrane signaling.

Like the interrelated nervous and vascular systems with which they are involved, these molecules have multiple and complex interrelationships that doubtless contribute to the molecular and chemical plasticity of the response capability of these two systems and to their involvement in the immune response system. They may release and be released by one another, may potentiate or inhibit the actions of one another, and may be augmented or diminished in action by the agonists or antagonists of one another.[7-11] Further plasticity and flexibility of action may be provided by multiple receptor types and subtypes, which receptor-specific and nonspecific agonists and antagonists and molecular cloning and amino acid sequencing may identify.

Recognition of the existence and significance of the neurovascular component of the cellular immune response was long delayed by attention fixation on the involvement of major molecules of this system (e.g., histamine) in the dramatic and awesome immediate hypersensitivity reactions of anaphylaxis and allergy. An appreciation of their more physiologic activities in delayed hypersensitivity and cellular immunity has taken somewhat longer.

II. WHY POSTULATE AN AXIS?

There are a number of major identified actions of the vasoactive neurotransmitter and neuromodulator molecules cited above that would appear to contribute to the purposeful function of an immune response defense system:

1. They function as local ''hormones'' of short-term and reversible action and are readily metabolized and inactivated.
2. They increase and decrease regional vascular permeability and regional blood and lymph flow and have their site of action at the anatomic level of postcapillary venules.
3. They act as neurotransmitters, neuroeffectors, and neuromodulators and may be produced and stored by neurons.
4. Their formation and release are affected by antigen-antibody interactions, and they play important roles in both immediate and delayed hypersensitivity reactions.
5. They are produced, stored, and released by important immune response cells — lymphocytes, macrophages, mast cells, and basophils. These cells also have surface receptors for these substances.
6. They exert important influences on immune response cell cellular cyclic nucleotide metabolism (cyclic AMP and cyclic GMP), and on PI (phosphatidyl inositol) transmembrane signaling in these cells.
7. They influence immune response cell chemotaxis and cell proliferation, transformation, and colony formation.
8. Their production and release are influenced by lymphokines, and they may be the effector molecules of some lymphokine actions. They also play important roles in the functions of interferon and other lymphokines and cytokines.
9. They are important products and play important roles in the functions of both the classical and alternate pathways of complement activation and may function as nonclassical growth factors.
10. They have multiple and intricate involvements in coagulation, anticoagulation, and fibrinolysis.
11. They influence lymphocyte traffic into and through regional lymph nodes and other lymphoid and nonlymphoid tissues.

III. MOLECULES OF AXIS

The mediator molecules of this axis are quite small. They are largely simple amino acids, decarboxylated amino acids (amines), or amino acids only slightly altered or extended in short chains (peptides) — or short-chain lipids. The molecules of this neurovascular immune response system axis have been divided by the author into small mediator molecules (as cited); transmembrane-signaling, cascade, and ancillary molecules; and receptor molecules (Table 1).

The mediator molecules may be stimulatory and excitatory, or inhibitory and modulatory. The sensory, defense, and memory molecules are on the stimulatory and excitatory side. Cascade molecules largely derive from the vascular system. These include the molecules of coagulation-fibrinolysis, the kinins (brady- and tachy-), EDRF (endothelium-derived relaxing factor), angiotensin-renin, complement cascade (classical and alternate pathways), and the arachidonic acid cascade (in which both thromboxane A_2 and prostacyclin [prostaglandin I_2]

TABLE 1
Molecules of the Neurovascular Immune Response Axis

Mediator molecules
- Amino acids
  - Excitatory amino acids: glutamate, aspartate
  - Inhibitory amino acids: GABA, glycine
- Amines (decarboxylated amino acids)
  - Histamine
  - Serotonin (5-hydroxytryptamine, 5-HT)
  - Dopamine and other catecholamines
- Short-chain peptides
  - Met-enkephalin (5 amino acids)
  - CCK_8 (8 amino acids)
  - Bradykinin (9 amino acids)
  - Substance P (11 amino acids)
  - Neurotensin (13 amino acids)
  - Bombesin (14 amino acids)
  - VIP (28 amino acids)
  - CGRP (37 amino acids)
- Short-chain lipids (20 carbons)
  - Prostaglandins
  - Thromboxane
  - Leukotrienes and lipoxins

Transmembrane signaling molecules
- Excitatory: phosphatidyl inositol (PI), cyclic GMP
- Inhibitory: cyclic AMP, adenosine

Cascade molecules: coagulation-fibrinolysis, kinins, EDRF/NO/nitric oxide angiotensin-renin, complement, arachidonic acid
Ancillary molecules: lymphokines, cytokines, growth factors, oncogenes
Receptor molecules: types, subtypes, gene products

are intimately and centrally involved in vascular system homeostasis and function). These mediator and cascade molecules are woven into a ''seamless web'' of biological defense and are interrelated in multiple and complex, as well as simple, ways, as cited earlier. Kallikrein, the bradykinin-forming enzyme, is also Hageman Factor (Factor XII) of the blood coagulation cascade. Kininase II, the bradykinin-degrading enzyme, also is antiotensin-converting enzyme of the angiotensin-renin cascade. Complement cascade C3a and C5a fragments, anaphylotoxins, are histamine releasers as well as chemotactic agents, etc.

Ancillary molecules, on the other hand, are largely immune response system cell derived. Lymphokines, cytokines, and interleukins are lymphocyte and macrophage derived. These molecules, along with mediator and cascade molecules, may also trigger the activation of both growth factors and oncogenes. Receptor and signaling molecules round out this remarkable panorama of extraordinary molecular plasticity integrating the nervous, vascular, and immune response systems and contributing to memory — immunological as well as central nervous system.

IV. AMMON, HIS HORNS, AND MEMORY

Ancient gods, as ancient man, had good times and bad, ups and downs, and shared human adversity as well as human triumph. The ancient Egyptian Theban god Ammon, King of the Egyptian gods, had humble origins as the village deity of Waset in the Fourth, or Scepter, Nome (district) of upper Egypt. Waset was located on the east bank of the Nile River and is the site of the modern city of Luxor.

FIGURE 1. Fragment from an ancient Egyptian limestone relief stela found at Deir el-Medina in Egypt and currently in the Fitzwilliam Museum at the University of Cambridge, England. It pictures the god Ammon as a ram with downturned horns.

At Waset, Ammon was the ram-headed god of virility, fertility, and pugnacity. The sheep of Ammon's ram's head was a peculiar kind of sheep. It was native to the area of Waset and the Scepter Nome of upper Egypt and was found nowhere else in Egypt. This sheep belonged to a flat-tailed species (*Ovis platyura aegyptiaca*). The ram of this sheep was also noted for qualities of virility, fertility, and pugnacity, as was Ammon, and had a characteristic peculiar to it in that its horns were particularly large, curved, and down turned. The Theban/Waset god Ammon adopted this ram with its down-turned horns and long-term survival characteristics (even in adversity) as his own symbol (Figures 1 and 2).

In the early part of the Middle Kingdom (Ca. 2040–1786 B.C.) a temple was built for Ammon at Waset, although this town was not yet the capital of the Nome. Waset shortly became the City of Ammon or *The City*. The ancient Greeks equated Ammon with Zeus, and the ancient Greek author Homer was one of the first to call Waset Thebes, perhaps as a compliment to his own Greek city of the same name.

Local deities played an important political and religious role in ancient Egypt, as local factions and local nobility and nonnobility leaders associated themselves with these individual local deities as faction identification markers. As a local faction or leader gained prominence and power, so did his associated local deity. This was established as the tradition of kingly (pharaohnic) legitimacy and enhancement, from the first king of the first dynasty (Ca. 3100 B.C.) to the death of the last native Egyptian Pharaoh in 343 B.C., and the initiation of the Ptolemaic Period of foreign dominance.

FIGURE 2. Ammon, presented as a ram-headed god with downturned horns, is featured on the bow of a funerary barge found in the tomb of the Pharaoh Seti I at Abydos, Egypt.

Ammon emerged from relative obscurity to major prominence as a result of a dynastic usurpation in which a powerful devotee of Ammon toppled his royal master to become the first king of the 12th dynasty, in 1991 B.C., as Amenemhet I (Ammon-emhet). The 12th dynasty (1991–1786 B.C.) is generally considered to be the most brilliant period in the whole of Egyptian history and gave symbolic importance to Ammon in these times and in the subsequent less-fortunate times of Hyksos (foreign) dominance, which followed the end of the 12th dynasty and the beginning of the second intermediate period of political troubles.

Native Egyptian pride and the desire to reunite the whole of Egypt under a Theban and Ammon-associated king led to the end of Hyksos rule by the Theban King Ahmose, who established (in 1570 B.C.) the second most powerful and significant dynasty in the history of ancient Egypt, the 18th dynasty (beginning of the New Kingdom). The queen of this new dynasty was given the title of "God's Wife of Ammon" and was accorded considerable respect and authority in her own right. King Ahmose's son and successor was named Amenhotep I (Ammon-hotep, favored of Ammon).

During this dynasty the great temples of Ammon were erected at Karnak and Luxor and served as the centers of a large and powerful priesthood that functioned as the administrative "brains" for an expanding Egyptian empire that included Libya, Palestine, Cyprus, and Syria and extended into the Euphrates valley.

The temples and priesthood of Ammon had vast landholdings and great economic power. The long papyrus of Ramses IV described the Ammon temple holdings during the reign of Ramses III and the large benefactions to it in land and personnel. At one time the temples and priesthood of Ammon controlled one third of the land and one fifth of the population of Egypt — a feat requiring brains, learning, and memory!

The collapse of the New Kingdom in 1085 B.C. sharply diminished the power, repute, and influence of Ammon, his temple, and his priesthood, as Egypt was invaded in succession by the Assyrians, Babylonians, and Persians.

A rather remarkable resurgence of interest in Ammon and in his stature occurred at the time of the expulsion of the hated Persians from Egypt in 332 B.C. and its conquest by Alexander the Great of Macedonia (and without a battle, thanks to Ammon). An astute Alexander, on arriving in Egypt, went promptly to the oasis of Siwa. This was the location of the oracle of Ammon, long famed in the ancient Greek world. At Siwa, Alexander was told by the oracle that he was the son of Ammon and quite entitled to his newly claimed position as Pharaoh. The Egyptians were overjoyed, as was Alexander. Newly minted coins had Alexander's head on one side and the ram's head of Ammon on the other.

More recently, an equally remarkable revival of interest in Ammon — and particularly in his horns — has occurred. Early gross anatomists of human and mammalian brains had noted a marked downward-curving configuration of the hippocampal formation of the forebrain which bore a striking resemblance to the horns of Ammon. This configuration and resemblance resulted in the naming of this central and core hippocampal area as the *Cornu Ammonis* (Horns of Ammon) (Figure 3). The sections of this horn have been labeled as *Cornu Ammonis* 1, 2, and 3 (abbreviated to CA1, CA2, and CA3).

Another strange twist (or curve) of Ammon's horn of the hippocampus occurred in the 1950s as surgical attempts to control otherwise uncontrollable epilepsy, by bilateral resections of the hippocampus, produced profound memory losses and defects.[12-14] This finding pointed to the hippocampus as a major memory acquisition area and stimulated a search for experimental animal models of memory. These efforts led to the important work of Lomo, Bliss, and associates[15-17] with hippocampal long-term potentiation (LTP). LTP may be produced by electric stimulation of the perforant pathway of the hippocampus leading to the CA3 and CA1 areas of the horn of Ammon. Numerous studies have suggested the CA3 area of the hippocampus to be the "pacemaker" of the functional anatomy of the hippocampus and an area of considerable importance for the tracking and study of the molecular mechanisms of memory.

The major vasoactive neurotransmitter serotonin recently has been identified as an important molecule in memory facilitation, in studies of two gastropod molluscs, *Aplysia* and *Hermissenda*.[18-21] A recent (1990) publication by Oleskevich and Descarries[22] presents a vivid delineation of the ram's head configuration of the whole hippocampus of the rat by light-microscopic radioautographic quantification of anatomical sectors of serotonin

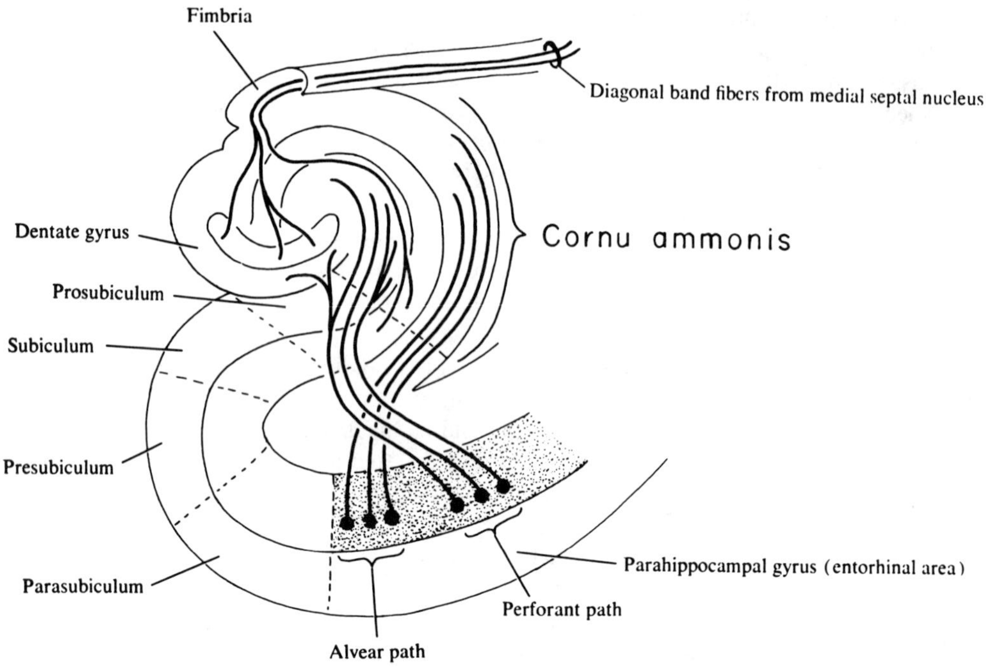

FIGURE 3. The main features of the memory-related hippocampus of the human brain, including the central location of the Cornu Ammonis (horns of Ammon). (After Pansky.[24])

innervation (Figure 4). Serotonin innervation was highest in Ammon's horn and in the subiculum outflow tract of the hippocampus — major memory areas.

V. CONCLUSION

Neurovascular immunology involves the study and understanding of the defense-survival-propagation and the learning and memory components of the vascular, nervous, and immune response systems of plants, invertebrates, and lower and higher vertebrates, including man, and of their basic and shared molecules (amino acids, amines, and short-chain peptides and lipids).

The vascular system was the first to appear in the evolutionary development of multicellular living organisms (vascular plants). The assumption of geographic motility by multicellular animals necessitated the addition of a nervous system with appropriate mediators and modulators. The lymphoid immune response system of vertebrates engrafted itself upon the more "primitive" existing vascular and nervous systems, yet retained the shared effector and modulator molecules of antiquity.

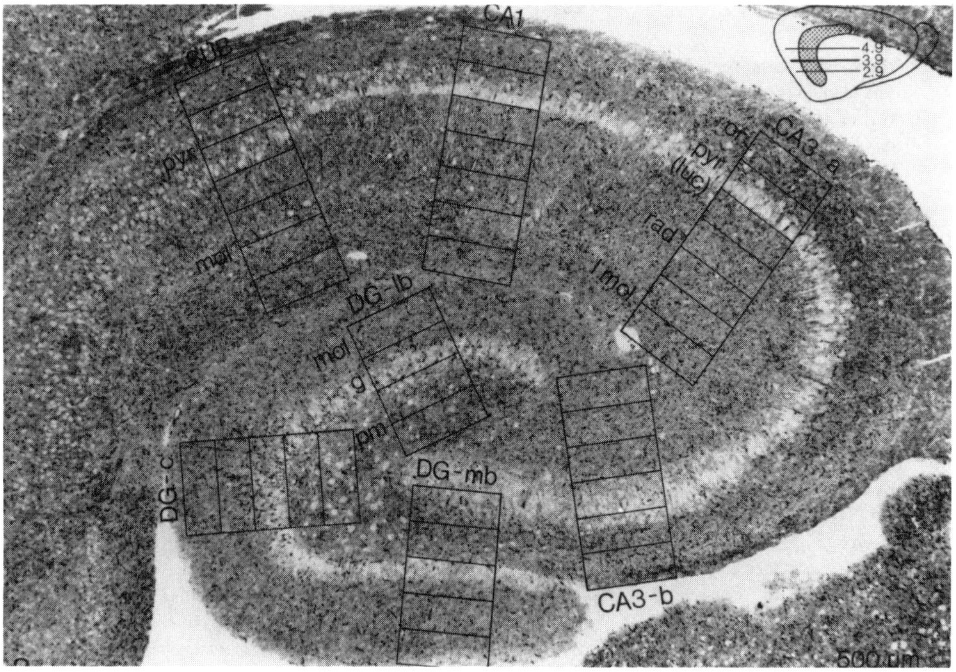

FIGURE 4. Photomicrograph showing radioautographic identification, location, and quantification of serotonin molecules (dark) in the whole hippocampus of the rat in its familiar ram's horn configuration. DG represents dentate gyrus impulse input area, and SUB represents the subiculum output area. Central areas CA3 (Cornu Ammonis 3) and CA1 (Cornu Ammonis 1) are the ''heart'' of the hippocampus. The CA3 area is considered the ''pacemaker'' of the hippocampus and a likely central area for the regulation of the laying down of memory and learning at the anatomical and molecular level. (From Oleskevich, S. and Descarries, L., *Neuroscience,* 34, 19, 1990. With permission.)

A commonality (homology) at the molecular level between central nervous system (CNS) memory and immunological memory was suggested by the author[23] in the last paragraph of the summary of his University of Cambridge (Trinity College) 1982 Ph.D. thesis, which includes the following:

> The relative vascular and neuronal contributions to the alterations in lymphocyte traffic observed in these studies are not identified. Both likely are of importance not only in contributing to on-site traffic regulation, but also in local (? ''axon reflex'') and more central (spinal and intracranial) coordination and integration and possibly even in memory. Immunologic memory may have something in common with other types of memory. A functioning immune response nervous system with considerable local action and integration may exist.

REFERENCES

1. **Majno, G., Gilmore, V., and Leventhal, M.,** On the mechanism of vascular leakage caused by histamine-type mediators, *Circ. Res.*, 21, 833, 1967.
2. **Majno, G., Palade, G. E., and Schoefl, G. I.,** Studies on inflammation. II. The site of action of histamine and serotonin along the vascular tree: a topographical study, *J. Biophys. Biochem. Cytol.*, 11, 607, 1961.
3. **Majno, G., Shea, S. M., and Leventhal, M.,** Endothelial contraction induced by histamine-type mediators. An electron microscopic study, *J. Cell Biol.*, 42, 647, 1969.
4. **Gowans, J. L.,** The effect of continuous reinfusion of lymph and lymphocytes on the output of lymphocytes from the thoracic duct of unanaesthetized rats, *Br. J. Exp. Pathol.*, 38, 67, 1957.
5. **Gowans, J. L.,** The recirculation of lymphocytes from blood to lymph in the rat, *J. Physiol. (London)*, 145, 54, 1959.
6. **Schoefl, G. I.,** The migration of lymphocytes across the vascular endothelium in lymphoid tissues, *J. Exp. Med.*, 136, 568, 1972.
7. **Moore, T. C., Normell, L., and Eiseman, B.,** Effect of serotonin loading on histamine release and blood flow of isolated perfused liver and lung, *AMA Arch. Surg.*, 87, 42, 1963.
8. **Moore, T. C., Normell, L., and Eiseman, B.,** Histamine metabolism in the isolated perfused canine liver, *Surg. Forum*, 14, 78, 1963.
9. **Eiseman, B., Moore, T. C., and Normell, L.,** Histamine metabolism in the isolated perfused pig liver, *Surg. Gynecol. Obstet.*, 118, 69, 1964.
10. **Moore, T. C., Normell, L., and Eiseman, B.,** Effect of serotonin and histamine on blood flow and tracheal resistance in isolated perfused lung, *Surgery*, 57, 730, 1965.
11. **Moore, T. C. and Eiseman, B.,** Serotonin metabolism in the isolated perfused canine liver, *Surgery*, 59, 531, 1966.
12. **Scoville, W. B. and Milner, B.,** Loss of recent memory after bilateral hippocampal lesions, *J. Neurol. Neurosurg. Psychiatry*, 20, 11, 1957.
13. **Milner, B.,** Amnesia following operation on the temporal lobes, in *Amnesia*, Whitty, C. W. M. and Zanglwill, O. L., Eds., Butterworths, London, 1966, 109.
14. **Milner, B.,** Disorders of learning and memory after temporal lobe lesions in man, *Clin. Neurosurg.*, 19, 421, 1972.
15. **Lomo, T.,** Frequency potentiation of excitatory synaptic activity in the dentate area of the hippocampal formation, *Acta Physiol. Scand.*, 68(suppl. 277). 128, 1966.
16. **Bliss, T. V. P. and Lomo, T.,** Long-lasting potentiation of synaptic transmission in the dentate area of the anesthetized rabbit following stimulation of the perforant path, *J. Physiol. (London)*, 232, 331, 1973.
17. **Bliss, T. V. P., Gardner-Medwin, A. R., and Lomo, T.,** Synaptic plasticity in the hippocampal formation, in *Macromolecules and Behaviour*, Ansell, G. B. and Bradley, P. B., Eds., MacMillan, London, 1973, 193.
18. **Montarolo, P. G., Goelet, P., Castellucci, V. F., Morgan, J., Kandel, E. R., and Schacher, S.,** A critical period for macromolecular synthesis in long-term heterosynaptic facilitation in *Aplysia*, *Science*, 234, 1249, 1986.
19. **Glanzman, D. L., Kandel, E. R., and Schacher, S.,** Target dependent structural changes accompanying long-term synaptic facilitation in *Aplysia* neurons, *Science*, 249, 799, 1986.
20. **Crow, T. and Forrester, J.,** Light paired with serotonin mimics the effect of conditioning a phototactic behavior of *Hermissenda*, *Proc. Natl. Acad. Sci. U.S.A.*, 83, 7975, 1986.
21. **Grover, L., Farley, J., and Auerbach, S. L.,** Serotonin involvement in in vitro conditioning of *Hermissenda crassicornis*, *Brain Res. Bull.*, 22, 363, 1989.
22. **Oleskevich, S. and Descarries, L.,** Quantified distribution of serotonin innervation in adult rat hippocampus, *Neuroscience*, 34, 19, 1990.
23. **Moore, T. C.,** The effects of inflammatory mediators and anaesthetic agents upon lymphocyte traffic of sheep *in vivo*, Ph.D. thesis, University of Cambridge (Trinity College), London, England, 1982.
24. **Pansky, B.,** in *Review of Neuroscience*, 2nd ed., Pansky, B., Allen, D. G., and Budd, G. C., Eds., Macmillan, New York, 1988, 269.

Chapter 2

EXCITATORY AMINO ACIDS

I. INTRODUCTION

The excitatory amino acids, aspartate and glutamate, are relative newcomers to the formerly exclusive club of established neurotransmitters. Their small size and the perception that they were too small to "count" and were largely substrates and metabolite building blocks for larger, more sophisticated, and more important molecules likely delayed the recognition of their true significance in the nervous system, and elsewhere as well. Difficulties in obtaining suitable antibodies for immunohistochemical study and tracking also delayed progress in this field.[1]

It is now recognized that the vast majority of synapses in the central nervous system use these excitatory amino acids as neurotransmitters.[2] Receptors for these excitatory amino acids are among the most abundant in the mammalian central nervous system.[3] L-Glutamate is generally believed to be the principal mediator of fast synaptic neurotransmission among vertebrate neurons.[4]

Glutamate is the most frequently encountered amino acid in the mammalian central nervous system.[5] It is stored in synaptic vesicles and is released from presynaptic terminals by neuronal membrane depolarization in a Ca^{++}-dependent manner.[6-8] The concentration of glutamate in the synaptic cleft may reach millimolar levels. The ability of glutamate to produce postsynaptic depolarization/excitation is influenced by both presynaptic and glial high-affinity uptake systems.[5]

Excitatory amino acid pathways are extensive throughout the brain and include cortical association and corticofugal and subcortical pathways.[1,9] Excitatory amino acids clearly are the predominant neurotransmitters in the mammalian forebrain. Virtually all neurons may be excited by glutamate.

The excitatory responses to glutamate are mediated by five major subtypes of receptors. Three of the receptor subtypes have been defined by the depolarizing/exciting actions of selective receptor agonists. They are *N*-methyl-D-aspartate (NMDA), kainate, and quisqualate/AMPA. A fourth receptor subtype, AP4, appears to function as an autoreceptor. A fifth subtype appears to function through inositol phosphate mobilization.[2]

The important involvement of the NMDA receptor in the initiation of the memory-involved hippocampal long-term potentiation (LTP) phenomenon brings these excitatory amino acids into a position of central importance as "memory" molecules in the molecular basis of learning and memory. Their involvement in other important physiological phenomena, including the immune response system, remains to be explored and established.

II. HISTORICAL BACKGROUND

1952. Hayashi[10,11] reports that topical application of glutamate and aspartate to mammalian cerebral cortex produces cortical seizure activity.

1959. Curtis, Phillis, and Watkins[12,13] describe iontophoretic application of glutamate and aspartate to spinal cord neurons as depolarizing (exciting) these neurons and as increasing the generation of action potentials in them.

1960. Curtis and Watkins[14] identify structural features required for excitatory activity as based on a "three-point-attachment" receptor model (Figure 1). Of the multiple compounds evaluated, excitatory activity was found in acidic amino acids with an

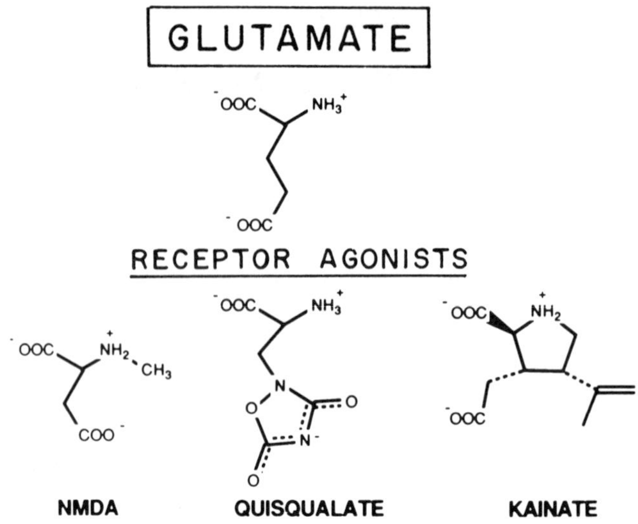

FIGURE 1. Basic structure of glutamate molecule and of major receptor agonists. (From Robinson, M. B. and Coyle, J. T., *FASEB J.*, 1, 446, 1987. With permission.)

acidic group in position alpha to an amino group (preferably primary), and with two or three carbon atoms distant to another group. The dicarboxylic amino acids (L-aspartate and L-glutamate), already shown to have excitatory activity in prior studies, satisfied these criteria and appeared to be prototypic.

1962. Watkins[15] reports the synthesis of some acidic amino acids that possess neuropharmacological activity. *N*-Methyl-D-aspartate (NMDA), an analogue of aspartic acid, was one of these synthetic molecules.

1963. Curtis and Watkins[16] report NMDA to be a potent neuronal excitant when applied to the mammalian spinal cord.

1968. McLennan et al.[17] describe different excitatory amino acid activities in different brain regions. This suggests more than one receptor.

1970– Shinozaki and associates[18,19] discover excitatory actions on rat cortical neurons and
1974. crayfish neuromuscular junctions by several antihelminthics and insecticides isolated from the seeds of the Cambodian nut plant *Quisqualis fructus* (quisqualic acid) and from the seaweed algae *Diginea simplex* (kainic acid). These observations also extend the evolutionary importance of these excitatory amino acids and their receptors into the plant kingdom as well as into Arthropoda of the invertebrate animal kingdom.

1974. Studies by Duggan[20] and by Johnston, Curtis, and associates[21] demonstrate differing spinal cord activities and potencies on different sections of the mammalian spinal cord: ventral vs. dorsal horn. L-Glutamate was found to be a more potent agonist upon dorsal-horn interneurons, while L-aspartate had greater potency of action on ventral-horn Renshaw interneurons. These findings provided additional strong support for the existence of multiple receptors for the excitatory amino acids.

1977– Synthesis of increasingly specific and potent receptor antagonists (by Watkins,
1981. McLennan, and their associates[22-24]) constitutes an important breakthrough in the study of these important excitatory amino acids.

1983. Monaghan, Cotman, and associates[25] employ quantitative autoradiologic techniques to identify and stratify excitatory amino acid binding sites in the memory-important hippocampus.

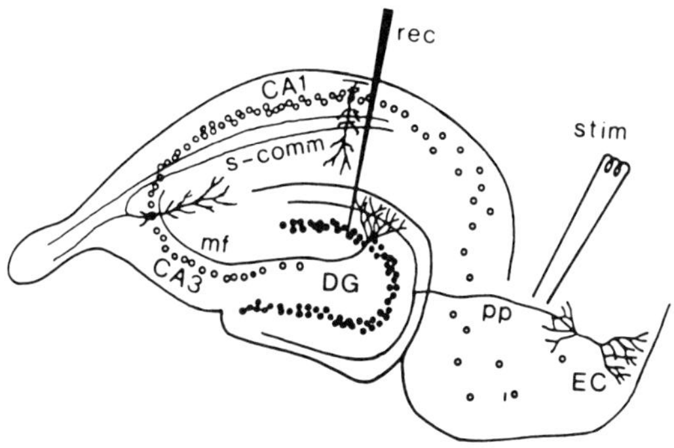

FIGURE 2. The means of studying long-term potentiation (LTP) of evoked potentials in the dentate gyrus (DG) of the hippocampus of the rat. Electrical tetanic stimuli (stim) are applied to the perforant path (pp) that originates in the entorhinal cortex (EC), and recordings (rec) are made from the granule cells of the dentate gyrus. This schematic drawing of the hippocampal formation also illustrates the major excitatory pathways of the trisynaptic circuit that, in addition to the perforant path from the entorhinal cortex to the dentate gyrus, also includes the mossy-fiber (mf) axons of the dentate gyrus granule cells that project to, and terminate on, proximal dendrites of CA3 pyramidal cells, and the Schaffer collateral axons of CA3 cell bodies that form a mixed pathway (S-COMM) with excitatory commissural fibers that originate in the contralateral hippocampal CA3 area to terminate on apical and basal dendrites of CA1 pyramidal cells. (From Bliss, T. V. P. and Lynch, M. A., in *Neurology and Neurobiology*, Vol. 35, Alan R. Liss, New York, 1988. With permission.)

1983. Collinridge, Kehl, and McLennan[26] demonstrate a direct involvement of the NMDA receptor in the hippocampal long-term potentiation (LTP) model of memory (Figure 2) and an associated increase in synaptic efficacy. They showed that AP5, a potent and selective NMDA antagonist, could block reversibly the induction of LTP. This study was carried out in the best-studied hippocampal pathway for LTP induction, the Schaffer/commissural-associational pathway to CA1 neurons (Cornu Ammonis 1).

1986. Harris and Cotman[27] identify a non-NMDA LTP pathway in the hippocampal mossy fiber pathway, which, from receptor levels,[1] is likely kinate receptor activated.

1987. Johnson and Ascher[28] describe marked potentiation of NMDA excitatory responses by nanomolar concentrations of glycine acting at a novel strychnine-insensitive binding site.

1988. CNQX, a potent and specific quisqualate/kainate (non-NMDA) receptor antagonist, is developed.[29,30]

1990. van den Pol, Wuarin, and Dudek[31] report that glutamate, through a non-NMDA receptor mechanism, plays an unexpectedly important role in neuroendocrine regulation in the hypothalamus, with a major and widespread role in the control of neuroendocrine neurons. They suggest that glutamate is the major fast excitatory neurotransmitter that controls not only the neuroendocrine system, but other hypothalamic regions as well. A potential role in neuroendocrine immunology may be seen here. If so, perhaps glutamate plays an important role in neurovascular immunology as well. These important potentials clearly are worthy of investigation in depth.

III. RECEPTOR TYPES AND ACTIONS

A. NMDA RECEPTOR

The NMDA glutamate receptor has been studied extensively. This receptor appears to be constituted as a receptor-ionophore complex. At least five pharmacologically distinct sites have been identified through which compounds may alter the activity of this receptor. They are (1) a transmitter binding site for L-glutamate binding, (2) a regulatory/stimulatory site for glycine that appears to be function as a speed-up recovery from desensitization factor,[32-34] (3) a channel site that binds phencyclidine,[35,36] (4) a voltage-dependent Mg^{++} binding site, and (5) an inhibitory divalent cation site that can bind Zn^{++}.[1,37]

Tonic activation of NMDA receptors by ambient glutamate has been reported, by Sah, Hestrin, and Nicoll,[38] to enhance the excitability of neurons. From voltage-clamp recordings and from noise analysis from pyramidal cells in hippocampal slices, they found evidence that NMDA receptors are tonically active. On the basis of the known concentration of glutamate in the extracellular fluid, they reasoned that this tonic action likely was produced by the ambient glutamate level. As NMDA receptors are voltage sensitive, the background activation of these receptors may be expected to impart a regenerative electrical property to the pyramidal cells, which facilitates the coupling between dendritic excitatory synaptic input and somatic action potential discharge in these neurons.

The Collingridge group[39] recently (1991) reported that a pharmacologically isolated NMDA-receptor-mediated synaptic response may undergo robust synapse-specific LTP. They considered that this observation added fundamentally to theories of synaptic plasticity because NMDA-receptor activation may, in addition to increasing synaptic efficiency, directly alter the plasticity of synapses. In addition, the NMDA receptor has been shown to regulate the expression of a class of neuronal proteins that underlie the morphological and physiological features of activity-dependent development.

An interesting neurovascular "connection" for the excitatory amino acids was made in 1989 in a report from the Moncada and Palmer group that had identified nitric oxide (NO) as the endothelium-derived relaxing factor (EDRF) of vascular physiology. This group, with the Garthwaites,[40] in 1989, reported that NMDA-receptor activation induced nitric acid synthesis from L-arginine in much the same manner as bradykinin and calcium ionophore induced its formation from L-arginine, by cultured endothelial cells in columns on microbeads.[41,42] With Knowles and Palacios, they have demonstrated the formation of nitric acid from L-arginine in the central nervous system, employing crude synaptosomes from rat brain.[43] On the immune response side, a similar generation of nitric oxide from L-arginine has been identified in macrophages[44] and in neutrophils.[45] A recent report in the October, 1991 issue of *Nature* described the ability of the stimulation of nitric oxide synthase formation of EDRF/NO/nitric oxide to lead to the production of LTP in the area of the mammalian hippocampus.[46] This suggests a pathway for all EDRF-stimulating molecules, such as bradykinin, substance P, serotonin, etc., to participate in LTP-type memory responses.

B. NON-NMDA RECEPTORS

1. Quisqualate

Quisqualate receptors differ from kainate receptors in both physiological and radioligand binding observations.[47-49] They also differ from kainate receptors rather significantly in anatomical distribution, as identified by autoradiography. Both receptors share a preferential telencephalic localization, however, with distinct localizations within this area. Quisqualate receptor sites are more numerous than kainate in the CA1 stratum radiatum, in the outer-cortical layers, in the lateral system, and in the molecular layer of the cerebellum.[50-52] Electrophysiological studies also have shown different anatomical distributions for quisqualate and kainate receptors.

In primary cultures of rat neocortical cells, excitatory amino acids stimulate inositol phosphate production, with a rank order of potency of quisqualate > ibotenate > glutamate > kainate > NMDA.[53] The individual responses were not additive. The results suggested that phosphoinositide mobilization activated by excitatory amino acids is mediated by a quisqualate-preferring receptor and that this activation is not blocked by any known antagonist.

In hippocampal neurons, activation of quisqualate "metabotropic" receptors has been observed to cause the release of Ca^{++} from intracellular stores[54-56] and to reduce voltage-dependent Ca^{++} currents.[57] Activation of a quisqualate ACPD "metabotropic" receptor has recently (1990) been found (by Charpak et al.[58]) to produce a slow type of excitation of hippocampal CA3 pyramidal cells. This slow-type excitation resulted from a depression of a Ca^{++}-dependent K^+ current and a depression of a voltage-gated K^+ current. In this 1990 study, combined voltage-clamp and microfluorometric recordings demonstrated that, although these quisqualate receptors can trigger an increase in intracellular Ca^{++} levels, suppression of K^+ currents is independent of changes in intracellular Ca^{++} concentration. These findings indicate that excitatory amino acids may not only act as fast inotropic transmitters, but also as slow neuromodulatory transmitters. An earlier (1987) report had identified this PI-stimulating glutamate receptor action as quisqualate preferring and as inhibited by pertussis toxin.[59]

In 1989, Tang et al.[60] reported that quisqualate-receptor-stimulated excitatory current may be carried by both high- and low-conductance channels by hippocampal neurons.

2. Kainate

The potency of agonists at the kainate receptors may be evaluated in spinal cord unmyelinated C-fiber afferents that have been found to possess a relatively pure population of kainate receptors. In telencephalic locations, kainate-receptor binding sites have high densities in the hippocampal CA3 stratum lucidum, in the deep cerebral cortical layers, in the striatum, in the reticular nucleus of the thalamus, and in the granular cell layer of the cerebellum.[61,62]

Electrophysiological studies have demonstrated the CA3 area of the hippothalamus to be unusually sensitive to kainic acid excitations.[63,64] Of the binding-site distributions for quisqualate and kainate, the kainate binding-site population has been observed to correspond to the brain areas that are unusually sensitive to kainate-receptor excitations and excitotoxicity.[65,66]

Excitatory amino acid agonists have been observed to increase the probability of the discharge of ventral horn cells in slices of spinal cord isolated from adult rats. The order of potency for faciliatory action was kainate > quisqualate > NMDA > L-glutamate > L-aspartate.[67]

Epileptic seizures, the hippocampus, and memory have had a long association. The use of bilateral surgical resection of the hippocampus to control intractable epilepsy led to the clinical observation that profound memory deficits result.[68-70] Experimental "kindling" of epileptiform seizures may be achieved by repeated subthreshold electrical stimulation to a local brain area[71-73] or by repeated applications of an excitatory amino acid to such an area.[74-76] The 1991 report of Gall et al.[76] is of particular interest, as the kainic-acid-induced seizures also stimulated an increased expression of a growth factor (nerve) in the hippocampus. This is of particular interest, as other vasoactive neurotransmitter "memory molecules," such as substance P, serotonin, bombesin, etc., also may act as growth factors.

"Kindling" differs from other models of synaptic plasticity or neuronal models of memory, such as LTP, in its extraordinary duration. Only a few low-level localized electrical stimuli will produce an augmented synaptic responsiveness of the stimulated circuit, which will persist for months without further reinforcement. In the usual "kindling" protocol, a subthreshold electrical stimulus is given repeatedly to a local brain area once or twice a day for a duration of 1 to 2 sec. When first given, the stimulation produces only a brief afterdischarge (AD) and no associated behavioral changes. Without any change in the intensity or duration of stimulation, and with the passage of time, the after-discharge gradually increases in duration and spreads from the stimulated area to increasingly distant and synaptically connected areas of the brain. Associated and progressive alterations in behavior accompany these electrophysical charges. The behavioral charge progresses from a momentary halting of ongoing locomotion, to localized twitching, and finally to a generalized seizure. Once generalized seizures have occurred, the halting of electrical stimulation for weeks, months, and even years will not eliminate the acquired electrophysiologically and behaviorally defined changes, and renewed application of the initial and behaviorally ineffective electrical stimulus will reliably and predictably result in a generalized seizure.

In some respects this artificial stimulus may be seen to impinge and ''piggyback'' itself on some normal set of pathways that are involved in the learning sensory processes that are involved in and basic to the laying down of memory and its molecular and structural bases.

The persisting enhancement of synaptic efficiency associated with the ''kindling'' process, its extension only along synaptically linked stations in the stimulated circuit,[77] and its strict dependence on protein synthesis[78,79] and axonal transport[77,79] have suggested to some observers[71-73] that ''kindling'' might be associated with some structural alteration in the synapse itself. A remodeling of synaptic architecture during hippocampal ''kindling'' has been described in a series of comparatively recent reports by Geinisman, Morrell, and deToledo-Morrell.[71,72] By employing a stereological technique, they observed that ''kindling'' produced a selective loss of nonperforated synapses, in contrast to perforated ones, with the ratio of perforated to nonperforated synapses increasing by 40 to 45% in ''kindled'' rats as compared with control animals. These findings suggested to the authors that synaptic efficacy may depend on the balance of the two synaptic types. They speculated that the selective elimination of nonperforated synapses may increase the potency of the remaining synaptic contacts in a manner reminiscent of synaptic remodeling during development. Reports by this group in 1990 described the increase in the relative proportion of perforated axospinous synapses following hippocampal ''kindling'' to be specific for the synaptic field of the stimulated axons[73] and to be associated with an increase structurally of the active zone in the hippocampal perforated synapses.[72] As the active zone is the site of intracellular transmission, the authors considered this active zone enlargement as a likely structural substrate of the synaptic plasticity that provides the brain's structural ''record'' of the ''kindling'' experience.

Gall et al.,[76] in 1991 reported that kainic-acid-induced seizures stimulate increased expression of nerve growth factor (NGF) mRNA in the rat hippocampus. NGF is differentially expressed anatomically in the adult central nervous system, with the highest level of NGF mRNA being found in neurons in the hippocampus, the olfactory cortex, and the neocortex.[76,80,81] Gall et al. investigated the influence of kainic-acid-induced limbic seizure activity on the expression of NGF on RNA in adult rat brain, employing *in situ* hybridization and S1 nuclease protection techniques with RNA probes complementary to murine and rat NGF mRNA. Intracerebroventricular injection of 0.5 μg of kainic acid (KA) produced a dramatic bilateral increase in the hybridization of the 35S-labeled mRNA within the stratum granulosum of the hippocampus. The increase was quite prompt, appearing 1 h post-KA injection and reaching a maximum 20-fold increase above control levels at 2 to 3 h post-injection. One may speculate concerning a possible involvement of a kainic-acid-induced increase in hippocampal NGF in the structural synaptic changes observed in experimental seizure activity, including the ratio of perforated and nonperforated synapses encountered in the Geinisman et al. studies cited above.

In addition to the above, kainic acid has been found to increase dopamine release in the striatum (employing *in vivo* microdialysis)[82,83] and to enhance GABA-dependent chloride uptake in the cortex.[84]

Kainic acid may be involved in mossy-fiber pathways in the rodent hippocampus, which displays LTP that is not blocked by NMDA-receptor antagonists.[27] This pathway has a paucity of NMDA receptors and a very high level of kainic acid receptors.[1]

In a 1990 report by Iino et al.,[85] a fraction of the kainate channels (type II) in cultured rat hippocampal neurons were discovered to have a very high Ca^{++} permeability — a finding consistent with the involvement of kainate channels in a mossy-fiber-generated LTP.

C. EVOLUTION AND RECEPTORS

Glutamate has been observed to activate an arthropod muscle channel that is fast desensitizing.[86]

The permeation of calcium ions through non-NMDA glutamate channels in salamander (*Ambystoma tigrinium*) retinal bipolar cells has been observed. L-Glutamate, and kainate and quisqualate, when rapidly applied to bipolar cells isolated from the salamander retina, produced a calcium conductance increase in 66% of the cells examined.[87] The quisqualate-induced currents decayed within a time course of tens of milliseconds. Kainate-induced currents demonstrated no desensitization. In these responding cells there was no response to NMDA. An absence of NMDA receptors was further confirmed by studies involving glycine, Mg^{++}, and CNQX (which blocked the non-NMDA responses observed). It was concluded that NMDA channels were absent from salamander retinal dipolar cells and that the calcium conductance observed was carried in non-NMDA channels. The authors interpreted these findings of high Ca^{++} permeability for some, indeed most, non-NMDA channels in this salamander retina bipolar cell preparation to raise the possibility that non-NMDA channels with calcium-conductance capability might be able to induce LTP in appropriate areas of the brain.

In nonmammalian lower vertebrates, glutamate receptors present a somewhat different representation, as the salamander findings above might suggest. In the Agnathan cyclostome lamprey, Dryer[88] found that the only excitatory amino acid stimulatory activity of the brain was produced through non-NMDA receptors. The application of glutamate to this brain preparation with voltage-clamp and current-clamp recording produced neuronal depolarization with a decrease in input resistance. The order of potency in these responses was quisqualate > kainate > glutamate > aspartate, with no NMDA-receptor response whatsoever. The inward currents evoked by glutamate as well as by kainate and quisqualate were attenuated reversibly by kynurenic acid (a non-NMDA antagonist), but not by 4-AP (a selective NMDA antagonist). The major conclusions from these studies are that excitatory synaptic events in the lamprey are mediated by non-NMDA-type receptors and that lamprey brain neurons do not express NMDA-type receptors. Interneurons inhibitory to Muller cells also appear to be driven by excitatory amino acids that act on non-NMDA receptors. The giant Muller cells are in the fast-conducting reticulospinal system of the lamprey. Brodin et al.[89] have described further evidence for excitatory amino acid transmission in lamprey reticulospinal neurons, by retrograde labeling with ^{3}H-D-aspartate. These studies were carried out after unilateral injections into the rostral spinal cord. The reticulospinal projections visualized are regarded as the oldest descending system controlling spinal motor functions, and its organization shows a comparatively high degree of conservation in vertebrate phylogeny.

Studies of retinotectal transmission in goldfish suggest a major role for quisqualate/kainate non-NMDA receptors.[90,91] An electrophysiological study of the action of glutamate on excitatory synaptic transmission in the optic tectum of the frog *Rana temporaria* showed the excitatory transmission to be blocked by CNQX (a non-NMDA receptor antagonist), but not by NMDA antagonists.[92]

A recent (1990) quantitative autoradiographic study for localization of L-glutamate binding sites in the chick brain revealed the highest concentration of binding sites in the molecular layer of the cerebellum, with high concentrations also found in the hippocampus, area parahippocampus, neostriatum, neostriatum intermedium, and hyperstriatum ventrale.[93]

IV. CONCLUSION

The excitatory amino acids are comparative newcomers to the big leagues of neurotransmission and neurotransmitters. For a long time they were considered too small to be important. That now has most decidedly changed. The major role of glutamate and aspartate and the glutamate/aspartate receptors in excitatory neurotransmission (including a major role in memory and the memory model, LTP) in the central nervous system is detailed in this chapter. The role of these molecules and receptors in peripheral actions involving the nervous, vascular, and immune response systems and neurovascular immunology remains a comparatively underinvestigated area. From the major role of these molecules and receptors in CNS memory, it is intriguing and challenging to consider their possible role and involvement in immunological memory and in the immune response system.

REFERENCES

1. **Cotman, C. W., Monaghan, D. T., Ottersen, O. P., and Storm-Mathisen, J.,** Anatomical organization of excitatory amino acid receptors and their pathways, *Trends Neurosci.*, 10, 273, 1987.
2. **Monaghan, D. T., Bridges, R. J., and Cotman, C. W.,** The excitatory amino acid receptors: their classes, pharmacology, and distinct properties in the function of the central nervous system, *Annu. Rev. Pharmacol. Toxicol.*, 29, 365, 1989.
3. **Wroblewski, J. T. and Danysz, W.,** Modulation of glutamate receptors: molecular mechanisms and functional implications, *Annu. Rev. Pharmacol. Toxicol.*, 29, 441, 1989.
4. **Collingridge, G. L. and Lester, R. A. J.,** Excitatory amino acid receptors in the vertebrate central nervous system, *Pharmacol. Rev.*, 40, 143, 1989.
5. **Fonnum, F.,** Glutamate: a neurotransmitter in mammalian brain, *J. Neurochem.*, 42, 1, 1984.
6. **Naito, S. and Ueda, T.,** Adenosine triphosphate-dependent uptake of glutamate into protein I-associated vesicles, *J. Biol. Chem.*, 258, 698, 1983.
7. **Potashner, S. J.,** The spontaneous and electrically evoked release from slices of guinea pig cerebral cortex of endogenous amino acids labeled via metabolism of D-[U-^{14}C] glucose, *J. Neurochem.*, 31, 177, 1978.
8. **Sandoval, M. E. and Cotman, C. W.,** Evaluation of glutamate as a neurotransmitter of cerebellar parallel fibers, *Neuroscience*, 3, 199, 1978.
9. **Fagg, G. E. and Foster, A. C.,** Amino acid neurotransmitters and their pathways in the mammalian central nervous system, *Neuroscience*, 9, 701, 1983.
10. **Hayashi, T. A.,** A physiological study of epileptic seizures following cortical stimulation in animals and its application to human clinics, *Jpn. J. Physiol.*, 3, 46, 1952.
11. **Hayashi, T. A.,** Effects of sodium glutamate on the nervous system, *Keio J. Med.*, 3, 183, 1954.
12. **Curtis, D. R., Phillis, J. W., and Watkins, J. C.,** Chemical excitation of spinal neurones, *Nature*, 183, 611, 1959.
13. **Curtis, D. R., Phillis, J. W., and Watkins, J. C.,** The chemical excitation of spinal neurones by certain acidic amino acids, *J. Physiol. (Lond.)*, 150, 656, 1960.
14. **Curtis, D. R. and Watkins, J. C.,** The excitation and depression of spinal neurons by structurally related amino acids, *J. Neurochem.*, 6, 117, 1960.
15. **Watkins, J. C.,** The synthesis of some acidic amino acids possessing neuropharmacological activity, *J. Med. Pharm. Chem.*, 5, 1187, 1962.
16. **Curtis, D. R. and Watkins, J. C.,** Acidic amino acids with strong excitatory actions on mammalian neurones, *J. Physiol. (Lond.)*, 166, 1, 1963.
17. **McLennan, H., Huffman, R. D., and Marshall, K. C.,** Patterns of excitation of thalamic neurones by amino acids and by acetylcholine, *Nature*, 219, 387, 1968.
18. **Shinozaki, H. and Konishi, S.,** Actions of several anti-helminthics and insecticides on rat cortical neurones, *Brain Res.*, 24, 368, 1970.

19. **Shinozaki, H. and Shibuya, I.**, A new potent excitant, quisqualic acid: effects on crayfish neuromuscular junction, *Neuropharmacology*, 13, 665, 1974.
20. **Duggan, A. W.**, The differential sensitivity to L-glutamate and L-aspartate of spinal interneurones and Renshaw cells, *Exp. Brain Res.*, 19, 522, 1974.
21. **Johnston, G. A. R., Curtis, D. R., Davies, J., and McCulloch, R. M.**, Spinal interneurone excitation by conformationally restricted analogues of L-glutamic acid, *Nature*, 248, 804, 1974.
22. **Biscoe, T. J., Evans, R. H., Francis, A. A., Martin, M. R., Watkins, J. C., Davies, J., and Dray, A.**, D-α-Aminoadipate as a selective antagonist of amino acid-induced and synaptic excitation of mammalian spinal neurones, *Nature*, 270, 743, 1977.
23. **Hall, J. G., McLennan, H., and Wheal, H. V.**, The actions of certain amino acids as neuronal excitants, *J. Physiol. (Lond.)*, 272, 52, 1977.
24. **Davies, J. D., Francis, A. A., Jones, A. W., and Watkins, J. C.**, 2-Amino-5-phosphonovalerate (2-APV), a potent and selective antagonist of amino acid-induced and synaptic excitation, *Neurosci. Lett.*, 21, 77, 1981.
25. **Monaghan, D. T., Holets, V. R., Toy, D. W., and Cotman, C. W.**, Anatomical distributions of four pharmacologically distinct ^{3}H-L-glutamate binding-sites, *Nature*, 306, 176, 1983.
26. **Collingridge, G. L., Kehl, S. J., and McLennan, H.**, Excitatory amino acids in synaptic transmission in the Schaffer collateral-commissural pathway of the rat hippocampus, *J. Physiol. (London)*, 334, 33, 1983.
27. **Harris, E. W. and Cotman, C. W.**, Long-term potentiation of guinea pig mossy fiber responses is not blocked by *N*-methyl-D-aspartate antagonists, *Neurosci. Lett.*, 70, 132, 1986.
28. **Johnson, J. W. and Ascher, P.**, Glycine potentiates the NMDA response in cultured mouse brain neurons, *Nature*, 325, 529, 1987.
29. **Honore, T., Davies, S. N., Drejer, J., Fletcher, E. J., Jacobsen, P., et al.**, Quinoxalinediones: potent competitive non-NMDA glutamate receptor antagonists, *Science*, 241, 701, 1988.
30. **Blake, J. F., Brown, M. W., and Collingridge, G. L.**, CNQX blocks acidic amino acid induced depolarizations and synaptic components mediated by non-NMDA receptors in rat hippocampal slices, *Neurosci. Lett.*, 89, 182, 1988.
31. **van den Pol, A. N., Wuarin, J.-P., and Dudek, F. E.**, Glutamate, the dominant excitatory transmitter in neuroendocrine regulation, *Science*, 250, 1276, 1990.
32. **Foster, A. C. and Kemp, J. A.**, Glycine maintains excitement, *Nature*, 338, 377, 1989.
33. **Mayer, M. L., Vyklicky, L., Jr., and Clements, J.**, Regulation of NMDA receptor desensitization in mouse hippocampal neurons by glycine, *Nature*, 338, 425, 1989.
34. **Kemp, J. A., Foster, A. C., Leeson, P. D., Priestley, T., Tridgett, R., Iversen, L. L., and Woodruff, G. N.**, 7-Chlorokynurenic acid is a selective antagonist at the glycine modulatory site of the *N*-methyl-D-aspartate receptor complex, *Proc. Natl. Acad. Sci. U.S.A.*, 85, 6547, 1988.
35. **Thomson, A. M., West, D. C., and Lodge, D.**, An *N*-methylaspartate receptor-mediated synapse in rat cerebral cortex: a site of action of ketamine? *Nature*, 313, 479, 1985.
36. **Kushner, L., Lerma, J., Zukin, S., and Bennett, M. V. L.**, Coexpression of *N*-methyl-D-aspartate and phencyclidine receptors in Xenopus oocytes injected with rat brain mRNA, *Proc. Natl. Acad. Sci. U.S.A.*, 85, 3250, 1988.
37. **Westbrook, G. L. and Mayer, M. L.**, Micromolar concentrations of Zn^{2+} antagonize NMDA and GABA responses of hippocampal neurons, *Nature*, 328, 640, 1987.
38. **Sah, P., Hestrin, S., and Nicoll, R. A.**, Tonic activation of NMDA receptors by ambient glutamate enhances excitability of neurons, *Science*, 246, 815, 1989.
39. **Bashir, Z. I., Alford, S., Davies, S. N., Randall, A. D., and Collingridge, G. L.**, Long-term potentiation of NMDA receptor-mediated synaptic transmission in the hippocampus, *Nature*, 349, 156, 1991.
40. **Garthwaite, J., Garthwaite, G., Palmer, R. M. J., and Moncada, S.**, NMDA receptor activation induces nitric oxide synthesis from arginine in rat brain slices, *Eur. J. Pharmacol.*, 172, 413, 1989.
41. **Palmer, R. M. J., Ferrige, A. G., and Moncada, S.**, Nitric oxide release accounts for the biological activity of endothelium-derived relaxing factor, *Nature*, 327, 524, 1987.
42. **Palmer, R. M. J., Rees, D. D., Ashton, D. S., and Moncada, S.**, L-Arginine is the physiological precursor for the formation of nitric oxide in endothelium-dependent relaxation, *Biochem. Biophys. Res. Commun.*, 153, 1251, 1988.
43. **Knowles, R. G., Palacios, L. M., Palmer, R. M. J., and Moncada, S.**, Formation of nitric oxide from L-arginine in the central nervous system: a transduction mechanism for stimulation of the soluble guanylate cyclase, *Proc. Natl. Acad. Sci. U.S.A.*, 86, 5159, 1989.
44. **Marletta, M. A., Yoon, P. S., Iyengar, R., Leaf, C. D., and Wishnok, J. S.**, Macrophage oxidation of L-arginine to nitrite and nitrate: nitric oxide is an intermediate, *Biochemistry*, 27, 8706, 1988.
45. **McCall, T. B., Boughton-Smith, N. K., Palmer, R. M. J., Whittle, B. J. R., and Moncada, S.**, Synthesis of nitric oxide from L-arginine by neutrophils. Release and interaction with superoxide anion, *Biochemistry*, 261, 293, 1989.

46. **O'Dell, T. J., Kandel, E. R., and Grant, S. G. N.,** Long-term potentiation in the hippocampus blocked by tyrosine kinase inhibitors, *Nature,* 353, 558, 1991.
47. **Fagg, G. E.,** L-Glutamate, excitatory amino-acid receptors and brain-functions, *Trends Neurosci.,* 8, 207, 1985.
48. **Mayer, M. L. and Westbrook, G. L.,** The physiology of excitatory amino-acids in the vertebrate central nervous system, *Prog. Neurobiol.,* 28, 197, 1987.
49. **Watkins, J. C. and Evans, R. H.,** Excitatory amino acid transmitters, *Annu. Rev. Pharmacol. Toxicol.,* 21, 165, 1981.
50. **Monaghan, D. T., Yao, D., and Cotman, C. W.,** Distribution of ^{3}H-AMPA binding sites in rat brain as determined by quantitative autoradiography, *Brain Res.,* 324, 160, 1984.
51. **Olsen, R. W., Szamraj, O., and Houser, C. R.,** [^{3}H]AMPA binding to glutamate receptor subpopulations in rat brain, *Brain Res.,* 402, 243, 1987.
52. **Rainbow, T. C., Wieczorek, C. M., and Halpain, S.,** Quantitative autoradiography of binding-sites for ^{3}H-AMPA, a structural analogue of glutamic-acid, *Brain Res.,* 309, 173, 1984.
53. **Patel, J., Moore, W. C., Thompson, C., Keith, R. A., and Salama, A. I.,** Characterization of the quisqualate receptor linked to phosphoinositide hydrolysis in neurocortical cultures, *J. Neurochem.,* 54, 1461, 1990.
54. **Furuya, S., Ohmori, H., Shigemoto, T., and Sugiyama, H.,** Intracellular calcium mobilization triggered by a glutamate receptor in rat cultured hippocampal cells, *J. Physiol. (Lond.),* 414, 539, 1989.
55. **Murphy, S. N. and Miller, R. J.,** A glutamate receptor regulates Ca^{2+} mobilization in hippocampal neurons, *Proc. Natl. Acad. Sci. U.S.A.,* 85, 8737, 1988.
56. **Murphy, S. N. and Miller, R. J.,** Two distinct quisqualate receptors regulate Ca^{2+} homeostasis in hippocampal neurons *in vitro, Mol. Pharmacol.,* 35, 671, 1989.
57. **Lester, R. A. J. and Jahr, C. E.,** Quisqualate receptor-mediated depression of calcium currents in hippocampal neurons, *Neuron,* 4, 741, 1990.
58. **Charpak, S., Gahwiler, B. H., Do, K. Q., and Knopfel, T.,** Potassium conductances in hippocampal neurons blocked by excitatory amino-acid transmitters, *Nature,* 347, 765, 1990.
59. **Sugiyama, H., Ito, I., and Hirono, C.,** A new type of glutamate receptor linked to inositol phospholipid metabolism, *Nature,* 325, 531, 1987.
60. **Tang, C.-M., Dichter, M., and Morad, M.,** Quisqualate activates a rapidly inactivating high conductance ionic channel in hippocampal neurons, *Science,* 243, 1474, 1989.
61. **Monaghan, D. T. and Cotman, C. W.,** Distribution of ^{3}H-kainic acid binding sites in rat CNS as determined by autoradiography, *Brain Res.,* 252, 91, 1982.
62. **Unnerstall, J. R. and Wamsley, J. K.,** Autoradiographic localization of high-affinity ^{3}H-kainic acid binding sites in the rat forebrain, *Eur. J. Pharmacol.,* 86, 361, 1983.
63. **Robertson, J. H. and Deadwyler, S. A.,** Kainic acid produces depolarization of CA3 pyramidal cells in the *in vitro* hippocampal slice, *Brain Res.,* 221, 117, 1981.
64. **Westbrook, G. L. and Lothman, E. W.,** Cellular and synaptic basis of kainic acid-induced hippocampal epileptiform activity, *Brain Res.,* 273, 97, 1983.
65. **Nadler, J. V., Perry, B. W., and Cotman, C. W.,** Intraventricular kainic acid preferentially destroys hippocampal pyramidal cells, *Nature,* 271, 676, 1978.
66. **Schwob, J. E., Fuller, T., Price, J. L., and Olney, J. W.,** Widespread patterns of neuronal damage following systemic or intracerebral injections of kainic acid: a histological study, *Neuroscience,* 5, 991, 1980.
67. **Hirayama, T., Ono, H., and Fukuda, H.,** Effects of excitatory and inhibitory amino acid agonists and antagonists on ventral horn cells in slices of spinal cord isolated from adult rats, *Neuropharmacology,* 29, 1117, 1990.
68. **Scoville, W. B. and Milner, B.,** Loss of recent memory after bilateral hippocampal lesions, *J. Neurol. Neurosurg. Psychiatry,* 20, 11, 1957.
69. **Milner, B.,** Amnesia following operation on the temporal lobes, in *Amnesia,* Whitty, C. W. M. and Zanglwill, O. L., Eds., Butterworths, London, 1966, 109.
70. **Milner, B.,** Disorders of learning and memory after temporal lobe lesions in man, *Clin. Neurosurg.,* 19, 421, 1972.
71. **Geinisman, Y., Morrell, F., and deToledo-Morrell, L.,** Remodeling of synaptic architecture during hippocampal "kindling", *Proc. Natl. Sci. U.S.A.,* 85, 3260, 1988.
72. **Geinisman, Y., Morrell, F., and deToledo-Morrell, L.,** Increase in the relative proportion of perforated axospinous synapses following hippocampal kindling is specific for the synaptic field of stimulated axons, *Brain Res.,* 507, 325, 1990.
73. **Geinisman, Y., deToledo-Morrell, L., and Morrell, F.,** The brain's record of experience: kindling-induced enlargement of the active zone in hippocampal perforated synapses, *Brain Res.,* 513, 175, 1990.

74. **Croucher, M. J. and Bradford, H. F.,** Kindling of full limbic seizures by repeated microinjections of excitatory amino acids into the rat amygdala, *Brain Res.,* 501, 58, 1989.
75. **Mori, N. and Wada, J. A.,** Bidirectional transfer between kindling induced by excitatory amino acids and electrical stimulation, *Brain Res.,* 425, 45, 1987.
76. **Gall, C., Murray, K., and Isackson, P. J.,** Kainic acid-induced seizures stimulate increased expression of nerve growth factor mRNA in rat hippocampus, *Mol. Brain Res.,* 9, 113, 1991.
77. **Morrell, F.,** *Epilepsy and the Corpus Callosum,* Reeves, A. G., Ed., Plenum Press, New York, 1985, 99–130.
78. **Morrell, F., Tsuru, N., Hoeppner, T. J., Morgan, D., and Harrison, W. H.,** Secondary eliptogenesis in frog forebrain: effect of inhibition of protein synthesis, *Can. J. Neurol. Sci.,* 2, 407, 1975.
79. **Jonec, V. and Wasterlain, C. G.,** Effect of inhibitors of protein synthesis on development of kindled seizures in rats, *Exp. Neurol.,* 66, 524, 1979.
80. **Korsching, S., Auburger, G., Heumann, R., Scott, J., and Thoenen, H.,** Levels of nerve growth factor and its mRNA in the central nervous system of the rat correlate with cholinergic innervation, *EMBO J.,* 4, 1389, 1985.
81. **Shelton, D. L. and Reichardt, L. F.,** Studies on the expression of the beta nerve growth factor (NGF) gene in the central nervous system: level and regional distribution of NGF mRNA suggest that NGF functions as a trophic factor for several distinct populations of neurons, *Proc. Natl. Acad. Sci. U.S.A.,* 83, 2714, 1986.
82. **Weskamp, G., Gasser, U. E., Dravid, A. R., and Otten, U.,** Fimbria-fornix lesion increases nerve growth factor content in adult rat septum and hippocampus, *Neurosci. Lett.,* 70, 121, 1986.
83. **Carrozza, D. P., Ferraro, T. N., Golden, G. T., Reyes, P. F., and Hare, T. A.,** Partial characterization of kainic acid-induced striatal dopamine release using in vivo microdialysis, *Brain Res.,* 543, 69, 1991.
84. **Schatzki, A., McMillian, M., and Miller, L. G.,** Glutamate, kainate and quisqualate enhance GABA-dependent chloride uptake in cortex, *Brain Res. Bull.,* 25, 239, 1990.
85. **Iino, M., Ozawa, S., and Tsuzuki, K.,** Permeation of calcium through excitatory amino acid receptor channels in cultured rat hippocampal neurons, *J. Physiol. (Lond.),* 424, 151, 1990.
86. **Dudel, J., Franke, C., Hart, H., Ramsey, R. L., and Usherwood, P. N. R.,** Rapid activation and desensitization by glutamate of excitatory, cation-selective channels in locust muscle, *Neurosci. Lett.,* 88, 33, 1988.
87. **Gilbertson, T. A., Scobey, R., and Wilson, M.,** Permeation of calcium ions through non-NMDA glutamate channels in retinal bipolar cells, *Science,* 251, 1613, 1991.
88. **Dryer, S. E.,** Excitatory amino acid-evoked membrane currents and excitatory synaptic transmission in lamprey reticulospinal neurons, *Brain Res.,* 443, 173, 1988.
89. **Brodin, L., Ohta, Y., Hokfelt, T., and Grillner, S.,** Further evidence for excitatory amino acid transmission in lamprey reticulospinal neurons: selective retrograde labeling with (^{3}H)-D-Aspartate, *J. Comp. Neurol.,* 281, 225, 1989.
90. **Langdon, R. B. and Freeman, J. A.,** Antagonists of glutaminergic neurotransmission block retinotectal transmission in goldfish, *Brain Res.,* 398, 169, 1986.
91. **Langdon, R. B. and Freeman, J. A.,** Pharmacology of retinotectal transmission in the goldfish: effects of nicotinic ligands, strychnine and kynurenic acid, *J. Neurosci.,* 7, 760, 1987.
92. **Nistri, A., Sivilotti, L., and Welsh, D. M.,** An electrophysiological study of the action of *N*-methyl-D-aspartate on excitatory synaptic transmission in the optic tectum of the frog *in vitro, Neuropharmacology,* 29, 681, 1990.
93. **Mitsacos, A., Dermon, C. R., Stassi, K., and Kouvelas, E. D.,** Localization of L-glutamate binding sites in chick brain by quantitative autoradiography, *Brain Res.,* 513, 348, 1990.

Chapter 3

INHIBITORY AMINO ACIDS

I. INTRODUCTION

The principal inhibitory amino acids long have been considered to be GABA (γ-amino butyric acid) and glycine. GABA is formed from glutamic acid (glutaminate) by the enzymatic action of glutamic acid decarboxylase (GAD). Glutamic acid is a polar, charged, acidic, hydrophilic amino acid that also functions as an excitatory amino acid. Glycine is an unaltered nonpolar hydrophobic amino acid.

Acceptance of a neurotransmitter (and modulator) role for these two amino acids was delayed significantly, quite like the situation with the excitatory amino acids, by their small and seemingly unimportant size and by the perception that their main purpose in ''life'' was to act as metabolic substrates and precursors for the synthesis of larger and more important molecules, such as peptides and proteins.

A major and important breakthrough for GABA came in 1963 with the publication of a discovery by Kravitz, Kuffler, and Potter[1] of a specific inhibitory effect of GABA on arthropod crustacean nerves. GABA was found to mimic the effects of inhibitory nerve stimulation and to be the only inhibitory amino acid found exclusively in the inhibitory nerve and in adequately stimulatory amounts. GABA release was correlated with nerve stimulation, and intracellular recordings from connective muscle showed that both nerve stimulation and application of GABA to an inhibitory neuron produced identical increases in Cl^- conductance. Picrotoxin, a Cl-channel-blocking agent blocked the effects of both inhibitory nerve stimulation and GABA application to inhibitory nerves.

Glycine's turn came 3 years later, in 1966, when Werman, Davidoff, and Aprison[2-4] published the first of three articles defining and documenting, by neurochemical and electrophysiological studies, the spinal cord postsynaptic neuronal inhibitory effects of glycine. Aprison and Werman,[5] a year earlier (1965), had published their findings relative to glycine distribution in the cat spinal cord. In 1972 Iversen and Bloom[6] described electron microscopic audiographic studies of the uptake of (^{3}H)-glycine by slices and homogenates of rat spinal cord and brain, and in 1974 Toung and Snyder[7] presented evidence that strychnine binding is associated with the tonic conductance mechanism at the glycine synaptic receptor.

In recent years the inhibitory amino acid reputation of both glycine and GABA has been tarnished somewhat by the identification of strong excitatory actions for both molecules. In 1987 Johnson and Ascher[8] of Paris, in a major *Nature* report, dropped a bombshell with the observation that glycine potentiates the excitatory amino acid response at the NMDA glutamate receptor in cultured mouse brain neurons. The action was not only potentiated, but *dramatically* potentiated and at a low glycine concentration. The action was clearly mediated by a new and non-strychnine-sensitive glycine receptor. This important and unexpected finding unleashed an avalanche of studies and publications, including a 1990 report by Tauch and Ashbeck[9] that showed that glycine synergistically potentiates the enhancement of a long-term potentiation (LTP) reponse induced by a sulfhydryl-reducing agent (DTT).

Next it was GABA's turn to fall from grace as an exclusive neuroinhibitory molecule. In 1990 and 1991 Mott and Lewis[10,11] and others[12,13] reported facilitation of the induction of LTP by $GABA_B$-receptor activation. These studies followed those demonstrating disinhibitory effects of $GABA_B$ receptors in multiple areas of the brain.[14-18]

These remarkable findings did not exhaust the capacity of ''inhibitory'' amino acids to surprise. In the September 13, 1990 issue of *Nature,* Baekkeskov et al.[19] identified the GABA-synthesizing enzyme GAD as the 64K autoantigen involved in the serious auto-

TABLE 1
Inhibitory Amino Acid Receptors

Amino acid	Receptor	Agonists	Antagonists	Actions
GABA	$GABA_A$[a]	Muscimol	Bicuculline	Inhibitory
	$GABA_B$	Baclofen	2-OH-saclofen	Excitatory
			Phaclofen	Enhances LTP
Glycine[b]	Strychnine sensitive	Glycine	Strychnine	Inhibitory
	Strychnine insensitive	Glycine	7-Chlorokynurenic acid	Excitatory
			(+)-HA-966	Enhances LTP

[a] $GABA_A$-receptor function also is modulated by picrotoxin and similar convulsants (inhibitory) and by barbiturates and benzodiazapines (stimulatory).

[b] Multiple glycine receptors have been shown to have both a tissue- and age-dependent distribution in the central nervous system. Light glycine-receptor mRNA is found predominantly in the neonatal rat spinal cord, and heavy glycine-receptor mRNA predominates in the adult rat spinal cord. Strychnine-sensitive neurons have been shown to be concentrated in the spinal cord and lower brain stem.

immune diseases insulin-dependent diabetes mellitus (IDDM) and stiff-man syndrome (SMS), a rare neurological disease that has been found to have a high association with IDDM. SMS does occur in association with IDDM, but with less frequency.

Reusens-Billen et al.,[20] in 1984, described localization of GABA high-affinity binding sites in the pancreas of neonatal rats. In a recent 1988 study Garry et al.[21] identified GABA immunoreactivity concentrated in the zymogen granules. Much earlier observations, in 1976, by Okada et al.[22] established the presence of GABA and its synthetic enzyme GAD at high levels in pancreatic islet cells. Subsequent immunocytochemical studies demonstrated that GABA, GAD, and GABA-transaminase (its metabolic-degrading enzyme) were present in a population of pancreatic B cells at the center of islets where these molecules were found to be colocalized with insulin.[23-25] These molecules, at the subcellular level, were observed to be concentrated in the mitochondrial and nuclear cytosol areas of the pancreatic B cells and not in secretory granules.[26,27]

When will the next shoe drop in this fascinating segment of central neurotransmitters and modulators?

II. GABA

A. RECEPTORS AND ACTIONS

1. $GABA_A$

The $GABA_A$ receptor site has been identified by Olsen et al.[28,29] and others as a GABA-receptor chloride-ion-channel protein complex with receptors for picrotoxin (a chloride-channel-blocking agent), barbiturates, and benzodiazopines, in addition to GABA. The major actions of GABA at this receptor site are inhibitory (Table 1).

Biochemical and molecular cloning studies have shown this receptor to be a hetero-oligometric protein that is composed of several distinct polypeptide types. The $GABA_A$ receptor appears to be a member of a gene superfamily of ligand-gated ion channels.[29] Molecular cloning has demonstrated that these receptor polypeptides share a 20 to 40% identity with each other[30] and a 10 to 20% identity with those of the nicotine acetylcholine receptors[31-35] and the strychnine-sensitive glycine receptors.[36-39]

The cloning and sequencing of subunits of the $GABA_A$ receptor by Schofield et al.[30] in 1987 was a major breakthrough in the understanding of the molecular and structural makeup of this receptor. Subsequent studies by this group have added to this understanding.[40]

2. $GABA_B$

Evidence for the existence of more than one GABA receptor is to be found in the 1970 report of DeGroat[41] which demonstrated that GABA may generate electrophysiological responses in sympathetic ganglia which differ from those through "classical" ($GABA_A$) receptors in the central nervous system. In 1979 Bowery et al.[42] described the effects of depolarizing agents on the efflux of GABA from sympathetic glial cells, and in 1981 this group[43] identified the presence of bicuculline-insensitive GABA receptors ($GABA_B$) on peripheral autonomic nerve terminals. An uptake of GABA by sympathetic ganglia had been reported earlier (1972) by this group (Bowery and Brown).[44]

The presence of both $GABA_B$- and $GABA_A$-receptor-mediated effects have been demonstrated in both the sympathetic and parasympathetic nervous systems.[45-51]

The discovery of a selective $GABA_B$-receptor antagonist, phaclofen, in 1987 by Kerr et al.,[52] was a major breakthrough in the study of the actions of GABA in the central and peripheral nervous systems as well as in areas outside the nervous system. They reported that phaclofen was both a central and peripheral antagonist of baclofen, the $GABA_B$-receptor agonist. Further studies by this group[53] resulted in the discovery (1988) of 2-OH-saclofen, the $GABA_B$-receptor antagonist used by Mott and Lewis[10] in their monumental 1991 report of the facilitation of the induction of the LTP (long-term potentiation) phenomenon by $GABA_B$ receptors.

Disinhibitory effects of $GABA_B$-receptor activation have been reported in several areas of the brain.[14-18] LTP enhancement also has been reported to be induced by the $GABA_B$ agonist baclofen.[11] Disinhibition also may be produced by the reduction in GABA release through $GABA_B$ autoreceptors on GABA-ergic nerve terminals.[54,55] Further disinhibition may result from the inhibition of the firing of GABA-ergic inhibitory interneurons by $GABA_B$ receptors on these interneurons.[56]

It has been suggested that $GABA_B$-receptor effects on LTP production may vary with the frequency of the stimulus used to induce LTP. Olpe and Karlsson,[12] in 1990, described the enhancement of high-frequency-induced LTP stimulation by $GABA_B$-receptor antagonists. The 1991 report and findings of Mott and Lewis[10] suggest that $GABA_B$-receptor modulation of synaptic plasticity may occur at electrical stimulatory frequencies in the range of the endogenous hippocampal theta rhythm that has been shown to modulate LTP *in vivo*. The Mott and Lewis findings[11] in the dentate gyrus (DG) area of the hippocampus, that $GABA_B$ receptors facilitate LTP induction, has been confirmed and duplicated recently (1991) in the CA1 region of the hippocampus, by Collingridge and associates.[13] These findings, with the disinhibitory effects of the $GABA_B$-receptor agonist baclofen in the cerebral cortex, have suggested to Mott and Lewis[10] that $GABA_B$ receptors may be important regulators of synaptic plasticity in many brain regions.

Whereas the $GABA_A$ receptor does not appear to be associated with any second messenger systems directly, the $GABA_B$ receptor is independent of chloride channels within membranes and appears to be negatively coupled to the adenylcyclase system and to function through a G-protein-involved mechanism.[57-60]

B. EVOLUTION

1. Annelida

GABA has been identified in, and implicated as, an inhibitory neurotransmitter in both the leech[61] and the earthworm.[62]

2. Arthropoda

An inhibitory effect on crustacean stretch receptors led to its functional identification as a neurotransmitter in the central nervous system.[63] A synaptic inhibitory effect had been described by Kuffler and Edwards[64] in 1958 prior to the careful dissection of this response by Kravitz, Kuffler, and Potter[1] (see "Introduction" to this chapter). Craelius and associates[65,66] have described the release of GABA by inhibitory interneurons in the crustacean crayfish, and the neurotrophic regulation of GABA receptors at a crayfish inhibitory synapse.

In insect arthropods, the GABA-synthesizing enzyme (GAD) has been shown by Emson et al.[67] to be present in inhibitory interneurons of the locust. Application of GABA has been found by Usherwood and Grundfest[68] to mimic the inhibitory postsynaptic potential in the extensor tibiae muscle of the locust, caused by stimulation of an axonal branch of the common inhibitor motor neuron. Kerkut and associates[69,70] have demonstrated an inhibitory (hyperpolarizing) effect of GABA on the iontophoretic application to central neurons of the cockroach, to the somata of the abdominal ganglion.

In a 1986 study of the distribution of GABA-like immunoreactivity in the brain of the honeybee, Schafer and Bicker[71] described GABA immunoreactivity in all major neuropile areas. Staining (PAP) in the optic lobes showed marked stratification. The receptor cells of the compound eyes, ocelli, and antennae were not found to be labeled. The dense reactivity in the optic and antennal neuropiles were interpreted as suggesting that GABA is more important in this arthropod insect in mediating local, rather than more distant, neural interactions. This group, in 1985, reported that mushroom body interneurons in the honeybee had GABA immunoreactivity.[72]

3. Mollusca

In 1986 and 1987 Richmond et al.,[73,74] of Calgary, Canada, presented pharmacological evidence for GABA-ergic neurotransmission of an excitatory nature in the pond snail *Helisoma trivolvis* (Mollusca, Pulmonata). GABA was shown to have pharmacological actions on identified buccal motorneurons, including activation of the central pattern generator (CPG) underlying feeding processes. This finding identified GABA as an important molecule in the functional regulation of feeding behavior in *Helisoma.* In 1991 these authors[75] reported in-depth studies of the presence, synthesis, immunoreactivity, and uptake of GABA in the nervous system of *Helisoma.* Employing chromatographic techniques, the authors demonstrated the presence of GABA and its differential distribution among the ganglia that constitute the central nervous system of *Helisoma.* Synthesis of ^{3}H-GABA *de novo* from ^{3}H-glutamate in the central nervous system also was shown. Levels of endogenous and newly synthesized GABA were found to be highest in the buccal, cerebral, and pedal ganglia. Indirect immunohistochemical study of whole amounts identified a central network of GABA-like immunoreactive neurons. Essentially all projections from GABA-immunoreactive neurons were confined to the central nervous system. These findings indicated a predominantly central role for GABA. Stained neurons were encountered on the dorsal surface of the buccal ganglia and throughout the cerebral and pedal ganglia. No GABA-immunoreactive cell bodies were found in the parietal, pleural, or visceral ganglia. An autoradiographic study of ^{3}H-GABA uptake in sectioned ganglia revealed a radiolabeled-cell pattern that closely resembled the distribution of GABA-immunoreactive neurons.

In other molluscs, GABA has been reported to have both excitatory (as in *Helisoma*) and inhibitory pharmacological actions on a number of central neurons.[76]

Detectable amounts of GABA have been found in other gastropods,[77-78] including *Aplysia*[79] and the terrestrial slug *Limax.* In a recent (1988) mapping of GABA immunoreactivity in *Limax maximus,* by Cooke and Gelperin,[80] the map of GABA neurons closely resembled that of *Helisoma.* In both molluscs, GABA-immunoreactive neurons were confined to the paired buccal, cerebral, and pedal ganglia. Both the presence and synthesis of GABA has been described in other molluscan classes: Bivalvia (clams, oysters, etc.) and Cephalopoda (octopi, squids, and cuttlefish).[81-83]

4. Vertebrata

a. Agnatha

Spinal cord neurons in the agnathan lamprey have been reported to contain $GABA_A$-receptor-activated Cl^- (chloride) channels[84-86] and to have both $GABA_A$ and $GABA_B$ receptors on presynaptic neurons.[87,88] $GABA_A$ and $GABA_B$ receptors on axons of lamprey spinal interneurons are phasically activated during fictive locomotion in the spinal preparation.[87] Intraspinal GABA neurons were considered to be the neurons most likely responsible for these locomotion-related effects. Both GABA and its synthesizing enzyme (GAD) have been shown to be present in the lamprey spinal cord.[85,89,90]

In a 1990 study of the location of GABA-ergic neurons in two species of lamprey, *Lampetra fluviatilis* and *Ichtyomyzon unicuspis,* Brodin et al.[91] described the presence of three types of GABA-immunoreactive cells in the spinal cord, employing both immunofluorescence and peroxidase-antiperoxidase (PAP) techniques. Three distinct morphological types of GABA-immunoreactive cell bodies were detected. Multipolar GABA-immunoreactive neurons were seen in the lateral gray cell column, while seemingly bipolar cells were found in the ventral part of the dorsal horn. Small liquor-contacting cells were encountered surrounding the central canal. A high density of immunoreactive fibers of spinal origin were observed in the lateral and ventral funiculi, while the dorsal column was found to have a comparatively low density. Dense GABA-immunoreactive fiber plexuses were identified in the lateral spinal margin and in the dorsal part of the dorsal horn (a sensory-pain-impulse-related area). Chronic lesions of the rostral spinal cord by the authors did not result in any observable loss of GABA-immunoreactive fibers below or above the lesion. This finding suggested that the three types of segmental GABA-immunoreactive neurons identified in this study were the main source of GABA-ergic innervation of the lamprey spinal cord.

The multipolar neurons in the Brodin et al.[91] study were the largest, with a perikaryal diameter of 15 to 25 μm. The bipolar-appearing cells were smaller (10 to 20 μm), with a rounded soma, and were distributed parasagitally and in the ventral aspect of the dorsal horn. Small spherical cells (10 to 15 μm) surrounded the central canal and had a club-like process that extended into the lumen of the spinal canal. These neuronal cells resembled the liquor-contacting neurons encountered in the spinal cord of various more advanced vertebrate species.

The GABA-ergic innervation of the lamprey spinal cord partly resembled that of other more advanced vertebrates, especially the high density of GABA-immunoreactive fibers in the dorsal horn of the spinal cord.[92-95] While differing morphological types of segmental GABA neurons, including liquor-contacting cells, have been identified in other more advanced vertebrates, amphibia,[96-98] and mammals,[92] brain stem GABA neurons in these more advanced vertebrates also appear to contribute to innervation of the spinal cord.[93,98,99]

C. NERVOUS SYSTEM

Involvement of GABA in the mammalian nervous system has already received considerable attention in the "Introduction" and "Receptors and Actions" sections of this chapter. Findings in several recent reports, however, are worthy of mention: largely mammalia.

GABA and glycine have been shown (by Todd and Sullivan[100] of Glasgow, U.K.) to coexist within cell bodies and axons in the rat spinal cord. A recent (1989) study of isolated synaptic vesicles by Kish et al.,[101] of the University of Michigan (Ann Arbor), has provided evidence that GABA, glycine, and glutamate have different and distinct uptake systems.

In a study of rat dorsal root neurons maintained in primary tissue culture, the effects of halothane, isoflurane, and enflurane general anesthetics on ionic currents induced by bath application of GABA to the neurons was investigated by Nakahiro et al.[102] of Northwestern

University. These general anesthetics were found to exert a dual effect on the GABA-receptor channel complex. They potentiated the nondesensitized current (both peak and sustained), and they suppressed the desensitized steady-state current. It is suggested that potentiation of the GABA-receptor channel response may be a primary action of general anesthetics, leading to surgical anesthesia.

Chronic exposure of developing cortical neurons (chick embryos 8.5 d *in ovo* and maintained *in vitro*) to GABA has been found to downregulate GABA/benzodiazepine receptors and GABA-gated chloride channels.[103]

Hippocampal fibers were observed (in a 1990 study by Meredith et al.[104] of Amsterdam) to make synaptic contacts with glutamate-decarboxylase-immunoreactive neurons in the rat nucleus accumbens. Axons arising in the hippocampal formation were found to make asymmetrical synapses with glutamate-decarboxylase-immunoreactive neurons in the nucleus accumbens. The postsynaptic targets included many spines and, less frequently, dendritic shafts and perikarya. Some of the immunolabeled neurons resembled medium-sized spiny-projection neurons, which suggested that the input from the hippocampus formation exerts a strong and direct influence on output pathways of nucleus accumbens of the important limbic system. The hippocampal formation projects widely and densely to the nucleus accumbens by way of the subiculum.[105,106] A major proportion of the subicular input to the nucleus accumbens has been found recently to be directed onto spines.[107,108] These spiny neurons, if GABA containing, may be the primary targets of the hippocampal formation input. This important 1990 study by Meredith et al.[104] provides direct evidence that the target neurons of the hippocampal fibers in nucleus accumbens are GABA-ergic. The finding of a strong monosynaptic relationship between hippocampal fibers and GABA-ergic neurons, which may be output neurons, not only provides evidence for a direct allocortical control of output pathways from the nucleus accumbens, but also poses some intriguing questions relative to this pathway, the hippocampus, GABA at the $GABA_B$-receptor LTP stimulatory locus, and "memory".

A 1991 report by Ericson et al.[109] of Sweden (Uppsala and Linkoping) established GABA immunoreactivity in the important tuberomammillary nucleus of the hypothalamus: major and sole location of the cell bodies of the widely projecting histaminergic nervous system. In this study, most of the neuronal perikarya of the ventral subgroup of the tuberomammillary nucleus were found to be GABA immunoreactive. The morphology of the GABA-immunoreactive perikarya was similar to the morphology of the histaminic neuronal perikarya. The GABA-immunoreactive perikarya were contacted by comparatively few neuronal terminals. The mean bouton-covering ratio of GABA-immunoreactive perikarya was 6.1%, while the mean bouton-covering ratio for GABA-immunoreactive dendrites in the tuberomammillary nucleus was established as 31%. It was particularly interesting that the GABA-immunoreactive perikarya and dendrites formed close contacts that never presented synaptic specializations. These findings indicate that the neurons of the histaminergic tuberomammillary nucleus contain the neurotransmitter GABA and that GABA, through its $GABA_A$ and $GABA_B$ receptors, may act as a modulator or enhancer of cellular processes in the tuberomammillary nucleus. It also is of interest that neurons of the tuberomammillary nucleus have few axosomatic contacts.[109]

D. VASCULAR SYSTEM

A variety of blood cells have been shown to contain GABA and its synthesizing and metabolizing enzymes.[60] They also have been shown to take up GABA specifically. GABA and GAD have been found in erythrocytes.[110] Platelets also have been shown to contain GABA and its related enzymes, as well as a high-affinity GABA uptake system.[110] Specific binding sites for GABA, the $GABA_A$ agonist muscimol, and various benzodiazepines also have been identified on platelet membranes.

Fugiwara,[111] in 1975, demonstrated the presence of GABA receptors on vascular smooth muscle of dog cerebral arteries. Krause[112] has explored involvement of local GABA mechanisms in vascular regulation. Napoleone et al.,[113] in 1987, described autoradiographic localization of the $GABA_A$-receptor agonist ^3H-mucimol in rat cerebral vessels.

E. IMMUNE RESPONSE SYSTEM

The GABA-metabolizing enzyme, GABA transaminase, has been found in lymphocytes.[60] Both GABA and GAD have been demonstrated to be present in the thymus of rodents.[110] In addition, Hall et al.,[114] in 1985, observed elevated levels of GABA in the thymus during the immune response. Recent evidence of a $GABA_A$-type (benzodiazopine) action on the proliferative responses of thymic cells of developing rats and upon proliferative responses of mouse spleen lymphocytes *in vitro* strongly suggests a role for GABA and its receptors in the immune response and with growth factor potentials.[115] More work clearly needs to be done in this important area.

III. GLYCINE

A. RECEPTORS AND ACTIONS

1. Strychnine-Sensitive Glycine Receptor

The strychnine-sensitive glycine receptor was the first glycine receptor to be identified and has been localized largely to the spinal cord and lower brain stem.[116-118] This receptor has been shown (by Betz and associates[119]) to be an oligomeric membrane glycoprotein with a molecular mass of 250 kDa made up of three polypeptides of 48, 58, and 93 kDa. These subunits of the strychnine-sensitive glycine receptor have been shown to have significant homology within their primary structure and may have evolved from a common ancestral receptor polypeptide and a single chloride-channel ancestor gene.

The strychnine-binding subunit of the rat glycine receptor has been cloned and sequenced by the Betz group[36] of Germany. The deduced polypeptide was found to have significant structural and amino acid sequence with nicotinic-acid-receptor proteins. This finding suggested the existence of a family of genes that encode neurotransmitter-gated ion channels.

The findings of Miledi and associates[38,39] suggest that, in the rat spinal cord, strychnine-sensitive glycine receptors have at least three different types of mRNAs encoding these functional receptors. These findings indicate that the glycine receptor and its mRNA are not homogeneous and that there are multiple glycine-receptor mRNAs in the central nervous system.

Monoclonal antibodies against the strychnine-sensitive glycine receptors have been employed in an immunocytochemical study to identify the anatomical localization of these receptors in the rat central nervous system.[117] In the diencephalon, a high density of immunoreactive structures was found in the lateral mammillary nucleus and in the caudal magnocellular nucleus. As the caudal magnocellular nucleus contains histamine and projects widely to the brain, this study suggested the possibility that histaminergic neurons may be influenced by glycine and its receptors of this type.

2. Strychnine-Insensitive Glycine Receptor

The agonists, antagonists, and basic actions involving the two major types of glycine receptors in mammals are presented in Table 1. The dramatic, watershed 1987 report of Johnson and Ascher[8] demonstrated the existence of a second excitatory and strychnine-insensitive glycine receptor. Recent (1990) reports suggest that there may be a heterogeneity of strychnine-insensitive glycine receptor sites.[120,121] Anatomical distribution of the glycine strychnine-insensitive receptor sites also was heterogeneous with respect to the rank order of binding densities, which was hippocampus > cerebral cortex > caudate putamen > thalamus > cerebellum > brain stem. The highest number and correlation of strychnine-insensitive glycine receptors and glutamate NMDA receptors was in the CA1 area of the hippocampus.[120] The D stereoisomers of serine and alanine bind strongly to this receptor site as agonists, even more strongly than L-glycine.[121] HA-966, a chiral molecule, is an antagonist for this receptor site. There is a stereoselectivity for the (R)-enantiomer of HA-966 at this glycine antagonist site.[122] Iversen and associates,[123] of the Merck, Sharpe, and Dohme Neuroscience Research Center of Harlow, Essex, U.K., in 1990, reported observations that enantiomers of HA-966 exhibit distinct central nervous system effects. They found that (+)-HA-966 is a selective glycine/NMDA-receptor antagonist, while (−)-HA-966 was a potent butyrolactone-like sedative with sedative/ataxic effects.

B. NERVOUS SYSTEM

The effects of glycine on the nervous system of mammals have been considered in both the "Introduction" and "Glycine Receptors and Actions" in this chapter. Nonetheless, some recently published observations in this area are of interest.

The administration of NMDA intraventricularly or intracisternally to rats has been found to increase cerebellar levels of cyclic GMP.[124] This response has been potentiated by glycine. Glycine also, in a concentration-dependent manner, has been observed to stimulate the release of noradrenalin from rat hippocampal brain slices.[125] Also in the hippocampal area, glycine has been shown to modulate NMDA-receptor desensitization in cultured mouse hippocampal neurons.[126]

Glycine synergistically has been shown to potentiate the enhancement of long-term potentiation in area CA1 of the hippocampal slice by a sulfhydryl reducing agent, dithiothreitol (DTT).[9] DTT has been found selectively to increase the responses of several central neurons to NMDA and to precipitate the occurrence of epileptiform activity in area CA1 of the hippocampus. It is of interest that the highest density of strychnine-insensitive glycine binding sites in the rat brain may be found in area CA1 of the hippocampus.[127]

The release of endogenous glycine from striatal neurons in primary culture by kainate has been reported recently.[128]

C. VASCULAR AND IMMUNE RESPONSE SYSTEMS

This is a grossly underinvestigated area. It is currently quite barren, with the exception of a 1989 report by Jain et al.[129] from India relating to modification of the immune response by glycine in animals.

IV. CONCLUSION

The "former" inhibitory amino acids GABA and glycine have experienced a reputational "sea change" with the discovery that agonist activation of $GABA_B$ and glycine strychnine-insensitive receptors may exert an excitatory response, leading to the stimulation of LTP, the memory model in the hippocampal area of the brain. Further explorations in this area, and in the periphery as well, offer a fruitful area for additional research of major significance to the nervous, vascular, and immune response systems. The involvement of these excitatory receptors in the immune response and in immunological memory should prove to be a most rewarding area for future research.

REFERENCES

1. **Kravitz, E. A., Kuffler, S. W., and Potter, D. D.,** Gamma-aminobutyric acid and other blocking compounds in Crustacea. Their relative concentrations in separated motor and inhibitory axons, *J. Neurophysiol.*, 26, 739, 1963.
2. **Werman, R., Davidoff, R. A., and Aprison, M. H.,** Glycine and postsynaptic inhibition in the cat spinal cord, *Physiologist*, 9, 318, 1966.
3. **Werman, R., Davidoff, R. A., and Aprison, M. H.,** Inhibition of motoneurons by iontophoresis of glycine, *Nature*, 214, 681, 1967.
4. **Werman, R., Davidoff, R. A., and Aprison, M. H.,** Inhibitory action of glycine on spinal neurons in the cat, *J. Neurophysiol.*, 31, 81, 1968.
5. **Aprison, M. H. and Werman, R.,** The distribution of glycine in the cat spinal cord, *Life Sci.*, 4, 2075, 1965.
6. **Iversen, L. L. and Bloom, F. E.,** Studies of the uptake of (^{3}H) glycine in slices and homogenates of rat brain and spinal cord by electron microscopic autoradiography, *Brain Res.*, 41, 131, 1972.
7. **Toung, A. B. and Snyder, S. H.,** The glycine synaptic receptor: evidence that strychnine binding is associated with the ionic conductance mechanism, *Proc. Natl. Acad. Sci. U.S.A.*, 71, 4002, 1974.
8. **Johnson, J. W. and Ascher, P.,** Glycine potentiates the NMDA response in cultured mouse brain neurons, *Nature*, 325, 529, 1987.
9. **Tauck, D. L. and Ashbeck, G. A.,** Glycine synergistically potentiates the enhancement of LTP induced by a sulfhydryl reducing agent, *Brain Res.*, 519, 129, 1990.
10. **Mott, D. D. and Lewis, D. V.,** Facilitation of the induction of long-term potentiation by $GABA_B$ receptors, *Science*, 252, 1718, 1991.
11. **Mott, D. D., Lewis, D. V., Ferrari, C. M., Wilson, W. A., and Swartzwelder, H. S.,** Baclofen facilitates the development of long-term potentiation in the rat dentate gyrus, *Neurosci. Lett.*, 113, 222, 1990.
12. **Olpe, H.-R. and Karlsson, G.,** The effects of baclofen and two $GABA_B$-receptor antagonists on long-term potentiation, *Naunyn-Schmiedeberg's Arch. Pharmacol.*, 342, 194, 1990.
13. **Davies, C. H., Starkey, S. J., Pozza, M. F., and Collingridge, G. L.,** $GABA_B$ autoreceptors regulate the induction of LTP, *Nature*, 349, 609, 1991.
14. **Mott, D. D., Bragdon, A. C., Lewis, D. V., and Wilson, W. A.,** Baclofen has a proepileptic effect in the rat dentate gyrus, *J. Pharmacol. Exp. Ther.*, 249, 721, 1989.
15. **Deisz, R. A. and Prince, D. A.,** Frequency-dependent depression of inhibition in guinea-pig neocortex *in vitro* by $GABA_B$ receptor feed-back on GABA release, *J. Pharmacol. Exp. Ther.*, 412, 513, 1989.
16. **Thompson, S. M. and Gahwiler, B. H.,** Activity-dependent disinhibition. III. Desensitization and GABA receptor-mediated presynaptic inhibition in the hippocampus *in vitro*, *J. Neurophysiol.*, 61, 524, 1989.
17. **Mott, D. D. and Lewis, D. V.,** *Soc. Neurosci. Abstr.*, 16, 945, 1990.
18. **Davies, C. H., Davies, S. N., and Collingridge, G. L.,** Paired-pulse depression of monosynaptic GABA-mediated inhibitory postsynaptic responses in the rat hippocampus, *J. Physiol. (Lond.)*, 424, 513, 1990.
19. **Baekkeskov, S., Aanstoot, H.-J., Christgau, A. R., Solimena, M., Cascalho, M., Folli, F., Richter-Olesen, H., and Camilli, P.-D.,** Identification of the 64K autoantigen in insulin-dependent diabetes as the GABA-synthesizing enzyme glutamic acid decarboxylase, *Nature*, 347, 151, 1990.
20. **Reusens-Billen, B., Pirlot, X., Remacle, C., Hoet, J. J., and deGasparo, M.,** Localization of GABA high-affinity binding sites in the pancreas of neonatal rat, *Cell Tissue Res.*, 235, 503, 1984.
21. **Garry, D. J., Garry, M. G., and Sorenson, R. L.,** Ultrastructural immunocytochemical localization of L-glutamate decarboxylase and GABA in rat pancreatic zymogen granules, *Cell Tissue Res.*, 252, 191, 1988.
22. **Okada, Y., Taniguchi, H., and Shimada, C.,** High concentration of GABA and high glutamate decarboxylase in pancreatic islets and in human insulinoma, *Science*, 194, 620, 1976.
23. **Vincent, S. R., Hokfelt, T., Wu, J. Y., Elde, R. P., Morgan, L. M., and Kimmel, J. R.,** Immunohistochemical studies on the GABA system in the pancreas, *Neuroendocrinology*, 36, 197, 1983.
24. **Okada, Y.,** Localization and function of GABA in the pancreatic islets, in *GABAergic Mechanisms in the Mammalian Periphery*, Erdo, S. L. and Bowery, N. G., Eds., Raven Press, New York, 1986, 223.
25. **Sakaue, M., Saito, N., and Tanaka, C.,** Immunohistochemical localization of gamma-aminobutyric acid (GABA) in the rat, *Histochemistry*, 86, 365, 1987.
26. **Garry, D. J., Coulter, H. D., McIntee, T. J., Wu, J.-Y., and Sorenson, R. L.,** Immunoreactive GABA transaminase within the pancreatic islet is localized in mitochondria of the B-cell, *J. Histochem. Cytochem.*, 35, 831, 1987.
27. **Garry, D. J., Sorenson, R. L., and Coulter, H. D.,** Ultrastructural localization of gamma amino butyric acid immunoreactivity in B cells of the rat pancreas, *Diabetologia*, 30, 115, 1987.

28. **Olsen, R. W., Bureau, M., Ransom, R. W., Deng, L., Dilber, A., Smith, G., Krestchatisky, M., and Tobin, A. J.,** The GABA receptor-chloride ion channel protein complex, *Ann. Rev.*, 1, 1976.
29. **Olsen, R. W. and Tobin, A. J.,** Molecular biology of $GABA_A$ receptors, *FASEB J.*, 4, 1469, 1990.
30. **Schofield, P. R., Darlison, M. G., Fujita, N., Burt, D. R., Stephenson, F. A., Rodriguez, H., Rhee, L. M., Ramachandran, J., Reale, V., Glencorse, T. A., Seeburg, P. H., and Barnard, E. A.,** Sequence and functional expression of the GABA-A receptor shows a ligand-gated receptor superfamily, *Nature*, 328, 221, 1987.
31. **Noda, M., Takahashi, H., Tanabe, T., Toyosato, M., Kikyotani, S., Furutani, Y., Hirose, T., Takashima, H., Iayama, S., Miyata, T., and Numa, S.,** Structural homology of *Torpedo california* acetylcholine receptor subunits, *Nature*, 302, 528, 1983.
32. **Boulter, J., Connolly, J., Deneris, E., Goldman, D., Heinemann, S., and Patrick, J.,** Functional expression of two neuronal nicotonic acetylcholine receptors from cDNA clones identifies a gene family, *Proc. Natl. Acad. Sci. U.S.A.*, 84, 7763, 1987.
33. **Schoepfer, R., Whiting, P., Esch, F., Blacher, R., Shimasaki, S., and Lindstrom, J.,** cDNA clones coding for the structural subunit of a chicken brain nicotinic acetylcholine receptor, *Neuron*, 1, 241, 1988.
34. **Changeux, J.-P., Giraudat, J., and Dennis, M.,** The nicotinic acetylcholine receptor: molecular architecture of a ligand-regulated ion channel, *TIBS*, 8, 459, 1987.
35. **Brisson, A. and Unwin, P. N. T.,** Quaternary studies of the acetylcholine receptor, *Nature*, 315, 414, 1985.
36. **Greeningloh, G., Rienitz, A., Schmitt, B., Methfessel, C., Zensen, M., Beyreuther, K., Gundelfinger, E. D., and Betz, H.,** The strychnine-binding subunit of the glycine receptor shows homology with nicotine acetylcholine receptors, *Nature*, 328, 215, 1987.
37. **Murakami, T., Araki, T., Yamano, M., Wanaka, A., Betz, H., and Tohyama, M.,** Localization of the glycine receptors in the rat central nervous system: an immunocytochemical analysis, *Adv. Exp. Med. Biol.*, 236, 71, 1988.
38. **Akagi, H. and Miledi, R.,** Heterogeneity of glycine receptors and their messenger RNAs in rat brain and spinal cord, *Science*, 242, 270, 1988.
39. **Akagi, H., Patton, D. E., and Miledi, R.,** Discrimination of heterogeneous mRNAs encoding strychnine-sensitive glycine receptors in Xenopus oocytes by antisense oligonucleotides, *Proc. Natl. Acad. Sci. U.S.A.*, 86, 8103, 1989.
40. **Levitan, E. S., Schofield, P. R., Burt, D. R., Rhee, L. M., Wisden, W., Kohler, M., Fujita, N., Rodriguez, H. F., Stephenson, A., Darlison, M. G., Barnard, E. A., and Seeburg, P. H.,** Structural and functional basis for $GABA_A$ receptor heterogeneity, *Nature*, 335, 76, 1988.
41. **DeGroat, W. C.,** The actions of γ-aminobutyric acid and related amino acids on mammalian autonomic ganglia, *J. Pharmacol. Exp. Ther.*, 172, 384, 1970.
42. **Bowery, N. G., Brown, D. A., and Marsh, S.,** γ-Aminobutyric acid efflux from sympathetic glial cells: effect of depolarizing agents, *J. Physiol. (Lond.)*, 293, 75, 1979.
43. **Bowery, N. G., Doble, A., Hill, D. R., Hudson, A. L., and Shaw, J. S., Turnbull, M. J., and Warrington, R.,** Bicuculine-insensitive GABA receptors on peripheral autonomic nerve terminals, *Eur. J. Pharmacol.*, 71, 53, 1981.
44. **Bowery, N. G. and Brown, D. A.,** γ-Aminobutyric acid uptake by sympathetic ganglia, *Nature*, 238, 89, 1972.
45. **Farkas, Z., Kasa, P., Balcar, V. J., Joo, F., and Wolff, J. R.,** Type A and type B GABA receptors mediate inhibition of acetylcholine release from cholinergic nerve terminals in the superior cervical ganglion of rat, *Neurochem. Int.*, 8, 565, 1986.
46. **Bowery, N. G. and Hill, D. R.,** GABA mechanisms in autonomic ganglia, in *GABAergic Mechanisms in the Mammalian Periphery*, Erdo, S. L. and Bowery, N. G., Eds., Raven Press, New York, 1986, 135.
47. **Balcar, V. J., Joo, F., Kasa, P., Dammasch, I. E., and Wolff, J. R.,** GABA receptor binding in rat cerebral cortex and superior cervical ganglion in the absence of GABAergic synapses, *Neurosci. Lett.*, 66, 269, 1986.
48. **Erdos, S. L.,** Peripheral GABAergic mechanisms, *Trends Pharmacol. Sci.*, 6, 205, 1985.
49. **Kerr, D. I. B. and Ong, J.,** GABAergic mechanisms in the gut: their role in the regulation of gut motility, in *GABAergic Mechanisms in the Mammalian Periphery*, Erdo, S. L. and Bowery, N. G., Eds., Raven Press, New York, 1986, 153.
50. **Cerubini, E. and North, R. A.,** Actions of γ-aminobutyric acid on neurons of guinea-pig myenteric plexus, *Br. J. Pharmacol.*, 82, 93, 1984a.
51. **Tanaka, C.,** γ-Aminobutyric acid in peripheral tissues, *Life Sci.*, 37, 2221, 1985.
52. **Kerr, D. I. B., Ong, J., Prager, R. H., Gynther, B. D., and Curtis, D. R.,** Phaclofen: a peripheral and central baclofen antagonist, *Brain Res.*, 423, 150, 1987.
53. **Kerr, D. I. B., Ong, J., Johnston, G. A. R., Abbenante, J., and Prager, R. H.,** 2-Hydroxy-saclofen: an improved antagonist at central and peripheral $GABA_B$ receptors, *Neurosci. Lett.*, 92, 92, 1988.

54. **Bonanno, G., Pellegrini, G., Asaro, D., Fontana, G., and Raiteri, M.,** $GABA_B$ autoreceptors in rat cortex synaptosomes: response under different depolarizing and ionic conditions, *Eur. J. Pharmacol.*, 172, 41, 1989.
55. **Harrison, N. L.,** On the presynaptic action of baclofen at inhibitory synapses between cultured rat hippocampal neurons, *J. Physiol. (Lond.),* 422, 433, 1990.
56. **Misgeld, U., Muller, W., and Brunner, H.,** Effects of (−) baclofen on inhibitory neurons in the guinea pig hippocampal slice, *Pfluegers Arch.*, 414, 139, 1989.
57. **Kamatchi, G. L. and Ticku, M. K.,** $GABA_B$ receptor activation inhibits Ca^{2+}-activated ^{86}Rb-efflux in cultured spinal cord neurons via G-protein mechanism, *Brain Res.*, 506, 181, 1990.
58. **Ticku, M. K. and Delgado, A.,** $GABA_B$ receptor activation inhibits Ca^{2+}-activated potassium channels in synaptosomes: involvement of G-proteins, *Life Sci.*, in press.
59. **Wojcik, W. J. and Neff, N. H.,** γ-Aminobutyric acid B receptors are negatively coupled to adenylate cyclase in brain and in the cerebellum. These receptors may be associated with granule cells, *Mol. Pharmacol.*, 25, 24, 1984.
60. **Erdo, S. L. and Wolff, J. R.,** γ-Aminobutyric acid outside the mammalian brain, *J. Neurochem.*, 54, 363, 1990.
61. **Cline, H.,** Evidence for GABA as a neurotransmitter in the leech, *J. Neurosci.*, 6, 2848, 1986.
62. **Ito, Y., Kuriyama, H., and Tashiro, N.,** Effect of gamma-aminobutyric acid and picrotoxin on the permeability of the longitudinal muscle of the earthworm to various anions, *J. Exp. Biol.*, 51, 363, 1969.
63. **Otsuka, M.,** GABA in the crustacean nervous system: a historical review, in GABA in Nervous System Function, Vol. 5, Roberts, E., Chase, T. N., and Tower, D. B., Eds., Raven Press, New York, 1976, 245.
64. **Kuffler, S. W. and Edwards, C.,** Mechanisms of gamma aminobutyric acid (GABA) action and its relation to synaptic inhibition, *J. Neurophysiol.*, 21, 589, 1958.
65. **Craelius, W. and Fricke, R. A.,** Release of ^3H-gamma-aminobutyric acid (GABA) by inhibitory neurons of the crayfish, *J. Neurobiol.*, 12, 249, 1981.
66. **Craelius, W., Tiberio, A., and Pomerantz, J.,** Neurotrophic regulation of GABA-receptors at a crayfish inhibitory synapse, *J. Neurobiol.*, 14, 123, 1983.
67. **Emson, P. C., Burrows, M., and Fonnum, F.,** Levels of glutamate decarboxylase, acetyltransferase, and acetylcholine esterase in identified motor neurons of the locust, *J. Neurobiol.*, 5, 33, 1974.
68. **Usherwood, P. N. R. and Grundfest, H.,** Peripheral inhibition in skeletal muscle of insects, *J. Neurophysiol.*, 28, 497, 1965.
69. **Kerkut, G. A., Pitman, R. M., and Walker, R. J.,** Sensitivity of neurones of the insect central nervous system to iontophoretically applied acetylcholine or GABA, *Nature,* 222, 1075, 1969.
70. **Pitman, R. M. and Kerkut, G. A.,** Comparison of the actions of iontophoretically applied acetylcholine and gamma-aminobutyric acid with EPSP and IPSP in cockroach central neurones, *Comp. Gen. Pharmacol.*, 1, 221, 1970.
71. **Schafer, S. and Bicker, G.,** Distribution of GABA-like immunoreactivity in the brain of the honeybee, *J. Comp. Neurol.*, 246, 287, 1986.
72. **Bicker, G., Schafer, S., and Kingan, T. G.,** Mushroom body feedback interneurons in the honeybee show GABA-like immunoreactivity, *Brain Res.*, 360, 394, 1985.
73. **Richmond, J. E., Bulloch, A. G. M., and Lukowiak, K. D.,** Electrophysiological and biochemical evidence for GABA as a neurotransmitter in *Helisoma trivolvis, Soc. Neurosci. Abstr.*, 13(2), 1070, 1987.
74. **Richmond, J. E., Murphy, A. D., Bulloch, A. G. M., and Lukowiak, K.,** Evidence for an excitatory effect of GABA on feeding patterned motor activity (PMA) of *Helisoma trivolvis, Soc. Neurosci. Abstr.*, 12, 792, 1986.
75. **Richmond, J. E., Bulloch, A. G. M., Bauce, L., and Lukowiak, K.,** Evidence for the presence, synthesis, immunoreactivity, and uptake of GABA in the nervous system of the snail *Helisoma trivolvis, J. Comp Neurol.*, 307, 131, 1991.
76. **Walker, R. J.,** Transmitters and modulators, in *The Mollusca, Vol. 9, Neurobiology and Behavior,* Part 2, Willows, A. O. D., Ed., Academic Press, New York, 1986, 279.
77. **Osborne, N. N.,** Occurrence of GABA and taurine in the nervous systems of the dogfish and some invertebrates, *Comp. Gen. Pharm.*, 2, 433, 1971.
78. **Osborne, N. N., Briel, G., and Neuhoff, V.,** Distribution of GABA and other amino acids in different tissues of the gastropod mollusc *Helix pomatia,* including *in vitro* experiments with [^{14}C] glucose and [^{14}C] glutamic acid, *Int. J. Neurosci.*, 1, 265, 1971.
79. **Cottrell, G. A.,** Serotonin and free amino acid analysis of ganglia and isolated neurones of *Aplysia dactylomela, J. Neurochem.*, 22, 557, 1974.
80. **Cooke, I. R. C. and Gelperin, A.,** Distribution of GABA-like immunoreactive neurons in the slug *Limax maximus, Cell Tissue Res.*, 253, 77, 1988.
81. **Vitellaro-Zuccarello, L. and De Biasi, S.,** GABA-like immunoreactivity in the pedal ganglia of *Mytilus galloprovincialis:* light and electron microscopic study, *J. Comp. Neurol.*, 267, 516, 1988.

82. **Bradford, H. F., Chain, E. B., Cory, H. T., and Rose, S. P. R.,** Glucose and amino acid metabolism in some invertebrate nervous systems, *J. Neurochem.,* 16, 969, 1969.
83. **Cory, H. T. and Rose, S. P. R.,** Glucose and amino acid metabolism in octopus optic and vertical lobes *in vitro, J. Neurochem.,* 16, 979, 1969.
84. **Gold, M. R. and Martin, A. R.,** Gamma-aminobutyric acid and glycine activate C1-channels with different characteristics in CNS neurons, *Nature,* 308, 639, 1984.
85. **Homma, S.,** Physiology and pharmacology of putative transmitters in lamprey central nervous system, *Prog. Neurobiol.,* 20, 287, 1983.
86. **Homma, S. and Rovainen, C. M.,** Conductance increases produced by gamma-aminobutyric acid and glycine in lamprey interneurones, *J. Physiol. (Lond.),* 279, 321, 1978.
87. **Alford, S., Christenson, J., and Grillner, S.,** Presynaptic $GABA_A$ and $GABA_B$ receptor-mediated phasic modulation in axons of spinal interneurons, *Eur. J. Neurosci.,* submitted.
88. **Christenson, J. and Grillner, S.,** Modulation of synaptic transmission in primary sensory neurons in the lamprey spinal cord, *Neuroscience,* 22 (suppl.), 1022, 1987.
89. **Brodin, L., Tossman, U., Ungerstedt, U., and Grillner, S.,** The effect of an uptake inhibitor (dihydrokainate) on endogenous excitatory amino acids in the lamprey spinal cord as revealed by microdialysis, *Brain Res.,* 458, 166, 1988.
90. **Wald, U., Selzer, M. E., and Krieger, N. E.,** Glutamic acid decarboxylase in sea lamprey (*Petromyzon marinus*): characterization, localization and developmental changes, *J. Neurochem.,* 36, 363, 1981.
91. **Brodin, L., Dale, N., Christenson, J., Storm-Mathisen, J., Hokfelt, T., and Grillner, S.,** Three types of GABA-immunoreactive cells in the lamprey spinal cord, *Brain Res.,* 508, 172, 1990.
92. **Barber, R. P., Vaughn, J. E., and Roberts, E.,** The cytoarchitecture of GABAergic neurons in rat spinal cord, *Brain Res.,* 238, 305, 1982.
93. **Fuji, K., Senba, E., Fujii, S., Nomura, I., Wu, J. Y., Ueda, Y., and Tohyama, M.,** Distribution, ontogeny and projections of cholecystokinin-8, vasoactive intestinal polypeptide, and gamma-aminobutyric acid containing neuron systems in the rat spinal cord, *Neuroscience,* 14, 881, 1985.
94. **Magoul, R., Onteniente, B., Geffard, M., and Calas, A.,** Anatomical distribution and ultrastructural organization of the GABAergic system in the rat spinal cord. An immunocytochemical study using anti-GABA antibodies, *Neuroscience,* 20, 1001, 1987.
95. **McLaughlin, B. J., Barker, R., Saito, K., and Wu, J. Y.,** Immunocytochemical localization of glutamate decarboxylase in the rat spinal cord, *J. Comp. Neurol.,* 164, 305, 1975.
96. **Dale, N., Robers, A., Ottersen, O. P., and Storm-Mathisen, J.,** The morphology and distribution of "Kolmer-Agduhr" cells, a class of cerebrospinal fluid-contacting neurons revealed in frog embryo spinal cord by GABA immunocytochemistry, *Proc. R. Soc. Lond.,* 232, 193, 1987.
97. **Dale, N., Roberts, A., Ottersen, O. P., and Storm-Mathisen, J.,** The development of a population of spinal cord neurons and their axonal projections revealed by GABA immunocytochemistry, *Proc. R. Soc. Lond.,* 232, 205, 1987.
98. **Roberts, A., Dale, N., Ottersen, O. P., and Storm-Mathisen, J.,** The early development of neurons with GABA immunoreactivity in the CNS of *Xenopus laevis* embryos, *J. Comp. Neurol.,* 261, 435, 1987.
99. **Millhorn, D. E., Hokfelt, T., Seroogy, K., Oertel, W., Verhofstad, A. A. J., and Wu, J. Y.,** Immunohistochemical evidence for colocalization of gamma-aminobutyric acid and serotonin in neurons of the ventral medulla projecting to the spinal cord, *Brain Res.,* 410, 179, 1987.
100. **Todd, A. J. and Sullivan, A. C.,** Light microscope study of the coexistence of GABA-like and glycine-like immunoreactivities in the spinal cord of the rat, *J. Comp. Neurol.,* 296, 496, 1990.
101. **Kish, P. E., Fischer-Bovenkerk, C., and Ueda, T.,** Active transport of γ-aminobutyric acid and glycine into synaptic vesicles, *Proc. Natl. Acad. Sci. U.S.A.,* 86, 3877, 1989.
102. **Nakahiro, M., Yeh, J. Z., Brunner, E., and Narahashi, T.,** General anesthetics modulate GABA receptor channel complex in rat dorsal root ganglion neurons, *FASEB J.,* 3, 1850, 1989.
103. **Hablitz, J. J., Tehrani, M. H. J., and Barnes, E. M., Jr.,** Chronic exposure of developing cortical neurons to GABA down-regulates GABA/benzodiazepine receptors and GABA-gated chloride currents, *Brain Res.,* 501, 332, 1989.
104. **Meredith, G. E., Wouterlood, F. G., and Pattiselanno, A.,** Hippocampal fibers make synaptic contacts with glutamate decarboxylase-immunoreactive neurons in the rat nucleus accumbens, *Brain Res.,* 513, 329, 1990.
105. **Groenewegen, H. J., Vermeulen-Van der Zee, E., te Kortschot, A., and Witter, M. P.,** Organization of the projections from the subiculum to the ventral striatum in the rat. A study using anterograde transport of *Phaseolus vulgaris*-leucoagglutinin, *Neuroscience,* 23, 103, 1987.
106. **Kelley, A. E. and Domesick, V.,** The distribution of the projection from the hippocampal formation to the nucleus accumbens in the rat: an anterograde- and retrograde-horse-radish peroxidase study, *Neuroscience,* 7, 2321, 1982.

107. **Meredith, G. E. and Wouterlood, F. G.,** Hippocampal and midline thalamic fibers and terminals in relation to the choline acetyltransferase-immunoreactive neurons in nucleus accumbens of the rat: a light and electron microscopic study, *J. Comp. Neurol.*, 296, 204, 1990.
108. **Totterdell, S. and Smith, A. D.,** Convergence of hippocampal and dopaminergic input onto identified neurons in the nucleus accumbens of the rat, *J. Chem. Neuroanat.*, 2, 285, 1989.
109. **Ericson, H., Kohler, C., and Blomqvist, A.,** GABA-like immunoreactivity in the tuberomammillary nucleus: an electron microscopic study in the rat, *J. Comp. Neurol.*, 305, 462, 1991.
110. **Oset-Gasque, M. J., Launay, J. M., and Gonzalez, M. P.,** GABAergic mechanisms in blood cells: their possible role, in *GABAergic Mechanisms in the Mammalian Periphery,* Erdo, S. L. and Bovery, N. G., Eds., Raven Press, New York, 1986, 305.
111. **Fugiwara, M., Muramatsu, I., and Shibata, S.,** γ-Aminobutyric acid receptors on vascular smooth muscle of dog cerebral arteries, *Br. J. Pharmacol.*, 55, 561, 1975.
112. **Krause, D. N.,** Involvement of local GABA mechanisms in vascular regulation, in *GABAergic Mechanisms in the Mammalian Periphery,* Erdo, S. L. and Bowery, N. G., Eds., Raven Press, New York, 1986, 193.
113. **Napoleone, P., Erdo, s. L., and Amenta, F.,** Autoradiographic localization of the $GABA_A$ receptor agonist ^{3}H-muscimol in rat cerebral vessels, *Brain Res.*, 423, 109, 1987.
114. **Hall, N. R., Suria, A., and Goldstein, A. L.,** Elevated levels of gamma amino butyric acid (GABA) in the thymus gland during the immune response, *Lymphokine Res.*, 4, 339, 1985.
115. **Pawlikowski, M., Lyson, K., Kunert-Radek, J., and Stepien, H.,** Effect of benzodiazepines on the proliferation of mouse spleen lymphocytes *in vitro, J. Neural Transm.*, 73, 161, 1988.
116. **Probst, A., Cortes, R., and Palacios, J. M.,** The distribution of glycine receptors in the human brain. A light microscopic autoradiographic study using [^{3}H]strychnine, *Neuroscience,* 17, 11, 1986.
117. **Murakami, T., Araki, T., Yamano, M., Wanaka, A., Betz, H., and Tohyama, M.,** Localization of the glycine receptors in the rat central nervous system: an immunocytochemical analysis, *Adv. Exp. Med. Biol.*, 236, 71, 1988.
118. **White, W. F., Brown, K. L., and Frank, D. M.,** Glycine binding to rat cortex and spinal cord: binding characteristics and pharmacology reveal distinct populations of sites, *J. Neurochem.*, 53, 503, 1989.
119. **Pfeiffer, F., Simler, R., Grenningloh, G., and Betz, H.,** Monoclonal antibodies and peptide mapping reveal structural similarities between the subunits of the glycine receptor of rat spinal cord, *Proc. Natl. Acad. Sci. U.S.A.*, 81, 7224, 1984.
120. **McDonald, J. W., Penney, J. B., Johnston, M. V., and Young, A. B.,** Characterization and regional distribution of strychnine-insensitive [^{3}H]glycine binding sites in rat brain by quantitative receptor autoradiography, *Neuroscience,* 35, 653, 1990.
121. **Danysz, W., Fadda, E., Wroblewski, J. T., and Costa, E.,** [^{3}H]D-serine labels strychnine-insensitive glycine recognition sites of rat central nervous system, *Life Sci.*, 46, 155, 1990.
122. **Pullan, L. M., Britt, M., Chapdelaine, M. J., Keith, R. A., LaMonte, D., Mangano, T. J., Patel, J., Powel, R. J., Stumpo, R. J., Warwick, P. J., Zinkand, W. C., and Salama, A. I.,** Stereoselectivity for the (R)-enantiomer of HA-966 (1-hydroxy-3-aminopyrrolidone-2) at the glycine site of the *N*-methyl-D-aspartate receptor complex, *J. Neurochem.*, 55, 1346, 1990.
123. **Singh, L., Donald, A. E., Foster, A. C., Hutson, P. H., Iversen, L. L., Iversen, S. D., Kemp, J. A., Leeson, P. D., Marshall, G. R., Oles, R. J., Priestley, T., Thorn, L., Tricklebank, M. D., Vass, C. A., and Williams, B. J.,** Enantiomers of HA-966 (3-amino-1-hydroxypyrrolid-2-one) exhibit distinct central nervous system effects: (+)-HA-966 is a selective glycine/*N*-methyl-D-aspartate receptor antagonist, but (−)-HA-966 is a potent γ-butyrolactone-like sedative, *Proc. Natl. Acad. Sci. U.S.A.*, 87, 347, 1990.
124. **Danysz, W., Wroblewski, J. T., Brooker, G., and Costa, E.,** Modulation of glutamate receptors by phencyclidine and glycine in the rat cerebellum: cGMP increase *in vivo, Brain Res.*, 479, 270, 1989.
125. **Mangano, T. J., Patel, J., Salama, A. I., and Keith, R. A.,** Glycine-evoked neurotransmitter release from rat hippocampal brain slices: evidence for the involvement of glutaminergic transmission, *J. Pharmacol. Exp. Ther.*, 252, 574, 1990.
126. **Vyklicky, L., Jr., Benveniste, M., and Mayer, M. L.,** Modulation of *N*-methyl-D-aspartic acid receptor desensitization by glycine in mouse cultured hippocampal neurones, *J. Physiol. (Lond.),* 428, 313, 1990.
127. **Bristow, D. R., Bowery, N. G., and Woodruff, G. N.,** Light microscopic autoradiographic localisation of [^{3}H]glycine and [^{3}H]strychnine binding sites in rat brain, *Eur. J. Pharmacol.*, 126, 303, 1986.
128. **Weiss, S., Kemp, D. E., Bauce, L., and Tse, F. W.,** Kainate evokes the release of endogenous glycine from striatal neurons in primary culture, *Neurosci. Lett.*, 107, 205, 1989.
129. **Jain, P., Khanna, N. K., Godhwani, J. L., and Solanki, A.,** Modification of immune response by glycine in animals, *Indian J. Exp. Biol.*, 27, 292, 1989.

Chapter 4

HISTAMINE

I. INTRODUCTION

A. GENERAL

Histamine was one of the first known and, for a long time, the most misunderstood and feared of all of the major vasoactive neurotransmitter substances. Like many of these substances, it is quite potent in its actions and short lived. In addition to its many and complex involvements of importance in the immune response system, it shares with many other of these substances a major involvement in the physiology of the nervous and vascular systems as well as in gut influences (stomach).

Although Sir Henry Dale, the major discoverer and early investigator of histamine as a naturally and physiologically occurring substance, received a Nobel Prize (1936), it was not for the discovery of this pesky substance but rather for the discovery of a more proper and useful substance, acetylcholine. The only histamine-research-related Nobel Prizes have been awarded for combatting its harmful and untoward actions and contributions to allergy and anaphylaxis (Bovet, 1957) and to peptic ulceration of the alimentary tract (Sir James Black, 1989).

Despite its long-time ''bad press'', histamine, at present and through evolution, must be regarded as one of the most important of the major vasoactive neurotransmitter mediator molecules, including a major and highly significant involvement in the cellular immune response.

B. STRUCTURE, FORMATION, STORAGE, RELEASE, AND METABOLISM

Histamine, an amine, is of a comparatively simple structure (Figure 1). It is formed intracellularly by the decarboxylation of the amino acid histidine by the action of the enzyme histidine decarboxylase (HDC).[1] Mast cells may both make and store histamine. They also may release histamine in controlled and physiological circumstances. Rarely and dramatically and in uniquely afflicted and sensitized individuals or animals, the release may be excessive or explosive, resulting in allergy or anaphylaxis — and death. Other cells may produce histamine as well. This was illustrated clearly in a 1980 study by Watanabe et al.,[2] who found near-normal levels of HDC in a strain of mast-cell-deficient mice.

Of considerable importance was the demonstration (in 1981 by Dy et al. of Paris[3]) of a T-cell-generated lymphokine that functioned as a histamine-producing cell stimulating factor, which they called HCSF. This HCSF also was found by them to be released by concanavalin A (ConA)-stimulated normal spleen cells. This group also reported that this putative lymphokine was capable of inducing in bone marrow cells the exponential growth and concentration of a mast-cell-like population. This was considered at the time to be a sort of ''mast-cell growth factor'' (personal communication from M. Dy to author in early 1980s). This factor now appears to be within the activities ascribed to interleukin 3.

In 1974 the author with Granger and Lemmi[4] had reported that both Con A and Con A-activated lymphokines *in vivo* had elevated rat splenic HDC levels and had produced translocation of histamine from nuclear to mitochondrial and microsomal subcellular areas, as determined by ultracentrifugation.

The disposal of HDC-formed histamine is by storage in granules in mast cells in tissues or in granules in basophils in blood or by metabolic degradation. Metabolic degradation of histamine has been shown by Schayer and Cooper[1] to follow two paths: oxidative deamination

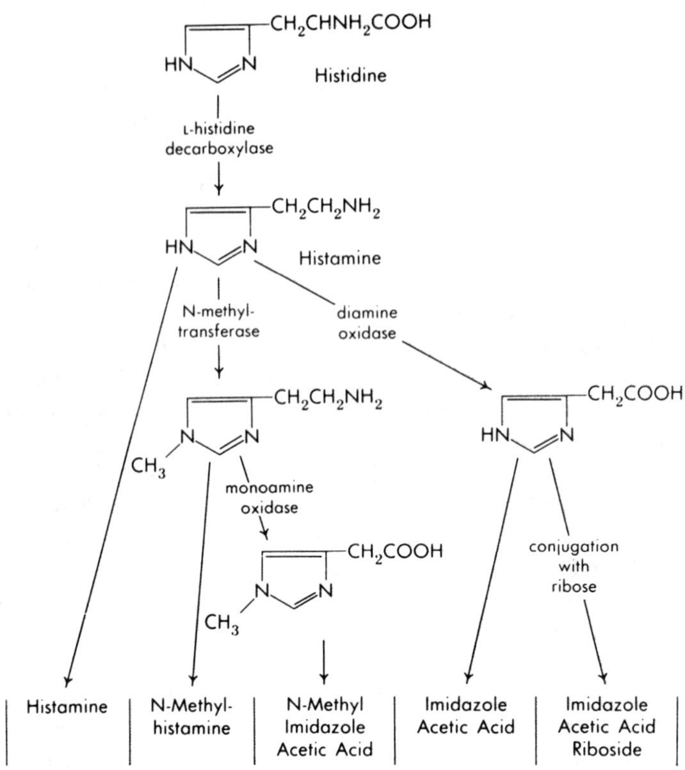

FIGURE 1. Metabolic synthesis of histamine from histidine by the enzymatic action of histidine decarboxylase and its metabolic degradation by methylation and by oxidative deamination. (After Schayer, R. W. and Cooper, J. A. D., *J. Appl. Physiol.*, 9, 481, 1956. With permission.)

by the enzyme diamine oxidase (histaminase), and methylation through the activity of the enzyme histamine methyltransferase. The relating levels of activity of these two enzymes vary from species to species and, in some species, from sex to sex. A small but predictable amount of histamine is excreted unchanged in urine, and this has been found to be a useful means of assessing histamine formation, as its volume corresponds closely to the total amount of histamine formed and released.

The arrangements of mast cells in the areas of blood vessels, and their recruitment to lymph nodes during immunological challenge, suggest that circulatory control, including both level of flow and vascular permeability of physiologically titratable order, likely are involved in levels of histamine release. The fortunately rare, but dramatic, excessive releases were cited earlier. Histamine release involves the need for Ca^{++} ions and metabolic energy with glycolytic substrates.[5] This energy and Ca^{++}-driven release involves the extrusion of histamine-containing granules by exostosis.

II. HISTORICAL BACKGROUND

1907. Histamine, like acetylcholine, was synthesized as a chemical curiosity before any occurrence in nature or any pharmacologic or physiologic activities were known.[6]

1910. Histamine was first encountered as a naturally occurring substance — initially in extracts of ergot. Its presence in ergot extracts later was found to be a product of bacterial contamination.[7,8]

1910. Production of histamine by decarboxylation of the amino acid histidine by bacterial putrefactive action was reported.[9]

1910–1911. Histamine (then referred to by its chemical name of β-imidazolyl-ethylamine) was subjected to intensive pharmacologic investigation by Dale and Laidlaw. These studies identified most of its major actions, including its potent vasodepressor activity and stimulation of multiple types of smooth muscle.[10,11]

1913. The capability of histamine to reproduce many of the responses of hypersensitivity allergic reactions and anaphylactic shock was recognized. It also was shown that hypersensitivity phenomena involved the reaction of antibody with cell-fixed antigen.[12,13]

1920. Histamine stimulation of gastric secretion was discovered.[14,15]

1927. The isolation of histamine from clearly fresh tissues (liver and lung) established it, at last, as a physiological substance and not just a product of putrefaction.[16] Its presence in a wide variety of tissues led to its being termed *histamine* by Dale, from the Greek word for tissue, "histos".

1927. Studies with the "triple response" of Lewis demonstrated that a histamine-like agent was liberated in the skin during anaphylactic-type actions and other types of injury, with both direct vascular and reflex nervous contributions, including an increase in vascular permeability. All of these factors contributed to the accompanying symptoms of local vasodilatation, itching, and edema formation.[17]

1929. Dale considers Lewis' H substance to be histamine, but warns that all of the events of anaphylaxis cannot be explained by histamine alone.[18-20]

1932. Release of histamine during antigen-antibody interactions is established.[21,22]

1937. Antihistamines are discovered. Histamine-blocking activity was discovered by Bovet and Staub[23] in one of a series of amine compounds with a phenolic ether function that was synthesized by Fourneau. Smooth-muscle spasms and anaphylactic shock were controlled. Toxic effects were gradually eliminated. In 1944 Bovet discovered pyrilamine maleata (mepyramine), a major antihistamine still in use.[23-25] Bovet, a Swiss-born and -educated chemist, was head of the therapeutic chemistry laboratory at the Institut Pasteur in Paris from 1937 to 1947. In 1957 he received the Nobel Prize for physiology or medicine for his discoveries of chemotherapeutic agents, which, in addition to antihistamines, included succinylcholine and other valuable curarimimetics.

1956. The development by Schayer of a radioisotopic method for the assay of histamine formation by the activity of the histamine-forming enzyme, histidine decarboxylase (HDC), opened a new era of inquiry into the physiologic role of histamine.[26] Kahlson et al.[27-29] of Lund (Sweden), employing Schayer's HDC assay, carried out a number of important studies that indicated that tissues undergoing rapid growth or regenerative repair had elevated levels of HDC and that inhibition of HDC enzyme activity could reverse or alter many of these processes.[27-29] Schayer speculated that the histamine newly formed by HDC activity, which was not stored and was immediately released (termed "induced histamine" by him), might play an important role in microcirculatory regulation.[30]

1957–1959. Gowans[31,32] of Oxford, U.K., discovers the recirculation of lymphocytes and the postcapillary venule of lymph nodes and lymphoid tissues as the anatomic site of the transit of lymphocytes from blood to lymphoid tissue and lymph.

1961. Majno, Palade, and associates,[33-36] of Harvard and the Rockefeller Institute, identify the site of action of histamine, serotonin, and histamine-type mediators (bradykinin) along the vascular tree to be the postcapillary venule. This site earlier had been identified by Gowans as the critical site in lymphocyte recirculation for the transit of small recirculating lymphocytes from blood to lymph node and lymphoid tissue and finally into lymph and the lymphatic circulation (see above).

1965–1970. An involvement of importance for histamine in the delayed hypersensitivity cellular immune response to the immunologic challenge of tissue (skin) allografting was established (in 1965–1966 by the author and Chang[37]) by the demonstration of an increase in the urinary excretion of histamine — a finding not encountered after skin allografting. These findings led to a series of studies to evaluate the effect of inhibition of the activity of the histamine-forming enzyme HDC on rat skin allograft survival. This inhibition was found (by the author, reported in *Nature* in 1967[38]) to significantly prolong allograft survival. Second-set skin allograft survival was found by the author[39] to be prolonged (at times indefinitely) only when HDC enzyme inhibition was utilized during both first- and second-set allografting. This finding suggested a possible (indeed probable) involvement of host lymphoid tissues in these findings. Direct isotopic measurement of skin allograft and host lymphoid (spleen) tissue HDC activity by the author and Schayer[40,41] revealed this activity to be progressively increased during allograft rejection, a finding not observed following skin isografting. These findings subsequently were extended to other organs and lymphoid tissues by the author and associates.[42,43] In time, an increase in the formation and urinary excretion of histamine was found after skin, kidney, and heart allografting in rats, dogs, and man, and after other immune challenges as well.[37,44-50] A theoretical paper regarding histamine and the cellular immune response (and "normal" physiology) was published by the author[51] in 1971.

1972. A second histamine receptor (H_2) was identified by the synthesis of an antagonist that blocked this receptor, by Black, Duncan, Darant, Ganellin, and Parson in 1972.[52] This was a truly major breakthrough in the understanding of the physiologic role of histamine. The failure of traditional antihistamines to block clearly important physiologic activities of histamine, such as gastric secretion, had been agonizingly perplexing and had stimulated speculation that a second receptor relating to these activities might exist.[53]

1972. The discovery of receptors for histamine on lymphocytes that appeared to function by triggering intracellular elevations in cyclic AMP activity (by Shearer, Melmon, Bourne, and associates[54-56]) thrust histamine more into the realm of a proper and "normal" physiologic molecule of significance. An especially important finding in this area was the 1975 report of Plaut, Lichtenstein, and Henney[57] showing that histamine receptors of the H_2 type on T cells, and in association with an increase in cellular cyclic AMP and a decrease in cellular cytotoxic activity, did not appear on sensitized cells until comparatively late in an immune response. This finding raises the intriguing possibility that histamine, like prostaglandin E_2, might serve as a "double agent" in the cellular immune response, initially triggering, facilitating, and amplifying a stimulatory immune response with its critical increase in blood flow and vascular permeability, and later functioning in a feedback and turn-off capability to modulate, limit, and halt the response it helped initiate.[58]

1975. The discovery of histamine metabolism and function in the rat brain and the exploration of an additional histamine receptor, H_3, in the central nervous system (by Schwartz and associates[59-62] of Paris) added an additional dimension to histamine as a major regulatory and modulatory molecule in the neurovascular normal function and defense armamentarium and enhanced its significance and function as a vasoactive neurotransmitter.

1978. A depression in malignant tumor HDC enzyme activity was observed by the author with Kopplemann and associates,[63,64] in experimental transplanted-tumor-bearing

TABLE 1
Histamine Receptor Types

Histamine receptor	Agonists	Antagonists
H_1	2-Methylhistamine	Mepyramine
	2-Pyridylethylamine	Pyrilamine
	Betahistidine	Chlorpheniramine
		Promethazine
		Terfenadine
H_2	4-Methylhistamine	Cimetidine
	Betazole	Ranitidine
	Dimaprit	Fanotidine
	Impromadine	Nizatidine
H_3	(R)-α-Methylhistamine	Thioperamine
		Impromadine
H_4 (intracellular)		DPPE[a]

[a] *N,N,*-diethyl-2-[4-(phenylmethyl)phenoxy]ethanamine-HCl

animals with rapid and progressive tumor growth — the more rapid the growth, the greater the decrease in HDC activity. This depression was reversed by intralesional injections with bradykinin and with an associated decrease in the rapidity of tumor growth and with injected tumor infiltration with small cells (lymphocytes).

1981. Dy, Lebel, Schneider, Hamburger, and associates,[3] also of Paris, report the discovery of new T cell HDC-inducing lymphokine that also appears to function as a mast-cell "growth factor". This finding further "improves" the reputation of histamine and brings it into the expanding age of lymphokines and cytokines.

III. RECEPTORS

A. H_1 RECEPTOR

The histamine H_1 receptor is the principal receptor for the stimulatory effects of histamine in the nervous, vascular, and immune response systems. The transmembrane signaling mechanism for this receptor is the phosphatidyl inositol (PI) dual second-messenger system. Representative agonists and antagonists for this and the other histamine receptors are shown in Table 1.

This receptor was the target of the first antihistamines, and its activation during anaphylaxis, allergy, and many other immediate hypersensitivity-type pathologic reactions has given histamine its bad name during most of the 20th century. These pathological reactions have resulted largely from the sudden, and at times explosive, release of histamine from storage sites in mast cells and basophils and have given rise to entire disciplines in clinical medicine (allergy, rheumatology, etc.).

The multiple and critical actions of histamine in normal physiology in the nervous, vascular, and immune response systems have been recognized and appreciated only slowly and only within quite recent years.

The importance of the discovery of this receptor and safe and effective antagonists for its blockage, control, and study, was recognized by the awarding of a Nobel Prize to Bovet of France in 1957.

B. H_2 RECEPTOR

The histamine H_2 receptor is the principal receptor for the inhibitory and downregulatory effects of histamine in the nervous, vascular, and immune response systems. Its transmembrane signaling is by the cyclic AMP second-messenger molecule in the nervous, vascular, and immune response systems.

This receptor was the second histamine receptor to be discovered. The failure of H_1-receptor antagonists to affect the long-known role of histamine in stimulating gastric acid secretion led to the discovery of antagonists to this receptor; the virtual elimination of surgery for peptic ulcerative disease of the esophagus, stomach, and duodenum, and to the awarding of a 1989 Nobel Prize to Sir James Black of England.

C. H_3 RECEPTOR

The histamine H_3 receptor, an autoreceptor, has been identified on the widely distributed histaminergic nerve bodies, fibers, and terminals of the central nervous system by the pioneering work of Schwartz and associates in France.[59-62,65-68] The observations that histamine inhibits its own synthesis release in brain slices by a negative feedback process operating at the level of histaminergic nerve endings led to the discovery of this receptor and to the discovery of agonist and antagonist drugs that bind to this receptor, with high specificity and selectivity and with high potency at nanomolar concentrations *in vitro,* by this highly productive Parisian group.[65] (R)-methylhistamine, a chiral agonist, is the most potent of the agonists, and thioperamide, an imidazolyl piperidine, has been found to be the most potent of the antagonists. These drugs have been shown to modify histamine synthesis and/or release from brain slices *in vitro* and in cerebral and peripheral tissues in the living rat. These drugs provide important new tools for the study and management of both physiological and pathological conditions in which histamine may be involved. These and related studies had led to a flurry of biochemical and immunohistochemical studies that have identified and mapped the anatomical location and distribution of this important histaminergic nervous system throughout the invertebrate and vertebrate phylogenetic trees.

D. H_4 RECEPTOR

The histamine H_4 receptor, also called $H_{I\text{-}C}$ (histamine receptor intracellular) is an intracellular histamine receptor that has been found to inhibit human platelet aggregation, by Brandes, Saxena, and associates[69-72] of Winnipeg, Canada. Intracellular antagonism of histamine at this receptor has been found to be produced by DPPE, a diphenylmethane derivative with antiproliferative action. These early findings suggest that histamine, through this receptor and as an intracellular messenger, may serve a critical role in cell function and as a possible growth factor.

IV. NERVOUS SYSTEM

A. GENERAL

The identification and study of histamine as an important neurotransmitter substance in both invertebrates and vertebrates is of comparatively recent vintage. The existence of a histaminergic neuronal system in the central nervous system is well established.[61,73-75] Antibodies against histamine and the histamine-forming enzyme (HDC) have been employed in the delineation of this system. Histamine may be synthesized in histaminergic neurons and stored in synaptosomes.

A human histamine-containing (histaminergic) neuronal system that resembles the serotoninergic and catecholaminergic systems in the primate cerebral cortex has been identified and tracked.[73] The widespread projections of histaminergic neurons has suggested a major role for this system in the maintenance of the homeostasis of the whole brain.[74]

The histaminergic system has been found to play important roles in growth and development, circadian rhythm regulation, thermoregulation, arousal and wakefulness, emotion, memory and learning, feeding and drinking, neuroendocrine regulation, auditory and vestibular regulation, and regulation of the cardiovascular system. A new histamine receptor, H_3, has been found to be located largely on nervous system cells. This receptor has been characterized as an autoreceptor, and specific agonist and antagonist molecules have been synthesized that bind specifically to this receptor, with essentially no binding to H_1 or H_2 histamine receptors.

Histamine-containing synaptosomes have been found in both the cerebral cortex and in the hypothalamus. The greatest concentration of histamine-containing synaptosomes and neuronal bodies is to be found in the hypothalamus.

In the human brain, the hypothalamus is below and immediately adjacent to the thalamus, the site of final subcortical neuronal relays of the major ascending pathways of the auditory, visual, and somatosensory systems. The thalamus plays a key role in sensory integration and in transforming information that reaches the cerebral cortex. Essentially all afferent sensory neuronal pathways that project to the cerebral cortex do so by means of a relay nucleus in the thalamus.

The hypothalamus is a small but mighty component of the central nervous system. The hypothalamus makes up less than 1% of the total volume of the brain, but nonetheless contains such a large number of the neuronal circuits concerned with vital functions and homeostasis that it has been characterized as the "visceral" brain. The neuronal cell bodies of the histaminergic nervous system are concentrated in the posterior hypothalamus, especially in the tuberomammillary nucleus area. Extensive interconnections of the histaminergic and other hypothalamic neurons with the limbic system, especially the hippocampal formation, and the cerebral cortex suggest a role of importance for the histaminergic neuronal system in memory and learning. In this regard, it is of interest that a conditioned-reflex release of histamine through an olfactory sensory neuronal mechanism has been reported and may represent a histamine-associated memory response.[76]

Brain cells other than neuronal may produce and store histamine. They include mast cells, circulating basophils, and vascular endothelial and smooth-muscle cells. Nonetheless, from 60 to 80% of histamine produced and stored in the brains of mammals is produced and stored by neuronal cells.

B. INVERTEBRATES

1. General

Histamine is a major neurotransmitter in the invertebrate nervous system. It has been found to mediate fast synaptic transmission in both arthropods[77-80] and molluscs.[81,82] It has been found directly to gate a chloride channel in lobster olfactory neurons.[83] It has also been found to activate a chloride channel involved in neurotransmission at a photoreceptor synapse in the blowfly.[77] In a single-cell animal, *Entamoeba histolytica,* histamine has been reported to stimulate phagocytic activity.[84]

2. Arthropoda

The evolutionary origin of the arthropods has been placed in the Precambrian period, more than 500 million years ago. The body plan is that of a segmented eucoelomate animal, with a pair of jointed appendages growing out of each segment. The evolutionary development of the arthropods involved specialization of these appendages and a condensation of the somites anteriorly into a unified region containing the organs. These developments were associated with a concentration of the primitive segmental ganglia and sensory organs in the cephalic area of these animals. Coelomic pouches of early embryogenesis later were fused into a common mixocoel. Arthropods have an open blood circulation system that is located in the hemocoel cavity and provides opportunity for a neurovascular "connection".

The circulating substance, hemolymph, contains agglutinable cells and coagulatable plasma, as well as a number of important lectins. In addition, non-self-defense capabilities include nodule formation, encapsulation, and phagocytosis.

a. Chelicerata

The chelicerates include the subgroups Merostomata and Arachnida and such animals as the horseshoe crab, spiders, scorpions, mites, and ticks. The horseshoe crab, *Limulus polyhemus,* has been found to use histamine as a neurotransmitter in visual processing. *Limulus* eyes have been studied extensively as models for basic information relating to visual transduction, light adaptation, retinal integration, and circadian rhythm in retinal function. In a 1991 study, biochemical and immunochemical studies identified histamine as a major afferent neurotransmitter in *Limulus* eyes.[85] The eyes and the visual regions of the brain to which their afferents projected both synthesized and stored histamine. Immunohistochemical studies revealed histamine-like immunoreactivity in the somata of photoreceptors in each of the eyes and in the regions of the brain where these photoreceptor terminated. Immunoreactivity also was intense in the cell bodies and axon terminals of eccentric cells in the lateral eye and in the projections of these cells to the brain. These eccentric cells mediate lateral inhibition, a major integrative process in the lateral eye, which increases contrast sensitivity, forming a fundamental mechanism for processing visual information in the lateral eye.

b. Crustacea

Crustaceans have a large population of circulating free hemocytes that possess a variety of defense-related capabilities and include such animals as crabs, crayfish, lobsters, shrimps, and barnacles.

Application of histamine to the somata of olfactory receptor neurons of the Caribbean spiny lobster (*Panulirus argus*) has been found to suppress both spontaneous and odor-evoking spiking in these neurons, by rapidly activating a chloride (Cl^-) current.[86] The rapid time course of this response, and other evidence that histamine mediates host synaptic transmission in both arthropods and molluscs, suggested that histamine might be acting by directly gating ion channels.

Evidence that lobster olfactory receptor neurons express a chloride channel that is directly gated by histamine was obtained by patch-clamp electrophysiological studies employing these neurons from the Caribbean spiny lobster and the American lobster (*Homarus americanus*) by McClintock and Ache of the University of Florida (St. Augustine).[83] Recordings were obtained from whole-cell somata as well as both outside-out and inside-out patches excised from neuron somata. Somata from these bipolar neurons were obtained by removing both the dendrite and the axon. Evidence from these studies indicates that this channel mediates the suppressive action of histamine on lobster olfactory neurons at the somata histamine-receptor level where histamine receptors are quite numerous. This inhibitory response was blocked by cimetidine, a vertebrate H_2 antagonist, and unaffected by pyrilamine, a vertebrate H_1 antagonist.

Histamine is a known odorant for lobster olfactory receptor neurons and may be synthesized and stored in the regions of the lobster central nervous system where the olfactory receptor-cell axons terminate, but not in the peripheral olfactory organ.[87] The excitatory dendritic responses of these olfactory neurons to histamine and the greater activity in histamine and somata chloride-channel-induced inhibition at neuronal depolarized potentials suggest a regulatory or feedback response to histamine at the somata level of these important sensory neurons.

Additional biochemical and immunohistochemical studies by this St. Augustine group have shown that histamine may be synthesized and stored by olfactory lobe tissue of the spiny lobster.[88]

c. *Insecta*

This arthropod assemblage is the largest monophyletic group of animals that have ever lived on Earth, with the number of individuals at any given moment being as high as 10^{18}.

Maxwell et al., in 1978,[89] employing the moth *Manduca sexta* to study regional CNS neurotransmitter synthesis, demonstrated the ability of the insect nervous system to synthesize histamine. In 1983 Elias and Evans[90] described the distribution of histamine in the nervous system of three insects by biochemical means. The insects were the cockroach (*Periplaneta*), the locust, and the moth. The retina was found to be a remarkably rich source of histamine. Very high levels of histamine also were found in the lamina, as compared with the other optic ganglia and with the cerebral ganglion as a whole. They also reported that the locust nervous system was able to metabolize as well as synthesize histamine. Large amounts of histamine were found to be synthesized by the retina.

Arthropod photoreceptors, particularly in the fly, have been found to respond to light with a rapid, chloride-mediated hyperpolarization. Histamine has been implicated as the neurotransmitter released by a variety of arthropod photoreceptors, and the rapid, chloride-mediated hyperpolarization of fly photoreceptor large monopolar cells responding to light has been shown to be mimicked by the application of histamine.

Hardie[77] of Cambridge University (U.K.), in 1989, reported documentation by electrophysiological patch-clamp studies of some basic properties of the histamine receptors present on large monopolar neuronal cells isolated from blowfly optic lobes. He employed both whole-cell and outside-out and inside-out patch-clamp recording studies. Studies were carried out on large monopolar cell (LMC) bodies enzymatically dissociated from the first visual neuropile, the lamina, of the blowfly. In the whole-cell recording configuration, most (93%) cell bodies studied responded to histamine with large currents, and it was confirmed that most of the current was carried by chloride. The number of channels opened and the increase in current conductance were comparable to the increase in conductance in intact LMC neurons generated by a saturating light stimulus. As with the lobster study cited above, histamine-induced current conductances increased with increasing induced neuronal depolarization. These studies indicated an involvement of importance of a histamine-activated chloride channel in neurotransmission at a photoreceptor synapse.

The first report of mapping the histaminergic nervous system in insects appeared in 1988. At that time, Pirvola et al.[91] described the combined use of biochemical and immunohistochemical measures to map the distribution of histamine and the histaminergic system in the brain and visual system of the cockroach *Blaberus craniifer*. They employed the improved carbodiimide method of tissue fixation for histamine immunohistochemistry.[92] These studies demonstrated widespread distribution of histamine-containing somata and fibers in the brain and particularly in the visual system. The most intense fluorescence immunolabeling was found in the retinal photoreceptors and in the first optic ganglion (the lamina), where the short visual fibers make synaptic connections with the monopolar neurons and where immunofluorescence also was found. Immunoreactive long visual fibers were shown to traverse the first optic ganglion, the lamina, and the outer chiasma to terminate in the distal medulla. Tracts of histamine-immunopositive fibers linked the optic fibers to the protocerebrum and to the central parts of the brain, including the central body, where thick immunoreactive fibers gave rise to intensely fluorescent varicose processes. Immunofluorescence also was found in the antennal lobes. The biochemical measurements of histamine levels correlated well with the histochemical data. The retinal and optic lobes were found to contain remarkably large amounts of histamine. The findings of these studies suggest an important involvement of histamine and the histaminergic nervous system in the visual processing of the cockroach at many different and critical levels.

Nässel et al., in 1990, reported a neurovascular "connection" in flies.[93] By immunohistochemical means, they demonstrated histamine-immunoreactive neurons innervating putative neurohemal areas and central neuropil in the thoracoabdominal ganglia of the flies *Drosophila melanogaster* and *Calliphora vomitoria.* Their findings suggested that histamine may be released from neurohemal terminals in the neural sheath of the thoracoabdominal ganglia. The neural sheath of flies is considered to be a formidable neurohemal organ.

By electron microscopic immunohistochemistry in *Calliphora,* Nässel et al. identified histamine-immunoreactive terminals containing small granular vesicles, outside the "blood-brain barrier" below the acellular basal lamina of the neural sheath. Previous studies by these and other investigators had shown the neurohemal areas of the neural sheath of these ganglia and some of their nerve roots in these flies to contain neurohemal terminals that reacted with antisera to serotonin and several neuropeptides, including substance P, enkephalin, and cholecystokinin.[94-101] The neuronal cellular source of the serotoninergic and peptidergic terminals and their anatomic distribution on the neural sheath differed from the histaminergic terminals. No other neurotransmitter substance was found to be colocalized in histaminergic neurons. The findings of these studies were interpreted as indicating that histamine-immunoreactive neurons are well adapted for coordinating large populations of follower neurons in diverse neurophil regions, including sensory, motor, and integrative centers.

3. Mollusca

Molluscs also are an ancient group of animals whose evolutionary origin has been traced back to the Precambrian period. Their body plan can be divided into a defined head, visceral mass, and foot. The primary body cavity of molluscs has been transformed into a large blood vascular system (hemocoel) in which there is a circulation of hemolymph. In most molluscs the circulatory system is open, without an endothelial cell layer, and with a systemic heart located in the peritoneal cavity. In the most advanced group of molluscs, Cephalopoda (squid, octopus, cuttlefish), beginning closure of the vascular system with capillary formation and blood vessel wall endothelium-type cell layering is found. Multiple cell types are found in the circulating hemolymph in which the principal respiratory protein is hemocyanin. Occasionally, hemoglobin may occur as in Gastropoda.

A histaminergic nervous system has been identified in the much-studied gastropod marine sea slug *Aplysia californica.*[102-104] Endogenous histamine has been shown to be present in single neurons isolated from the central nervous system of *Aplysia.*[103] Histaminergic neuronal activity has been reported to be involved in *Aplysia* in arousal, modulation, and coordination of feeding behavior.[105-109] Histamine in this marine mollusc also has been shown to mediate slow hyperpolarizing (inhibitory) synaptic potentials[110] and to be involved both directly and through a histaminergic neuron in presynaptic inhibition and conductance changes.[111-113] A histaminergic neuron also has been found to modulate synaptic activity of a serotininergic metacerebral cell in *Aplysia.*[114]

C. NONHUMAN VERTEBRATES

Vertebrates constitute the major and overwhelmingly dominant subphylum vertebrata of the phylum chordata, whose principal body plan characteristics include a supporting notochord (a dorsal hollow neural tube, body segmented, and bilaterally symmetrical) and a closed and endothelialized blood vascular system with the heart below the pharynx. In vertebrata, a bony or cartilaginous vertebra replaces the notochord, and the brain and sense organs are considerably enlarged.

1. Agnatha, Amphibia, and Reptilia

Although the presence of histamine in fish and amphibian brains has been known by biochemical assay for a number of years,[115,116] it is only quite recently that the improved carbodiimide tissue fixative method has permitted the tracking of histaminergic nervous systems in the brains of lampreys, frogs, and turtles.[117-119]

Immunohistochemical tracking of the histaminergic nervous system of the agnathan cyclostone lamprey, *Lampetra fluviatilis,* has identified numerous histamine-immunoreactive cell bodies in the dorsal and ventral hypothalamic nuclei and in the adjacent high-density postfundibular commissural nucleus.[118] Immunoreactive neuronal fibers of high density from these cell bodies were traced dorsally to the corpus striatum and the septal nucleus, where they terminated in dense plexuses. A smaller group of cell bodies were seen in the border area between the mesencephalon and thrombencephalon, the midbrain and medulla. Their fibers were sparcely distributed in the mesencephalon. These findings and similarities to corresponding histaminergic neuronal systems in amphibians, reptiles, and mammals suggested that this neuronal system is phylogenetically old and conserved in basic form and structure during evolution. The principal differences between the histaminergic nervous system of cyclostome brains and higher vertebrates was the presence of some cell bodies outside the otherwise, restricted posterior ventral hypothalamus area, and the restriction of hypothalamus-originating fibers to the basal forebrain.

Study of the histaminergic nervous system of the adult amphibian South African clawed frog, *Xenopus laevis,* showed the cell bodies to be located in, and restricted to, a small area of the posterolateral hypothalamus.[119] The number of cell bodies was relatively low as compared with the mammalian brain — approximately 4% of that seen in the rat. A thick process from each cell terminated in the ventricle. The densest fiber networks were in the medial basal forebrain, particularly the medial amygdala and septum.

Study of the brain histaminergic nervous system of adult Reeve's turtles, *Chinemys reevesii,* revealed a concentration of histaminergic neuronal cell bodies in the hypothalamus, where they were found only in the posterior part of the basal hypothalamus with wide but concentrated distribution (Figure 2).[117] The dense concentration of fiber bundles differed from the more diffuse pattern of neuronal fiber distribution encountered in the mammalian central nervous system. The neuronal cell bodies were smaller than observed in mammals. In addition, the number of histaminergic cell bodies was smaller, and these cell bodies did not form clusters as found in mammals. These cell bodies usually bore only one dendrite-like process, whereas mammalian histaminergic neurons have multiple dendrites.

2. Mammalia

Major progress has been made in recent years in tracking the histaminergic nervous system in mammals ranging from from rodent rats and guinea pigs to protoprimate tree shrews at the primate/nonprimate interface. Although an insectivore group, tree shrews are frequently classed as primates because of their remarkably advanced brains and versatile limbs. The rat and the guinea pig provide a unique opportunity to study the histaminergic nervous system of wild and domesticated rodents. The guinea pig, a South American rodent, has been domesticated since pre-Incan times.

The tracking of the histaminergic nervous system, as in the above studies, has been carried out by immunohistochemical identification of the presence and concentration (fluorescence intensity) of histamine and its forming enzyme, histidine decarboxylase (HDC), in nerve bodies, nerve fibers, and nerve terminals in the brains of the animals studied. The carbodiimide tissue fixation method of the Panula et al.[92,120] group has greatly facilitated these recent studies.

Histamine immunoreactivity has been identified in rat brain neuronal cell bodies to be located exclusively in the tuberomammillary (TM) nucleus of the posterior hypothalamus,

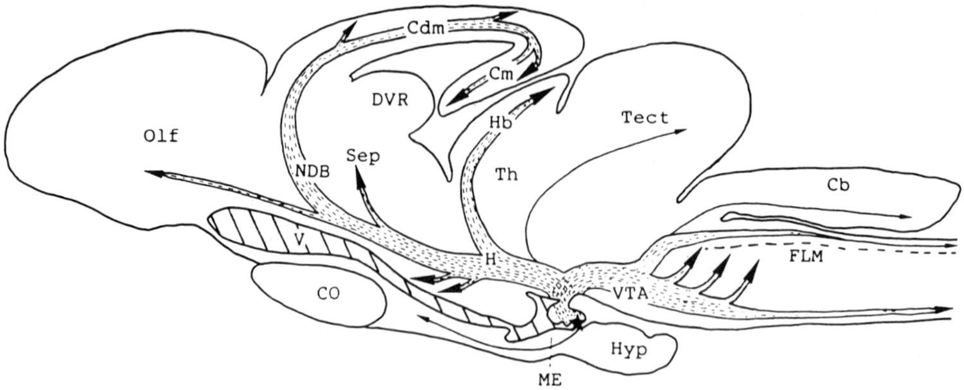

FIGURE 2. Projections of the histaminergic nervous system from the posterior part of the ventral hypothalamus (star) of the turtle brain. The hypothalamus (H), median eminence (ME), thalamus (Th), septum (Sep), ventral tegmental area (VTA), hypophysis (Hyp), and olfactory bulb (Olf) are identified for reference. (From Inagaki, N., et al., *J. Comp. Neurol.*, 297, 132, 1990. With permission.)

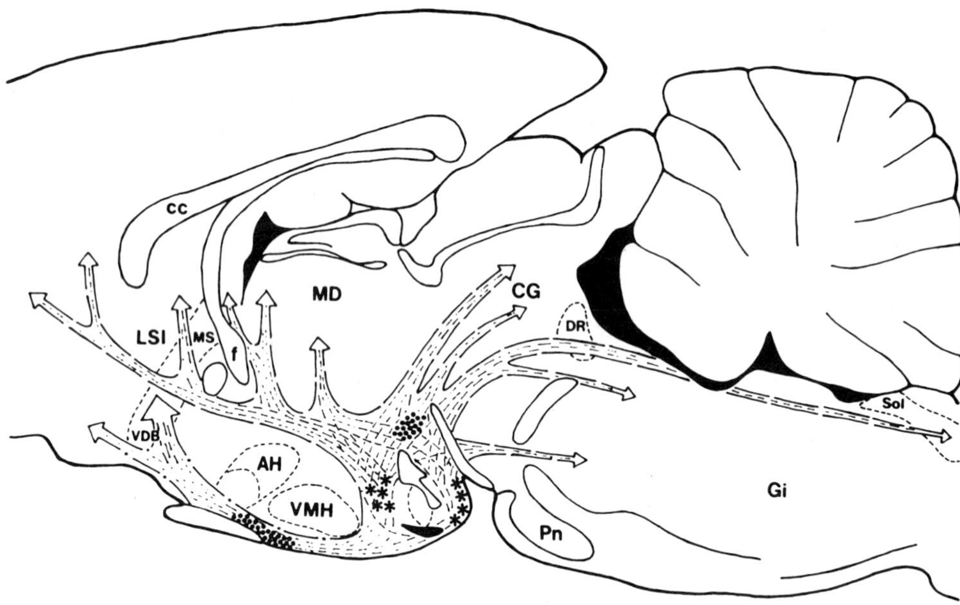

FIGURE 3. Histaminergic nervous system of the rat brain shown in sagittal section. The stars represent histaminergic neuronal cell bodies that were found exclusively in the tuberomammillary nucleus of the hypothalamus. The small round dots represent histaminergic neuronal fiber projections that cross the midline. Dashes identify ipsolateral fibers and projections, with arrows indicating the direction of the projections. For orientation: VMH = ventromedial hypothalamic nucleus, MD = mediodorsal thalamic nucleus, DR = dorsal raphe nucleus (5-HT-ergic cell bodies), and CG = central (periaqueductal) gray (? pain transmission). (From Panula, P., et al., *Neuroscience,* 28, 585, 1989. With permission.)

as in other vertebrates.[120] There was a widespread distribution of histamine-containing nerve fibers and terminals throughout the rat brain (Figure 3). Dense networks of immunoreactive fibers were seen in the medial septum, the nucleus of the diagonal band, and the ventral tegmental area (VTA), and moderate density fibers were seen throughout the cerebral cortex, olfactory bulb and tubercles, bed nucleus of the striaterminalis, amygdala, hippocampus (basal parts), colliculi, substantia nigra, parabrachial nuclei, and nucleus of the solitary tract. These distributions suggest a role of importance for histamine in the functions of the limbic system and in visual and hearing functions as well as in olfactory sensory modulation and responses. The major projection of fibers was rostral (to higher areas of the brain) rather than caudal. The descending pathway appeared to consist of a major dorsal and a minor ventral pathway. Serial frontal, saggital, and horizontal tissue sections indicated that histaminergic nerve fibers crossed the midline at several areas of the brain. These findings with histamine-content tracking were in agreement with those of HDC (histidine decarboxylase)-tracking studies.[121-124] These tract-tracing experiments permitted the identification of the bilateral projection patterns of the histaminergic nervous system.

Autoradiographic studies with radioactive mepyramine as a ligand for histamine H_1 receptors showed a good correlation between a high density of binding sites in the cortex and the dense cortical network of varicose histaminergic fibers and terminals found in the immunohistochemical studies.[120] A high density of H_1-binding sites was found in the rostral hippocampus.

Tracking of the histaminergic nervous system in the guinea pig brain revealed the neuronal cell bodies to be large and limited to the posterior hypothalamus, as in the rat, but more numerous and distributed more widely than in the rat.[125] Most histaminergic fibers followed a ventromedial route to the forebrain. They reached the amygdala ventral to the medial forebrain bundle, the hippocampus (zone CA1) via the subiculum, and the hindbrain structures via the medial central (periaqueductal) gray. Fiber projections of the guinea pig brain were denser to the hippocampus, thalamus, pons medulla, and neurohypophysis than were found in the rat.

On autoradiographic study and as compared with the rat, histamine and immunoreactive-histamine and histamine H_1-binding sites in nerve fibers and terminals in guinea pigs were higher in most areas, especially in such important survival and adjustment areas of the brain as the hippocampus, the thalamus, the periaqueductal grey, and the pons, vestibular, raphe, and inferior olivary nuclei. These divergent findings in wild and long-domesticated rodents raise interesting questions regarding this nervous system, its functions, and the potential for and effects of domestication.

Tracking of the histaminergic nervous system of the brain of the tree shrew (*Tupaia bolangeri*) is of particular interest because this animal, with a remarkably advanced brain and versatile limbs, is kept in the taxonomic gray interface between primates and nonprimates, largely because of bad teeth (primitive dentition). More histaminergic neuronal fibers were present in the tree shrew than were found in the guinea pig.[126] The difference was greatest in the globus pallidus, thalamus, and deep cerebellar nuclei. A second descending histaminergic neuronal pathway was found in the tree shrew, with a dense fiber network around the locus ceruleus, a site of major importance in the catecholaminergic nervous system and location of a high concentration of recently cloned noradrenergic receptor sites. Of particular interest, and perhaps of significance memory- and learning-wise, was the laminar organization of the histaminergic fibers in the hippocampus. The striking feature of the histaminergic fibers in the tree shrew hippocampus was their concentration in distinct cell layers, but evenly in these areas. In the guinea pig these hippocampal histaminergic neuronal fibers were distributed unevenly in the different areas and without apparent laminar organization.

The rat and mouse have been the major experimental animals for other nontracking studies of the histaminergic nervous system in mammals. Cultured mouse neuroblastoma cells (N1E-115) have been used to demonstrate histamine H_1-receptor-mediated cyclic GMP formation by these cells.[127] An involvement of arachidonic acid and lipoxygenase in this cyclic GMP-stimulatory process also has been observed by these investigators.[128] An H_1-receptor-stimulated increase in cyclic GMP also involved the presence of extracellular Ca^{++}

as an absolute requirement. This group of Richelson and associates also found that histamine causes an accumulation of inositol triphosphates in these cells. In view of the above tree shrew locus ceruleus observations and the intercommunications of the histaminergic and catecholaminergic nervous systems, it is of further interest that these N1E-115 murine neuroblastoma cells are an adrenergic clone. An additional observation of interest with respect to a possible neurovascular "connection" is that histamine, through H_1 receptors, also stimulates phosphoinositol (PI) activation in cultured endothelial cells.[129,130]

The 1967 identification of histamine in rat brain synaptosomes and synaptic vesicles[131] and the 1973 subcellular localization of its forming enzyme, HDC (histidine decarboxylase), in rat brain synaptosomes were important stimuli for the in-depth study of the role of brain histamine in normal physiology and the role of the histaminergic nervous system and brain neuronal histamine receptors in these processes.

The stimulation of histamine H_1 receptors generally has been found, in the rodent nervous system, to have an excitatory and depolarizing effect. Endogenous histamine has been reported to excite neurons of the guinea pig superior cervical ganglion *in vitro*.[132]

Histamine-immunoreactive nerve fibers has been found in the spinal cords of rats, guinea pigs, pigs, and tree shrews.[133] These fibers were found in the pain-sensation-related superficial lamina area of the dorsal horn of the spinal cord, with densest fiber locations in lamina X and laminae I and II. These findings suggested to the authors a possible involvement of these histamine-containing (histaminergic) fibers in pain sensation. An additional histaminergic neuron and pain-sensation report appeared in 1990. Histaminergic neurons were found to receive substance P-ergic inputs in the posterior hypothalamus of the rat. Through the use of an immunoelectron microscopic mirror method, Tamiya et al.[134] of Japan, in 1990, were able to demonstrate synaptic connections between histaminergic neurons and substance P afferents in the caudal magnocellular nucleus of the hypothalamus. Substance P-immunoreactive terminals made synaptic contacts with the somata, the somatic spines, and the dendrites of HDC-immunoreactive neurons. These findings suggest that substance P afferents exert monosynaptic influence on the central histaminergic neuronal system. They also bring substance P, neuronal influences, and its heat, pain, and memory potentials more rostral in the mammalian central nervous system and provide an important "connection" with the histaminergic nervous system and with histamine and its multiple involvements in the nervous, vascular, and immune response systems.

In rats, immunosuppressive influences, such as anesthesia or corticosteroid administration, appear to have a downregulatory effect on brain histamine turnover.[135-137] In thipental-anesthetized rats, a marked reduction in histamine turnover in the brain was found and was associated with marked depressions in both the synthesis of histamine by HDC activity and in the metabolic degradation of histamine by methylation. These marked changes occurred without alteration in the endogenous levels either of histamine or of histidine.[137] By analogy with other brain amines,[138,139] it was suggested that histamine turnover depends on nerve impulse flow. Anesthesia with barbiturates has been reported to cause a marked reduction in the turnover of other vasoactive neurotransmitter amines, and it has been speculated that anesthesia produces a rather nonspecific (or at least cause unknown) diminution in neuronal electrical activity.[140,141] This decrease in neuronal activity has been associated with decreased transmitter turnover, since this turnover has been found, in multiple studies, to depend on nerve impulse flow. In a more recent study, administration of sodium pentobarbital to rats produced a 62 to 85% reduction in histamine turnover throughout nine brain regions, with no effects on the steady-state levels of either histamine or its brain metabolite, *tele*-methylhistamine.[135] These findings were regarded by the author as strengthening the hypothesis that neuronal activity is responsible for histamine in the brain. They also supported a role for brain histamine in arousal and sleep mechanisms and in pain perception. In another study histamine was found in pentobarbital-treated rats to cause a dose-related decrease in narcosis duration.[137] Prolonged (6 weeks) treatment of rats with a glucocorticoid was found to decrease the ability of brain synaptosomes to accumulate histidine and to depress the activity of the histamine-catabolizing enzyme, *N*-methyltransferase, by 57%. Levels of brain HDC and histamine were unchanged.[142] These findings suggest a decreased turnover of the neuronal pool of histamine, similar to that observed with anesthesia and anesthetic agents, and raise the question of a neuronal- and histamine-turnover-related influence in the immunosuppressive actions of corticosteroids and anesthesia.

Histamine stimulation of rat neuronal depolarization (excitation) has been reported to occur in medial vestibular nucleus neurons, as recorded intracellularly *in vitro*.[143] The depolarization was concentration dependent and of rapid onset, with a slow recovery (return to resting potential) and with depolarization lasting up to 15 min (? LTP, long-term potentiation). Electrophysiological studies of the effects of histamine on neurons of the hypothalamic arcuate nucleus demonstrated an excitatory response, with a histamine-increased spontaneous neuronal firing rate in 63% of the arcuate neuronal cells tested.[144] The response was dose dependent and was stable during synaptic blockade. It was found to be mediated through histamine H_1 receptors. A significant and most interesting correlation was found in the responses to histamine and substance P, in that all units excited by substance P also were excited by histamine — an additional histamine and substance P "connection" in neuronal excitation.

An excitatory effect of histamine on electroencephalograms (EEGs) of the thalamus and cortex of rats has been observed by Tasaka et al.[145] of Okayama, Japan. The effect was blocked by antagonists of the histamine H_1 receptor. An excitatory effect of histamine in the memory and learning-critical CA1 area of the rat hippocampus by inhibition of an inhibitory response also has been reported by Springfield and Geller.[146] The histamine-inhibited inhibitory response was paired-pulse induced. The paired-pulse inhibitory response was achieved by presenting two electrical stimuli at various intervals to the Schaffer commissural pathway to hippocampal CA1 pyramidal neurons. Pulses of 0.2 msec were presented at 1/60 Hz. When two closely spaced stimuli are presented, the amplitude of the second evoked (test) spike has been found to differ from the amplitude of the first (conditioning) spike. The ratio of these amplitudes has been found to be a function of the interstimulus interval. Shorter interstimulus intervals have been observed to produce "paired-pulse" inhibition, while longer interstimulus intervals produce facilitation,[147] and facilitation in the hippocampal CA1 area is the essence of learning and memory.

Memory-molecule-involved neuronal tracts and their anatomical projections in the peripheral and central nervous systems are of particular interest in probing neurovascular components of the immune response and their collective and integrated contributions to neurovascular immunology, and at the molecular level, to central nervous system and immunological memory. As indicated above, a memory-molecule (histamine and substance P) "connection" has been tracked rostrally from the superficial laminae of the dorsal horn of the spinal cord to the posterior hypothalamus. In this regard, it is of particular interest that histamine-containing neuronal cell bodies in the posterior hypothalamus are the source of a discrete histaminergic nerve pathway that travels along the median forebrain bundle to innervate the hippocampus and the cerebral cortex.[148,149]

Histamine stimulatory effects *in vivo* have been observed in arousal actions, including increased grooming and exploratory behavior, in rats and cats.[136,150] These arousal actions appear to be mediated through histamine H_1 receptors. Histamine-H_1-receptor stimulation has been found to suppress food intake in the rat, while H_1-receptor antagonists elicit feeding but not drinking.[151,152] In addition, evidence has been provided to suggest that diurnal fluctuation in feeding behavior of rats may reflect circadian variations of hypothalamic neuronal histamine levels.

The discovery of neuronal H_3 autoreceptors and their study in-depth by Schwartz and associates[59-62,65-68,153-156] in Paris has been a major breakthrough in the study of the role of histamine in normal physiology and in the identification of histamine as an important neurotransmitter and neuromodulator within the nervous system. Most of these studies were carried out in rats. Histamine-induced depression of histamine formation and release by K^+-depolarized histaminergic neurons led to the discovery of a K^+-unique autoreceptor and, subsequently, to important and specific agonists and antagonists to this H_3 receptor. The 1972 finding of Atack and Carlsson[157] that endogenous histamine was released from brain slices depolarized by potassium was an important early finding and stimulus. Potassium (K^+)-induced depolarization of histaminergic neurons from rat hypothalamic brain slices was reported (in 1975 by the Schwartz group[156]) to increase the role of histamine formation from HDC enzyme activity in hypothalamic cell bodies and nerve terminals. Exogenous histamine was found by this group to depress both the level of histamine release and histamine formation by depolarized histaminergic neurons and neuronal terminals throughout the vast areas of histaminergic neuronal distribution.

Histamine-H_3-receptor ligands have been found to have important effects on various biochemical indices of histaminergic neuronal activity in the rat brain. The specific H_3-receptor antagonist, thioperamide, has enhanced markedly the turnover of histamine in the rat cerebral cortex, with an associated long-lasting decrease in synaptosomal histamine and an increase in the methylation metabolite, N^T-methylhistamine. (R)d-methylhistamine, a chiral agonist of histamine, had the opposite and equally long-lasting effect of reducing markedly cortical histamine synthesis, with a decrease in *tele*-methyl-histamine levels.[65] Thioperamide antagonizes the actions of both histamine and (R)d-methylhistamine at H_3 receptors in both rat and human brain. These selective agonists and antagonists hold considerable promise for both the study and modulation of brain histamine metabolism and its consequences in experimental animals as well as in man.

D. HUMANS

The existence of a histaminergic neuronal population in the human brain was first demonstrated in 1990 by Panala et al.[158] of Finland. The location of the histaminergic system cell bodies corresponded to the tuberomammillary nucleus, and these cells represented a prominent part of all cells in the basal hypothalamus. Some cell bodies were found outside this area, in the lateral tuberal nucleus, in the supermammillary nucleus, and along the third ventricle. These histaminergic system cell bodies in humans, thus, were found in more widespread areas in the human hypothalamus than in the hypothalami of rodents and other nonhuman mammals and vertebrates.

The organization of the human histamine-containing neuronal system was observed to be quite similar to the catecholaminergic and serotoninergic systems in the primate cerebral cortex, with nerve fibers originating from neuronal cell bodies located in distant small nuclei (e.g., the locus ceruleus for the catecholaminergic system) and distributed to all cortical laminae. Histamine is stored in nerve fibers in the human cerebral cortex in sites that are compatible with its function as a neurotransmitter, and the majority of these fibers are unmyelinated. Both thick and thin histaminergic fibers were seen. A well-organized network of varicose fibers was found throughout the frontal and temporal cortices — areas of importance for sensory information receipt and analysis and for associated memory and learning. Numerous histaminergic neuronal types were seen throughout the frontal cortex, with an accumulation of histamine-containing fibers on the pial surface of lamina I. In this lamina, many long fibers were seen to run parallel to the surface of the brain. Some histaminergic fibers surrounded cortical cell bodies in a basket-like manner.

In another recent (1990) report by Kendall and Firth,[159] histamine has been found to increase inositol phospholipid hydrolysis in human cortical slices obtained at surgery. This effect was shown to be inhibited specifically by adenosine and one of its analogues. This inhibition was antagonized by theophylline.

The findings of these still quite limited and quite recent studies of the human histaminergic nervous system suggest a role of major importance for this system and an expansion of the phylogenetic tree for the histaminergic nervous system.

The large concentration of histaminergic system cell bodies in the tuberomammillary nucleus of the human basal hypothalamus and the identified degeneration of tuberomammillary neurons in Alzheimer's disease[160] suggest a possible involvement of histamine and the histaminergic neuronal system in the debilitating process in which severe memory loss is such a major component and in which the substance P "connection" may have some real significance.

V. VASCULAR SYSTEM

A. POSTCAPILLARY VENULES AND ENDOTHELIUM

In 1961 Majno, Palade, and Schoefl[33,34] identified the postcapillary venule as the site along the vascular tree of the action in promoting vascular permeability of histamine and serotonin (as well as bradykinin in other studies). Electron microscopic studies were employed in these investigations. Subsequent studies explored the role of other histamine-like vasoactive neurotransmitter substances and the mechanisms of this mediator-induced leakage. The mediator-stimulated endothelial cell separation first demonstrated by them in 1961 was later attributed to endothelial cell contraction. These mediators, in addition to histamine and serotonin, also have been found to include bradykinin.[35,36] The endothelial cell contraction hypothesis has been based on the observations that endothelial cells contain contractile elements and that these contractile elements are involved with the cellular cytoskeleton.[161-168] These contractile elements have been found to modulate the dimensions of the junctional gaps of the postcapillary venules. It is of further interest in this regard that the contractile elements in venular endothelial cells are located at cell junctions.

In 1982 Heltianu et al.[169] reported that histamine receptors are most dense in venular endothelial cells and that these receptors are strategically localized near endothelial cell junctions and in close proximity to the contractile elements. In 1986 Horan et al.[170] of Houston presented evidence, based on intravital light microscopy and fluorimetric measurements of plasma suffusate tracer, that prolonged histamine suffusions *in vivo* produced transient increases in macromolecular efflux, which were dependent on the formation of discrete venular macromolecular leakage sites. They considered these findings to be "proof" of the Majno-Palade hypothesis.

It is also of interest that histamine has been shown to stimulate phosphatidyl inositol (PI) hydrolysis in cultured venous endothelial cells.[129,171-173] A calcium-dependent link in histamine-induced increases in postcapillary venule permeability has been described.[174]

There is considerable evidence to suggest that Ca^{++} modulates endothelial cell metabolic and morphologic responses to histamine and to other mediators of inflammation. The finding of endothelial cell cytoskeleton actin and myosin fragments that are immunochemically indistinguishable from those of smooth muscle[167] has suggested that endothelial cells and smooth-muscle cells may share a similar contractile mechanism.[175] These filaments in the perijunctional filament web have been localized to regions of interendothelial contact.[176,177] It is, accordingly, of particular interest that histamine receptors have been found to be concentrated in the plasmalemma overlying regions of interendothelial contact.

Rotrosen and Gallin,[175] in 1986, reported a histamine-induced cytoskeletal alteration to underlie enhanced cultured human venous endothelial monolayer permeability to albumin. In this study endothelial cells were stained with a fluorescent marker of F-actin and extracted with methanol to yield a quantitative index of actin polymerization. The F-actin content of endothelial cells exposed to histamine was found to be decreased significantly. A prior study had documented a decrease in F-actin content of endothelial cells that were exposed to agents known to increase vascular permeability.[178] These findings were associated, in the Rotrosen-Gallin studies, with an increase in endothelial cell cytosolic calcium, as monitored with a fluorescent Ca^{++} indicator, quin 2. Both the calcium and the endothelial cell actin and cell shape changes were mediated through a histamine-H_1-receptor activation and mimicked by calcium ionophores.

In electrophysiologic studies, endothelial cell depolarization (stimulation) was found to be associated with the passage of an inward calcium current.[179] In these studies, histamine-stimulated cells remained depolarized as long as the agonist and extracellular Ca^{++} were present.

B. SKIN MICROVASCULATURE AND MICROCIRCULATION *IN VIVO*

The microcirculation of the skin has long been a major site and resource for the investigation of the role of histamine and related vasoactive neurotransmitter molecules in inflammation and immediate hypersensitivity. The well-known and classical "triple response" of wheal and flare of Sir Thomas Lewis[17] was the first demonstration of this histamine-induced phenomenon.

The local administration of histamine to subcutaneous tissues elicits the classical vascular responses found and known to be associated with acute inflammation. These include an increase in blood flow and blood volume (hyperemia/flare), as well as an increase in microvascular permeability and edema formation (wheal). Associated phenomena of pain and itching may also be produced by histamine.[180] The permeability triggering effect of histamine is quite prompt, but also quite limited in time, as cited above.[170]

The involvement of histamine in the microvascular and microcirculatory alterations found in acute inflammation in the popular subcutaneous administration model also is associated and interrelated with actions of other inflammogenic vasoactive neurotransmitter and modulator molecules, such as bradykinin, substance P, and PGE_2. A comediator role for histamine in a two-mediator hypothesis was proposed by Williams in 1977.[181,182] His proposal was that the initiation and continuation of an acute inflammatory process was likely based on an interaction of mediators and a synergism between vasodilators, such as the prostaglandins, and microvascular-permeability-inducing agents such as histamine. The histamine-induced vascular permeability increase appears to be through a histamine H_1 receptor.[170,183,184]

Histamine-induced venular endothelial cell hydrolysis of inositol phosphate (PI-transmembrane-signaling cycle) and arachidonic acid release and regulation, as cited above, and in association with Ca^{++} mobilization, and associated alteration in endothelial cell cytoarchitecture, would appear to provide an attractive and reasonable molecular basis for these processes. The histamine-stimulated release of substance P and involvement of substance P in an axon reflex "neurogenic" inflammatory response help complete this picture of the dynamic early events involved in the establishment of an acute inflammatory response.

The prompt, controlled, appropriate, and physiologic local-site release of histamine from stored sites in mast cells, basophils, neuronal cells, endothelial cells, and vascular smooth-muscle cells would appear to be the initial and triggering event in the initiation of an acute inflammatory response. The contribution of histamine in the transformation from an acute to a chronic or delayed-type hypersensitivity reaction involves an accelerated intracellular formation of histamine by accelerated HDC enzyme activity. The extremely important 1981 observations of Mertin and Stackpoole[185] regarding the obligatory involvement of PGE_2 in the early minutes of the initiation of an inflammatory process in the critical early period events of a cellular immune response are similar to those of histamine. Recent observations by the author and associates[186] of the prompt appearance of arachidonic acid metabolites, PGE_2 and thromboxane B_2 (TXB_2), in popliteal node efferent lymph of sheep within minutes of nodal drainage-area immunization with killed *Salmonella muenchen* bacteria are in additional support of an extremely early involvement of arachidonic acid metabolites, including PGE_2, in the initiation of an inflammatory process. An early period release of histamine from perivascular mast cells is also likely involved.

Some recent reports add information of interest. A 1988 report by Harper et al.[187] of Dundee, U.K., described the use of laser Doppler velocimetry to study the effects of injected histamine and PGE_2 on the microcirculation of human skin. The laser Doppler flowmeter is of particular value, in contrast to other methods, because of its ability to detect dynamic changes in blood-flow velocity in skin microvasculature. Four concentrations of histamine and PGE_2 were used. Both mediators produced prompt and marked acceleration of cutaneous blood flow. The PGE_2-induced response differed from that of histamine in that the times to maximum response and its duration and the period of decay were longer with PGE_2. The PGE_2-induced cutaneous erythema was localized and restricted to the injection site. That of histamine extended out from the injection site in a flare, with central edema around the injection site. This difference is due to the actions of histamine in producing both an increase in microvascular blood flow and an increase in microvascular permeability, whereas PGE_2 produces only an increase in microvascular blood flow.

In a 1990 study (by Pentland et al.[188] of Washington University in St. Louis) of human skin explants exposed to ultraviolet (UV) injury, a synergism of histamine and PGE_2 in the associated UV erythema encountered *in vivo* was observed. These studies demonstrated that endogenous histamine, by a UV-triggered enhanced histamine sensitivity, stimulates PGE_2 synthesis and release in human skin after UV injury by potentiation of a histamine-induced prostaglandin response.

In another recent (1989) report concerning the sequence of events in substance P-mediated plasma extravasation in rat skin (blister model of inflammation), an early histamine-independent plasma-extravasation response was identified and followed by a late histamine-dependent response.[189] In other studies, both histamine and prostaglandins have been found to be implicated in the cutaneous inflammatory response to substance P.

C. EARLY VASCULAR BLOOD-FLOW RESPONSE TO IMMUNIZATION AND ORGAN ALLOGRAFTING

Rapid sequential scintiscan studies by the author and associates,[47] following canine renal allografting and employing intravenous ^{131}I-hippurate administration and with probes over the heart (central venous area), the allograft, and the recipient urinary bladder, have permitted a correlation of renal allograft blood flow with the release of histamine into allograft urine and into the recipient urinary bladder. Two phases of elevations in the urinary excretion of histamine were observed. The initial and early elevation in histamine release into allograft urine was associated with a marked and sharp increase in allograft blood flow, as determined by scintiscan data (see "Immune Response System; Transplantation" section of this chapter). Isotope-labeled microsphere studies, by Hay and Hobbs[190] of Toronto, following drainage-area immunization have identified a prompt and marked increase in the blood flow to rabbit popliteal lymph nodes.

D. LARGER VESSELS: ARTERIES, ARTERIOLES, VEINS, AND LYMPHATICS

The mechanisms of action of histamine on arteries vary with different species and with the anatomical location and sizes of the arteries investigated. Cerebral artery responses to histamine have been studied in monkeys,[191] cats,[192,193] dogs,[194] rabbits,[195] and man.[196] In a 1990 report by Toda of Japan,[196] histamine *in vitro* predominantly caused contraction of autopsy-obtained human middle cerebral arteries. On the other hand, third and fifth branches of these middle cerebral arteries responded exclusively with a dose-dependent relaxation. The removal of the endothelium suppressed the relaxation or converted it into a contraction. It appeared from agonist and antagonist studies that the histamine-induced relaxation was mediated by endothelial H_1 receptors whose activation led to the release of endothelium-derived relaxing factor (EDRF/nitric oxide). The contraction appeared to be associated with activation of smooth-muscle H_1 receptors. Monkey cerebral arteries (strips) and receptor subtypes also were investigated in this report and found to differ quantitatively in response from responses observed above with human cerebral arteries. The distal relaxations were similar to those found in dog middle cerebral arteries.[194]

The study of the isolated basilar artery from the rat cerebral circulation, by Chang et al.[197] of the University of Lund (Sweden), revealed histamine to be a potent vasodilator agent acting through a combination of H_1 and H_2 receptors. A 1990 report from Richmond, VA, of an *in vivo* study of histamine action on mouse cerebral arteries described competing endothelium-dependent and endothelium-independent dilatation. This study by Rosenblum et al.[198] involved the use of television microscopy and an image-splitting technique to monitor changes produced in diameter by histamine applied locally to mouse pial arterioles *in vivo*. A high dose of histamine constricted the arterioles, whereas lower doses relaxed them. Constriction was blocked and dilatation occurred when selective *in situ* microvascular endothelial damage was produced by light from a helium-neon laser in the presence of intravascular Evans blue. From these findings it was concluded that the constriction was endothelium dependent and resulted from the production and release of an endothelium-derived constricting factor (EDCF). Constriction also was blocked by histamine-H_1-receptor antagonists and by treatment of the arterioles with indomethacin, an arachidonic acid cyclooxygenase inhibitor. Histamine-H_1-receptor blockade unmasked a dilating effect of a histamine dose too low to affect the diameter of arterioles not treated with the histamine-H_1-receptor blockers. A histamine-H_2-receptor blocker interfered with the relaxation produced by low-dose histamine. These findings were interpreted by the authors as indicating for mouse pial arterioles that histamine can interact with histamine H_1 receptors on the endothelium to release an endothelium-derived constricting factor that causes constriction of the underlying vascular smooth muscle while simultaneously interacting with histamine H_2 receptors in the smooth muscle which mediates relaxation of the vessel.

A 1988 study (by Watanabe et al.[199] of Japan) of the action of histamine on lymphatic vessels was carried out by employing isolated bovine mesenteric lymphatics and investigating histamine actions on the spontaneous contractions of these vessels by recording isometric tensions. Histamine at lower concentrations produced a dose-dependent deceleration of the rhythm of spontaneous contractions. Higher concentrations of histamine caused a dose-related acceleration of the rhythm in association with a slight elevation of basal tone. Studies with histamine-receptor agonists and antagonists suggest that both H_1 and H_2 receptors are located on the plasma membrane of smooth-muscle cells in bovine mesenteric lymphatics and that excitations of H_1 and H_2 receptors, respectively, produce an acceleration and a deceleration of the rhythm of spontaneous contractions in lymphatic smooth muscles.

VI. IMMUNE RESPONSE SYSTEM

A. CELLS

1. Lymphocytes

When human cytolytic T lymphocytes (CTL) were generated in the presence of histamine, histamine was found (in 1989 by Kahn et al.[200] of Stanford University) to suppress the generation of MHC Class I-specific CTL and, simultaneously, to stimulate and increase the generation of MHC Class II-specific cytolysis. As histamine was not present in the cytolytic assays, it was determined that histamine acted at the level of CTL generation. The lytic ability of effector cells was the same as that of controls when histamine was added to the lytic assay with CTL grown without histamine. The histamine-induced suppression of MHC Class I-specific cytolysis was blocked by continuous culture with the histamine-H_2-receptor antagonist. It was not blocked by continuous culture with the histamine-H_1-receptor antagonist pyrilamine. Continuous culture of human CTL with histamine had no effect either on T-cell proliferation or on the expression of the cell surface molecules tested. Addition of phytohemagglutinin (PHA) to the cytolytic assay reversed the suppression of MHC Class I-specific cytolysis, a finding that indicated that the cytolytic machinery was intact. These data suggested that the cytolytic suppression was mediated by the histamine H_2 receptor, with a cytolytic-stimulation enhancement mediated by the histamine H_1 receptor. Others have observed histamine-H_1-receptor stimulation of lymphocyte cytolytic activity.[201-203]

Histamine-H_1-receptor activation also has been found to have an immunostimulatory effect on the proliferation of lymphocytes[204] and on the production of a histamine-induced migration-inhibitory factor by lymphocytes.

A subset of human rosetted lymphocytes containing high concentrations of histamine has been isolated. The function of this subset, however, remains obscure.[205]

When the spleen cells of normal and mast-cell-deficient mice were cultured, their HDC activity was found (in 1989 by Aoi and associates[206] of Nagoya and Osaka, Japan) to be increased, with increases in the histamine concentration in the cells and in the medium. The removal of adherent cells reduced the control HDC activity as well as that increased by mitogen stimulation. Purified lymphocytes were found to have an increase in HDC activity in response to Con A, but not to LPS. Neither mitogen had any effect on B lymphocytes. Anti-T-lymphocyte antibodies and complement blocked the HDC activity responses. Con A-induced T-cell increases in HDC activity were found to be stimulated either by the addition of conditioned culture medium from peritoneal adherent cells activated by the LPS mitogen, or by the addition of paraformaldehyde-fixed peritoneal adherent cells. It also was found to be increased by the addition of recombinant interleukin-1 (rIL-1). These observations strongly suggest that histamine may be synthesized by T lymphocytes through HDC enzyme activity and that this synthesis may be enhanced by soluble factors released from macrophages. Medium conditioned with mouse spleen cells, in this study, also was found to contain a large amount of HDC enzyme activity. Others had reported that spleen cells produce histamine in response to T-cell mitogen.[207,208] These observations concerning an enhanced T-lymphocyte production of histamine and HDC enzyme activity in the T-lymphocyte macrophage system provide additional support for an additional immunoregulatory and stimulatory component of histamine and HDC in the cellular immune response. It is of further interest and that their stimulatory responses may be inhibited by glucocorticoids.[209]

Infected mice challenged with the antigen of the infecting organism have been found (in 1990 by Lebel, Schneider, Dy, and associates[210] of Paris) to have marked increases *in vivo* in HDC enzyme activity in both spleen and bone marrow cells. The HDC-stimulating activity was quite rapid (marked at 1 h, with a maximum of 4 h). This finding, along with those of others, suggests a role of importance for histamine and its synthesis by HDC in hematopoietic organs in the stimulation of multipotent stem-cell cycling.[208,211-218] The Dy and Hamburger group[219] of Paris earlier had reported the identification in *in vitro* studies of a histamine synthesis and HDC-enhancing lymphokine associated with tissue allograft rejection *in vivo* and cited even earlier studies by the author and associates[40-43] concerning increases in tissue and organ allograft levels of HDC during allograft rejection *in vivo* in rats.

Enhanced histamine synthesis by HDC enzyme activity in murine hematopoietic cells by granulocyte-macrophage colony-stimulating factor (GM-CSF) has been found to be further enhanced and synergized by IL-1 or recombinant tumor necrosis factor alpha (rTNFα), with these synergistic effects being mediated by prostaglandin E_2 (PGE_2).[220] Increased HDC activity produced by interleukin-3 (IL-3) was not found to be synergized in this manner.

On the histamine-associated stimulatory side, histamine has been reported (in 1990 by Hellstrand and Hermodsson[221] of Goteborg, Sweden) to augment natural killer-cell cytotoxicity of human $CD16^+$ non-T lymphocytes *in vitro* against NK-sensitive target cells. This augmented activity was by a prostaglandin- and IL-1-independent accessory function of monocytes. This appeared to involve a cell-to-cell mediated interaction. The findings of this important study suggest that histamine may have an important role in nonspecific tumor defense by regulating and enhancing an immunostimulatory interplay between monocytes and natural killer (NK) cells.[222]

2. Mast Cells, Basophils, and Platelets

Mast cells, basophils, and platelets are being recognized increasingly as major players in the neurovascular immune response. These cells and structures are major sites for the synthesis, storage, and release of major neurovascular and immune-response-system mediator and modulator molecules, including histamine, serotonin, and others. Recent (1989) observations (by Plaut et al.[223] of Baltimore and Bethesda, MD) of considerable importance demonstrate that these cells, such as mast cells, may also synthesize and secrete a number of lymphokines, such as IL-3, IL-4, IL-5, and IL-6, which are derived principally from T cells.

To add to the current excitement, the origins, growth stimuli, and fate of mast cells and basophils remain shrouded in mystery. Major studies of these "mysteries" (by the Guy-Grand, Dy, and Hamburger group[224-230] of Paris and the Ginsburg group[231-239] of Israel, as well as by Schrader et al.[240,241]) have explored and presented some intriguing possibilities. Antigen-stimulated lymph node cells (likely lymphocytes), in the presence of stimulating antigen and of fibroblasts and histamine, appear to be key factors in the "genesis" of mast cells, and likely also of basophils (circulating "mast cells"). The Guy-Grand group report of 1978 presented evidence that the lymphocyte Thy-1 antigen was present on the cell membranes of precursors of murine mucosal mast cells in helminth-infected animals.[225]

In the 1989 report by Plaut et al.,[223] murine mast cell lines were obtained by culturing fetal liver from mice with IL-3, with propagation of homogeneous cell populations. These fetal-liver-derived mast cells possessed properties similar to intestinal mucosa, but differed from connective-tissue-type (peritoneal) mast cells. Stimulation of these cells in culture by calcium ionophores or by cross-linkage of high-affinity plasma membrane Fc/epsilon (IgE receptors) receptors (Fc RI) resulted in the stimulation of increased messenger RNA (mRNA) levels and the secretion of a series of lymphokines that generally are associated with a subset of murine T cells. These stimulus-resulting lymphokines included IL-3 (a mast cell growth factor), IL-4 (an IgE switch and production factor), IL-5 (an eosinophil differentiation and activation factor), and IL-6 (a factor controlling immunoglobulin secretion) — a rather formidable output. The authors suggest that mast cells may play an important, and heretofore unrecognized, role by amplifying strongly signals normally mediated by T cells and by ensuring access to a wider variety of target cells.

Recent reports by Brandes, Saxena, and associates[69-72,242] have greatly expanded our knowledge and understanding of normal platelet and mast cell function and physiology and of the role of histamine and a newly identified histamine receptor, an intracellular receptor (H_4), in these processes. This group has shown that histamine newly formed and synthesized in platelets by HDC enzyme activity acts as a second messenger in human platelets, through a novel intracellular histamine receptor (H_4) of micromolar affinity. Histamine acts as a second messenger in mediating platelet aggregation in response to a variety of aggregation agonists, including phorbol ester (PMA), collagen, platelet-activating factor (PAF), and thrombin. Both platelet aggregation and the accompanying secretion of platelet dense granules containing serotonin have been shown by them to be blocked by inhibitors of HDC and by an antagonist of the H_4 receptor. Alpha-fluoromethylhistidine (α-FMH) was employed as the HDC inhibitor and DPPE, N_1N-diethyl-2-[4-(phenylmethyl)phenoxy] ethanamine. HCl was used as an antagonist for the novel histamine H_4 receptor.

Further studies by this group employing both a reversible HDC inhibitor, alpha-methyl histidine (α-MH) and the irreversible HDC inhibitor (α-FMH), found that HDC inhibition with both inhibitors suppressed stimulated platelet aggregation and a parallel increase in elevation of platelet histamine content. Platelet permeabilization with saponin revealed most (75%) of the histamine increase to be cytoplasmic. Both reversible and irreversible HDC inhibitors accumulate within cells by active transport through the histidine transporter. Saponin is a detergent that has some preferential ability to solubilize plasma membrane, as compared to intracellular membranes.

DPPE, an arylalkamine, also is a potent antiestrogen-binding-site ligand and an antiproliferative agent. DPPE is a selective ligand for the microsomal antiestrogen binding site (AEBS); Ki = 65 *M* for [^{3}H] tamoxifen and antagonizes the binding of [^{3}H] histamine (Ki = 1 μM) at an AEBS-associated site distinct from H_1-, H_2-, and H_3-receptor sites. It has been suggested that arylalkylamines may act to block a "final common pathway" for mast cell histamine release.[243]

It has further been identified by the above investigators that the inhibitory effects of these agents (α-FMH and DPPE) are reversed in saponin-permeabilized, but not intact, platelets by micromolar amounts of histamine and that in phorbol ester (PMA)-stimulated platelets, DPPE inhibits and histamine reverses key platelet-aggregation-associated ultrastructural changes. These ultrastructural changes include the swelling and fusion of platelet granular membranes with each other and with those of the surface-connected canalicular system. These findings indicate an important, indeed, major, role for histamine in the processes essential for platelet granule secretion.

As platelets are nonnucleated, the rapid (30 sec) rise in histamine levels in response to stimulation suggests a nongenomic activation of platelet HDC. This is further confirmed by the failure of the inhibition of protein synthesis (cyclohexamide) or of RNA synthesis (actinomycin D) to block HDC-enzyme-activity-induced increases in stimulated platelet histamine levels and synthesis.

Collagen, a more physiologic mediator of platelet function, promoted platelet aggregation and stimulated histamine formation comparable to the other platelet aggregation stimulatory agents. As with the other agents, DPPE inhibited collagen-stimulated platelet aggregation, and this inhibition, as with the other agents, was reversed by added histamine.

The Brandes, Saxena, et al. group has explored the platelet intracellular location of histamine in stimulated cells and its location with respect to that of serotonin within the same stimulated platelet cells. These authors emphasize that while histamine is present in the platelets of many species, certain species, such as rabbits and pigs, have rather high levels of platelet histamine that are largely and primarily located in dense granules. They reported in 1989 that resting human platelets contain relatively small amounts of histamine; they may synthesize additional histamine by elevated agonist-stimulated HDC enzyme activity. In addition to phorbol esters, the agonists studies included collagen, thrombin, PAF, and a combination of phorbol ester and U46619 (a thromboxane mimetic). Their studies showed that serotonin in stimulated human platelets is largely (90%) stored in platelet-dense granules, whereas histamine is largely (80%+) cytoplasmic in location.

The addition of histamine to rabbit and human platelets recently (1990) has been reported to stimulate arachidonic acid release from these platelets by way of a histamine H_1 receptor and to enhance phorbol ester (PMA)-stimulated serotonin secretion. This histamine-induced activation of phospholipase A_2 and the resulting arachidonic acid release by histamine appeared to be mediated by pertussis-toxin-sensitive GTP-binding G proteins with particular sensitivity to beta-gamma ($\beta\gamma$) subunits of G proteins.

Mast cell granule secretion, in most respects, is similar to that encountered in platelets. As with platelets, most cells of humans vary with those of rats, mice, and other rodents (rabbits) with respect to histamine location, synthesis, and storage within the cells and the relative amounts of histamine and serotonin. While murine mast cells contain relatively larger amounts of serotonin, histamine is the predominant vasoactive neurotransmitter substance found in human mast cells. As with platelet granule secretion, mast cell granule secretion is both energy and calcium dependent and requires granule and cell membrane fusion. In both mast cells and platelets, granule secretion may be triggered by intracellular histamine production by increased HDC enzyme activity stimulators, such as phorbol esters and Con A.

The histamine-H_4-receptor antagonist DPPE has been found to block Con A-stimulated rat mast cell histamine release. The most potent inhibition of histamine release was from nonpermeable Con A-stimulated rat mast cells. The effects on serotonin secretion were not monitored in these studies. It has been reported that nonstimulated mast cells, especially the less mature forms, secrete significant extracellular amounts of histamine. This released histamine may be of importance in normal physiology, as contrasted with the massive releases in anaphylaxis and allergy. From an evolutionary standpoint, anaphylaxis and its causes are of comparatively recent origin and largely primate dependent.

Proton nuclear magnetic resonance studies of rat mast cell histamine have provided evidence suggesting that histamine in rat mast cells is relatively labile, with rapid changes between bound histamine and pools of free histamine in water compartments confined in the granule matrix.[244] These studies were carried out on peritoneal-cavity-derived rat mast cells.

Mammalian mast cells vary within species (rats and guinea pigs to man) and anatomical locations within a single species with respect to their relative contents of histamine and serotonin and with respect to the uptake and synthesis patterns of these potent amines and the effects of agonist and antagonist drugs on these processes.[245]

The rat peritoneal mast cell is obtained and concentrated with relative ease and has now been the mammalian mast cell of choice for most mast cell studies. Rat peritoneal mast cells contain more histamine than serotonin, whereas rat mast cells from lung and mesenteric tissues contain amounts of serotonin in excess of histamine. Rat mast cells, regardless of location, appear to have different release and uptake patterns of histamine release and a preference for serotonin uptake. Release agonists, such as compound 48/80, Con A, and the calcium ionophore A23187, are more effective liberators of histamine than serotonin from rat mast cells.[243,245]

Rat peritoneal mast cells have been found (by Beaven, WoldeMussie, and associates[246,247] of the N.I.H., Bethesda, MD) to have subpopulations of cells at different stages of maturity, ranging from immature (early postmitotic) to mature. These mast cells may be separated by size, with the smaller cells being immature and the larger ones being mature. The smaller "immature" cells contain no granules and store little histamine, but have high levels of histamine synthetic activity from HDC action. These cells release histamine spontaneously and at a steady state. Their histamine release levels are unaffected by either compound 48/80 or the calcium ionophore A23187.

The critical steps in histamine synthesis and storage in rat mast cells have been observed to include the uptake of histidine (histidine transporter system), the decarboxylation of histidine (HDC), and the translocation of newly formed histamine into storage granules. The uptake/transporter system has a high affinity for histidine.[246,248,249] HDC enzyme activity in rat mast cells has been shown by immunohistochemistry studies to be present in discrete patches on the inner surface of the plasma membrane of these cells.[250]

The use of the HDC inhibitor α-fluoromethylhistidine in the above studies was found to produce a more rapid decline of intracellular histamine levels in small, ''immature'' rat mast cells than in the larger, ''mature'', and granule-containing cells.[246] These findings indicated that histamine turnover is more rapid in the small, ''immature'' mast cells. Although termed ''immature'', these small mast cells appear to be doing the bulk of the important histamine-forming and controlled-release work, while the large, bloated, and granule-containing ''mature'' cells seem to function largely as histamine sponges whose most important work is to provide a livelihood for allergists and rheumatologists. The possible controlled and physiological release of histamine from large and ''mature'' mast cell granules in vascular, lymphoid, and nonlymphoid tissue homeostasis, and appropriate and controlled defense, would appear to be worthy of considerable study.

Studies of the rates of histamine synthesis during the growth cycle of transformed rat basophil cells (2H3 cells) has revealed findings similar to those encountered in rat mast cells, with histamine synthesis maturity greatest in the small and ''immature'' cells. Interleukin-1 beta ($IL\text{-}1_\beta$) has been reported to cause histamine release from human basophils.[251] A considerably less potent effect on histamine release from these human basophils was produced by recombinant tumor necrosis factor (rTNF).

Human bone marrow, cord blood, and peripheral blood have been found to contain progenitor cells that, during culture, mature into histamine-containing basophilic cells. Different proportions of progenitor cells were found from these three sites, and all with different abilities to produce histamine and become granulated basophilic cells.[252] The basophilic cells from the peripheral blood contained the highest levels of histamine. Mature basophilic cells from peripheral blood were highly unstable in culture. In a human bone marrow suspension-culture system, recombinant human IL-3 (rhIL-3) has been reported to support the formation of blasts and eosinophils and to induce the differentiation of basophils and the formation of intracellular histamine during a 14-d culture period. In these studies by Mayer et al.[253] of Vienna, Austria, recombinant human IL-3 has also been observed to have similar stimulating effects on simian bone marrow precursors. In this regard, it is of interest that IL-3 has been found to share a 96% sequence homology with gibbon IL-3, but only a 29% sequence homology with murine IL-3. In an *in vivo* study employing nonhuman primates (Rhesus monkeys), rhIL-3 has been reported to stimulate basophil differentiation and histamine production and to prime GM-CSF-responsive progenitors.

3. Macrophages

In 1985, Carlsson, Dohlsten, and Sjogren[254] of the University of Lund (Sweden) reported evidence that histamine modulates the production of interferon gamma and interleukin-2 by mitogen-activated human mononuclear blood cells by means of a histamine-H_2-receptor activation mechanism. The human mononuclear blood cells were obtained from peripheral blood, and the mitogen employed by these investigators was staphylococcus enterotoxin A (SEA). The inhibitory effects could be reversed by the histamine-H_2-receptor antagonist cimetidine, but not by the histamine-H_1-receptor antagonist diphenhydramine.

In 1986, Endo et al.[255] of Sendai, Japan, published in *Cellular Immunology* an important *in vivo* study in which mice injected with culture supernatants of a murine macrophage line (P338D1 cells) experienced a rapid increase in the levels of the histamine-forming HDC enzyme activity in their spleens, livers, and lungs. Attempts to purify these supernatants HDC-enzyme-inducing factors produced low-purity preparations whose *in vivo* activity suggested the HDC-inducing factor(s) to be interleukin-1, which produced a similar HDC-inducing activity in spleens, livers, and lungs. Injections with these semipurified fractions into mice *in vivo* increased tissue HDC levels within 2 h, with peak elevations at 2 to 4 h postinjection. The increases in tissue HDC levels were accompanied by marked (up to tenfold) elevations in tissue histamine levels.

A 1987 study by Parker and associates[256] at Washington University (St. Louis) described histamine (and dopamine)-induced augmentation of interferon-gamma-induced phagocytosis by murine macrophages *in vitro*. This was identified as a histamine-H_1-receptor-mediated immunostimulator activity, as it was blocked by the histamine-H_1-receptor antagonists pyrilamine and chlorpheniramine and was not blocked by the histamine-H_2-receptor antagonist cimetidine. These actions were described as being in a dose-related manner and at physiological concentrations.

Falus and Meretey[257] of Budapest, Hungary, reported in *Molecular Immunology* (1988) an interesting and significant study that histamine, by way of histamine-H_1-receptor stimulation, considerably upregulated the crucial complement C_3 gene expression and C_3 secretion in murine macrophages. A downregulation was produced by histamine-H_2-receptor activation. The authors interpreted this important data as indicating a distinct signal-processing mechanism acting on histamine H_1 and histamine H_2 receptors, which resulted in opposite effects on the transcriptions of the C_3 gene in macrophages.

A 1990 publication from Nagoya, Japan, in *Immunology* (by Okamoto et al.[258]) described a histamine-induced enhancement of interleukin-1 release from a cultured macrophage cell line (P388D1) to be largely by way of histamine-H_1-receptor activation.

B. TRANSPLANTATION

An increase in the urinary excretion of histamine has been found (by the author and associates[37,46-50]) to occur following skin, kidney, and heart allografting in rats, dogs, and man and following other immune response challenges as well. The urinary excretion of histamine has been found to be a useful and reliable indicator of the amount of histamine formed and released by cells and tissues. Although the great bulk of formed and released histamine is metabolized by methylation and oxidative deamination, a small but reliably predictable percent is excreted unchanged in mammalian urine.[1]

In the rat skin allografting and autografting studies, adult female rats of the Sprague-Dawley and Long-Evans strains were used. Female rats were used because of a higher urinary excretion of histamine than is found in the male.[27-29] The elevation in urinary histamine excretion following skin allografting generally started shortly after allografting and peaked at the time of the observed allograft rejection.[37] The pattern of urinary histamine excretion elevation after skin allografting of a Sprague-Dawley rat is shown in Figure 4, and that after skin allografting of a Long-Evans rat is shown in Figure 5. The elevation in urinary histamine excretion elevation following allografting was substantially greater after skin allografting than following autografting, and despite two major incisions and wound healings in the autografted rats (Figure 6). In second-set skin allografting, the accelerated rejection was associated with a prompt and marked increase in the urinary excretion of histamine (Figure 7).

In renal allografting, the allografts are immediately vascularized after completion of the arterial and venous anastomoses, whereas in skin allografting revascularization of the allograft requires a period of time for recipient vascular ingrowth. Following renal allografting between mongrel dogs, two sharp elevations in the recipient urinary excretion of histamine were found (Figure 8).[47] The first occurs immediately postallografting, and the second at 6 to 7 d postallografting. The first sharp elevation was found to be associated time-wise with a sharp increase in renal allograft blood flow as monitored by daily postallografting

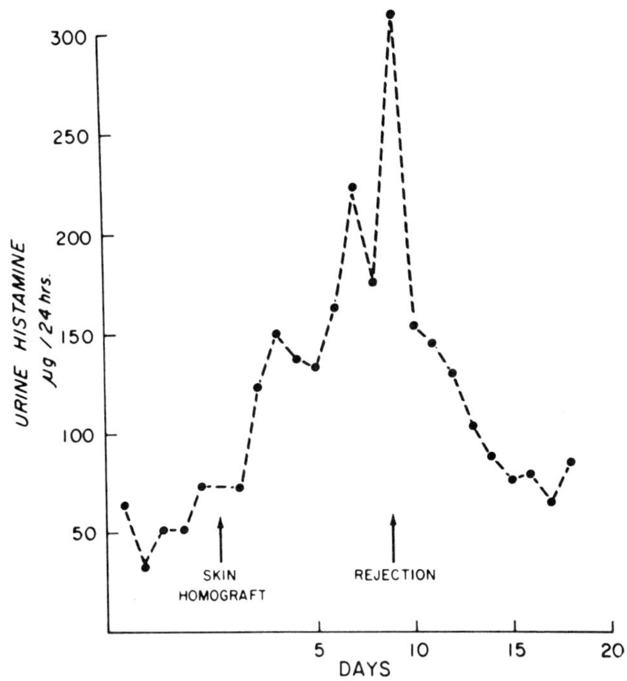

FIGURE 4. Urinary excretion of histamine by a Sprague-Dawley rat during rejection of a skin homograft/allograft. (From Moore, T. C. and Chang, J. K., *Ann. Surg.*, 167, 232, 1968. With permission.)

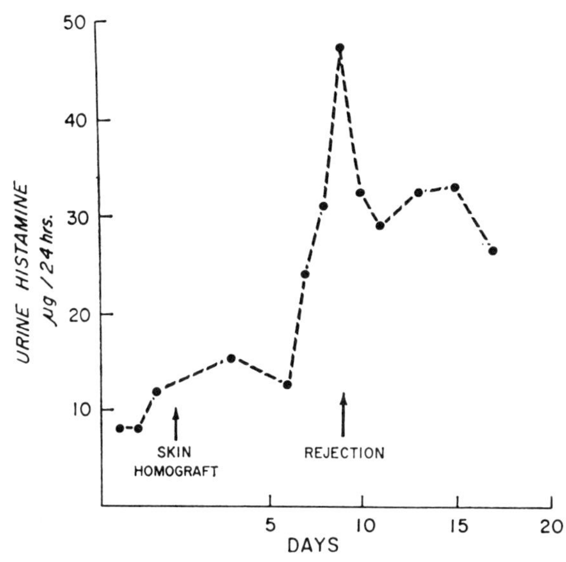

FIGURE 5. Urinary excretion of histamine by a Long-Evans rat before and after skin homografting/allografting. (From Moore, T. C. and Chang, J. K., *Ann. Surg.*, 167, 232, 1968. With permission.)

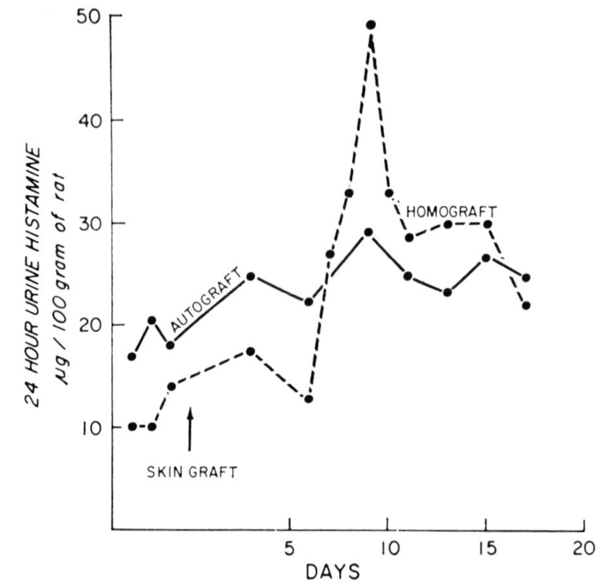

FIGURE 6. Pattern of urinary histamine excretion following skin autografting and homografting/allografting in rats of comparable size and age. (From Moore, T. C. and Chang, J. K., *Ann. Surg.*, 167, 232, 1968. With permission.)

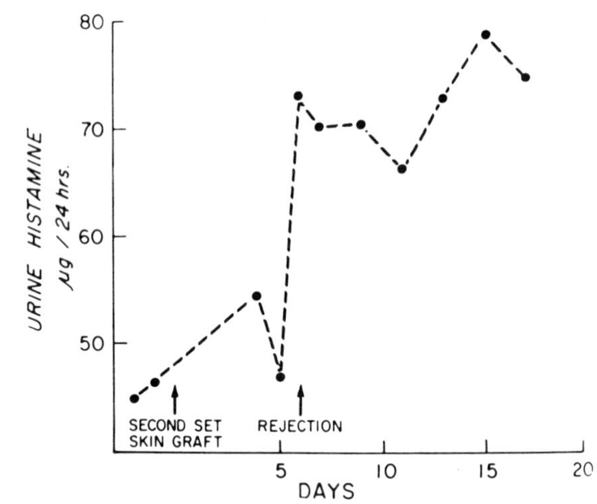

FIGURE 7. Urinary excretion of histamine before and after second-set skin allografting of a Sprague-Dawley rat. The second-set grafting was carried out 19 d after the initial allografting and from the same donor. (From Moore, T. C. and Chang, J. K., *Ann. Surg.*, 167, 232, 1968. With permission.)

scintiscans. Rapid sequential renal scans with a time sequence scintillation camera were obtained as serial 2-min scans of the renal allograft and the recipient urinary bladder for 30 min after intravenous injection of radioisotopic (^{131}I) hippuran. The renal blood flow, as a renal-blood-flow index, was derived by comparison of this data with the slope of an external heart-isotope clearance curve. The end of the second sharp elevation in the urinary excretion of histamine occurred at the time of a sharp and progressive rise in recipient dog serum creatinine levels. Both renal allograft blood flows and urinary histamine excretion levels

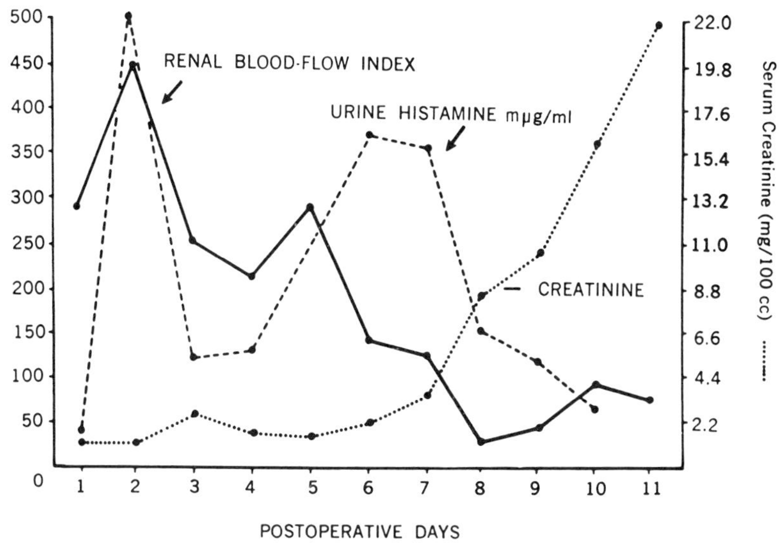

FIGURE 8. Correlation of renal allograft blood-flow indices (isotope scintiscan determined) with urinary histamine excretion during renal allograft rejection in a bilaterally nephrectomized dog. (From Moore, T. C., et al., *Arch. Surg.*, 101, 45, 1970. With permission.)

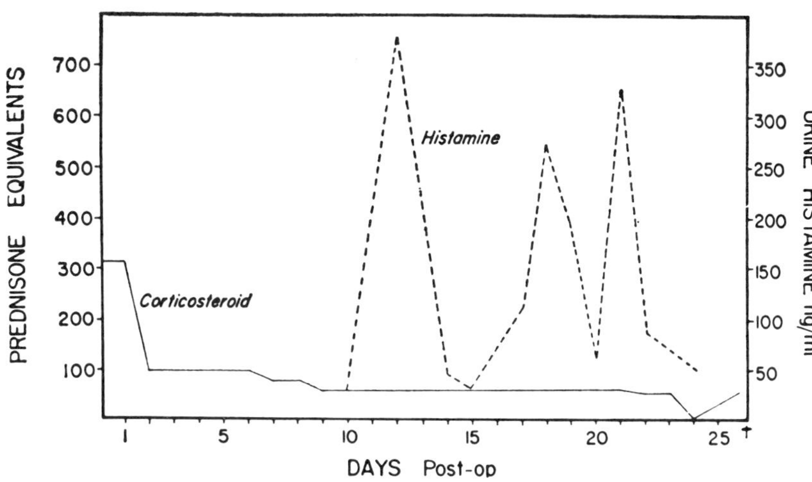

FIGURE 9. Repeated high spikes in the urinary excretion of histamine are associated with low levels of corticosteroid administration following canine heart allografting (orthotopic) with immunosuppression. (From Moore, T. C., et al., *Ann. Surg.*, 174, 51, 1971. With permission.)

fell sharply with progressing allograft rejection. The initial and early event rise in the urinary excretion of histamine suggested an immediate poststimulation release of histamine from blood-vessel-associated mast cells, as commented upon earlier in this chapter (''Vascular System'').

Heart allografting in dogs with azathioprine and corticosteroid immunosuppression also was reported (by the author and associates[48]) to be associated with marked elevations in the urinary excretion of histamine. Low doses of corticosteroids were accompanied by repeated high elevations in the urinary excretion of histamine (Figure 9). A sudden and sharp lowering of corticosteroid dosage was associated with a sharp elevation in the levels of urinary

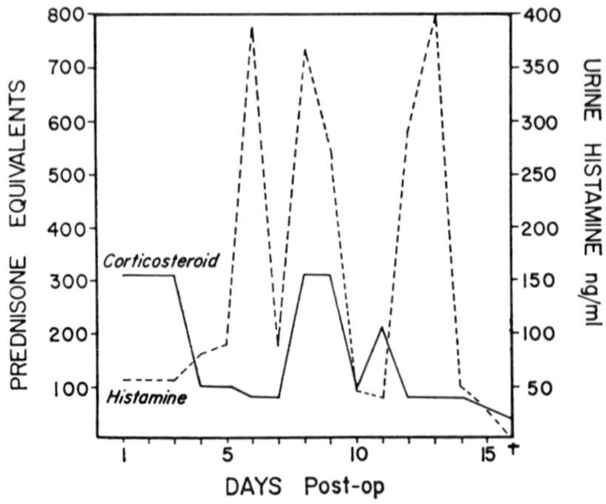

FIGURE 10. Marked depression of urinary histamine excretion in association with high levels of corticosteroid administration following canine orthotopic heart transplantation with immunosuppression. A marked outpouring of histamine in the urine occurred following a sharp reduction in the level of corticosteroid administration. (From Moore, T. C., et al., *Ann. Surg.*, 174, 51, 1971. With permission.)

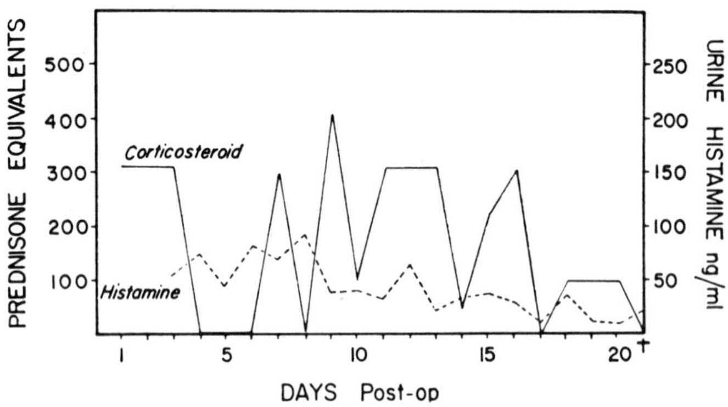

FIGURE 11. Comparatively low levels of urinary excretion of histamine following heart allografting of a dog on immunosuppression that received repeated high doses of corticosteroids. (From Moore, T. C., et al., *Ann. Surg.*, 174, 51, 1971. With permission.)

histamine excretion by the recipient animal (Figure 10). Frequently repeated high doses of corticosteroid were found to maintain low levels of urinary histamine excretion and to prevent rejection episodes (Figure 11). Very high levels of corticosteroid administration to a human heart transplant recipient produced marked depressions of urinary histamine excretion and also prevented the appearance of rejection episodes during this period (Figure 12).

An increase in the urinary excretion of histamine has been observed (by the author and associates[259]) to occur following sheep red blood cell (SRBC) immunization of rats. Kahlson et al.[27-29] of the University of Lund (Sweden) observed in the 1960s that accelerated histamine formation occurs in animals undergoing hypersensitivity reactions, which are protected from anaphylactic death by antihistamines.

The development by Schayer[26] of a radioisotopic method of measuring the level of intracellular histamine from stimulated HDC enzyme activity (in 1956) was a major breakthrough in the study of the role of histamine and its intracellular formation in normal physiological reactions. In 1969, the author with Schayer[40] reported a marked increase in

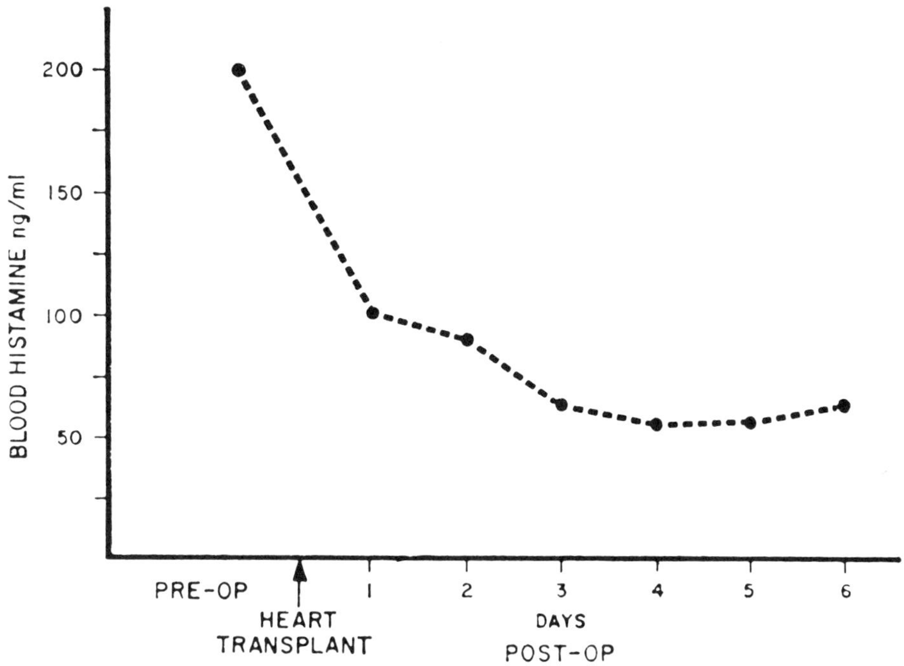

FIGURE 12. Progressive depressions in the levels of blood histamine excretion in the early days following human heart transplantation when the patient was receiving very high levels of corticosteroids. (From Moore, T. C., et al., *Ann. Surg.*, 174, 51, 1971. With permission.)

rat skin allograft HDC enzyme activity during allograft rejection, with no elevation in skin autograft HDC activity during the period of the marked increase in allograft HDC activity (Figure 13). Skin allografting in the rat was also found by the author and Schayer,[41] to be associated with an increase in recipient central lymphoid tissue (spleen) HDC activity (Figures 14 and 15).

Renal allografting studies between inbred Brown Norway and Lewis strains of rats also have shown (by the author and associates[44,45]) to result in marked elevations in HDC activity in recipient (Lewis) rat allografts, thymuses, and spleens (Figure 16). The peak elevation in recipient splenic HDC activity occurred at 4 d postallografting. These elevations have not been encountered in renal-isografted animals (Figure 17).

In these studies of renal allografting between inbred rat strains, of particular interest was the difference between tissues with extensive and with limited exposure to host immunologically directed defense-reaction capabilities. Renal capsules that are immediately revascularized and exposed over a large area per gram of tissue to non-self-''hostile'' host tissue experienced an immediate and marked elevation in capsular HDC enzyme activity that is not shared by isografted renal capsules (Figure 18). Allograft renal parenchyma that is exposed to only a very small bulk of circulating host blood nucleated cells, largely lymphocytes, does not experience this marked and immediate rise in tissue HDC activity. Its increase in HDC does not begin until day 5 after allografting and rises progressively thereafter as rejection progresses (Figure 19).

The increase in allograft size during rejection across this histocompatibility barrier is associated with an increased in allograft infiltration of lymphocytes, especially large lymphocytes. This response was observed (by the author with Granger[260]) to be associated with an increase in the amount of lymphotoxin (LT) produced per cell in this infiltrate in the heart heterotopic allograft model (Figure 20). When organ allografts across this same

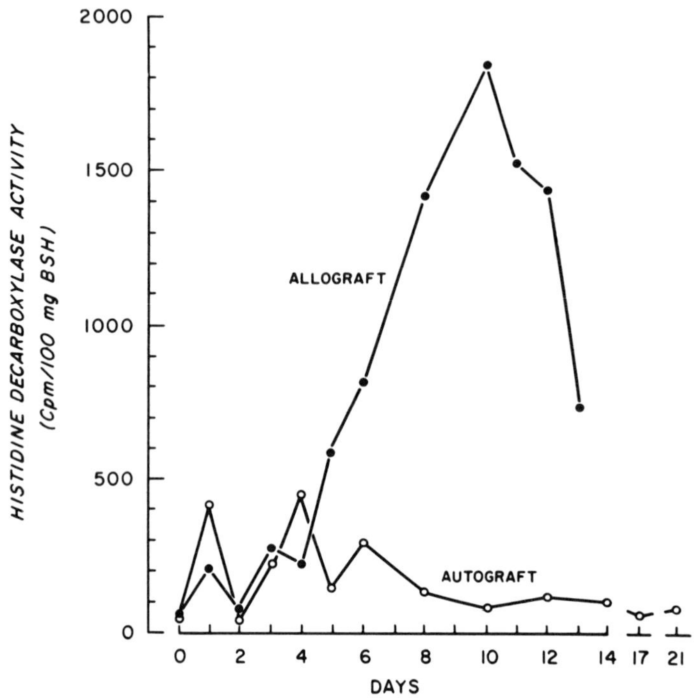

FIGURE 13. Alterations in histidine decarboxylase histamine-forming enzyme activity of autografted and allografted full-thickness rat skin at various intervals following grafting. Allografted skin was from inbred Lewis to inbred PA rats. No immunosuppression was used. (From Moore, T. C. and Schayer, R. W., *Transplantation*, 7, 99, 1969. With permission.)

histocompatibility barrier (BN to Lewis) are compared, the time and degree of increases in allograft HDC and allograft-infiltrating lymphocyte LT activity are quite similar, almost identical.

Heart allografting (BN to Lewis rats) also was found, by the author and associates,[45] to be accompanied by an increase in recipient thymic HDC activity, starting at day 4 after allografting (Figure 21). This, interestingly, was the day of the peak increase in recipient splenic HDC after renal allografting. The early post-heart-allografting decrease in thymic HDC activity likely was operative-stress induced. It was accompanied by a sharp decrease in thymic weight during this period (Figure 22). This also was likely operative-stress induced. The placement of the heart allografts was heterotopic into the recipient abdominal aorta, with no operative invasion of the recipient thoracic cavity.

Rat immune system central challenge by sheep red blood cell (SRBC) immunization also was observed, by the author and associates,[261] to be accompanied by an increase in immunized rat splenic HDC activity. As with organ-allografted rats, the peak increase in splenic HDC activity occurred at 4 d post-immune system challenge. The peak increase in antibody formation, by the Jerne blood agar plaque assay, occurred in these animals on the following day, day 5 postimmunization (Figure 23).

Inhibition of recipient HDC enzyme activity in rats was reported to prolong significantly the survival of skin allografts. This report in the August 19, 1967 issue of *Nature*[38] was the author's first report concerning the role of vasoactive neurotransmitter substances, such as histamine, in cellular-immunity- and delayed-hypersensitivity-type immune responses. Second-set rat skin survival was found by the author[39] to be prolonged only when HDC activity inhibition was employed during both first- and second-set allografting. Inhibition of rat HDC

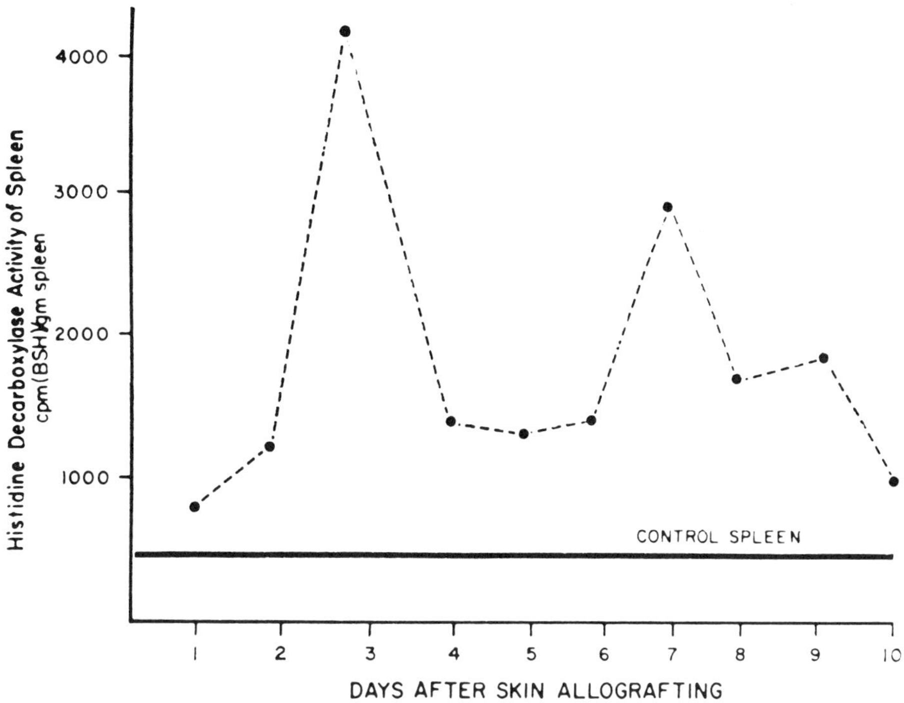

FIGURE 14. Elevations in rat splenic histidine decarboxylase histamine-forming enzyme activity above normal, control values at various intervals following skin allografting between inbred rat strains (Lewis to PA). In multiple studies the initial peak in splenic histidine decarboxylase enzyme activity occurred at 3 and 4 d following allografting. No immunosuppression was used. (From Moore, T. C. and Schayer, R. W., *Ann. Surg.*, 171, 609, 1970. With permission.)

enzyme activity also has been observed, by the author and Lawrence,[262] to produce a marked reduction in antibody formation following immunization with SRBC and with *Salmonella* flagellar antigen. This inhibition also blocked the occurrence of an immunization-associated increase in spleen sizes of the immunized animals.

C. CANCER

Rapid malignant tumor growth in hamsters was reported, in 1978 by the author and associates,[63] to be accompanied by marked depressions in tumor-tissue levels of HDC enzyme activity, with the greatest depressions in HDC found in those tumors with the most rapid growth (Figures 24 and 25). Repeated intralesional tumor injections with bradykinin slowed and halted tumor growth in a manner that was both dose and frequency of injection related[64,263] and was accompanied by marked tumor infiltration with small round cells (lymphocytes). Small round cells rarely were found in noninjected tumors. Tumors with the greatest growth inhibition by bradykinin injections also were found to have the greatest elevations in tumor HDC levels above those of noninjected tumors.[63] Bradykinin-injection-induced increases in malignant tumor HDC levels had been reported earlier, in 1974 and 1975.[64,264]

The spleens of hamsters with rapidly growing and large tumors were found, by the author and associates,[265] to have markedly and significantly ($p = 0.00005$) depressed HDC enzyme levels in comparison to the spleens of normal and nontumor-bearing animals (Figure 26). In addition, hamsters that received the rejected tumor cells had significantly higher splenic HDC levels ($p = 0.000025$) than animals that received tumor cells and developed tumors (Figure 27).

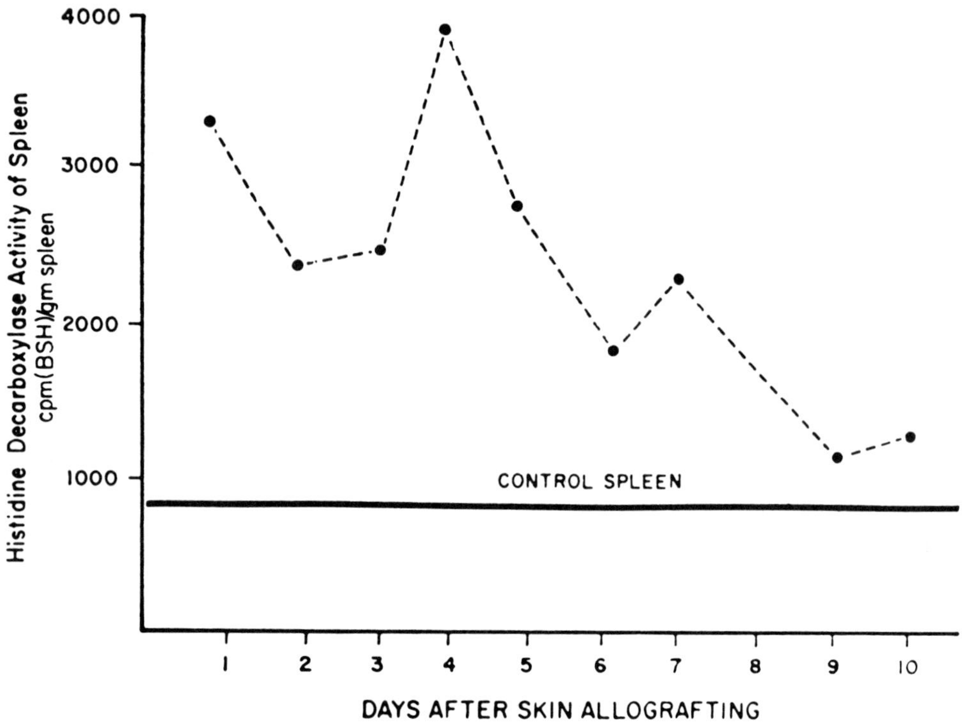

FIGURE 15. A second series of splenic histidine decarboxylase enzyme activity following skin allografting between inbred rat strains (Lewis to PA) and above control, normal levels for recipient PA rats. (From Moore, T. C. and Schayer, R. W., *Ann. Surg.*, 171, 609, 1970. With permission.)

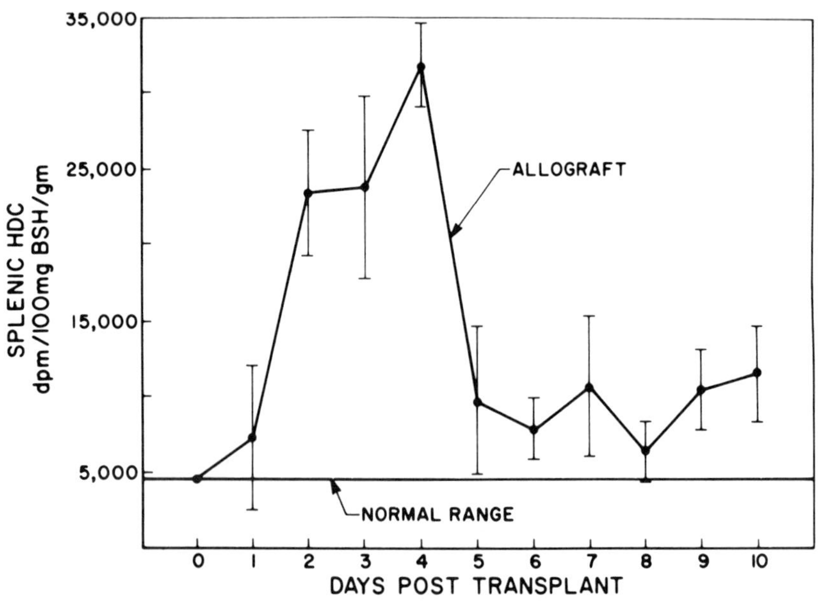

FIGURE 16. Splenic HDC enzyme activity at various intervals following renal allografting between inbred rat strains (Brown-Norway/BN to Lewis rats). As in second series of skin allografting (Lewis to PA), splenic HDC levels peaked at 4 d after allografting. In this case the allografted tissue was an immediately revascularized organ (kidney). (From Moore, T. C., et al., *Surgery*, 75, 203, 1974. With permission.)

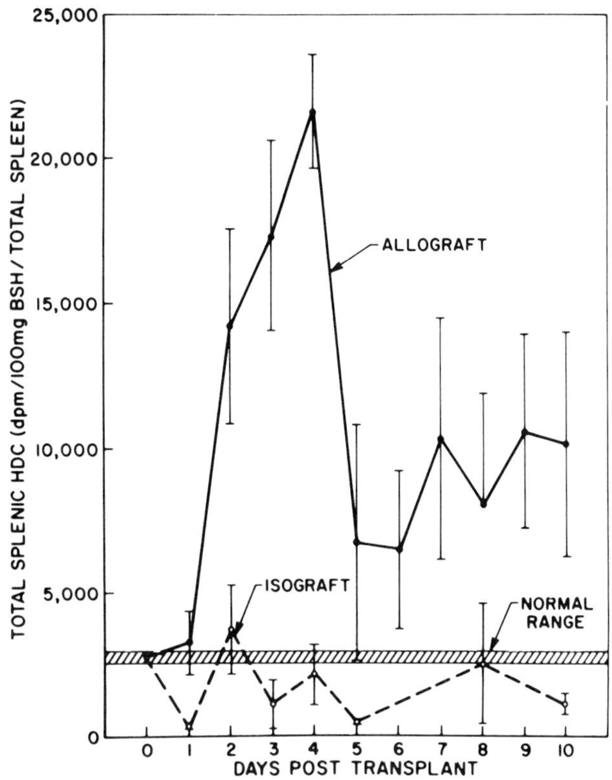

FIGURE 17. Splenic HDC enzyme activity for total spleens following renal allografting and isografting between and within inbred rat strains (BN to Lewis, and Lewis to Lewis). (From Moore, T. C., et al., *Surgery,* 75, 203, 1974. With permission.)

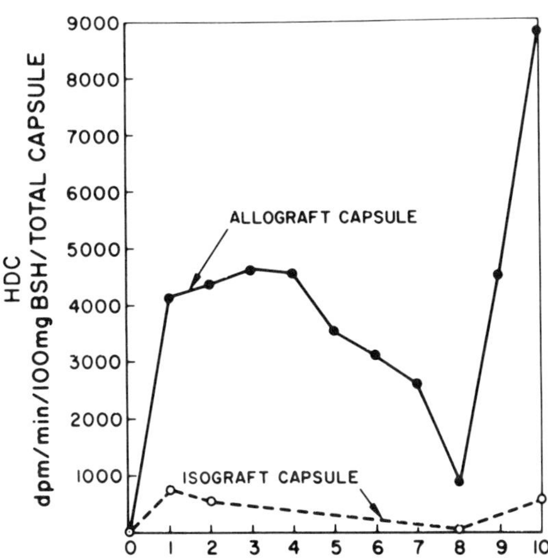

FIGURE 18. Comparison of rat kidney capsule allograft and isograft HDC histamine-forming activity at intervals following renal allografting between inbred rat strains (BN to Lewis). There is a marked and early increase in allograft capsule HDC levels. The capsule, immediately revascularized, has an extensive contact with host "foreign" tissue per volume and weight, as compared with allograft parenchyma that is exposed only to circulating host blood. (From Moore, T. C., et al., *Surgery,* 75, 203, 1974. With permission.)

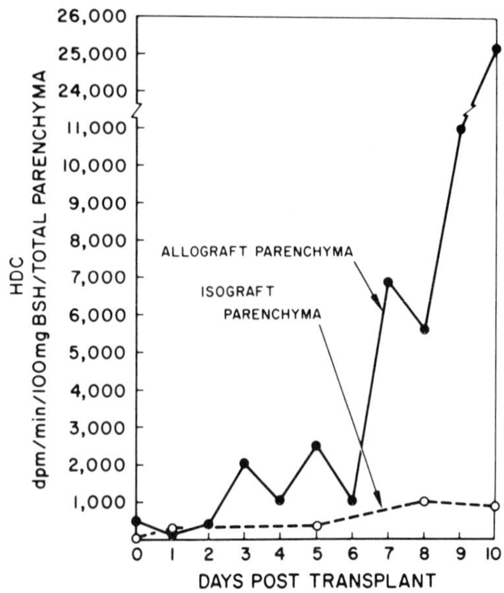

FIGURE 19. Marked increases in total renal allograft parenchymal HDC histamine-forming enzyme activity at intervals postgrafting between inbred rat strains (BN to Lewis), as compared with that of isograft parenchymas (Lewis to Lewis). (From Moore, T. C., et al., *Surgery,* 75, 203, 1974. With permission.)

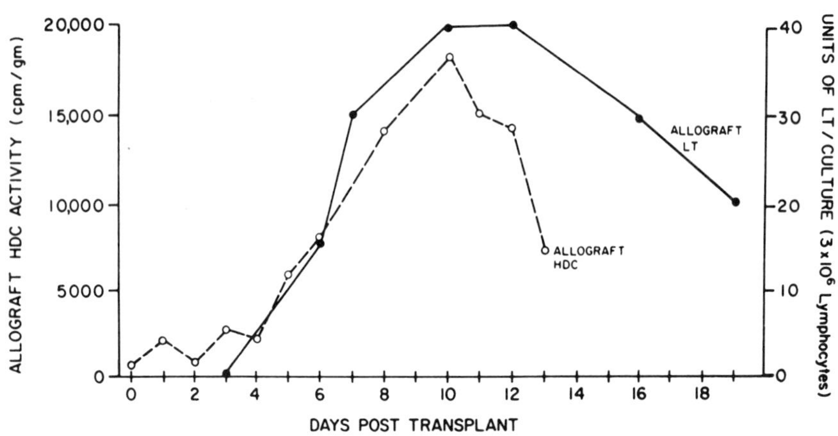

FIGURE 20. Comparison of lymphotoxin (LT) levels and HDC levels, with organ allografting across the same histocompatibility barrier in rats at intervals postallografting. The LT levels were from heart allografts in a study by G. A. Granger and the author. The HDC levels were from kidney allografting studies by the author and associates. (From Moore, T. C., *Bull. Int. Chir.,* 24, 467, 1975. With permission.)

In an important 1983 clinical study, Burtin et al.[266] of Paris reported that 163 patients with solid malignant tumors had significantly lower blood histamine levels than normal noncancer-bearing subjects. The study group included patients with unresected primary tumors, metastases, or both. Stability of the blood histamine levels was associated with the stability of the disease — they observed that progressive decreases in blood histamine levels preceded relapse, detection of metastasis, or both. Patients with blood histamine levels less than 15 mg ml^{-1} (normal levels of 52 + 18.4 ng ml^{-1}) had survivals of less than 2 months. In this report and in a subsequent 1989 report by this group,[267] successful tumor resection

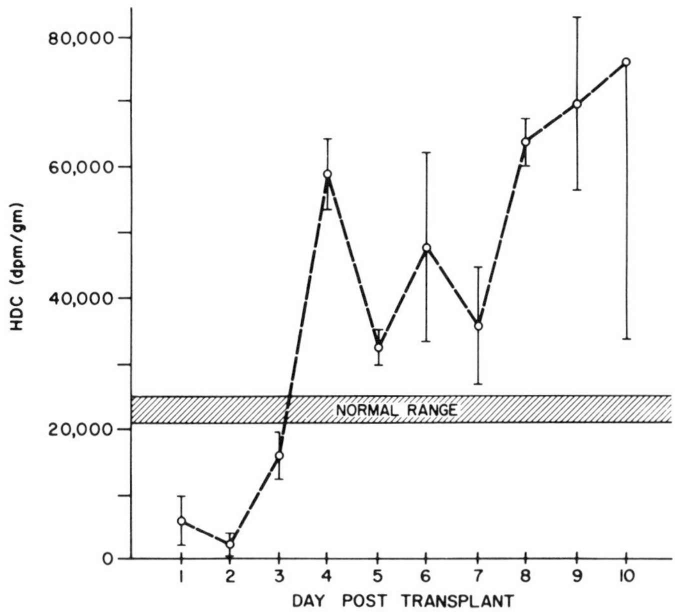

FIGURE 21. Recipient rat thymic histidine decarboxylase histamine-forming enzyme activity at intervals following heterotopic (abdominal) heart allografting (BN to Lewis), as compared with normal levels of thymic HDC enzyme activity of ungrafted animals. (From Moore, T. C., et al., *Surgery,* 76, 733, 1974. With permission.)

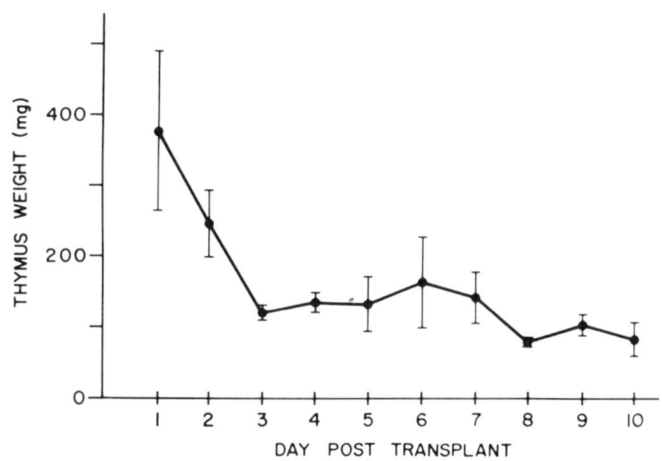

FIGURE 22. Recipient rat thymic weights at intervals postheart allografting (heterotopic into abdominal aorta). (From Moore, T. C., et al., *Surgery,* 76, 733, 1974. With permission.)

(without metastasis) restored blood histamine levels to normal. Daily intraperitoneal injections of histamine also have been reported, by this group,[268,269] to inhibit tumor growth. A similar effect was observed in this study with serotinin injections. In this important study by Burtin et al.,[268] stimulation of histamine type 1 or serotonin receptors and inhibition of histamine type 2 receptors were found to play beneficial roles in the host's defense against malignant and cancerous tumors in fibrosarcoma-bearing mice. Genetically mast-cell-deficient mice that also have markedly depressed HDC enzyme activity[270] have been found, by

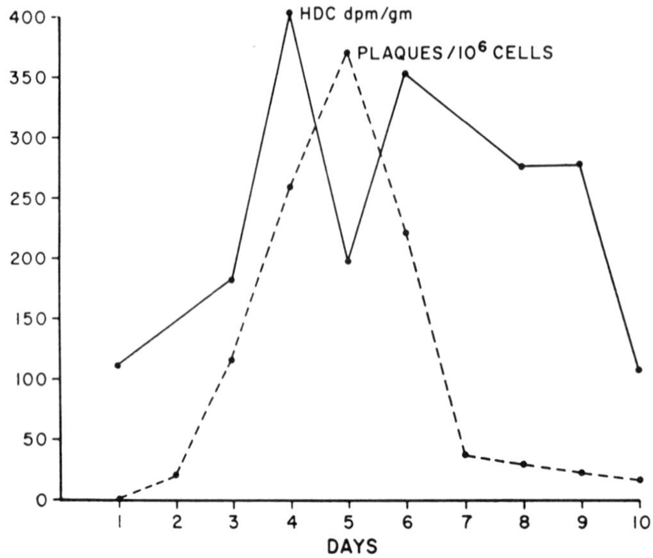

FIGURE 23. Elevation in splenic histamine-forming HDC enzyme activity in the initial 10 d following sheep red blood cell (SRBC) immunization, as compared with the number of antibody-forming spleen cells (plaques per 10^6 nucleated cells). HDC histamine-forming enzyme activity increased and peaked 1 d before the peak elevation in the number of antibody-forming cells. (From Moore, T. C., et al., *Transplantation,* 11, 347, 1971. With permission.)

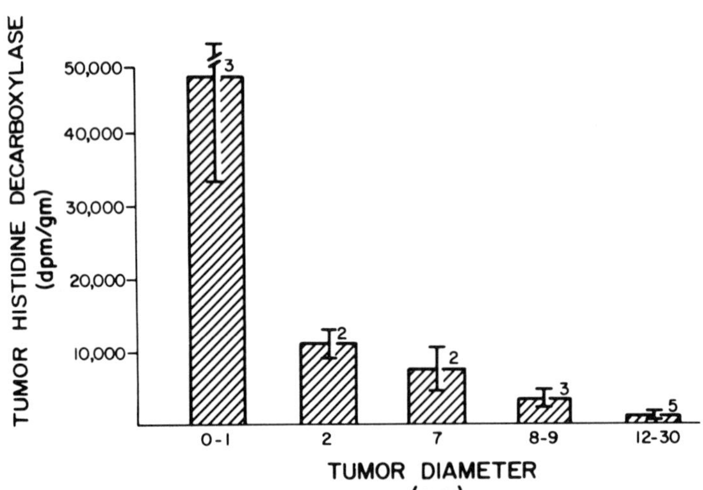

FIGURE 24. Bar graph showing the effects of progressive increases in tumor size on levels of HDC in the tumors. Marked decreases in the levels of the histamine-forming enzyme (HDC) activity was found in the largest and most rapidly growing tumors. (From Moore, T. C., et al., *Ann. Surg.,* 188, 175, 1978. With permission.)

Burtin et al.,[271] to have depressed resistance to tumor incidence and metastasis. The authors interpret these results as confirming the protective role of mast cells (and histamine) against malignant tumors.

A role for histamine at the H_1 receptor for the antitumor activity of endotoxin has been reported in 1984 by Bloksma et al.[272] of Utrecht, the Netherlands. Their study involved the role of histamine in the antitumor activity of endotoxin against solid syngeneic Meth-A

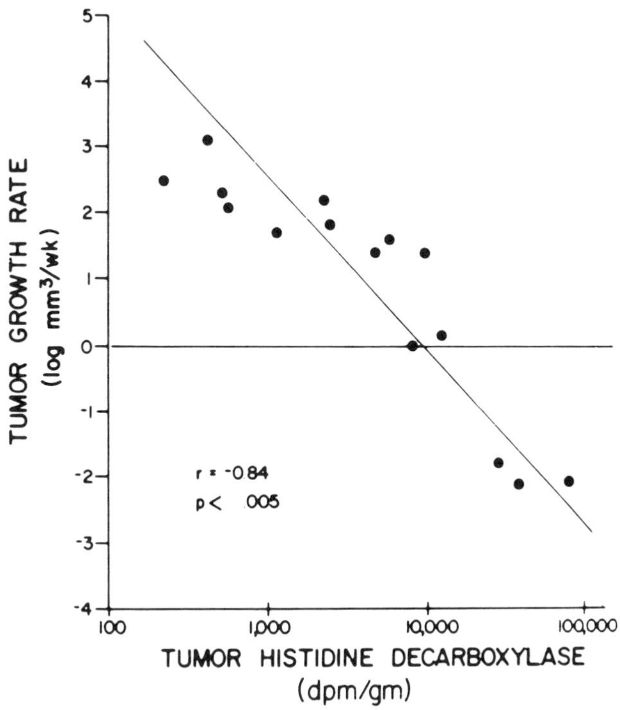

FIGURE 25. A statistically significant ($p < 0.005$) negative correlation of tumor growth rates (expressed as log mm^3/week) and tumor HDC enzyme activity for 15 animals studied (the slower the growth rate, the higher the tumor HDC). The coefficience of correlation (r) for the 15 tumors analyzed by this method was -0.84. (From Moore, T. C., et al., *Ann. Surg.*, 188, 175, 1978. With permission.)

sarcoma in BALB/c mice. Endotoxin was found to induce hemorrhagic necrosis and regression of these tumors. Histamine and the selective histamine-H_1-receptor agonist, 2-pyridylethylamine, mimicked the induction of malignant tumor necrosis but did not produce regression. This effect on tumor necrosis could be inhibited by histamine-H_1-receptor antagonists, but not by a histamine-H_2-receptor antagonist, cimetidine. The incidence of endotoxin-induced tumor regression was reduced by both histamine-H_1-receptor antagonists, diphenhydramine and promethazine, and was potentiated by the histamine-H_2-receptor antagonist, cimetidine. The authors interpreted this interesting and important data as indicating that intact histamine H_1 receptors are required for the induction of malignant tumor regression and the production of antitumor factors by endotoxin. They also make the significant observation that concomitant histamine-H_2-receptor blockade may facilitate this process.

An important 1987 report by Pribluda, Heberman, and Gorelick,[273] of N.I.H.-N.C.I., Fredrick, MD, indicated that histamine *in vivo* both stimulated NK cell activity and inhibited metastasis in mice. The effect of histamine in these studies had its maximum effect after intraperitoneal administration. It was of particular interest that this *in vivo* response differed from *in vitro* findings — again suggesting the importance of *in vivo* data in evaluating physiological *in vivo* responses to these vasoactive neurotransmitter molecules that require intact vascular and nervous systems for their physiologic roles. They observed that histamine significantly increased NK cell activity of spleen cells of C57BL/6 mice. This augmentation was most marked at 24 h after intraperitoneal administration of histamine and had disappeared by 72 h after histamine administration. This observation again buttresses the importance of histamine (like prostaglandin E_2) in the early phases and induction of the cellular immune response.

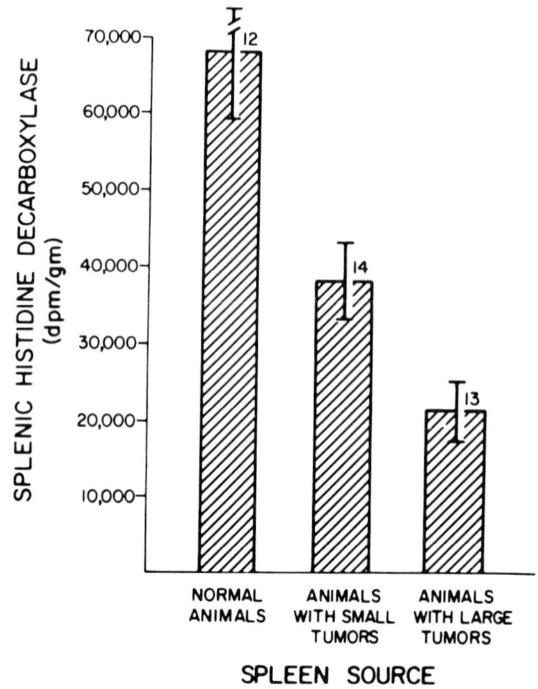

FIGURE 26. Progressive decreases in splenic HDC histamine-forming enzyme activity with progressive tumor growth from small to large size. The significance for the decrease was $p < 0.005$ for small tumors and $p < 0.00005$ for large tumors. The number of animals studied is indicated that the top of each bar. (From Moore, T. C., et al., *Surg. Gynecol. Obstet.*, 146, 559, 1978. With permission.)

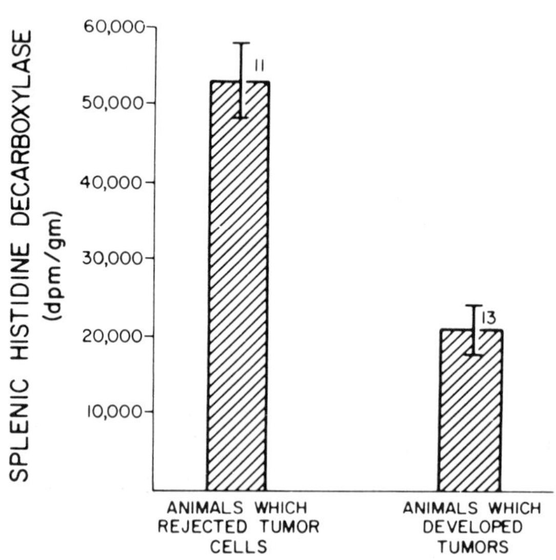

FIGURE 27. Significantly higher ($p < 0.00025$) splenic histamine-forming HDC enzyme levels are shown for animals that rejected inoculated malignant tumor cells and did not develop tumors, as compared with animals that developed tumors following an injection of tumor cells. These animals were studied at comparable periods of time after tumor cell inoculation. (From Moore, T. C., et al., *Surg. Gynecol. Obstet.*, 146, 559, 1978. With permission.)

VII. CONCLUSION

Histamine, the first vasoactive neurotransmitter molecule to be discovered, was synthesized before its existence in nature was identified. Although Nobel Prizes have been awarded twice in the past 35 years for histamine research, neither prize was for the discovery of histamine nor for the dilineation of its particularly important role in the normal functioning of the nervous, vascular, or immune response systems. Both prizes were for antiactions and related to the overwhelmingly bad reputation of histamine as a pesky and trouble-making molecule (allergy, anaphylaxis, and peptic ulceration). Although the discoverer of histamine, Sir Henry Dale, did receive a Nobel Prize, it was for the discovery of the good molecule, acetylcholine, not the bad molecule, histamine.

In recent years more attention has been devoted to histamine in normal physiology and as a ''good'' and important molecule. Greater knowledge of its forming enzyme, histidine decarboxylase (HDC), and its isotopic measurement and study in tissues and its immunohistological tracking have provided much important information, including its role in defense against tissue and organ allografts and its involvement in untreated malignant tumor growth. The identification of additional histamine receptors and the development of selective and potent agonists and antagonist to these new receptors have greatly expanded the scope of histamine research, with particular reference to the delineation of the histaminergic nervous system in the central nervous system. The brilliant work of J.-C. Schwartz and associates of Paris has been of major importance and significance in this area.

Histamine appears to function somewhat like prostaglandin E_2 (PGE_2) as an obligatory molecule early in a cellular immune response. Immuno-, neuro-, and vascular stimulatory actions of histamine seem largely to be generated through activation of histamine H_1 receptors, while inhibitory actions tend to be through histamine-H_2-receptor activation.

The work of Majno and associates in identifying the site of action of histamine (and bradykinin and serotonin) along the vascular tree at the postcapillary venule level (site of transit of small recirculating lymphocytes into lymph nodes) brought histamine center stage in the immune response. It also should be emphasized that histamine, like the other major vasoactive neurotransmitter ''memory molecules'', also has impressive evolutionary credentials.

REFERENCES

1. **Schayer, R. W. and Cooper, J. A. D.,** Metabolism of C^{14} histamine in man, *J. Appl. Physiol.,* 9, 481, 1956.
2. **Watanabe, T., Maeyama, K., Yamatodani, A., Yamada, M., Kitamura, Y., and Wada, H.,** Histadine decarboxylase activities in mutant mice deficient in mast cells, *Life Sci.,* 26, 1569, 1980.
3. **Dy, M., Lebbel, B., Kamoun, P., and Hamburger, J.,** Histamine production during the allograft response: demonstration of a new lymphokine enhancing histamine synthesis, *J. Exp. Med.,* 153, 293, 1981.
4. **Moore, T. C., Granger, G. A., and Lemmi, C. A. E.,** *In vivo* elevation in rat splenic histidine decarboxylase activity by intra-splenic injection of Con A and Con A activated lymphokines, *IRCS Med. Sci.,* 2, 1458, 1974.
5. **Sullivan, T. J. and Parker, C. W.,** Pharmacologic modulation of inflammatory mediator release by rat mast cells, *Am. J. Pathol.,* 85, 437, 1976.
6. **Windaus, A. and Vogt, W.,** Syntheses des Imidazolyläthylamins, *Ber d. deutsch chem. Gessell,* 40, 3691, 1907.
7. **Barger, G. and Dale, H. H.,** Chemical structure and sympathomimetic action of amines, *J. Physiol. (Lond.),* 41, 19, 1910.
8. **Kutscher, F.,** Die physiologische Wirkung einer Secalebase und des Imidazolyathylamins, *Zbl. Physiol.,* 24, 163, 1910.

9. **Ackermann, D.,** Uber den bakteriellen Abbau des Histidine, *Zeitschr. Physiol. Chem.*, 65, 504, 1910.
10. **Dale, H. H. and Laidlaw, P. P.,** The physiological action of β-imidazolylethylamine, *J. Physiol. (Lond.)*, 41, 318, 1910.
11. **Dale, H. H. and Laidlaw, P. P.,** Further observations on the action of β-imidazolylethylamine, *J. Physiol. (Lond.)*, 43, 182, 1911.
12. **Dale, H. H.,** The anaphylactic reaction of plain muscle of the guinea pig, *J. Pharmacol. Exp. Ther.*, 4, 167, 1913.
13. **Dale, H. H. and Laidlaw, P. P.,** Histamine shock, *J. Physiol. (Lond.)*, 52, 355, 1919.
14. **Keeton, R. W., Luckhardt, A. B., and Koch, F. C.,** Gastrin studies. IV. The response of the stomach mucosa to food and gastrin bodies as influenced by atropine, *Am. J. Physiol.*, 51, 469, 1920.
15. **Popielski, L.,** β-Imidazolyathylamin und die Organextrakte. Erster Teil: β-imidazolylathylamin als machtiger Erreger der Magendrusen, *Pflugers Arch. Ges. Physiol.*, 178, 214, 1920.
16. **Best, C. H., Dale, H. H., Dudley, H. W., and Thorpe, W. V.,** The nature of the vasodilator constituents of certain tissue extracts, *J. Physiol. (Lond.)*, 62, 397, 1927.
17. **Lewis, T.,** *The Blood Vessels of the Human Skin and Their Responses*, Shaw & Sons, London, 1927.
18. **Dale, H. H.,** Some chemical factors in the control of the circulation, *Lancet*, 1, 1179, 1929a.
19. **Dale, H. H.,** Some chemical factors in the control of the circulation, *Lancet*, 1, 1233, 1929b.
20. **Dale, H. H.,** Some chemical factors in the control of the circulation, *Lancet*, 1, 1285, 1929c.
21. **Dragstedt, C. A. and Gebauer-Fuelnegg, E.,** Studies in anaphylaxis. I. The appearance of a physiologically active substance during anaphylactic shock, *Am. J. Physiol.*, 102, 512, 1932.
22. **Bartosch, R., Feldberg, W., and Nagel, E.,** Das Freiwerden eines histaminahnlichen stoffes beider Anaphylaxie des Maeschweinchens, *Pfleugers Arch.*, 230, 120, 1932.
23. **Bovet, D. and Staub, A.,** Action protectrice des ethers phenoliques au cours de l'intoxication histaminique, *C.R. Soc. Biol. (Paris)*, 124, 547, 1937.
24. **Staub, A.-M. and Bovet, D.,** Action de la thymoxyethyldiethylamine (929F) et des ethers phenoliques sur le choc anaphylactique du cobaye, *C.R. Soc. Biol. (Paris)*, 125, 818, 1937.
25. **Bovet, D., Horclois, R., and Walthert, F.,** Properties anti-histaminique de la *N-p*-methoxy-benzyl-*N*-dimethyl aminoethyl and amino-pyridine, *C.R. Soc. Biol. (Paris)*, 138, 99, 1944.
26. **Schayer, R. W.,** Formation and binding of histamine by rat tissues *in vivo*, *Am. J. Physiol.*, 187, 63, 1956.
27. **Kahlson, G.,** New approaches to the physiology of histamine, *Perspect. Biol. Med.*, 5, 179, 1962.
28. **Kahlson, G., Rosengren, E., and Thunberg, R.,** Observations on the inhibition of histamine formation, *J. Physiol. (Lond.)*, 169, 467, 1963.
29. **Kahlson, G. and Rosengren, E.,** New approaches to the physiology of histamine, *Physiol. Rev.*, 48, 155, 1968.
30. **Schayer, R. W.,** Induced synthesis of histamine, microcirculatory regulation and the mechanism of glucocorticoid hormones, *Prog. Allergy*, 7, 187, 1963.
31. **Gowans, J. L.,** The effect of continuous reinfusion of lymph and lymphocytes on the output of lymphocytes from the thoracic duct of unanesthetized rats, *Br. J. Exp. Pathol.*, 38, 67, 1957.
32. **Gowans, J. L.,** The recirculation of lymphocytes from blood to lymph in the rat, *J. Physiol. (Lond.)*, 145, 54, 1959.
33. **Majno, G. and Palade, G. E.,** Studies on inflammation. I. The effect of histamine and serotonin on vascular permeability: an electron microscopic study, *J. Biophys. Biochem. Cytol.*, 11, 571, 1961.
34. **Majno, G., Palade, G. E., and Schoefl, G. I.,** Studies on inflammation. II. The site of action of histamine and serotonin along the vascular tree: a topographical study, *J. Biophys. Biochem. Cytol.*, 11, 607, 1961.
35. **Majno, G., Gilmore, V., and Leventhal, M.,** On the mechanism of vascular leakage caused by histamine-type mediators, *Circ. Res.*, 21, 833, 1967.
36. **Majno, G., Shea, S. M., and Leventhal, M.,** Endothelial contraction induced by histamine-type mediators. An electron microscopic study, *J. Cell Biol.*, 42, 647, 1969.
37. **Moore, T. C. and Chang, J.,** Urinary histamine excretion in the rat following skin homografting and autografting, *Ann. Surg.*, 167, 232, 1968.
38. **Moore, T. C.,** Histidine decarboxylase inhibitors and the survival of skin homografts, *Nature*, 215, 871, 1967.
39. **Moore, T. C.,** Histidine decarboxylase inhibitors and second-set allograft survival, *Arch. Surg.*, 99, 470, 1969.
40. **Moore, T. C. and Schayer, R. W.,** Histidine decarboxylase activity of autografted and allografted rat skin, *Transplantation*, 7, 99, 1969.
41. **Moore, T. C. and Schayer, R. W.,** Histidine decarboxylase activity of the rat spleen following skin allografting, *Ann. Surg.*, 171, 609, 1970.

42. **Moore, T. C., Sinclair, M. C., and Lemmi, C. A. E.,** Elevation in host splenic and allograft parenchymal and capsular histidine decarboxylase activity and weights following renal allografting in rats, *Surgery,* 75, 203, 1974.
43. **Moore, T. C., Sinclair, M. C., McAlpin, C. D., and Lemmi, C. A. E.,** Effect of aging, of organ allografting and of surgical stress on rat thymus histidine decarboxylase activity, *Surgery,* 76, 733, 1974.
44. **Moore, T. C., Sinclair, M. C., and Lemmi, C. A. E.,** Elevation in host splenic and allograft parenchymal and capsular histidine decarboxylase activity and weights following renal allografting in rats, *Surgery,* 75, 203, 1974.
45. **Moore, T. C., Sinclair, M. C., McAlpin, C. D., and Lemmi, C. A. E.,** Effect of aging, of organ allografting and of surgical stress on rat thymus histidine decarboxylase activity, *Surgery,* 76, 733, 1974.
46. **Moore, T. C., Chang, J., and Hume, D. M.,** Urinary excretion of histamine in renal transplant patients, *Surg. Gynecol. Obstet.,* 127, 1023, 1968.
47. **Moore, T. C., Thompson, D. P., and Hayes, M.,** Correlation of renal blood flow indices and histamine metabolism data after canine renal allografting, *Arch. Surg.,* 101, 45, 1970.
48. **Moore, T. C., Cleveland, R. J., Thompson, D. P., and Nelson, R. J.,** Vasoactive amine metabolism following canine and human heart transplantation, *Ann. Surg.,* 174, 51, 1971.
49. **Moore, T. C., Thompson, D. P., and Glassock, R. J.,** Elevation in urinary and blood histamine following clinical renal transplantation, *Ann. Surg.,* 173, 381, 1971.
50. **Sinclair, M. C., Lemmi, C. A. E., and Moore, T. C.,** Elevation in urinary excretion of histamine following renal allografting in rats, *J. Surg. Res.,* 17, 43, 1974.
51. **Moore, T. C.,** A theory of the role of histamine metabolism in transplant rejection, *Surg. Gynecol. Obstet.,* 132, 489, 1971.
52. **Black, J. W., Duncan, W. A. M., Durant, C. J., Ganellin, C. R., and Parsons, E. M.,** Definition and antagonism of histamine H_2-receptors, *Nature,* 236, 385, 1972.
53. **Ash, A. S. F. and Schild, H. O.,** Receptors mediating some actions of histamine, *Br. J. Pharmacol.,* 27, 427, 1966.
54. **Bourne, H. R., Melmon, K. L., and Lichtenstein, L. M.,** Histamine augments leukocytes adenosine 3′,5′ monophosphate and blocks antigenic histamine release, *Science,* 173, 743, 1971.
55. **Shearer, G. M., Melmon, K. L., Weinstein, Y., and Sela, M.,** Regulation of antibody response by cells expressing histamine receptors, *J. Exp. Med.,* 136, 1302, 1972.
56. **Shearer, G. M., Weinstein, Y., and Melmon, K. L.,** Enhancement of immune response potential of mouse lymphoid cells fractionated over insolubilized conjugated histamine columns, *J. Immunol.,* 113, 597, 1974.
57. **Plaut, M., Lichtenstein, L. M., and Henney, C. S.,** Properties of a sub-population of T-cells bearing histamine receptors, *J. Clin. Invest.,* 55, 856, 1975.
58. **Moore, T. C.,** The effects of inflammatory mediators and anaesthetic agents upon lymphocyte traffic of sheep *in vivo,* Ph.D. thesis, University of Cambridge (Trinity College), London, 1982.
59. **Schwartz, J.-C., Lampart, C., and Rose, C.,** Histamine formation in rat brain *in vivo;* effects of histidine loads, *J. Neurochem.,* 19, 801, 1975.
60. **Schwartz, J.-C.,** Histaminergic mechanisms in brain, *Ann. Rev. Pharmacol. Toxicol.,* 17, 325, 1977.
61. **Schwartz, J.-C.,** Histamine receptors in brain, *Life Sci.,* 25, 895, 1979.
62. **Schwartz, J.-C., Pollard, H., and Tam Quoch, T.,** Histamine as a neurotransmitter in mammalian brain: neurochemical evidence, *J. Neurochem.,* 35, 26, 1980.
63. **Moore, T. C., Koppelmann, L. E., and Lemmi, C. A. E.,** Decrease in histamine forming enzyme activity of non-metastisizing fibrosarcomas in hamsters with progressive tumor growth, *Ann. Surg.,* 188, 175, 1978.
64. **Koppelmann, L. E., Moore, T. C., Lemmi, C. A. E., and Porter, D. D.,** Accumulation of mononuclear cells in tumors with growth slowing and elevation in host splenic histidine decarboxylase activity following repeated tumor injections with bradykinin, *Surgery,* 78, 181, 1975.
65. **Arrang, J.-M., Garbarg, M., Lancelot, J.-C., Lecomte, J.-M., Pollard, H., Robba, M., Schunack, W., Schwartz, J.-C.,** Highly potent and selective ligands for histamine H_3-receptors, *Nature,* 327, 117, 1987.
66. **Schwartz, J.-C., Arrang, J.-M., Garbarg, M., Gulat-Marnay, C., and Pollard, H.,** Modulation of histamine synthesis and release in brain via presynaptic autoreceptors and heteroreceptors, *Ann. N.Y. Acad. Sci.,* 604, 40, 1990.
67. **Arrang, J.-M., Garbarg, M., and Schwartz, J.-C.,** Autoinhibition of brain histamine release by a novel class (H_3) of histamine receptor, *Nature,* 302, 832, 1983.
68. **Arrang, J.-M., Garbarg, M., and Schwartz, J.-C.,** Autoregulation of histamine release in brain by presynaptic H_3-receptors, *Neuroscience,* 15, 553, 1985.

69. **Brandes, L. J., Gerrard, J. M., Bogdanovic, R. P., Lint, D. W., Reid, R. E., and LaBella, F. S.,** Correlation of the antiproliferative action of diphenylmethane-derivative antiestrogen binding site ligands with antagonism of histamine binding but not of protein kinase C-mediated phosphorylation, *Cancer Res.*, 48, 3954, 1988.
70. **Saxena, S. P., McNicol, A., Brandes, L. J., Becker, A. B., and Gerrard, J. M.,** A role for intracellular histamine in collagen induced platelet aggregation, *Blood,* 75, 407, 1990.
71. **Saxena, S. P., Brandes, L. J., Becker, A. B., Simons, K. J., LaBella, F. S., and Gerrard, J. M.,** Histamine is an intracellular messenger mediating platelet aggregation, *Science,* 243, 1596, 1989.
72. **Brandes, L. J. and Hermonat, M. W.,** A diphenylmethane derivative specific for the antiestrogen binding site found in rat liver microsomes, *Biochem. Biophys. Res. Comm.*, 123, 724, 1984.
73. **Wada, H., Yamatodani, A., Inagaki, N., Itowi, N., Wang, N. P., and Fukui, H.,** Histaminergic neuron system and its function, *Adv. Exp. Med. Biol.*, 236, 343, 1988.
74. **Panula, P., Airaksinen, M. S., Pirvola, U., and Kotilainen, E.,** A histamine-containing neuronal system in human brain, *Neuroscience,* 34, 127, 1990.
75. **Russell, W., Henry, D. P., Phebus, L. A., and Clemens, J. A.,** Release of histamine in rat hypothalamus and corpus striatum in vivo, *Brain Res.*, 512, 95, 1990.
76. **Russell, M., Dark, D. A., Cummins, R. W., Ellman, G., Calloway, E., and Peek, H. V. S.,** Learned histamine release, *Science,* 225, 733, 1984.
77. **Hardie, R. C.,** A histamine-activated chloride channel involved in neurotransmission at a photo receptor synapse, *Nature,* 339, 704.
78. **Pirvola, V., Tumomisto, L., Yamatodani, A., and Panula, P.,** Distribution of histamine in the cockroach brain and visual system: an immunocytochemical and biochemical study, *J. Comp. Neurol.*, 276, 514, 1988.
79. **Nassel, D. R., Pirvila, U., and Panula, P.,** Histamine like immunoreactive neurons innervating putative neurohaemal areas and central neuropil in the thoracoabdominal ganglia of the flies *Drosophila* and *Calliphora, J. Comp. Neurol.*, 297, 525, 1990.
80. **Elias, M. S. and Evans, P. D.,** Autoradiographic localization of ^{3}H-histamine accumulation by the visual system of the locust, *Cell Tissue Res.*, 232, 221, 1984.
81. **Weinreich, D., Weiner, C., and McCaman, R.,** Endogenous levels of histamine in single neurons isolated from CNS of *Aplysia californica, Brain Res.*, 84, 341, 1976.
82. **McCaman, R. E. and Weinreich, D.,** Histaminergic synaptic transmission in the cerebral ganglion of *Aplysia, Neurophysiology,* 53, 1016, 1985.
83. **McClintock, T. S. and Ache, B. W.,** Histamine directly gates a chloride channel in lobster olfactory neurons, *Proc. Natl. Acad. Sci. U.S.A.*, 86, 8137, 1989.
84. **Khan, M. A. and Sen, P. C.,** Effect of histamine and serotonin on phagocytic activity of *Entamoeba histolytica, Indian J. Exp. Biol.*, 26, 730, 1988.
85. **Battelle, B. A., Calman, B. G., Andrews, A. W., Grieco, F. D., Mileziva, M. B., Callaway, J. C., and Stuart, A. E.,** Histamine: a putative afferent neurotransmitter in Limulus eyes, *J. Comp. Neurol.*, 305, 527, 1991.
86. **Bayer, T. A., McClintock, T. S., Grunert, U., and Ache, B. W.,** Histamine induced modulation of olfactory receptor neurons in two species of lobsters. *P. argus* and *H. americanus, J. Exp. Biol.*, 145, 133, 1989.
87. **Orona, E., Battelle, B.-A., and Ache, B. W.,** Immunohistochemical and biochemical evidence for histamine and GABA in the olfactory lobes of the spiny lobster, *Chem. Senses,* 14, 681, 1989.
88. **Orona, E., Battelle, B.-A., and Ache, B. W.,** Immunohistochemical and biochemical evidence for the putative inhibitory neurotransmitters histamine and GABA in lobster olfactory lobes, *J. Comp. Neurol.*, 294, 633, 1990.
89. **Maxwell, G. D., Tait, J. F., and Hildebrand, J. G.,** Regional synthesis of neurotransmitter candidates in the CNS of the moth *Manduca sexta, Comp. Biochem. Physiol.*, 61C, 109, 1978.
90. **Elias, M. S. and Evans, P. D.,** Histamine in the insect nervous system: distribution, synthesis and metabolism, *J. Neurochem.*, 41, 562, 1983.
91. **Pirvola, U., Tuomisto, L., Yamatodani, A., and Panula, P.,** Distribution of histamine in the cockroach brain and visual system: an immunocytochemical and biochemical study, *J. Comp. Neurol.*, 276, 514, 1988.
92. **Panula, P., Happola, O., Airaksinen, M. S., Auvinen, S., and Virkamaki, A.,** Carbodiimide as a tissue fixative in histamine immunohistochemistry and its application to developmental biology, *J. Histochem. Cytochem.*, 36, 259, 1988.
93. **Nässel, D. R., Pirvola, U., and Pertti, P.,** Histamine-like immunoreactive neurons innervating putative neurohaemal areas and central neurophil in the thoraco-abdominal ganglia of the flies *Drosophila* and *Calliphora, J. Comp. Neurol.*, 297, 525, 1990.

94. **Nässel, D. R. and Elekes, K.,** Serotonergic terminals in the neural sheath of the blowfly nervous system: ultrastructural immunocytochemistry and 5-7-dihydroxytryptamine labelling, *Neuroscience,* 15, 293, 1985.
95. **Veenstra, J. A.,** Diversity in neurohaemal organs for homologous neurosecretory cells in different insect species as demonstrated by immunocytochemistry with an antiserum to molluscan cardioexcitatory peptide, *Neurosci. Lett.,* 73, 33, 1987.
96. **Nässel, D. R. and O'Shea, M.,** Proctolin-like immunoreactive neurons in the blowfly central nervous system, *J. Comp. Neurol.,* 265, 437, 1987.
97. **Nässel, D. R., Ohlsson, L. G., and Cantera, R.,** Metamorphosis of identified neurons innervating thoracic neurohaemal organs in the blowfly: transformation of cholecystokinin-like immunoreactive neurons, *J. Comp. Neurol.,* 267, 343, 1988a.
98. **Nässel, D. R., Holmqvist, B. I., and Moverus, B. J. A.,** Vasopressin- and proctolin-like immuno-reactive efferent neurons in blowfly abdominal ganglia: development and ultrastructure, *J. Comp. Neurol.,* 283, 450, 1989a.
99. **Nässel, D. R., Lundquist, T., Hoog, A., and Grimelius, L.,** Substance P-like immunoreactive neurons in the nervous system of *Drosophila, Brain Res.,* 507, 225, 1990.
100. **Dove, H., Thorpe, A., and Nässel, D. R.,** Light and electron microscopic immunocytochemistry of peptidergic neurons innervating thoracico-abdominal neurohaemal areas in the blowfly, *Cell Tissue Res.,* 253, 583, 1988.
101. **Lundquist, T. and Nässel, D. R.,** Substance P, FMRFamide- and gastrin/cholecystokinin-like immunoreactive neurons in the thoraco-abdominal ganglia of the flies *Drosophila* and *Calliphora, J. Comp. Neurol.,* 294, 161, 1990.
102. **Weinreich, D.,** Synaptic responses mediated by identified histamine-containing neurones, *Nature,* 267, 854, 1977.
103. **Weinreich, D., Weiner, C., and McCaman, R.,** Endogenous levels of histamine in single neurons isolated from CNS of *Aplysia californica, Brain Res.,* 84, 341, 1976.
104. **McCaman, R. E. and Weinreich, D.,** Histaminergic synaptic transmission in the cerebral ganglion of *Aplysia, Neurophysiology,* 53, 1016.
105. **Chiel, H. J., Weiss, K. R., and Kupfermann, I.,** An identified histaminergic neuron acts pre- and postsynaptically to inhibit the outputs of identified buccal-cerebral interneurons, *Soc. Neurosci. Abstr.,* 9, 913, 1983.
106. **Chiel, H. J., Weiss, K. R., and Kupfermann, I.,** An identified histaminergic neuron modulates feeding motor circuitry in *Aplysia, J. Neurosci.,* 6, 2427, 1986.
107. **Weiss, K. R., Chiel, H. J., Koch, U., and Kupfermann, I.,** Activity of an identified histaminergic neuron, and its possible role in arousal of feeding behavior in semi-intact *Aplysia, J. Neurosci.,* 6, 2403, 1986a.
108. **Weiss, K. R., Chiel, H. J., and Kupfermann, I.,** Sensory function and gating of histaminergic neuron C2 in *Aplysia, J. Neurosci.,* 6, 2416, 1986b.
109. **Chiel, H. J., Kupfermann, I., and Weiss, K. R.,** An identified histaminergic neuron can modulate the outputs of buccal-cerebral interneurons in *Aplysia* via presynaptic inhibition, *J. Neurosci.,* 8(1), 49, 1988.
110. **McCaman, R. E. and Weinreich, D.,** On the nature of histamine-mediated slow hyperpolarizing synaptic potentials in identified molluscan neurones, *J. Physiol. (Lond.),* 328, 485, 1982.
111. **Hinzen, D. H. and Riehl, J.,** Presynaptic inhibition and histamine in cerebral ganglion of *Aplysia, Arzneim. Forsch. (Drug Res.),* 35, 292, 1985.
112. **Kretz, R., Shapiro, E., and Kandel, E. R.,** Presynaptic inhibition produced by an identified presynaptic inhibitory neuron. I. Physiological mechanisms, *J. Neurophysiol.,* 55, 113, 1986a.
113. **Kretz, R., Shapiro, E., Bailey, C. H., Chem, M., and Kandel, E. R.,** Presynaptic inhibition produced by an identified presynaptic inhibitory neuron. II. Presynaptic conductance changes caused by histamine, *J. Neurophysiol.,* 55, 131, 1986b.
114. **Weiss, K. R., Shapiro, E., and Kupfermann, I.,** Modulatory synaptic actions of an identified histaminergic neuron on the serotonergic metacerebral cell of *Aplysia. J. Neurosci.,* 6, 2393, 1986.
115. **Airaksinen, M. S. and Panula, P.,** The histaminergic system in the guinea pig central nervous system: an immunocytochemical mapping study using an antiserum against histamine, *J. Comp. Neurol.,* 273, 163, 1988.
116. **Woodruff, G. N., Oniwinde, A. B., and Kerfut, G. A.,** Histamine in tissues of the snail, crab, goldfish, frog and mouse, *Comp. Biochem. Physiol.,* 31, 599, 1969.
117. **Inagaki, N., Panula, P., Yamatodani, A., and Wada, H.,** Organization of the histaminergic system in the brain of the turtle *Chinemys reevesii, J. Comp. Neurol.,* 297, 132, 1990.
118. **Brodin, L., Hokfelt, T., Grillner, S., and Panula, P.,** Distribution of histaminergic neurons in the brain of the lamprey *Lampetra fluviatilis* as revealed by histamine-immunohistochemistry, *J. Comp. Neurol.,* 292, 435, 1990.

119. **Airaksinen, M. S. and Panula, P.,** Comparative neuroanatomy of the histaminergic system in the brain of the frog *Xenopus laevis, J. Comp. Neurol.,* 292, 412, 1990.
120. **Panula, P., Pirvola, U., Auvinen, S., and Airaksinen, M. S.,** Histamine-immunoreactive nerve fibers in the rat brain, *Neuroscience,* 28, 585, 1989.
121. **Ericson, H., Watanabe, T., and Kohler, C.,** Morphological analysis of the tuberomammillary nucleus in the rat brain: delineation of subgroups with antibody against L-histidine decarboxylase as a marker, *J. Comp. Neurol.,* 263, 1, 1987.
122. **Inagaki, N., Yamatodani, A., Ando-Yamamoto, M., Tohyama, M., Watanable, T., and Wada, H.,** Organization of histaminergic fibers in rat brain, *J. Comp. Neurol.,* 273, 283, 1988.
123. **Takeda, N., Inagaki, S., Taguchi, Y., Tohyama, M., Watanable, T., and Wada, H.,** Origins of histamine-containing fibers in the cerebral cortex of rats studied by immunohistochemistry with histidine decarboxylase as a marker and transection, *Brain Res.,* 323, 55, 1984.
124. **Watanable, T., Taguchi, Y., Shiosaka, S., Tanaka, J., Kubota, H., Terano, Y., Tohyama, M., and Wada, H.,** Distribution of the histaminergic neuron system in the central nervous system of rats, a fluorescent immunohistochemical analysis with histidine decarboxylase as a marker, *Brain Res.,* 295, 13, 1984.
125. **Airaksinen, M. S. and Panula, P.,** The histaminergic system in the guinea pig central nervous system: an immunocytochemical mapping study using an antiserum against histamine, *J. Comp. Neurol.,* 273, 163, 1988.
126. **Airaksinen, M. S., Flugge, G., Fuchs, E., and Panula, P.,** Histaminergic system in the tree shrew brain, *J. Comp. Neurol.,* 286, 289, 1989.
127. **Oakes, S. G., Iaizzo, P. A., Richelson, E., and Powis, G.,** Histamine-induced intracellular free Ca^{++}, inositol phosphates and electrical changes in murine N1E-115 neuroblastoma cells, *J. Pharmacol. Exp. Ther.,* 247, 114, 1988.
128. **Snider, R. M., McKinney, M., Forray, C., and Richelson, E.,** Neurotransmitter receptors mediate cyclic GMP formation by involvement of arachidonic acid and lipoxygenase, *Proc. Natl. Acad. Sci. U.S.A.,* 81, 3905, 1984.
129. **Lo, W. W. Y. and Fan, T.,** Histamine stimulates inositol phosphate accumulation via the H_1-receptor in culture human endothelial cells, *Biochem. Biophys. Res. Commun.,* 148, 47, 1987.
130. **Resink, T. J., Grigorian, G. Y., Moldavaeva, A. K., Danilov, S. M., and Buhler, F. R.,** Histamine-induced phosphoinositide metabolism in cultured umbilical vein endothelial cells, *Biochem. Biophys. Res. Commun.,* 144, 438, 1985.
131. **Kataoka, K. and DeRobertis, E.,** Histamine in isolated small nerve endings and synaptic vesicles of rat brain cortex, *J. Pharmacol. Exp. Ther.,* 156, 114, 1967.
132. **Christian, E. P., Undem, B. J., and Weinreich, D.,** Endogenous histamine excites neurones in the guinea-pig superior cervical ganglion *in vitro, J. Physiol. (Lond.),* 409, 297, 1989.
133. **Panula, P., Flugge, G., Fuchs, E., Pirvola, U., Auvinen, S., and Airaksinen, M. S.,** Histamine-immunoreactive nerve fibers in the mammalian spinal cord, *Brain Res.,* 484, 234, 1989.
134. **Tamiya, R., Hanada, M., Narita, N., Inagaki, S., Tohyama, M., and Takagi, H.,** Histaminergic neurons receive substance P-ergic inputs in the posterior hypothalamus of the rat, *Exp. Brain Res.,* 79, 261, 1990.
135. **Houch, L. B.,** Barbiturate-induced inhibition of rat brain histamine turnover, *Biochem. Pharmacol.,* 36, 3321, 1987.
136. **Kalivas, P. W.,** Histamine-induced arousal in the conscious and pentobarbital-pretreated rat, *J. Pharmacol. Exp. Ther.,* 222, 37, 1982.
137. **Pollard, H., Bischoff, S., and Schwartz, J.-C.,** Turnover of histamine in rat brain and its decrease under barbiturate anesthesia, *J. Pharmacol. Exp. Ther.,* 190, 88, 1974.
138. **Corrodi, H., Fuxe, K., and Hokfelt, T.,** The effect of barbiturates on the activity of catecholamine neurons in rat brain, *J. Pharm. Pharmacol.,* 18, 556, 1966.
139. **Corrodi, H., Fuxe, K., and Hokfelt, T.,** The effect of some psychoactive drugs on central monoamine neurons, *Eur. J. Pharmacol.,* 1, 363, 1967.
140. **Schuberth, J., Sparf, B., and Sundwall, A.,** On the turnover of acetylcholine in the brain, in *Drugs and Cholinergic Mechanisms in the Central Nervous System,* Heilbronn, E. and Winter, A., Eds., Research Institute of National Defense, Stockholm, 1970, 177.
141. **Persson, T. and Waldeck, B.,** A reduced rate of turnover of brain noradrenaline during pentobarbitone anaesthesia, *J. Pharm. Pharmacol.,* 23, 377, 1971.
142. **Walajtys-Rode, E., Waskiewicz, J., and Rafalowska, U.,** Effect of *in vivo* glucocorticoid administration on histamine metabolism in rat brain synaptosomes, *Neurosci. Lett.,* 97, 340, 1989.
143. **Phelan, K. D., Nakamura, J., and Gallagher, J. P.,** Histamine depolarizes rat medial vestibular nucleus neurons recorded intracellularly *in vitro, Neurosci. Lett.,* 109, 287, 1990.
144. **Jorgenson, K. L., Kow, L.-M., and Pfaff, D. W.,** Histamine excites arcuate neurons *in vitro* through H_1 receptors, *Brain Res.,* 502, 171, 1989.

145. **Tasaka, K., Chung, Y. H., and Sawada, K.**, Excitatory effect of histamine on EEGs of the cortex and thalamus in rats, *Agents Actions,* 27, 1, 1989.
146. **Springfield, S. A. and Geller, H. M.**, Histamine modulates local inhibition in the rat hippocampal slice, *Cell. Mol. Neurobiol.*, 8, 431, 1988.
147. **Anderson, P., Eccles, J. C., and Loyning, Y.**, Location of postsynaptic inhibitory synapses on hippocampal pyramids, *J. Neurophysiol.*, 27, 592, 1964.
148. **Garbarg, M., Barbin, G., Bischoff, S., Pollard, H., and Schwartz, J. C.**, Dual localization of histamine in an ascending neuronal pathway and in non-neuronal cells evidenced by lesions in the lateral hypothalamic area, *Brain Res.*, 106, 333, 1976.
149. **Sastry, B. S. R. and Phillis, J. W.**, Evidence for an ascending histaminergic pathway to the cerebral cortex, *Can. J. Physiol. Pharmacol.*, 54, 782, 1976.
150. **Lin, J.-S., Sakai, K., and Jouvet, M.**, Evidence for histaminergic arousal mechanisms in the hypothalamus of cat, *Neuropharmacology,* 27, 111, 1988.
151. **Ookuma, K., Yoshimatsu, H., Sakata, T., Fujimoto, K., and Fukagawa, K.**, Hypothalamic sites of neuronal histamine action on food intake by rats, *Brain Res.*, 490, 268, 1989.
152. **Fukagawa, K., Sakata, T., Shiraishi, T., Yoshimatsu, H., Fujimoto, K., Ookuma, K., and Wada, H.**, Neuronal histamine modulates feeding behavior through H_1-receptor in rat hypothalamus, *Am. Phys. Soc.*, 605, 1989.
153. **Garbarg, M., Tuong, M. D. T., Gros, C., and Schwartz, J. C.**, Effects of histamine H_3-receptor ligands on various biochemical indices of histaminergic neuron activity in rat brain, *Eur. J. Pharmacol.*, 164, 1, 1989.
154. **Arrang, J.-M., Garbarg, M., and Schwartz, J.-C.**, Autoinhibition of histamine synthesis mediated by presynaptic H_3-receptors, *Neuroscience,* 23, 149, 1987.
155. **Schwartz, J. C.**, Histamine as a transmitter in brain, *Life Sci.*, 17, 503, 1975.
156. **Verdiere, M., Rose, C., and Schwartz, J.-C.**, Synthesis and release of histamine studied on slices from rat hypothalamus, *Eur. J. Pharmacol.*, 34, 157, 1975.
157. **Atack, C. and Carlsson, A.**, *In vitro* release of endogenous histamine together with noradrenaline and 5-hydroxytryptamine, from slices of mouse cerebral hemispheres, *J. Pharm. Pharmacol.*, 24, 990, 1972.
158. **Panula, P., Airaksinen, M. S., Pirvola, U., and Kotilainen, E.**, A histamine-containing neuronal system in human brain, *Neuroscience,* 34, 127, 1990.
159. **Kendall, D. A. and Firth, J. L.**, Inositol phospholipid hydrolysis in human brain; adenosine inhibition of the response to histamine, *Br. J. Pharmacol.*, 100, 37, 1990.
160. **Saper, C. B. and German, D. C.**, Hypothalamic pathology in Alzheimer's disease, *Neurosci. Lett.*, 74, 364, 1987.
161. **Joris, I., Majno, G., and Ryan, G. B.**, Endothelial contraction *in vivo:* a study in the rat mesentery, *Virchows Arch. (Cell Pathol.)*, 12, 73, 1972.
162. **Svensjo, E., Arfors, K.-E., Raymond, R. M., and Grega, G. J.**, Morphological and physiological correlation of bradykinin-induced macromolecular efflux, *Am. J. Physiol.*, 236, 600, 1979.
163. **Renkin, E. M., Carter, R. D., and Joyner, W. L.**, Mechanism of the sustained action of histamine and bradykinin on transport of large molecules across capillary walls in dog paw, *Microvasc. Res.*, 7, 49, 1974.
164. **Svensjo, E., Persson, C. G. A., and Rutili, G.**, Inhibition of bradykinin induced macromolecular leakage from post-capillary venules by a B_2-adrenoreceptor stimulant, terbutaline, *Acta Physiol. Scand.*, 101, 504, 1977.
165. **Joyner, W. L., Svensjo, E., and Arfors, K.-E.**, Simultaneous measurements of macromolecular leakage and arteriolar blood flow as altered by PGE_1 and B_2-stimulation in the hamster cheek pouch, *Microvasc. Res.*, 18, 301, 1979.
166. **Becker, C. G. and Nachman, R. L.**, Contractile proteins of endothelial cells, platelets, and smooth muscle, *Am. J. Pathol.*, 71, 1, 1973.
167. **Drenckhan, D.**, Cell motility and cytoplasmic filaments in vascular endothelium, *Prog. Appl. Microcirc.*, 1, 53, 1983.
168. **Drenckhan, D.**, Cell motility and cytoplasmic filaments in vascular endothelium, *Prog. Appl. Microcirc.*, 1, 53, 1983.
169. **Heltianu, C., Simionescu, M., and Simionescu, N.**, Histamine receptors of the microvascular endothelium revealed in situ with a histamine-ferritin conjugate: characteristic high-affinity binding sites in venules, *J. Cell Biol.*, 93, 357, 1982.
170. **Horan, K. L., Aldamski, S. W., Ayele, W., Langone, J. J., and Grega, G. J.**, Evidence that prolonged histamine suffusions produce transient increases in vascular permeability subsequent to the formation of venular macromolecular leakage sites: proof of the Majno-Palade hypothesis, *Am. J. Pathol.*, 123, 570, 1986.

171. **Resick, T. J., Grigorian, G. Y., Moldabaeva, A. K., Danilov, S. M., and Buhler, F. R.,** Histamine-induced phosphoinositide metabolism in cultured human umbilical vein endothelial cells, association with thromboxane and prostacyclin release, *Biochem. Biophys. Res. Commun.*, 144, 438, 1987.
172. **Brock, T. A. and Capasso, E. A.,** Thrombin and histamine activate phospholipase C in human endothelial cells via a phorbol ester-sensitive pathway, *J. Cell. Physiol.*, 136, 54, 1988.
173. **Voyno-Yasenetskaya, T. A., Panchenko, M. P., Nupenko, E. V., Rybin, V. O., and Tkachuk, V. A.,** Histamine and bradykinin stimulate the phosphoinositide turnover in human umbilical vein endothelial cells via different G-proteins, *FEBS Lett.*, 259, 67, 1989.
174. **Liddell, R. H. A., Scott, A. R. W., and Simpson, J. G.,** Histamine-induced changes in the endothelium of post-capillary venules: effects of chelating agents and cytochalasin B., *Bibl. Anat.*, 20, 109, 1981.
175. **Rotrosen, D. and Gallin, J. I.,** Histamine type I receptor occupancy increases endothelial cytosolic calcium, reduces F-actin, and promotes albumin diffusion across cultured endothelial monolayers, *J. Cell Biol.*, 103, 2379, 1986.
176. **Simionescu, M., Simionescu, N., and Palade, G. E.,** Segmental differentiation of cell junctions in the vascular endothelium. The microvasculature, *J. Cell Biol.*, 67, 863, 1975.
177. **Simionescu, M., Simionescu, N., and Palade, G. E.,** Structural basis of permeability in sequential segments of the microvasculature of the diaphragm. II. Pathways followed by microperoxidase across the endothelium, *Microvasc. Res.*, 15, 17, 1978.
178. **Shepro, D. and Hechtman, H. B.,** Endothelial serotonin uptake and mediation of prostanoid secretion and stress fiber formation, *Fed. Proc.*, 44, 2616, 1985.
179. **Northover, B. J.,** The membrane potential of vascular endothelial cells, *Adv. Microcirc.*, 9, 135, 1980.
180. **Keele, C. A. and Armstrong, D.,** Substances producing pain and itch, Philosophical Soc. Monogr. No. 12, Arnold, London, 1964, 33.
181. **Williams, T. J.,** Chemical mediators of vascular responses in inflammation: a two mediator hypothesis, *Br. J. Pharmacol.*, 61, 447, 1977.
182. **Williams, T. J.,** Prostaglandin E_2, prostaglandin I_2 and the vascular changes of inflammation, *Br. J. Pharmacol.*, 65, 517, 1979.
183. **Owen, D. A. A., Poy, E., and Woodward, D. F.,** Evaluation of the role of histamine H_1- and H_2-receptors in cutaneous inflammation in the guinea-pig produced by histamine and mast cell degranulation, *Br. J. Pharmacol.*, 69, 615, 1980.
184. **Mortillaro, N. A., Branger, D. N., Kvietys, P. R., Rutili, G., and Taylor, A. E.,** Effects of histamine and histamine antagonists on intestinal capillary permeability, *Am. J. Physiol.*, 240, G381, 1981.
185. **Mertin, J. and Stackpoole, A.,** Anti-PGE antibodies inhibit *in vivo* development of cell-mediated immunity, *Nature*, 294, 456, 1981.
186. **Moore, T. C., Spruck, C. H., Lami, J. L., and Said, S. I.,** Prompt elevations of PGE_2 and thromboxane B_2 in peripheral node efferent lymph of sheep following drainage area immunization, *Immunopharmacology*, 17, 73, 1989.
187. **Harper, E. I., Beck, J. S., Spence, V. A., and Brown, R. A.,** Effect of histamine and prostaglandin E_2 on the microcirculation in the skin, *Agents Actions*, 24, 102, 1988.
188. **Pentland, A. P., Mahoney, M., Jacobs, S. C., and Holtzman, M. J.,** Enhanced prostaglandin synthesis after ultraviolet injury is mediated by endogenous histamine stimulation, *J. Clin. Invest.*, 86, 566, 1990.
189. **Khalil, Z. and Helme, R. D.,** Sequence of events in substance P-mediated plasma extravasation in rat skin, *Brain Res.*, 500, 256, 1989.
190. **Hay, J. B. and Hobbs, B. B.,** The flow of blood to lymph nodes and its relations to lymphocyte traffic and the immune response, *J. Exp. Med.*, 145, 31, 1977.
191. **Duckworth, J. W., Lance, J. W., Lord, G. D., and Mylecharane, E. J.,** Histamine receptors in the cranial circulation of the monkey, *Br. J. Pharmacol.*, 58 (Abstr.), 444, 1976.
192. **DeLey, G., Weyne, J., Demeester, G., and Leusen, I.,** Response of local blood flow in the caudate nucleus of the cat to intraventricular administration of histamine, *Stroke*, 13, 499, 1982.
193. **Edvinsson, L., Gross, P. M., and Mohamed, A.,** Characterization of histamine receptors in cat cerebral arteries *in vitro* and *in situ, J. Pharmacol. Exp. Ther.*, 225, 168, 1983.
194. **Toda, N. T., Okamura, T., and Miyazaki, M.,** Heterogeneity in the response to vasoconstrictors of isolated dog proximal and distal middle cerebral arteries, *Eur. J. Pharmacol.*, 106, 291, 1984.
195. **Sercombe, R., Verrecchia, C., Philipson, V., Oudart, N., Dimitriadou, V., Bouchaud, C., and Seylaz, J.,** Histamine-induced constriction and dilatation of rabbit middle cerebral arteries *in vitro:* role of the endothelium, *Blood Vessels*, 23, 137, 1986.
196. **Toda, N.,** Mechanism underlying responses to histamine of isolated monkey and human cerebral arteries, *Am. Physiol. Soc.*, 311, 1990.
197. **Chang, J.-Y., Hardebo, J. E., and Owman, CH.,** Differential vasomotor action of noradrenaline, serotonin, and histamine in isolated basilar artery from rat and guinea-pig, *Acta Physiol. Scand.*, 132, 91, 1988.

198. **Rosenblum, W. I., Nelson, G. H., and Weinbrecht, P.,** Histamine elicits competing endothelium-dependent constriction and endothelium-independent dilatation *in vivo* in mouse cerebral arterioles, *Stroke*, 21, 305, 1990.
199. **Watanabe, N., Kawai, Y., and Ohhashi, T.,** Dual effects of histamine on spontaneous activity in isolated bovine mesenteric lymphatics, *Microvasc. Res.*, 36, 239, 1988.
200. **Khan, M. M., Keaney, K. M., Melmon, K. L., Clayberger, C., and Krensky, A. M.,** Histamine regulates the generation of human cytolytic T lymphocytes, *Cell. Immunol.*, 121, 60, 1989.
201. **Schwartz, A., Sutton, S. L., and Gershon, R. K.,** Regulation of in vitro cytotoxic T lymphocyte generation. I. Evidence that killer cell precursors differentiate to effector cells in two steps, *J. Exp. Med.*, 155, 783, 1982.
202. **Nair, M. P. N. and Schwartz, S. A.,** Effect of histamine and histamine antagonists on natural and antibody-dependent cellular cytotoxicity of human lymphocytes *in vitro, Cell. Immunol.*, 81, 45, 1983.
203. **Siegel, J. N., Schwartz, A., Askenase, P. W., and Gershon, R. K.,** T-cell suppression and contrasuppression induced by histamine H2 and H1 receptor agonists, respectively, *Proc. Natl. Acad. Sci. U.S.A.*, 79, 5052, 1982.
204. **Hallberg, T. and Dohlsten, M.,** Binding of histamine-albumin conjugates to human lymphocytes: evidence for labelling of histamine H-1 but not H-2 receptors, *Acta Pathol. Microbiol. Scand. Sect. C*, 95, 199, 1987.
205. **Khan, M. M., Fishwild, D. M., Melmon, K. K., Coutre, St., Engleman, E. G., and Bauer, J. A.,** A subset of human rosetted lymphocytes contains previously unidentified histamine, *Clin. Immun. Immunopathol.*, 52, 147, 1989.
206. **Nakashima, I., Kitamura, Y., Asai, H., and Nakano, K., and Aoi, R.,** Histamine synthesis by mouse T lymphocytes through induced histidine decarboxylase, *Immunology*, 66, 219, 1989.
207. **Dy, M. and Lebel, B.,** Skin allografts generate an enhanced production of histamine and histamine-producing cell-stimulating factor (HCSF) by spleen cells in response to T cell mitogens, *J. Immunol.*, 130, 2343, 1983.
208. **Schneider, E., Pollard, H., Lapault, F., Guy-Grand, D., Minkowski, M., and Dy, M.,** Histamine-producing cell stimulating activity, *J. Immunol.*, 139, 3710, 1987.
209. **Oh, C. and Nakano, K.,** Inhibition by glucocorticoids of mitogen-dependent histamine biosynthesis caused by histidine decarboxylase in cultured mouse spleen cells and periotoneal adherent cells, *Immunology*, 65, 433, 1988.
210. **Lebel, B., Schneider, E., Piquet-Pellorce, C., Machavoine, F., Kindler, V., Luffau, G., and Dy, M.,** Antigenic challenge of immunized mice induces endogeneous production of IL-3 that increases histamine synthesis in hematopoietic organs, *J. Immunol.*, 145, 1222, 1990.
211. **Byron, J. W.,** Mechanism for histamine H_2 receptor induced cell-cycle changes in the bone marrow stem cell, *Agents Actions*, 7, 409, 1977.
212. **Byron, J. W.,** Pharmacodynamic basis for the interaction of cimetidine with bone marrow stem cells (CFU-S), *Exp. Hematol.*, 8, 256, 1980.
213. **Tang, Z. and Xu, Y.,** Comparison of the effects of 4-methyl histamine on the cell cycle response between CFU-S and it subpopulation, *Int. J. Cell Cloning*, 5, 511, 1987.
214. **Schneider, E., Piquet-Pellorce, C., and Dy, M.,** New role for histamine in interleukin 3-induced proliferation of hematopoietic stem cells, *J. Cell. Physiol.*, 143, 337, 1990.
215. **Shounan, Y. and Xu, Y. H.,** The influence of histamine at various concentrations on the cell cycle state of hematopoietic stem cells (CFU-S), *Int. J. Cell Cloning*, 6, 290, 1988.
216. **Xu, Y. H., Wang, Q. R., Pang, Z. J., Li, M. F., and Huang, X. K.,** Histamine H_1 and H_2 receptors and proliferation of hemopoietic progenitors, *Exp. Hematol.*, 17, 666, 1989.
217. **Nakaya, N. and Tasaka, K.,** The influence of histamine on precursors of granulocytic leukocytes in murine bone marrow, *Life Sci.*, 42, 999, 1988.
218. **Nakaya, N. and Tasaka, K.,** Histamine incorporation into murine myeloblasts and promyelocytes, *Biochem. Pharmacol.*, 37, 4523, 1988.
219. **Dy, M., Lebel, B., Kamoun, P., and Hamburger, J.,** Histamine production during the anti-allograft response. Demonstration of a new lymphokine enhancing synthesis, *J. Exp. Med.*, 153, 293, 1981.
220. **Piquet-Pellorce, C., Homo-Delarche, F., and Dy, M.,** Interleukin 1 and/or tumor tumor necrosis factor-a synergize with granulocyte-macrophage colony-stimulating factor to enhance histamine synthesis in hematopoietic cells: role of prostaglandin E_2, *Eur. J. Immunol.*, 19, 1999, 1989.
221. **Hellstrand, K. and Hermodsson, S.,** A cell-to-cell mediated interaction involving monocytes and non-T/CD16$^+$ natural killer (NK) cells is required for histamine H_2-receptor-mediated NK-cell activation, *Scand. J. Immunol.*, 31, 631, 1990.
222. **Pribluda, C., Herberman, R., and Gorelik, E.,** Effect of histamine on NK cell activity and metastasis formation, *Fed. Proc.*, 46, 1500, 1987.

223. **Plaut, M., Pierce, J. H., Watson, C. J., Hanley-Hyde, J., Nordan, R. P., and Paul, W. E.,** Mast cell lines produce lymphokines in response to cross-linkage of FcεRI or to calcium ionophores, *Nature,* 339, 64, 1989.
224. **Guy-Grand, D., Dy, M., Luffau, G., and Vassalli, P.,** Gut mucosa mast cells. Origin, traffic and differentiation, *J. Exp. Med.,* 160, 12, 1984.
225. **Guy-Grand, D., Griscelli, C., and Vassalli, P.,** The mouse gut T lymphocyte: a novel type of T cell. Nature, origin and traffic in mice in normal and graft versus host conditions, *J. Exp. Med.,* 148, 1661, 1978.
226. **Filho, M. A., Dy, M., Lebel, B., Luffau, G., and Hamburger, J.,** *In vitro* and *in vivo* histamine-producing cell-stimulating factor (or IL3) production during *Nippostrongylus brasiliensis* infection: coincidence with self-cure phenomenon, *Eur. J. Immunol.,* 13, 841, 1983.
227. **Piquet-Pellorce, C., Homo-Delarche, F., and Dy, M.,** Interleukin 1 and/or tumor necrosis factor-a synergize with granulocyte-macrophage colony-stimulating factor to enhance histamine synthesis in hematopoietic cells: role of prostaglandin E_2, *Eur. J. Immunol.,* 19, 1999, 1989.
228. **Abbud, M., Dy, M., Lebel, B., Luffau, G., and Hamburger, J.,** *In vitro* and *in vivo* histamine producing cell stimulating factor production during *Nippostrongylus brasiliensis* infection: coincidence with self-cure phenomenon, *Eur. J. Immunol.,* 13, 841, 1983.
229. **Dy, M., Schneider, E., Gastinel, L. N., Auffray, C., Mermod, J. J., and Hamburger, J.,** Histamine-producing cell-stimulating activity. A biological activity shared by interleukin 3 and granulocyte-macrophage colony stimulating factor, *Eur. J. Immunol.,* 17, 1243, 1987.
230. **Dy, M., Schneider, E., Guy-Grand, D., and Lebel, B.,** Histamine-producing cell-stimulating factor (HCSF) and interleukin 3 (IL-3): their effects on histidine and ornithine decarboxylases, in *Lymphokines,* Vol. 15, Schrader, J. W., Ed., Academic Press, New York, 1988, 39.
231. **Davidson, S., Mansour, A., Gallily, R., Smolarski, M., Rofolovitch, M., and Ginsburg, H.,** Mast cell differentiation depends on T cells and granule synthesis on fibroblasts, *Immunology,* 48, 439, 1983.
232. **Ginsburg, H.,** The *in vitro* differentiation and culture of normal mast cells from the mouse thymus, *Ann. N.Y. Acad. Sci.,* 103, 20, 1963.
233. **Ginsburg, H.,** Graft versus host reaction-tissue culture. I. Lysis of monolayers of embryo mouse cells from strains differing in the H-2 histocompatibility locus by rat lymphocytes sensitized *in vitro, Immunology,* 14, 621, 1968.
234. **Ginsburg, H., Ben-Shahar, D., Hammel, I., and Ben-David, E.,** Degranulation capacity of mast cells grown in cell culture in the presence of histamine releaser, *Nature,* 280, 151, 1979.
235. **Ginsburg, H., Ben-Shahar, D., and Ben-David, E.,** Mast cell growth on fibroblast monolayers: two cell entities, *Immunology,* 45, 371, 1982.
236. **Ginsburg, H. and Lagunoff, D.,** The *in vitro* differentiation of mast cells. Culture of cells from immunized mouse lymph nodes and thoracic duct lymph on fibroblast monolayers, *J. Cell Biol.,* 35, 685, 1967.
237. **Ginsburg, H., Naot, Y., and Hollander, N.,** Lysis and necrosis: analysis of two cytotoxic phenomena mediated by lymphocytes, *Israel J. Med. Sci.,* 12, 435, 1976.
238. **Ginsburg, H., Nir, I., Hammel, I., Eren, R., Weissman, B. A., and Naot, Y.,** Differentiation and activity of mast cells following immunization in cultures of lymph node cells, *Immunology,* 35, 485, 1978.
239. **Ginsburg, H., Olson, E. C., Huff, T., Okudaira, H., and Ishizaka, T.,** Enhancement of mast cell differentiation *in vitro* by T cell factor(s), *Int. Arch. Allergy Appl. Immunol.,* 66, 447, 1981.
240. **Schrader, J. W.,** The *in vitro* production and cloning of the P cell, a bone marrow-derived null cell that expresses H-2 and I antigens has mast cell-like granules and is regulated by a factor released by activated T cells, *J. Immunol.,* 126, 452, 1981.
241. **Schrader, J. W., Clark-Lewis, I., Crapper, R. M., Leslie, K. B., Schrader, S., Varigos, G., and Ziltener, H. J.,** The panspecific hemopoietic interleukin 3: physiology and pathology, in *Lymphokines,* Vol. 15, Schrader, J. W., Ed., Academic Press, New York, 1988, 281.
242. **Brandes, L. J., Sukhu, B., and Bogdanovic, R. P.,** Does intracellular histamine mediate mast cell histamine release? *Biochem. Biophys. Res. Commun.,* 167, 665, 1990.
243. **Read, G. W. and Lagunoff, D.,** Antagonism of the final common pathway of mast cell histamine secretion by arylalkylamines, *J. Pharmacol. Exp. Ther.,* 237, 357, 1986.
244. **Rabenstein, D. L., Ludowyke, R., and Lagunoff, D.,** Protein nuclear magnetic resonance studies of mast cell histamine, *Biochemistry,* 26, 6923, 1987.
245. **Purcell, W. M., Cohen, D. L., and Hanahoe, T. H. P.,** Comparisons of histamine and 5-hydroxytryptamine content and secretion in rat mast cells isolated from different anatomical locations, *Int. Arch. Allergy Appl. Immunol.,* 90, 382, 1989.
246. **WoldeMussie, E., Aiken, D., and Beaven, M. A.,** Evidence for rapid histamine turnover and loss of histamine from immature rat mast cells, *Biochem. Pharmacol.,* 35, 903, 1986.

247. **Beaven, M. A., Aiken, D. L., WoldeMussie, E., and Soll, A. H.,** Changes in histamine synthetic activity, histamine content and responsiveness to compound 48/80 with maturation of rat peritoneal mast cells, *J. Pharmacol. Exp. Ther.,* 224, 620, 1983.
248. **WoldeMussie, E., Aiken, D. L., and Beaven, M. A.,** Changes in histidine uptake and histamine synthesis during the growth cycle of rat basophilic leukemia (2H3) cells, *J. Pharmacol. Exp. Ther.,* 232, 20, 1985.
249. **WoldeMussie, E. and Beaven, M. A.,** Alpha-fluoromethylhistidine. Kinetics of uptake and inhibition of histamine synthesis in basophil (2H3) cell cultures, *Mol. Pharmacol.,* 28, 191, 1985.
250. **Taguchi, Y., Watanabe, T., Kubota, H., Hayashi, H., and Wada, H.,** Purification of histidine decarboxylase from the liver of fetal rats and its immunochemical and immunohistochemical characterization, *J. Biol. Chem.,* 259, 5214, 1984.
251. **Haak-Frendscho, M., Dinarello, C., and Kaplan, A. P.,** Recombinant human interleukin-1 beta causes histamine release from human basophils, *J. Allergy Clin. Immunol.,* 82, 218, 1988.
252. **Ahlstedt, S., Hammarstrom, I., Into-Malmberg, M.-B., and Lehtonen, P.,** Differences in the kinetics of histamine formation and granulation of human basophilic cells from bone marrow, peripheral blood, and cord blood, *Scand. J. Immunol.,* 26, 631, 1987.
253. **Mayer, P., Valent, P., Schmidt, G., Liehl, E., and Bettelheim, P.,** The *in vivo* effects of recombinant human interleukin-3: demonstration of basophil differentiation factor, histamine-producing activity, and priming of GM-CSF-responsive progenitors in nonhuman primates, *Blood,* 74, 613, 1989.
254. **Carlsson, R., Dohlsten, M., and Sjogren, H. O.,** Histamine modulates the production of interferon-γ and interleukin-2 by mitogen-activated human mononuclear blood cells, *Cell. Immunol.,* 96, 104, 1985.
255. **Endo, Y., Suzuki, R., and Kumagai, K.,** Macrophages can produce factors capable of inducing histidine decarboxylase, a histamine-forming enzyme, *in vivo* in the liver, spleen, and lung of mice, *Cell. Immunol.,* 97, 13, 1986.
256. **Sternberg, E. M., Wedner, H. J., Leung, M. K., and Parker, C. W.,** Effect of serotonin (5-HT) and other monoamines on murine macrophages: modulation of interferon-y induced phagocytosis, *J. Immunol.,* 138, 4360, 1987.
257. **Falus, A. and Meretey, K.,** Effect of histamine on the gene expression and secretion of complement components C2, factor B and C3 in murine macrophages — an opposite signal processing via H1 and H2 receptors, *Mol. Immunol.,* 25, 1093, 1988.
258. **Okamoto, H., Oh, C. K., and Nakano, K.,** Possible involvement of adenosine 3′:5′-cyclic monophosphate and extracellular calcium ions in histamine stimulation of interleukin-1 release from macrophage-like P388D1 cells, *Immunology,* 70, 186, 1990.
259. **Sinclair, M. C., Moore, T. C., Lemmi, C. A. E., and Orlando, J. C.,** Increase in the urinary excretion of histamine following SRBC immunization of rats, *Am. J. Surg.,* 126, 624, 1973.
260. **Granger, G. A. and Moore, T. C.,** Accumulation of lymphotoxin secreting lymphoid cells in rat heart allografts, *Surgery,* 76, 542, 1974.
261. **Moore, T. C., Lemmi, C. A. E., Orlando, J. C., and Pinkerton, W.,** Correlation of splenic histidine decarboxylase activity and antibody formation following SRBC immunization, *Transplantation,* 11, 347, 1971.
262. **Moore, T. C. and Lawrence, W., Jr.,** Suppression of antibody formation by histidine decarboxylase inhibitors, *Transplantation,* 8, 224, 1969.
263. **Koppelmann, L. E., Moore, T. C., and Porter, D. D.,** Increased plasma kallikrein activity and tumor growth suppression associated with intralesional bradykinin injections in hamsters, *J. Pathol.,* 126, 1, 1978.
264. **Koppelman, L., Moore, T. C., Lemmi, C. A. E., and Porter, D. D.,** Effects of bradykinin injection into syngenic SV-40 hamster tumors on rate of tumor growth and splenic histidine decarboxylase activity, *IRCS Med. Sci.,* 2, 1272, 1974.
265. **Moore, T. C., Koppelmann, L. E., and Lemmi, C. A. E.,** An experimental study of tumor growth and splenic histamine enzymes, *Surg. Gynecol. Obstet.,* 146, 559, 1978.
266. **Burtin, C., Noirot, C., Paupe, J., and Scheinmann, P.,** Decreased blood histamine levels in patients with solid malignant tumours, *Br. J. Cancer,* 47, 367, 1983.
267. **Galoppin, L., Noirot, C., Wastiaux, J. P., Scheinmann, P., Paupe, J., and Burtin, C.,** Comparison between number of basophils, blood histamine, and histamine release in cancer and noncancer patients, *J. Allergy Clin. Immunol.,* 84, 501, 1989.
268. **Burtin, C., Scheinmann, P., Salomon, J. C., Lespinats, G., and Canu, P.,** Decrease in tumour growth by injections of histamine or serotonin in fibrosarcoma-bearing mice: influence of H_1 and H_2 histamine receptors, *Br. J. Cancer,* 45, 54, 1982.
269. **Burtin, C., Scheinmann, P., Salomon, J. C., Lespinats, G., and Canu, P.,** The influence of intraperitoneal injections of histamine on tumour growth in fibrosarcoma-bearing mice, *Cancer Lett.,* 12, 195, 1981.

270. **Watanabe, T., Kitamura, Y., Maeyama, K., Go, S., Yamatodani, A., and Wada, H.,** Absence of increase of histidine decarboxylase activity in mast cell-deficient W/W mouse embryos before parturition, *Proc. Natl. Acad. Sci. U.S.A.*, 78, 4209, 1981.
271. **Burtin, C., Ponvert, C., Fray, A., Scheinmann, P., Lespinats, G., Loridon, B., Canu, P., and Paupe, J.,** Inverse correlation between tumor incidence and tissue histamine levels in w/w^v, w^v/t, and t/t mice, *JNCI,* 74, 671, 1985.
272. **Bloksma, N., van de Wiel, P., Hofhuis, F., Kuper, F., and Willers, J.,** Role of histamine in the antitumour activity of endotoxin, *Cancer Immunol. Immunother.*, 17, 33, 1984.
273. **Pribluda, C., Herberman, R.,and Gorelik, E.,** Effect of histamine on NK cell activity and metastasis formation, Fed. Proc., 46, 1500, 1989.

Chapter 5

SEROTONIN (5-HT)

I. INTRODUCTION

A. GENERAL

Serotonin, like histamine, is a major vasoactive neurotransmitter and neuromodulator molecule with a long evolutionary history of involvement in biological defense and communication. Also like histamine, it is a decarboxylated amino acid product and in its early history was a pest. A further similarity to histamine is to be found in its wide distribution in the animal and plant kingdoms. In the plant kingdom, it is found in fruits, nuts, beans, and the stings of stinging nettles.

It is both a defense and a communications molecule in the most primitive of multicellular animals, the cnidarian coelenterates, a phylum that includes animals such as corals, the common jellyfish (*Aurelia*), Portuguese man-of-war (*Physalia*), *Hydra*, and sea anemone (*Epiactis*) and, as recently described by Descarries and associates[1] of Montreal, is present in the pennatulid anthozoan *Renilla koellikeri*. These are the most primitive organisms endowed with a nervous system. The cnidarian coelenterates are considered the "pioneer" phylum with respect to the evolutionary emergence and development of nervous systems. The cnidarian phylum name derives from the Greek word for the stinging nettle.

Serotonin has been identified biochemically in all three classes of cnidarians: anthozoans (*Metridium*[2]), hydrozoans (*Hydra littoralis*[3]), and scyphozoans (medusa *Cynea lamarcki*[4]). In *Hydra oligactis* and *Hydra vividis*, serotonin has been detected histochemically in the stenotele nematocysts of the nematocytes of these animal's tentacles.[5] These nematocysts are specialized synaptosome-like urticant organelles that are found singly in nematocytes. In *Renilla*, serotonin has been found in both the neurons of the nerve-net nervous system and in the nematocyst-containing nematocytes of the tentacles.[1] Serotonin, thus, appears to be a significant constituent and to play a significant food-gathering and defense role in the chemical weaponry of these most primitive of the animal carnivores.

The mind-boggling physiological effects of naturally occurring congeners of serotonin of plant origin have been known since prehistory. These "tryptamines" of plant origin, with their potent effects on brain function, have long been used by aboriginal tribes in the Americas in tribal and religious rites to induce mental changes considered appropriate to these rites. The cahobe bean of Caribbean shores had both DMT (*N,N*-dimethyltryptamine) and bufotenine (its 5-hydroxy derivative) as active principles. Aboriginal rites of the Central Americas also included a variety of hallucinogenic mushrooms (*Psilocybe mexicana*) whose active hallucinogenic molecules included such 4-substituted tryptamines as psilocine (4-hydroxy-*N,N*-methyltryptamine). The synthetic hallucinogen LSD (lysergic acid diethelamide) is also a 4-substituted tryptamine. The observation that LSD is structurally similar to serotonin (5-HT) and that it could inhibit smooth-muscle responses to 5-HT heightened interest in this component of 5-HT activity in the early days of its postdiscovery study.

Serotonin was originally discovered in enterochromaffin cells of the gastrointestinal tract and has subsequently been found in platelets and vascular tissue and in the central and peripheral nervous systems.

Interest of the biomedical scientific community in 5-HT has fluctuated wildly since its initial discovery and early study as "enteramine" by Vitorio Erspamer of Italy in the late 1930s and early 1940s.[6] Enteramine was a potent vasoactive substance extracted from the enterochromaffin cells of the gastrointestinal mucosa of the rabbit. In 1952 Erspamer and

Asero[7] identified the chemical nature and structure of enteramine as 5-hydroxytryptamine (5-HT). The vascular actions of 5-HT, hence, were its first identified biological activity. Erspamer also was the first to demonstrate the widespread distribution in nature of this important amine.[8-11] Identified involvements in the nervous and immune response systems came later.

B. STRUCTURE, FORMATION, STORAGE, RELEASE, AND METABOLISM

5-HT is formed by the decarboxylation of the 5-hydroxy form of the amino acid L-tryptophan (Figure 1). The rate-limiting step in its endogenous synthesis is the hydroxylase of tryptophan by the enzyme tryptophan 5-hydroxylase. The 5-hydroxytryptophan is decarboxylated to 5-HT by the nonspecific L-amino acid decarboxylase. Formed 5-HT is taken up into secretory granules and stored.

Platelets actively accumulate 5-HT during their passage through intestinal blood vessels. This anatomical area has comparatively high levels of 5-HT resulting from its secretion by enterochromaffin cells.

The uptake and storage of 5-HT in granules of platelets and synaptosomes is similar to that of the catecholamines, with similar transport, release, and reuptake mechanisms, including reserpine disruption of storage and depletions of brain concentrations. An amount of 5-HT equal to the total body content of this amine is synthesized endogenously each day. Brain turnover of serotonin in the brain is estimated at 1 h and at 17 h in the gastrointestinal tract. These findings suggest a role of considerable importance for 5-HT in normal physiology at a variety of levels.

The metabolic degradation of serotonin is by oxidative deamination by monoamine oxidase (MAO) to 5-hydroxyindoleacetaldehyde. This is further degraded by oxidation to 5-hydroxyindoleacetic acid (5-HIAA) or by reduction by alcohol dehydrogenase to 5-hydroxytryptophol (5-HTOL). 5-HIAA is the principle metabolite and is excreted in the urine, along with 5-HTOL and its metabolites.

II. HISTORICAL BACKGROUND

Pre-1946. Since the middle of the 19th century, physiologists had known of a constrictor material that appeared in serum when blood was allowed to clot. It went by a variety of names, including ''vasotonin'' and ''serotonin''. It also was found to be present in defibrinated blood as a nuisance substance to physiologists who wished to use this blood in perfusion experiments. These resourceful physiologists discovered that this substance could be removed by passing the blood through the lungs.

1946. Erspamer,[6] who had been working in Italy for years on the properties of intestinal enterochromaffin cells and had obtained in acetone extracts of intestine a gut-stimulating factor of basic nature that he believed to be an amine and had named ''enteramine'', reported his suggestion that this substance (enteramine) might be an indole alkylamine.[8] He had accumulated a considerable amount of pharmacological data on the activity of enteramine and had found it to be present in many vertebrate and invertebrate tissues.[11]

1948. The vasoconstrictor substance called ''serotonin'' was isolated in pure form as a crystalline complex as a part of an intensive research effort into the causes of hypertension, at the Cleveland Clinic in Cleveland, OH (Rapport, Green, and Page,[12] 1948; Page,[13] 1954).

1949. The active principle of ''serotonin'' is characterized as 5-hydroxytryptamine and is called serotonin (Rapport,[14] 1949).

1951. Serotonin (5-hydroxytryptamine) is synthesized (Hamlin and Fischer,[15] 1951).

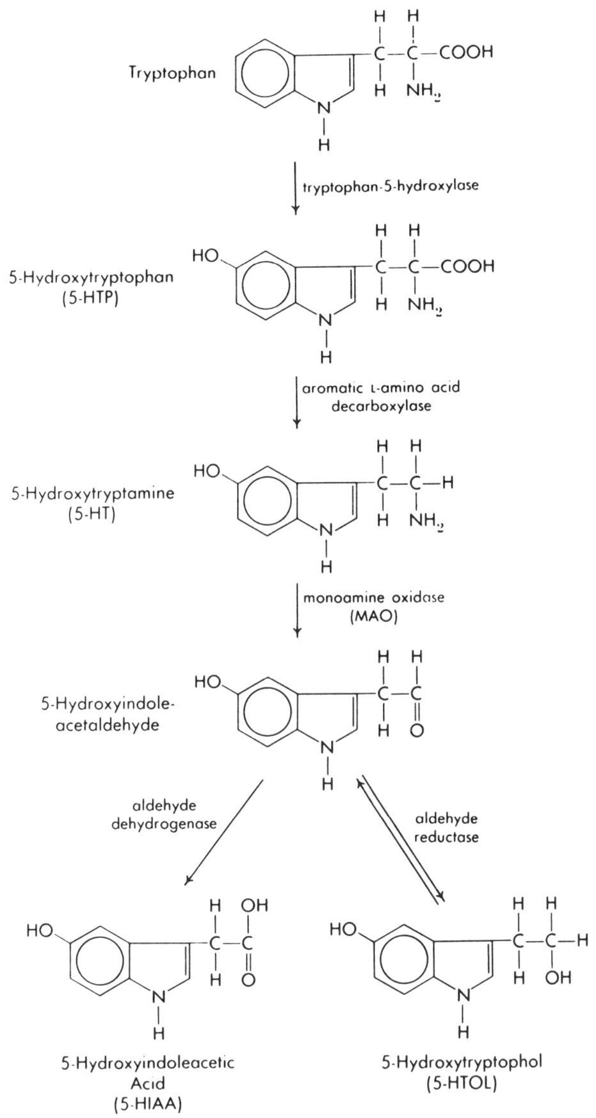

FIGURE 1. The formation of serotonin/5-hydroxytryptamine from the amino acid tryptophan by decarboxylation of its 5-hydroxy products and the metabolic degradation of 5-hydroxytryptamine by monoamine oxidase (MAO) enzyme activity and subsequent steps. (From Douglas, W. W., *Goodman and Gilman's The Pharmacological Basis of Therapeutics,* 6th ed., Macmillan, New York, 1980. With permission.)

1952. Enteramine is identified as 5-hydroxytryptamine (Erspamer and Asero,[7] 1952).

1952–1982. An explosion in serotonin research occurred that was restrained by competing enthusiasm for prostaglandin research. Serotonin had been found to be widely distributed in nature and to have a multiplicity of pharmacological actions — often confusing and contradictory. Its roles as a gut substance of importance and as a major neurotransmitter with strong psychopharmacologic involvements attracted the most continuing interest.

III. RECEPTORS

A. GENERAL

The identification and sorting out of 5-HT receptor types and subtypes, as well as those of other vasoactive neurotransmitters, has involved the use of a wide spectrum of methodologies and techniques. These have included pharmacological, biochemical, electrophysiological, radiolabeled-ligand binding, and molecular biological techniques.

The identification of 5-HT as the chemical nature of both serotonin and enteramine, in the late 1940s and early 1950s, triggered a search for its receptor or receptors. The exciting advances relating to histamine receptor antagonist molecules doubtlessly spurred this search.

Studies by Rocha e Silva, Valle, and Picarelli[16] published in 1953 and by Gaddum and Picarelli[17] in 1957 were the first to establish the existence of a 5-HT receptor. These studies demonstrated that cocaine had antagonist activity against 5-HT-induced (and neuronally mediated) contractions of guinea pig isolated ileum. It was of particular interest that cocaine at low concentrations specifically antagonized the 5-HT effect without affecting contractions induced by pilocarpine or histamine. This receptor was characterized as an ''M'' receptor. The 1978 and 1979 investigations by Fozard et al.[18,19] identified the receptor that mediated the neuronally mediated 5-HT-induced tachycardia response in the isolated rabbit heart as belonging to the same M-type receptor described above for the isolated guinea pig ileum.

These studies and the 1979 publication by Peroutka and Snyder[20] that identified two distinct 5-HT receptors in the central nervous system by radioligand binding techniques triggered the 1980s avalanche of studies and publications relating to 5-HT (serotonin) and its multiple, varied, widely distributed, and likely quite important receptor types and subtypes.

B. 5-HT$_1$ RECEPTORS

1. 5-HT$_{1A}$ Receptor

The 5-HT$_{1A}$ subtype of the 5-HT-receptor ''family'' was the first 5-HT-receptor type or subtype to be cloned and sequenced. This was achieved and reported in mid-1988 by Fargin et al.[21] They found that a receptor genomic clone (G-21) that resembled a beta-adrenergic receptor sequence encoded the 5-HT$_{1A}$ receptor. The demonstration of the existence of more than one class of 5-HT$_1$ receptor (by Pedigo et al. in 1981[22]) and the development of 8-OH-DPAT in 1982 (by Gozlan et al.[23]) as a highly selective radioligand for the specific labeling of 5-HT$_{1A}$ sites in brain membranes were important steps in this process. The discovery in 1987 that several 5-HT$_{1A}$ ligands (such as buspirone, isapirone, and gepirone) had anxiolytic and antidepressant effects in animal models and man gave added significance and impetus to these efforts.[24,25] The studies of Fargin et al.,[21] cited above, also demonstrated high-affinity binding of 8-OH-DPAT to a G-21 genomic-clone-transfected monkey kidney cell line.

The 5-HT$_{1A}$ receptor is a G-protein-linked receptor, with its greatest amino acid sequence variability from other cloned biogenic amines being found in its third cytoplasmic loop. Autoradiographic studies have shown the highest densities of 5-HT$_{1A}$ binding sites to be in the CA1 and dentate gyrus (DG) regions of the hippocampus and in the raphe nuclei, home of the cell bodies of the serotonin nervous system. The key and distinguishing features of this 5-HT-receptor subtype, in comparison with other 5-HT receptors and receptor subtypes, are shown in Table 1.

2. 5-HT$_{1B}$ (5-HT$_{1B}$ and 5-HT$_{1D}$)

The 5-HT$_{1B}$ and 5-HT$_{1D}$ receptors appear to be homologous, with significant pharmacological, anatomical, and biochemical similarities. Their principal differences are species

TABLE 1
5-HT Receptors

Receptor		Cloned	Second messenger	Selective agonists	Selective antagonists	Regional distribution
5-HT_1	5-HT_{1A}	Yes	Inhibition of adenylate cyclase	8-OH-DPAT Busprone MDL 72832	None	Hippocampus CA1 Dentate gyrus Raphe nuclei
	5-HT_{1B} $(1_B + 1_D)$	No	Inhibition of adenylate cyclase	RU 24969	None	Rat and mouse ($5\text{-HT}1_B$) Globus pallidus Dorsal subiculum Substantia nigra Bovine ($5\text{-HT}1_D$) Caudate Basal ganglia
5-HT_2	5-HT_2	Yes[a]	PI	DOB	Ketanserin Ritanserin	Mesolimbic brain areas Frontal cortex Nucleus accumbens Tuberum olifactorium Smooth muscle cells Platelets
	5-HT_{1C}	Yes[a]	PI	None	Mianserin Mesulergine LSD	Cerebral cortex Choroid plexus Basal ganglia Hippocampus Hypothalamus Spinal cord
	5-HT_3	No	PI Cyclic GMP	2-Methyl-5HT	Odansetron Zacopride Granisetron MDL 72222 ICS 205-930	Peripheral NS Spinal cord substantia gelatinosa Cerebral cortex limbic areas Lymphocytes NK cells Monocytes/macrophages
	5-HT_4	No	Cyclic AMP	5-Methoxytryptamine 5-CT	None	Mouse embryo colliculi Adult guinea pig hippocampus

[a] These receptors share a 51% amino acid homology, including identical 5th transmembrane domains.

specificities with 5-HT_{1B} being found in pigs, cows, chickens, turtles, frogs, and man, but not in mice and rats. These receptors share a low affinity for 8-OH-DPAT and are here grouped and combined as 5-HT_{1B}. This receptor functions as an "autoreceptor" in the rat, guinea pig, and human brain cortex. It is negatively coupled to adenylate cyclase in the substantia nigra of both rats and calves. This receptor has not as yet been cloned.

C. 5-HT_2 RECEPTORS

1. 5-HT_2

5-HT_2 receptors have been shown to have specific locations in mesolimbic brain areas, including the frontal cortex, the nucleus accumbens, and the tuberum olfactorium.[26] They also have been identified on smooth-muscle cells and on platelets. Their initial detection was obtained in radioligand binding studies employing selective agonists and antagonists. DOB was an agonist ligand. Ketanserin and vitanserin were the principal antagonist ligands that were employed in the receptor-identification studies.[20,26-28] The 5-HT_2 receptor has been observed to function through the PI (phosphatidyl inositol) second-messenger system with stimulation of the turnover of phosphatidyl inositol.

The cloning and sequencing of a 5-HT_2 receptor, encoded by 449 amino acids, was reported in 1988 by Pritchett et al.[29] They isolated a functional 5-HT_2 receptor cDNA from a cloned cDNA library, employing probes generated from the amino acid sequences 88–104 and 134–149 of the membrane-spanning regions 2 and 3 of the cloned 5-HT_{1C} receptor (currently classified as a 5HT_2-type receptor). The cDNA library was from rat forebrain, an area known to be rich in 5-HT receptors. The cloned and sequenced 449 amino acid 5-HT_2 receptor shared a 51% homology with the 460 amino acid 5-HT_{1C} receptor. When expressed in a mammalian cell line, the 5-HT_2 sequence displayed a pharmacological profile consistent with a 5-HT_2-receptor assignment. Radioligand studies revealed spiperone binding with high affinity, which was potently inhibited by ketanserin and mianserin, but not by 8-OH-DPAT or MDL-72222 (5-HT_1 and 5-HT_3 antagonists). The expressed clone also was shown to mediate 5-HT stimulation of intracellular calcium levels as well as phosphoinositide hydrolysis. The amino acid sequence of the human 5-HT_2 receptor has been found to differ 10% from that of the rat 5-HT_2 receptor.[30]

Administration of a 5-HT_2 receptor agonist to rats produces head-twitch responses and clonic seizures. In humans, 5-HT_2-agonist administration results in hallucinations. 5-HT_2 antagonists may block these effects, but when given by themselves produce remarkably few effects. This is one of many serotonin (5-HT) mysteries that clearly need further study.

2. 5-HT_{1C}

The 5-HT_{1C} receptor was discovered in the porcine choroid plexus by mesulergine radioligand binding studies by Pazos et al. in 1984.[31,32] Binding of this receptor in the rat choroid plexus by LSD was reported in 1985.[33] The choroid plexus of both pigs and rats is a particularly useful area for the study of 5-HT_2 receptors, due to its high 5-HT_{1C}-receptor density (10^6 per cell) and to the lack of significant levels of other 5-HT receptors. Like the 5-HT_2 receptor, the 5-HT_{1C} receptor functions through the PI second-messenger system.

The cloning and molecular characterization of the 5-HT_{1C} receptor was reported in 1988 by Julius et al.[34] The cloning strategy involved sucrose gradient fractionation of adult rat choroid plexus total RNA for enrichment of 5-HT receptor mRNA. A cDNA library was constructed from the choroid-plexus-size fractioned mRNA population in an expression vector (λZAP), permitting the *in vitro* transcription of cDNA inserts. RNA transcribed from populations of cDNA clones were screened for their ability to induce 5-HT-induced inward currents in injected and patch-clamped *Xenopus* oocytes. A single functional 5-HT_{1C} receptor was purified from a large population of clones by procedures of sib selection (the progressive subdivision of a pool of cDNA clones, giving positive responses into smaller pools until a single positive clone is obtained). The amino acid sequence of this 5-HT_{1C} receptor reveals that it belongs to the rhodopsin family of G protein-coupled receptors that have been identified as traversing cellular cytoplasmic membranes seven times.

Injection of RNA transcribed *in vitro,* from the above isolated 5-HT_{1C}-receptor clone, into *Xenopus* oocytes was observed to elicit serotonin sensitivity of the 5-HT_{1C}-receptor type. Superfusion of injected oocytes with mianserin (a 5-HT_{1C} blocker) blocked inward current evoked by 5-HT. Similarly, membranes from mouse 3T3 fibroblasts transformed with this clone were found to exhibit high-affinity binding sites for ligands that interact with the 5-HT_{1C} receptor on choroid plexus cells, including radiolabeled LSD. The relative ability of specific antagonists to inhibit radiolabeled-LSD binding to the transformed 3T3 fibroblast cell membranes paralleled their potency on choroid plexus cell membranes. Mianserin was about two orders of magnitude more effective in displacing LSD binding than was spiperone. These transformed 3T3 fibroblasts exhibit an increase in intracellular Ca^{++} concentrations in response to added serotonin.

In situ hybridization and RNA blot analysis revealed that the 5-HT_{1C} receptor is widely distributed throughout the central nervous system anatomically. *In situ* hybridization demonstrated a heavy silver grain density associated with epithelial cells of the choroid plexus in the third, fourth, and lateral ventricles as well as in numerous neuronal cell groups throughout the central nervous system. RNA blot analysis revealed 5-HT_{1C}-receptor presence at significant levels in the choroid plexus and in other regions of the brain, including the cortex, basal ganglia, hypothalamus, hippocampus, olfactory bulb, pons medulla, and the spinal cord. 5-HT_{1C}-receptor mRNA was not detected in the cerebellum, liver, kidney, intestine, heart, or lung.

The high density (10^6 per cell) of 5-HT_{1C} receptors in the choroid plexus is likely of great importance in central nervous system physiology and function and indicates an important neurovascular role and "connection" for this 5-HT receptor. In this regard, it is of interest that serotoninergic nervous system innervation of the choroid plexus is largely from fibers that terminate in the vasculature.[33,35] The choroid plexus produces and secretes the bulk of the cerebrospinal fluid of the brain and acts as a bidirectional transport system between blood and brain compartments.[36] The secretory epithelial cells of the choroid plexus, which are responsible for these activities, have been found to form a monolayer at the surface of the choroid plexus, with their apical surfaces being in direct contact with the cerebrospinal fluid of the ventricles. The blood-brain barrier of the choroid plexus is somewhat unusual in that the capillaries contain leaky junctions between individual endothelial cells. This arrangement permits components of blood plasma to leak out freely and to bathe the basal and lateral surfaces of these secretory epithelial cells. The epithelial cells are joined by tight junctions that prevent the direct transfer of plasma components into the cerebrospinal fluid. Accordingly, the function of this unique blood-brain barrier is controlled and influenced by the transport, productive, and secretory properties and capabilities of these epithelial cells.

The concentration of 5-HT_{1C} receptors in these choroid plexus epithelial cells, as identified above by mRNA heavy grain density in *in situ* hybridization,[34] confirms earlier studies localizing the 5-HT_{1C} receptor in these cells by high-resolution autoradiographic-labeling study of these and adjacent cells,[33] including receptors in choroid plexus tumors of epithelial origin,[37] and in studies on cultured choroid plexus cells in tissue culture.[38] Choroid plexus tumors of epithelial cell origin are of interest because these tumors, which are encountered in certain transgenic mice, are composed of epithelial cells that have the highest density of 5-HT_{1C} receptors encountered in the central nervous system.

Serotonin has been found (by Faraci et al.[39]) to exert strong effects on blood flow to the choroid plexus, and these efforts are quite different from serotonin effects on the blood flow to other vascular beds. There is clearly much more to be learned in this important area of critical and interrelated serotonin neuro- and vascular activity.

D. 5-HT_3 RECEPTOR

The 5-HT_3 serotonin receptor was the first to be identified and was initially termed the "M receptor", as cited above in the general introduction to this section, and subsequently (1986) reclassified as 5-HT_3 receptors.[40]

Although initially identified in peripheral tissues, 5-HT_3 receptors have been found to be widely distributed throughout the central and peripheral nervous systems and in areas of importance in both the cardiovascular and immune response system areas, including lymphocytes, NK cells, monocytes, and macrophages (see Table 1).[41] In the rat forebrain, the highest densities of 5-HT_3-receptor sites have been shown to be within subnuclei of the amygdala (including the amygdalo-hippocampal area), cortical areas (primary olfactory and entorhinal), and the hippocampus.[42]

Serotonin functions through 5-HT_3 receptors by both the PI and cyclic GMP second-messenger stimulatory systems. In a 1989 report, serotonin was shown to raise cyclic-GMP levels in a widely used neuronal cell line (neuroblastoma $\times$ glioma hybrid cells, clone 108CC15) by stimulation of 5-HT_3 receptors.[43] In 1975 two reports described serotonin as stimulating monocyte and leukocyte chemotaxis and increasing the cyclic GMP levels of these cells.[44,45]

5-HT_3 receptors also have been shown to be involved in pain-response mediation, when applied to exposed blister bases in man.[46] These 5-HT_3 receptors also are found in the important pain impulse and transmission area of lamina I of the dorsal horn of the spinal cord, the substantia gelatinosa.[47,48] These substantia gelatinosa receptors are reduced markedly in rats treated neonatally with capsaicin. Capsaicin-sensitive afferent nerve fibers are known to conduct specific nociceptive information.[49] These findings suggest that 5-HT_3 receptors occur presynaptically on both peripheral and central primary afferent terminals.

E. 5-HT_4 RECEPTOR

Bockaert and associates,[50,51] in 1988 and 1989, described a "nonclassical" 5-HT_4 serotonin receptor that stimulated cyclic AMP production in a dose-dependent manner in cell cultures from mouse embryo colliculi and in adult guinea pig hippocampal membranes. This may turn out to be an important feedback-inhibitory receptor for serotonin.

IV. NERVOUS SYSTEM

A. GENERAL

Like the histaminergic nervous system, the neuronal cell bodies of the serotoninergic nervous system have a rather limited and discrete anatomical location in the central nervous system of mammals. This system has been most widely studied in the rat and the cat of mammalian species. The mammalian serotoninergic neuronal cell bodies are located largely in the raphe nuclei and reticular formation of the brain stem, with the raphe nuclei being the principal sites of neuronal cell body location (Figure 2).

Serotonin, also like histamine, may be synthesized and stored in neuronal cells, including their synaptosomes. Antibodies to conjugated serotonin and to its rate-limiting forming enzyme, tryptamine hydroxylase, have been of value in identifying and tracking the serotoninergic nervous system in both vertebrates and invertebrates.

In the mammalian brain stem, the major raphe nuclei of the serotoninergic nervous system have a rather sharp anatomical delineation with respect to those largely projecting their fibers rostrally (cephalad) and those projecting their fibers caudally (Figures 3 and 4).

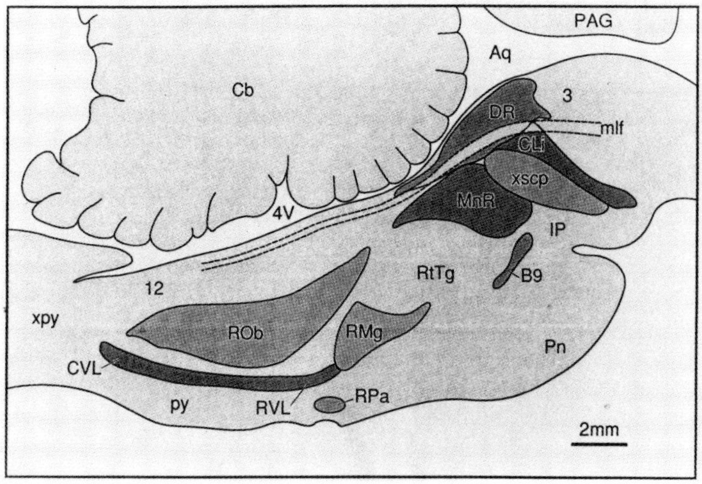

FIGURE 2. Schematic presentation of a sagittal projection through a cat brain stem of the serotoninergic neuronal cell body groups of the raphe nuclei and the reticular formation. The rostral division (to the right) that includes the dorsal raphe (DR) nucleus, the median raphe (MnR) nucleus, the caudal linear (CLi), and B9 supralemniscal cell group are clearly separate from the caudal divisions (left) that include the raphe obscurus (ROb), raphe magnus (RMg), and raphe pallidus (RPa) nuclei, as well as the collections of serotoninergic cell bodies in the rostral and caudal reticular formation of the medulla (CVL and RVL). (From Törk, I., *Ann. N.Y. Acad. Sci.*, 600, 9, 1990. With permission.)

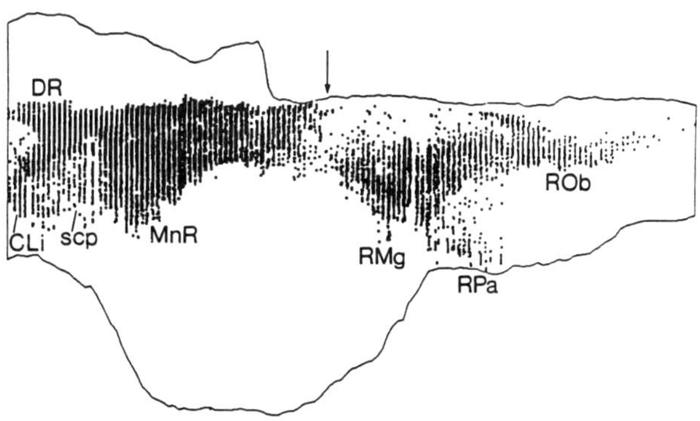

FIGURE 3. Computer-generated sagittal section of the serotoninergic raphe nuclei of the human brain stem, showing the clear anatomical separation of the rostral and caudal cell body groups (arrow at top). (From Törk, I., *Ann. N.Y. Acad. Sci.*, 600, 9, 1990. With permission.)

The large and important dorsal and medial raphe nuclei are in the rostral group, with the obscurus, magnus, and pallidus nuclei in the caudal group. The rostral group is located in the midbrain and rostral pons, and the caudal group largely in the medulla oblongata. The extensive projections of the serotoninergic cell body groups of the brain stem to the forebrain, brain stem, and spinal cord are pictured schematically in Figure 5.

A dual serotoninergic system innervating the forebrain has been suggested by Törk,[52] with different projections of differing axonal types from the dorsal and median raphe nuclei (Figure 6). He also has suggested differential spinal cord projections of the caudal group, with the principal target of the raphe magnus nucleus being the superficial areas of lamina I of the dorsal horn, the substantia gelatinosa, which is so important in thermal-heat sensory pain impulse transmission and associated substance P release in this area (Figures 7 and 8).

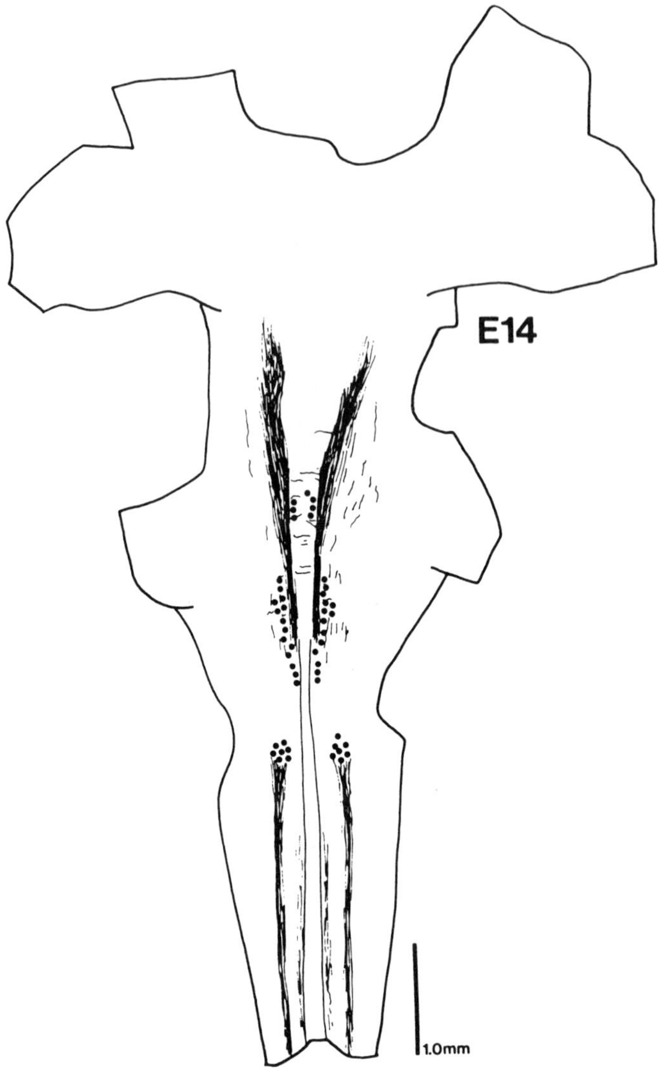

FIGURE 4. Cameralucida drawing showing the rostral and caudal cell bodies (black dots) and fiber projections (black lines) of the serotoninergic system of the embryonic rat brain stem at embryonic day 14 (E14). Some caudal fiber projections from the rostral cell body groups may be seen. (From Törk, I., *Ann. N.Y. Acad. Sci.*, 600, 9, 1990. With permission.)

B. INVERTEBRATES

1. General

Like histamine, serotonin is a major neurotransmitter in the invertebrate nervous system, with major involvements in both biological defense and in communication. It has been tracked back phylogenetically to the most primitive of nervous systems in the most primitive of multicellular animals, the cnidarian coelenterates. It also has been identified as an important learning and memory molecule, in extensive studies of two much-investigated marine molluscs, *Aplysia* and *Hermissenda.*

2. Cnidaria/Coelenterata

The cnidarian coelenterates are the most primitive of multicellular animals and the most primitive to possess a nervous system. Contemporary cnidarians have only two body layers,

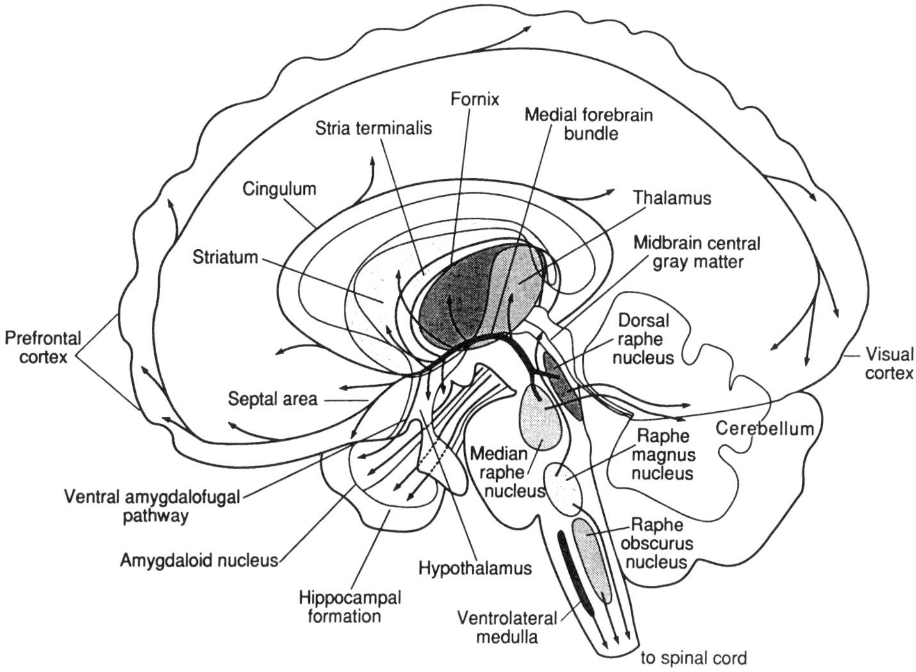

FIGURE 5. Schematic portrayal of the extensive forebrain, brain stem, and spinal cord projections of fibers from the serotoninergic neuronal cell body groups of the raphe nuclei and the ventrolateral medulla. (From Törk, I., *Ann. N.Y. Acad. Sci.*, 600, 9, 1990. With permission.)

endoderm and ectoderm, and are considered contemporary representatives of the earliest evolutionary organisms possessing a nervous system.

The phylum Cnidaria is divided into three classes (Anthozoa, Scyphozoa, and Hydrozoa) and includes such diverse marine animals as the common jellyfish, the Portuguese man-of-war, the sea anemone, sea fans, and reef-forming sea coral. Most of these animals are carnivores and use their stinging cells to capture food, as well as in self defense.

The cnidarian nervous system consists of bipolar or multipolar neurons that are linked by chemical synapses or gap junctions to form one or more network-type integrating units that are characterized as ''nerve nets''. Neuronal processes from these neuronal cell bodies are unmyelinated and not identified as either axons or dendrites. These processes appear to share features of both axons and dendrites and are called ''neurites''.

Serotonin has been identified by biochemical assay in all three classes of cnidarians: *Metridium* (Anthozoa),[2] *Hydra littoralis* (Hydrazoa),[3] and the medusa form of *Cyanea lamarcki* (Scyphozoa).[4] An association of serotonin with cnidarian nerve cells (*Hydra littoralis*) was first reported in 1964 by Wood and Lentz.[53] The presence of serotonin in the stenotele nematocysts of cnidarian nematocytes (*Hydra oligactis* and *Hydra viridis*) was reported in 1978 by Castano and Rossi.[5] The nematocysts are specialized stinging and urticant organelles that are found singly in the nematocytes.

An important observation by Anctil[54] in 1989 that serotonin enhanced spontaneous motor activities of the pennatulid anthoazoan *Renilla koellikeri* prompted an immunohistochemical search for the existence of 5-HT neurons in this cnidarian coelenterate, by Umbriaco, Anctil, and Descarries.[1] Serotonin also has been shown to augment the responsiveness of sea anemone muscle preparations.[55,56]

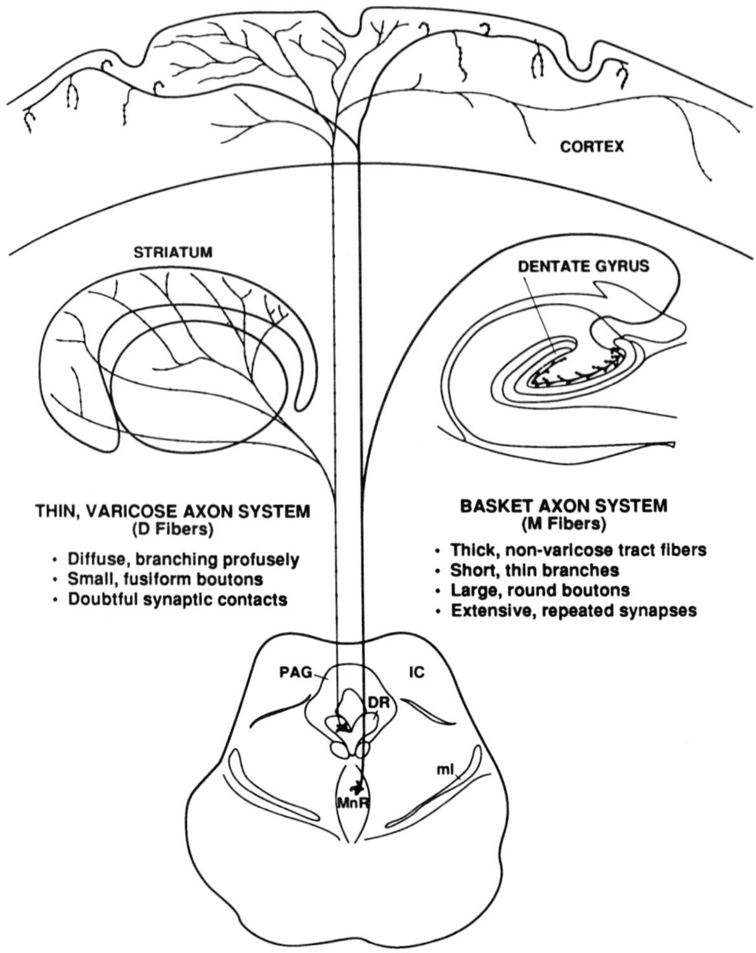

FIGURE 6. Dual serotoninergic nervous system innervation of the forebrain as envisioned by Törk. (From Törk, I., *Ann. N.Y. Acad. Sci.*, 600, 9, 1990. With permission.)

The study of Umbriaco et al.[1] identified 5-HT-immunostaining cellular elements in both neurons and in nematocyst-containing nematocytes. Neuronal 5-HT immunostaining involved all but the cell nucleus, while in the nematocytes it was confined to the nematocysts.

5-HT-immunoreactive neurons had distinctive features depending on their anatomical location. Neurons in the mesoglea, an extracellular matrix sandwiched between the ectoderm and endoderm, were invariably bipolar, with two short beaded processes. Ectodermal neurons, on the other hand, were largely pseudounipolar and their cell bodies generally spanned the whole ectoderm. When 5-HT-immunoreactive neurons were found in the oral complex, 5-HT-immunostained nematocysts also were seen in this region. The appearance of the *Renilla* 5-HT-immunostaining "nerve net" is shown in Figure 9. It is of interest that in the cnidarian *Hydra,* nerve cells and nematocytes have been shown to come from a common multipotent stem cell. The authors of these studies speculated that the extensive 5-HT-immunoreactive nerve net of the cnidarian *Renilla* might contribute to the coordination of its behavioral adaptation to turbulent subtidal waters. Serotonin was found by Anctil[54] to selectively enhance rhythmic activity of turgescent *Renilla.* This effect was mimicked by allowing water to flow over turgescent *Renilla* colonies.

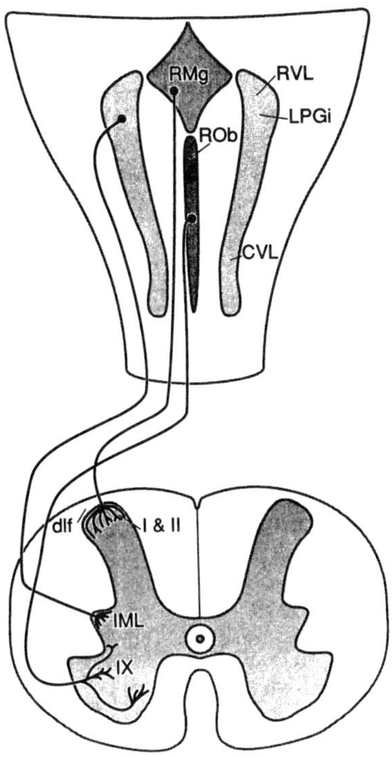

FIGURE 7. Schematic presentation of serotoninergic neuronal system fiber projections from the brain stem to the spinal cord. This presentation emphasizes three principal targets of these projections. Projections from the raphe magnus nucleus (RMg) are to laminae I and II of the dorsal horn, those from the raphe obscurus (ROb) nucleus are to the ventral horn, and those from the rostral ventrolateral medulla (RVL) are to the interomediolateral cell column (IML). (From Törk, I., *Ann. N.Y. Acad. Sci.*, 600, 9, 1990. With permission.)

3. Platyhelminthes (Flatworms)

Flatworms are the lowest invertebrate phylum to possess three embryonic body layers (endoderm, mesoderm, and ectoderm) and to have a head region that contains concentrated sense organs and nervous tissue (''brain''). Flatworms are bilaterally symmetrical, flattened, and they lack respiratory, circulatory, and skeletal systems. There is no body cavity or coelom, and the body is not divided into true segments.

The phylum is divided into five classes: Turbellaria (planarians), Monogenea, Aspidocotylea, Cestoda (tapeworms), and Trematoda (flukes). With the exception of Turbellaria, all classes are parasitic during part or all of their life cycle.

Serotonin-immunoreactive neurons and pathways had been found in both parasitic flatworms (cestodes)[57,58] and nonparasitic flatworms (Turbellaria).[59] 5-HT has been shown to stimulate muscle contractions in flatworms.[60,61]

4. Annelida

Annelids are segmented worms with transverse body rings (annulations), a body cavity (coelom), and movable bristles (setae). The body wall is covered with a cuticle that is secreted by the epidermis and consists of an outer circular and inner longitudinal muscle layer. The circulatory system is closed. The nervous system is made up of a dorsal, bilobed brain and a pair of connective nerves that encircle the digestive tract and unite to form a ventral nerve cord with one ganglion per segment. There are three classes of annelids: Polychaeta (marine worms), Oligochaeta (freshwater or terrestrial worms), and Hirudinea (leeches).

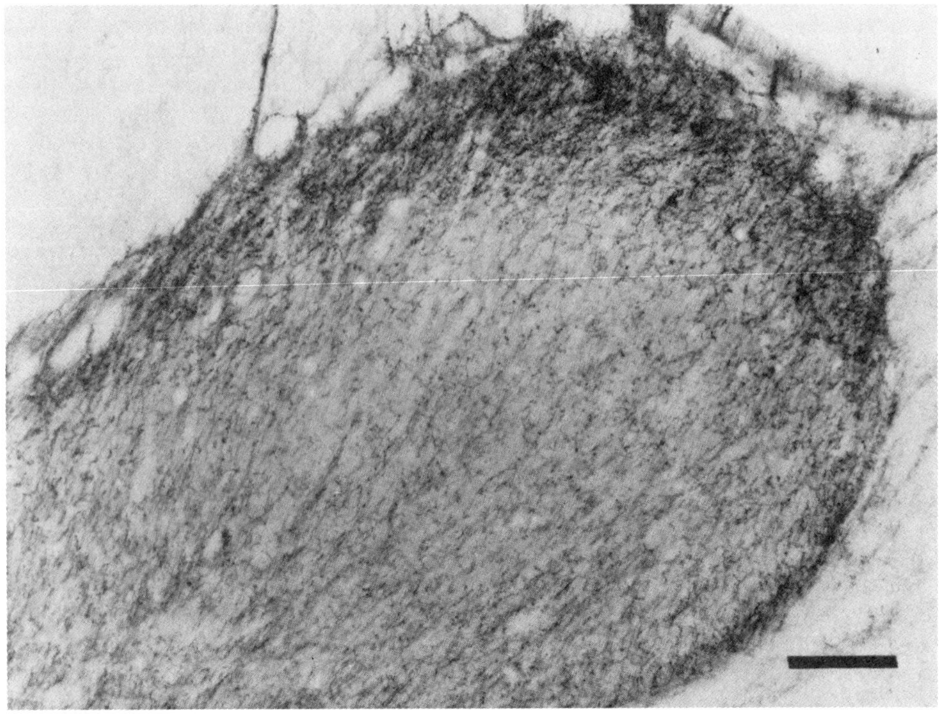

FIGURE 8. Photomicrograph showing serotonin axon immunoperoxidase reactivity in the spinal cord dorsal horn, with highest fiber density in the area closest to the surface of lamina I. (From Törk, I., *Ann. N.Y. Acad. Sci.*, 600, 9, 1990. With permission.)

Serotoninergic neurons have been identified in both the central and peripheral nervous systems of the Polychaeta scale worm *Harmothoe imbricata*.[62] 5-HT-immunoreactive neuronal somata were found to be located in discrete cell masses of the dorsally situated cerebral ganglion and in the segmental ganglion of the ventral nerve cord. Large and small 5-HT-immunoreactive varicose fibers were found in the neuropile of these ganglia and in the interganglionic connective tissue and segmental nerves as well. The cerebral ganglion of *Harmothoe* is the most highly organized of all annelid brains and is considered to be a forerunner of the arthropod brain.[63] It is suggested that the serotoninergic nerve circuits in this worm may be involved in sensorimotor integration and in intersegmental coordination. The widespread 5-HT immunoreactivity in the parapodal ganglion, which was associated with both neuronal somata and neuropile, suggests a key role for serotonin in peripheral integrative activity associated with locomotion.

Serotonin-containing neurons have been identified in the central nervous system of both glossiphoniid and hirudinid leeches.[64,65] Five types of serotonin-containing neurons have been found in the ganglia of midbody segments of both of these types of leeches. One of these neuron cell types, the Retzius cell, has been found to synthesize, store, release, and take up serotonin and to exert physiological behavioral influences that can be mimicked by serotonin.[66-73] These Retzius cells also have been cited as neuroeffector cells that control the release of mucus by the leech.[68] Willard,[74] in 1981, described the effects of serotonin on the generation of the motor program for swimming by the medicinal leech. In 1984 Lent and Dickinson[70] observed that serotonin integrates the feeding behavior of the medicinal leech *Hirudo medicinalis*. Serotonin also have been reported to increase potassium conductance in neuropile ganglion cells of the medicinal leech.

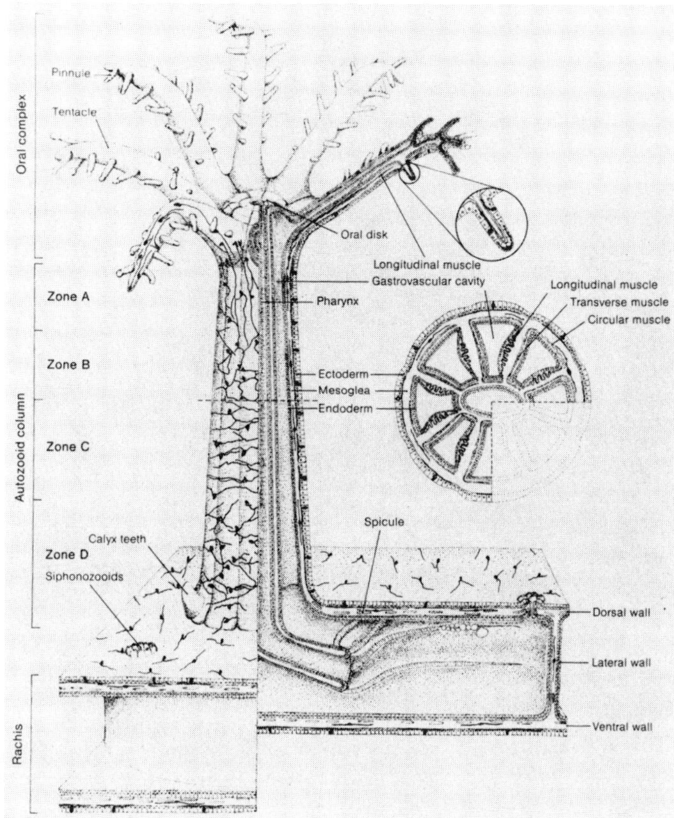

FIGURE 9. Schematic drawing presenting the 5-HT-immunoreactive neurons in the cnidarian coelenterate *Renilla koellikeri.* This drawing presents a single autozooid as it emerges from a piece of rachis. A transverse section of the autozooid also is presented and shows the histological distribution of the 5-HT-immunoreactive nerve cells bodies. The outer ectoderm contains the majority of 5-HT-immunoreactive neuronal cell bodies that are interconnected by projections/"neurites" that form the characteristic "nerve net" of this primitive phylum — the earliest multicellular animals to possess a nervous system. (From Umbriaco, D., Anctil, M., and Descarries, L., *J. Comp. Neurol.,* 291, 167, 1990. With permission.)

In 1987 Kuffler, Nicholls, and Drapeau,[75] of the University of Basel (Switzerland) reported studies relating to transmitter localization and vesicle turnover at a serotoninergic synapse between identified leech neurons in culture. Identified leech neurons, isolated from the central nervous system and maintained in tissue culture, have been found to provide important opportunities for the study of chemical and electrical synaptic transmission. With the chemical synapses that develop in culture, it has been possible to study the quantal release of transmitter and its modulation, to a remarkably full extent.[76-80] In addition, one may voltage-clamp the presynaptic cell and demonstrate the relationship between the membrane potential of the terminals and transmitter release. This valuable model also permits study of the structure of the synaptic junctions that develop between these leech neuronal cells in culture. With careful dissection from fresh leeches, chemical synaptic transmission between neuronal cells in culture becomes established in approximately 7 d. Retzius cells are employed as presynaptic neuronal cells in this culture model, and P sensory cells as postsynaptic cells. The neurotransmitter used by Retzius cells has been shown unambiguously to be serotonin.[76,78-80]

Employing this leech neuron culture model, Kuffler et al.,[75] in an electron microscopic study, investigated the ultrastructure of the synaptic junctions that develop between leech

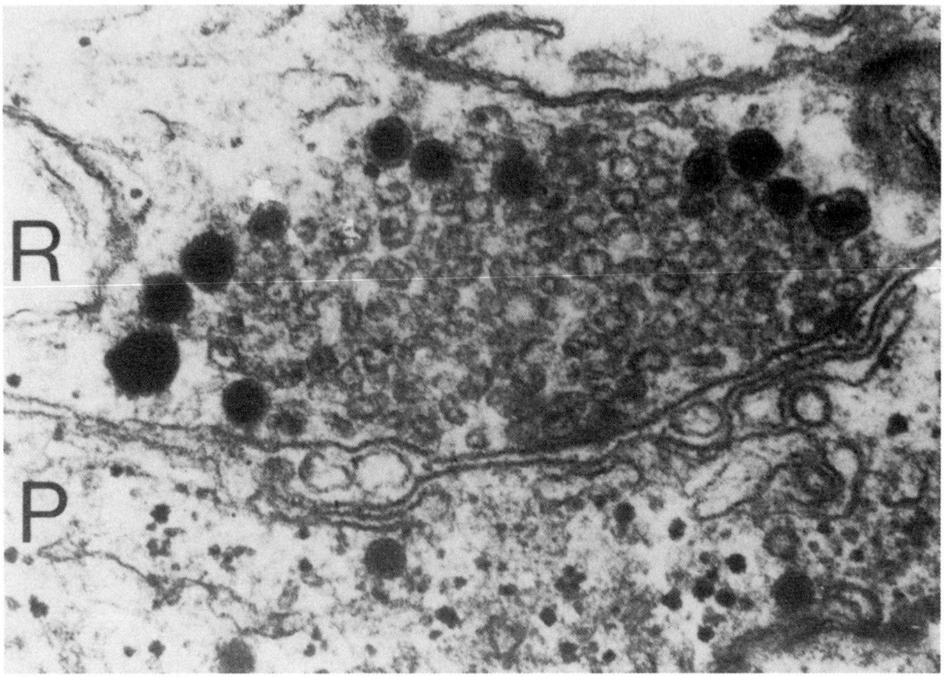

FIGURE 10. Electron photomicrograph showing a serotoninergic synapse between identified leech neurons in culture. This ultrathin section presents the synapse of a presynaptic Retzius (R) cell onto a postsynaptic P sensory cell (P) cocultured for 5 d. The presynaptic and serotonin-containing Retzius cell is shown to have a cluster of clear vesicles, which is opposed to the presynaptic membrane and presynaptic thickenings. The clear vesicles are capped by dense-core serotonin-containing vesicles. The synaptic cleft is wider opposite the presynaptic densities, and extracellular material has been identified within the synaptic cleft in many sections. The presynaptic Retzius cells, as shown here, are known to liberate serotonin. (From Kuffler, D. P., Nicholls, J., and Drapeau, P., *J. Comp. Neurol.*, 256, 516, 1987. With permission.)

neuronal cells in culture. Only preparations in which the synaptic potentials in the postsynaptic P sensory cells were greater than 5 mV were used for electron microscopy. The structures of the junctional synapses that developed bore a striking resemblance to *in vivo* synapses both in invertebrates and in vertebrates (Figure 10). The membrane separation between the presynaptic cell (Retzius) and the postsynaptic cell (P sensory) was wider in the synaptic than in the extrasynaptic regions. The presynaptic Retzius cell contained clusters of clear vesicles that were opposed to thickenings of the presynaptic membrane. The clear-vesicle clusters were capped by a layer of dense-core serotonin-containing vesicles that did not contact the presynaptic-membrane thickenings. Subsynaptic cisternae were identified in the postsynaptic cell opposite the presynaptic-cell-membrane thickening. Extracellular material was seen in the synaptic cleft. Similar synaptic structures were observed to develop between pairs of Retzius cells in culture, and a single Retzius cell was found to be able to form synapses upon itself (autapses) (Figure 11). Structures having the appearance of transmitter release sites were observed at a distance from any postsynaptic membrane, and these areas were presumed to be sites for the diffuse release of transmitter. Antibody to 5-HT labeled with colloidal gold showed 5-HT to be localized in the dense-core vesicles in Retzius cells (Figure 12). Vesicle recycling was identified by following the uptake of extracellular markers, following depolarization stimulation, by raising the culture media concentrations of K^+ in the presence of Ca^{++}. The extracellular markers used were colloidal gold, horseradish

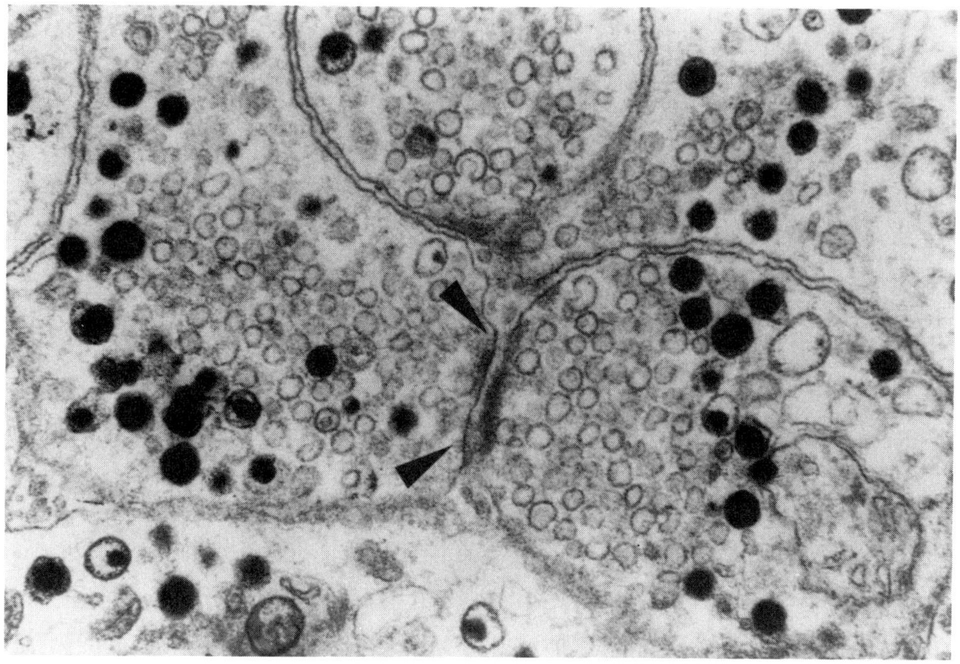

FIGURE 11. Electron photomicrograph demonstrating a self-to-self synapse (autapse) of a serotonin-containing leech Retzius cell (shown by pointers) with presynaptic membrane thickenings and associated synaptic cleft widening. The caps of serotonin-containing dense-core vesicles surround the clusters of clear vesicles that are opposed to the presynaptic membranes. (From Kuffler, D. P., et al., *J. Comp. Neurol.*, 256, 516, 1987. With permission.)

peroxidase (HRP), and ferritin. Ferritin was encountered in dense-core and clear vesicles and in cisternae.

These important and basic studies demonstrate that serotoninergic synaptic structures develop between pairs of identified leech neuronal cells in tissue culture in which chemical transmission occurs and in which transmitter release leads to vesicle recycling, with turnover of dense-core and clear vesicles.

The presynaptic cluster of clear vesicles surrounded by dense-core vesicles, the presence of active zones and widened intercellular synaptic clefts, and the occurrence of cisternae under the postsynaptic membranes observed in the above leech studies provided pictures with striking similarity to those encountered at serotoninergic synapses in vertebrates and mammals and in their brains.[81-83]

It is of interest that serotonin has been identified in neuronal-cell dense-core vesicles in the molluscan *Aplysia* as well as in mammalian brain.[84-90]

5. Arthropoda

General body and related data for arthropods and their major subgroups has been presented in Chapter 4 (''Histamine'') and will not be repeated here.

a. Chelicerata

Serotonin has been found to be synthesized and stored in the central nervous system of the horseshoe crab, *Limulus polyphemus*. 5-HT immunoreactivity has been observed in neuronal fibers in the neuropile of the central body, in the neuropiles of the visual centers (the lamina and medulla of the visual lobes and in the ocellar ganglion), in the optic tract that connects the ocellar ganglion with the posterior medial medulla, and in the central

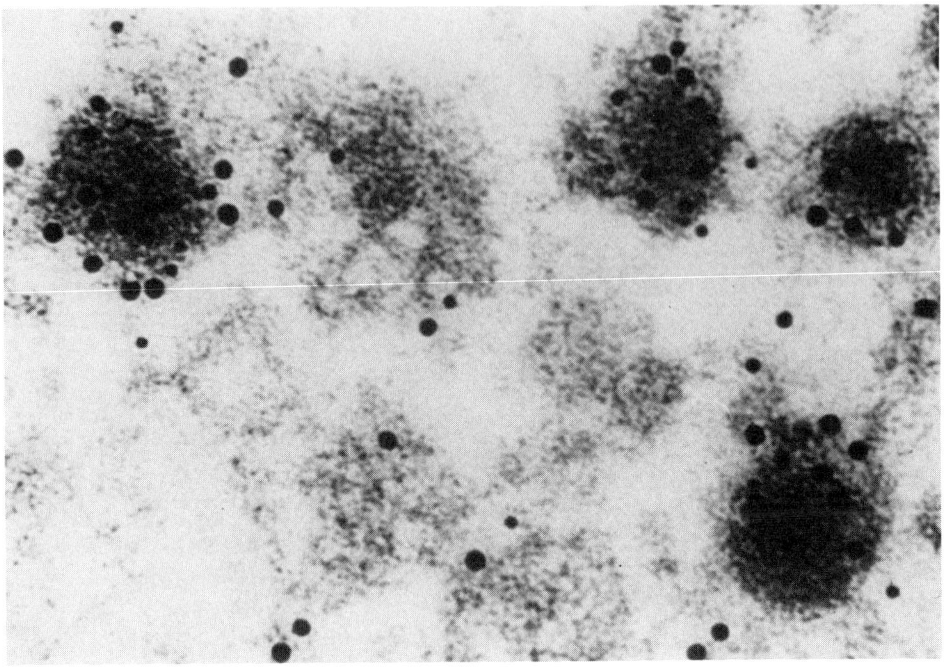

FIGURE 12. Ultrathin section electron photomicrograph presenting antibody gold labeling of serotonin in dense-core vesicles of a leech presynaptic Retzius cell. In this study 95% of the dense-core vesicles became labeled with gold particles per vesicle ranging from 1 to 20. (From Kuffler, D. P., et al., *J. Comp. Neurol.*, 256, 516, 1987. With permission.)

neuropile of the brain.[91] These findings suggest that serotonin functions as a neurotransmitter in the *Limulus* brain and that it plays a role in central visual processing.

b. Crustacea

Kravitz and associates[92,93] have reported interesting studies relating to the role of serotonin and a serotoninergic nervous system in the lobster, which mediate aggressive and dominant behavior. On injection into freely moving lobsters, serotonin and octopamine (a phenol analogue of norepinephrine) produce stable and stereotyped postures that resemble those in dominant (serotonin) and subordinate (octopamine) lobsters. Serotonin caused the animals to stand tall on the tips of their walking legs with claws open and forward in an aggressive posture. Their abdomen was loosely tucked downward. Octopamine produced the opposite posture, with a low stand and subordinate positioning of claws and legs. It was suggested that these amines reach lobster muscles by circulation in the hemolymph from peripheral release sites.

Serotonin-containing neurons have been demonstrated (by Sandeman et al.[94,95] of New South Wales, Australia) in the optic and olfactor lobes and in proto-, deutero-, and trito-cerebrum in two species of crayfish, *Cherax destructor* and *Orconectes virilis*. The architecture of olfactory interneurons suggested that the input from a large number of chemoreceptors may be summed and conveyed to terminals along the axons of the olfactory globular tract, the only established output pathway from the olfactory neuropiles.

The application of serotonin to a walking-leg opener muscle of the freshwater crayfish *Procambarus clarkii* (by Dixon and Atwood[96] of Toronto) caused presynaptic depolarization and long-lasting facilitation of excitatory postsynaptic potentials (EPSPs). The frequency of spontaneously released transmitter quanta also was increased. These results suggested an involvement of the PI second-messenger system in these observations. Presynaptic injection of inositol triphosphate (IP_3) mimicked most of these findings, as did a phorbol ester that activates protein kinase C. Presynaptic injection of a protein kinase C inhibitor blocked the serotonin-induced facilitation in this crayfish model. The protein kinase C activation appeared to be the mediator of the long-lasting phase of the observed facilitation (? LTP).

c. *Insecta*

The relative simplicity of the insect nervous system, the presence of identifiable neurons (location, shape, axonal projection, synaptic connections, electrochemical properties, etc.), the superficial location of large cell bodies, and the comparative ease of preparation for intracellular recodings has made the insect a popular animal for neurophysiological and functional anatomical study and as models for neural integration and neurochemical mechanisms. Extensive studies in this area relating to serotonin and serotinin-immunoreactive neurons in insects have been reviewed recently (1988) and in considerable depth by Nässel[97] of the University of Lund in Sweden.

The nervous system of insects is composed of a central nervous system formed by segmental ganglia, and a peripheral nervous system that includes neurohemal organs. The ganglia of the central nervous system may fuse to form a cephalic supraesophageal ganglion mass (brain), a cephalic subesophageal ganglion mass, and separate or fused thoracic and abdominal ganglia. The basic organization of an insect ganglion involves a central neuropile (where synapses occur) and axonal tracts and a peripheral cortex made up of neuronal and glial cell bodies where synapses do not occur.

Serotonin and serotonin-immunoreactive neurons are widely distributed and much studied in insects. In all of 22 insect species studied, all neuropile regions of the optic lobes contained 5-HT-immunoreactive processes.[97] The morphology of 5-HT neurons in insect optic lobes has suggested that they act over extensive areas and likely over the entire projected retinoptic mosaic.

Insect 5-HT neuron cell bodies tend to aggregate in clusters in each segmental ganglion and in different regions of the brain. The largest number of 5-HT neurons in the insect central nervous system appear to be interneurons. In the insect peripheral nervous system, 5-HT neurons have been found to supply extensive neurohemal areas in the neural sheath of the nervous system.[98]

Immunohistochemical study of the distribution of serotonin in the visual system of the honeybee (by Schaefer and Bicker[99] of Berlin) has revealed a distinct pattern of labeled layers in the three visual ganglia: the lamina (the outermost ganglia), the medulla (the second visual ganglion), and the lobula. In the lamina C layer, 5-HT- and GABA-immunoreactive wide-field tangential cells of striking similarity have been found. These have been shown to be separate cells with projections and arborizations in close apposition to each other. The common projection of these fibers in the C layer of the lamina suggests that serotoninergic and GABA-ergic pathways may interact in visual processing in this insect.

Study of 5-HT-containing neurons in the cockroach supraesophageal (cerebral) ganglion, by Klemm et al.,[100] demonstrated complex and wide distribution, with extensive arborization and innervation of many brain regions. Clusters of 5-HT fibers were demonstrated in the neuropile of all brain regions (protocerebrum, deuterocerebrum, and tritocerebrum).

In the nerves of the mouth parts of the cockroach *Periplaneta americana,* serotoninergic axons have been shown (by Davis[101] of Storrs, CT), to give rise to a large number of secondary branches that have been found to form an extensive neurohemal system at the surface of these nerves. Both the large surface and the cephalic location of this system suggest that serotonin release in appropriate amounts may play a role in the physiology of feeding in this insect. Lange et al.,[102] of Toronto, have demonstrated the calcium-dependent release of endogenous serotonin from neurohemal areas of the blood-feeding bug *Rhodnius prolixus*.

In a study of learning performance in a procedure of Pavlovian conditioning of the fly *Drosophila melanogaster* (by Vaysse et al.[103] of Toulouse, France), alimentary-tract elevation of body serotonin levels enhanced the learning performance relative to a proboscis-extension reflex. Prior studies by this group[104,105] had shown that serotonin depletion by inhibition of serotonin synthesis had produced a decrease in learning performance in this model, which was correlated with a reduction in serotonin concentrations.

6. Mollusca

a. Gastropoda

Serotonin has been suggested to play an important role in the molecular and cellular mechanisms of both associative and nonassociative learning in gastropods.[106] The marine mollusc sea slug or sea hare *Aplysia californica,* of the subclass Opisthobranchia and the order Anaspidea, has been studied extensively with respect to the induction, storage, and retrieval of both short-term and long-term memory. In *Aplysia* the shell has been reduced to a flat plate. Simple defensive reflexes in *Aplysia* may exhibit both short-term (minutes) and long-term (days) sensitization. The siphon and gill defensive withdrawal reflex in *Aplysia* was first demonstrated (in 1973 by Pinsker, Kandel, and associates[107] of Columbia University, New York) to be sensitized in a long-lasting manner by repeated noxious stimuli to the tail. This long-term sensitization may be produced in a training session as short as 1.5 h.[108] Monosynaptic connections from sensory neurons were shown, by the Kandel group of investigators, to participate in the storage of long-term memory relating to this defensive gill- and siphon-withdrawal reflex.[109] It was shown that a part of the mechanism underlying this long-term behavioral sensitization involves long-lasting changes in the efficiency of synapses between the sensory and motor neurons responsible for this withdrawal reflex. Sensory neurons and noxious stimuli are recurrent themes in the laying down of memory and are likely involved in its physiological and molecular basis.

These same *Aplysia* sensory and motor synapses may also be facilitated in a short-lasting manner during short-term behavioral sensitization of the gill withdrawal reflex.[110] This short-term synaptic facilitation of the sensory neuron-to-motor neuron synapse appears to be caused largely by the sensory neuron and in association with depression in a special serotonin-sensitive K^+ current (S current).[110-114] Depression of this S current produces a broadening of the action potential, with more Ca^{++} entering the sensory neuron during each action potential, resulting in the release of more transmitter. These events contribute to a short-lasting increase in the input resistance of the sensory neuron and an increase in its membrane excitability. A given intracellular sensory neuronal stimulus, accordingly, generates a longer train of action potentials in these sensory neurons. Modulation of this serotonin-sensitive potassium channel has been observed in sensory nerve processes, such as the growth cone, as well as in the nerve body. This observation increases the likelihood that this type of modulation may also occur at the small and difficult-to-study presynaptic sensory nerve terminal.[115] Growth cones are the immediate precursors to the synaptic terminals which are too small to study. They do share a number of properties with the nerve terminals, including the capability of releasing transmitter.

The mechanisms underlying long-term facilitation have been explored by means of cell culture methodology in which isolated sensory and motor neurons of the gill withdrawal reflex are cocultured. In this *in vitro* environment, the sensory neurons readily form synapses with gill motoneurons.[116-118] These synaptic connections may be facilitated in a short-lasting manner by single applications of 5-HT,[116] or in a long-lasting way by repeated applications of 5-HT.[119] The repeated application methodology involved four 5-min 1-μM 5-HT applications, each separated by 15 min of washing. It was found that short-term facilitation in this nerve culture model did not require protein synthesis, whereas new protein synthesis was required for long-term facilitation. Further study demonstrated that the long-term facilitation of sensory neurons of *Aplysia* could occur in isolated sensory neurons cultured in the absence of their postsynaptic motor neuron partners and that both isolated and nonisolated long-term facilitations required new protein synthesis.[120] These findings imply that expression of gene products is required for the modulation of ion channels in the establishment of long-term facilitation — a most important observation relating to long-term potentiation and memory.

These findings have triggered a search for the genes and gene products involved in the long-term facilitation process. In order to identify genes specific for the sensory neurons of *Aplysia,* Kandel and associates,[121] in May 1991, reported in *Science* their employment of a miniaturized differential screening method based on the polymerase chain reaction. Pleural ganglia homogenous clusters of neurons, which are the primary mechanoreceptors for the tail, foot, parapodium, and body wall of *Aplysia* and which constitute the afferent pathway of the monosynaptic component of the tail withdrawal reflex and which show both short-term and long-term behavioral plasticity, were employed to endeavor to identify molecules specific to these sensory neurons, in comparison with an identified excitatory motor neuron (R2) to the mucous cells of the foot of this molluscan gastropod. One mRNA was isolated which was expressed in every sensory neuron cluster. This mRNA encodes a peptide that appears to function as an inhibitory cotransmitter on a subpopulation of follower cells and selectively inhibits certain postsynaptic cells, but not others. This function would appear to allow the sensory neurons to achieve target-specific synaptic actions.

Serotonin also has been reported to regulate the phase of a biological clock in the eye of *Aplysia.*[122] A serotonin-induced phase shift in the regulation of a circadian rhythm in the *Aplysia* eye has also been found to be protein-synthesis dependent. A 34-kDa protein has been reported to be directly involved in the phase shift produced by 5-HT.

Another gastropod mollusc of the order Nudibranchia (shell lacking) of the subclass Opisthobranchia, *Hermissenda crassicornis,* has been subjected to considerable study with respect to a possible role for serotonin in associative learning and memory.[123-131] *Hermissenda* belongs to the Opisthobranch order Nudibranchia, whereas *Aplysia* belongs to the order Aplysiomorpha.

Serotonin has been shown to be involved in associative learning and conditioning responses to light of Type-B photoreceptors of *Hermissenda* eyes, both in intact animals and in single-cell *in vitro* photoreceptor preparations. Alkon and associates at the National Institutes of Health (N.I.H., Bethesda, MD) have demonstrated the presence of 5-HT in cell bodies and processes of the neuropile of the *Hermissenda* pedal and cerebropleural ganglia.[130-132] Crow and associates have reported 5-HT enhancement of the Type-B photo response of *Hermissenda* and the presence of 5-HT-immunoreactive fibers and varicosities on the optic nerve and in the synaptic region of the cerebropleural ganglia neuropile near the Type B photoreceptor synaptic terminals of this marine mollusc.[123,129] They also have reported that the pairing of light with 5-HT applied directly to the surgically exposed circumesophageal nervous system of otherwise-intact *Hermissenda* mimics the effects of conditioning on long-term changes in phototactic behavior.[125]

In three major 1989 publications, Farley and associates,[126-128] of Princeton University, made important contributions to a better understanding of a role of importance for serotonin in the mechanisms underlying associative learning in *Hermissenda.* They demonstrated the presence of 5-HT-containing nerve fibers in the *Hermissenda* central nervous system cerebropleural ganglia neuropile, which are likely presynaptic to Type B photoreceptors and the release of 5-HT in this area in a calcium-dependent manner.[126] They also reported the prevention of *in vitro* conditioning-produced short-term neural correlates of learning in Type B photoreceptors of *Hermissenda,* by drugs known to interfere with 5-HT neurotransmission. Their studies of 5-HT effects on single-cell Type B photoreceptors in *in vitro* preparations were of particular importance and significance.[127] The circumesophageal nervous system was removed from the animals by dissection and prepared for intracellular recording as in prior studies. The single-cell photoreceptors were synaptically isolated by razor removal of the axodendritic process. This left the cell capable of responding to light with depolarizing generator potentials, but without complicating impulse activity or synaptic potentials. Peak and steady-state light-induced generator potentials were obtained following 8 to 10 min of dark adaptation. The light-stimulation protocol called for three consecutive 30-sec light steps delivered every 2.0 min after the 8 to 10 min of dark adaptation. Light and 5-HT responses in terms of ionic currents and long-lasting depolarization values were measured. Serotonin was found to mimic many of the persistent effects of associative training upon *Hermissenda* Type B photoreceptor light responses and ionic conductances. Voltage-clamp analysis identified five separate and distinct effects of 5-HT. These multiple-conductance changes resulted in a decreased membrane conductance and enhancement of the light-induced generator potential of these Type B photoreceptor cells and, thus, paralleled the effects of associative training. Lipophilic analogues of cyclic AMP produced no changes in this *in vitro* model in these studies. In an earlier report by these authors,[133] protein kinase C (PKC) activation was found to induce conductance changes in *Hermissenda* photoreceptors, similar to those encountered in associative learning.

Serotonin also has been found to play an important role in the growth, development, and function of the nervous system of another and more advanced subclass of Gastropoda: Pulmonata. Pulmonates, as terrestrial and freshwater gastropods, are the largest subclass of molluscs. These land snails and slugs have a mantle cavity that generally is vascularized as a lung, and are well adapted to terrestrial life and water economy.

Posthatching, the pulmonate pond snail *Lymnaea stagnalis* experiences a 20-fold increase in the linear dimensions of the ganglia that make up the central nervous system. Posthatching, the population of 5-HT-immunoreactive neurons in *Lymnaea* increases both by doubling the number of nerve cells, largely in identified discrete clusters, and by growth of individual cells.[134] Direct-developing pulmonate gastropods, such as *Lymnaea,* develop through larval forms directly within the egg and leave the egg on hatching with a functional nervous system capable of producing most of the same complex behaviors as the adult nervous system. Any neuronal changes, accordingly, must be integrated into an existing system, with a minimal disruption of functionality.

An identified excitatory monosynaptic connection between pairs of serotonin-containing neurons in the brain of the European terrestrial snail *Helix pomatia* has been reported recently by Vehovszky et al.[135] of Hungary. The presynaptic neuron was always the giant LP_3 cell in the left pedal ganglion, and the postsynaptic cell was one of the visceral or right parietal serotonin cells. These serotonin-containing and monosynaptically connected neuronal cells appeared, in this terrestrial pulmonate, to take part in important cardiorenal and other visceral regulation and integration of functions.

Serotonin has been reported to induce an increase of Ca^{++} current in a group of identified ventral neurons of the European garden snail, *Helix aspera.*[136] This effect was mimicked by intracellular injection of cyclic GMP. Cyclic GMP-dependent protein kinase was found to enhance the Ca^{++} current of these identified ventral *Helix* neurons and to potentiate the 5-HT-induced Ca^{++} current increase in these same neurons. The vasoactive neurotransmitter peptide cholecystokinin-8 (CCK-8) has been found to be colocalized in a giant serotonin-containing neuron in the circumesophageal ganglia of the same *Helix aspersa.*[137]

b. Cephalopoda

Cephalopods are the most advanced class of Mollusca and include animals such as squids, octopi, and cuttlefish. Substantial phylogenetic advances are to be found in cephalopod nervous and vascular systems. The central nervous system is highly developed, with the major ganglionic centers being concentrated in the head. Cephalopods are unique among invertebrates in the degree of cephalization and cerebralization that has been achieved. The uniting of the major ganglion centers of the central nervous system has produced a brain of considerable complexity. The true brain thus formed is encased in a cartilaginous cranium. The eyes also are highly developed and most closely resemble the eyes of vertebrates. The octopus is capable of learning and of considerable intelligence. Cephalopods appear to be sociable animals and may be found together in droves and schools.

Less is known about squids because of difficulties in maintaining them in confinement. In certain squids the mantle is innervated by giant paired dorsal axons that are the largest nerve fibers known. These giant axons have been subjected to considerable study, and much of current knowledge of the mechanism of nerve impulse conduction obtained by Hodgkin, Huxley, and others has come from studies on these giant-squid neurons and fibers. The giant squids (*Archyteuthis*) are the largest living invertebrates and may attain lengths in excess of 60 ft with tentacles extended.

Most cephalopods possess pigment cells in the skin, called chromatophores. These chromatophores are controlled by brain-connected nerves and play important roles in protective behavior patterns, including attack, alarm, camouflage, rest, and defense. Additional defense behavioral patterns include skin sculpture changes and tentacle curling. These highly intelligent cephalopods use their heads and demonstrate considerable skill in hunting, stalking, and luring. These defense measures have developed with considerable survival-generated stimulus, as extinct forms of cephalopoda outnumber the living. With the exception of the nautilus (subclass Mautiloidea), the majority of cephalopods (subclass Coleoidea — squid, octopus, cuttlefish) show marked reduction in the size and extent of the characteristic molluscan shell.

Marked improvement in the structure and organization of the vascular system also has been achieved by the cephalopods. Members of Coleoidea all possess a closed circulatory system with blood vessels. There is a capillary network with a strong microvascular. Larger blood vessels have an incomplete endothelial layer and a more or less three-layered wall structure. Hemocyanin respiratory pigment-containing hemolymph is propelled in this closed circulatory system by one systemic heart and two branchial hearts at the base of each gill. Gas exchange takes place in the gills.

Serotonin has been identified in the optic lobe of squid, by biochemical assay.[138,139] This lobe also has been shown to contain large populations of serotonin-containing neurons.[140,141] A 5-HT-metabolizing enzyme, monoamine oxidase (MAO), has been identified in the squid central nervous system. Inhibition of this 5-HT-degrading enzyme has increased 5-HT levels in the squid central nervous system, from comparatively low resting levels.[142]

Localization of 5-HT-immunoreactive neurons in the central nervous system of *Octopus vulgaris* has been achieved (by Uemura et al.[143] of Japan) by reliable and nonfading peroxidase-antiperoxidase complex (PAP) immunohistochemical methodology.[143] This expanded on previous fluorescence histochemical studies of 5-HT neurons in the cephalopods.[138,144] 5-HT-containing neuronal cells were identified in areas of the subesophageal ganglia (chromatophore, vasomotor, pedal, branchial, and palliovisceral lobes) and in areas of the supraesophageal ganglia (buccal and basal lobes).

The finding of 5-HT-containing neurons in the vasomotor lobe and in chromatophore lobes, which are the final common pathway to the periphery for the skins and chromatophore muscles, provide a neurovascular and defense "connection" for serotonin in this most advanced segment of the invertebrate family of animals and strongly suggests the value of more in-depth study in this area where dramatic phylogenetic changes have occurred and where so much extraordinary in-depth study has been done on the immediately less-advanced class of molluscs, the gastropods, including the memory and learning mechanisms and the responses and biochemical changes in the "heat, pain, and memory" scenario.

C. VERTEBRATES

1. Agnatha

The central nervous system of the agnathan river lamprey (*Lampetra fluviatilis*) is comparatively simple, yet shares a number of features with the central nervous systems of higher vertebrates, despite the fact that the cyclostomes diverged from main vertebrate line from 450 million years ago. Serotonin and an intrinsic 5-HT system have been identified in the lamprey spinal cord and brain stem.[145-150] A few of these 5-HT-immunoreactive cells also contain tachykinin immunoreactivity.[148] Cell bodies along the spinal cord midline have been found to give rise to a dense ventromedial 5-HT plexus.[150] 5-HT also was found to coexist with calcitonin gene-related peptide (CGRP) and bombesin-like peptides in a dorsal-root-column system of the spinal cord. This fiber system was unchanged after degeneration of descending fibers and was presumed to originate from cell bodies in the dorsal-root ganglia that demonstrated the same immunoreactivities. A cholecystokinin-like peptide also has been found to be present in 5-HT neurons in the lamprey spinal cord.[147]

The lamprey spinal cord is considered to represent a favorable model for the study of the physiological and molecular bases of neuronal events, including neurotransmitters and their mechanisms of action.[151] These uses of the lamprey nervous system involve detailed study of the neurobiological bases of rhythmic motor acts in vertebrates. The serotonin-containing spinal cord interneurons of the lamprey occur in a midline row below the central canal, forming a dense ventromedial plexus of terminals in which there are dendrites of motor neurons and premotor interneurons. It is speculated that the serotonin systems may play an important role, from their anatomical positions and connections, in influencing the integrating the motor pattern of effective locomotion.

2. Chondrichthyes

Serotonin immunoreactivity has been found to be widely distributed in the spinal cord of four species of elasmobranch fish: the Atlantic and brown sting rays (*Dasyatis sabina* and *Dasyatis fluviorum*), and eagle ray (*Aetobatis narinari*), the shovel-nose ray (*Rhinobatis battilum*), and the black-tip shark (*Carcharhinus melanopterus*).[152,153] The concentration of 5-HT immunoreactivity in varicosities and fibers (found by Cameron et al.[152] of the University of Queensland, Brisbane, Australia) was highest in the outer substantia gelatinosa of the spinal cord dorsal horn. The high concentration of 5-HT varicosities in the outer substantia gelatinosa, and its colocalization in this area with substance P, CGRP, and bombesin, suggest that serotonin in elasmobranches may be employed as a modulator and integrator of activity in second-order dorsal-horn neurons and in peptide-containing primary afferent sensory terminals. Although the amount of sensory information reaching the elasmobranch spinal cord is comparatively limited with respect to that of mammalian species, the distribution of these major vasoactive neurotransmitter substances in the dorsal horns of these two vertebrate groups, primitive and advanced, is strikingly similar. 5-HT fibers in the lateral and ventral funiculi indicate that some of the 5-HT immunoreactivity in the dorsal horn of elasmobranches may be of supraspinal origin.

3. Osteichthyes

Serotonin-containing neuronal cell bodies and fibers have been encountered in a number of teleost fish, including goldfish,[154,155] trout,[156] sickelback,[157,158] platyfish,[159] and a morymid electric fish, *Gnathonemus petersii*.[160,161] Serotonin-containing nerve cell bodies were found in the raphe region of the brain stem, in the hypothalamus, and in the transition zone between the dorsal thalamus and the pretectum. Serotoninergic cells in the thalamopretectal zone have not been encountered in other vertebrates and appear to be a teleostean specialization. Many of the hypothalamic neurons have cerebrospinal-fluid-contacting processes. As with other vertebrates, no serotoninergic cell bodies were found in the telencephalon.

Serotoninergic nerve-fiber density has been found to be quite high in the medial forebrain bundle and surrounding preoptic and hypothalamic regions as well as in several telencephalic areas.

In the mormyrid electric fish *Gnathonemus*, special serotoninergic processes are involved with the control and integration of the electric organ activity. The electric organ in mormyrid fish is derived from muscles at the base of the tail. In all Mormyridae, a short length of body musculature toward the tail is modified to form the electric organ. These muscles no longer have the ability to contract and have undergone considerable reorganization to form thin, flat electroplates. These are the structures in which the electricity is produced. There are from 300 to 400 electroplates in mormyrid electric organs. Electrical discharges from these organs are in the form of pulses. The frequency and nature of these pulses are different in each species and may range in output from 120 to 300 pulses per second. The electric organ is innervated by electromotoneurons which are located in the spinal cord. The initiation and irregular rhythm of this all-or-none electrical pulse discharge depend on descending neuronal activity that originates in the electric-organ-discharge (EOD) command and medullary relay nuclei. These nuclei are located in the ventral brain stem, immediately rostral to the inferior olive and close to the caudal serotonin-containing raphe nuclei.[162]

The electric organs and the electrical field produced around these mormyrid fish are of considerable survival importance. The flow pattern of the established electrical field may be distorted by the differing conductivity of the objects that pass through it, both animate and inanimate. These variations in conductivity are detected by modified nerve cells (mormyromasts) in the skin. These sensory inputs enable the mormyrids to distinguish between prey, predators, and obstacles in the turbid water they often inhabit. The discharges from these electric organs may also function as signals to other mormyrids, as well as aiding in navigation and detection.

In the recent (1989) study of Grant et al.,[160] from France (Gif sur Yvette), both the EOD medullary relay and command nuclei contained serotoninergic varicose fibers that surrounded the neuron somata and coursed through the interstitial neuropile. Some of these axons could be traced to neuron cell bodies of origin immediately ventral to the EOD command nucleus, at a distance of only 200 μm. These anatomical approximations may be of added significance, since serotonin may also be released at nonjunctional sites. It was of importance that the same serotoninergic axons in this Grant et al. study projected to both the EOD command and EOD relay nuclei, as the command nucleus is the site of decision for the initiation of the descending electromotor signal.[163,164] The EOD premotor relay nucleus contains strongly electronically coupled neurons and serves to synchronize the efferent message before transmitting it to the spinal electromotoneurons.[163,165,166]

4. Amphibia

Study of the 5-HT neuronal system of the brain of the urodele crested newt, *Triturus cristatus carnifex* (L), has shown the bulk of the serotoninergic system cell bodies to extend from the raphe region of the medulla oblongata through the isthmus toward the mesencephalic tegmentum.[167] Serotoninergic fiber innervation was extensive in the preoptic area and in the habenular complex. In the telencephalon, serotonergic fibers were found in the dorsal pallium, the primordium hippocampus, the striatum, and the olfactory bulbs. The serotoninergic periventricular cerebrospinal-fluid-contacting system found prominently throughout submammalian classes, and absent in mammals, was found in this crested newt.

Dorsal-root ganglion cells of the bullfrog *Rana catesbiana* have been found to be sensitive to 5-HT,[168] and 5-HT has been found in unique sets of retinal neurons of the African clawed frog, *Xenopus laevis*.[169]

5. Reptilia

Identification of the distribution of the cell bodies and fibers of the serotoninergic nervous system in the lizard *Gekko gecko* showed cell bodies in the diencephalon in the hypothalamic periventricular organ and the ependymal wall of the infundibular recess, while in the midbrain they were observed in the raphe and reticular nuclei.[170] There also was extensive distribution of 5-HT fibers to the forebrain and midbrain. In the phylogenetic series of fishes–amphibians–reptiles–birds–mammals, there is a gradual decrease in the number of cerebrospinal-fluid-contacting serotoninergic cells in the hypothalamic periventricular layer and a remarkable increase in the number of serotoninergic cells in the midbrain tegmentum. During phylogeny the role of the diencephalic serotoninergic neurons becomes less important. Amniotes, as compared to fishes and amphibians, show a remarkable increase in the number of serotoninergic cells in their midbrain and isthmic regions. This development culminates in mammals with a strongly developed dorsal raphe nucleus. Mammals, which lack cerebrospinal-fluid-contacting cells completely, are the only group of vertebrates in which a dense supraependymal plexus of serotoninergic neuronal elements has been identified in cerebral ventricles.

In this lizard, as in mammals, a strong serotoninergic innervation of sensory pathways, especially visual, could be recognized. In *Gekko,* as in the turtle *Clemmys japonica,*[171] a well-laminated serotoninergic innervation of the midbrain tectum is found, as well as a strong serotoninergic innervation of visual relay centers.

Morphometric analysis of the entire population of retinal serotoninergic bipolar cells in the turtle *Pseudemys scripta elegans* has been obtained, and the findings have provided important implications for retinal image processing and for motion detection.[172] Intensification of 5-HT immunofluorescence was achieved by blockage of serotonin degradation. This permitted unusually complete staining of the entire population of serotoninergic bipolar cells within the retina and made possible a detailed morphological analysis of single-labeled cells that functionally constituted a homogeneous neuronal population. Bipolar cells in this homogeneous population all contain serotonin, are hyperpolarized by light, and release serotonin in darkness.[173] Bipolar cells are the retinal neurons that connect the outer plexiform layer of the retina with the inner plexiform layer. Their axons cross the inner nuclear layer. Because of circuitous axonal paths, the output location of the bipolar cells is shifted relative to the input location in the retina. This shift varies with the retinal location of the cell. Despite functional homogeneity, the morphology of the serotoninergic bipolar cells was found to undergo dramatic changes depending on retinal location.[172] The morphometric data obtained revealed a distorted projection of serotoninergic bipolar neurons from the outer plexiform layer into the inner plexiform layer, with excessive representation in the visual streak area. This anatomical configuration was interpreted as helping to increase visual acuity of the visual streak area, with additional likely assistance in motion detection.

6. Aves

5-HT immunoreactivity has been found to occur very early in the avian central nervous system.[174-178] By chick embryo day 4, 5-HT-immunoreactive neuronal cells appeared in both the mesencephalon and rhombencephalon.[178] There were two cell groups in the rhombencephalon: rostral and caudal. The rostral group of cell bodies was quite numerous and sent axons toward more rostral parts of the brain; the cells in the caudal group were small in number and scattered. The number of serotoninergic cell bodies increased with embryonic development. By day 8, virtually all of the serotoninergic cell bodies found in the adult chick were present. By day 10, 5-HT-immunoreactive cells were found in the paraventricular organ of the diencephalon. Serotoninergic neuronal-fiber projections took somewhat longer. Nonetheless, serotoninergic neuronal fibers were found in almost all brain areas by day 16. Soon after the appearance of 5-HT-immunoreactive neuronal cell bodies in the medullary raphe nuclei (days 5 and 6), serotoninergic fibers may be identified in the marginal layer of the cervical and lumbar spinal cord (days 6 and 8). A longer period is required for axonal arbors of serotoninergic fibers to differentiate in the gray matter.[175,177] This process may continue even after hatching. In the posthatched chick and adult chicken, serotonin-containing neuronal fibers are widely distributed in the chick brain in a manner comparable to mammals. The distribution of serotonin-containing nerve fibers to brain stem motor nuclei is similar to that encountered in mammals studied.

7. Mammalia

a. General

The majority of studies concerning the anatomical and functional aspects of serotonin in mammals have used rats, mice, guinea pigs, and cats. The rapid metabolic degradation of serotonin in the human central nervous system has limited in-depth studies in man.

b. Peripheral Nervous System

Extracellular recordings from the isolated rat cervical vagus nerve have shown it to be stimulated (depolarized) by 5-HT.[179] This stimulation was mimicked by a 5-HT_3-receptor agonist (2-methyl-5-HT) and was blocked by a 5-HT_3-receptor antagonists (metoclopramide, MDL 72222, and ICS 205-930).

Intracellular recordings from guinea pig coeliac ganglion cells maintained *in vitro* and exposed to 5-HT by pressure injection showed three types of response.[180] A fast, transient depolarization occurred in 43% of the cells, a fast transient followed by a slow depolarization (biphasic response) occurred in 30%, and a slow sustained in 25%. 5-HT_3-receptor antagonists blocked the fast depolarization, indicating that this response was 5-HT_3-receptor mediated. The receptor mediating the slow depolarization was not identified. In an earlier study by this group,[181] serotonin was applied to guinea pig coeliac ganglion cells by superfusion. In this earlier study, slow depolarization occurred in 60% of the neurons exposed to serotonin and subjected to single-cell recording. Substance P was found to depolarize 15% of these neurons. This was a sharp reduction from substance P depolarization of inferior mesenteric ganglion cells (80%).[182]

Intracellular recordings from rabbit nodose ganglion neurons *in vitro* (and separated from an attached vagus nerve) have shown serotonin to increase the excitability of C-fiber neurons by two distinct mechanisms.[183] Higashi and Nishi[184] earlier (1982) had shown that 5-HT produces a strong depolarization, with associated conductance increase of the somal membrane of C-fibers located in the rabbit nodose ganglion.

A role of importance for serotonin in retinal function and vision, cited above in the "Reptilia" section for turtles with an ancestry and origin dating back more than 200 million years, appears to be shared by mammals, which are regarded as descended from this remarkable Reptilia class. In addition to their impressive lineage, turtles also are the longest living of all vertebrates — perhaps good vision, a good brain, and a conservative approach to evolutionary change help. "And the voice of the turtle is heard in the land" (Song of Solomon, *Old Testament*).

Cutliffe and Osborne[185] of Oxford, in 1987, reported significant serotonin-induced stimulation of inositol phosphate formation and accumulation in rabbit retinas and showed that at least a part of this response was mediated by 5-HT_2 receptors. Earlier studies had described the presence of serotonin, serotonin-containing and serotonin-binding cells, and cell membranes in mammalian retinas, bovine as well as rabbit.[186-190] A 1989 report from Harvard[191] describes connections of serotonin (indoleamine)-accumulating cells in the rabbit retina. The use of an important photooxidization technique permitted easy identification of the dendrites of the serotonin- and indoleamine-accumulating cells, seen with electron microscopy.[192-194] This study was carried out following injection of 5,7-dihydrooxytryptamine into the vitreous body of the rabbit eye. Dendrites of the indoleamine-accumulating cells could be easily identified forming a dense plexus near the junction of the inner plexiform and ganglion cell layer. In this area they exhibit large synaptic endings that occupy a substantial fraction of the surface of the rod bipolar terminals. These dendrites receive input from rod bipolars at dyad synapses where the other postsynaptic partner is a dendrite of a narrow-field bistratified amacrine cell, and also receive amacrine cell input throughout the inner plexiform layer. The only outputs found by the authors were reciprocal synapses onto rod bipolar endings. These findings suggest that these amacrine (unipolar nerve) cells exert important effects on the transmission of scotopic (vision in dim light with dark-adapted eyes) information through the retina. The cell bodies of the indoleamine-accumulating cells were located predominantly in the innermost nuclear row of the inner nuclear layer. The most common synapse established by these cells is with axonal endings of bipolar cells. The labeled processes were postsynaptic to bipolar endings at dyad synapses.

c. *Spinal Cord*

Serotonin and substance P have been found to be colocalized in spinal cord neurons and in neurons projecting to the spinal cord. In the medullar raphe pallidus and raphe obscurus nuclei of the cat, virtually all of the spinally projecting neurons contained substance P and/or serotonin.[195] Substance P and serotonin were colocalized in half of these spinal raphe neurons. Similar localizations were encountered in the caudal VLM (ventrolateral medulla). No substance P neurons, however, were found in the rostral VLM projecting to the spinal cord. Substance P also has been reported to enhance the release of endogenous serotonin from the rat ventral spinal cord where most of the substance P and serotonin are stored in the same nerve endings.[196] Pharmacological depletion of 5-HT from the spinal cord dorsal-horn neurons of the cat increases significantly the proportion of dorsal-horn neurons excited by substance P and by serotonin.

Electronmicroscopic study of the cat lumbar ventral spinal cord revealed that 75% of both substance P- and 5-HT-immunoreactive boutons were present in the type of bouton that contained a high number of large dense-core vesicles.[197] The large number of boutons containing substance P and 5-HT making synaptic contacts with the proximal dendrites in the motor nucleus suggest the importance of this pathway in the function of motoneurons. An earlier study employing immunogold methodology had demonstrated colocalization of substance P and 5-HT in large dense-core vesicles at the electron microscopic level.[198] A later (1989) study by members of this group, employing subcellular fractionation, reported that 5-HT and substance P are, to a large extent, colocalized in nerve endings of descending bulbospinal neurons in the rat ventral spinal cord, but are stored differentially, with 5-HT in both small and large synaptic vesicles and substance P found only in large vesicles.[199] In the dorsal horn, substance P and CGRP were colocalized in central branches in large synaptic vesicles. The amount of 5-HT in the ventral spinal cord was approximately two times that found in the dorsal spinal cord, on differential centrifugation.

Serotonin has been found to have a powerful excitatory effect on cells in the superficial dorsal horn of the spinal cord of cats.[200-202] This excitatory action could be blocked by methysergide. In the rat spinal cord dorsal-root ganglion neurons, there appear to be two distinct depolarizing and excitatory responses to 5-HT on these primary afferent neurons, which are mediated by 5-HT_2 and 5-HT_3 receptors.[203] This response did not show desensitization. These two 5-HT receptors may not only occur in the same tissue, but also on the same cell. In most tissues these receptors have faciliatory actions, including many neural systems.[204-207]

It has been suggested that rhythmic generators in the spinal cord may be involved in mammalian (and lamprey, see above) locomotion and respiratory activity[151,208-211] and that serotonin may be involved in the activation, integration, and control of these rhythmic generators.[212] Serotonin has been reported (by Legendre et al.[212] of Bordeaux, France) to have a prolonged excitatory effect on fetal mouse endogenous pacemaker neurons in spinal cord cell primary culture. These endogenous pacemaker neurons are distinguished from other cells in the cultures by their large somata and thick neurites.[213,214] In this model, serotonin evoked a response of longer duration (? LTP) than was produced by noradrenaline. The response of these in-culture pacemaker neurons to 5-HT at the resting membrane potential level was characterized by a long-lasting depolarization that was accompanied by an increase in output resistance. Suprathreshold depolarizations produced by 5-HT induced rhythmic activity in both beating pacemaker neurons and in bursting pacemaker neurons. These responses to 5-HT were dose dependent and appeared to be due to a closure of K^+ channels. The input-resistance variation produced by 5-HT was voltage dependent, and it increased exponentially when membrane potential was depolarized.

Peripheral nociceptive stimulation has been found (by Weil-Fagazza et al.[215] of Paris) to markedly alter 5-HT in the central nervous system. A prolonged and intense nociceptive electrical stimulus to the tails of anesthetized rats stimulated an increase in 5-HT synthesis, which depended on the part of the central nervous system studied. The dorsal spinal cord was the most sensitive, with a 25% increase in 5-HT synthesis. The increase in synthesis was 14% in the ventral spinal cord and 16% in the brain stem. 5-HT synthesis in the forebrain was not affected significantly. A nonnociceptive stimulus did not alter 5-HT synthesis in the areas studied, nor did the intravenous administration of morphine. Morphine did block the nociceptive-stimulus-induced increase in 5-HT synthesis in the areas studied, with the exception of the ventral spinal cord, an action blocked by naloxone. Naloxone alone failed to influence the nociceptive-stimulus-induced increases in 5-HT synthesis.

Morphological and physiological characterization of spinal cord dorsal-horn neurons (by Hylden et al.[216] at the N.I.H., Bethesda, MD) has identified lamina I cells as projecting to the midbrain or thalamus. The differential distribution of 5-HT fibers in spinal cord laminae VII and X indicate that descending 5-HT fibers preferentially innervate those regions of lamina VII that contain sympathetic and parasympathetic neurons. Sympathetic nuclei are more richly innervated with 5-HT than are spinal parasympathetic nuclei. Spinal cord transections demonstrate that the vast majority of spinal 5-HT-immunoreactive neurons descend from the brain stem. Nonetheless, some 5-HT fibers are of intrinsic and local origin.

d. Brain Stem

The mammalian brain stem is the location site of the serotoninergic nervous system cell bodies and the site of origin of both caudal- and rostral-projecting nerve fibers from this system. It also is the site of multiple connections and important communicative pathways between the brain and the periphery. Included in these pathways and connections are those relating to pain and stimulatory sensory impulses, which likely are of importance in the laying down of the molecular basis for learning and memory. Many of these areas receive serotoninergic projections. One such area is the parafasicularis (PF) nucleus of the thalamus. Noxious tail-pinch stimulation of the rat has been shown to cause 68% of PF neurons to increase their firing rates to 242% of their initial baseline level of activity.[217] In other studies, this area has been shown to be responsive to painful stimuli in cats, monkeys, and rats, as is the thalamus.[218-222] The PF also is known to receive convergent inputs from the dorsal raphe nucleus (the major site of serotoninergic system nerve bodies) and from pain stimulation sites.[221]

The dorsal raphe (DR) nucleus is the most rostral (forward) of the brain stem serotoninergic nuclei. Serotonin-containing neurons are the most numerous and the most important neuronal population of this nucleus.[223-225] In this nucleus in the rat, significant inter- and intracellular relationships between 5-HT, substance P, and GABA have been identified.[223] Synaptic contacts by 5-HT nerve terminals on substance-P-containing nerve somata and dendrites have been shown.

Acute and chronic systemic administration of GABA and GABA-mimetics and direct injection into raphe nuclei, including the DR nucleus, were found (by Nishikawa and Scatton[226,227] of Bagneux, France) to reduce 5-HT synthesis and 5-hydroxytryptophan (5-HT-precursor) in major cerebral serotoninergic neurons, including the hippocampus, the striatum, the prefrontal cortex, and the olfactory tubercule. This action appeared to be mediated through the brain stem raphe nuclei. Electron microscopic study of the DR nucleus has identified the existence of serotoninergic dendrodendritic synaptic contacts and axon terminals in this area, and these findings suggest direct local interactions of importance between serotoninergic neurons in the nucleus raphe dorsalis (DR).[228]

Serotonin facilitation of a dendritic calcium conductance in the substantia nigra of the guinea pig has been reported by Nedergaard et al. at Oxford, U.K.[229] Within the substantia nigra, dendrites of dopaminergic neurons that project to the striatum play an active role in transmitter and protein release. 5-HT in fibers afferent to the substantia nigra is present in terminals that make direct synaptic contact with dopaminergic neurons. This 5-HT has a site-dependent, receptor-mediated, faciliatory effect on a specific dendritic calcium-dependent potential in nigrostriatal neurons *in vitro*. This important observation suggests that 5-HT may produce a significant alteration in a local dendritic calcium-dependent potential, without causing inhibition or stimulation in the cell as a whole. It also may indicate an ionic mechanism by which 5-HT and other transmitters trigger the dendritic release of transmitter and protein.

The above observations and the presence of serotoninergic terminals on dopamine cells in the substantia nigra[230] and in the ventromedial mesencephalic tegmentum (VMT)[231] suggest the likelihood of important interactions between these two monoamines, in these and other segments of the central nervous system. High levels of 5-HT have been detected in the VMT,[232] and both the dorsal and median raphe nuclei send a large number of projections to this area, through the transegmental serotoninergic fiber pathway and on the way to the medial forebrain bundle.[233] These axons likely innervate the ventral striatum and the cortex.[233-236] Approximately 50% of all VMT terminals with synapses can be labeled with 5-HT.[236,237] Careful study by Halliday and Törk[237] of Sydney, Australia, has revealed heterogeneous distribution of 5-HT cells and fibers throughout the five-component nuclei of the rat VMT, suggesting multiple and varied 5-HT influences in this area. An additional monoamine "connection" has been observed, as 5-HT-immunoreactive fibers have been identified in the histaminergic system cell body area, the tuberomammillary nucleus. Study of the 5-HT-immunoreactive fibers showed vesicle-containing terminal boutons that formed synapses with dendrites of various sizes.

The mammalian visual system receives dense innervation by noradrenergic fibers from the locus ceruleus and by serotoninergic fibers from the raphe nuclei.[238] Visual hallucinogenic effects in man of serotonin-receptor ligands, such as psilocine and lysergide, support the importance of serotoninergic sensory inputs on visual function. The superior colliculus plays an important role in the integration of visual inputs.[239] Serotonin has been found to be both accumulated and released in rabbit superior collicular slices.[238,240]

In an important 1989 study by Eaton and Salt,[241] of London, serotonin, on iontophoric injection at low currents, has been found to produce a marked facilitation of responses to excitatory amino acids, as recorded extracellularly from single-neuron preparations from the rat ventrobasal thalamus. This is an especially important observation, since excitatory amino acids are thought to mediate sensory input to the ventrobasal thalamus as well as to play an important role in the establishment of long-term memory. The findings of this study demonstrated that 5-HT may facilitate responses to all glutamate receptors (NMDA, quisqualate, and kainate), confirming earlier observations of Nedergaard et al.,[242] and added kainate to the glutamate receptors facilitated. The facilitation of glutamate, but not nicotonic or muscarinic cholinergic receptors, indicates that these faciliatory responses are relatively selective for excitatory amino-acid-receptor-mediated responses. These findings confirm a similar selectivity with respect to cholinergic receptors, as encountered in the rat neocortex.[243]

e. Brain

The involvement of serotonin and the fibers and projections of the serotoninergic neuronal system in the basal ganglia, hippocampus, and the cerebral cortex will be considered in this section.

In 1991, serotonin was reported (by Walker et al.,[244] of Rutgers University) to play a role in the biosynthesis of tachykinins in the rat neostriatum. Pharmacological depletion of 5-HT by subchronic pCPA (*p*-chlorophenylalanine) treatment *in vivo* decreased preprotachykinin mRNA levels in the neostriatum. Raising extracellular 5-HT levels with zimelidine (a 5-HT-uptake inhibitor) or clorgyline (a monoamine oxidase inhibitor) raised levels of preprotachykinin mRNA. DOI, a 5-HT_2-receptor agonist, increased significantly both preprotachykinin mRNA and substance P immunoreactivity in the rat neostriatum. These findings indicate that neostriatal tachykinin synthesis is influenced by 5-HT levels and provide another link in the 5-HT-substance P "connection".

Activation of 5-HT_3 receptors in the rat neostriatum (striatal slices *in vitro*) has been found to increase the release of dopamine from this area.[245] *In vivo* intraventricular administration of a 5-HT_3 agonist (2-Me-5HT) has been shown to release dopamine from the nucleus accumbens (a part of the striatum).[246] These findings, and the substantia nigra observations from Oxford cited above[229,242] in the "Brain Stem" section, provide support for an important "dopamine" connection for 5-HT and may link this "connection" with the 5-HT_3 receptor.

An important involvement of 5-HT in hippocampal long-term potentiation (LTP) produced by brief, high-frequency stimulation of the perforant path of the hippocampus was reported in rats in 1983 by Bliss, Goddard, and Riives.[247] 5-HT-depleted rats experienced severe depletion of the LTP phenomenon, an experimental phenomenon widely regarded as an important model of memory and learning.[248-250] The duration of LTP was reduced by 5-HT depletion, and the threshold required to achieve LTP was increased by 5-HT depletion. Neither of these responses were affected by noradrenalin (NA) depletion. Long-term potentiation of dentate gyrus granule-cell excitability was reduced by 5-HT depletion. The perforant path of the hippocampus is an excitatory fiber system that originates in the entorhinal cortex and terminates on granule cells in the dentate gyrus. The LTP of synaptically evoked responses has been described for a number of other excitatory pathways of the hippocampus. This phenomenon has received a massive amount of research attention. These are important observations, as serotoninergic fibers from the raphe nuclei in mammals project widely to the hippocampus, with the greatest density of terminals in the hilus of the dentate gyrus.[251-257] They also recall memory-stimulating and -facilitating effects of serotonin in the substantially more-primitive marine molluscs *Aplysia* and *Hermissenda* (see above).

A serotonin-induced late depolarization (excitation) was recently (1989) reported as a serotonin effect on hippocampal dentate granule cells.[258] Prolonged facilitation (30 to 40 min) responses to mossy-fiber stimulation of the CA3 (Cornu Ammonis 3) area of the guinea pig and rat hippocampus was reported from Russia in 1982 by Otmakhov and Bragin.[259]

Two 1989 publications of significance dealt with effects of 5-HT on rat hippocampal CA1 neurons. Colino and Halliwell,[260] of London, described differential modulation of three separate potassium conductances in hippocampal CA1 neurons by serotonin: one inhibitory and two simulatory. One of the stimulatory responses was slow developing and long lasting. It led to CA1 neuronal depolarization and excitation. Andrade and Nicoll,[261] of San Francisco, described serotonin-induced actions on hippocampal CA1-region pyramidal cells. The initial response was hyperpolarization (inhibition) followed by a longer-lasting depolarization (excitation). The initial response appeared to be 5-HT_{1A}-receptor mediated. No receptor was identified for mediation of the longer-lasting depolarization response. Extensive serotonin localization in the Horns of Ammon in the hippocampus is beautifully illustrated in Figure 4 of Chapter 1.

Clark, Sharp, and associates,[262] of Oxford, reported in 1989 a series of studies involving the effects of manipulating the activity of central 5-HT neurons on extracellular 5-HT levels (microdialysis-determined) in the ventral hippocampus of anesthetized rats. Levels were reduced by 57% in rats pretreated for 14 d by intracerebroventricular administration of a 5-HT neurotoxin (5,7-dihydroxytryptamine). Electrical stimulation of the dorsal raphe (DR) nucleus for 20 min produced a rapid increase in 5-HT levels, which declined promptly on cessation of stimulation and were frequency dependent. Injection of a 5-HT_{1A} agonist into the DR nucleus resulted in a dose-related decrease in 5-HT levels. It was concluded that the spontaneous output of 5-HT into hippocampal extracellular dialysates originated predominantly from central 5-HT neurons and changed in accordance with their electrical activity.

In rats, grafted 5-HT neurons have been found (by Daszuta, Bjorklund, et al. of France and Sweden) to maintain 5-HT release in the normal range[263] and to contribute in an essential and required way in the amelioration by grafting of spatial memory impairment produced by combined cholinergic and serotoninergic denervation of the forebrain.[264]

Serotonin innervation of the mammalian cerebral cortex was first demonstrated (in 1965 by Fuxe,[265] of Stockholm) by means of a fluorescence histochemical technique. Ultrastructural delineation of 5-HT axon terminals in the cerebral cortex (by Descarries et al.[266] in 1975) by means of a high-resolution radioautography produced the unexpected observation that these 5-HT terminals seldom were involved in morphologically differentiated synaptic junctions. This finding led these investigators to speculate regarding the ''nonsynaptic'' release of 5-HT in the cortical area of the brain, and a particular and unique mode of action of 5-HT in this brain area. The demonstration of the widespread intracortical 5-HT varicosities (by Beaudet and Descarries[267,268]) led to the extension of this speculation to the concept that 5-HT, through this predominantly nonjunctional innervation, might possess the capability of acting on a much larger and more numerous cellular target in the cortical area than the range of action permitted by a limited number of postsynaptic connections of the more traditional and orthodox kind. A recent (1989) study by the Descarries group,[269] employing PAP immunocytochemistry, concluded that the adult rat neocortical 5-HT innervation is predominantly nonjunctional in the three major cortical areas investigated: frontal (primary motor), parietal (primary sensory), and occipital (primary visual). The 5-HT axon terminals (varicosities) were uniform in size and shape in the four cortical layers studied. The synaptic incidence ranged from 28 to 46%, and the junctions were always asymmetrical and exclusively found on dendritic spines and shafts, with a greater frequency on spines. Non-5-HT dendritic spines and shafts and axon varicosities frequently were identified in the immediate microenvironment of 5-HT varicosities. These findings reinforced this group's prior views concerning a highly divergent 5-HT afferent system to the mammalian brain cortex, with the potentials for widespread, global, and sustained influences in this most advanced and sophisticated of the areas of the central nervous system.

The cerebral neocortex of the macaque monkey was reported (in 1988 by a San Diego group of Bloom, Morrison, and associates[270]) to be densely innervated by serotonin-containing nerve fibers. The greatest density of these serotoninergic fibers occurs in the primary sensory regions, including the primary visual cortex. In the primary visual cortex (area 17) of this monkey, 5-HT-immunoreactive fibers were found in all cortical layers with labeled varicosites. A prior study by this group[271] had compared serotoninergic and noradrenergic (NA) innervation of cortical and subcortical visual structures in Old World (macaque) and New World (squirrel) monkeys. Few species differences were encountered for either monoamine. 5-HT fiber density and uniformity of distribution was greater. NA fibers (identified by antisera against human dopamine-β-hydroxylase) appeared preferentially to innervate areas involved in spatial analysis and visuomotor response, rather than those areas of 5-HT innervation involving feature extraction and pattern analysis. In another 1988 report from Paris by Berger et al.,[272] the regional and laminar distribution of both serotonin and dopamine (DA) innervation of the entire macaque cerebral cortex were investigated by radioautographic means. In all but the motor areas, 5-HT fibers exhibited a much higher density than did DA fibers. The greatest density of 5-HT innervation was in the visual areas, with the density in cortical area 17 greater than in area 18. A considerably expanded DA cortical innervation was observed in this monkey in comparison with rat and rodent studies. In the cat cerebral cortex small-caliber 5-HT varicose axons have been found (by Mulligan and Törk) to predominate.[273]

Important brain-cortical-area interrelationships of 5-HT and other important non-5-HT "memory molecules" of a vasoactive neurotransmitter type (NMDA and substance P) have been reported. Agnati, Fuxe, Hokfelt et al.,[274] of Stockholm, reported in 1983 that *in vitro* additions of substance P to spinal cord membrane preparations increased the number of 5-HT binding sites. Ogren, Fuxe, and Agnati[275] described an effect of intraventricular-injected substance P in increasing significantly the rat head-twitch behavior induced by low doses of the 5-HT precursor, 5-HTP, given subcutaneously. This 5-HT head-twitch behavior in rats appears to be controlled by 5-HT_2 receptors.[276]

Serotonin has been found (by Nedergaard et al.,[277] of Aarhus, Denmark) to facilitate the depolarization (excitation) of cat neocortical neurons induced by glutamate-receptor agonist "memory molecule" NMDA (*N*-methyl-D-aspartate). The effects of 5-HT on neuronal responses to the excitatory amino acid agonist NMDA has been explored in rat neocortical slices by Reynolds et al.,[278] of Toronto, using current-clamp and single-electrode voltage-clamp techniques. Prior perfusion of these neocortical slices with 5-HT significantly increased both the amplitude and the duration of the responses of these neurons to NMDA. The response was long lasting and could not be reversed during the time course of the experiment. 5-HT did not alter the neuronal response to quisqualic acid, another glutamate-receptor agonist. Substance P has been found to enhance basal and potassium-evoked release of 5-HT from slices of rat cerebral cortex.[279]

In a recent report in *Nature* (November 7, 1991), Blakely et al.,[280] from Duke and Emory Universities, described cloning and expression of a functional serotonin transporter from rat brain. The authors found sequence similarities with human noradrenaline and rat

GABA transporters, and *in situ* hybridization to the dorsal and medial subdivisions of the rat serotoninergic midbrain raphe complex. They identified a single cDNA (from rat brain) which was sufficient to form a fully functional 5-HT transporter in nonneuronal cells.

V. VASCULAR SYSTEM

Despite its name, the role of serotonin in the vascular system is by no means clear. In a 1988 review of the involvement of serotonin in the vascular system and in vascular responses, Hollenberg, of Harvard[281] states, ''Given the myriad of actions, many of which were not blocked by serotonin antagonists, it is not surprising that serotonin's role in normal circulatory physiology and in disease has remained obscure.''

In vascular system areas of importance to the immune response system, the influence of serotonin is toward vasodilatation with increases in blood flow and vascular permeability. In 1986 Orlandi, Blackshear, and Hollenberg[282] reported a unique microvascular response to serotonin in dogs in a serotonin-induced visualization of the popliteal lymph node by angiography. Infusion of serotonin into the femoral artery induced striking visualization of the popliteal lymph node in 19 of 20 dogs. This particularly important and significant response was unique to serotonin and was not duplicated by other vasodilator or vasoconstrictor substances. The timing of appearance and disappearance of the contrast medium during 5-HT infusion suggested that an increase in capillary blood flow secondary to arteriolar dilatation in the lymph node was responsible for the augmented angiographic visualization. This augmented visualization was not blocked by either methysergide or ketanserin.

Both blood flow and vascular permeability in lymph nodes and lymphoid tissue play important and critical roles in the initiation and propagation of immune responses. Virtually all of the small recirculating lymphocytes that reach lymph nodes arrive in arterial blood and enter the lymphoid tissues by way of passage between the cells of the postcapillary venule area.[283-294]

In studies on inflammation, Majno, Palade, and associates[295-297] have identified the postcapillary venule as the anatomical location in the vascular tree of serotonin-induced increase in vascular permeability and ''leakage'' with macromolecular passage between endothelial cells and out of the intravascular system.

The most likely source of 5-HT in inflammation and immunoinflammatory reactions is the platelet, which appears to act as a serotonin ''sponge''. In rodents, serotonin also is found in mast cells.

In the supported-sheep lymphocyte traffic model, the author has reported that serotonin infused acutely into cannulated afferent lymphatics of popliteal lymph nodes has produced prompt and sharp increases in lymphocyte traffic, as monitored by the output of small recirculating lymphocytes into nodal efferent lymphatics, chronically cannulated.[298-301] This was interpreted as an immunostimulative response.

Cerebrovascular endothelial cells in culture have been shown to take up, synthesize, and metabolize serotonin.[302] Vascular endothelial cells have been reported to mediate endothelium-dependent relaxation of arteries.[303,304]

A perivascular serotoninergic innervation of pial arteries and arterioles in the rat has been demonstrated.[305] The fibers appeared to have a central neuronal origin. They appeared to arise from the median and dorsal raphe nuclei in the brain stem. In the gerbil, the major cerebral arteries of the circle of Willis appear to have a dual or combined serotoninergic innervation, with fibers arising in the superior cervical ganglion as well as in the raphe nuclei.[306]

VI. IMMUNE RESPONSE SYSTEM

Serotonin is a major platelet product released at sites of immunoinflammation. Circulating platelets contain high concentrations of serotonin. In various immunological and immunoinflammatory reactions, substantial amounts of serotonin have been detected locally.[307-309] These local elevations in 5-HT levels are presumed to be the result of platelet aggregation and subsequent serotonin release, or of mast cell degranulation. A serotonin-induced increase in vascular permeability may play an important role in the recruitment of immune response effector cells to inflammatory and immunoinflammatory foci.[310] A serotonin enhancement in lymphocyte chemotactic activity would augment the effectiveness of this response.[311]

The 1990 demonstration of 5-HT receptors on T-lymphocytes and an immunostimulative T-cell response to 5-HT activation was a major breakthrough in dissecting and delineating the effective role of serotonin in the cellular immune response. The authors of this study, Aune et al.[312] of the Miles Research Center, West Haven, CT, used the Jurkat cell model system. These cells are a transformed human T-cell line. These studies showed that Jurkat cells bind 5-HT, with an average dissociation constant of 90 n*M*, and that 5-HT stimulates an increase in inositol phosphate and intracellular-Ca^{++} levels in these cells. These findings indicate a clearly immunostimulative effect of 5-HT on these important lymphocytes of the T-cell class.

B-cells also have been shown to have 5-HT receptors. Choquet and Korn,[313] of the Pasteur Institute (Paris), used a pre-B-cell line with whole-cell patch-clamp technique to show that 5-HT in physiological doses regulated a voltage-gated potassium channel by activating different subsets of 5-HT receptors. 5-HT induced in the recorded cells both an increase in the maximum potassium conductance and an acceleration of the inactivation process. The increase in maximum conductance was shown to be due to activation of 5-HT_1-like receptors. Acceleration of the inactivation process was shown to be due to activation of 5-HT_3 receptors, and this acceleration was shown to be mimicked by the 5-HT_3 agonist 2-methyl-5-HT. The findings of these studies indicate important common features of 5-HT effects on both the nervous and immune response effector cells and further emphasize the significance of the neurovascular immunoresponse "connection" relating to serotonin.

In other immune response actions, serotonin has been shown to augment murine bone-marrow macrophage phagocytosis at low concentrations of interferon-gamma.[314]

Hellstrand and Hermodsson in 1987 and 1990[315,316] reported important observations relating to a serotonin immunostimulative activity. They found that serotonin, in concentrations ranging from 10^{-4} to 10^{-7} *M*, augmented human natural killer (NK) cell cytotoxicity against four types of NK-sensitive target cells by a mechanism of action that involved a prostaglandin- and IL-1-independent accessory function of monocytes. Monocytes obtained by countercurrent centrifugal elutration, not by adherence, reconstituted the effect of serotonin when added to nonadherent NK cells. The NK-enhancing effects of serotonin were mimicked by two 5-HT_{1A}-receptor agonists. The development of the NK response involved an induction phase that required the presence of both serotonin and accessory monocytes, and an effector phase that did not require the presence of either monocytes or serotonin. Serotonin-activated NK cells continued to demonstrate enhanced cytotoxicity for at least 8 h after removal of the serotonin and monocytes. This long-term potentiation of an immune-type response calls to mind the long-term potentiation of a nervous system response by serotonin in the hippocampus, reported in 1983 by Bliss et al.[247] and cited earlier in this chapter in the "Brain" component of the "Nervous System" section.

In several experiments in the Hellstrand and Hermodsson report, serotonin-activated NK cells killed more than 75% of one of the target-cell types (K562 cells) even at low effector-to-target-cell ratios and low-baseline natural-killer-cell cytotoxicity. The NK cells used in this study were CD3−/CD16+ lymphocytes with LGL morphology that did not require interaction with the CD3/ti antigen-receptor complex for target-cell recognition (Hercend and Schmidt definition). An interesting observation was that NK-insensitive Daudi target cells were lysed by serotonin-activated NK cells. Since resting human NK cells may conjugate with Daudi targets, but do not lyse them, it was concluded by the authors that serotonin both activates human NK cells and expands their target spectrum. In the 1987 report by these authors, the NK-enhancing effect of serotonin was found to be interferon and IL-2 independent. The authors have expanded their studies to show that serotonin-activated monocytes recovered by countercurrent centrifugation, rather than adherence, may enhance the cytotoxicity of small lymphocytes with high buoyant density and low spontaneous cytotoxicity. They also have found that this serotonin-enhancement of monocyte/NK cell cytolytic activity is operable for histamine as well.

The authors, in their 1990 report, have suggested that serotonin may play an important role in nonspecific tumor defense by influencing a critical interplay between monocytes and NK cells. With A. Asea, they have explored the potential of a peripheral blood *in vivo* serotonin-activation of circulating NK cells as an antimetastatic effect. They have treated mice with serotonin (25 mg/kg) for 12 to 24 h prior to intravenous injection of NK-sensitive B16 melanoma cells and have monitored the mice for pulmonary metastases 10 to 14 d later. They found that the serotonin treatment markedly (by more than 50%) reduced the number of pulmonary metastases.[316]

In two 1975 reports, Sandler et al.[44,45] observed that serotonin increased human monocyte levels of cyclic GMP and stimulated leukocyte and monocyte chemotaxis.

Cyproheptadine, with antiserotonin and antihistamine action, has been reported to prolong the survival of renal allografts in rats[317] and in man.[318]

In a 1979 study of 52 cancer patients and 76 normal age-matched controls, cancer patients had higher platelet levels of the 5-HT-degrading enzyme (MAO), and lower urinary excretion of 5-HT, 5-H1AA, and tyramine.[319] Burtin et al.,[320] of Paris, in 1982, reported that the best results in controlling MCA-induced malignant tumor growth in mice and prolonging animal survival were obtained with daily intraperitoneal injections of serotonin. Animal survival was decreased by the antiserotonin agent methysergide. The serotonin-releasing agent reserpine has found to decrease fibrosarcoma growth *in vivo*.[321] Depletion of 5-HT has been shown to increase tumor growth in melanoma-bearing mice.[322]

As cited in the "Vascular System" section of this chapter, the author[298-301] has reported the production of prompt, sharp, and prolonged increases in the traffic of small recirculating lymphocytes through the popliteal lymph nodes of sheep *in vivo,* by the acute infusion of serotonin into cannulated popliteal lymph node afferent lymphatics (Figure 13). The increase in lymphocyte traffic was monitored by sequential evaluations of the output of these small recirculating lymphocytes into popliteal efferent lymph of animals, with chronic cannulation of the popliteal node efferent lymphatic. The serotonin-stimulated increase in lymphocyte traffic was prolonged (? LTP) and involved both small recirculating and blast lymphocytes. The increased traffic and small recirculating lymphocyte outputs into efferent lymph were largely targeted to the T-cell outputs, as there was little change in the level of Ig^+-cell (B-cell) outputs into efferent lymph (Figure 14).

VII. GROWTH FACTOR ACTIONS

A 1989 commentary article by Berk and Alexander[323] that appeared in *Biochemical Pharmacology* was devoted to considerations of vasoactive effects of growth factors. A year later (1990), another commentary article appeared in *Biochemical Pharmacology* (by Seuwen

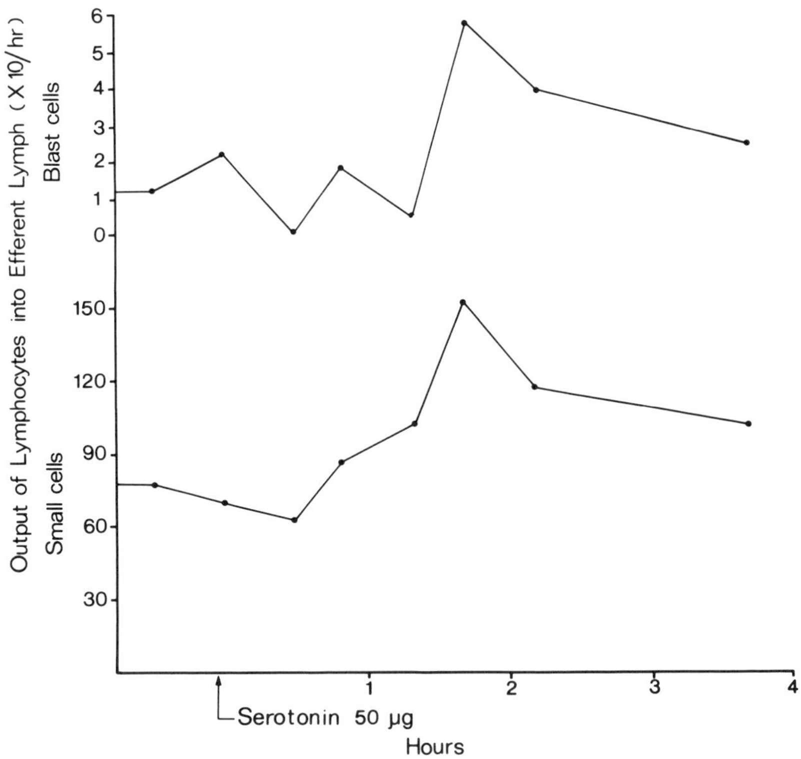

FIGURE 13. Acute infusion of serotonin (50 μg) into a cannulated afferent lymphatic of a popliteal lymph node of a sheep produces a sharp, prompt, and prolonged increase in the output of both small recirculating and blast lymphocytes into the nodal efferent lymph and lymphatic, chronically cannulated. (From Moore, T. C., *Immunology,* 52, 511, 1984. With permission.)

and Pouyssegur,[324] of the University of Nice, France) concerning an important role for serotonin, a major vasoactive neurotransmitter, as a growth factor. This important article also considered other potent vasoactive neurotransmitter molecules, such as bradykinin and and bombesin, which may function as growth factors and promote mitogenesis and DNA synthesis. These "nonclassical" growth factors have in common the ability to activate phosphoinositide (PI) breakdown and/or inhibit adenylate cyclase. These nonclassical growth factors activate a PI-specific phospholipase C (PLC) and act through G-protein-coupled receptors that may be pertussis toxin sensitive. The "classical" growth factors, such as EGF, PDGF, FGF, and IGF, on the other hand, function through receptor tyrosine kinases.

Interest in serotonin as a growth factor was triggered by a 1986 article by Nemecek et al.[325] (in *Proc. Natl. Acad. Sci. U.S.A.*) that described serotonin-induced stimulation of aortic smooth-muscle-cell mitogenesis. In 1988 Kavanaugh, Williams, and associates[326] reported vascular smooth-muscle-cell DNA synthesis to be induced by serotonin. Also in 1988 Seuwen et al.[327] published an article in *Nature* concerning serotonin-induced stimulation of DNA by fibroblasts, with action being mediated through G-protein-coupled receptors.

VIII. CONCLUSION

Serotonin (5-HT), is, indeed, a most remarkable molecule with an extraordinary scope of biological activity, which ranges from vasoactive neurotransmission (including pain stimulus and transmission), to immunostimulation, to mediation of inflammation, to action as

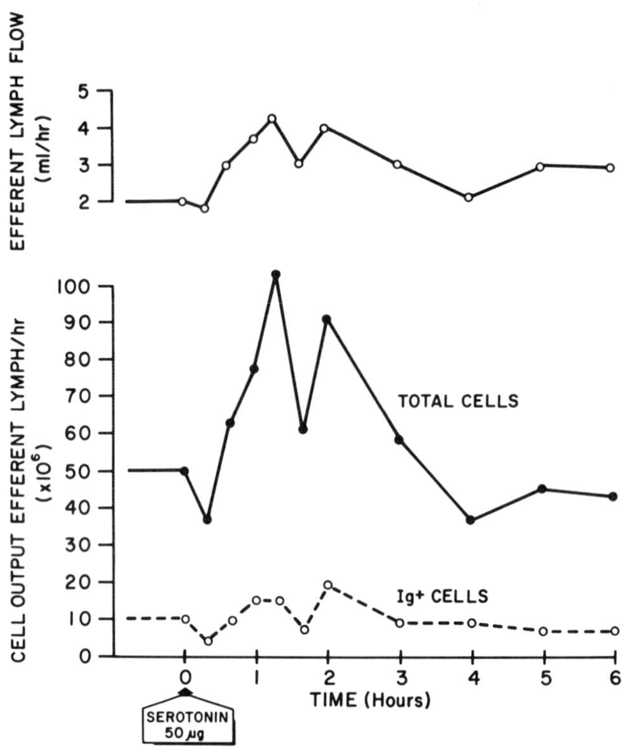

FIGURE 14. The serotonin-induced increase in lymphocyte traffic in this sheep model and on the small recirculating population of lymphocytes was found to be targeted largely toward the T cells, as there was little change in the output into efferent lymph of Ig^+ small lymphocytes. (From Moore, T. C., et al., *IRCS Med. Sci.*, 8, 272, 1980. With permission.)

a growth factor, to function as a major "memory molecule" (CNS as well as immunological). It appears to be one of the most ancient, phylogenetically, of all of the currently known vasoactive neurotransmitter molecules. It plays an important role in the most primitive of nervous systems (*Renilla*), and in this cnidarian coelenterate also functions in both food gathering and stinging defense. It clearly is one of the "big four" vasoactive neurotransmitter "memory molecules", along with substance P, bradykinin, and histamine.

5-HT neurons constitute the most widely and ubiquitously distributed transmitter-defined system in the vertebrate brain. No region of the central nervous system lacks 5-HT innervation, including each of the circumventricular organs (blood-brain interface) and the cerebroventricular cavity.[328-330] Cell bodies of the 5-HT neuronal system are strategically located in the brain stem, with major projections both rostrally and caudally and with extensive fiber interplay and interdigitations and synaptic contacts with the critical dopaminergic, as well as with the substance P and histamine, nervous systems.

The important contribution of serotoninergic neuronal system innervation of the mammalian (rat) hippocampus is dramatically presented in the whole-hippocampus light microscope radioautograph (silvergrains) presented as an illustration in the 1990 *Neuroscience* publication by Oleskevich and Descarries[90] (see Figure 4 in Chapter 1). This radioautograph also delineates vividly the Horns of Ammon (Cornu Ammonis), which are such a central hippocampal anatomical component. The Cornu Ammonis are designated CA1, CA2, and CA3 and are central to memory laying down. The 5-HT innervation of the hippocampus is rich and widespread, with midbrain raphe-issued 5-HT fibers reaching the hippocampus by three major pathways: the cingulum, the fimbria-fornix, and the amygdaloid.[90] Greater 5-HT innervation densities in the hippocampus are observed in the subiculum and in Ammon's horn than in the dentate gyrus.[90] The influence of this extensive anatomical innervation of the hippocampus by the serotoninergic neuronal system is further amplified by the large number of varicosities (130) in critical areas such as the CA3.[90] The pyramidal cells of CA3 show spontaneous bursting and are thought of as pacemaker cells in the hippocampus.[331,332] Disinhibition of these CA3 "pacemaker" cells causes epileptic activity in CA1, the hippocampal Cornu Ammonis area innervated by CA3. Destruction of the CA3-CA1 connections has been shown to prevent abnormal activity in CA1 cells.[331,332] As disinhibitory effects of 5-HT have been demonstrated in the hippocampus,[333,334] the dense 5-HT innervation of the CA3 area of the hippocampus, documented in the 1990 study of Oleskevich and Descarries,[90] suggested to these authors that actions of 5-HT in the stratum oriens, either directly on dendrites of CA3 pyramidal cells or indirectly on interneurons, could play an important role in influencing neuronal activity throughout the Cornu Ammonis. Oleskevich and Descarries further speculate, and are intrigued by the possibility, that the 5-HT innervation might exert its major action in the "heart" of the hippocampus, while the noradrenaline (NA) innervation (inhibitory) has been found by them (Oleskevich, Descarries, and Lacaille[335]) to predominate in the hippocampal input (dentate gyrus) and output (subiculum) zones.

The role of serotonin/5-HT as the sting/pain-producing molecule of both the stinging nettles (*Urtica dioica, Urtica urens,* and *Urtica pilulifera*) and the cnidarian coelentrate *Renilla koellikeri* raises the interesting question of the relationship between the words cnidarian and nettle. Nettle is a Teutonic/Anglo-Saxon word for the stinging plant, and the Latin and Greek words for the stinging nettle are cnida and κνιδη. As the Romans lacked the letter "k" in their usual alphabet, the pronunciation of cnidarian, cnidoblast, and cnidocyst (serotonin-containing nematocyst) follow the phonetic Greek κ (kappa). The dramatic stinging power and extreme pain produced by stinging nettles has led to the extensive use of this pesky plant in literature. An association between the pain-inducing nettle and learning and memory is suggested in Hotspur's speech in Shakespeare's *Henry IV,* Part 1 (Act 2, Scene 3): "Out of this this Nettle, Danger; we pluck this Flower, Safety." From science to literature, serotonin has a legitimate, important, and major role in the "heat, pain, and memory" scenario.

Serotonin and related molecules clearly are of immense importance in many key and critical biological processes, including neurovascular immunology, and doubtless have many more biological "secrets" to yield, including those of immunological and central nervous system memory.

REFERENCES

1. **Umbriaco, D., Anctil, M., and Descarries, L.,** Serotonin-immunoreactive neurons in the cnidarian *Renilla koellikeri, J. Comp. Neurol.,* 291, 167, 1990.
2. **Welsh, J. A.,** 5-Hydroxytryptamine in coelenterates, *Nature,* 186, 811, 1960.
3. **Kline, E. S. and Weissbach, H.,** Hydroxyindoleamines in the nematocyst of *Hydra littoralis, Life Sci.,* 4, 63, 1965.
4. **Carlberg, M. and Rosengren, E.,** Biochemical basis for adrenergic neurotransmission in coelenterates, *J. Comp. Physiol.,* 155, 251, 1985.
5. **Castano, P. and Rossi, S.,** Cytochemical, ultrastructural and fluorescence study of the nervous net of *Hydra* sp., *J. Submicr. Cytol.,* 10, 381, 1978.
6. **Erspamer, V.,** Pharmakologische Studien uber Enteramin. I. Mitteilung: Uber die Wirkun von Acetonextrakten der Kaninchenmagenschleimhaut auf den Blutdruck und auf isolierte ube lebende Organe, *Arch. Exp. Pathol. Pharmakol.,* 196, 343, 1940.

7. **Erspamer, V. and Asero, B.,** Identification of enteramine, the specific hormone of the enterochromaffin cell system, as 5-hydroxytryptamine, *Nature,* 169, 800, 1952.
8. **Erspamer, V.,** Richerche farmacoligiche sull'enteramina, VII, Enteramina indolalchilamine del velena di rospo, *Arch. Sci. Biol. (Bologna),* 31, 86, 1946.
9. **Erspamer, V.,** Active substances in the posterior salivary gland of octopoda, *Acta Pharmacol. Toxicol.,* 4, 213, 1948.
10. **Erspamer, V.,** Pharmacology of indolealkamines, *Pharmacol. Rev.,* 6, 425, 1954.
11. **Erspamer, V.,** Occurrence of indolealkylamines in nature, in 5-hydroxytryptamine and related indolealkylamines, Erspamer, V., Ed., *Handbo. Exp. Pharmacol.,* 19, 132, 1966.
12. **Rapport, M. M., Green, A. A., and Page, I. H.,** Serum vasoconstrictor (serotonin). IV. Isolation and characterization, *J. Biol. Chem.,* 176, 1243, 1948.
13. **Page, I. H.,** Serotonin (5-hydroxytryptamine), *Physiol. Rev.,* 34, 563, 1954.
14. **Rapport, M. M.,** Serum vasoconstrictor (serotonin). V. The presence of creatinine in the complex: a proposed study of the vasoconstrictor principle, *J. Biol. Chem.,* 180, 961, 1949.
15. **Hamlin, K. E. and Fischer, F. E.,** Synthesis of 5-hydroxytryptamine, *J. Am. Chem. Soc.,* 73, 5007, 1951.
16. **Rocha e Silva, M., Valle, J. R., and Picarelli, Z. P.,** A pharmacological analysis of the mode of action of serotonin (5-HT) upon the guinea-pig ileum, *Br. J. Pharmacol.,* 8, 378, 1953.
17. **Gaddum, J. H. and Picarelli, Z. P.,** Two kinds of tryptamine receptor, *Br. J. Pharmacol.,* 12, 323, 1957.
18. **Fozard, J. R. and Mobarek Ali, A. T. M.,** Blockade of neuronal tryptamine receptors by metoclopramide, *Eur. J. Pharmacol.,* 49, 109, 1978.
19. **Fozard, J. R., Mobarok Ali, A. T. M., and Newgrosh, G.,** Blockade of serotonin receptors on autonomic neurones by (–)cocaine and some related compounds, *Eur. J. Pharmacol.,* 59, 195, 1979.
20. **Peroutka, S. J. and Snyder, S. H.,** Multiple serotonin receptors: differential binding of ^{3}H-serotonin, ^{3}H-lysergic acid diethylamide and ^{3}H-spiroperidol, *Mol. Pharmacol.,* 16, 687, 1979.
21. **Fargin, A., Raymond, J. R., Lohse, M. J., Kobilka, B. K., Caron, M. G., and Lefkowitz, R. J.,** The genomic clone G-21 which resembles a beta-adrenergic receptor sequence encodes the 5-HT_{1A} receptor, *Nature,* 335, 358, 1988.
22. **Pedigo, N. W., Yamamura, H. I., and Nelson, D. L.,** Discrimination of multiple [^{3}H]5-hydroxytryptamine binding sites by the neuroleptic spiperone in rat brain, *J. Neurochem.,* 36, 220, 1981.
23. **Gozlan, H., El Mestikawy, S., Pichat, L., Glowinski, J., and Hamon, M.,** Identification of presynaptic serotonin autoreceptors using a new ligand: ^{3}H-PAT, *Nature,* 305, 140, 1983.
24. **Traber, J. and Glaser, T.,** The 5-HT_{1A} receptor-related anxiolytics, *Trends Pharmacol. Sci.,* 8, 432, 1987.
25. **Cervo, L. and Samanin, R.,** Potential antidepressant properties of 8-hydroxy-2-(di-n-propylamino) tetralin, a selective $serotonin_{1A}$ receptor agonist, *Eur. J. Pharmacol.,* 144, 223, 1987.
26. **Leysen, J. E. and Pauwels, P. J.,** 5-HT_2 receptors, roles and regulation, *Ann. N.Y. Acad. Sci.,* 600, 183, 1990.
27. **Leysen, J. E., Niemegeers, C. J. E., Van Nueten, J. M., and Laduron, P. M.,** [^{3}H]ketanserin (R 41 468), a selective ^{3}H-ligand for $serotonin_2$ receptor binding sites. Binding properties, brain distribution, and functional role, *Mol. Pharmacol.,* 21, 301, 1982.
28. **Titeler, M., Herrick, K., and Lyon, R. A.,** [^{3}H]DOB: a specific agonist radioligand for 5-HT_2 serotonin receptors, *Eur. J. Pharmacol.,* 117, 145, 1985.
29. **Pritchett, D. B., Bach, A. W. J., Wozny, M., Taleb, O., Dal Toso, R., Shih, J. C., and Seeburg, P. H.,** Structure and functional expression of cloned rat serotonin 5-HT-2 receptor, *EMBO J.,* 7, 4135, 1988.
30. **Leysen, J. E., Niemegeers, C. J. E., Tollenaere, J. P., and Laduron, P. M.,** Serotoninergic component of neuroleptic receptors, *Nature,* 272, 168, 1978.
31. **Pazos, A., Hoyer, D., and Palacios, J. M.,** Binding of serotoninergic ligands to the porcine choroid plexus: characterization of a new type of 5-HT recognition site, *Eur. J. Pharmacol.,* 106, 539, 1984.
32. **Pazos, A., Hoyer, D., and Palacios, J. M.,** Mesulergine, a selective serotonin-2 ligand in the rat cortex, does not label these receptors in porcine and human cortex: evidence for species differences in brain serotonin-2 receptors, *Eur. J. Pharmacol.,* 106, 531, 1984.
33. **Yagalogg, N. A. and Hartig, P. R.,** ^{125}I-Lysergic acid diethylamide binds to a novel serotoninergic site on rat choroid plexus epithelial cells, *J. Neurosci.,* 5, 3178, 1985.
34. **Julius, D., MacDermott, A. B., Axel, R., and Jessell, T. M.,** Molecular characterization of a functional cDNA encoding the serotonin 1c receptor, *Science,* 24, 558, 1988.
35. **Hartig, P. R., Hoffman, B. J., Kaufman, M. J., and Hirata, F.,** The 5-HT_{1C} receptor, *Ann. N.Y. Acad. Sci.,* 600, 149, 1990.
36. **Johanson, C.,** Choroid plexus, in *Encyclopedia of Neuroscience,* Vol. 1, Adelman, G., Ed., Birkhouser Publishers, Boston, 1987, 236.

37. **Yagaloff, K. A., Lozano, G., Van Dyke, T., Levine, A. J., and Hartig, P. R.,** Sertonin 5-HT_{1C} receptors are expressed at high density on choroid plexus tumors from transgenic mice, *Brain Res.*, 385, 389, 1986.
38. **Tsutsumi, M., Skinner, M. K., and Sanders-Bush, E.,** Transferrin gene expression and synthesis by cultured choroid plexus epithelial cells, *J. Biol. Chem.*, 264, 9626, 1989.
39. **Faraci, F. M., Mayhan, W. G., and Heistad, D. D.,** Effect of serotonin on blood flow to the choroid plexus, *Brain Res.*, 478, 121, 1989.
40. **Bradley, P. B., Angel, G., Feniuk, W., Fozard, J. R., Humphrey, P. P. A., Middlemiss, D. N., Mylecharane, E. J., Richardson, B. P., and Saxena, P. R.,** Proposals for the classification and nomenclature of functional receptors for 5-HT, *Neuropharmacology*, 25, 563, 1986.
41. **Tyers, M. B.,** 5-HT_3 receptors, *Ann. N.Y. Acad. Sci.*, 600, 194, 1990.
42. **Schmidt, A. W. and Peroutka, S. J.,** 5-Hydroxytryptamine receptor "families", *FASEB J.*, 3, 2242, 1989.
43. **Reiser, G. and Hamprecht, B.,** Serotonin raises the cyclic GMP level in a neuronal cell line via 5-HT_3 receptors, *Eur. J. Pharmacol.*, 172, 195, 1989.
44. **Sandler, J. A., Clyman, R. L., Manganiello, V. C., and Vaughn, M.,** The effect of serotonin (5-hydroxytryptamine) and derivatives on guanosine 3′–5′ monophosphate in human monocytes, *J. Clin. Invest.*, 55, 431, 1975.
45. **Sandler, J. A., Gallin, J. I., and Vaughn, M.,** Effects of serotonin carbamylcholine and ascorbic acid on leukocyte cyclic GMP and chemotaxis, *J. Cell Biol.*, 67, 480, 1975.
46. **Richardson, B. P., Engel, G., Donatsch, P., and Stadler, P. A.,** Identification of serotonin M-receptor subtypes and their specific blockade by a new class of drugs, *Nature*, 316, 126, 1985.
47. **Waeber, C. K., Dixon, K., Hoyer, D., and Palacios, J. M.,** Localisation by autoradiography of neuronal 5-HT_3 receptors in the mouse CNS, *Eur. J. Pharmacol.*, 151, 351, 1988.
48. **Hamon, M., Galissot, M. C., Menard, F., Gozlan, H., Bourgoin, S., and Verge, D.,** 5-HT_3 bindings sites are on capsaicin-sensitive fibers in the rat spinal cord, *Eur. J. Pharmacol.*, 164, 315, 1989.
49. **Hayes, A. G. and Tyers, M. B.,** Effects of capsaicin on nociceptive heat, pressure and chemical thresholds and on substance P levels in the rat, *Brain Res.*, 189, 561, 1980.
50. **Dumuis, A., Bouhelal, R., Sebben, M., Cory, R., and Bockaert, J.,** A nonclassical 5-hydroxytryptamine receptor positively coupled with adenylate cyclase in the central nervous system, *Mol. Pharmacol.*, 34, 880, 1988.
51. **Dumuis, A., Sebben, M., and Bockaert, J.,** Benzamide derivatives are agonists of nonclassical 5-HT receptors (5-HT_4) positive coupled with adenylate cyclase in the brain, in *Abstracts of the International Symposium on Serotonin: From Cell Biology to Pharmacology and Therapeutics*, 1989.
52. **Törk, I.,** Anatomy of the serotoninergic system, *Ann. N.Y. Acad. Sci.*, 600, 9, 1990.
53. **Wood, J. G. and Lentz, T. L.,** Histochemical localization of amines in *Hydra* and in the sea anemones, *Nature*, 201, 88, 1964.
54. **Anctil, M.,** Modulation of a rhythmic activity by serotonin via cyclic AMP in the coelenterate *Renilla kollikeri, J. Comp. Physiol.*, in press.
55. **Ross, D. M.,** The effects of ions and drugs on neuromuscular preparations of sea anemones. I. On preparation of the column of *Calliactis* and *Metridium, J. Exp. Biol.*, 37, 732, 1960a.
56. **Ross, D. M.,** The effects of ions and drugs on neuromuscular preparations of sea anemones. II. On sphincter preparations of *Calliactis* and *Metridium, J. Exp. Biol.*, 37, 753, 1960b.
57. **Sukhdeo, S. C. and Sukhdeo, M. V. K.,** Immunohistochemical and electrochemical detection of serotonin in the nervous system of *Fasciola hepatica,* a parasitic flatworm, *Brain Res.*, 463, 57, 1988.
58. **Webb, R. A. and Mizukawa, K.,** Serotonin-like immunoreactivity in the cestode *Hymenolepsis diminuta, J. Comp. Neurol.*, 234, 431, 1985.
59. **Reuter, M., Wikgren, M., and Lehtonen, M.,** Immunocytochemical demonstration of 5-HT-like and FMRFamide-like substances in whole mounts of *Microstomum lineare* (Turbellaria), *Cell Tissue Res.*, 246, 7, 1986.
60. **Mansour, T. E.,** The pharmacology and biochemistry of parasitic helminths, *Adv. Pharmacol.*, 3, 129, 1964.
61. **Anctil, M., Boulay, D., and Lariviere, L.,** Monoaminergic mechanisms associated with control of luminescence and contractile activities in the coelenterate, *Renilla kollikeri, J. Exp. Zool.*, 223, 11, 1982.
62. **Miron, M.-J. and Anctil, M.,** Serotoninlike immunoreactivity in the central and peripheral nervous system of the scale worm *Harmothoe imbricata* (Polychaeta), *J. Comp. Neurol.*, 275, 429, 1988.
63. **Bullock, T. H. and Horridge, G. A.,** *Structure and Functions of the Nervous System of Invertebrates,* Vol. 1, W. H. Freeman, San Francisco, 1965.
64. **Glover, J. C.,** Serotonin storage and uptake by identified neurons in the leech *Haementeria ghilianii, J. Comp. Neurol.*, 256, 117, 1987.

65. **Walz, W. and Schlue, W. R.,** Ionic mechanism of a hyperpolarizing 5-hydroxytryptamine effect on leech neuropile glial cells, *Brain Res.*, 250, 111, 1982.
66. **Coggeshall, R. E.,** Autoradiographic and chemical localization of 5-hydroxytryptamine in identified neurons in the leech, *Anat. Rec.*, 172, 489, 1972.
67. **Henderson, L. P.,** The role of 5-hydroxytryptamine as a transmitter between identified leech neurones in culture, *J. Physiol. (Lond.)*, 339, 309, 1983.
68. **Lent, C. M.,** Retzius cells: neuroeffector cells controlling mucuous release by the leech, *Science,* 179, 693, 1973.
69. **Lent, C. M.,** Morphology of neurons containing monoamines within leech segmental ganglia, *J. Exp. Zool.*, 216, 311, 1981.
70. **Lent, C. M. and Dickinson, M. H.,** Serotonin integrates the feeding behavior of the medicinal leech, *J. Comp. Physiol. A,* 154, 457, 1984.
71. **Mason, A. and Kristan, W. B., Jr.,** Neuronal excitation, inhibition and modulation of leech longitudinal muscle, *J. Comp. Physiol.*, 146, 527, 1982.
72. **Wallace, B. G.,** Neurotransmitter chemistry, in *Neurobiology of the Leech,* Muller, K. J., Nicholls, J. G., and Stent, G. S., Eds., Cold Spring Harbor, New York, 1981, 147.
73. **Zipser, B.,** Horseradish peroxidase nerve backfilling in leech, *Brain Res.*, 182, 441, 1980.
74. **Willard, A. L.,** Effects of serotonin on the generation of the motor program for swimming by the medicinal leech, *J. Neurosci.*, 1, 936, 1981.
75. **Kuffler, D. P., Nicholls, J., and Drapeau, P.,** Transmitter localization and vesicle turnover at a serotoninergic synapse between identified leech neurons in culture, *J. Comp. Neurol.*, 256, 516, 1987.
76. **Dietzel, I. D., Drapeau, P., and Nicholls, J. G.,** Voltage dependence of 5-hydroxytryptamine release at a synapse between identified leech neurons in culture, *J. Physiol. (Lond.),* 382, 101, 1986.
77. **Fuchs, P., Nicholls, J. G., and Ready, D.,** Membrane properties and selective connexions of identified leech neurones in culture, *J. Physiol. (Lond.),* 316, 203, 1981.
78. **Fuchs, P. A., Henderson, L. P., and Nicholls, J. G.,** Chemical transmission between individual Retzius and sensory neurones of the leech in culture, *J. Physiol. (Lond.),* 323, 195, 1982.
79. **Henderson, L.,** The role of 5-hydroxytryptamine as a transmitter between identified leech neurons in culture, *J. Physiol. (Lond.),* 339, 311, 1983.
80. **Henderson, L. P., Kuffler, D. P., Nicholls, J., and Zhang, R.-J.,** Structural and functional analysis of synaptic transmission between identified leech neurones in culture, *J. Physiol. (Lond.),* 340, 347, 1983.
81. **Arluison, M. and De La Manche, I. S.,** High resolution radioautographic study of the serotonin innervation of the rat corpus striatum after intraventricular administration of ⟨3H⟩ 5-hydroxytryptamine, *Neuroscience,* 5, 229, 1980.
82. **Chan-Palay, V.,** Serotonin axons in the supra- and subependymal plexuses and the leptomeninges; their roles in local alterations of cerebrospinal fluid and vasomotor activity, *Brain Res.*, 102, 103, 1976.
83. **Ruda, M. A. and Gobel, S.,** Ultrastructural characterization of axonal endings in the substantia gelatinosa which take up [3H] serotonin, *Brain Res.*, 184, 57, 1980.
84. **Descarries, L., Beaudet, A., and Watkins, K. C.,** Serotonin nerve terminals in adult rat neocortex, *Brain Res.*, 100, 563, 1975.
85. **Goldman, J. E., Shin Kim, K., and Schwartz, J. M.,** Axonal transport of ^{3}H serotonin in an identified neuron of Aplysia Californica, *J. Cell Biol.*, 70, 304, 1976.
86. **Howes, E. A., McLaughlin, B. J., and Heslop, J. P.,** The autoradiographic association of fast transported material with dense core vesicles in the central nervous system of Anadenta cygnea (L.), *Cell Tissue Res.*, 153, 545, 1974.
87. **Maley, B. and Elde, R.,** The ultrastructural localization of serotonin immunoreactivity within the nucleus of the solitary tract of the cat, *J. Neurosci.*, 2, 1499, 1982.
88. **Ruda, M. A., Coffied, J., and Steinbusch, H. W. M.,** Immunocytochemical analysis of serotoninergic axons in laminae I and II of the lumbar spinal cord of the cat, *J. Neurosci.*, 2, 1660, 1982.
89. **Descarries, L., Audet, M. A., Doucet, G., Garcia, S., Oleskevich, S., Seguela, P., Soghomonian, J.-J., and Watkins, K. C.,** Morphology of central serotonin neurons. Brief review of quantified aspects of their distribution and ultrastructural relationships, *Ann. N.Y. Acad. Sci.*, 600, 81, 1990.
90. **Oleskevich, S. and Descarries, L.,** Quantified distribution of the serotonin innervation in adult rat hippocampus, *Neuroscience,* 34, 19, 1990.
91. **Chamberlain, S. T., Pepper, J., Battelle, B.-A., Wyse, G. A., and Lewandowski, T. J.,** Immunoreactivity in *Limulus.* II. Studies of serotoninlike immunoreactivity, endogenous serotonin, and serotinin synthesis in the brain and lateral eye, *J. Comp. Neurol.*, 251, 363, 1986.
92. **Kravits, E. A.,** Hormonal control of behavior: amines and the biasing of behavioral output in lobsters, *Science,* 241, 1775, 1988.

93. **Livingstone, M. S., Harris-Warrick, R. M., and Kravitz, E. A.,** Serotonin and octopamine produce opposite postures in lobsters, *Science,* 208, 76, 1980.
94. **Sandeman, D. C., Sandeman, R. E., and Aitken, A. R.,** Atlas of serotonin-containing neurons in the optic lobes and brain of the crayfish, *Cherax destructor, J. Comp. Neurol.,* 269, 465, 1988.
95. **Sandeman, R. E. and Sandeman, D. C.,** Serotonin-like immunoreactivity of giant olfactory interneurons in the crayfish brain, *Brain Res.,* 403, 371, 1987.
96. **Dixon, D. and Atwood, H. L.,** Phosphatidylinositol system's role in serotonin-induced facilitation at the crayfish neuromuscular junction, *J. Neurophysiol.,* 62, 239, 1989.
97. **Nässel, D. R.,** Serotonin and serotonin-immunoreactive neurons in the nervous system of insects, *Prog. Neurobiol.,* 30, 1, 1988.
98. **Nässel, D. R. and Elekes, K.,** Serotoninergic terminals in the neural sheath of the blowfly nervous system: electron microscopical immunocytochemistry and 5,7-dihydroxytryptamine labeling, *Neuroscience,* 15, 293, 1985.
99. **Schafer, S. and Bicker, G.,** Common projection areas of 5-HT- and GABA-like immunoreactive fibers in the visual system of the honeybee, *Brain Res.,* 380, 368, 1986.
100. **Klemm, N., Steinbusch, H. W. M., and Sundler, F.,** Distribution of serotonin-containing neurons and their pathways in the supraoesophageal ganglion of the cockroach *Periplaneta americana* (L.) as revealed by immunocytochemistry, *J. Comp. Neurol.,* 225, 387, 1984.
101. **Davis, N. T.,** Neurosecretory neurons and their projections to the serotonin neurohemal system of the cockroach *Periplaneta americana* (L.), and identification of mandibular and maxillary motor neurons associated with this system, *J. Comp. Neurol.,* 259, 604, 1987.
102. **Lange, A. B., Orchard, I., and Lloyd, R. J.,** Immunohistochemical and electrochemical detection of serotonin in the nervous system of the blood-feeding bug, *Rhodnius prolixus, Arch. Insect Biochem. Physiol.,* 8, 187, 1988.
103. **Vaysse, G., Galissie, M., and Corbiere, M.,** Induced variation of serotonin in *Drosophila melanogaster* and its relation to learning performance, *J. Comp. Psychol.,* 102, 225, 1988.
104. **Marenco, M. J., Galissie, M., and Vaysse, G.,** Experimental modifications of the larval nutritive medium in *Drosophila melanogaster,* and learning process of the imago, *Drosophila Inf. Serv.,* 60, 221, 1984.
105. **Vaysse, G., Sanz, J. G., Marenco, M. J., Galissie, M., and Abravanel, G.,** Repercussions sur le comportement et l'apprentissage de *Drosophila melanogaster,* d'une carence alimentaire en DL-tryptophane et d'une inhibition de synthese de la serotonine [Consequences for behavior and learning in *Drosophila melanogaster,* of a DL-tryptophan alimentary deficiency and of the inhibition of serotonin synthesis], *Bull. Soc. d'Histoire Nat. Toulouse,* 119, 27, 1983.
106. **Byrne, J. H.,** Cellular analysis of associative learning, *Physiol. Rev.,* 67, 329, 1987.
107. **Pinsker, H. M., Hening, W. A., Carew, T. J., and Kandel, E. R.,** Long-term sensitization of a defensive withdrawal reflex in *Aplysia, Science,* 1821, 1039, 1973.
108. **Eskin, A., Garcia, K. S., and Byrne, J. H.,** Information storage in the nervous system of *Aplysia:* specific proteins affected by serotonin and cAMP, *Proc. Natl. Acad. Sci. U.S.A.,* 86, 2458, 1989.
109. **Frost, W. N., Catellucci, V. F., Hawkins, R. D., and Kandel, E. R.,** The monosynaptic connections from the sensory neurons participate in the storage of long-term memory for sensitization of the gill- and siphon-withdrawal reflex in *Aplysia, Proc. Natl. Acad. Sci. U.S.A.,* 82, 8266, 1985.
110. **Castellucci, V. and Kandel, E. R.,** Presynaptic facilitation as a mechanism for behavioral sensitization in *Aplysia, Science,* 194, 1176, 1976.
111. **Klein, M. and Kandel, E. R.,** Presynaptic modulation of voltage-dependent Ca^{2+} current: mechanism for behavioral sensitization in *Aplysia californica, Proc. Natl. Acad. Sci. U.S.A.,* 75, 3512, 1978.
112. **Klein, M., Camardo, J. S., and Kandel, F. R.,** Serotonin modulates a specific potassium current in the sensory neurons that show presynaptic facilitation in *Aplysia, Proc. Natl. Acad. Sci. U.S.A.,* 79, 5713, 1982.
113. **Siegelbaum, S. A., Camardo, J. S., and Kandel, E. R.,** Serotonin and cyclic AMP close single K^+ channels in *Aplysia* sensory neurones, *Nature,* 299, 413, 1982.
114. **Klein, M., Hochner, B., and Kandel, F. R.,** Facilitatory transmitters and cAMP can modulate accommodation as well as transmitter release in *Aplysia* sensory neurons: evidence for parallel processing in a single cell, *Proc. Natl. Acad. Sci. U.S.A.,* 83, 7994, 1986.
115. **Siegelbaum, S. A., Belardetti, F., Camardo, J. S., and Shuster, M. J.,** Modulation of the serotonin-sensitive potassium channel in *Aplysia* sensory neurone cell body and growth cone, *J. Exp. Biol.,* 124, 287, 1986.
116. **Rayport, S. G. and Schacher, S.,** Synaptic plasticity *in vitro:* cell culture of identified *Aplysia* neurons mediating short-term habituation and sensitization, *J. Neurosci.,* 6, 759, 1986.
117. **Schacher, S.,** Differential synapse formation and neurite outgrowth at two branches of the metacerebral cell of *Aplysia* in dissociated cell culture, *J. Neurosci.,* 5, 2028, 1985.

118. **Schacher, S. and Proshansky, E.,** Neurite regeneration by *Aplysia* neurons in dissociated cell culture: modulation by *Aplysia* hemolymph and the presence of the initial axon segment, *J. Neurosci.*, 3, 2403, 1983.
119. **Montarolo, P. G., Goelet, P., Castellucci, M. V., Morgan, J., Kandel, E. R., and Schacher, S.,** A critical time window for macromolecular synthesis in long-term heterosynaptic facilitation in *Aplysia, Science,* 234, 1249, 1986.
120. **Dale, N., Kandel, E. R., and Schacher, S.,** Serotonin produces long-term changes in the excitability of *Aplysia* sensory neurons in culture that depend on new protein synthesis, *J. Neurosci.*, 7, 2232, 1987.
121. **Brunet, J.-F., Shapiro, E., Foster, S. A., Kandel, E. R., and Iino, Y.,** Identifications of a peptide specific for *Aplysia* sensory neurons by PCR-based differential screening, *Science,* 252, 856, 1991.
122. **Yeung, S. J. and Eskin, A.,** Involvement of a specific protein in the regulation of a circadian rhythm in *Aplysia* eye, *Proc. Natl. Acad. Sci. U.S.A.*, 84, 279, 1987.
123. **Crow, T. J. and Bridge, M. S.,** Serotonin modulates photoresponses in *Hermissenda* Type B photoreceptors, *Neurosci. Lett.*, 60, 83, 1985.
124. **Jacklet, J. W. and Acosta-Urquidi, J. A.,** Serotonin decreases a background current and increases calcium and calcium-activated current in pedal neurons of *Hermissenda, Cell Mol. Neurobiol.*, 5, 407, 1985.
125. **Crow, T. and Forrester, J.,** Light paired with serotonin mimics the effect of conditioning on phototactic behavior of *Hermissenda, Proc. Natl. Acad. Sci. U.S.A.*, 83, 7975, 1986.
126. **Auerbach, S. L., Grover, L., and Farley, J.,** Neurochemical and immunocytochemical studies of serotonin in the *Hermissenda* central nervous system, *Brain Res. Bull.*, 22, 353, 1989.
127. **Farley, J. and Wu, R.,** Serotonin modulation of *Hermissenda* type B photoreceptor light responses and ionic currents: implications for mechanisms underlying associative learning, *Brain Res. Bull.*, 22, 335, 1989.
128. **Grover, L., Farley, J., and Auerbach, S. L.,** Serotonin involvement in in vitro conditioning of *Hermissenda, Brain Res. Bull.*, 22, 363, 1989.
129. **Land, P. W. and Crow, T.,** Serotonin immunoreactivity in the circumesophageal nervous system of *Hermissenda Crassicornis, Neurosci. Lett.*, 62, 199, 1985.
130. **Jacklet, J., Acosta-Urquidi, J., and Alkon, D.,** Inward currents and calcium dependent potassium currents are increased and other potassium currents decreased by serotonin in pedal neurons of *Hermissenda, Soc. Neurosci. Abstr.*, 10, 145, 1984.
131. **Heldman, E. and Alkon, D.,** Neurotransmitter synthesis in the nervous system of the mollusc *Hermissenda, Exp. Biochem. Physiol.*, 590, 117, 1978.
132. **Heldman, E., Grossman, Y., Jerussi, T. P., and Alkon, D. L.,** Cholinergic features of photoreceptor synapses in *Hermissenda, J. Neurophysiol.*, 42, 153, 1979.
133. **Farley, J. and Auerbach, S. L.,** Protein kinase C activation induces conductance changes in *Hermissenda* photoreceptors like those seen in associative learning, *Nature,* 319, 220, 1986.
134. **Croll, R. P. and Chiasson, B. J.,** Postembryonic development of serotoninlike immunoreactivity in the central nervous system of the snail, *Lymnaea stagnalis, J. Comp. Neurol.*, 280, 122, 1989.
135. **Vehovszky, A., Kemenes, G., and S.-Rozsa, K.,** Monosynaptic connections between serotonin-containing neurones labeled by 5,6-dihydroxytryptamine-induced pigmentation in the snail *Helix pomatia* L, *Brain Res.*, 484, 404, 1989.
136. **Paupardin-Tritsch, D., Hammond, C., Gerschenfeld, H. M., Nairn, A. C., and Greengard, P.,** cGMP-dependent protein kinase enhances Ca^{2+} current and potentiates the serotonin-induced Ca^{2+} current increase in snail neurones, *Nature,* 323, 812, 1986.
137. **Osborne, N. N., Cuello, A. C., and Dockray, G. J.,** Substance P and cholecystokinin-like peptides in *Helix* neurons and cholecystokinin and serotonin in a giant neuron, *Science,* 216, 409, 1982.
138. **Tansey, E. M.,** Aminergic fluorescence in the cephalopod brain, *Phil. Trans. R. Soc. Lond. Ser. B.*, 291, 127, 1980.
139. **Welsch, J. H. and Morehead, M.,** Identification and assay of 5-hydroxy-tryptamine in molluscan tissue by fluorescence method, *Science,* 129, 1491, 1959.
140. **Feldman, S. C.,** Distribution of immunoreactive somatostatin (ISRIF) in the nervous system of the squid, *Loligo paelei, J. Comp. Neurol.*, 245, 238, 1986.
141. **Feldman, S. C.,** Neuromodulators and neurotransmitters in the molluskan nervous system, *Soc. Neurosci.*, (Abstr.) 1983.
142. **Youdim, M. B. H., Feldman, S. C., Pappas, G. D., and Pollard, H. B.,** Serotonin metabolism and the nature of monoamine oxidase in squid central nervous system, *Brain Res.*, 381, 300, 1986.
143. **Uemura, T., Yamashita, T., Haga, C., Miyazaki, N., Kondo, H., and Matsushita, M.,** Localization of serotonin-immunoreactivity in the central nervous system of *Octopus vulgaris* by immunohistochemistry, *Brain Res.*, 406, 73, 1987.
144. **Matus, A. I.,** Histochemical localization of biogenic monoamines in the cephalic ganglia of *Octopus vulgaris, Tissue Cell,* 5, 591, 1973.

145. **Honma, S.,** Presence of monoaminergic neurons in the spinal cord and intestine of the lamprey, *Arch. Histol. Jpn.,* 32, 383, 1970.
146. **Harris-Warrick, R. M., McPhee, J. C., and Philler, J. A.,** Distribution of serotoninergic neurons and processes in the lamprey spinal cord, *Neuroscience,* 14, 1127, 1985.
147. **Van Dongen, P. A. M., Hokfelt, T., Grillner, S., Rehfeld, J. F., and Verhofstad, A. A. J.** A cholecystokinin-like peptide is present in 5-hydroxytryptamine neurons in the spinal cord of the lamprey, *Acta Physiol. Scand.,* 125, 557, 1985a.
148. **Van Dongen, P. A. M., Hokfelt, T., Grillner, S., Verhofstad, A. A. J., Steinbusch, H. W. M., Cuello, A. C., and Terenius, L.,** Immunohistochemical demonstration of some putative neurotransmitters in the lamprey spinal cord and spinal ganglia: 5-hydroxytryptamine-, tachykinin-, and neuropeptide Y-immunoreactive neurons and fibers, *J. Comp. Neurol.,* 234, 501, 1985b.
149. **Van Dongen, P. A. M., Hokfelt, T., Grillner, S., Verhofstad, A. A. J., and Steinbusch, H. W. M.,** Possible target neurons of 5-hydroxytryptamine fibers in the lamprey spinal cord, *J. Comp. Neurol.,* 234, 523, 1985c.
150. **Brodin, L., Buchanan, J. T., Hilfelt, T., Grillner, S., Rehfeld, J. F., Frey, P., Verhofstad, A. A. J., Dockray, G. J., and Walsh, J. H.,** Immunohistochemical studies of cholecystokininlike peptides and their relation to 5-HT, CGRP, and bombesin immunoreactivities in the brainstem and spinal cord of lampreys, *J. Comp. Neurol.,* 271, 1, 1988.
151. **Grillner, S.,** Neurobiological bases of rhythmic motor acts in vertebrates, *Science,* 228, 143, 1985.
152. **Cameron, A. A., Plenderleith, M. B., and Snow, P. J.,** Organization of the spinal cord in four species of elasmobranch fish: cytoarchitecture and distribution of serotonin and selected neuropeptides, *J. Comp. Neurol.,* 297, 201, 1990.
153. **Ritchie, T. C. and Leonard, R. B.,** Immunocytochemical demonstration of serotoninergic cells, terminals and axons in the spinal cord of the stingray, *Dasyatis sabina, Brain Res.,* 240, 334, 1982.
154. **Kah, O. and Chambolle, P.,** Serotonin in the brain of the goldfish, *Carassius auratus.* An immunocytochemical study, *Cell Tissue Res.,* 234, 319, 1983.
155. **Yoshida, M., Nagatsu, I., Kawakami-Kondo, Y., Karasawa, N., Spatz, M., and Nagatsu, T.,** Monoaminergic neurons in the brain of goldfish as observed by immunohistochemical techniques, *Experientia,* 39, 1171, 1983.
156. **Frankenhuis-van de Heuvel, T. H. M. and Nieuwenhuys, R.,** Distribution of serotonin-immunoreactivity in the diencephalon and mesencephalon of the trout, *Salmo gairdneri*: cell bodies, fibers and terminals, *Anat. Embryol.,* 169, 193, 1984.
157. **Ekstrom, P., Nyberg, L., and van Veen, T.,** Ontogenetic development of serotoninergic neurons in the brain of a teleost, the three spined stickleback. An immunohistochemical analysis, *Dev. Brain Res.,* 17, 209, 1985.
158. **Ekstrom, P. and van Veen, T.,** Distribution of 5-hydroxytriptamine (serotonin) in the brain of the teleost *Gasterosteus aculeatus, J. Comp. Neurol.,* 226, 307, 1984.
159. **Margolis-Kazan, H., Halpern-Sebold, L. R., and Schriebman, M. P.,** Immunocytochemical localization of serotonin in the brain and pituitary gland of the platyfish, *Xiphophorus maculeates, Cell Tissue Res.,* 240, 311, 1985.
160. **Grant, K., Clausse, S., Libouban, S., and Szabo, T.,** Serotoninergic neurons in the mormyrid brain and their projection to the preelectromotor and primary electrosensory centers: immunohistochemical study, *J. Comp. Neurol.,* 281, 114, 1989.
161. **Meek, J. and Joosten, H. W. J.,** Distribution of serotonin in the brain of the mormyrid teleost *Gnathonemus petersii, J. Comp. Neurol.,* 281, 206, 1989.
162. **Sambo, T., Ravaille, M., Libouban, S., and Enger, P. S.,** The mormyrid rhombencephalon. 1. Light and EM investigations on the structure and connections of the lateral line lobe nucleus with HRP labeling, *Brain Res.,* 26, 1, 1983.
163. **Elekes, K. and Szabo, T.,** The mormyrid brainstem. III. Ultrastructure and synaptic organization of the medullary "pacemaker" nucleus, *Neuroscience,* 15, 431, 1985.
164. **Grant, K., Bell, C. C., Clausse, S., and Ravaille, M.,** Morphology and physiology of the brainstem nuclei controlling the electric organ discharge in mormyrid fish, *J. Comp. Neurol.,* 245, 514, 1986.
165. **Bennett, M. V. L., Pappas, G. D., Aljure, E. F., and Nakajima, Y.,** Physiology and ultrastructure of electrotonic junctions. II. Spinal and medullary electromotor nuclei in mormyrid fish, *J. Neurophysiol.,* 30, 180, 1967.
166. **Elekes, K., Ravaille, M., Bell, C. C., Libouban, S., and Szabo, T.,** The mormyrid brainstem. II. The medullary relay nucleus: an ultrastructural HRP study, *Neuroscience,* 15, 417, 1985.
167. **Fasolo, A., Franzoni, M. F., Gaudino, G., and Steinbusch, H. W. M.,** The organization of serotonin-immunoreactive neuronal systems in the brain of the crested newt, *Triturus cristatus carnifex* Laur., *Cell Tissue Res.,* 243, 239, 1986.

168. **Morita, K. and Katayama, Y.,** 5-Hydroxytryptamine effects on the somata of bullfrog primary afferent neurons, *Neuroscience,* 21, 1007, 1987.
169. **Frederick, J. M., Rayborne, M. E., and Hollyfield, J. G.,** Serotoninergic neurons in the retina of *Xenopus laevis:* selective staining, identification, development, and content, *J. Comp. Neurol.,* 281, 516, 1989.
170. **Smeets, W. J. A. J. and Steinbusch, H. W. M.,** Distribution of serotonin immunoreactivity in the forebrain and midbrain of the lizard *Gekko gecko, J. Comp. Neurol.,* 271, 419, 1988.
171. **Ueda, S., Yakeuchi, Y., and Sano, Y.,** Immunohistochemical demonstration of serotonin neurons in the central nervous system of a turtle *(Clemmys japonica), Anat. Embryol.,* 168, 1, 1983.
172. **Schutte, M. and Weiler, R.,** Morphometric analysis of serotoninergic bipolar cells in the retina and its implications for retinal image processing, *J. Comp. Neurol.,* 260, 619, 1987.
173. **Weiler, R. and Schutte, M.,** Kainic acid induced release of serotonin from OFF-bipolar cells in the turtle retina, *Brain Res.,* 360, 379, 1985b.
174. **Wallace, J. A.,** An immunocytochemical study of the development of central serotoninergic neurons in the chick embryo, *J. Comp. Neurol.,* 236, 443, 1985.
175. **Sako, H., Kojima, T., and Okado, N.,** Immunohistochemical study on the development of serotoninergic neurons in the chick. II. Distribution of cell bodies and fibers in the spinal cord, *J. Comp. Neurol.,* 253, 79, 1986.
176. **Sako, H., Kojima, T., and Okado, N.,** Immunohistochemical study on the development of serotoninergic neurons in the chick. I. Distribution of cell bodies and fibers in the brain, *J. Comp. Neurol.,* 253, 61, 1986.
177. **Kojima, T., Homma, S., Sako, H., Shimizu, I., Okada, A., and Okado, N.,** Developmental changes in density and distribution of serotoninergic fibers in the chick spinal cord, *J. Comp. Neurol.,* 267, 580, 1988.
178. **Sako, H., Kojima, T., and Okado, N.,** Immunohistochemical study on the development of serotoninergic neurons in the chick. I. Distribution of cell bodies and fibers in the brain, *J. Comp. Neurol.,* 253, 61, 1986.
179. **Ireland, S. J. and Tyers, M. B.,** Pharmacological characterization of 5-hydroxytryptamine-induced depolarization of the rat isolated vagus nerve, *Br. J. Pharmacol.,* 90, 229, 1987.
180. **Wallis, D. I. and Dun, N. J.,** A comparison of fast and slow depolarizations evoked by 5-HT in guinea-pig coeliac ganglion cells *in vitro, Br. J. Pharmacol.,* 93, 110, 1988.
181. **Dun, N. J., Kiraly, M., and Ma, R. C.,** Evidence for a serotonin-mediated slow excitatory potential in the guinea-pig coeliac ganglia, *J. Physiol. (Lond.),* 351, 61, 1984.
182. **Dun, N. J. and Minota, S.,** Effects of substance P on neurones of the inferior mesenteric ganglia of the guinea pig, *J. Physiol. (Lond.),* 321, 259, 1981.
183. **Christian, E. P., Taylor, G. E., and Weinreich, D.,** Serotonin increases excitability of rabbit C-fiber neurons by two distinct mechanisms, *J. Appl. Physiol.,* 67, 584, 1989.
184. **Higashi, H. and Nishi, S.,** 5-Hydroxytryptamine receptors of visceral primary afferent neurones on rabbit nodose ganglia, *J. Physiol. (Lond.),* 323, 543, 1982.
185. **Cutcliffe, N. and Osborne, N. N.,** Serotonergic and cholinergic stimulation of inositol phosphate formation in the rabbit retina. Evidence for the presence of serotonin and muscarinic receptors, *Brain Res.,* 421, 95, 1987.
186. **Mitchell, C. K. and Redburn, D. A.,** Analysis of pre- and postsynaptic factors of the serotonin system in rabbit retina, *J. Cell Biol.,* 100, 64, 1985.
187. **Osborne, N. N.,** In vitro experiments on the metabolism uptake and release of 5-hydroxytryptamine in bovine retina, *Brain Res.,* 184, 283, 1980.
188. **Osborne, N. N.,** *Biology of Serotonergic Transmission,* Pergamon Press, Oxford, England, 1982.
189. **Osborne, N. N. and Beaton, D. W.,** Studies on serotonin-accumulating cells in rabbit retinal cultures, *Brain Res.,* 362, 287, 1986.
190. **Thomas, T. N. and Redburn, D.,** 5-Hydroxytryptamine — a transmitter of bovine retina, *Exp. Eye Res.,* 28, 55, 1979.
191. **Sandell, J. H., Masland, R. H., Raviola, E., and Dacheux, R. F.,** Connections of indoleamine-accumulating cells in the rabbit retina, *J. Comp. Neurol.,* 283, 303, 1989.
192. **Maranto, A. R.,** Neuronal mapping: a photooxidation reaction makes Lucifer Yellow useful for electron microscopy, *Science,* 217, 953, 1982.
193. **Sandell, J. H. and Masland, R. H.,** A system of indoleamine-accumulating neurons in the rabbit retina, *J. Neurosci.,* 6, 3331, 1986.
194. **Sandell, J. H. and Masland, R. H.,** Photoconversion of some fluorescent markers to a diaminobenzidine product, *J. Histochem. Cytochem.,* 36, 555, 1988.
195. **Marson, L.,** Evidence for colocalization of substance P and 5-hydroxytryptamine in spinally projecting neurons from the cat medulla oblongata, *Neurosci. Lett.,* 96, 54, 1989.

196. **Franck, J., Fried, G., and Brodin, E.,** Substance P enhances the release of endogenous serotonin from rat ventral spinal cord, *Eur. J. Pharmacol.*, 174, 85, 1989.
197. **Ulfhake, B., Arvidsson, U., Cullheim, S., Hokfelt, T., Brodin, E., Verhofstad, A., and Visser, T.,** An ultrastructural study of 5-hydroxytryptamine-, thyrotropin-releasing hormone- and substance P-immunoreactive axonal boutons in the motor nucleus of spinal cord segments L7–S1 in the adult cat, *Neuroscience*, 23, 917, 1987.
198. **Pelletier, G., Steinbusch, H. W. M., and Verhofstad, A. A. J.,** Immunoreactive substance P and serotonin present in the same dense-core vesicles, *Nature*, 293, 71, 1981.
199. **Fried, G., Franck, J., Brodin, E., Born, W., Fischer, J. A., Hiort, W., and Hokfelt, T.,** Evidence for differential storage of calcitonin gene-related peptide, substance P and serotonin in synaptosomal vesicles of rat spinal cord, *Brain Res.*, 499, 315, 1989.
200. **Todd, A. J. and Millar, J.,** Antagonism of 5-hydroxytryptamine-evoked excitation in the superficial dorsal horn of the cat spinal cord by methysergide, *Neurosci. Lett.*, 48, 167, 1984.
201. **Belcher, G., Ryall, R. W., and Schaffner, R.,** The differential effects of 5-hydroxytryptamine, noradrenaline and raphe stimulation on nociceptive and non-nociceptive dorsal horn interneurones in the cat, *Brain Res.*, 151, 307, 1978.
202. **Todd, A. J. and Millar, J.,** Receptive field and responses to ionophoretically applied noradrenaline and 5-hydroxytryptamine of units recorded in laminae I–III of cat dorsal horn, *Brain Res.*, 288, 159, 1983.
203. **Todorovic, S. and Anderson, E. G.,** 5-HT_2 and 5-HT_3 receptors mediate two distinct depolarizing responses in rat dorsal root ganglion neurons, *Brain Res.*, 511, 71, 1990.
204. **Araneda, R. C. and Andrade, R.,** 5-HT_2 receptors in rat prefrontal cortex, *Soc. Neurosci. Abstr.*, 14.2, 845, 1988.
205. **Davies, M. F., Deisz, R. A., Prince, D. A., and Peroutka, S. J.,** Two distinct effects of 5-hydroxytryptamine on single cortical neurons, *Brain Res.*, 423, 347, 1987.
206. **Davies, M., Wilkinson, L. S., and Roberts, M. H. T.,** Evidence for excitatory 5-HT_2 receptors on rat brainstem neurons, *Br. J. Pharmacol.*, 94, 483, 1988.
207. **Uchimura, N. and North, R. A.,** 5-Hydroxytryptamine acts at 5-HT_2 receptors to reduce Gk in rat nucleus accumbens neurons, *Soc. Neurosci. Abstr.*, 14.2, 845, 1988.
208. **Grillner, S.,** Locomotion in vertebrates: central mechanisms and reflex interaction, *Physiol. Rev.*, 55, 247, 1975.
209. **Grillner, S. and Zangger, P.,** On the central generation of locomotion in the low spinal cat, *Exp. Brain Res.*, 34, 241, 1979.
210. **Viola, D. and Freton, E.,** Evidence for respiratory and locomotor generators in the rabbit cervicothoracic cord and their interactions, *Exp. Brain Res.*, 49, 247, 1983.
211. **Viola, D. and Vidal, C.,** Evidence for distinct spinal locomotion generators supplying respectively fore and hindlimbs in the rabbit, *Brain Res.*, 155, 182, 1978.
212. **Legendre, P., Guzman, A., Dupouy, B., and Vincent, J. D.,** Excitatory effect of serotonin on pacemaker neurons in spinal cord cell culture, *Neuroscience*, 28, 201, 1989.
213. **Legendre, P., Dupouy, B., and Vincent, J. D.,** Excitatory effect of noradrenaline on pacemaker cells in spinal cord primary cultures, *Neuroscience*, 24, 647, 1988.
214. **Legendre, P., McKenzie, S. S., Dupouy, B., and Vincent, J. D.,** Evidence for bursting pacemaker neurones in cultured spinal cord cells, *Neuroscience*, 16, 753, 1985.
215. **Weil-Fugazza, J., Godefroy, F., and Le Bars, D.,** Increase in 5-HT synthesis in the dorsal part of the spinal cord, induced by a nociceptive stimulus: blockade by morphine, *Brain Res.*, 297, 247, 1984.
216. **Hylden, J. L. K., Hayashi, H., Ruda, M. A., and Dubner, R.,** Serotonin innervation of physiologically identified lamina I projection neurons, *Brain Res.*, 370, 401, 1986.
217. **Andersen, E. and Dafny, N.,** An ascending serotoninergic pain modulation pathway from the dorsal raphe nucleus to the parafascicularis nucleus of the thalamus, *Brain Res.*, 269, 57, 1983.
218. **Albe-Fessard, D. and Kruger, L.,** Duality of unit discharges from cat centrum medianum in response to natural and electrical stimulation, *J. Neurophysiol.*, 25, 3, 1962.
219. **Casey, K. L.,** Unit analysis of nociceptive mechanisms in the thalamus of the awake squirrel monkey, *J. Neurophysiol.*, 29, 727, 1966.
220. **Emmers, R.,** Thalamic mechanisms that process a temporal pulse code for pain, *Brain Res.*, 103, 425, 1976.
221. **McClung, R. E. and Dafny, N.,** The parafasicularnus of the thalamus exhibits convergence input from dorsal raphe and the spinal tract of the trigeminal, *Brain Res.*, 197, 525, 1980.
222. **Peshanski, M., Guilbaud, G., and Gautron, M.,** Postintralaminar nuclei in the rat-neuronal responses to noxious and non-noxious cutaneous stimuli, *Exp. Neurol.*, 72, 226, 1981.
223. **Magoul, R., Onteniente, B., Oblin, A., and Calas, A.,** Inter- and intracellular relationship of substance P-containing neurons with serotonin and GABA in the dorsal raphe nucleus: combination of autoradiographic and immunocytochemical techniques, *J. Histochem. Cytochem.*, 34, 735, 1986.

224. **Chan-Palay, V.,** Indoleamine neurons and their processes in the normal rat brain and in chronic diet induced thiamine deficiency demonstrated by uptake of ^{3}H-serotonin, *J. Comp. Neurol.,* 174, 467, 1977.
225. **Descarries, L., Beaudet, A., Watkins, K., and Garcia, S.,** The serotonin neurons in nucleus raphe dorsalis of adult rat, *Anat. Rec.,* 183, 520, 1979.
226. **Nishikawa, T. and Scatton, B.,** Inhibitory influence of GABA on central serotoninergic transmission. Involvement of the habenulo-raphe pathways in the GABAergic inhibition of ascending cerebral serotoninergic neurons, *Brain Res.,* 331, 81, 1985.
227. **Nishikawa, T. and Scatton, B.,** Inhibitory influence of GABA on central serotoninergic transmission. Raphe nuclei as the neuroanatomical site of the GABAergic inhibition of cerebral serotoninergic neurons, *Brain Res.,* 331, 91, 1985.
228. **Chazal, G. and Ralston, H. J., III,** Serotonin-containing structures in the nucleus raphe dorsalis of the cat: an ultrastructural analysis of dendrites, presynaptic dendrites, and axon terminals, *J. Comp. Neurol.,* 259, 317, 1987.
229. **Nedergaard, S., Bolam, J. P., and Greenfield, S. A.,** Facilitation of a dendritic calcium conductance by 5-hydroxytryptamine in the substantia nigra, *Nature,* 333, 174, 1988.
230. **Azmitia, E. C. and Segal, M.,** An autoradiographic analysis of the differential ascending projections of the dorsal and median raphe nuclei in the rat, *J. Comp. Neurol.,* 179, 641, 1978.
231. **Herve, D., Pickel, V. M., Joh, T. H., and Beaudet, A.,** Serotonin axon terminals in the ventral tegmental area of the rat: fine structure and synaptic input to dopaminergic neurons, *Brain Res.,* 435, 71, 1987.
232. **Saavedra, J., Brownstein, M., and Palkovits, M.,** Serotonin distribution in the limbic system of the rat, *Brain Res.,* 79, 437, 1974.
233. **Parent, A., Descarries, L., and Beaudet, A.,** Organization of ascending serotonin systems in the adult rat brain. A radioautographic study after intraventricular administration of [^{3}H]5-hydroxytryptamine, *Neuroscience,* 6, 115, 1981.
234. **Conrad, L., Leonard, C., and Pfaff, D.,** Connections of the median and dorsal raphe nuclei in the rat: an autoradiographic and degeneration study, *J. Comp. Neurol.,* 156, 179, 1974.
235. **Loughlin, S. E. and Fallon, J. W.,** Substantia nigra and ventral tegmental area projection to cortex: topography and collaterlization, *Neuroscience,* 11, 425, 1984.
236. **Törk, I.,** Raphe nuclei and serotonin containing systems, in *The Rat Nervous System, Vol. 2., Hindbrain and Spinal Cord,* Paxinos, G., Ed., Academic Press, Sydney, 1985, 43.
237. **Halliday, G. M. and Törk, I.,** Serotonin-like immunoreactive cells and fibres in the rat ventromedial mesencephalic tegmentum, *Brain Res. Bull.,* 22, 725, 1989.
238. **Wichmann, T., Limberger, N., and Starke, K.,** Release and modulation of release of serotonin in rabbit superior colliculus, *Neuroscience,* 32, 141, 1989.
239. **Kassel, J.,** Somatotopic organization of SI corticotectal projections in rats, *Brain Res.,* 231, 247, 1982.
240. **Wichmann, T. and Starke, K.,** Uptake, release, and modulation of release of noradrenaline in rabbit superior colliculus, *Neuroscience,* 26, 621, 1988.
241. **Eaton, S. A. and Salt, T. E.,** Modulatory effects of serotonin on excitatory amino acid responses and sensory synaptic transmission in the ventrobasal thalamus, *Neuroscience,* 33, 285, 1989.
242. **Nedergaard, S., Engberg, I., and Flatman, J. A.,** The modulation of excitatory amino acid responses by serotonin in the cat neocortex *in vitro, Cell. Mol. Neurobiol.,* 7, 367, 1987.
243. **Reynolds, J. N., Baskys, A., and Carlen, P. L.,** The effects of serotonin on N-methyl-D-aspartate and synaptical evoked depolarizations in rat neocortical neurons, *Brain Res.,* 456, 286, 1988.
244. **Walker, P. D., Riley, L. A., Hart, R. P., and Jonakait, G. M.,** Serotonin regulation of tachykinin biosynthesis in the rat neostriatum, *Brain Res.,* 546, 33, 1991.
245. **Blandina, P., Goldfarb, J., and Green, J. P.,** Activation of a 5-HT$_3$ receptor release dopamine from rat striatal slice, *Eur. J. Pharmacol.,* 155, 349, 1988.
246. **Jiang, L. H., Ashby, C. R., Jr., Kasser, R. J., and Wang, R. Y.,** The effect of intraventricular administration of the 5-HT$_3$ receptor agonist 2-methylserotonin on the release of dopamine in the nucleus accumbens: an in vivo chronocoulometric study, *Brain Res.,* 513, 156, 1990.
247. **Bliss, T. V. P., Goddard, G. V., and Riives, M.,** Reduction of long-term potentiation in the dentate gyrus of the rat following selective depletion of monoamines, *J. Physiol. (Lond.),* 334, 475, 1983.
248. **Bliss, T. V. P.,** Synaptic plasticity in the hippocampus, *Trends Neurosci.,* 2, 42, 1979.
249. **Bliss, T. V. P. and Gardner-Medwin, A. R.,** Long-lasting potentiation of synaptic transmission in the dentate area of the unanesthetized rabbit following stimulation of the perforant path, *J. Physiol. (Lond.),* 232, 357, 1973.
250. **Bliss, T. V. P. and Lomo, T.,** Long-lasting potentiation of synaptic transmission in the dentate area of the anaesthetized rabbit following stimulation of the perforant path, *J. Physiol. (Lond.),* 232, 231, 1973.
251. **Azmita, E. C.,** The serotonin-producing neurons of the midbrain median and dorsal raphe nuclei, in *Handbook of Psychopharmacology,* Vol. 9, *Chemical Pathways in the Brain,* Iversen, L. L., Iversen, S. D., and Snyder, S. H., Eds., Plenum, New York, 1978, 233.

252. **Jones, B. E. and Moore, R. Y.,** Ascending projections of the locus coeruleus in the rat. II. Autoradiographic study, *Brain Res.,* 127, 23, 1977.
253. **Lindvall, O. and Bjorklund, A.,** The organization of the ascending catecholamine neuron systems in the rat brain, *Acta Physiol. Scand.,* 412 (Suppl.), 1, 1974.
254. **Loy, R., Koziell, D. A., Lindsey, J. D., and Moore, R. Y.,** Noradrenergic innervation of the adult rat hippocampal formation, *J. Comp. Neurol.,* 189, 699, 1980.
255. **Moore, R. Y. and Halaris, A. E.,** Hippocampal innervation by serotonin neurons of the midbrain raphe in the rat, *J. Comp. Neurol.,* 164, 171, 1975.
256. **Segal, M.,** Physiological and pharmacological evidence for a serotoninergic projection to the hippocampus, *Brain Res.,* 94, 115, 1975.
257. **Ungerstedt, U.,** Stereotaxic mapping of the monoamine pathways in the rat brain, *Acta Physiol. Scand.,* 367 (Suppl.), 1, 1971.
258. **Baskys, A., Niesen, C. E., Davies, M. F., and Carlen, P. L.,** Modulatory actions of serotonin on ionic conductances of hippocampal dentate granule cells, *Neuroscience,* 29, 443, 1989.
259. **Otmakhov, N. A. and Bragin, A. G.,** Effects of norepinephrine and serotonin upon spontaneous activity and responses to mossy fiber stimulation of CA_3 neurons in hippocampal slices, *Brain Res.,* 253, 173, 1982.
260. **Colino, A. and Halliwell, J. V.,** Differential modulation of three separate K-conductances in hippocampal CA1 neurons by serotonin, *Nature,* 328, 73, 1987.
261. **Andrade, R. and Nicoll, R. A.,** Pharmacologically distinct actions of serotonin on single pyramidal neurones of the rat hippocampus recorded in vitro, *J. Physiol. (Lond.),* 394, 99, 1987.
262. **Sharp, T., Bramwell, S. R., Clark, D., and Grahame-Smith, D. G.,** In vivo measurement of extracellular 5-hydroxytryptamine in hippocampus of the anaesthetized rat using microdialysis: changes in relation to 5-hydroxytryptaminergic neuronal activity, *J. Neurochem.,* 53, 234, 1989.
263. **Daszuta, A., Kalen, P., Strecker, R. E., Brundin, P., and Bjorklund, A.,** Serotonin neurons grafted to the adult rat hippocampus. II. 5-HT release as studied by intracerebral microdialysis, *Brain Res.,* 498, 323, 1989.
264. **Nilsson, O. G., Brundin, P., and Bjorklund, A.,** Amelioration of spatial memory impairment by intra-hippocampal grafts of mixed septal and raphe tissue in rats with combined cholinergic and serotoninergic denervation of the forebrain, *Brain Res.,* 515, 193, 1990.
265. **Fuxe, K.,** Evidence for the existence of monoamine neurons in the central nervous system. IV. Distribution of monoamine nerve terminals in the central nervous system, *Acta Physiol. Scand.,* 64, 39, 1965.
266. **Descarries, L., Beaudet, A., and Watkins, K. C.,** Serotonin nerve terminals in adult rat neocortex, *Brain Res.,* 100, 563, 1975.
267. **Beaudet, A. and Descarries, L.,** Quantitative data on serotonin nerve terminals in adult rat neocortex, *Brain Res.,* 111, 301, 1976.
268. **Beaudet, A. and Descarries, L.,** The monoamine innervation of rat cerebral cortex: synaptic and non-synaptic axon terminals, *Neuroscience,* 8, 851, 1979.
269. **Seguela, P., Watkins, K. C., and Descarries, L.,** Ultrastructural relationships of serotonin axon terminals in the cerebral cortex of the adult rat, *J. Comp. Neurol.,* 289, 129, 1989.
270. **de Lima, A. D., Blood, F. E., and Morrison, J. H.,** Synaptic organization of serotonin-immunoreactive fibers in primary visual cortex of the macaque monkey, *J. Comp. Neurol.,* 247, 280, 1988.
271. **Morrison, J. H. and Foote, S. L.,** Noradrenergic and serotoninergic innervation of cortical, thalamic, and tectal visual structures in old and new world monkeys, *J. Comp. Neurol.,* 243, 117, 1986.
272. **Berger, B., Trottier, S., Verney, C., Gaspar, P., and Alvarez, C.,** Regional and laminar distribution of the dopamine and serotonin innervation in the macaque cerebral cortex: a radioautographic study, *J. Comp. Neurol.,* 273, 99, 1988.
273. **Mulligan, K. A. and Törk, I.,** Serotoninergic innervation of the cat cerebral cortex, *J. Comp. Neurol.,* 270, 86, 1988.
274. **Agnati, L. F., Fuxe, K., Benfenati, F., Zini, I., and Hokfelt, T.,** On the functional role of coexistence of 5-HT and substance P in bulbospinal 5-HT neurons. Substance P reduces affinity and increases density of ^{3}H-5-HT binding sites, *Acta Physiol. Scand.,* 177, 299, 1983a.
275. **Ogren, S.-O., Fuxe, K., and Agnati, L. F.,** Evidence for a modulatory influence of centrally administered substance P on a 5-hydroxytryptamine-mediated behavioural response in the male rat, *Acta Physiol. Scand.,* 125, 743, 1985.
276. **Ogren, S.-O., Fuxe, K., Berge, O. G., and Agnati, L. F.,** Effects of chronic administration of antidepressant drugs on central serotoninergic receptor mechanism, in *Frontiers in Neuropsychiatric Research,* Usdin, E., Goldstein, M., Firedhoff, A. J., and Georgotas, A., Eds., Macmillan, London, 1983, 93.
277. **Nedergaard, S., Engberg, I., and Flatman, J. A.,** Serotonin facilitates NMDA responses of cat neocortical neurones, *Acta Physiol. Scand.,* 128, 323, 1986.

278. **Reynolds, J. N., Baskys, A., and Carlen, P. L.,** The effects of serotonin on N-methyl-D-aspartate and synaptically evoked depolarizations in rat neocortical neurons, *Brain Res.*, 456, 286, 1988.
279. **Solti, M. and Bartfai, T.,** Tachykinin regulation of serotonin release: enhancement of [^{3}H]serotonin release from rat cerebral cortex by neuromedin K and substance P acting at distinct receptor sites, *Brain Res.*, 401, 377, 1987.
280. **Blakely, R. D., Berson, H. E., Fremeau, R. T., Jr., Caron, M. G., Peek, M. M., Prince, H. K., and Bradley, C. C.,** Cloning and expression of a functional serotonin transporter from rat brain, *Nature*, 354, 66, 1991.
281. **Hollenberg, N. K.,** Serotonin and vascular responses, *Ann. Rev. Pharmacol. Toxicol.*, 28, 41, 1988.
282. **Otlandi, C., Blackshear, J. L., and Hollenberg, N. K.,** Serotonin-induced visualization of the popliteal lymph node by angiography, *Invest. Radiol.*, 21, 201, 1986.
283. **Gowans, J. L.,** The effect of continuous reinfusion of lymph and lymphocytes on the output of lymphocytes from the thoracic duct of unanesthetized rats, *Br. J. Exp. Pathol.*, 38, 67, 1957.
284. **Gowans, J. L.,** The recirculation of lymphocytes from blood to lymph in the rat, *J. Physiol. (Lond.)*, 145, 54, 1959.
285. **Gowans, J. L. and Knight, E. J.,** The route of recirculation of lymphocytes in the rat, *Proc. R. Soc. Lond. B. Biol. Sci.*, 159, 257, 1964.
286. **Gowans, J. L., McGregor, D. D., Cowen, D. M., and Ford, C. E.,** Initiation of immune responses to small lymphocytes, *Nature*, 196, 651, 1962.
287. **Hall, J. G. and Morris, B.,** The output of cells in lymph from the popliteal nodes of sheep, *Q. J. Exp. Physiol.*, 47, 360, 1962.
288. **Hall, J. G. and Morris, B.,** The lymph-born cells of the immune response, *Q. J. Exp. Physiol.*, 48, 235, 1963.
289. **Hall, J. G. and Morris, B.,** The origin of cells in the efferent lymph from a single lymph node, *J. Exp. Med.*, 121, 901, 1965a.
290. **Hall, J. G. and Morris, B.,** The immediate effect of antigens on the cell output of a lymph node, *Br. J. Exp. Pathol.*, 46, 450, 1965b.
291. **Hall, J. G., Morris, B., Moreno, G. D., and Bessis, M. C.,** The ultrastructure and function of cells in lymph following antigenic stimulation, *J. Exp. Med.*, 125, 91, 1967.
292. **Hay, J. B. and Hobbs, B. B.,** The flow of blood to lymph nodes and its relations to lymphocyte traffic and the immune response, *J. Exp. Med.*, 145, 31, 1977.
293. **Schoefl, G. I.,** The migration of lymphocytes across the vascular endothelium in lymphoid tissue, *J. Exp. Med.*, 136, 568, 1972.
294. **Ottaway, C. A. and Parrott, D. M. V.,** Regional blood flow and its relationship to lymphocyte and lymphoblast traffic during a primary immune reaction, *J. Exp. Med.*, 150, 218, 1979.
295. **Majno, G., Gilmore, V., and Leventhal, M.,** On the mechanism of vascular leakage caused by histamine-type mediators, *Circ. Res.*, 21, 833, 1967.
296. **Majno, G., Palade, G. E., and Schoefl, G. I.,** Studies on inflammation. II. The site of action of histamine and serotonin along the vascular tree: a topographical study, *J. Biophys. Biochem. Cytol.*, 11, 607, 1961.
297. **Majno, G., Shea, S. M., and Leventhal, M.,** Endothelial contraction induced by histamine-type mediators. An electron microscopic study, *J. Cell Biol.*, 42, 647, 1969.
298. **Moore, T. C., Hopkins, J., McConnell, I., and Lachmann, P. J.,** Efferent lymph output of lymphocytes following infusions of prostaglandin E_2 and other vasoactive mediators into cannulated afferent popliteal lymphatics in sheep, *IRCS Med. Sci.*, 8, 272, 1980.
299. **Moore, T. C.,** The effects of inflammatory mediators and anaesthetic agents upon lymphocyte traffic of sheep in vivo, Ph.D. thesis, University of Cambridge (Trinity College), England 1982.
300. **Moore, T. C.,** Modification of lymphocyte traffic by vasoactive neurotransmitter substances, *Immunology*, 52, 511, 1984.
301. **Moore, T. C.,** Mediator and physiological influences on lymphocyte traffic in sheep, in *Immunology of the Sheep*, Morris, B. and Miyasaka, M., Eds., Editiones Roche, Basel, 1985, 327.
302. **Maruki, C., Spatz, M., Ueki, Y., Nagatsu, I., and Bembry, J.,** Cerebrovascular endothelial cell culture: metabolism and synthesis of 5-hydroxytryptamine, *J. Neurochem.*, 43, 316, 1984.
303. **Cohen, R. A., Shepherd, J. T., and Vanhoutte, P. M.,** 5-Hydroxytryptamine can mediate endothelium-dependent relaxation of coronary arteries, *Am. J. Physiol.*, 245, H1077, 1983.
304. **Cocks, T. M. and Angus, J. A.,** Endothelium-dependent relaxation of coronary arteries by noradrenaline and serotonin, *Nature*, 305, 627, 1983.
305. **Edvinsson, L., Degueurce, A., Duverger, D., MacKenzie, E. T., and Scatton, B.,** Central serotoninergic nerves project to the pial vessels of the brain, *Nature*, 306, 55, 1983.
306. **Griffith, S. G., Lincoln, J., and Burnstock, G.,** Serotonin as a neurotransmitter in cerebral arteries, *Brain Res.*, 247, 388, 1982.

307. **Henson, P. M.,** Mechanisms of release of constituents from rabbit platelets by antigen-antibody complexes and complement, *J. Immunol.,* 105, 476, 1970.
308. **Essman, W. B.,** Serotonin distribution in tissues and fluids, in *Serotonin in Health and Disease,* Essman, W. B., Ed., Spectrum, New York, 1978, 15.
309. **Olson, P. S., Ljungqvist, U., and Bergentz, S.-E.,** Analysis of platelet, red cell and fibrin content in experimental arterial and venous thrombi, *Thromb. Res.,* 5, 1, 1974.
310. **Schwartz, A. and Askenase, P. W.,** The effect of locally injected vasoactive amines on the elicitation of delayed-type hypersensitivity, *J. Immunol.,* 118, 159, 1977.
311. **Paegelow, I. and Lange, P.,** Pharmacological studies on lymphocytes. 2. Effects of indole derivatives and their antagonists on the lymphocyte migration, *Agents Actions,* 15, 361, 1984.
312. **Aune, T. M., Kelley, K. A., Ranges, G. E., and Bombara, M. P.,** Serotonin-activated signal transduction via serotonin receptors on Jurkat cells, *J. Immunol.,* 145, 1826, 1990.
313. **Choquet, D. and Korn, H.,** Dual effects of serotonin on a voltage-gated conductance in lymphocytes, *Proc. Natl. Acad. Sci. U.S.A.,* 85, 4557, 1988.
314. **Sternberg, E. M., Wedner, H. J., Leung, M. K., and Parker, C. W.,** Effect of serotonin (5-HT) and other monoamines on murine macrophages: modulation of interferon-γ induced phagocytosis, *J. Immunol.,* 138, 4360, 1987.
315. **Hellstrand, K. and Hermodsson, S.,** Role of serotonin in the regulation of human natural killer cell cytotoxicity, *J. Immunol.,* 139, 869, 1987.
316. **Hellstrand, K. and Hermodsson, S.,** Enhancement of human natural killer cell cytotoxicity by serotonin: role of non-T/CD16+ NK cells, accessory monocytes, and 5-HT_{1A} receptors, *Cell. Immunol.,* 127, 199, 1990.
317. **Schanzer, H., Bramis, J., Nirmul, G., Barrows, L., Severin, C., and Taub, T.,** The effect of cyproheptadine (periactin) on rejection of rat kidney allografts, *Transplantation,* 19, 291, 1975.
318. **Rattazzi, R., Simmons, R. L., and Najarian, J. S.,** Use of cyproheptadine in human cadaver transplantation; a controlled randomized prospective study, *Transplant. Proc.,* 9, 985, 1977.
319. **Feldman, J. M.,** Effect of metastatic cancer on platelet monoamine oxidase activity and serotonin metabolism, *Cancer,* 44, 1751, 1979.
320. **Burtin, C., Scheinmann, P., Salomon, J. C., Lespinats, G., and Canu, P.,** Decrease in tumour growth by injections of histamine or serotonin in fibrosarcoma-bearing mice: influence of H_1 and H_2 histamine receptors, *Br. J. Cancer,* 45, 54, 1982.
321. **Nelson, M., Nelson, D. S., and Hopper, K. E.,** Inflammation and tumor growth, *Am. J. Pathol.,* 104, 114, 1981.
322. **Van Loveren, H., Den Otter, W., Meade, R., Terheggan, P. M. A. B., and Askenase, P. W.,** A role for mast cells and the vasoactive amine serotonin in T cell-dependent immunity to tumors, *J. Immunol.,* 134, 1292, 1985.
323. **Berk, B. C. and Alexander, R. W.,** Vasoactive effects of growth factors, *Biochem. Pharmacol.,* 38, 219, 1989.
324. **Seuwen, K. and Pouyssegur, J.,** Serotonin as a growth factor, *Biochem. Pharmacol.,* 39, 985, 1990.
325. **Nemecek, G. M., Coughlin, S. R., Handley, D. A., and Moskowitz, M. A.,** Stimulation of aortic smooth muscle cell mitogenesis by serotonin, *Proc. Natl. Acad. Sci. U.S.A.,* 83, 674, 1986.
326. **Kavanaugh, W. M., Williams, L. T., Ives, H. E., and Coughlin, S. R.,** Serotonin-induced deoxyribonucleic acid synthesis in vascular smooth muscle cells involves a novel, pertussis toxin-sensitive pathway, *Mol. Endocrinol.,* 2, 599, 1988.
327. **Seuwen, K., Magnaldo, I., and Pouyssegur, J.,** Serotonin stimulates DNA synthesis in fibroblasts acting through 5-HT_{1B} receptors coupled to a G-protein, *Nature,* 335, 254, 1988.
328. **Descarries, L., Audet, M. A., Doucet, G., Garcia, S., Oleskevich, S., Seguela, P., Soghomonian, J.-J., and Watkins, K. C.,** Morphology of central serotonin neurons. Brief review of quantified aspects of their distribution and ultrastructural relationships, *Ann. N.Y. Acad. Sci.,* 600, 81, 1990.
329. **Steinbusch, H. V. M.** Serotonin-immunoreactive neurons and their projections in the CNS, in *Handbook of Chemical Neuroanatomy,* Vol. 3, *Classical Transmitters and Transmitter Receptors in the CNS,* Part II, Bjorklund, A., Hokfelt, T., and Kuhar, M. J., Ed., Elsevier, Amsterdam, 1984, 68.
330. **Takeuchi, Y.,** Distribution of serotonin neurons in the mammalian brain, in *Neuronal Serotonin,* Osborne, N. N. and Hamon, M., Eds., John Wiley & Sons, New York, 1988, 25.
331. **Traub, R. D., Miles, R., and Wong, R. K. S.,** Model of the origin of rhythmic population oscillations in the hippocampal slice, *Science,* 243, 1319, 1989.
332. **Wong, R. K. S., Traub, R. D., and Miles, R.,** Cellular basis of neuronal synchrony in epilepsy, in *Advances in Neurology,* Vol. 44, Delgado-Escueta, A. V., Ward, A. A., Jr., Woodbury, D. M., and Porte, R. J., Eds., Raven Press, New York, 1986, 583.

333. **Andrade, R. and Nicoll, R. A.,** Pharmacologically distinct actions of serotonin on single pyramidal neurones of the rat hippocampus recorded *in vitro, J. Physiol. (Lond.),* 394, 99, 1987.
334. **Ropert, N.,** Inhibitory action of serotonin in CA1 hippocampal neurons *in vitro, Neuroscience,* 26, 69, 1988.
335. **Oleskevich, S., Descarries, L., and Lacaille, J.-C.,** Quantified distribution of the noradrenaline innervation in the hippocampus of adult rat, *J. Neurosci.,* 9, 3803, 1989.

Chapter 6

DOPAMINE AND OTHER CATECHOLAMINES

I. INTRODUCTION

Dopamine is the principal catecholamine neurotransmitter of the central nervous system of mammals. Dopamine consitutes more than one half of all the central nervous system catecholamine content, and 80% of the central nervous system dopamine is located in the extremely important basal ganglia area. These ganglia comprise three large subcortical gray masses which are derived from the telencephalon and which contain neurons and intermixed, finely myelinated axons. These ganglia are physically separated from the diencephalon by the internal capsule, and they are intimately related to the cerebral cortex, the thalamus, the subthalamus, and the substantia nigra (site of dopaminergic nervous system nerve bodies). The major anatomical divisions of the basal ganglia are the caudate nucleus, the putamen, the globus pallidus, and the amygdala. In addition to the caudate nucleus, extremely high concentrations of dopamine are found in the nucleus accumbens (site of Parkinson's disease pathology), the olfactory tubercule, the central nucleus of the amygdala, the medium eminence, and restricted fields of the frontal cortex.

Recognition of dopamine as a major neurotransmitter was delayed because of the prolonged perception of dopamine largely or solely as a metabolic precursor of norepinephrine. This perception was brought into serious question by analytical studies that demonstrated marked anatomical differences in the central nervous system distributions of dopamine and norepinephrine.

Dopamine is formed by decarboxylation of the hydroxylated form of the amino acid tyrosine (Figure 1). The rate-limiting step in the biosynthesis of dopamine and the other catecholamines is the hydroxylation of tyrosine by the enzymatic action of tyrosine hydroxylase. The hydroxylated product, dopa, is transformed into dopamine by decarboxylation through the enzymatic action of the L-aromatic amino acid decarboxylase (also known as "dopa decarboxylase"). This is the same enzyme that forms 5-hydroxytryptamine (serotonin) from 5-hydroxytryptophan.

Norepinephrine is formed in the central and peripheral nervous systems and in certain cells of the medulla of the adrenal gland from dopamine by the enzymatic action of dopamine β-hydroxylase (DBH). Other cells of the adrenal medulla also contain the enzyme phenylethanolamine-*N*-methyl transferase, which converts norepinephrine into epinephrine.

Norepinephrine in the central nervous system is found in comparatively large amounts in the hypothalamus and in areas of the limbic system, such as the amygdala central nucleus and the dentate gyrus of the hippocampus. The cell bodies of the noradrenergic nervous system arise largely in the locus ceruleus of the pons, with some cell bodies located in the lateral tegmental area of the reticular formation.

Epinephrine is a circulating hormone that arises from special cells in the medulla of the adrenal gland. Its involvement in the central nervous system is unclear at present. Epinephrine-containing neurons have been identified in the medullary reticular formation, with restricted protections to a few nuclei in the pons and diencephalon.

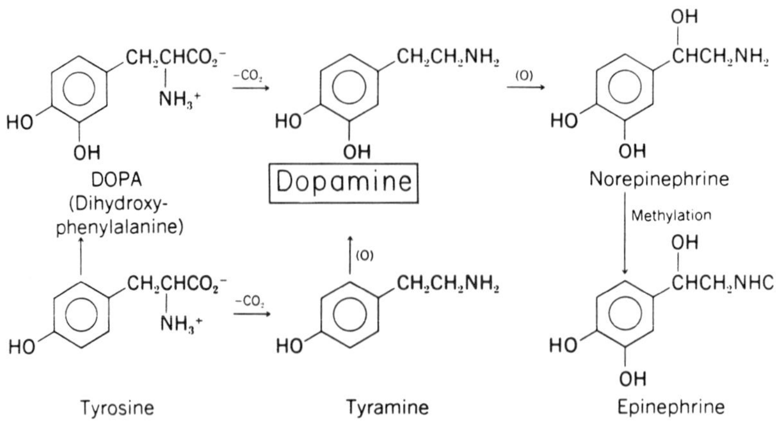

FIGURE 1. Central location of dopamine and metabolic pathways of catecholamine metabolism. (After Holum.[329])

TABLE 1
Dopamine and Norepinephrine Receptor Types and Subtypes

Catecholamine	Receptor types	Agonists	Antagonists
Dopamine	D_1	SKF 38393	SCH 23390
	D_2	Apomorphine	Butyrophenones
		Quinpirole	S-sulpride
			Haloperidol
	D_3	TL 99	UH 232
		Pergolide	AJ 76
Norepinephrine	α_1	Phenylephrine	Prazosin
	α_2	Clonidine	Rauwolscine
			Yohimbine
	β_1	Dobutamine	Metoprolol
			Practolol
	β_2	Terbutaline	Butoxamine

II. RECEPTOR TYPES AND SUBTYPES

A. DOPAMINE

1. D_1

The highest concentration of D_1 dopamine receptors is in the caudate nucleus of the neostriatum, where they are involved in the stimulation of adenyl cyclase activity.[1] In both rodents and humans, the cerebral cortex contains a large number of D_1 receptors, where the dopamine receptor is largely, if not exclusively, of the D_1 type. In the human neocortex, D_1 receptors do not appear to be involved in the stimulation of adenyl cyclase activity.[2] Selective D_1-receptor agonists and antagonists have become available comparatively recently and may help clear up some of the functional activity of this receptor (Table 1). In rodents, concurrent stimulation of D_1 and D_2 receptors in the basal ganglia appears to be required for normal motor activity.

In animals, D_1-receptor activation has been observed to cause self-stimulation,[3] grooming,[4] behavioral arousal,[5] and desynchronization of the electroencephalogram. The significance of these animal findings for humans is completely unknown.

2. D_2

The highest concentration of D_2-dopamine-receptor subtypes also is in the caudate nucleus area of the neostriatum in the dorsal striatum area. Activation of this receptor inhibits adenyl cyclase.[6] Butyrophenomes, the most potent of neuroleptics, bind at the D_2-receptor site. There are two D_2-receptor subtypes, D_{2A} ($D_{2(444)}$) and D_{2B} ($D_{2(415)}$), which are formed by alternate splicing, but display the same pharmacology and drug specificity.[7] These receptors are synthesized by both dopamine and postsynaptic neurons.[8-10] The D_2 receptor cDNA was cloned in 1988.[11] Activation of D_2 receptors has been found to produce a 95% inhibition of the stimulation of cyclic AMP by forskolin, a potent cyclic AMP stimulatory agent. This finding suggests a coupling of D_2 receptor with Gi protein. Each antipsychotic neuroleptic studied has been found to be from 10 to 10,000 times more potent at the D_2 receptor than on the D_1 receptor.[12,13]

3. D_3

Cloning and characterization of the D_3 receptor was reported in the September 13, 1990 issue of *Nature* by Schwartz and associates of Paris.[7] This receptor belongs to the superfamily of G protein-coupled receptors and is encoded by a gene homologous to the D_2-receptor gene.

This receptor differs in both its pharmacology and signaling system from the D_1 and D_2 receptors. It has been observed to be both an autoreceptor and a postsynaptic receptor and is localized to limbic areas of the brain which are known to be associated with cognitive, emotional, and endocrine functions.[7] It also has been found to mediate some of the effects of antipsychotic drugs and of drugs employed in the management of Parkinson's disease that were thought to interact only with D_2 receptors.

D_3-receptor mRNA was found by the Schwartz group[7] to overlap D_2 mRNA in location, being concentrated in the paleostriatum, as opposed to the neostriatum. D_3-receptor mRNA was shown to be expressed at high levels in the olfactory tubercule islands of Calleja complex and in the nucleus accumbens, with highest levels in the islands of Calleja. These areas, with the ventral and venteromedial parts of the caudate putamen, form the ventral striatum. The ventral striatum area receives afferents from the prefrontal or allocortex, and the amygdala and its inputs from the ventral tegmental area. D_3-receptor signals also were found in the hippocampus.

B. NOREPINEPHRINE

There are four subtypes of adrenergic receptors: α_1, α_2, β_1, and β_2. Their agonists and antagonists are presented in Table 1. These receptors are found in both the peripheral and central nervous systems. The β_1 adrenergic receptors are largely associated with neurons, whereas β_2 receptors have been encountered more frequently in glial and vascular cells and areas. Both the α_1 and α_2 receptors have been identified in central nervous system areas that are targets for noradrenergic neurons. The functions of these receptors in the central nervous system remain poorly understood. Adrenergic β_1 receptors are found in the myocardium of the heart, and β_2 receptors are encountered in noncardiac smooth muscle and at other sites. The human gene encoding another β adrenergic receptor, β_3, was isolated recently (1989) by Emorine et al.[14] It appears to be involved with atypical catecholamine responses such as those involving adipose tissue.

Postsynaptic and adrenergic receptors are excitatory and are designated as α_1 while presynaptic receptors are labeled α_2. Clonidine is a potent agonist at α_2 receptors, while phenylephrine selectively activates postsynaptic α_1 receptors. Additional heterogeneity of both α_1 and α_2 receptors has been identified in recent years, although the functional roles of these receptors remains unclear.

III. DOPAMINE

A. NERVOUS SYSTEM

1. General

The dopaminergic nervous system possesses considerable morphological and topographical diversity. There are three major morphological classes of dopaminergic neurons. Ultrashort neurons are found within the amacrine cells of the retina and periglomercular cells of the olfactory bulb. Intermediate-length dopaminergic neurons are found within the hypothalamus. Long-projection dopaminergic neurons transverse between major dopamine-containing neurons in the substantia nigra and ventral tegmentum (VTA) and their targets in the striatum, the limbic areas of the cerebral cortex, and other major areas of the limbic system.

Early anatomical descriptions of the dopaminergic neuronal system divided the cell bodies into two main subgroups: A9, with cell bodies in the substantia nigra compacta and projecting in the nigrostriatal pathway, and the other (A10) in the ventral tegmental area (VTA) with mesolimbic and mesocorticolimbic projections and with projections to the nucleus accumbens and the other five brain areas, such as the amygdala and the septum. A recent classification divides the mesotelencephalic dopamine nervous system into three parts: (1) the mesostriatal, with neuronal site terminations in the caudate nucleus, putamen, globus pallidus, and the nucleus accumbens; (2) the mesoallocortical, with site terminations in the amygdala, olfactory tubercle, and the septal area over the piriform cortex; and (3) the mesoneocortical, with site terminations in all neocortical areas.[15]

Despite rather extensive anatomical dopaminergic system tracking studies, comparatively little is known about the functional role of dopamine in the cerebral cortex. The recent report of Berger et al.[16] describing the regional and laminal distribution of dopamine innervation in the macaque monkey brain identified that dopaminergic innervation of the whole cerebral cortex of the macaque brain was of considerable importance, as it indicated that the mesoneocortic dopamine neuronal pathways were considerably more widespread than previously thought. The studies of Berger et al.[16] showed that dopamine innervation of the cerebral cortex displays major differences between rodents and primates, which were characterized by expanded cortical targets and by a highly differentiated laminar distribution. These findings suggested a greater functional importance and a direct role of dopamine, at the cortical level, in the control of motor and higher order brain functions and that, with phylogenetic development, dopamine appears to play a considerably more important role at the cortical level.

The importance of the mesoneocortical dopaminergic projection can be shown with lesions of the rat ventral tegmental area, which contains the majority of dopamine neuronal cell bodies projecting to the cerebral cortex. Lesions in this area produce a syndrome including defective adjustment to new situations, enhanced distractability to environmental stimuli, disturbed avoidance responses, impaired hoarding behavior, hypoemotivity, hyperactivity, and repetitive behavior.[17,18] The general disruptions of many sorts of behavioral actions produced by a variety of drugs acting on dopaminergic transmissions provide further emphasis that the dopaminergic systems provide some sort of general and fundamental "function" in all of the behaviors concerned and tested.[19,20] The whole dopaminergic network, including all of its systems, appears to modulate function as a single set with functional interdependence between systems, which appears to be necessary for normal integrative processes, harmonious behavior, and general homeostasis.[21] Enormous plasticity and compensatory mechanisms are operating within the dopaminergic system, with considerable response to external stimuli and considerable potential for major involvement in learning and memory.

Dopaminergic neurons do not appear to have specific functions as such. Their roles seem to be homeostatic and regulatory, as they allow forebrain and cortical neuronal systems to functional normally.[21] Lesions of the dopaminergic neurons have been shown to disturb important brain integrative functions such as learning and memory, cognitive functions, and reinforcement processes. Major deficits such as these demonstrate clearly why organisms cannot survive normally without dopamine and why dopaminergic regulation is required for adoptive psychobiological capabilities to develop in a harmonious manner. Behavioral responses necessary for survival are almost impossible without dopamine.[21] These findings and observations suggest a role of considerable importance for the dopaminergic system in its interactions with other vasoactive neurotransmitter substances that appear to have more specific actions.

A 1990 report by Lindefors et al.,[22] from the Karolinska Institute of Stockholm, stands as a major breakthrough in experimental efforts to identify and characterize the extremely important, yet ellusive, role and function of the dopaminergic nervous system in modulating and influencing brain function and related events. The significant study by this group involved the use of *in situ* hybridization technology to study the expression of preprotachykinin (PPT), preprocholecystokinin (CCK), preprosomatostatin (SOM), and preproneuropeptide Y (NPY) mRNA in the caudate putamen and in the frontoparietal cortex of the rat brain, with unilateral 6-hydroxydopamine injection-lesioning of midbrain dopamine neurons in this study with oligodeoxyribonucleotide probes. The results of this study indicated that midbrain dopamine neurons differentially control expression of the neuropeptides investigated in a complex and likely quite-important fashion. There was no change in CCK mRNA expression following dopamine deafferentation in either the frontoparietal cortex (normally high amounts of CCK) or the caudate putamen (normally quite low amounts of CCK). A predeafferentation study of PPT mRNA — the substance P precursor gene — in the caudate-putamen area showed a dense signal density in the caudate putamen and nucleus accumbens, with a few intense dots. It was of interest that from 40 to 60% of all neurons in the caudate putamen were PPT mRNA positive. Dopamine deafferentation of the caudate putamen decreased the numerical density of PPT mRNA neurons in this area by one fifth and decreased the number of grains over PPT mRNA-positive neurons by one half. These findings show that dopamine deafferentation of the caudate putamen clearly decreased PPT mRNA expression and provide strong evidence for a tonic stimulatory effect by striatal dopamine on PPT expression in the caudate putamen and, by inference, a role of importance for substance P in this important area. NPY and SOM mRNA appeared linked and were increased in the caudate putamen and decreased in the frontoparietal cortex by dopamine deafferentation.

2. Invertebrates

a. Platyhelminthes (Flatworms)

In 1983 Lent et al.[23] reported the identification of dopamine within a neuron of the central nervous system of the medicinal leech *Hirudo medicinalis*. In 1991, Friesen and associates,[24] at the University of Virginia, extended their prior studies[25-29] and described important involvements of dopamine in the survival-important swimming and biting actions of this same species of leech. These actions were similar to those observed for serotonin by these and other authors.[30-40]

Single injections of reserpine into the crop of this medicinal leech, in the 1991 study by Friesen et al.,[24] were found to reduce dopamine levels in both the cephalic and midbody ganglia to less than 1% of control values within 3 d, with the levels remaining maximally depressed for approximately 1 month. Acute exposure of isolated ganglia to reserpine for periods of up to 6 h was observed to reduce dopamine levels to 20 to 30% of control levels. Reserpine treatment of intact leeches eliminated biting behavior within 4 d of crop injection. This biting behavior remained depressed below control levels for approximately 4 months. During this period of reserpine-induced depression of biting behavior, biting behavior could be restored to normal by adding dopamine to the pond water bathing these intact leeches. These authors observed that the minimum levels of dopamine required for the expression of normal swimming and biting behaviors are lower for swimming than for biting.

b. Arthropoda

A dopaminergic nervous system has been identified recently (1989) in a crustacean arthropod, the lobster *Homarus gammarus*, by Barthe, Geffard, and associates.[41] These authors employed a highly specific monoclonal antibody directed against conjugated dopamine (dopamine-gluteraldehyde-bovine serum albumin).[42-45] Dopaminergic structures were identified in the 5th abdominal ganglion where the important swimmeret beating is elicited. The swimmeret is a crustacean abdominal limb or appendage that is adapted for swimming (a ''swimming foot''). The swimmeret system of crustaceans is one of the best studied and defined with respect to the important central pattern generator (CPG) and rhythmic motor behavior of these animals.[46-49]

In their immunohistochemical study, Barthe et al.[41] reported the identification of dopaminergic cell bodies near the midline in the lateroventral area of the 5th abdominal ganglion of *Homarus gammarus*. A large number of immunoreactive processes also were encountered. Large processes were stained in the anterior part of the 5th abdominal ganglion, which could be followed into the anterior connectives. This observation suggested dopaminergic connections between ganglia. In addition, numerous dopaminergic processes were identified that remained within the 5th abdominal ganglion. The majority of these appeared to be very thin processes and varicosities which were located in the dorsal part of the neuropile, an area in which most synaptic controls are believed to occur in the swimmeret system of the lobster. A few larger dopaminergic processes without varicosities were located in the lateral neuropile, posterior to the previously described thin processes. It was suggested by the authors that these larger processes might correspond to the integrative segment or to the axons of dopaminergic cells.

In addition to the important immunohistochemical studies, the authors also investigated the action of dopamine on the 5th abdominal ganglion of *Homarus* by two different approaches, one *in vivo* and one *in vitro*. For the *in vivo* study, a behavioral preparation was employed involving ventral insertion of a chronic syringe needle near the 5th abdominal ganglion. This needle was used for the injection of dopamine into the hemolymph of a freely moving animal. Three muscles were chosen for electromyographical recordings to analyze behavioral modifications: a claw levator muscle, a depressor muscle of a 4th leg, and a remotor muscle of a 5th swimmeret. Injection of dopamine produced a rapid and vigorous response. This response consisted of an elevation of the whole animal, as shown by leg-depressor activity, a projection of the claws, and a strong abdominal extension that was correlated with rhythmic beating of the swimmerets at high frequency. These findings *in vivo* were highly reproducible and were similar to a previously described agonistic behavior.

The *in vitro* studies were carried out by superfusion of segments of the abdominal nervous cord (4th, 5th, and 6th ganglia), with associated nerve roots pinned down, dorsal side up, in Petri dishes. Both extracellular and intracellular recordings of motoneuronal activity were obtained. Dopamine superfusion of this *in vitro* preparation always was found to induce an increase in motoneuron activity. In addition, in 40% of the experiments (10 of 25), dopamine superfusion also elicited a 1-Hz rhythmic bursting activity in the swimmeret nerves.

In *in vitro* studies on operated animals in which the abdominal neuronal chain had been severed between the last thoracic ganglion and the 1st abdominal ganglion, dopamine injection near the 5th ganglion was found to restore lost vigorous swimmeret beating activity.

In the above studies, dopamine appeared to exert a more powerful swimmeret beating activation in *in vivo* than in *in vitro* preparations. The authors concluded that dopamine is an especially important neuromodulator in the triggering and maintenance of essential "life-support" swimmeret activity in *Homarus gammarus*.

Dopamine also has been found to be present and of importance in the Insecta assemblage of arthropods. It has been reported to elicit the flight pattern of the moth *Manuda sexta*[50] and to be active in olfactory pathway and antennal lobes of the honeybee *Apis mellifera*.[51,52] In a correlative light and electronmicroscopy study by Nässel et al.,[53] dopamine-immunoreactive neurons in the visual system of the blowfly, *Calliphora erythrocephala*, have been mapped in detail. Three types of dopamine-immunoreactive neurons were identified in the optic lobes. One type consisted of a population of several thousand columnar small-field amacrine neurons in the second neuropile region, the medulla. The term "amacrine" designates a type of small nerve cell in the inner nuclear layer of the retina, which has neurites with characteristics of both axons and dendrites (Oxford English Dictionary, 2nd ed., 1989). The other two types of dopamine-immunoreactive neurons were large-field-projection neurons that innervated the next and more central synaptic region that comprised the lobula and lobula plate as well as centers in the blowfly midbrain. The cell bodies of these large-field-projection neurons of the dopaminergic system were located lateroventrally in the brain. The findings of this study suggest that dopamine may function as a neurotransmitter of importance in functionally different classes of neurons in the blowfly visual system: both the amacrine neurons and the projection neurons.

c. *Mollusca*

Dopamine has been reported to occur in comparatively high concentrations in the central nervous systems of a number of molluscan gastropods.[54-57] A neurotransmitter role for dopamine has been suggested and explored in several species of this phylum.[58-60] Dopamine has been observed to play an important part in the activation of the feeding motor pattern in *Helisoma trivolvis*,[61] *Limax maximus*,[62] and *Lymnaea stagnalis*.[63] In *Lymnaea* and *Planorbis*, two snail species, a giant dopamine-containing neuron in the pedal ganglion has been observed to form monosynaptic connections with a sizeable number of visceral and parietal neurons.[64,65] It is of particular interest that both presynaptic and postsynaptic dopamine receptors have been identified in *Lymnaea stagnalis*.[66,67] An apparent involvement of dopamine in the electrical activity of neuroendocrine growth-hormone-producing cells also has been reported.[68] A recent (1990) report by Werkman et al.[69] suggests a possible involvement of dopamine in hormonal communications within the central nervous system of *Lymnaea stagnalis*. Studies of this group of investigators and others have indicated the difficulties and pitfalls of trying to identify vertebrate-type dopamine receptors in invertebrates and molluscs.[70-72]

Gill muscle contractions in *Aplysia* have been reported to be initiated and modulated by dopamine.[73] Dopamine also has been observed to produce spike narrowing and presynaptic inhibition of siphone sensory neurons in *Aplysia*.[74,75]

A 1991 report by Elekes, Geffard, and associates, of France, Hungary, and England, presented detailed mapping of dopamine-containing neurons in the central nervous system of the pond snail *Lymnaea stagnalis*.[76] This study included the mapping of inter- and intraganglionic axon tracts. Earlier and preliminary studies had suggested the presence of dopamine immunoreactivity in the buccal and in the pedal and cerebral ganglia of *Lymnaea*. In the 1991 report, the number of dopamine-immunoreactive neurons was found to be similar to the number of serotonin-immunoreactive neurons described in a prior study.[77] Nonetheless, the locations and ranges of sizes of dopamine- and serotonin-immunoreactive neuronal cells were quite different. Dopamine-immunoreactive neuronal cells ranged in size from small to medium, whereas serotonin-immunoreactive cells in this same animal species were numerous and of a large and giant type in size. These studies also appeared to rule out colocalization of dopamine and serotonin in most neuronal cell clusters of *Lymnaea*. These and the above findings clearly suggest a role of importance for dopamine, as well as serotonin, in the defense- and survival-related central nervous system functions of gastropod molluscs.

d. Echinodermata

A report by Capasso et al.,[78] from Naples, Italy, has described an adenylcyclase-stimulating effect of catecholamines, especially dopamine, on the eggs of sea urchins. The responses to dopamine were found to be mediated by guanine nucleotide regulatory proteins, G proteins.

3. Vertebrates

Significant and dramatic advances in the sophistication and circuitry of the neuronal nervous system clearly played a major role in the emergence of our phylogenetic vertebrate ancestors from the mud and muck of primordial waters and in their climb onto the safety of dry land and sunny skies, as well as their productive survival and propagation in this attractive new environment.

Fossil evidence has indicated that, of all living reptiles, turtles and crocodilians are the most closely related to the stem reptiles from which birds and mammals evolved.[79] Accordingly, it is to these and other reptiles that we must look for those critical evolutionary changes in the extremely important neuronal nervous system.

In all studies of catecholaminergic nervous systems in bony fish, dopamine-immunoreactive neurons have not been encountered in the midbrain tegmentum.[80] In bony fish there is a striking absence of differentiated mesencephalic neuronal nuclei, such as ''substantia nigra'' and ''ventral tegmental area''.

A 1988 study (by Smeets[81]) of dopamine immunoreactivity in the forebrain and midbrain of the snake *Python regius* has established that this representative of a previously unstudied reptilian radiation shares many key features of the dopaminergic nervous systems of lizards and turtles that have been investigated with the same antibodies.[82-84] From this and prior studies by Smeets and associates, it was clear that a basic reptilian dopaminergic nervous system exists. All reptiles studied had in common the finding that, in the olfactory bulbs, the majority of dopamine-immunoreactive cell bodies surround the glomeruli, with only a small number of dopamine-immunoreactive cells being found in the the external plexiform layer. In addition, dopamine-immunoreactive cell bodies were not encountered in the telencephalon proper in any of the reptile studies. Furthermore, only a limited number of cerebrospinal-fluid-containing dopaminergic neurons were found in the hypothalamus — a significant change from the bony fish. A particularly important reptilian finding is that the mesencephalic tegmentum contains a large number of dopamine-immunoreactive neurons that can, more or less, be subdivided into a medial-ventral tegmental area (VTA) and a lateral substantia nigra (SN). The important 1988 study by Smeets[81] also provided evidence that the structure previously labeled rostral septal nucleus by Halpern[85] is actually the nucleus accumbens.

In 1984 Parent and associates[86,87] drew attention to the distribution of dopaminergic neurons as following a trend through phylogeny from fish to mammals. This trend involved a reduction in the number of catecholaminergic and dopaminergic neurons in the hypothalamus — especially cerebrospinal-fluid-contacting neurons. This change was accompanied by an increase in the number of dopaminergic neurons in the midbrain, with a corresponding increase in the complexity and number of fibers innervating the basal ganglia. These investigators suggested that the midbrain dopaminergic cell groups arose or appeared coincident ''with the appearance of the terrestrial vertebrates — the reptiles.''

The location and distribution of midbrain dopaminergic neuronal populations in reptiles are quite similar to those of mammals and appear to be a basic feature of brain organization in terrestrial vertebrates. The progressive disappearance of catecholaminergic and dopaminergic cerebrospinal-fluid-contacting neurons also appears to be a significant feature of terrestrial life. Dopaminergic cerebrospinal-fluid-contacting neurons are abundant in fish and amphibians, less numerous in reptiles and birds, and virtually absent in mammals.[88-93]

4. Other Molecules

a. Substance P

The heat, pain, memory, and immunostimulating molecule, substance P,[94,95] has many and important connections with dopamine and the dopaminergic nervous system.

High concentrations of substance P have been identified in pathways from the striatum (caudate putamen) to the substantia nigra.[96] In addition, binding studies for substance P have been found in both the striatum and substantia nigra.[97] Multiple reciprocal relationships between substance P-containing and dopaminergic neurons in the nigrostrial system have been observed. Dopaminergic neurons of the substantia nigra have been observed to be excited by iontophoretic administration of substance P.[98]

Dopaminergic neuronal axon terminals in the neostriatum have been found to form synapses with substance P-containing neurons.[99] Members of this same group of investigators,[100] in Osaka and Kagawa, Japan, recently (1990) presented direct evidence at the synaptic level that dopamine-containing neurons directly receive synaptic input from substance P neuronal afferents in the VTA. This evidence was obtained by use of an immunoelectron microscopic mirror method. The importance of this information is evidenced by the physiological and pharmacological studies that have demonstrated a close relationship between dopamine and substance P in the VTA[101-105] and the known projection of dopaminergic neurons from the VTA to such important memory, learning, and survival areas as the hippocampus, limbic areas, and the cerebral cortex.[106] On light microscopy study, substance P-immunoreactive nerve terminals have been observed frequently to surround VTA neurons.[107]

Dopamine-containing neuronal cell bodies in the VTA also richly innervate the nucleus accumbens, a neuronal population central to the organization of the mesolimbic dopaminergic nervous system.[108,109] Activation of this pathway has been shown to produce an increase in locomotor activity that is thought to be dependent on dopamine releases in the nucleus accumbens.[110] The importance of dopamine and the dopaminergic nervous system in this behavioral response has been established both by 6-hydroxydopamine lesions either of the dopaminergic neuronal cell body (A10) or the terminal region of this mesolimbic dopaminergic pathway, and by intranucleus accumbens infusions of dopamine antagonists that have been shown to attenuate amphetamine-induced locomotor activity.[110-113] Both substance P immunoreactivity and receptors have been identified in the nucleus accumbens.[114-116] It is of further interest that *in vivo* injections of substance P into the nucleus accumbens increases dopamine metabolites in this area.[117] *In vitro* investigations have established that substance P can enhance the release of dopamine from brain slices.[118,119] In 1986 Elliot et al.,[120] from Duke University, reported *in vivo* evidence for a tonic facilitory influence of substance P on dopamine release in the nucleus accumbens.

High concentration of substance P-precursor-gene mRNA in the striatum (caudate putamen) was cited earlier in this chapter.[22] Abundant substance P-precursor-gene mRNA (PPT) labeling in the rat caudate putamen also has been reported by others.[121-122] The investigators also have shown unilateral dopamine deafferentation to inhibit PPT substance P-precursor-gene mRNA in the caudate-putamen area, as revealed both by *in situ* hybridization and mRNA blots. The 1990 study by Lindefors et al.[22] demonstrated that dopamine deafferentation in the caudate putamen induces a reduction in substance P precursor (PPT) gene mRNA expression in the number of hybridization-positive neurons. Earlier studies had suggested an involvement of both dopamine D_1 and D_2 receptors in dopamine regulation of substance P precursor synthesis in rat caudate putamen.[123-125]

Substance P is encountered in the substantia nigra of the rat in fibers that originate in the ipsilateral caudate-putamen area.[126-128] Substance P fibers in the substantia nigra are concentrated in the pars reticulata area where they surround the dendrites of substantia nigra dopaminergic neurons, as cited earlier.[107] In this area synaptic contacts between substance P-positive neuronal endings and dopaminergic neuronal cell bodies and dendrites have been identified by electron microscopy.[129,130]

In 1978 Cheramy et al.[131] made the important observation that the immunoneutralization of substance P in the cat substantia nigra by intranigral administration of substance P antiserum caused a decrease in dopamine release from the dendrites and terminals of dopaminergic neurons in the ipsilateral striatum (caudate putamen). This significant finding suggests that substance P exerts a tonic excitatory modulation of the nigrostriatal dopaminergic nervous system projection. Subsequent studies have confirmed an excitatory input of substance P on dopaminergic neurons.[132-137] The use of *in vivo* microdialysis has been of considerable use in further defining the effects of intranigral injections of substance P on the extracellular levels of dopamine in the ipsilateral striatum (caudate putamen). This important technique has also provided an opportunity to investigate the effects of substance P antagonists on these responses.[133,138,139] Initial substance P antagonists were found to have neurotoxic effects.[140,141]

The 1990 development of a nontoxic and substance P-specific substance P antagonist (Spanide II), by Folkers and associates[142] at the Karolinska Institute in Stockholm, has been an important advance in the study, control, and manipulation of potent substance P activities *in vivo* and *in vitro*. Also, in a 1990 report, Reid, Hokfelt, Folkers, and associates,[143] of Stockholm, described the inhibition of intranigral substance P-stimulated release of dopamine in the striatum (caudate putamen) by Spanide II. In additional recent reports, Spanide II has been observed to produce antinociceptive effects on the hot-plate test and to block facilitation of the spinal flexor reflex stimulated by intrathecal substance P administration or by conditioning stimulation of C fibers.[144-145] These results were achieved without complicating "neurotoxic" effects.[142,146]

These dopaminergic neurons that project from the substantia nigra to the striatum (caudate putamen) are critical and of immense importance to the normal and survival-related function of the basal ganglia. The outputs of the cortico-striato-palladial and the cortico-striato-nigral pathways are influenced by the dopaminergic neurons that make direct synaptic contact with the projection neurons of the striatum (caudate putamen).[137] The release of dopamine from dopaminergic neuronal terminals in the striatum (and influenced by substance P, as described above) constitutes a major modulatory and/or excitatory feedback and feed-forward mechanism for information flow through the basal ganglia. The functional importance of this system is dramatically presented in the marked motor and occasional cerebrocortical dementia symptoms that occur in Parkinson's disease and in models of Parkinson's disease which are produced by the selective destruction of striatum nigral dopaminergic neurons.[147] In 1983 Mauborgne et al.[148] reported decreased levels of substance P in striatonigral neurons and, including the pallidium, in the brains of Parkinson's disease patients observed by Tenovuo et al.[149] Dysfunction in the striatonigral substance P pathway recently (1986) has been suggested by Barker[150] as a primary site in the pathology of idiopathic Parkinsonism. It also has been proposed by Barker[151] that substance P may be the active molecule in the adrenal tissue that has been employed for grafting in Parkinsonian patients.

The release of dopamine within the striatum (caudate putamen), substance P-influenced (see above), is from terminals that are in synaptic contact with strial output and input neurons and are influenced by the activity of nigral axons as well as by the activity of afferent synaptic inputs to the dendrites and perikarya of the parent neurons in the substantia nigra.

Accordingly, it is of considerable importance to know the chemistry, origin, and pattern of synaptic input to dopaminergic neurons to achieve a clearer appreciation and understanding of the role of the dopaminergic neuronal system in the functional organization of the basal ganglia and of other areas of dopaminergic neuronal projection.

The substantia nigra is known to receive many afferent inputs of diverse chemical nature and origin. These include excitatory inputs from the cerebral cortex,[152] from striatum (caudate putamen) by neurons containing GABA and peptides (including substance P),[153] from GABA-containing neurons of the globus pallidus,[154-156] and from the dorsal raphe nucleus as a serotoninergic input.[157,158] All of these molecules have been identified as playing important roles in sensory stimulation and excitation and in the establishment of memory and "learning" in the central nervous system, the immune response system, or both.

In 1990 Bolam and Smith,[159] of Oxford University, reported the results of their studies that demonstrated that dopaminergic neurons in the substantia nigra receive symmetrical synaptic input from GABA-ergic and substance P-containing terminals and that showed that a portion of these terminals contain both GABA and substance P. These findings suggested to the authors that the major synaptic input to dopaminergic neurons is from GABA-ergic (as well as substance P-ergic) terminals and that a part of this innervation is derived from the striatum.

In a combined immunohistochemistry and autoradiographic-axon-tracing study, Beckstead,[160] of Charlottesville, VA reported in 1987 that striatal substance P neuronal cell clusters coincide with high-density terminal zones of the discontinuous nigrostriatal dopaminergic projection system in the cat. These findings suggested to the author that striatal substance P neuronal cells that send their axons selectively to the entopeduncular nucleus and the substantia nigra are a major target of nigrostriatal dopamine transmission.

The data presented here document in some depth the extent of substance P and dopamine interrelatedness in adult and mature mammalian nervous systems. Recent (1989 and 1990) reports also extend this interrelatedness to developing mammalian nervous systems.[161-163] The patchy infrastructure of the striatum (caudate putamen) is established prior to birth and serves postnatally as a substrate for the progression of separate "waves" of elevated neuronal peptide content, including a substance P wave.[162] These "waves" in rats have been observed to proceed along rostromedial wave trajectories to involve interdigitating neuronal domains. At birth, patches of dense dopaminergic immunoreactivity were present across the breadth of the rostral striatum. At birth, patches displaying substance P immunoreactivity were present only in the lateral half of the striatum. Substance P did not appear in the medial half of the striatum until the postnatal day 3.

Sivam[161] of Indiana University–Gary, in 1989, reported findings of a study that examined the influences of dopamine-receptor stimulation on substance P (and met-enkephalin) systems of the basal ganglia of rats that were lesioned as neonates with 6-hydroxydopamine. It had been proposed that 6-hydroxydopamine-lesioned rats might serve as an experimental model for the dopamine deficiency and self-injurious behavior responses encountered in the childhood neurological disease Lesch-Nyhan syndrome.[164,165] The self-injurious behavior responses seen clinically in Lesch-Nyhan syndrome children is associated with choreoathetoid movements, mental retardation, and reduced brain dopamine levels.[166-168] It was suggested in 1988, by Criswell et al.,[169] that the self-injurious behavior reponses encountered in the Lesch-Nyhan syndrome might result from supersensitive dopamine D_1 receptors and an associated reduced adenosine activity that might, in concert, facilitate a hyperactive dopaminergic function.

Prior studies by Breese and associates[164,165,170-174] had shown that the destruction of nigrostriatal dopamine-containing neurons with 6-hydroxydopamine during the early postnatal period in rats produces a severe dopamine loss and an increased susceptibility for self-injurious behavior when the animals are challenged with a mixed dopamine D_1/D_2 agonist, such as L-dopa, or a selective D_1 agonist, such as SKF-38393. The self-injurious behavior in these animals was evidenced by self biting associated with lacerations of the skin. Dopamine agonists did not elicit this response in adult 6-hydroxydopamine-lesioned rats in related studies by the Breese group. A number of recent studies have shown substance P and the substance P neuronal system to be influenced by nigrostriatal dopaminergic neurons.[121,175-180]

In his 1989 report, Sivam[161] observed neonatal (3 d of age) dopaminergic denervation of rats by 6-hydroxydopamine to produce a dopaminergic-D_1-receptor-mediated depletion of substance P in the striatonigral neurons when the animals were tested as adults. Administration of the dopamine precursor, L-dopa, to lesioned animals induced self-injurious behavior, increased dopamine levels, and produced a greater decrease in substance P levels in the striatum and substantia nigra (64% reduction) than was found in animals with lesioning only (28% reduction). The dopa-induced decreases in substance-P levels and the stimulated self-injurious behavior of lesioned animals were prevented by pretreatment with the dopamine-D_1-receptor antagonist SCH-28890. Administration of the dopamine-D_1-receptor agonist SKF-38393, but not the dopamine-D_2-receptor agonist LY-171555, to lesioned animals mimicked the L-dopa responses in all respects, with the exception that the agonists did not alter dopamine levels. The findings of this study demonstrated for the first time that the self-injurious behavior precipitated by dopamine agonists in neonatal denervated animals is accompanied by a marked and selective decrease in substance P in striatonigral substance P neurons. This process appeared to the author to have two components: first, a retarded development of the substance P neuronal system due to neonatal dopamine denervation, and then a depletion of much of the remaining substance P associated with enhanced substance P release triggered by dopamine-D_1-receptor-mediated activated of striatonigral substance P neurons.

In a 1990 report of a collaborative study, Sivam with Krause, Breese, and Hong[163] described their findings in a series of experiments that explored the effects of neonatal (3 d of age) lesioning of dopaminergic neurons with 6-hydroxydopamine on dopamine, met-enkephalin, substance P, and serotonin metabolism and an expression of preproenkephalin and preprotachykinin mRNA in the striatum at various days postlesioning up to 35 d of age. The dopaminergic lesion led to a marked depletion of dopamine and its metabolites in the striatum (caudate putamen) and stimulated an increase in 5-HT levels and those of its metabolite, 5-HIAA, in this area. The lesioning did not effect met-enkephalin or preproenkephalin mRNA levels in the striatum up to 25 d of age, but did increase them at 35 d of age. In contrast to the enkephalin response, dopaminergic lesioning resulted in a decreased developmental expression of substance P and its genetic precursor PPT mRNA throughout the 35-d observation period. The results of this important study indicate clearly and impressively that the normal development of the met-enkephalin, substance P, and 5-HT neuronal systems of the striatum (caudate putamen) is dependent on the availability of dopamine and the integrity of the dopaminergic nervous system.

b. Serotonin (5-HT)

Like substance P, serotonin also possesses good credentials both as a memory and as an immunostimulative molecule. An excitatory effect of serotonin by way of its input into the substantia nigra from the dorsal raphe nucleus was cited above.[157,158] An increase in 5-HT and its 5-HIAA metabolite in the striatum of newborn rats following chemical dopamine deafferentation also is cited above.

A most interesting dopamine-substance P-serotonin "connection" has been reported recently (1990) by Hagan et al.[181] These investigators from Glaxo, England, described blocking, by 5-HT_3-receptor antagonists, of a behavioral hyperactivity response resulting from rat brain VTA injection of a long-lasting synthetic analogue of substance P (DiMe-C7) and associated with raised mesolimbic dopaminergic activity. This study brings together two important lines of investigation. Hyperactivity behavioral responses of elevated mesolimbic dopamine activity in the rat and marmoset have been reported to be blocked by serotonin 5-HT_3-receptor antagonists (by Hagan et al.[182] and Costall et al.[183,184]). Blandina et al.,[185] in 1988, observed that the 5-HT_3-receptor antagonist ICS-205-930 could reduce the 5-HT-induced decrease in a spontaneous overflow of endogenous dopamine from superfused slices of rat brain striatum. This finding provided experimental support for the concept that 5-HT_3-receptor antagonists can modulate dopamine release. These findings and this concept are given added importance when one considers the strategic localizations of 5-HT_3 receptors in the mammalian central nervous system: predominantly in the limbic and cortical area of the forebrain and including the entorhinal cortex, the frontal cortex, the amygdala, and the nucleus accumbens/olfactory tubercules. It is, accordingly, of particular importance that serotonin 5-HT_3-receptor blockade also has been found to reduce both the hyperactivity response and the increase in dopamine metabolism produced by DiMe-C7 treatment in other critical areas of the brain, including the nucleus accumbens and the amygdala.[181,186,187]

The other related line of investigation contributing to the 1990 report of Hagan et al. derives from the important studies of Iversen and associates[188-191] in the early and mid-1980s. These studies described the behavioral effects produced by the infusion of the stable analogue of substance P (DiMe-C7) into VTA and substantia nigra of rat brains, and associated these behavioral responses with selective activation of mesolimbic and mesocortical dopamine metabolism. This model of dopamine-mediated behavioral hyperactivity in the rat has been widely used in experimental investigations of functional components of the dopaminergic nervous system in animals. The hyperactive behavioral effects are readily monitored and include increased locomotion, rearing, and wet-dog shakes. At times, these responses were preceded by marked lacrimation.

It is of further interest that the Blandina group[192] at the Mount Sinai Medical School in New York, in 1988, provided direct and unambiguous evidence that 5-HT_3 receptors in the rat striatum modulate the release of endogenous dopamine.

In 1990 Jiang et al.,[193] of Stony Brook, NY, described the effects of the serotonin 5-HT_3-receptor agonist, 2-Me-5HT, on dopamine release in the nucleus accumbens of rats, employing an *in vivo* chronocoulometric recording technique. Intraventricular administration of the receptor agonist was found to increase the release of endogenous dopamine from the nucleus accumbens, in a dose-dependent manner. This response was blocked by the selective 5-HT_3 antagonist, granisetron. The increase in dopamine release produced by the 5-HT_3 agonist was blocked completely in animals that had received acute bilateral injections of 6-hydroxydopamine into the medial forebrain bundle. The results suggested to the authors that the 2-Me-5HT_3-receptor-agonist-induced dopamine release in the nucleus accumbens was mediated by 5-HT_3 receptors and that this effect is dependent upon the impulse flow of dopamine cells. As the dopamine hypothesis of schizophrenia considers that this severe psychotic disorder may be associated with a hyperactive tone in the dopaminergic nervous system, the authors suggest that alterations in 5-HT_3 receptors and their function might contribute to the pathophysiology of schizophrenia. In this regard, they cite interesting evidence from Sorensen et al.[194] that 5-HT_3 antagonists exhibit antipsychotic properties.

In 1988 Nedergaard et al.[195] reported that iontophoretic administration of 5-HT *in vitro* onto distal, but not proximal, dendrites of dopaminergic neurons in the substantia nigra resulted in a facilitation of dendritic calcium conductance. In 1990 Kelland et al.,[196] of

Detroit, reported studies identifying serotoninergic afferent regulation of the basic physiology and pharmacological responsiveness of nigrostriatal dopamine neurons. They interpreted their findings as suggesting that 5-HT systems exert subtle influences on the activity and responsiveness of nigrostriatal dopamine neurons.

c. Excitatory Amino Acids

Kalivas et al.,[197] in 1989, reported their results in a study that involved the microinjection of excitatory amino acids into the ventral tegmental area (VTA/A10) of the brains of rats. Glutamate produced an increase in motor behavior that was antagonized and blocked by pretreatment with a dopamine-D_2-receptor antagonist, haloperidol. In an investigation of the role glutamate/aspartate excitatory amino-acid-receptor subtypes play in this response, these investigators observed that the motor stimulant effect also was produced by kainic acid, but not by *N*-methyl-D-aspartate (NMDA) or by quisqualic acid. Through the use of *in vivo* dialysis, the authors found that glutamate injection into the A10/VTA region enhanced dopamine release in the nucleus accumbens as well as increasing locomotor activity. They also observed that glutamate injection into the A10/VTA region increased postmortem-determined concentrations of dopamine metabolites in the medial prefrontal cortex, the nucleus accumbens, and the A10/VTA region. NMDA injection selectively increased dopamine metabolism in the prefrontal cortex, while the kainic-acid-produced increases were localized to the nucleus accumbens and the A10/VTA region. The authors interpreted these findings as indicating that the excitatory amino acid glutamate is a regulatory neurotransmitter of A10/VTA dopaminergic neurons, with the NMDA-receptor-subtype-modulating dopaminergic neurons projecting to the prefrontal cortex, while the kainic-acid-receptor subtype exerts its regulatory/modulatory effects on mesoaccumbens-projecting dopaminergic neurons arising from the A10/VTA area. An additional finding of interest was that NMDA, but not glutamate or kainic acid, increased the concentration of serotonin metabolites in all of the brain areas studied — further documentation of the interrelationship of these important ''memory molecules''.

In 1990 Imperato et al.[198] related that perfusion *in vivo*, in awake and freely moving rats, of nucleus accumbens and caudate nucleus with a Ringer solution containing quisqualic acid or kainic acid increased the release of dopamine in both areas in a dose-dependent manner. The potency in stimulating dopamine release was higher for kainic acid than for quisqualic acid. The stimulation produced by kainic acid was similar in both areas perfused, whereas quisqualic acid stimulated dopamine release more in the nucleus accumbens than in the caudate nucleus. The maximum increase in stimulated dopamine release of 150% of basal release was similar for both receptor-subtype agonists. NMDA did not affect dopamine release except at very high concentrations. Dopamine release in these studies was measured by transcerebral microdialysis.

Other studies also have failed to establish dopaminergic nervous system release of dopamine by stimulation of the NMDA-receptor subtype. The potent and selective noncompetitive NMDA antagonist, MK-801, has been employed in these studies. *In vitro* studies with synaptosomes and *in vivo* experiments with push-pull cannulae implanted above the striatum (caudate putamen) failed to demonstrate any effect of MK-801 on dopamine release.[199,200] Although MK-801 has been found to cause locomotor stimulation in monoamine-depleted mice, it has been established that this effect was independent of catecholamine release.[201] In a 1990 publication by Kashihara et al.,[202] of Okayama, Japan, additional evidence was presented that the locomotor and behavioral effects of MK-801 in rats was not due to the release of dopamine in the striatal area of these animals.

A modulatory role for nigrostriatal dopaminergic fibers in the control of excitatory amino acid excitotoxicity produced by NMDA and kainic acid has been described from London by Chapman et al.[203] Prior unilateral lesioning of the nigrostriatal dopaminergic pathway by focal injection of 6-hydroxydopamine has been found to decrease the excitotoxic effects of both kainic acid and NMDA.

In 1990, Lindefors and Ungerstedt,[204] from the Karolinska Institute in Stockholm, reported significant observations concerning the interplay of two of the central nervous system's giant (in effects) small molecules, dopamine and glutamate. The great importance of dopamine and the dopaminergic nervous system has been rather extensively documented throughout this chapter, as has that of glutamate and the excitatory amino acids in Chapter 2. Glutamate, as an excitatory neurotransmitter, has a global function, and glutamate neurons are found in virtually every part of the mammalian central nervous system. The significance of glutamate is exemplified by the multiple glutaminergic pathways, the observations that cortical neuronal input to basal ganglia are carried largely through glutamate neurons, and the massive cortiostriatal glutaminergic input.[205-212] Lindefors and Ungerstedt[204] measured tissue levels and *in vivo* release of glutamate in the caudate putamen (striatum) with and without unilateral 6-hydroxydopamine lesioning of mesostriatal/mesocorticolimbic dopamine neurons. The unilateral dopaminergic lesioning produced a bilateral increase in tissue levels of glutamate (45% ipsilateral and 39% contralateral to the lesion). Extracellular levels of glutamate, determined by microdialysis, also were elevated bilaterally (107% ipsilateral and 94% contralateral to the lesion). These results clearly establish that mesencephalic dopamine neurons regulate neuronal glutamate bilaterally in the caudate putamen/striatum areas and provide further impressive and dramatic evidence for the global nature and influence of glutamate in the nervous system of mammals. These important interrelationships of dopamine and glutamate can hardly be without significance in the sensory input molecular phenomena underlying the establishment of learning and memory, in the nervous system as well as the immune response system.

d. Enkephalins

A number of studies have localized enkephalins to the striatum of the mammalian brain.[213-219] An interrelationship with substance P also has been identified in the 1983 report by Del Rio et al.[213] that described the substance P-induced release of met-enkephalin from slices of striatum (caudate putamen).

Catecholamine (dopamine and norepinephrine) innervation at the synaptic level of enkephalinergic neurons in the guinea pig hypothalamus has been demonstrated (in a 1988 report by Mitchell et al.,[220] from Lille, France) employing a novel histochemical/cytochemical technique. An earlier report (1986) by Kubota et al.[221] had identified a modulatory action on enkephalinergic neurons in the rat neostriatum by dopaminergic nerve endings.

A 1990 report from Rouen and London by Dourmap et al.[222] describes an increase in dopamine release in the striatum of rats, by microdialysis infusion of an inhibitor of enkephalin-degrading enzymes in this area.

e. Calcium

Mochida and Libet,[223] from Tokyo and San Francisco, reported in 1990 that long-term enhancement (LTE) by dopamine may be mediated by Ca^{++} and calmodulin.

B. VASCULAR SYSTEM

Vasodilatation mediated by dopamine D_1 receptors has been identified in the renal, mesenteric, splenic, and cerebral vascular beds of the systemic vascular circulation.[224] Efforts to clarify the actions of dopamine on vascular beds of intact animals have been complicated by associated alterations in cardiac output, the concomitant induced release of other vasoactive neurotransmitter molecules, and concurrent neurogenic reflexes.

Gorman[225,226] has effectively employed the isolated perfused lung of the ferret in order to "ferret out" the role of dopamine-receptor subtypes that mediate the dopamine-induced vasodilatation in this useful pulmonary circulation model. His studies established a dopamine-D_1-receptor mediation role in this vasodilation and provided no evidence for the existence or influence of dopamine D_2 receptors contributing to this vasodilatory response. A 1989 report from Australia, by Wanstall and O'Donnell,[227] suggested that the vasodilator responses to dopamine in rat perfused mesentery vasculature were age dependent as well as mediated by dopamine D_1 receptors. The vasodilator response in this preparation of isolated perfused mesentery vasculature was abolished by the dopamine-D_1-receptor selective antagonist, SCH-23390. Responses to all of the vasodilator drugs investigated in this study decreased in magnitude between the ages of 1 and 2 months. Vasodilatory responses to dopamine declined further in 4-month-old rats and were negligible at 6 and 22 to 24 months of age. The authors concluded that, in the rat mesenteric vasculature, there is a nonspecific decline in responses to vasodilator drugs during development (1 to 2 months). Following this period, there is a specific decline in dopamine-D_1-receptor-mediated vasodilatory responses. They speculated that this different time course suggests that animal age influences receptor numbers or their response in terms of second-messenger activation.

An important investigation of the influence of dopamine and its receptor-subtype neurohypophyseal blood flows in sheep was reported by Page et al.[228] in 1990. Radiolabeled microspheres were used to determine blood flows. Intravenous dopamine infusion did not change median eminence blood flow or neural-lobe blood flow. It did, however, increase renal blood flow and blood flow in the choroid plexus. Intravenous infusion of haloperidol caused a significant decrease in blood flow to the regional neurohypophysis. The sheep in these studies were under pentobarbitol sodium anesthesia, an agent that slightly but uniformly depresses blood flow throughout the brain of sheep. Infusion of the dopaminergic D_2 agonist, bromocriptine, did not alter blood flow to any of the areas studied. The number of the infusions and the doses of the D_1 agonists were too small to produce useful data.

Although the synthesis of dopamine was first reported by Barger and Ewins[229] in 1910, it remained a synthetic and metabolic way-station curiosity for many, many years. The first hint of a possible physiological role and a difference from other catecholamines appeared in 1942 in the publication by Holtz and Credner.[230] This report concerned a vascular effect of dopamine. The authors observed that dopamine in guinea pigs and rabbits produced a vasodepressor effect. This was in sharp and dramatic contrast to the blood-pressure-elevating effects of norepinephrine and epinephrine. In these studies vasopressor effects could be achieved only with very large doses of dopamine. The passage of 21 years and the conclusion of a world war led to additional studies of this phenomenon by the Holtz group and others. Early studies are well described in a 1982 minireview concerning vascular dopamine receptors by Brodde[231] of Essen, Germany. The failure to inhibit this dopamine-induced vasodepressor effect by atropine, antihistamines, or β-adrenergic-receptor blocking agents suggested a possible dopamine-specific receptor as the cause of the observed vasodepressor and vasodilator effects. Injection of dopamine into the renal arteries of dogs has been found to produce vasodilatation in small doses and vasoconstriction in large doses. The vasoconstriction could be prevented by prior treatment with phenoxybenzamine, an α-adrenergic-receptor-blocking agent. Dopamine-containing neuronal elements have been identified in the canine kidney.[232]

In 1974 von Essen,[233] of Goteborg, Sweden, found that dopamine stimulates specific dopamine receptors in cerebral vessels, with vasodilatation resulting. Vasodilatation occurred largely in response to doses of dopamine in the medium range. In 1983 Toda[234] described dopaminergic-receptor function in primates (monkeys) to be greater in cerebral, renal, and mesenteric arteries than in femoral and coronary arteries. A dopaminergic influence in cerebral choroid plexus blood flow was investigated in sheep by the radioactive microsphere technique by Townsend et al.[235] Dopamine infusion in this sheep model was found to significantly increase both choroid plexus and renal blood flows. The increase in choroid plexus blood flow was 55% of control flows and that of renal blood flow was 49%.

A blood-flow-increasing and vasodilator effect of dopamine appears to be most potently expressed in the spleens of rabbits. Brodde[231] considers the isolated rabbit splenic artery to be the most suitable tool for the investigation of dopamine vascular receptors. This is of particular interest, as the spleen is a major component of the immune response system, and lymphoid organ blood flow is a major factor in lymphocyte traffic into and through these areas. Hall and Morris[236] have shown that 95 to 98% of lymphocytes in the efferent lymph of peripheral lymph nodes of sheep entered these lymph nodes from their arterial blood supply. Hay and Hobbs[237] have demonstrated, with the radioactive microsphere technique, that the arterial blood flow to a single primary peripheral lymph node of a sheep may be increased as much as sixfold by an antigenic challenge in its drainage area.

C. IMMUNE RESPONSE SYSTEM

From the previous section it is apparent that one of the important influences of dopamine on the immune response system is its effect in increasing blood flow to lymphoid organs and tissues of the immune response system, and the resulting increase in lymphocyte traffic. Both Gowans and Ford have emphasized the importance of lymphocyte traffic in the initiation and maintenance of the immune response.[238-242]

The author and associates[243] have observed a prompt and sharp increase in lymphocyte traffic through primary peripheral (popliteal) lymph nodes of sheep to be produced by the systemic pump infusion of dopamine into the central venous (superior vena cava) areas of sheep (Figure 2). The increase in lymphocyte traffic was monitored by the output of small recirculating lymphocytes into the nodal efferent lymphatic that was chronically cannulated. The increase in lymphocyte traffic was maintained throughout the period of dopamine infusion. This was not encountered with infusion of phenylephrine, an α-adrenergic agonist. With phenylephrine, cannula clotting and severe hemorrhaging into efferent lymph also were produced. From other sheep lymphocyte traffic studies by the author,[244,245] these results were considered immunostimulative.

Dopamine receptors have been described as present on lymphocytes by several investigators.[246-248] In 1987, Ovadia and Abramsky[249] of Jerusalem described the binding of radiolabeled dopamine to isolated membranes of rat lymphocytes as specific, saturable, and of high affinity. Their finding that the specific binding in membranes obtained from Con A-activated lymphoblasts was greater than the binding in unactivated lymphoid cells was of particular interest.

In 1988, Devoino et al.[250] from Siberia described the stimulation of antibody production by increasing endogenous dopamine levels through administration of L-dopa. They employed sheep red blood cell (SRBC) immunization and the SRBC plaque-forming cell assay in these studies. In other publications, dopamine-receptor blockade was found to decrease the number of plaque-forming cells and to exert an immunosuppressive effect. Peripheral dopamine depletion by intraperitoneal 6-hydroxydopamine administration has been observed to have an inhibitory/depressant effect on the response to SRBC immunization in this same plaque assay.[251,252]

Destructive lesions placed in the central nervous system have been demonstrated to alter lymphoid tissue architecture,[253,254] to diminish the immune response to immunization,[254,255] to impair the expression of delayed-type hypersensitivity,[256,257] and to suppress NK (natural killer) cell activity.[258]

Cross and associates,[259] at the University of Kentucky, reported in 1987 a series of important studies relating to the immunologic deficits following in the wake of central nervous system depletion of dopamine, produced by the injection of 6-hydroxydopamine into the systerna magna of mice. They found the primary antibody immune response to SRBC

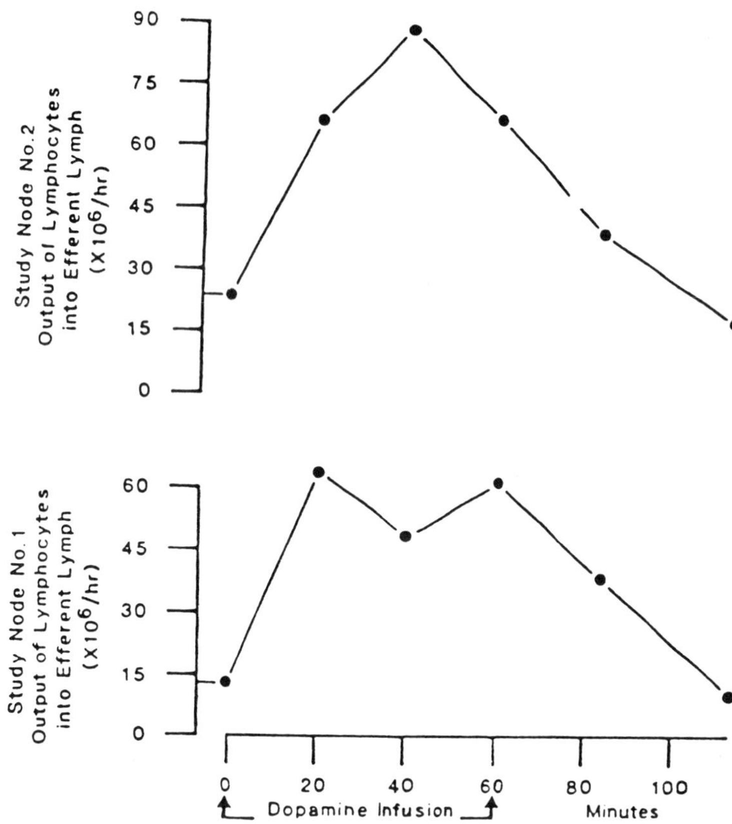

FIGURE 2. The results of two studies of systemic dopamine infusion intravenously (central) on the prompt elevation and sustained elevation in the output into sheep popliteal lymph node efferent lymph (chronically cannulated) of small recirculating lymphocytes. (From Moore, T. C., et al., *Immunopharmacology,* 14, 151, 1987. With permission.)

immunization to be impaired, as well as the establishment of immunological ''memory''. These impairments were not accompanied by an increase in serum corticosteroid levels or by alterations in specific lymphoid cell subpopulations. Extension of these studies provided interesting evidence that these immunosuppressive effects of central nervous system dopamine depletion were associated also with a nonspecific increase in splenic suppressor-cell activity and a requirement for concomitant immunization for the expression of the enhanced suppressor-cell activity. It was of further interest that these splenic suppressor cells were not antigen specific. The authors interpret these data as suggesting that central nervous system depletion of dopamine (and other catecholamines) increases suppressor-cell activity, but does not alter the number of suppressor cells present in the splenic T cell compartment. Furthermore, it appears that the central depletion does not recruit additional suppressor cells to the spleen, nor does it induce the expansion by proliferation of resident suppressor cells. Uniquely, these immunoinhibitory responses are associated with suppressor-cell activity only in a subpopulation of lymphocytes (and possibly other cells) that are antigen primed. These authors further speculated that this data might indicate that dopamine and other catecholamine depletion in the brain might interface with the immune system through alterations in the homeostatic humoral milieu, rather than by altering and acting directly on lymphocyte function.

In 1987 Dasgupta and Lahiri[260] reported from Calcutta, India, that intraperitoneal administration of dopamine to mice bearing transplantable Ehrlich ascites carcinoma resulted in a significant inhibition of tumor growth, with an appreciable increase in host-animal survival time. Enzyme assays indicated an inhibition of tumor growth as well as active lysis of tumor cells. The direct effect of intraperitoneal dopamine injection on tumor-cell proliferation also was shown by a marked associated inhibition in DNA synthesis. Prior studies by others had shown dopamine and its analogues to have selective antitumor activity in a restricted group of malignancies, melanomas, and murine leukemias.[261-265] Prior studies by Dasgupta and Lahiri also had revealed an inverse relationship between brain dopamine levels and the growth of transplanted murine sarcoma-180, as well as an antitumor effect of dopamine on sarcoma-180 tumors.[266,267] In 1990 this group reported additional studies supporting prior findings with use of a dopamine analogue (DHBA).[268] In this study tumor inhibition was accompanied by marked improvements in hemoglobin concentration, red blood cell count, and bone marrow cellularity.

IV. OTHER CATECHOLAMINES

A. NOREPINEPHRINE

1. General

In 1951 Professor U. S. von Euler, of the Karolinska Institute of Stockholm, discoverer of substance P and the prostaglandins, added to an already extraordinary productive and distinguished career the correct identification of noradrenaline (norepinephrine) as the catecholamine-group molecule responsible for neurotransmission in the sympathetic nervous system. In addition to its role in neurotransmission, Professor von Euler also demonstrated that noradrenaline could be, and was, stored in neurons. Professor von Euler, son of a Nobel Laureate, received a much-deserved Nobel Prize in 1970.

In the mid- and late 1950s, noradrenaline also was identified in the brain by biochemical studies by Vogt,[269] Montagu,[270] Carlsson et al.,[271] and Bertler and Rosengren.[272] From these high points, the descent into progressive chaos was prompt, complete, and persistent (to the present). Despite an avalanche of publications, the definitive physiological and biochemical roles of norepinephrine and epinephrine in the functions of the central nervous system and related systems remain a mystery and an enigma. From childhood we have been taught, and have come to believe, that a "flow of adrenaline" means "action". While that may have limited validity peripherally and related principally to the adrenal medulla and as a consequence of sudden action rather than producing it, there is increasing evidence that this concept is both untrue and misleading as it relates to the central nervous system where noradrenaline appears to function in selective areas largely as an inhibitory and downregulatory molecule, with epinephrine being largely a "fellow-traveling" metabolite.

Two 1990 publications by leading international authorities in the neurosciences highlight the current state of affairs concerning the roles of norepinephrine and epinephrine in the central nervous system. In his section on "Central Neurotransmitters" in "Neurohumoral Transmission and the Central Nervous System" in the 1990 edition of the prestigious and classic Goodman and Gillman's pharmacology textbook, *The Pharmacological Basis of Therapeutics,* Floyd E. Bloom,[273] of La Jolla in southern California, does not attempt to hide a significant level of frustration regarding information relating to central nervous system roles for norepinephrine and epinephrine. Qualifying statements in his article include: "a number of features to the mode of action of this biogenic amine have complicated the acquisition of convincing evidence", "the pharmacological properties of such synapses have been complex", "although the mechanisms underlying these effects are not clear", "the mechanisms underlying the inhibitory effects of α-adrenergic receptors are not understood", and, finally, "clearly, further study is necessary" and "the physiological properties of these connections have not as yet been studied."

These sentiments are mirrored rather closely in the 1990 publication from the University of Cambridge, by Selden, Robbins, and Everitt.[274] Their article concerns central noradrenergic function and contains such phrases as "the complex and seemingly opposing effects", "these important neurophysiological studies do not, in themselves, argue for a particular functional role of noradrenergic transmission", "also compromises any simple role for NA" (noradrenaline), "more equivocal", "conflicting results", and "but the results of these studies have also been equivocal and often contradictory." They also state that "the hypothesis that memory consolidation depends on telencephalic NA transmission is weakened by the well-documented lack of effect of central NA depletion on a wide variety of learning tasks (Mason and Iversen,[275] 1975, Roberts,[276] 1981, Robbins and Everitt,[277] 1982, Robbins et al.,[278] 1985). The observation that NA depletion primarily affects the acquisition, but not the performance, of adversely motivated learning (Cole and Robbins,[279] 1987) also compromises a simple role for NA in anxiety."

The author of this book, a novice in the neurosciences, after triaging some 20,000 recent (last 10 years) articles relating to norepinephrine and epinephrine and cited in the *Cumulated Index Medicus*, and after photocopying a small forest worth of articles of seeming relevance to this *Neurovascular Immunology* volume, can only say a heartfelt "Amen" to the above observations. Not only are the results confusing and conflicting, even the assays and their interpretations have been subject to increasing recent questioning and concern.[280,281]

2. Nervous System

a. General

Whereas dopamine and its fibers are diffusely projecting, norepinephrine molecule concentrations and fiber projections are more limited and less diffuse anatomically within the central nervous system. The largest amounts of norepinephrine are concentrated in specific brain areas, such as the hypothalamus, the central nucleus of the amygdala, and the dentate gyrus of the hippocampus. Most noradrenergic neuron cell bodies are located in the locus ceruleus (LC) of the pons, with some cell bodies also found in the lateral tegmental portion of the reticular formation.

b. Peripheral

An important 1988 publication from Chile and Brazil, by Lara and Bastos-Ramos,[282] identified an important peripheral action of an excitatory amino acid, kainic acid, and an interrelationship with norepinephrine. They observed that a single dose of kainic acid to the peripheral ganglia innervating the vas deferens of Sprague-Dawley rats induces a time-dependent decrease of norepinephrine in the organ. These conclude from their experimental data that noradrenergic neurons of the rat vas deferens are under glutamergic control and that this might be important in the motor control of the organ. This organ is of importance in the reproductive survival of this and related mammalian species of animals.

c. Central: Hypothalamus and Corticosteroid Secretion

An important 1989 study from the Rockefeller Institute of New York (Leibowitz et al.[283]) explored the role of norepinephrine in the control of the hypothalamo-pituitary-adrenal axis by testing the effects of direct hypothalamic infusion of norepinephrine on the release of corticosterone in awake and freely moving rats. Infusion of norepinephrine into the paraventricular nucleus (PVN) induced a dose-dependent increase in the circulating levels of corticosterone. In a significant mapping study, the stimulatory effect of norepinephrine was found to be anatomically localized. The greatest increase in circulating-corticosteroid levels (up to 12 μg%) occurred with PVN infusions. A somewhat smaller increase was produced by infusions into the dorsomedial (DM) nucleus of the hypothalamus. No response whatsoever occurred following norepinephrine injections dorsal to the PVN into the ventromedial (VM) or supraoptic (SO) nuclei or into the lateral or the posterior hypothalamus. It was also of particular interest and significance that injection of dopamine into the PVN had no observable effect on circulating-corticosteroid levels. Other recent articles support a significant involvement of norepinephrine in corticosteroid secretion.[284-287]

This same Rockefeller University group, in another 1989 report,[286] described corticosterone-dependent alterations of norepinephrine in seven discrete areas of the rat hypothalamus following chronic adrenalectomy and subsequent replacements and depletions. An increase in the turnover of norepinephrine, especially in the PVN, was observed. Dopamine utilization in these studies and in these areas was unaltered. The authors interpreted these findings as indicating that circulating corticosteroids may have a modulatory role in the regulation of norepinephrine in discrete hypothalamic areas and, as a result, have an impact on the control of various related physiological responses.

Corticotropin-releasing factor-41 (CRF-41) is located within the hypothalamus, largely within the parvocellular neurons of the paraventricular nucleus (PVN).[288,289] Neuronal axons project from this area to the important zona externa of the medial eminence. In a 1988 publication from St. Bartholomew's Hospital in London, Tsagarakis et al.[287] report that norepinephrine stimulates the release of CRF-41 from rat hypothalamus PVN *in vitro*. These authors interpreted their findings as suggesting that the norepinephrine effect was mediated through beta-adrenoreceptors. Norepinephrine stimulation of CRF-41 from the hypothalamus also has been reported from France[285] and from Japan.[290]

The potential central-nervous-system-stimulating effect of norepinephrine on the elevation of circulating-corticosteroid levels in the systemic vascular system is a demonstration of a major inhibitory and suppressive effect of norepinephrine on total body responses, including immunological. To the author of this volume, as a long-time clinical kidney transplant surgeon, the awesome immunosuppressive powers of corticosteroids need no further delineation.

d. Central: Inhibitory in Nonmammalians

In a 1987 amphibian study, Wohlberg et al.,[291] of Miami, FL, reported that by sucrose gap recordings from ventral roots, exposure of motoneurons of the isolated superfused frog spinal cord to norepinephrine prior to the application of the excitatory amino acids L-glutamate or L-aspartate inhibited the excitation/depolarization of the motoneurons produced by either of these excitatory amino acids alone — an inhibitory response.

In 1990, Elmslie and Cohen[292] observed that iontophoresis of norepinephrine into a group of neurons in the pigeon brain, considered to be the equivalent of the lateral geniculate nucleus, produced an *inhibitory* response relating to maintained activity.

e. Central: Inhibition of Kindling-Induced Seizure Activity in the Hippocampus

Excitation of hippocampal input areas by sensory stimuli, by excitatory molecules such as excitatory amino acids, and by electrical stimulation, as of the perforant path in LTP production, represent essential components of the establishment of learning and memory at the molecular and cellular level. On the down side, stimulation may be excessive and may result in the kindling of seizures, as in epilepsy, and may have permanent and fatal consequences. Norepinephrine inputs and concentrations in the hippocampus may act as an important inhibitory brake in these excitatory and possibly excitotoxic stimulatory inputs.

In a 1990 report Weiss et al.[293] described mediation by ascending noradrenergic projections of antikindling effects resulting from electrical stimulation of the locus ceruleus (LC). The kindling was generated in the amygdala, an area rich in norepinephrine from projection from cell bodies in the LC.

Kokaia et al.,[294] from Lund, Sweden, reported in 1989 that by an intracerebral dialysis technique in the hippocampus of rats during both focal and generalized hippocampal seizures evoked by electrical stimulation, marked increases in released norepinephrine occurred. There was no increase in 5-HT release in this area during the induced seizures. These findings, with those above, suggest a seizure (? and memory) inhibitory effect of norepinephrine in the hippocampus, with no inhibitory effect of 5-HT implied.

A 1988 report from Denver by Mynlieff and Dunwiddie[295] presented evidence that a noradrenergic depression of synaptic responses in the rat was mediated by noradrenergic α_1 receptors. There was no evidence of 5-HT or dopamine-receptor involvement.

A 1987 publication, by Barry, Bjorklund, and associates,[296] of Copenhagen and Lund, presented significant documentation of norepinephrine-induced inhibition of kindling-induced epilepsy. In their study, norepinephrine-rich cell suspensions were prepared from the LC region of fetal rats. These norepinephrine-rich cell suspensions were grafted bilaterally into the hippocampus of rats made hypersensitive to hippocampal kindling by a neurotoxic lesion of the central catecholamine system. Animals with grafts demonstrated a marked suppression of the onset and progression of kindling-induced epilepsy. This response was correlated with the degree of graft-derived innervation of the host hippocampal formation. The authors concluded that grafted norepinephrine-rich neurons from the LC area of fetal rats can modulate the excitability of epileptic brain regions of rats.

These "kindling" studies and their inhibition by norepinephrine and the noninvolvement in inhibition by "memory" molecules, such as 5-HT and dopamine, are of particular importance. Kindling is one of the most important and extensively studied models of epilepsy. The term "kindling" doubtlessly derives from the maneuvers involved in the setting, making, and starting of a fire — a component of the heat, pain, and memory saga — a staple from Prometheus to the present.

Kindling in experimental epilepsy involves repeated administrations of electrical stimuli, initially subconvulsive, to an area such as the hippocampus or amygdala, in progressive intensification of stimulus-inducing seizure-type activity until a generalized eliptiform seizure results. Excitability triggered or kindled by electrical stimulation or excitatory amino acid administration (especially kainic acid) in the limbic system has been considered to be analogous to complex-partial/temporal-lobe epilepsy, the most common type of epilepsy in adult humans. It was for this type of epilepsy, in its most severe and relentless form, that the bilateral surgical removal of the hippocampus was first suggested and first carried out. The resulting severe and disabling memory deficits first brought attention to the hippocampus and its important and critical requirement for the establishment of learning and memory.

The ascending noradrenergic projection system originating in the LC has appeared to provide a tonic inhibitory or suppressive action on the development of kindling and kindling-induced seizures. Damage to this system and projections has been shown to greatly facilitate the kindling process and the resulting development of seizure activity. Intracerebral grafting of fetal brain stem neurons from the LC have been demonstrated, by Bjorklung of Lund and associates,[296] to restore transmitter release and turnover in dennervated brain regions. Data provided by this group include electrophysiological evidence to indicate the reestablishment of tonic inhibitory neurotransmission in the grafted areas.

f. Pain, PAG, and LC: More Inhibition

Recent reports by Aran and Proudfit[297] (1990) and by Millar and Williams,[298] (1989) from Chicago and London, provide impressive evidence from their work, and that of others, that norepinephrine and norepinephrine-containing noradrenergic neurons projecting from the LC and related midbrain areas are major inhibitors of nociceptive/pain transmission from the substantia gelatinosa area in the superficial lamina I of the dorsal horn of the spinal cord of mammals. The substantia gelatinosa is a major sensory-impulse reception area in the central nervous system for pain sensations from the periphery, especially for thermal pain.

In 1987 Duggan et al.,[299] from Australia, found that the noxious heating of skin releases immunoreactive substance P in the substantia gelatinosa of the cat. In this regard and respect to the norepinephrine "connection", it is of particular interest that norepinephrine acts on primary afferent terminals in the spinal cord dorsal horn to diminish substance P release.[300-302] In one of these studies, morphine produced a response similar to norepinephrine. Jones and Gebhart,[303] in 1986, observed noradrenergic inhibition of heat-evoked responses recorded from neuronal units responding to noxious stimuli. The substantia gelatinosa and adjacent superficial areas of the dorsal horn have received massive attention in studies concerned with pain and its control.

Anatomical studies to determine the origins and cell body locations of noradrenergic neurons that terminate in the substantia gelatinosa and adjacent superficial lamina I areas of the dorsal horn of the spinal cord have utilized a variety of tract-tracing methodologies and a combination of lesions and histochemistry. The most reliably demonstrated origins are the ventral LC and the nucleus subceruleus. Autoradiographic investigations have demonstrated the presence of noradrenergic terminals in the dorsal horn, with the highest density encountered in the superficial laminae.[304]

Iontophoretic application of norepinephrine to nociceptive dorsal horn neurons has been found to reduce their spontaneous activity. It also has been observed to inhibit their activation by noxious peripheral stimuli.[305-308]

Intrathecal administration of norepinephrine or norepinephrine agonists has been shown to produce potent and dose-dependent antinociception/pain inhibition.[309-311] Further confirmation of these findings was presented by Aran and Proudfit in their 1990 publication,[297] with the additional finding that the norepinephrine nociceptive action on intrathecal application was synergistically enhanced by adenosine agonists. It is of further interest that intrathecal administration of noradrenergic antagonists has been observed to produce hyperalgesia.[312,313]

Electrical stimulation of LC neurons has been shown to produce stimulation-induced antinociception. Electrophysiological studies also have demonstrated antinociceptive effects on dorsal horn neurons. This was produced both by electrical stimulation of LC neurons and by microinjection of the excitatory amino acid, glutamic acid, into the LC.

The midbrain periaqueductal gray (PAG) has also received a considerable amount of attention as our important pain modulation area of the central nervous system. Stimulation of the PAG has been reported to induce behavioral analgesia in rats, cats, and man.[314-317] This analgesia is considered in neurons of the spinal cord dorsal horn that have ascending projections.

The 1989 report of Millar and Williams[298] for PAG-stimulation-induced antinociception by way of descending noradrenergic neuronal projections utilized both electrical stimulation of the PAG and stimulation by iontophoretically applied norepinephrine. Only units that could be excited by iontophoresis of glutamate were analyzed. The inhibition affected both glutamate-evoked activity and responses to nociceptive stimuli. A major and significant finding of this study was that electrical stimulation of the PAG produced very similar effects to iontophoresis of norepinephrine, both in terms of excitation of small inhibitory interneuronal units and in inhibition of the nociceptive component of the input to large units. The authors interpreted their findings as giving strong support to the view that the PAG exerts its antinociceptive actions through a descending noradrenergic pathway. An additional observation of importance in these studies of Millar and Williams was that both norepinephrine and PAG stimulation could block glutamate-evoked activity.

There is also pharmacological evidence that norepinephrine exerts its antinociceptive actions by way of an opiate link.[318] Adrenergic antagonists such as phentolamine can reduce or block analgesia produced by intrathecal or medullary microinjection of opiates.[319-321]

3. Vascular System

Vasoconstriction is a major effect of the application of norepinephrine to the vascular system. In a 1990 study of the effects of phenylephrine and dopamine on cerebral blood flow, blood volume, and oxygenation in young rabbits, Koyama et al.[322] observed that infusion of phenylephrine, an α_1 noradrenergic-receptor agonist, led to an increase in blood pressure and cerebral blood flow without changes in cerebral blood volume, and oxygenation. These findings suggested to the authors that the phenylephrine-induced increases in arterial blood flow and vascular resistence might lead to the rupture of weak cerebral vessels in premature infants — an actual finding of the author and associates[243] with systemic administration of phenylephrine to explore its effects in lymphocyte traffic through primary peripheral lymph nodes of sheep *in vivo*. They also found that low-dose dopamine had no adverse effects on cerebral vessel integrity or cerebral hemodynamics — a finding again comparable to the study by the author and co-workers[243] concerning dopamine-infusion effects on lymph node lymphocyte traffic.

A 1984 report by Randich and Maixner[323] is of particular interest here in that they found that noradrenergic agonists, such as phenylephrine, in systemic doses that elevated arterial blood pressure, also induced antinociception.

4. Immune Response System

In the author's 1987 report[243] of the effects of systemic dopamine and phenylephrine infusion into central venous (superior vena cava) areas on lymphocyte traffic in chronically cannulated and in awake and unsedated sheep, phenylephrine infusion produced an immediate first 20-min sample increase in the output of small recirculating lymphocytes into peripheral nodal efferent lymphatics, chronically cannulated, followed by a sharp fall in output despite a continued elevation in systemic arterial blood pressure (Figure 3). This may be interpreted as an immunosuppressive effect, especially as compared with the dopamine infusion responses cited earlier in this chapter. Hemorrhage into efferent lymph and lymphatic was a frequent occurrence, as was silastic efferent lymphatic occlusion by blood clotting. These findings did not occur with dopamine infusions.

In a 1988 report, Dabney et al.[324] describe hemodynamic alterations produced by perfusion of a lymphatic vessel in the paw of an anesthetized dog with norepinephrine. The addition of norepinephrine to the lymphatic perfusate significantly increases the lymphatic perfusion pressure. Perfusion of prenodal lymphatics disconnected from downstream vessels and nodes indicated that this increase in perfusion pressure occurred primarily in prenodal lymph vessels. The observed increase in lymphatic pressure was blocked by phentolamine.

Hellstrand and Hermodsson,[325] of Goteborg, Sweden, reported in 1989 that adrenergic stimulation induced an inhibition of human natural-killer-cell cytotoxicity. This response was mimicked by the β_2-adrenergic-receptor agonist, terbutaline, and was completely and potentially antagonized by the mixed β_1/β_2-receptor antagonist, propranolol. Pretreatment with IL-2 also virtually abolished the response to adrenaline.

B. EPINEPHRINE

Epinephrine, as an endogenous molecule of physiological significance, appears to be largely produced in, and released from, the adrenal medula through the metabolic activity of phenylethanolamine-*N*-methyl transferase (PNMT). Its involvement in central nervous system function, beyond being a catecholaminergic systems metabolic end product or at best a cotransmitter, has been questioned.

A 1989 publication by Sved,[326] of the University of Pittsburgh, suggests that PNMT-containing catecholaminergic neurons are not necessarily adrenergic. In 1987 Routledge and Marsden[327] observed that none of a series of drugs tested could differentiate between adrenergic and noradrenergic neurons. A 1987 review by Mefford[328] seriously questioned the

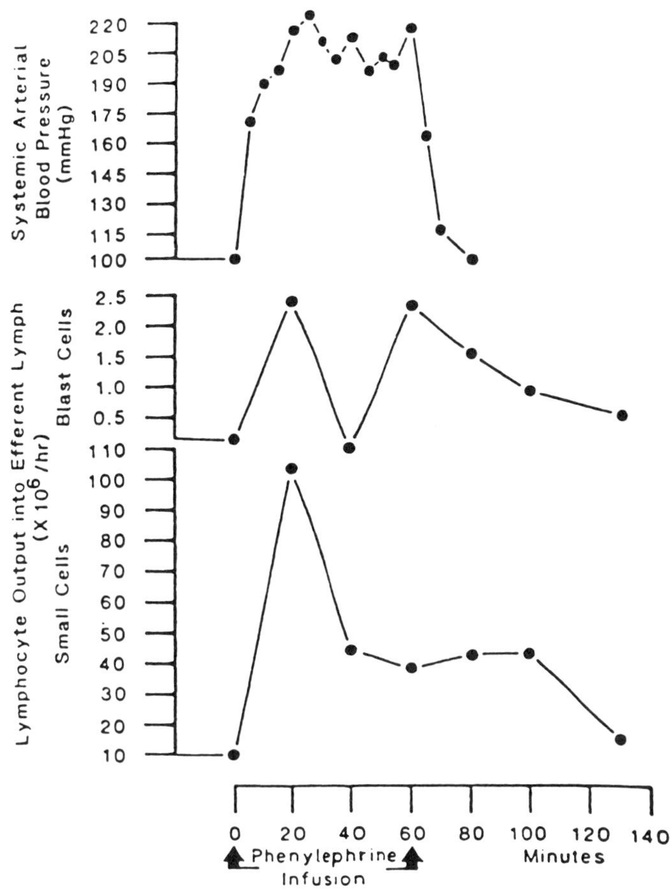

FIGURE 3. A brief, but unsustained, increase in the output of small recirculating lymphocytes into sheep popliteal lymph node efferent lymph (chronically cannulated) in association with central venous infusion of phenylephrine despite a steady and sustained increase in arterial blood pressure. (From Moore, T. C., et al., *Immunopharmacology,* 14, 151, 1987. With permission.)

existence of epinephrine neurons in the rat brain and reviewed a considerable amount of experimental findings in support of this proposition.

V. CONCLUSION

From the standpoint of central nervous system normal function, dopamine is *the* catecholamine and the most important single molecule in the monitoring, directing, and protection of central nervous system homeostasis. As the text of this chapter indicates, there are many and close relationships between dopamine and the dopaminergic nervous system and major immuno- and neurostimulatory "memory molecules" such as substance P, serotonin, and histamine, and the substance P-ergic, serotoninergic, and histaminergic nervous systems of the brain, brain stem, and spinal cord. In a sense, dopamine is a "mystery" molecule, and while much has been learned about its biological actions and activities, much more remains to be learned. Actions of dopamine in the periphery relating to the nervous, vascular, and immune response systems and to neurovascular immunology remain a grossly underinvestigated area. It is hoped that future research efforts will help close this gap.

REFERENCES

1. **Clement-Cormier, Y. C., Kebabian, C. J. W., Petzold, G. L., and Greengard, P.,** Dopamine-sensitive adenylate cyclase in mammalian brain: a possible site of action of antipsychotic drugs, *Proc. Natl. Acad. Sci. U.S.A.*, 71, 1113, 1974.
2. **DeKeyser, J., Dierckx, R., Vanderheyden, P., Ebinger, G., and Vauquelin, G.,** Dopamine D_1 receptors in human putamen, frontal cortex and calf retina: differences in guanine nucleotide regulation of agonist binding and adenylate cyclase stimulation, *Brain Res.*, 443, 77, 1988.
3. **Ferrer, J. M. R., Sanguinetti, A. M., Vives, T., and Mora, F.,** Effects of agonists and antagonists of D_1 and D_2 dopamine receptors on self-stimulation of the medial prefrontal cortex in the rat, *Pharmacol. Biochem. Behav.*, 19, 211, 1983.
4. **Molloy, A. G. and Waddington, J. L.,** Dopaminergic behavior stero-specifically promoted by the D_1 agonist R-SK & F 38393 and selectively blocked by the D1 antagonist SCH 23390, *Psychopharmacology,* 82, 409, 1984.
5. **Ongini, E., Caporali, M. G., and Massotti, M.,** Stimulation of dopamine D-1 receptors by SKF 38393 induces EEG desynchronisation and behavioral arousal, *Life Sci.*, 37, 2327, 1985.
6. **Snyder, S. H.,** The dopamine connection, *Nature,* 347, 121, 1990.
7. **Sokoloff, P., Giros, B., Martres, M.-P., Bouthenet, M.-L., and Schwartz, J.-C.,** Molecular cloning and characterization of a novel dopamine receptor (D_3) as a target for neuroleptics, *Nature,* 347, 146, 1990.
8. **Gros, B., Sokoloff, P., Martres, M.-P., Riou, J.-F., Emorine, L. J., and Schwartz, J.-C.,** Alternate splicing directs the expression of two D_2 dopamine receptor isoforms, *Nature,* 342, 923, 1989.
9. **Mosma, F. J., McVittie, L. D., Gerfen, C. R., Mahan, L. C., and Sibley, D. R.,** Multiple D_2 dopamine receptors produced by alternate RNA splicing, *Nature,* 342, 926, 1989.
10. **Selbie, L. A., Wayes, G., and Shine, J.,** The major dopamine D_2 receptor: molecular analysis of the human D_{2a} subtype, *DNA,* 8, 683, 1989.
11. **Bunzow, J. R., VanTol, H. H. M., Grandy, D. K., Albert, P., Salon, J., Christie, M., Machida, C. A., Neve, K. A., and Civelli, O.,** Cloning and expression of rat D_2 dopamine receptor cDNA, *Nature,* 336, 783, 1988.
12. **Seeman, P., Lee, T., Chau-Wong, M., and Wong, K.,** Antipsychotic drug doses and neuroleptic/dopamine receptors, *Nature,* 261, 717, 1976.
13. **Creese, I., Burt, D. R., and Snyder, S. H.,** Dopamine receptor binding predicts clinical and pharmacological potencies of antipsychotic drugs, *Science,* 192, 481, 1976.
14. **Emorine, L., Marullo, S., Delavier-Klutchko, C., Kaveri, S. V., Durieu-Trautmann, O., and Strosberg, A. D.,** Structure of the gene for human B_2-adrenergic receptor: expression and promoter characterization, *Proc. Natl. Acad. Sci. U.S.A.*, 84, 6995, 1987.
15. **DeKeyser, J., Herregodts, P., and Ebinger, G.,** The mesoneocortical dopamine neuron system, *Neurology,* 40, 1660, 1990.
16. **Berger, B., Trottier, S., Verney, C., Gaspar, P., and Alvarez, C.,** Regional and laminar distribution of the dopamine and serotonin innervation in the macaque cerebral cortex: a radioautographic study, *J. Comp. Neurol.*, 273, 99, 1988.
17. **LeMoal, M., Stinus, L., and Galey, D.,** Radiofrequency lesion of the ventral mesencephalic tegmentum: neurological and behavioral considerations, *Exp. Neurol.*, 50, 521, 1976.
18. **Simon, H., Scatton, B., and LeMoal, M.,** Dopaminergic A10 neurones are involved in cognitive functions, *Science,* 286, 150, 1980.
19. **Wise, R. A.,** The dopamine synapse and the notion of ''pleasure centers'' in the brain, *Trends Neurosci.*, 3, 91, 1980.
20. **Wise, R. A.,** Brain dopamine and reward, in *Theory in Psychopharmacology,* Cooper, S. J., Ed., Academic Press, London, 1, 103, 1981.
21. **LeMoal, M. and Simon, H.,** Mesocorticolimbic dopaminergic network: functional and regulatory roles, *Physiol. Rev.*, 71, 155, 1991.
22. **Lindefors, N., Brene, S., Herrera-Marschitz, M., and Persson, H.,** Neuropeptide gene expression in brain is differentially regulated by midbrain dopamine neurons, *Exp. Brain Res.*, 80, 489, 1990.
23. **Lent, C. M., Mueller, R. L., and Haycock, D. A.,** Chromatographic and histochemical identification of dopamine within an identified neuron in the leech nervous system, *J. Neurochem.*, 41, 481, 1983.
24. **O'Gara, B. A., Chae, H., Latham, L. B., and Friesen, W. O.,** Modification of leech behavior patterns by reserpine-induced amine depletion, *J. Neurosci.*, 11, 96, 1991.
25. **Friesen, W. O., Cometa, A. K., and Hashemzadeh, H.,** Serotonin modifies synaptic transmission in neuronal circuits of the leech *(Hirudo medicinalis), Soc. Neurosci. Abstr.*, 14, 384.
26. **Hashemzadeh-Gargari, H. and Friesen, W. O.,** Modulation of swimming activity in the medicinal leech by serotonin and octopamine, *Comp. Biochem. Physiol.*, 94C, 295, 1989.

27. **O'Gara, B. A. and Friesen, W. O.,** Termination of swimming activity by stimulation of a neuron in subesophageal ganglion of the leech, Hirudo medicinalis, *Soc. Neurosci. Abstr.*, 13, 1060.
28. **O'Gara, B. A. and Friesen, W. O.,** Reserpine depletes serotonin from the CNS of the leech *(Hirudo medicinalis)* and drastically alters behavior, *Soc. Neurosci. Abstr.*, 14, 383, 1988.
29. **O'Gara, B. A., Chae, H., and Friesen, W. O.,** Aminergic neurotransmitters reverse the effect of reserpine on biting behavior but not swimming activity in the leech, *Hirudo medicinalis, Soc. Neurosci. Abstr.*, 15, 27, 1989.
30. **Willard, A. L.,** Effects of serotonin on the generation of the motor program for swimming by the medicinal leech, *J. Neurosci.*, 1, 936, 1981.
31. **Marsden, C. A. and Kerkut, G. A.,** Fluorescent microscopy of the 5HT- and catecholamine-containing cells in the central nervous system of the leech *Hirudo medicinalis, Comp. Biochem. Physiol.*, 31, 851, 1969.
32. **Nusbaum, M. P.,** Synaptic basis of swim initiation in the leech. III. Synaptic effects of serotonin-containing interneurones (cells 21 and 61) on swim CPG neurones (cells 18 and 208), *J. Exp. Biol.*, 122, 303, 1986.
33. **Nusbaum, M. P. and Kristan, W. J., Jr.,** Swim initiation in the leech by serotonin-containing interneurons, cells 21 and 61, *J. Exp. Biol.*, 122, 277, 1986.
34. **Lent, C. M.,** Fluorescent properties of monoamine neurons following glyoxylic acid treatment of intact leech ganglia, *Histochemistry*, 75, 77, 1982.
35. **Lent, C. M.,** Quantitative effects of a neurotoxin upon serotonin levels within tissue compartments of the medicinal leech, *J. Neurobiol.*, 15, 309, 1984.
36. **Lent, C. M.,** Serotonergic modulation of the feeding behavior of the medicinal leech, *Brain Res. Bull.*, 14, 643, 1985.
37. **Lent, C. M. and Dickinson, M. H.,** Serotonin integrates the feeding behavior of the medicinal leech, *J. Comp. Physiol. A*, 154, 457, 1984.
38. **Lent, C. M., Ono, J., Keyser, K. T., and Karten, H. J.,** Identification of serotonin within vital-stained neurons from leech ganglia, *J. Neurochem.*, 321, 1559, 1979.
39. **Lent, C. M., Fliegner, K. H., Freedman, E., and Dickinson, M. H.,** Ingestive behaviour and physiology of the medicinal leech, *J. Exp. Biol.*, 137, 513, 1988.
40. **Lent, C. M., Dickinson, M. H., and Marshall, C. G.,** Serotonin and leech feeding behavior: obligatory neuromodulation, *Am. Zool.*, 29, 1241, 1989.
41. **Barthe, J. Y., Mons, N., Cattaert, D., Geffard, M., and Clarac, F.,** Dopamine and motor activity in the lobster *Homarus gammarus, Brain Res.*, 497, 368, 1989.
42. **Mons, N. and Geffard, M.,** Specific antisera against the catecholamines; L-3,4-dihydroxyphenylalanine, dopamine, noradrenaline, and octopamine tested by an enzyme-linked-immunosorbent-assay, *J. Neurochem.*, 48, 1826, 1987.
43. **Geffard, M., Kah, O., Chambelle, P., LeMoal, M., and Delaage, M.,** Premiere application immunocytochimique d'un anticorps anti-dopamine a l'etude du system nerveaux central, *C.R. Acad. Sci. (Paris)*, 295, 797, 1982.
44. **Geffard, M., Buijs, R. M., Seguela, P., Pool, W. C., and LeMoal, M.,** First demonstration of highly specific and sensitive antibodies against dopamine, *Brain Res.*, 294, 161, 1984.
45. **Chagnaud, J. L., Mons, N., Tuffet, S., Grandier-Vazeilles, X., and Geffard, M.,** Monoclonal antibodies against glutaraldehyde-conjugated dopamine, *J. Neurochem.*, 49, 487, 1987.
46. **Hughes, G. M. and Wiersma, C. A. G.,** The coordination of swimmeret movement in the crayfish *Procambarus clarkii* (girard), *J. Exp. Biol.*, 37, 657, 1960.
47. **Mulloney, B., Acevedo, L. D., and Bradbury, A. G.,** Modulation of the crayfish swimmeret rhythm by octopamine and the neuropeptide proctolin, *J. Neurophysiol.*, 58, 584, 1987.
48. **Paul, D. H. and Mulloney, B.,** Local interneurons in the swimmeret system of the crayfish, *J. Comp. Physiol.*, 156, 489, 1985.
49. **Cattaert, D. and Clarac, F.,** Rami motor neurons and motor control of the swimmeret system of *Homarus gammarus, J. Comp. Physiol.*, 160, 55, 1987.
50. **Maxwell, G. D., Tait, J. F., and Hildebrand, J. G.,** Regional synthesis of neurotransmitter candidates in the CNS of the moth *Manduca sexta, Comp. Biochem. Physiol.*, 61C, 109, 1978.
51. **Macmillan, C. S. and Mercer, A. R.,** An investigation of the role of dopamine in the antennal lobes of the honeybee, *Apis mellifera, J. Comp. Physiol.*, 160, 359, 1987.
52. **Mercer, A. R. and Menzel, R.,** The effects of biogenic amines on conditioned and unconditioned responses to olfactory stimuli in the honeybee *Apis mellifera, J. Comp. Physiol.*, 145, 363, 1982.
53. **Nässel, D. R., Elekes, K., and Johansson, K. U. I.,** Dopamine-immunoreactive neurons in the blowfly visual system: light and electron microscopic immunocytochemistry, *J. Chem. Neuroanat.*, 1, 311, 1988.
54. **Dahl, E., Falck, B., Lindquist, M., and von Mecklenburg, C.,** Monoamines in molluscan neurones, *K. Fysiogr. Saellsk Lund Faerh.*, 32, 89, 1962.

55. **Guthrie, P. B., Neuhoff, V., and Osborne, N. N.,** Dopamine, noradrenaline, octopamine and tyrosine hydroxylase in the gastropod *Helix pomatia, Comp. Biochem. Physiol.,* 52C, 109, 1975.
56. **McCaman, M. W., Ono, J. K., and McCaman, R. E.,** Dopamine measurements in molluscan ganglia and neurons using a new, sensitive assay, *J. Neurochem.,* 32, 1111, 1979.
57. **Straub, H. and Kuhlmann, D.,** Identification and quantitative measurements of biogenic monoamines in the central nervous tissue of some gastropods, *Comp. Biochem. Physiol.,* 78C, 319, 1984.
58. **Gospe, S. M.,** Minireview: studies of dopamine pharmacology in mollusca, *Life Sci.,* 33, 1945, 1983.
59. **S.-Rozsa, K.,** The pharmacology of molluscan neurons, *Prog. Neurobiol.,* 23, 79, 1984.
60. **Walker, R. J.,** Transmitters and modulators, in *The Mollusca, Vol. 9, Neurobiology and Behavior,* Willows, A. O. D., Ed., Academic Press New York, 1986, 279.
61. **Trimble, D. L. and Barker, D. L.,** Activation by dopamine of patterned motor output from the buccal ganglia of *Helisoma trivolvis, J. Neurobiol.,* 15, 37, 1984.
62. **Wieland, S. J. and Gelperin, A.,** Dopamine elicits feeding motor program in *Limax maximus, J. Neurosci.,* 3, 1735, 1983.
63. **Kyriakides, M. A. and McCrohan, C. R.,** Effect of putative neuromodulators on rhythmic buccal motor output in *Lymnaea stagnalis, J. Neurobiol.,* 20, 635, 1989.
64. **Berry, M. S. and Cottrell, G. A.,** Excitatory, inhibitory and biphasic synaptic potentials mediated by an identified dopamine-containing neurone, *J. Physiol. (Lond.),* 244, 589, 1975.
65. **Winlow, W., Haydon, P. G., and Benjamin, R. P.,** Multiple postsynaptic actions of the giant dopamine-containing neurone R.Pe.D.1 of *Lymnaea stagnalis, J. Exp. Biol.,* 94, 137, 1981.
66. **Stoof, J. C., Werkman, T. R., Lodder, J. C., and DeVlieger, T. A.,** Growth hormone producing cells in the snail *Lymnaea stagnalis* as a model for mammalian dopamine receptors? *Trends. Pharmacol. Sci.,* 7, 7, 1986.
67. **Audesirk, T. E.,** Characterization of pre- and postsynaptic dopamine receptors in *Lymnaea, Comp. Biochem. Physiol.,* 93C, 115, 1989.
68. **DeVlieger, T. A., Lodder, J. C., Stoof, J. C., and Werkman, T. R.,** Dopamine receptor stimulation induces a potassium dependent hyperpolarizing response in growth hormone producing cells of the gastropod mollusc *Lymnaea stagnalis, Comp. Biochem. Physiol.,* 83C, 429, 1986.
69. **Werkman, T. R., DeVlieger, T. A., and Stoof, J. C.,** Indications for a hormonal function of dopamine in the central nervous system of the snail, *Lymnaea stagnalis, Neurosci. Lett.,* 108, 167, 1990.
70. **Kemenes, G., Elekes, K., Hirpi, L., and Benjamin, P. R.,** A comparison of four techniques for mapping the distribution of serotonin and serotonin-containing neurons in fixed and living ganglia of the snail, *Lymnaea, J. Neurocytol.,* 18, 193, 1989.
71. **Straub, H. and Kuhlmann, D.,** Identification and quantitative measurements of biogenic monoamines in the central nervous tissue of some gastropods, *Comp. Biochem. Physiol.,* 78C, 319, 1984.
72. **Gerschenfeld, H. M.,** Chemical transmission in invertebrate central nervous system and neuromuscular junctions, *Physiol. Rev.,* 53, 1, 1973.
73. **Schwann, J. W., Sinback, C. N., Kebabian, P. R., and Carpenter, D. O.,** Motoneurons which may utilize dopamine as their neurotransmitter, *Cell. Mol. Neurobiol.,* 2, 309, 1982.
74. **Billy, A. J. and Walters, E. T.,** Modulation of mechanosensory threshold in *Aplysia* by serotonin, small cardioactive peptide$_B$ (SCP$_B$), FMRFamide, acetylcholine, and dopamine, *Neurosci. Lett.,* 105, 200, 1989.
75. **Abrams, T. W., Castellucci, V. F., Camardo, J. S., Kandel, E. R., and Lloyd, P. E.,** Two endogenous neuropeptides modulate the gill and siphone withdrawal reflex in *Aplysia* by presynaptic facilitation involving cAMP-dependent closure of a serotonin-sensitive potassium channel, *Proc. Natl. Acad. Sci. U.S.A.,* 81, 7956, 1984.
76. **Elekes, K., Kemenes, G., Hiripi, L., Geffard, M., and Benjamin, P. R.,** Dopamine-immunoreactive neurones in the central nervous system of the pond snail *Lymnaea stagnalis, J. Comp. Neurol.,* 307, 214, 1991.
77. **Kemenes, G., Elekes, K., Hiripi, L., and Benjamin, P. R.,** A comparison of four techniques for mapping the distribution of serotonin and serotonin-containing neurons in fixed and living ganglia of the snail, *Lymnaea, J. Neurocytol.,* 18, 193, 1989.
78. **Capasso, A., Creti, P., DePetrocellis, B., DePrisco, P. P., and Parisi, E.,** Role of dopamine and indolamine derivatives in the regulation of the sea urchin adenylate cyclase, *Biochem. Biophys. Res. Commun.,* 154, 758, 1988.
79. **Romer, A. S. and Parsons, T. S.,** *The Vertebrate Body,* Saunders, Philadelphia, 1986, 624.
80. **Ekstrom, P., Honkanen, T., and Steinbusch, H. W. M.,** Distribution of dopamine-immunoreactive neuronal perikarya, and fibres in the brain of a teleost, *Gasterosteus aculeatus* L. Comparison with tyrosine hydroxylase- and dopamine-β-hydroxylase-immunoreactive neurons, *J. Chem. Neuroanat.,* 3, 233, 1990.
81. **Smeets, W. J. A. J.,** Distribution of dopamine immunoreactivity in the forebrain and midbrain of the snake *Python regius*: a study with antibodies against dopamine, *J. Comp. Neurol.,* 271, 115, 1988.

82. **Smeets, W. J. A. J.,** The monoaminergic systems in the forebrain and midbrain of reptiles investigated with specific antibodies against serotonin, dopamine and noradrenaline, in *The Forebrain of Reptiles: Current Concepts of Structure and Function,* Schwerdtfeger, W. K. and Smeets, W. J. A. J., Eds., Karger, Basel, 1988.
83. **Smeets, W. J. A. J., Hoogland, P. V., and Voorn, P.,** The distribution of dopamine immunoreactivity in the forebrain and midbrain of the lizard *Gekko gecko*: an immunohistochemical study with antibodies against dopamine, *J. Comp. Neurol.,* 253, 46, 1986b.
84. **Smeets, W. J. A. J., Jonker, A. J., and Hoogland, P. V.,** Distribution of dopamine in the forebrain and midbrain of the red-eared turtle, *Pseudemys scripta elegans,* reinvestigated using antibodies against dopamine, *Brain Behav. Evol.,* 30, 121, 1987.
85. **Halpern, M.,** The telencephalon of snakes, in *Comparative Neurology of the Telencephalon,* Ebbesson, S. O. E., Ed., Plenum Press, New York, 1980, 257.
86. **Parent, A.,** Distribution of monoamine-containing neurons in the brain stem of the frog, *Rana temporaria, J. Morphol.,* 139, 67, 1973.
87. **Parent, A., Poitras, D., and Dube, L.,** Comparative anatomy of central monoaminergic system, in *Handbook of Chemical Neuroanatomy,* Vol. 2, *Classical Transmitters in the CNS,* Part I, Bjorklund, A. and Hokfelt, T., Eds., Elsevier, Amsterdam, 1984, 409.
88. **Brauth, S. E.,** Catecholamine neurons in the brainstem of the reptile *Caiman crocodilus, J. Comp. Neurol.,* 270, 313, 1988.
89. **Corio, M. and Doerr-Schott, J.,** The monoaminergic system in the diencephalon of the newt tadpole, *Triturus alpestris* (Mert). A histofluorescence study, *J. Hirnforsch.,* 29, 377, 1988.
90. **Franzoni, M. F., Thibault, J., Fasolo, A., Martinoli, M. G., Scaranari, F., and Calas, A.,** Organization of tyrosine-hydroxylase immunopositive neurons in the brain of the crested newt, *Triturus cristatus carnifex, J. Comp. Neurol.,* 251, 121, 1986.
91. **Meredith, G. E. and Smeets, W. J. A. J.,** Immunocytochemical analysis of the dopamine system in the forebrain and midbrain of *Raja radiata*. Evidence for a substantia nigra and ventral tegmental area in cartilaginous fish, *J. Comp. Neurol.,* 265, 530, 1987.
92. **Smeets, W. J. A. J., Hoogland, P. V., and Voorn, P.,** The distribution of dopamine immunoreactivity in the forebrain and midbrain of the lizard *Gekko gecko:* an immunohistochemical study with antibodies against dopamine, *J. Comp. Neurol.,* 253, 46, 1986.
93. **Smeets, W. J. A. J., Jonker, A. J., and Hoogland, P. V.,** Distribution of dopamine in the forebrain and midbrain of the red-eared turtle, *Pseudemys scripta elegans,* reinvestigated using antibodies against dopamine, *Brain Behav. Evol.,* 30, 121, 1987.
94. **Moore, T. C., Whitley, G. A., Lami, J. L., and Said, S. I.,** Substance P increases and prolongs increased output of T4 (CD4) lymphocytes from lymph nodes of sheep *in vivo*: is it a mediator of immunological memory? *Immunopharmacology,* 20, 207, 1990.
95. **Duggan, A. W., Morton, C. R., Zhao, Z. Q., and Hendry, I. A.,** Noxious heating of skin releases immunoreactive substance P in the substantia gelatinosa of the cat: a study with antibody probes, *Brain Res.,* 403, 345, 1987.
96. **Kohno, J., Shiosaka, S., Shinoda, K., Inagaki, S., and Tohyama, M.,** Two distinct strio-nigral substance P pathways in the rat: an experimental immunohistochemical study, *Brain Res.,* 308, 309, 1984.
97. **Quirion, R., Shults, C. W., Moody, T. W., Pert, C. B., Chase, T. N., and O'Donohue, T. L.,** Autoradiographic distribution of substance P receptors in the rat central nervous system, *Nature,* 303, 714, 1983.
98. **Pernow, B.,** Substance P, *Pharmacol. Rev.,* 35, 85, 1983.
99. **Kubota, Y., Inagaki, S., and Kito, S.,** Innervation of substance P neurons by catecholaminergic terminals in the neostriatum, *Brain Res.,* 375, 163, 1986.
100. **Tamiya, R., Hanada, M., Kawai, Y., Inagaki, S., and Takagi, H.,** Substance P afferents have synaptic contacts with dopaminergic neurons in the ventral tegmental area of the rat, *Neurosci. Lett.,* 110, 11, 1990.
101. **Bannon, M. J., Elliott, P. J., Alpert, J. E., Goedert, M., Iversen, S. D., and Iversen, L. L.,** Role of endogenous substance P in stress-induced activation of mesocortical dopamine neurones, *Nature,* 306, 791, 1983.
102. **Cador, M., Rivet, J.-M., Kelley, A. E., Moal, M. L., and Stinus, L.,** Substance P, neurotensin and enkephalin injections into the ventral tegmental area: comparative study on dopamine turnover in several forebrain structures, *Brain Res.,* 486, 357, 1989.
103. **Kelley, A. E., Stinus, L., and Iversen, S. D.,** Behavioural activation induced in the rat by substance P infusion into ventral tegmental area: implication of dopaminergic A10 neurones, *Neurosci. Lett.,* 11, 335, 1979.
104. **Kelley, A. E., Cador, M., and Stinus, L.,** Behavioral analysis of the effect of substance P injected into the ventral mesencephalon on investigatory and spontaneous motor behavior in the rat, *Psychopharmacology,* 85, 37, 1985.

105. **Stinus, L., Kelley, A. E., and Iversen, S. D.,** Increased spontaneous activity following substance P infusion into A10 dopaminergic area, *Nature,* 276, 616, 1978.
106. **Bjorklund, A. and Lindvall, O.,** Dopamine-containing systems in the CNS, in *Handbook of Chemical Neuroanatomy*, Vol. 2, *Classical Transmitters in the CNS,* Part I, Bjorklund, A. and Hokfelt, T., Eds., Elsevier, Amsterdam, 1984, 55.
107. **Ljungdahl, A., Hokfelt, T., Nilsson, G., and Goldstein, M.,** Distribution of substance P-like immunoreactivity in the central nervous system of the rat. II. Light microscopic localization in relation to catecholamine-containing neurons, *Neuroscience,* 3, 945, 1978.
108. **Dahlstrom, A. and Fuxe, K.,** Evidence for the existence of monoamine-containing neurons in the CNS. I. Demonstration of monoamines in the cell bodies of brainstem neurons, *Acta Physiol. Scand.*, 232 (Suppl.), 1, 1964.
109. **Lindvall, O. and Bjorklund, A.,** The organization of the ascending catecholamine neuron systems in the rat brain as revealed by the glyoxylic acid fluorescence method, *Acta Physiol. Scand.*, 412 (Suppl.), 1, 1974.
110. **Kelly, P. H., Seviour, P. W., and Iversen, S. D.,** Amphetamine and apomorphine responses in the rat following both 6-OHDA lesions of the nucleus accumbens septi and or striatum, *Brain Res.*, 94, 507, 1975.
111. **Creese, I. and Iversen, S. D.,** Blockade of amphetamine-induced motor stimulation in the adult rat following neonatal treatment with 6-hydroxydopamine, *Brain Res.*, 55, 369, 1973.
112. **Joyce, E. M. and Iversen, S. D.,** The effect of 6-OHDA lesions to mesolimbic terminals on spontaneous behaviors in the rat, *Neurosci. Lett.*, S289 (Suppl.), 1981.
113. **Koob, G. F., Stinus, L., and LeMoal, M.,** Hyperactivity and hypoactivity produced by lesions to the mesolimbic dopamine system, *Behav. Brain Res.*, 3, 341, 1981.
114. **Gale, K., Costa, E., Toffano, G., Hong, J. S., and Guidotti, A.,** Evidence for a role of nigral GABA and substance P in the haloperidol-induced activation of striatal tyrosine hydroxylase, *J. Pharmacol. Exp. Ther.*, 206, 29, 1978.
115. **Kanazawa, I. and Jessell, T.,** Post-mortem changes and regional distribution of substance P in the rat and mouse nervous system, *Brain Res.*, 117, 362, 1976.
116. **Shults, C. W., Quirion, R., Chronwall, B., Chase, T. N., and O'Donohue, T. L.,** A comparison of the anatomical distribution of substance P and substance P receptors in the rat central nervous system, *Peptides,* 5, 1097, 1984.
117. **Kalivas, P. W. and Miller, J. S.,** Substance P modulation of dopamine in the nucleus accumbens, *Neurosci. Lett.*, 48, 55, 1984.
118. **Shalby, I. A., Kotake, C., Hoffman, P. C., and Heller, A.,** Release of dopamine from coaggregate cultures of mesencephalic tegmentum and corpus striatum, *J. Neurosci.*, 3, 1565, 1983.
119. **Silbergeld, E. K. and Walters, J. R.,** Synaptosomal uptake and release of dopamine in substantia nigral effects of GABA and substance P, *Neurosci. Lett.*, 12, 119, 1979.
120. **Elliott, P. J., Nemeroff, C. B., and Kilts, C. D.,** Evidence for a tonic facilitatory influence of substance P on dopamine release in the nucleus accumbens, *Brain Res.*, 385, 379, 1986.
121. **Young, W. S., Bonner, T. I., and Brann, M. R.,** Mesencephalic dopamine neurons regulate the expression of neuropeptide mRNAs in the rat forebrain, *Proc. Natl. Acad. Sci. U.S.A.*, 83, 9827, 1986.
122. **Gerfen, C. R. and Young, W. S.,** Distribution of striatonigral and striatopallidal peptidergic neurons in both patch and matrix compartments: an *in situ* hybridization histochemistry and fluorescent retrograde tracing study, *Brain Res.*, 460, 161, 1988.
123. **Oblin, A., Zivkovic, B., and Bartholini, G.,** Involvement of the D-2 dopamine receptor in the neuroleptic-induced decrease in nigral substance P, *Eur. J. Pharmacol.*, 105, 175, 1984.
124. **Sonsalla, P. K., Gibb, J. W., and Hanson, G. R.,** Opposite responses in the striato-nigral substance P system to D_1 and D_2 receptor activation, *Eur. J. Pharmacol.*, 105, 185, 1984.
125. **Haverstick, D. M., Rubenstein, A., and Bannon, M. J.,** Striatal tachykinin gene expression regulated by interaction of D-1 and D-2 dopamine receptors, *J. Pharmacol. Exp. Ther.*, 248, 858, 1989.
126. **Brownstein, M. J., Mroz, E. A., Tappaz, M. L., and Leeman, S. E.,** On the origin of substance P and glutamate decarboxylase (GAD) in the substantia nigra, *Brain Res.*, 135, 315, 1977.
127. **Gale, K., Hong, J. S., and Guidotti, A.,** Presence of substance P and GABA in separate striatonigral neurons, *Brain Res.*, 136, 371, 1977.
128. **Kanazawa, I., Emson, P. C., and Cuello, C.,** Evidence for the existence of substance P containing fibers in striatonigral and pallidonigral pathways in rat brain, *J. Pharm. Pharmacol.*, 29, 447, 1977.
129. **Kawai, Y., Takagi, H., Kumoi, Y., Shiosaka, S., and Tohyama, M.,** Nigrostriatal dopamine neurons receive substance P-ergic inputs in the substantia nigra: application of the immunoelectron microscopic mirror technique to the fluorescent double staining for transmitter selective projections, *Brain Res.*, 401, 371, 1987.
130. **Somogyi, P., Priestly, J. V., Cuello, A. C., Smith, A. D., and Bolam, J. P.,** Synaptic connections of substance P-immunoreactive nerve terminals of substantia nigra of the rat, *Cell. Tissue Res.*, 223, 469, 1982.

131. **Cheramy, A., Michelot, R., Leviel, V., Glowinski, J., and Kerdelhue, B.,** Effect of the immunoneutralization of substance P in the cat substantia nigra on the release of dopamine from dendrites and terminals of dopaminergic neurons, *Brain Res.*, 155, 404, 1978.
132. **Baruch, P., Artaud, F., Godeheu, G., Barbeito, L., Glowinski, J., and Cheramy, A.,** Substance P and neurokinin A regulate by different mechanisms dopamine release from dendrites and nerve terminals of the nigrostriatal dopamine neurons, *Neuroscience*, 25, 889, 1986.
133. **Herrera-Marschitz, M., Christensson-Nylander, I., Sharp, T., Staines, W., Reid, M., Hokfelt, T., Terenius, L., and Ungerstedt, U.,** Striato-nigral dynorphin and substance P pathways in the rats. II. Functional analysis, *Exp. Brain Res.*, 64, 193, 1986.
134. **Innis, R. B., Andrade, R., and Aghajanian, G. K.,** Substance K excites dopaminergic and non-dopaminergic neurons in rat substantia nigra, *Brain Res.*, 335, 381, 1985.
135. **James, T. A. and Starr, M. S.,** Behavioral and biochemical effects of substance P injected into the substantia nigra of the rat, *J. Pharm. Pharmacol.*, 29, 181, 1977.
136. **Pinnock, R. D. and Dray, A.,** Differential sensitivity of presumed dopaminergic and non-dopaminergic neurons in rat substantia nigra to electrophoretically applied substance P, *Neurosci. Lett.*, 29, 153, 1982.
137. **Kelly, A. F. and Inversen, S. D.,** Substance P infusion into the substantia nigra of the rat: behavioral analysis and involvement of striatal dopamine, *Eur. J. Pharmacol.*, 60, 171, 1979.
138. **Reid, M., Herrera-Marschitz, M., Hokfelt, T., Terenius, L., and Ungerstedt, U.,** Differential modulation of striatal dopamine release by intranigral injection of gamma-aminobutyric acid (GABA), dynorphin A and substance P, *Eur. J. Pharmacol.*, 147, 411, 1988.
139. **Reid, M. S., Herrera-Marschitz, M., Hokfelt, T., Ohlin, M., Valentino, K., and Ungerstedt, U.,** Effects of intranigral substance P and neurokinin A injections on striatal dopamine release. I. Interactions with substance P antagonists, *Neuroscience*, 36, 643, 1990.
140. **Hokfelt, T., Vincent, S., Hellsten, L., Rosell, S., Folkers, K., Markey, K., Goldstein, M., and Cuello, C.,** Immunohistochemical evidence for a "neurotoxic" action of (D-Pro^2, D-TRP^{7,9})-substance P, an analogue with substance P antagonist activity, *Acta Physiol. Scand.*, 113, 571, 1981.
141. **Post, C. and Paulson, I.,** Antinociceptive and neurotoxic actions of substance P analogues in the rat spinal cord after intrathecal administration, *Neurosci. Lett.*, 57, 159, 1985.
142. **Folkers, K., Feng, D.-M., Asano, N., Hakanson, R., Wiesenfeld-Hallin, Z., and Leander, S.,** Spantide II, an effective tachykinin antagonist having high potency and negligible neurotoxicity, *Proc. Natl. Acad. Sci. U.S.A.*, 87, 4833, 1990.
143. **Reid, M. S., Hokfelt, T., Herrera-Marschitz, M., Hakanson, R., Feng, D. M., Folkers, K., Goldstein, M., and Ungerstedt, U.,** Intranigral substance P stimulation of striatal dopamine release is inhibited by Spantide II: a new tachykinin antagonist without apparent neurotoxicity, *Brain Res.*, 532, 175, 1990.
144. **Wiesenfeld-Hallin, Z., Xu, X.-J., Hakanson, R., Feng, D.-M., and Folkers, K.,** The specific antagonistic effect of intrathecally injected Spantide II on substance P and C-fiber conditioning stimulation induced facilitation of the nociceptive flexor reflex in the rat, *Brain Res.*, 526, 284, 1990.
145. **Wiesenfeld-Hallin, Z., Xu, X.-J., Hakanson, R., Feng, D.-M., and Folkers, K.,** Plasticity of the peptidergic mediation of spinal reflex facilitation after peripheral nerve section in the rat, *Neurosci. Lett.*, 116, 293, 1990.
146. **Wiesenfeld-Hallin, Z., Xu, X.-J., Kristensson, K., Hakanson, R., Feng, D.-M., and Folkers, K.,** Antinociception and substance P antagonistic effects of intrathecally injected spantide II in rat: no signs of motor impairment or neurotoxicity, *Regul. Pept.*, 29, 1, 1990.
147. **Yahr, M. D. and Bergmann, K. J., Eds.,** *Parkinson's Disease, Advances in Neurology,* Vol. 45, Raven Press, New York, 1987.
148. **Mauborgne, A., Javoy-Agid, F., Legrand, J. C., Agid, Y., and Cesselin, F.,** Decrease of substance P-like immunoreactivity in the substantia nigra and pallidum of parkinsonian brains, *Brain Res.*, 268, 167, 1983.
149. **Tenovao, O., Runne, V. K., and Viljanen, M. K.,** Substance P immunoreactivity in the post-morten Parkinsonian brain, *Brain Res.*, 303, 113, 1984.
150. **Barker, R. A.,** Substance P and Parkinson's disease: a causal relationship, *J. Theor. Biol.*, 120, 253, 1986.
151. **Barker, R. A.,** Adrenal grafting for Parkinson's disease: a role for substance P, *Int. J. Neurosci.*, 46, 47, 1989.
152. **Gariano, R. F. and Groves, P. M.,** Burst firing induced in midbrain dopamine neurons by stimulation of the medial prefrontal and anterior cingulate cortices, *Brain Res.*, 462, 194, 1988.
153. **Graybiel, A. M. and Ragsdale, C. W.,** Biochemical anatomy of the striatum, in *Chemical Neuroanatomy,* Emson, P. C., Ed., Raven Press, New York, 1983, 427.
154. **Fisher, R. S.,** GABA-ergic pallindonigral and "accessory" striatonigral connections demonstrated in cats by double peroxidase labeling, *Synapse,* 4, 154, 1989.

155. **Schmued, L., Phermsangngam, P., Lee, H., Thio, S., Chen, E., Truong, P., Colton, E., and Fallon, J.,** Collateralization and GAD immunoreactivity of descending pallidal efferents, *Brain Res.*, 487, 131, 1989.
156. **Smith, Y. and Bolam, J. P.,** The output neurones and the dopaminergic neurones of the substantia nigra receive a GABA-containing input from the globus pallidus in the rat, *J. Comp. Neurol.*, 296, 47, 1990.
157. **Steinbusch, H. W. M.,** Serotonin-immunoreactive neurons and their projections in the CNS, in *Handbook of Chemical Neuroanatomy*, Vol. 3, Bjorklund, A., Hokfelt, T., and Kuhar, M., Eds., Elsevier, Amsterdam, 1984, 68.
158. **Wirtshafter, D., Stratford, T. R., and Asin, K. E.,** Evidence that serotonergic projections to the substantia nigra in the rat arise in the dorsal, but not the median, raphe nucleus, *Neurosci. Lett.*, 77, 261, 1987.
159. **Bolam, J. P. and Smith, Y.,** The GABA and substance P input to dopaminergic neurones in the substantia nigra of the rat, *Brain Res.*, 529, 57, 1990.
160. **Beckstead, R. M.,** Striatal substance P cell clusters coincide with the high density terminal zones of the discontinuous nigrostriatal dopaminergic projection system in the cat: a study by combined immunohistochemistry and autoradiographic axon-tracing, *Neuroscience*, 20, 557, 1987.
161. **Sivam, S. P.,** D_1 dopamine receptor-mediated substance P depletion in the striatonigral neurons of rats subjected to neonatal dopaminergic denervation: implications for self-injurious behavior, *Brain Res.*, 500, 119, 1989.
162. **Zahm, D. S., Eggerman, K. W., Sprung, R. F., Wesche, D. E., and Payne, E.,** Postnatal development of striatal neurotesin immunoreactivity in relation to clusters of substance P immunoreactive neurons and the "dopamine islands" in the rat, *J. Comp. Neurol.*, 296, 403, 1990.
163. **Sivam, S. P., Krause, J. E., Breese, G. R., and Hong, J.-S.,** Dopamine-dependent postnatal development of enkephalin and tachykinin neurons of rat basal ganglia, *J. Neurochem.*, 56, 1499, 1991.
164. **Breese, G. R., Baumeister, A. A., McCown, T. J., Emerick, S. G., Frye, G. D., Crotty, K., and Mueller, R. A.,** Behavioral differences between neonatal and adult 6-hydroxy-dopamine treated rats to dopamine agonists: relevance to neurological symptoms in clinical syndromes with reduced brain dopamine, *J. Pharmacol. Exp. Ther.*, 21, 343, 1984.
165. **Breese, G. R., Baumeister, A. A., McCown, T. J., Emerick, S. G., Frye, G. D., and Mueller, R. A.,** Neonatal 6-hydroxy-dopamine treatment: model of susceptibility for self-mutilation in the Lesch-Nyhan syndrome, *Pharmacol. Biochem. Behav.*, 21, 459, 1984.
166. **Lesch, M. and Nyhan, W. D.,** A familial disorder of uric acid metabolism and central nervous system function, *Am. J. Med.*, 36, 561, 1964.
167. **Lloyd, K. G., Hornykiewicz, O., Davidson, L., Shannak, K., Farley, I., Goldstein, M., Shibuya, M., Kelley, W. N., and Fox, I. H.,** Biochemical evidence of dysfunction of brain neurotransmitters in the Lesch-Nyhan syndrome, *N. Engl. J. Med.*, 305, 1106, 1981.
168. **Wilson, J. M., Young, A. B., and Kelley, W. N.,** Hypoxanthine-guanine phosphoribosyl transferase deficiency: the molecular basis of the clinical syndromes, *N. Engl. J. Med.*, 309, 100, 1983.
169. **Criswell, H., Mueller, R. A., and Breese, G. R.,** Assessment of purine-dopamine interactions in 6-hydroxydopamine lesioned rats: evidence for pre- and post-synaptic influences by adenosine, *J. Pharmacol. Exp. Ther.*, 244, 493, 1988.
170. **Breese, G. R., Baumeister, A., Napier, T. C., Frye, G. D., and Mueller, R. A.,** Evidence that D1 dopamine receptors contribute to the supersensitive behavioral responses induced by L-dihydroxyphenylalanine in rats treated neonatally with 6-hydroxydopamine, *J. Pharmacol. Exp. Ther.*, 235, 287, 1985.
171. **Breese, G. R., Napier, T. C., and Mueller, R. A.,** Dopamine agonist-induced locomotor activity in rats treated at differing ages: functional supersensitivity of D1 dopamine receptors in neonatally lesioned rats, *J. Pharmacol. Exp. Ther.*, 234, 447, 1985.
172. **Breese, G. R., Duncan, G. E., Napier, T. C., Bondy, S. C., Iomo, L. C., and Mueller, R. A.,** 6-Hydroxydopamine treatments enhance behavioral responses to intracerebral microinjection of D1 and D2 dopamine agonists into nucleus accumbens and striatum without changing dopamine antagonist binding, *J. Pharmacol. Exp. Ther.*, 240, 167, 1987.
173. **Breese, G. R., Hulebak, K. L., Napier, T. C., Baumeister, A., Frye, G., and Mueller, R. A.,** Enhanced muscimol-induced behavioral responses after 6-OHDA lesions: relevance to susceptibility for self-mutilation behavior in neonatally lesioned rats, *Psychopharmacology*, 91, 356, 1987.
174. **Breese, G. R., Mueller, R. A., Napier, T. C., and Duncan, G. E.,** Neurobiology of D1 dopamine receptors after neonatal 6-hydroxydopamine treatments: relevance to Lesch-Nyhan disease, *Adv. Exp. Med. Biol.*, 204, 197, 1986.
175. **Bannon, M. J., Lee, J.-M., Girand, P., Young, A., and Affolter, H.-U.,** Dopamine antagonist haloperidol decreases substance P, substance K and preprotachykinin mRNAs in rat striatonigral neurons, *J. Biol. Chem.*, 261, 6640, 1986.
176. **Bannon, M. J., Elliott, P. J., and Bunney, E. B.,** Striatal tachykinin biosynthesis: regulation of mRNA and peptide levels by dopamine agonists and antagonists, *Mol. Brain Res.*, 3, 31, 1987.

177. **Hanson, G., Alphs, L., Wolf, W., Pradhan, S., and Lovenberg, W.,** Response of striatonigral substance P systems to a dopamine receptor agonist and antagonist, *Neuropharmacology,* 20, 541, 1981.
178. **Hong, J.-S., Yoshikawa, K., Kanamatsu, T., and Sabol, S. L.,** Modulation of striatal enkephalinergic neurons by antipsychotic drugs, *Fed. Proc.,* 44, 2535, 1985.
179. **Li, S. J., Sivam, S. P., McGinty, J. F., Huang, Y. S., and Hong, J. S.,** Dopaminergic regulation of tachykinin metabolism in the striatonigral pathway, *J. Pharmacol. Exp. Ther.,* 243, 792, 1987.
180. **Sivam, S. P., Strunk, C., Smith, D. R., and Hong, J. S.,** Proenkephalin A gene regulation in the rat striatum: influence of lithium and haloperidol, *Mol. Pharmacol.,* 30, 186, 1986.
181. **Hagan, R. M., Jones, B. J., Jordan, C. C., and Ryers, M. B.,** Effect of 5-HT_3 receptor antagonists on responses to selective activation of mesolimbic dopaminergic pathways in the rat, *Br. J. Pharmacol.,* 99, 227, 1990.
182. **Hagan, R. M., Butler, A., Hill, J. M., Jordan, C. C., Ireland, S. J., and Tyers, M. B.,** Effect of the 5-HT_3 receptor antagonist, GR38032F, on responses to injection of a neurokinin agonist into the ventral tegmental area of the rat brain, *Eur. J. Pharmacol.,* 138, 303, 1987.
183. **Costall, B., Domeney, A. M., Naylor, R. J., and Tyers, M. B.,** Effects of the 5-HT_3 receptor antagonist, GR38032F, on raised dopaminergic activity in the mesolimbic system of the rat and marmoset brain, *Br. J. Pharmacol.,* 92, 881, 1987a.
184. **Costall, B., Domeney, A. M., Naylor, R. J., and Tyers, M. B.,** Inhibition by 5-HT_3 antagonists of hyperactivity caused by dopamine infusion into the rat accumbens, *Br. J. Pharmacol.,* 93, 194P, 1987b.
185. **Blandina, P., Goldfarb, J., and Green, J. P.,** Activation of 5-HT_3 receptor releases dopamine from rat striatal slice, *Eur. J. Pharmacol.,* 155, 349, 1988.
186. **Costall, B., Domeney, A. M., Naylor, R. J., and Tyers, M. B.,** Effects of the 5-HT_3 receptor antagonist, GR38032F, on raised dopaminergic activity in the mesolimbic system of the rat and marmoset brain, *Br. J. Pharmacol.,* 92, 881, 1987a.
187. **Costall, B., Domeney, A. M., Gerrard, P. A., Kelly, M. E., Naylor, R. J., and Tyers, M. B.,** Inhibition by ondansetron of behavioral effects in the rat which follow withdrawal from treatment with drugs of abuse, *Br. J. Pharmacol.,* 96, 340P, 1989.
188. **Eison, A. S., Eison, M. S., and Iversen, S. D.,** The behavioral effects of a novel substance P analogue following infusion into the ventral tegmental area or substantia nigra of the rat brain, *Brain Res.,* 238, 137, 1982b.
189. **Eison, A. S., Iversen, S. D., Sandberg, B. E. B., Watson, S. P., Hanley, M. R., and Iversen, L. L.,** Substance P analog, DiMe-C7: evidence for stability in rat brain and prolonged central actions, *Science,* 215, 188, 1982a.
190. **Elliott, P. J., Alpert, J. E., Bannon, M. J., and Iversen, S. D.,** Selective activation of mesolimbic and mesocortical dopamine metabolism in rat brain by infusion of a stable substance P analogue into the ventral tegmental area, *Brain Res.,* 363, 145, 1986.
191. **Elliott, P. J. and Iversen, S. D.,** Behavioural effects of tachykinins and related peptides, *Brain Res.,* 381, 68, 1986.
192. **Blandina, P., Goldfarb, J., Craddock-Royal, B., and Green, J. P.,** Release of endogenous dopamine by stimulation of 5-hydroxytryptamine$_3$ receptors in rat striatum, *J. Pharmacol. Exp. Ther.,* 251, 803, 1989.
193. **Jiang, L. H., Ashby, C. R., Jr., Kasser, R. J., and Wang, R. Y.,** The effect of intraventricular administration of the 5-HT_3 receptor agonist 2-methylserotonin on the release of dopamine in the nucleus accumbens: an *in vivo* chronocoulometric study, *Brain Res.,* 513, 156, 1990.
194. **Sorensen, S. M., Humphreys, T. M., and Palfreyman, M. G.,** Effect of acute and chronic MDL73,147EF, a 5-HT_3 receptor antagonist on A9 and A10 dopamine neurons, *Eur. J. Pharmacol.,* 163, 115, 1989.
195. **Nedergaard, S., Bolam, J. P., and Greenfield, S. A.,** Facilitation of a dendritic calcium conductance by 5-hydroxytryptamine in the substantia nigra, *Nature,* 333, 174, 1988.
196. **Kelland, M. D., Freeman, A. S., and Chiodo, L. A.,** Serotonergic afferent regulation of the basic physiology and pharmacological responsiveness of nigrostriatal dopamine neurons, *J. Pharmacol. Exp. Ther.,* 253, 803, 1990.
197. **Kalivas, P. W., Duffy, P., and Barrow, J.,** Regulation of the mesocorticolimbic dopamine system by glutamic acid receptor subtypes, *J. Pharmacol. Exp. Ther.,* 251, 378, 1989.
198. **Imperato, A., Honore, T., and Jensen, L. H.,** Dopamine release in the nucleus caudatus and in the nucleus accumbens is under glutamatergic control through non-NMDA receptors: a study in freely-moving rats, *Brain Res.,* 530, 223, 1990.
199. **Clineschmidt, B. V., Martin, G. E., and Bunting, P. R.,** Anticonvulsant activity of (+)-5-methyl-10,11,-dihydro-5H-dibenzo[a,d]cyclohepten-5,10-imine (MK-801), a substance with potent anticonvulsant, central sympathomimetic, and apparent anxiolytic properties, *Drug. Dev. Res.,* 2, 134, 1982.
200. **Martin, G. E.,** Catecholamine release within the striatum of the freely moving rat, *Ann. N.Y. Acad. Sci.,* 473, 151, 1986.

201. **Carlsson, M. and Carlsson, A.,** The NMDA antagonist MK-801 causes marked locomotor stimulation in monoamine-depleted mice, *J. Neural Transm.*, 75, 221, 1989.
202. **Kashihara, K., Hamamura, T., Okumura, K., and Otsuki, S.,** Effect of MK-801 on endogenous dopamine release in vivo, *Brain Res.*, 528, 80, 1990.
203. **Chapman, A. G., Durmuller, N., Lees, G. J., and Meldrum, B. S.,** Excitotoxicity of NMDA and kainic acid is modulated by nigrostriatal dopaminergic fibres, *Neurosci. Lett.*, 107, 256, 1989.
204. **Lindefors, N. and Ungerstedt, U.,** Bilateral regulation of glutamate tissue and extracellular levels in caudate-putamen by midbrain dopamine neurons, *Neurosci. Lett.*, 115, 248, 1990.
205. **Cheramy, A., Romo, R., Godehue, G., Baruch, P., and Glowinski, J.,** *In vivo* presynaptic control of dopamine release in the cat caudate nucleus. II. Facilitatory or inhibitory influence of L-glutamate, *Neuroscience,* 19, 1081, 1986.
206. **Divac, I., Fonnum, F., and Storm-Mathisen, J.,** High affinity uptake of glutamate in terminals of corticostriatal axons, *Nature,* 266, 377, 1977.
207. **Fonnum, F., Soreide, A., Kvale, I., Walker, J., and Wallaas, I.,** Glutamate in cortical fibers, in *Glutamate as a Neurotransmitter,* DiChiara, G. and Gessa, G. L., Eds., Raven Press, New York, 1981, 29.
208. **Fonnum, F.,** Glutamate: a neurotransmitter in mammalian brain, *J. Neurochem.*, 42, 1, 1984.
209. **McGeer, P. L., McGeer, E. G., Shiferer, U., and Singh, K.,** A glutamatergic corticostriatal path? *Brain Res.*, 128, 369, 1977.
210. **Roberts, P. J. and Anderson, S. D.,** Stimulatory effect of L-glutamate and related amino acids on ^{3}H-dopamine release from rat striatum: an *in vitro* model for glutamate actions, *J. Neurochem.*, 32, 1539, 1979.
211. **Roberts, P. J., McBean, G. J., Sharif, N. A., and Thomas, E. M.,** Striatal glutamatergic function: modifications following specific lesions, *Brain Res.*, 235, 83, 1982.
212. **Rowlands, G. J. and Roberts, P. J.,** Specific calcium-dependent release of endogenous glutamate from rat striatum is reduced by destruction of the cortio-striatal tract, *Exp. Brain Res.*, 30, 239, 1980.
213. **Del Rio, J., Naranjo, J. R., Yang, H.-Y. T., and Costa, E.,** Substance P-induced release of Met^5-enkephalin from striatal or periaqueductal gray slices, *Brain Res.*, 279, 121, 1983.
214. **Pickel, V. M., Sumal, K. K., Beckley, S. C., Miller, R. J., and Reis, D. J.,** Immunocytochemical localization of enkephalin in the neostriatum of rat brain: a light and electron microscopic study, *J. Comp. Neurol.*, 189, 721, 1980.
215. **Hong, J. S., Yang, H. Y. T., Fratta, W., and Costa, E.,** Chronic treatment with haloperidol accelerates the biosynthesis of enkephalins in rat brain, *Brain Res.*, 160, 192, 1979.
216. **Hong, J.-S., Yoshikawa, K., Kanamatsu, T., and Sabol, S. L.,** Modulation of striatal enkephalinergic neurons by antipsychotic drugs, *Fed. Proc.*, 44, 2535, 1985.
217. **Sivam, S. P., Breese, G. R., Krause, J. E., Napier, T. C., Mueller, R. A., and Hong, J.-S.,** Neonatal and adult 6-hydroxydopamine-induced lesions differentially alter tachykinin and enkephalin gene expression, *J. Neurochem.*, 49, 1623, 1987.
218. **Sivam, S. P., Strunk, C., Smith, D. R., and Hong, J. S.,** Proenkephalin A gene regulation in the rat striatum: influence of lithium and haloperidol, *Mol. Pharmacol.*, 30, 186, 1986.
219. **Tang, F., Costa, E., and Schwartz, J. P.,** Increase of proenkephalin mRNA and enkephalin content of rat striatum after daily injections of haloperidol for two to three weeks, *Proc. Natl. Acad. Sci. U.S.A.*, 80, 3841, 1983.
220. **Mitchell, V., Beauvillain, J.-C., Poulain, P., and Mazzuca, M.,** Catecholamine innervation of enkephalinergic neurons in guinea pig hypothalamus: demonstration by an *in vitro* autoradiographic technique combined with a post-embedding immunogold method, *J. Histochem. Cytochem.*, 36, 533, 1988.
221. **Kubota, Y., Inagaki, S., Kito, S., Takagi, H., and Smith, A. D.,** Ultrastructural evidence of dopaminergic input to enkephalinergic neurons in rat neostriatum, *Brain Res.*, 367, 374, 1986.
222. **Dourmap, N., Michael-Titus, A., and Costentin, J.,** Local enkephalins tonically modulate dopamine release in the striatum: a microdialysis study, *Brain Res.*, 524, 153, 1990.
223. **Mochida, S. and Libet, B.,** Postsynaptic long-term enhancement (LTE) by dopamine may be mediated by Ca^{2+} and calmodulin, *Brain Res.*, 513, 144, 1990.
224. **Lokhandwala, M. F. and Barrett, R. J.,** Cardiovascular dopamine receptors: physiological, pharmacological and therapeutic implications, *J. Auton. Pharmacol.*, 3, 189, 1982.
225. **Gorman, P.,** Vasodilator response to dopamine in the ferret pulmonary circulation, *Br. J. Pharmacol.*, 94, 212, 1988.
226. **Gorman, P.,** Inhibition by sulpride of dopamine induced pulmonary vasodilatation in the ferret, *J. Physiol. (Lond.),* 382, 56P, 1986.
227. **Wanstall, J. C. and O'Donnell, S. R.,** Vasodilator responses to dopamine in rat perfused mesentery are age-dependent, *Br. J. Pharmacol.*, 98, 302, 1989.

228. **Page, R. B., Gropper, M., Woodward, E., Townsend, J., Davis, S., and Bryan, R. M.,** Role of catecholamines in regulating ovine median eminence blood flow, *Am. J. Physiol.,* 258, 1242, 1990.
229. **Barger, G. and Ewins, A. J.,** *J. Chem. Soc. (Lond.),* 97, 2253, 1910.
230. **Holtz, P. and Credner, K.,** *Naunyn-Schmiedeberg's Arch. Exp. Pathol. Pharmakol.,* 200, 356, 1942.
231. **Brodde, O.-E.,** Vascular dopamine receptors: demonstration and characterization by *in vitro* studies, *Life Sci.,* 31, 289, 1982.
232. **Dinerstein, R. J., Vannice, J., Henderson, R. C., Roth, L. J., Goldberg, L. I., and Hoffman, P. C.,** Histofluorescence techniques provide evidence for dopamine-containing neuronal elements in canine kidney, *Science,* 205, 497, 1979.
233. **Von Essen, C.,** Effects of dopamine on the cerebral blood flow in the dog, *Acta Neurol. Scand.,* 50, 39, 1974.
234. **Toda, N.,** Alpha adrenergic receptor subtypes in human, monkey and dog cerebral arteries, *J. Pharmacol. Exp. Ther.,* 226, 861, 1983.
235. **Townsend, J. B., Ziedonis, D. M., Bryan, R. M., Brennan, R. W., and Page, R. B.,** Choroid plexus blood flow: evidence for dopaminergic influence, *Brain Res.,* 290, 165, 1984.
236. **Hall, J. G. and Morris, B.,** The origin of cells in the efferent lymph from a single lymph node, *J. Exp. Med.,* 121, 901, 1965a.
237. **Hay, J. B. and Hobbs, B. B.,** The flow of blood to lymph nodes and its relations to lymphocyte traffic and the immune response, *J. Exp. Med.,* 145, 31, 1977.
238. **Gowans, J. L.,** The effect of continuous reinfusion of lymph and lymphocytes on the output of lymphocytes from the thoracic duct of unanaesthetized rats, *Br. J. Exp. Pathol.,* 38, 67, 1957.
239. **Gowans, J. L.,** The recirculation of lymphocytes from blood to lymph in the rat, *J. Physiol. (Lond.)* 145, 54, 1959.
240. **Gowans, J. L. and McGregor, D. D.,** Immunological activities of lymphocytes, *Prog. Allergy,* 9, 1, 1965.
241. **Ford, W. L. and Simmonds, S. J.,** The tempo of lymphocyte circulation from blood to lymph in the rat, *Cell Tissue Kinet.,* 5, 175, 1972.
242. **Ford, W. L.,** Lymphocyte migration and immune responses, *Prog. Allergy,* 19, 1, 1975.
243. **Moore, T. C., Lippman, M., Spruck, C. H., and Gamal, R.,** Increased output of lymphocytes in lymph efferent from the lymph nodes of sheep during systemic arterial hypertension induced by phenylephrine or dopamine, *Immunopharmacology,* 14, 151, 1987.
244. **Moore, T. C.,** Mediator and physiological influences on lymphocyte traffic in sheep, in *Immunology of the Sheep,* Morris, B. and Miyasaka, Eds., Editiones Roche, Basel, 1985, 327–354.
245. **Moore, T. C.,** Neurovascular immunology: a 25 year odyssey, *Ann. N.Y. Acad. Sci.,* 650, 30, 1992.
246. **Lefur, G., Phan, T., and Uzan, A.,** Identification of sterospecific ^{3}H-spiroperidol binding sites in mammalian lymphocytes, *Life Sci.,* 26, 1139, 1980.
247. **Uzan, A., Phan, T., and LeFur, G.,** Selective labeling of murine B lymphocytes by ^{3}H-spiroperidol, *J. Pharm. Pharmacol.,* 33, 102, 1981.
248. **Shaskan, E. G., Ballow, M., Lederman, M., Margoles, S. L., and Melchreit, R.,** Spiroperidol binding sites on mouse lymphoid cells. Effects of ascorbic acid and psychotropic drugs, *J. Neuroimmunol.,* 6, 59, 1984.
249. **Ovadia, H. and Abramsky, O.,** Dopamine receptors on isolated membranes of rat lymphocytes, *J. Neurosci. Res.,* 18, 70, 1987.
250. **Devoino, L., Alperina, E., and Idova, G.,** Dopaminergic stimulation of the immune reaction: interaction of serotoninergic and dopaminergic systems in neuroimmunomodulation, *Int. J. Neurosci.,* 40, 271, 1988.
251. **Kasahara, K., Tanaka, S., Ito, T., and Hamashima, Y.,** Suppression of the primary immune response by chemical sympathectomy, *Res. Commun. Chem. Pathol. Pharmacol.,* 16, 687, 1977.
252. **Hall, N. R., McClure, J. E., Hu, S., Tave, N. S., Seals, C. M., and Goldstein, A. L.,** Effects of 6-hydroxydopamine upon primary and secondary immune responses, *Immunopharmacology,* 5, 39, 1982.
253. **Isakovic, K. and Jankovic, B. D.,** Neuroendocrine correlates of the immune response. II. Changes in the lymphatic organs of brain lesioned rats, *Int. Arch. Allergy,* 45, 373, 1973.
254. **Cross, R. J., Markesbery, W. R., Brooks, W. H., and Roszman, T. L.,** Hypothalamic-immune interactions. I. The acute effect of anterior hypothalamic lesions on the immune response, *Brain Res.,* 196, 79, 1980.
255. **Cross, R. J., Jackson, J. C., Brooks, W. H., Sparks, D. L., Markesbery, W. R., and Roszman, T. L.,** Neuroimmunomodulation: impairment of humoral immune responsiveness by 6-hydroxydopamine treatment, *Immunology,* 57, 145, 1986.
256. **Macris, N. J., Schiavi, R. C., Camerino, R. C., and Stein, M.,** Effect of hypothalamic lesions on immune processes in the guinea pig, *Am. J. Physiol.,* 219, 1205, 1970.
257. **Stein, M., Schiavi, R. C., and Camerino, M.,** Influence of brain and behavior on the immune system, *Science,* 191, 435, 1976.

258. **Cross, R. J., Markesbery, W. R., Brooks, W. H., and Roszman, T. L.,** Hypothalamic-immune interactions: neuromodulation of natural killer cell activity by lesioning of the anterior hypothalamus, *Immunology,* 51, 399, 1984.
259. **Cross, R. J., Brooks, W. H., and Roszman, T. L.,** Modulation of T-suppressor cell activity by central nervous system catecholamine depletion, *J. Neurosci. Res.,* 18, 75, 1987.
260. **Dasgupta, P. S. and Lahiri, T.,** Antitumor effect of i.p. dopamine in mice bearing Ehrlich ascites carcinoma, *J. Cancer Res. Clin. Oncol.,* 113, 363, 1987.
261. **Wick, M. M.,** Dopamine: novel antitumor agent active against B-16 melanoma in vivo, *J. Invest. Dermatol.,* 71, 163, 1978.
262. **Wick, M. M.,** Levodopa and dopamine analogs, melanin precursors as antitumor agents in experimental human and murine leukemia, *Cancer Treat. Rep.,* 63, 991, 1979.
263. **Wick, M. M.,** Levodopa and dopamine as therapy for human malignant melanoma, *Clin. Res.,* 28 (Abstr.), 254, 1980a.
264. **Wick, M. M.,** Levodopa and dopamine analogs, DNA polymerase inhibitors and antitumor agents in human and murine melanoma, *Cancer Res.,* 40, 1414, 1980b.
265. **Wick, M. M. and Mul, A.,** Synthesis and biologic evaluation of dopamine analog *N*-acetyldopamine in experimental leukemia in mice, *J. Natl. Cancer Inst.,* 66, 351, 1981.
266. **Dasgupta, P. and Lahiri, T.,** Dopamine inhibits Sarcoma 180 tumor growth *in vivo, Indian J. Cancer Chemother.,* 7, 27, 1985.
267. **Dasgupta, P. and Lahiri, T.,** Brain dopamine concentration and inhibition of Sarcoma 180 tumor growth following chronic restraint stress, *IRCS Med. Sci. Biochem.,* 14, 112, 1986.
268. **Lahiri, T., Banerjee, S., Dasgupta, P. S., and Ray, M. R.,** Tumor inhibition and hematological improvements by dopamine analog 3,4-dihydroxybenzylamine in mice bearing transplantable carcinoma, *Neoplasma,* 37, 387, 1990.
269. **Vogt, M.,** The concentration of sympathin in different parts of the central nervous system under normal conditions and after the administration of drugs, *J. Physiol. (Lond.),* 123, 451, 1954.
270. **Montagu, K. A.,** Adrenaline and noradrenaline concentrations in rat tissues, *Biochem. J.,* 63, 559, 1956.
271. **Carlsson, A., Lindqvist, M., Magnusson, I., and Waldeck, B.,** On the presence of 3-hydroxytyramine in brain, *Science,* 127, 471, 1958.
272. **Bertler, A. and Rosengren, E.,** Occurrence and distribution of catecholamines in brain, *Acta Physiol. Scand.,* 47, 350, 1959.
273. **Bloom, F. E.,** Neurohumoral transmission and the central nervous system, in *Goodman and Gillman's The Pharmacological Basis of Therapeutics,* Pergamon Press, New York, 1990, chap. 12.
274. **Selden, N. R. W., Robbins, T. W., and Everitt, B. J.,** Enhanced behavioral conditioning to context and impaired behavioral and neuroendocrine responses to conditioned stimuli following ceruleocortical noradrenergic lesions: support for an attentional hypothesis of central noradrenergic function, *J. Neurosci.,* 10, 531, 1990.
275. **Mason, S. T. and Iversen, S. D.,** Learning in the absence of forebrain noradrenaline, *Nature,* 258, 422, 1975.
276. **Roberts, D. C. S.,** An evaluation of the role of noradrenaline in learning, in *Theory in Psychopharmacology,* Vol. 1, Cooper, S. J., Ed., Academic Press, London, 1981, 123.
277. **Robbins, T. W. and Everitt, B. J.,** Functional studies of the central catecholamines, *Int. Rev. Neurobiol.,* 32, 303, 1982.
278. **Robbins, T. W., Everitt, B. J., Cole, B. J., Archer, T., and Mohammed, A.,** Functional hypotheses of the ceruleocortical noradrenergic projection: a review of recent experimentation and theory, *Physiol. Psychol.,* 13, 127, 1985.
279. **Cole, B. J. and Robbins, T. W.,** Dissociable effects of lesions to the dorsal or ventral noradrenergic bundle on the acquisition, performance, and extinction of aversive conditioning, *Behav. Neurosci.,* 101, 476, 1987.
280. **Henriksen, J. H. and Christensen, N. J.,** Plasma norepinephrine in humans: limitations in assessment of whole body norepinephrine kinetics and plasma clearance, *Am. J. Physiol.,* 257, 743, 1989.
281. **Liggett, S. B., Shah, S. D., and Cryer, P. E.,** Human tissue adrenergic receptors are not predictive of responses to epinephrine *in vivo, Am. J. Physiol.,* 256, 600, 1989.
282. **Lara, H. and Bastos-Ramos, W.,** Glutamate and kainate effects on noradrenergic neurons innervating the rat vas deferens, *J. Neurosci. Res.,* 19, 239, 1988.
283. **Leibowitz, S. F., Diaz, S., and Tempel, D.,** Norepinephrine in the paraventricular nucleus stimulates corticosterone release, *Brain Res.,* 496, 219, 1989.
284. **Feldman, S., Conforti, N., and Melamed, E.,** Involvement of ventral noradrenergic bundle in corticosterone secretion following neural stimuli, *Neuropharmacology,* 27, 129, 1988.

285. **Szafarczyk, A., Guillaume, V., Conte-Devolx, B., Alonso, G., Malaval, F., Pares-Herbute, N., Oliver, C., and Assenmacher, I.,** Central catecholaminergic system stimulates secretion of CRH at different sites, *Am. J. Physiol.*, 255, 463, 1988.
286. **Jhanwar-Uniyal, M., Renner, K. J., Bailo, M. T., Luine, V. N., and Leibowitz, S. F.,** Corticosterone-dependent alterations in utilization of catecholamines in discrete areas of rat brain, *Brain Res.*, 500, 247, 1989.
287. **Tsagarakis, S., Holly, J. M. P., Rees, L. H., Besser, G. M., and Grossman, A.,** Acetylcholine and norepinephrine stimulate the release of corticotropin-releasing factor-41 from the rat hypothalamus *in vitro*, *Endocrinology*, 123, 1962, 1988.
288. **Swanson, L. W., Sawchenko, P. E., Rivier, J., and Vale, W.,** Organisation of ovine corticotrophin-releasing factor immunoreactive cells and fibers in the rat brain: an immunocytochemical study, *Neuroendocrinology*, 36, 165, 1983.
289. **Kruseman, A. C. N., Linton, E. A., Ackland, J., Besser, G. M., and Lowry, P.,** Heterogeneous immunocytochemical reactivities of oCRF-41 like material in the human hypothalamus, pituitary and gastrointestinal tract, *Neuroendocrinology*, 38, 212, 1984.
290. **Suda, T., Yajima, F., Tomori, N., Sumitomo, T., Nakagani, Y., Ushiyama, T., Demura, H., and Shizume, K.,** Inhibitory effect of norepinephrine on immunoreactive corticotropin-releasing factor release from the rat hypothalamus *in vitro*, *Life Sci.*, 40, 1645, 1987.
291. **Wohlberg, C. J., Hackman, J. C., and Davidoff, R. A.,** Epinephrine and norepinephrine modulate neuronal responses to excitatory amino acids and agonists in frog spinal cord, *Synapse*, 1, 202, 1987.
292. **Elmslie, K. S. and Cohen, D. H.,** Iontophoresis of norepinephrine onto neurons of the pigeon's lateral geniculate nucleus: characterization of an inhibition response, *Brain Res.*, 517, 134, 1990.
293. **Weiss, G. K., Lewis, J., Jimenez-Rivera, C., Vigil, A., and Corcoran, M. E.,** Antikindling effects of locus coeruleus stimulation: mediation by ascending noradrenergic projections, *Exp. Neurol.*, 108, 136, 1990.
294. **Kokaia, M., Kalen, P., Bengzon, J., and Lindvall, O.,** Noradrenaline and 5-hydroxytryptamine release in the hippocampus during seizures induced by hippocampal kindling stimulation: an *in vivo* microdialysis study, *Neuroscience*, 32, 647, 1989.
295. **Mynlieff, M. and Dunwiddie, T. V.,** Noradrenergic depression of synaptic responses in hippocampus of rat: evidence for mediation by alpha$_1$-receptors, *Neuropharmacology*, 27, 391, 1988.
296. **Barry, D. I., Kikvadze, I., Brundin, P., Bolwig, T. G., Bjorklund, A., and Lindvall, O.,** Grafted noradrenergic neurons suppress seizure development in kindling-induced epilepsy, *Proc. Natl. Acad. Sci. U.S.A.*, 84, 7812, 1987.
297. **Aran, S. and Proudfit, H. K.,** Antinociception produced by interactions between intrathecally administered adenosine agonists and norepinephrine, *Brain Res.*, 513, 255, 1990.
298. **Millar, J. and Williams, G. V.,** Effects of iontophoresis or noradrenaline and stimulation of the periaqueductal gray on single-unit activity in the rat superficial dorsal horn, *J. Comp. Neurol.*, 287, 119, 1989.
299. **Duggan, A. W., Morton, C. R., Zhao, Z. Q., and Hendry, I. A.,** Noxious heating of skin releases immunoreactive substance P in the substantia gelatinosa of the cat: a study with antibody probes, *Brain Res.*, 403, 345, 1987.
300. **Kuraishi, Y., Hirota, N., Sato, Y., Kaneko, S., Satoh, M., and Takagi, H.,** Noradrenergic inhibition of the release of substance P from the primary afferents in the rabbit spinal dorsal horn, *Brain Res.*, 359, 177, 1986.
301. **Pang, I. H. and Vasko, M. R.,** Morphine and norepinephrine but not 5-hydroxytryptamine and gamma-amino butyric acid inhibit the potassium-stimulated release of substance P from rat spinal cord slices, *Brain Res.*, 6, 268, 1986.
302. **Hylden, J. L. K. and Wilcox, G. L.,** Pharmacological characterization of substance P-induced nociception in mice: modulation by opioid and noradrenergic agonists at the spinal level, *J. Pharmacol. Exp. Ther.*, 226, 398, 1983.
303. **Jones, S. L. and Gebbart, G. F.,** Quantitative characterization of coeruleospinal inhibition of nociceptive transmission in the rat, *J. Neurophysiol.*, 56, 1397, 1986.
304. **Seybold, V. S. and Elde, R. P.,** Receptor autoradiography in thoracic spinal cord: correlation of neurotransmitter binding sites with sympathoadrenal neurons, *J. Neurosci.*, 4, 2533, 1984.
305. **Belcher, G., Ryall, R. W., and Schaffner, R.,** The differential effects of 5-hydroxytryptamine, noradrenaline and raphe stimulation on nociceptive and non-nociceptive dorsal horn interneurons in the cat, *Brain Res.*, 151, 307, 1978.
306. **Fleetwood-Walker, S. M., Mitchell, R., Hope, P. J., Molony, V., and Iggo, A.,** An alpha-2 receptor mediates the selective inhibition by noradrenaline of nociceptive responses of identified dorsal horn neurons, *Brain Res.*, 334, 243, 1985.
307. **Headley, P. M., Duggan, A. W., and Griersmith, B. T.,** Selective reduction by noradrenaline and 5-hydroxytryptamine of nociceptive responses of cat dorsal horn neurons, *Brain Res.*, 145, 185, 1978.

308. **Howe, J. R. and Zieglgansberger, W.,** Responses to rat dorsal horn neurons to natural stimulation and to iontophoretically applied norepinephrine, *J. Comp. Neurol.*, 255, 1, 1987.
309. **Kuraishi, Y., Hirota, N., Sato, M., and Takagi, H.,** Antinociceptive effects of intrathecal opioids, noradrenaline and serotonin in rats: mechanical and thermal algesic tests, *Brain Res.*, 326, 168, 1985.
310. **Reddy, S. V., Maderdrut, J. L., and Yaksh, T. L.,** Spinal cord pharmacology of adrenergic agonist-mediated antinociception, *J. Pharmacol. Exp. Ther.*, 213, 525, 1980.
311. **Reddy, S. V. R. and Yaksh, T. L.,** Spinal noradrenergic terminal system mediates antinociception, *Brain Res.*, 189, 391, 1980.
312. **Proudfit, H. K. and Hammond, D. L.,** Alterations in nociceptive threshold and morphine-induced analgesia produced by intrathecally administered amine antagonists, *Brain Res.*, 218, 393, 1981.
313. **Sagen, J. and Proudfit, H. K.,** Effect of intrathecally administered noradrenergic antagonists on nociception in the rat, *Brain Res.*, 310, 295, 1984.
314. **Mayer, D. J., Wolfe, T. L., Akil, H., Carder, B., and Liebskind, J. C.,** Analgesia from electrical stimulation in the brainstem of the rat, *Science*, 174, 1351, 1971.
315. **Liebeskind, J. C., Guilbaud, G., Besson, J. M., and Oliveras, J. L.,** Analgesia from electrical stimulation of the periaqueductal grey matter in the cat: behavioural observations and inhibitory effects on spinal cord interneurones, *Brain Res.*, 50, 441, 1973.
316. **Richardson, D. E. and Akil, H.,** Pain reduction by electrical stimulation in man. I. Acute administration in periaqueductal and periventricular sites, *J. Neurosurg.*, 47, 178, 1977.
317. **Fardin, V., Oliveras, J. L., and Besson, J. M.,** A reinvestigation of the analgesic effects induced by stimulation of the periaqueductal gray matter in the rat. I. The production of behavioural side effects together with analgesia, *Brain Res.*, 306, 105, 1985.
318. **Wilcox, G. L., Carlsson, K.-H., Jochim, A., and Jurna, I.,** Mutual potentiation of antinociceptive effects of morphine and clonidine on motor and sensory responses in rat spinal cord, *Brain Res.*, 405, 84, 1987.
319. **Akaike, A., Shibata, M., Satoh, M., and Takagi, H.,** Analgesia induced by microinjection of morphine into and electrical stimulation of the nucleus reticularis paragigantocellularis of rat medulla oblongata, *Neuropharmacology*, 17, 775, 1978.
320. **Satoh, M., Akaike, A., Takahiro, N., and Takagi, H.,** Evidence for involvement of separate mechanisms in the production of analgesia by electrical stimulation of the nucleus reticularis paragigantocellularis and nucleus raphe magnus in the rat, *Brain Res.*, 194, 525, 1980.
321. **Proudfit, H. K. and Hammond, D. L.,** Alterations in nociceptive threshold and morphine-induced analgesia produced by intrathecally administered amine antagonists, *Brain Res.*, 218, 393, 1981.
322. **Koyama, K., Mito, T., Takashima, S., and Suzuki, S.,** Effects of phenylephrine and dopamine on cerebral blood flow, blood volume, and oxygenation in young rabbits, *J. Pediatr. Neurol.*, 6, 87, 1990.
323. **Randich, A. and Maixner, W.,** Interactions between cardiovascular and pain regulatory systems, *Neurosci. Biobehav. Rev.*, 8, 343, 1984.
324. **Dabney, J. M., Buehn, M. J., and Dobbins, D. E.,** Constriction of lymphatics by catecholamines, carotid occlusion, or hemorrhage, *Am. J. Physiol.*, 255, 514, 1988.
325. **Hellstrand, K. and Hermodsson, S.,** An immunopharmacological analysis of adrenaline-induced suppression of human natural killer cell cytotoxicity, *Arch. Allergy Appl. Immunol.*, 89, 334, 1989.
326. **Sved, A. F.,** PNMT-containing catecholaminergic neurons are not necessarily adrenergic, *Brain Res.*, 481, 113, 1989.
327. **Routledge, C. and Marsden, C. A.,** Comparison of the effects of selected drugs on the release of hypothalamic adrenaline and noradrenaline measured in vivo, *Brain Res.*, 426, 103, 1987.
328. **Mefford, I. N.,** Are there epinephrine neurons in the rat brain? *Brain Res. Rev.*, 12, 383, 1987.
329. **Holum, J. R.,** *Organic and Biological Chemistry*, John Wiley & Sons, New York, 1978, 110.

Chapter 7

MET-ENKEPHALIN

I. INTRODUCTION

Met-enkephalin is a five-amino-acid-residue vasoactive neurotransmitter peptide with stimulative actions in both the nervous and immune response systems. It was discovered in 1975 by Hughes, Kosterlitz, and associates,[1] at the end of a diligent and intensive search for endogenous agonists for long-known opiate receptors.

Three basic classes of endogenous substances identified in this successful search were the enkephalins, the dynorphins, and the endorphins. These three distinct families of endogenous peptides all derive from different large precursor molecules. In the central nervous system of vertebrates, these three families of peptides are almost exclusively found in different neurons, have different distributions, and may be presumed to play differing roles in the functions of these central nervous systems.[2]

Preproenkephalin is the 263 amino acid residue enkephalin precursor. By proteolytic cleavage it gives rise to one leu-enkephalin and six met-enkephalins, five-amino-acid-sequence molecules. Met-enkephalin precursors and receptors are widely distributed in the nervous, vascular, and immune resposne system. Like many of these potent vasoactive neurotransmitter substances of a stimulatory nature, met-enkephalin is rapidly metabolized and has a short half-life in blood and tissues.

At least four opoid-receptor subtypes have been identified by ligand binding and pharmacological studies: mu, kappa, delta, and epsilon. Beyond this, there is a great confusion and contradiction.[3] Problems of receptor solubilization, precipitation, cloning, and sequencing have impeded molecular characterization of the receptor types and subtypes. In addition, subtype specificity has been observed to vary between different organs and animal species. To complicate matters further, ligand specificity is only relative, and cross reactivity often occurs. Truly selective antagonists have been found to be even more rare than agonists. To add further confusion, selectivity also varies among species of animals.

One important area free of confusion and contradiction relates to the phylogenetic antiquity of met-enkephalin. In September 1990 Shipp, Stephano, and associates reported in *Nature* that CD10/NEP-related structures regulate enkephalin-mediated inflammatory responses in organisms whose ancestors diverged some 500 million years ago: man and the bivalve marine mollusc *Mytilus edulis*.

II. EVOLUTION

A. PLATYHELMINTHES/FLATWORMS

In 1983 Venturini et al.[4] employed radioimmunological and immunocytochemical techniques to demonstrate the presence of met-enkephalin in the planarian flatworm *Dugesia gonocephala*. The radioimmunoassay showed high concentrations of met-enkephalin. The immunocytochemical studies localized met-enkephalin to neurons and the neuropile. An earlier (1980) report by Zipser[5] had identified met-enkephalin in specific leech neurons.

B. ANNELIDA

In 1979 Alumets et al.[6] described the neuronal localization of immunoreactive met-enkephalin in annelid earthworms, *Lumbricus terrestris*. In this study, many neuronal cell bodies of different sizes, and nerve fibers displaying intense enkephalin immunoreactivity, were found in the cerebral ganglion. These cell bodies were located in the periphery of the cerebral ganglion, while the nerve fibers predominated in the central part of the ganglion. Numerous enkephalin-immunoreactive nerves extended from the cerebral ganglion into the circumpharyngeal connectives. In addition, enkephalin-immunoreactive neuronal cell bodies and a few immunoreactive nerves also were encountered in the subpharyngeal ganglion and in this abdominal ganglion of this annelid earthworm.

C. ARTHROPODA

Stephano and associates[7-9] have identified the presence of enkephalins in the nervous tissue and gut of the blattarian insect *Leucophaea maderae*. They also have observed a behavioral role for enkephalins in regulating locomotor activity. Application of enkephalin agonists and antagonists to the cerebral ganglia of *Leucophaea* have increased and decreased locomotor activity.

Duve et al.[10] have reported the presence of met-enkephalin immunoreactivity in the blowfly *Calliphora vomitoria*. Met-enkephalin immunoreactivity also has been found in the cerebral ganglion of the locust *Locust migratoria*[11] and in the midgut of the cockroach *Periplaneta*.[12]

An enkephalin-degrading activity in arthropod hemolymph was reported in 1985 by Coletti-Previero et al.[13] Their studies included the celicerata horseshoe crab *Limulus polyphemus* and the crustacean crayfish *Astacus fluviatilis*. At least three different enzymes were found to be involved in this inactivation: an aminopeptidase, a carboxypeptidase, and a peptidyldipeptidase. They considered these enzymes to be acting concomitantly.

D. MOLLUSCA

Met-enkephalin has been found in both the nervous system and the gut of the bivalve marine mollusc mussel *Mytilus edulis*. The Stephano group has identified met-enkephalin in the nervous system of this marine bivalve, including the pedal ganglion, and have suggested multiple binding sites.[14-17] Andersen et al.[18] detected met-enkephalin in both the digestive tract and nervous system of *Mytilus*, by immunohistological studies.

Schot et al.,[19] in 1981, employed immunocytochemical methodology to establish the presence of met-enkephalin in the nervous system of the pond snail *Lymnae stagnalis*.

Met-enkephalin has been shown to be present in the central nervous system of the opisthobranch gastropod *Aplysia kurodai* and the pulmonate gastropod *Achatina fulica*.[20] In this study, met-enkephalin-immunoreactive neurons were abundant in the cerebral ganglion of *Achatina*, and the authors speculated that met-enkephalin is likely to act as a neurotransmitter on neurons in the subesophageal ganglion of this gastropod mollusc. It was of interest that met-enkephalin and serotonin were found to coexist in a giant neuron of the right parietal ganglion of *Achatina*.

In 1982 Doble and Greenberg[21] observed that met-enkephalin relaxes the rectums of the clams *Mercenaria mercenaria* and *Macrocallista nimbosa*.

In 1981 Martin et al.[22] established the presence of enkephalin-like material in the vena cava neuropile of the cephalopod mollusc *Octopus vulgaris*.

E. NONMAMMALIAN VERTEBRATES

In 1984 Georges and Dubois[23] described the identification of met-enkephalin immunoreactivity in the nervous ganglion and the ovary of a protochordate, *Ciona intestinalis*. It also has been reported to be present in the brains of agnathan lampreys.[24]

In 1985 Nozaki and Gorbman[25] described immunoreactivity for both met-enkephalin and substance P in cells of the adenohypothesis of larvae and adult sea lampreys, *Petromyzon marinus*.

In 1981 Finger[26] reported the finding of met-enkephalin immunoreactivity in the gustatory lobes and visceral nuclei in the brains of bony fish, goldfish, and catfish. In 1982 and 1983

Levina and Gordon,[27-29] in a series of articles, presented evidence of an interesting and unexpected, and possibly defense-related, biological activity of met-enkephalin in the Teleost zebra fish, *Brachydanio rerio*. They demonstrated that met-enkephalin could induce the dispersion of pigment granules in the melanophores of this bony fish, as measured video-densitometrically. Their findings suggested a central nature of the met-enkephalin-induced dermal darkening.

Implications for the evolution of proenkephalins were investigated by Lindberg and White of New Orleans.[30] They selected three different orders of reptiles (turtles, alligators, and lizards) for study. These orders represented lineages that diverged more than 250 million years ago. Reptilian brains contained large quantities of met-enkephalin. Of the three reptile orders studied, turtles had the highest levels of brain met-enkephalin: two times the levels encountered in lizards. Alligator levels of brain met-enkephalin were intermediate. The levels of turtle brain met-enkephalin were approximately five times higher than those of rat brains. Turtle brains also were found to contain a higher ratio of met-enkephalin to met-enkephalin-Arg-Phe (MERF) than was identified in the brains of alligators and lizards. Reptilian brains, like avian brains, differ from mammalian brains in that they do not contain the met-enkephalin-Arg-Gly-Leu (MERGL) enkephalin sequence. This is of particular interest, as birds are considered to be descended from a reptilian ancestor not shared by mammals. The authors interpreted their findings as suggesting that the leu-enkephalin and MERGL sequences followed the evolutionary appearance of met-enkephalin and MERF and that met-enkephalin was the original duplicating unit.

In a 1986 publication, Wolters et al.,[31] of the Netherlands, described the distribution of met-enkephalin (and substance P) in the brainstem and spinal cord of the lizard *Varanus exanthematicus*. Highest concentration of met-enkephalin were encountered in cell bodies in the caudal hypothalamus and in the spinal cord in the superficial layers of the dorsal horn (lamina I) and around the central canal. The ventral horn of the spinal cord, including the motoneuron area, was found to receive only a comparatively sparse enkephalinergic innervation.

In 1987 Reiner,[32] of the University of Michigan, presented a detailed study of the distribution of proenkephalin-derived peptides in the central nervous system of two species of turtles, *Pseudemys scripta* (red-eared turtles) and *Chyremys picta* (painted turtles). Met-enkephalin was found to be present in considerably higher concentrations in the basal ganglia than in the overlying telencephalic regions. In the basal ganglia, met-enkephalin was identified in striatal neurons and fibers and in pallidal fibers. The hypothalamus was also found to be rich in enkephalinergic neurons and fibers, which also were present in the midbrain central gray. Met-enkephalin-containing neurons were abundant in the entry zones of the trigeminal nerve and in dorsal root fibers of the spinal cord. As in the Lindberg and White study (above), Reiner did not find MERGL enkephalin sequences in this order of reptiles. Reiner also observed, as did Lindberg and White, a greater overall similarity between the brains of reptiles and birds, with respect to their met-enkephalin content and distribution, than between the brains of reptiles and mammals. He considered the principal difference in brain enkephalinergic innervation between birds and reptiles, on one hand, and mammals, on the other, to relate largely to the larger size and somewhat different cytoarchitectural organization of the midbrain roof in birds and reptiles, as compared to that of mammals, and to an attendant hypertrophy of cell groups related to the midbrain roof in birds and reptiles, as contrasted with mammals. He concluded that the major features of enkephalin distribution in birds and mammals had already arisen in the reptilian ancestors of living birds, reptiles, and mammals.

Met-enkephalin immunoreactivity has been detected in the chicken spinal cord as early as embryonic stage 29–30 (incubation day 6).[33] In the embryo, as in the adult, met-enkephalin immunoreactivity of cell perikarya in the spinal cord was mainly concentrated in lamina I of the dorsal horn and in the nucleus of the dorsolateral funiculus throughout the spinal cord. Met-enkephalin (and substance P) also have been identified in preganglionic terminals of avian ciliary ganglia.[34] It is of further interest that met-enkephalin immunoreactivity has been encountered in vocal control regions of the zebra finch brain.[35]

III. NERVOUS SYSTEM

A. SPINAL CORD

Met-enkephalin-containing neuronal cells have been demonstrated in the dorsal horn and the intermediate gray and around the central canal of the mammalian spinal cord. The overlapping distributions of met-enkephalin-containing cells and long, ascending projection neurons have suggested that some of the met-enkephalin-containing neurons may be of the long-projection type. The 1983 observations of Pickel et al.[36] suggest that this may be so. They observed the accumulation of met-enkephalin in axons and cell bodies of neurons caudal to the sites of bulbar or spinal cord transection. In a major 1986 publication on this topic, Nahin and Micevych[37] employed a combination of retrograde transport of True Blue fluorescent dye and immunohistochemical localization of met-enkephalin in spinoreticular-tract neurons. Their findings provided the important evidence that a population of these long-ascending spinoreticular-tract neurons contain met-enkephalin.

Another major spinal cord and met-enkephalin study appeared in 1989. Cesselin et al.[38] observed that noxious thermal stimuli in rats would stimulate the segmental release of met-enkephalin from the spinal cords of study animals. Their investigation involved changes induced by noxious thermal stimulation on the activity of met-enkephalin-containing neurons at various levels of the spinal cord. The cord levels studied were the whole cord, the lumbar area, and the cervicotrigeminal area. By immersing the muzzle of intact rats in water heated to 52°C, they were able to demonstrate marked increases in the discharge of met-enkephalin into artificial cerebrospinal perfusates from the whole cord and from the cervicotrigeminal area, but not from the lumbar spinal cord. Heating the tail by immersion in water heated to 52°C stimulated an increase in met-enkephalin release into the perfusate from the whole spinal cord of intact as well as cervically transected rats. The noxious thermal stimulating technique used in these studies consisted of immersing depilated (hairless) muzzel or tail in water at 52°C for 10 sec for three times per minute for 30 min. These important findings demonstrated that noxious thermal stimulation may induce a segmental release of met-enkephalin and that this release occurs only within those spinal cord zones that receive the noxious-stimulation input. These extremely important observations bring met-enkephalin, a stimulatory molecule, very much into the "heat, pain, and memory" scenario as a major "memory molecule" candidate.

B. BRAIN STEM

Evidence for a major role for met-enkephalin and the enkephalins as neurotransmitters in the mammalian brain stem was presented in a particularly significant and important report in *Nature* in 1978 by the Iversens and co-workers.[39] They reported that enkephalin may be released from rat brain globus pallidus slices *in vitro* in response to depolarization caused by elevation of external potassium ion concentrations, and that this evoked release is completely dependent on the presence of calcium ions in the external medium. This response was considered to be characteristic of a number of other neurotransmitters known to be released as neurotransmitters in the central nervous system. The authors, furthermore, considered these findings to strongly support a neurotransmitter role for enkephalin in the mammalian brain. The globus pallidus of the brain stem is an important anatomical area for the enkephalingergic nervous system. Globus pallidus enkephalin concentrations are quite high, some six to eight times higher than observed in other areas of the rat brain,[40,41] and with especially high concentrations in nerve terminals.

Another met-enkephalin landmark publication appeared in 1983. In this report, Haber and Nauta[42] redefined the borders of the globus pallidus and brought into close anatomical proximation two major ''memory molecules'' (substance P and met-enkephalin), and in an important anatomical area. Their redefinition employed the aid of the unique pattern of enkephalin and substance P immunoreactivity encountered in pallidum of both the monkey and the rat. Their study employed the PAP (peroxidase-antiperoxidase) histochemical method. The unique pattern identified by them appeared in the form of ribbon-like fibers (termed ''woolly fibers'' by the authors) which are enkephalin or substance P immunoreactive and that enmesh the dendrites of the large pallidal neurons. These enmeshing woolly fibers constitute a plexus of thin enkephalin- or substance P-immunoreactive striatopallidal fibers. A dense enkephalin-immunoreactive woolly fiber plexus was found to fill the entire external pallidal segment (termed ''dorsal pallidum'' by the authors) and extended from there in various directions, generally ventral. The most massive rostral extension of these enkephalin-immunoreactive fibers defined the subcommissural or ''ventral pallidum''. From this anatomical site, the enkephalin-immunoreactive fibers expanded ventralward into the olfactory tubercle. This finding supported the suggestion that many of the large cells of the tubercle are pallidal neurons. Additional (and important) extensions from the enkephalin-immunoreactive ''dorsal pallidum'' plexus were found to invade the ventral strial region (including the nucleus accumbens), a dorsal region of the amygdala, and the bed nucleus of the striaterminalis — all important ''limbic'' areas of major importance in memory and learning.

Substance P-immunoreactive woolly fibers in this Haber and Nauta[42] study, like their enkephalin-immunoreactive counterparts, filled the ''dorsal pallidum'' and invaded the olfactory tubercle. A dense substance P-immunoreactive woolly fiber plexus also was observed to fill the internal pallidal segment (entopeduncular nucleus), with medialward expansion into the lateral hypothalamus. A loose plexus of enkephalin-immunoreactive woolly fibers also was involved in entopeduncular nucleus invasion of the hypothalamus. In Chapter 10 the role of substance P as another major ''memory molecule'' will be developed more fully.

These extremely important observations suggested to the authors that woolly fibers extending outward beyond the conventionally recognized borders of the pallidum represent pallidal elements that are innervated by enkephalin- or substance P-immunoreactive fibers that arise from the ventromedial striatal regions which, in turn, are innervated by limbic structures. These findings also bring these two critical ''memory molecules'' of the nervous and immune response systems into anatomical contact and involvement with the memory-important limbic and hippocampal systems.

The ventromedial region of the rat globus pallidus includes most cholinergic neurons that, from the basal nucleus of Meynert, project in turn to the cerebral cortex.[43,44] The forebrain regions of the rat also contain cholinergic neurons that are characterized by the presence of nerve fibers that are immunoreactive for substance P.[45,46] A dense network of enkephalin-immunoreactive fibers from the ventromedial region of the globus pallidus and from the subpallidal substantia innominata also provides a direct input to cholinergic neurons in the forebrain region of the cerebral cortex.[45,47]

In a 1990 report of major significance, Arluison et al.,[48] of Paris, employed a combination of the retrograde transport of gold-labeled wheat germ agglutinin with immunoperoxidase methodology (G-WGA-HRP) to determine parent cell body origin of met-enkephalin-immunoreactive nerve fibers that innervate the ventral and/or medial regions of the rat globus pallidus. On the injected side, nerve cell bodies that were found to be labeled retrogradely or which were double labeled were numerous in the central nucleus of the amygdala, scattered in the bed nucleus striaterminalis, and scarce in the fundus striati. In the central nucleus of the amygdala, 40% of the retrogradely labeled cell bodies were immunoreactive for met-enkephalin, while in the basal nucleus striaterminals 30% of the cell bodies were so labeled. From these observations, the authors concluded the met-enkephalinergic afferents of the ventromedial region of the globus pallidus, which contains the basal nucleus of Meynert in the rat, are largely of limbic origin. The authors considered that their findings supported the important concept that the ventral region of the rat globus pallidus is located at the crossroad of the striatopallidal system (the extended amygdala and the magnocellular corticopetal cell complex). They concluded with the critical observation that ''this unique situation allows the integration of informations from very diverse sources in order to regulate the most elaborate behaviors'' — including (the author of this volume suggests) a major role at the molecular level of the ''memory molecules'', substance P and met-enkephalin, in the establishment of memory (and learning) in both nervous and immune response systems.

In 1985 Frey and Hoffman,[49] of San Antonio, TX, reported that microiontophoresis of met-enkephalin and morphine onto single globus pallidus neurons produced opposite effects — *excitation* by met-enkephalin and *inhibition* for morphine — on spontaneous and/or glutamate-evoked activity of single neurons in locally anesthetized and paralyzed rats.

In 1986 Kalivas and Richardson-Carlson[50] reported that endogenous enkephalin in the rat ventral tegmental area (VTA) can increase the activity of A10 dopamine neurons. They concluded that these observations supported a physiological role for enkephalins in mesolimbic and mesocortical dopamine-mediated behaviors.

An additional met-enkephalin-dopamine ''connection'' of importance was reported from France by Normand et al.[51] In an *in situ* hybridization study, they observed that dopaminergic striatonigral neurons tonically inhibit preproenkephalin biosynthesis in the ipsolateral striatum and influence preproenkephalin gene expression in a subpopulation of enkephalin neurons in the contralateral striatum. They interpreted these results as indicating that permanent alteration of an endogenous neurotransmitter system (as by 6-hydroxydopamine-induced destruction of all dopaminergic neurons in the right substantia nigra, in this study) can affect gene transcription for neuropeptides in its target neurons.

Another important ''memory molecule'', serotonin, was brought into the picture in a 1988 publication by Pretel and Ruda[52] of Rochester, NY. They explored the anatomical basis for interactions between serotonin-immunoreactive neurons in nuclei raphe magnus and pallidus and substance P- and met-enkephalin-immunoreactive axonal varicosities, by means of a double-labeled immunocytochemical method. They found that both met-enkephalin and substance P could be localized to axons that contacted serotonin-immunoreactive neurons. The frequency and the location of these inputs differed in nuclei raphe magnus and pallidus. The greatest number of contacts with serotonin-immunoreactive neurons was by substance P-immunoreactive varicosities.

Also in the memory area and on the clinical side, Parkinson's disease in man, which is known to have a dementia component, has been described (by Taquet et al.,[53] of Paris) to affect met-enkephalin and leu-enkephalin levels of the human brain differently. In Parkinsonian brains a significant reduction in the levels of both met-enkephalin and leu-enkephalin were encountered in both the globus pallidus and the putamen. Met-enkephalin only was found to be decreased in the substantia nigra and VTA. These findings also suggested that met-enkephalin and leu-enkephalin may be located in different neuronal populations in the human brain — a very important observation with respect to met-enkephalin as a stimulatory molecule in both the central nervous system and in the immune response system.

C. HIPPOCAMPUS AND SEIZURE KINDLING

In 1977 Nicoll et al.[54] described neuronal excitatory effects produced by the microiontophoretic application of met-enkephalin to the rat hippocampus. In the hippocampus all responsive cells were excited by met-enkephalin. These results suggested to the authors that met-enkephalin may belong to a new class of central neurotransmitters (excitatory/stimulatory).

In 1980 Lee et al.[55] presented evidence to suggest that the met-enkephalin stimulatory and excitatory effects in the hippocampus result from a depression in the firing of inhibitory neurons that produce enkephalin-induced excitations of hippocampal pyramidal cells by disinhibition. A similar response had been recorded earlier (1979) by Zieglgansberger et al.[56] A 1982 publication by French and Zieglgansberger[57] suggests that the excitatory effect of met-enkephalin on hippocampal cells was exerted through a receptor population that differs from morphine-receptor effects — again pulling met-enkephalin out of the morphine/inhibitory "trap".

In 1988 Madison and Nicoll[58] described met-enkephalin-induced inhibition (hyperpolarization) of interneurons in the hippocampus through receptors that were coupled to potassium channels. These authors commented on an earlier (1986) report by their group[59] in which $GABA_B$ receptors (LTP inducing; see Chapter 2) on pyramidal cells are coupled to the same potassium channels as serotonin receptors. They speculated that it would be interesting to determine if the $GABA_B$ receptors and the met-enkephalin receptors on interneurons converge into the same class of potassium channel. They made the important point that there is an economy in the selective localization of met-enkephalin receptors on inhibitory interneurons in that, while pyramidal cells in the hippocampus greatly outnumber interneurons, some of these interneurons may make synaptic contact with as many as 200 pyramidal cells. Accordingly, activation of the relatively few met-enkephalin receptors in the hippocampus on interneurons may have an unusually widespread and powerful influence on the stimulation of hippocampal pyramidal cell excitability and resulting and related responses, *including memory* (this author's speculation).

In 1987 Haigler et al.[60] observed that microiontophoretic application of met-enkephalin to rat hippocampal cells in the CA3 and CA4 areas produced a greater increase in firing than it did when applied to cells in the CA1 area. This observation is of particular importance with respect to the role of the CA3 area of the hippocampus as the "pacemaker" of the hippocampus.

Also in 1987 Weisner and Henriksen[61] described the effects of *in vivo* electrophoretic application of a met-enkephalin analogue to a dentate gyrus of the hippocampus. This application had a stimulative and excitatory effect on evoked responsiveness to stimulation of the perforant-path input, as produced by stimulation of the afferent perforant path and as monitored in increased amplitude of population spikes evoked by this stimulation of the perforant path. Population spikes were increased in amplitude at all three stimulus levels employed. These findings confirmed that exogenously applied met-enkephalin or an analogue can enhance the excitatory response of the dentate gyrus to perforant-path stimulation. These observations also have important implications relating to LTP and potential memory-type responses at the molecular level.

In 1988 Dunwiddie and Su[62] observed that evoked-population-spike increases in the CA1 region of the hippocampus produced by met-enkephalin analogues given in association with activation by Schaffer collateral and commissural afferents was blocked by pretreatment with pertussis toxin. These findings suggest that the excitatory responses in the hippocampus to met-enkephalin and its analogues are the result of interactions with receptors that are coupled by means of pertussis-toxin-sensitive GTP-binding proteins (G proteins).

Electrical and chemical "kindling" of seizures in seizure-prone areas of the mammalian brain, such as the hippocampus and the amygdala, has been used extensively in the study of the mechanisms of epileptic seizures of various types. In 1986 Kanamatsu et al.[63] reported that chemical kindling with kainic-acid-induced seizures in the rat caused an increase in met-enkephalin release, which was followed by a compensatory increase in met-enkephalin biosynthesis in both the hippocampus and the entorhinal cortex.

Amygdaloid kindling, which has been considered to represent an experimental model of temporal lobe epilepsy, was described (in 1986 by McGinty et al.[64]) to produce an increase in met-enkephalin immunoreactivity in the rat hippocampus and a decrease in dynorphin A immunoreactivity in this area. The met-enkephalin increase was by 43%, and the dynorphin A decrease was by 71%. This also suggests a divergence in met-enkephalin and dynorphin responses.

In a 1988 publication from Irvine, CA, Gall[65] described the influence of recurrent limbic seizure activity on mouse hippocampal mossy fiber axonal-system neuropeptide levels. Seizures were induced by placement of a small unilateral electrolytic lesion in the dentate gyrus hilus or by intraventricular infusion of kainic acid. Both types of seizure induction produced marked elevations in hippocampal mossy fiber met-enkephalin levels and reductions in dynorphin levels, in these fibers. In a 1990 study of the role of neuropeptides in the electrical kindling model of epileptogenesis, Wanscher et al.,[66] of Denmark, found that hippocampal kindling also produced marked elevations in met-enkephalin immunoreactivity in rat hippocampal mossy fiber terminals. Electrical stimulation of rat hippocampal granule cells *in vivo* has been found (by *in situ* hybridization histochemistry) to increase the expression of proenkephalin mRNA and to decrease markedly the levels of prodynorphin mRNA, on the stimulated sides. These findings provided additional evidence that distinct and different mechanisms regulate the expression of proenkephalin mRNA and prodynorphin mRNA in the rat hippocampus and, again, separate stimulatory and excitatory enkephalin responses from those of dynorphin.

In an important 1989 publication, Tanaka et al.,[67] of Japan, studied chemical kindling with met-enkephalin and identified a potent stimulatory/excitatory and epileptogenic effect of met-enkephalin on the rat brain. They described the development of generalized convulsions after repeated met-enkephalin microinjections into the right amygdala of rats and the development of the transfer phenomenon between chemical kindling with met-enkephalin and amygdaloid electrical kindling. In this study, 1 week after the completion of met-enkephalin kindling, one or two electrical stimulations of the right amygdala of met-enkephalin-kindled rats resulted in a generalized convulsion. The mean number of electrical stimulations needed to produce generalized convulsions in the test group after completion of met-enkephalin kindling was significantly less than the number of electrical stimulations required in the saline control group. The duration of after-discharge in the first generalized convulsion induced by electrical amygdaloid stimulation in met-enkephalin-kindled rats was significantly longer than that in the first generalized convulsion induced by electrical stimulation of the amygdala in the saline control group of rats (? LTP). A prior report by this same Tanaka group[68] demonstrated that the concentration of met-enkephalin-binding sites increased in rat amygdala one week after the completion of amygdala electrical kindling. As kindling also involves new protein synthesis, these met-enkephalin-induced kindling findings would appear to place met-enkephalin in the growth factor category as well as in the ''memory molecule'' category.

D. OTHER MOLECULES

1. Substance P

Substance P appears to be the most important ''other'' molecule in the met-enkephalin molecular interrelationship scheme in addition to serotonin (see ''Nervous System — Brain Stem'' section of this chapter). There is extensive overlapping of the distribution of these two (met-enkephalin and substance P) very important ''memory molecules'', as well as colocalization in single neurons.

In a detailed 1986 report, Penny et al.[69] presented evidence of a substantial population overlap of met-enkephalin and substance P neurons in the neostriatum of the cat and the rat. Met-enkephalin immunoreactivity was encountered in 41% of the neostriatal neurons of the cat. Substance P immunoreactivity was detected in 38% of rat neostriatal neurons and in 39% of those of the cat. This study also provided documentation of the coexistence of met-enkephalin and substance P within single neurons of the mammalian central nervous system — an important observation for synergistic, rather than antagonistic, activities of these important stimulatory and excitatory sensory-impulse related ''memory molecules''.

In a 1987 publication[70] extensive colocalization of met-enkephalin and substance P also was found in the central nervous system of the primate Old World Rhesus monkey *Macaca mulatta*. A 1983 report[71] identifies colocalization of met-enkephalin with substance P in the nucleus tractus solitarii of the cat, and it was suggested by the authors that this association might involve a cooperative involvement in the regulation of the autonomic nervous system.

A 1982 report from Yale University, by de Lanerolle and LaMotte,[72] provides additional important connectional information related to met-enkephalin and substance P in the human spinal cord. They studied the distribution of substance P and met-enkephalin immunoreactivity in the human sacral, lumbar, and thoracic spinal cord, by an indirect peroxidase-antiperoxidase (PAP) method. The laminal distributions of met-enkephalin and substance P immunoreactivity were similar, with immunoreactive axons and terminal-like processes identified in the substantia gelatinosa and in the marginal zone. The distribution of substance P fibers and terminal profiles in the human spinal cord was found to be similar to those encountered in other mammals. With respect to met-enkephalin distribution in man in comparison with other mammals, some differences were encountered, particularly with respect to the pain reception areas of laminae I, II, and III.

In 1987 Tashiro et al.,[73] from Kyoto University in Japan, described the coexistence of met-enkephalin and substance P immunoreactivities in axonal components in lumbar segments of the cat spinal cord. Axonal components double-labeled with these two important memory-related vasoactive neuropeptides were widely distributed throughout the spinal gray. A 1988 report, also from Japan,[74] focuses on colocalization of met-enkephalin and substance P in the rat lumbar spinal cord. Colocalization was found to be most prominent within the substantia gelatinosa in the superficial layers of lamina I of the dorsal horn. This is an area of particular significance and importance for sensory reception of impulses of thermal pain. In 1987 Duggan et al.,[75] of Canberra, Australia, established by an antibody probe methodology that tissue-destructive thermal heat (50 to 52°C) in the periphery of cats would produce the release of substance P from axon terminals in the dorsal horn substantia gelatinosa.

The 1988 report of Senba et al.[74] referred to above also confirmed that 95% of the substance P-immunoreactive neurons in the substantia gelatinosa were also immunoreactive for met-enkephalin. Also in this area, 45% of the met-enkephalin-immunoreactive neuronal cells were immunoreactive for substance P. This remarkable colocalization of these two major ''memory molecules'' in this critical thermal-heat reception area in the central nervous system has immense implications for their cooperative and collaborative involvement at the molecular level in the laying down of sensory-input-triggered memory and learning and for the ''heat, pain, and memory'' scenario — the Prometheus Thesis of Moore.

A 1983 report from the National Institute of Mental Health in Washington, D.C., by Tana et al.,[76] described the release of met-enkephalin from the rat spinal cord, produced by the injection of substance P into fluid perfusing the spinal cord and subarachnoid spaces, both *in vivo* and *in vitro*. The authors interpreted these observations as indicating that central nervous system receptors for substance P in this area of the rat spinal cord were involved in the triggering of the release of met-enkephalin.

A 1985 publication from the Netherlands, by Huitinga et al.,[77] is of particular importance with respect to the enkephalin and substance P "connection" in neuronal-pathway projections. Their study involved a combination of fluorescent retrograde tracing and immunohistochemical staining for substance P and enkephalin to trace, in the rat, the projections from the interpeduncular nucleus (a distinct, but relatively small, nucleus in the ventral mesencephalon) to the dorsal tegmental area (DTA), which includes the caudal extension of the serotonin-involved dorsal raphe nucleus. Their findings established the existence of both enkephalin and substance P-containing projections from this important interpeduncular nucleus to the dorsal tegmental area. This is an especially important observation because, from the connectional point of view, the interpeduncular nucleus serves as an integration center through which limbic forebrain structures, by way of the habenula, can exert their influences on the midbrain serotoninergic raphe nuclei and on the dorsal tegmental area or region. In the dorsal tegmental area, interpeduncular neuronal efferents are largely directed to the caudal part of the serotonin-related dorsal raphe nucleus and to the dorsolateral tegmental nucleus. These two important and highly significant nuclei give rise, respectively, to serotoninergic and cholinergic ascending pathways that project to and reach widespread areas of the forebrain. What could be more important with respect to the interrelationships of these two extremely important and critical "memory molecules"?

Many terminal-like structures of met-enkephalin and substance P-immunoreactive terminals were observed to be in close apposition to neurons in several laminae. By correlation of three-dimensional organization of neurons, their neuronal cell types, in apposition to immunoreactive met-enkephalin and substance P-immunoreactive processes, could be tentatively identified. It was of considerable interest, and likely significance, that frequently, and in many of the laminae, the same neuron was observed to be surrounded by both met-enkephalin- and substance P-immunoreactive processes. From the large variety of neuronal cell types found to be supplied by met-enkephalin- and substance P-immunoreactive processes in this study and from the wide distribution of these two important vasoactive neuropeptides in the spinal cord, the authors speculated that met-enkephalin and substance P might be involved in and play important and significant roles in a number of critical spinal cord and related central and peripheral nervous system functions, including spinothalamic, autonomic, and motor.

2. Serotonin

In 1986, Leger et al.,[78] from Lyon, France, described the distribution of enkephalin-immunoreactive cell bodies in relation to serotonin-containing neurons in the raphe nuclei of the cat. These raphe nuclei are made up of neuronal cell groups that are located in the midline from the caudal medulla oblongata to the midlevel of the mesencephalon. From a structural and functional standpoint, they belong to the reticular formation of the brain stem. An important feature of these raphe nuclei in the cat is that they contain a majority of the central nervous system 5-HT neuronal cell bodies: 77% as established by histochemical studies.[79] In all raphe nuclei the 5-HT cell bodies are intermingled with non-5-HT cells and processes. In their study, Leger et al.[78] examined five of the nine raphe nuclei and found that all the raphe nuclei examined contained met-enkephalin-immunoreactive cell bodies along their whole caudorostral extent. By comparing consecutive sections treated separately with anti-5HT and antienkephalin antiserum, it was shown that certain 5-HT cells in each raphe nucleus contained enkephalin immunoreactivity as well. In a quantitative examination, the authors demonstrated that in the nucleus raphe pallidus (NRP) and the nucleus raphe obscurus (NRO), approximately one half of the total immunoreactive neuronal population was immunoreactive for both 5-HT and met-enkephalin. In the nucleus raphe magnus (NRM) approximately one third of the total immunoreactive population was immunoreactive to both

5-HT and met-enkephalin, while in the raphe nucleus pontis and dorsalis (NRD) the proportion of immunoreactivity involving both molecules was almost negligible. Among the 5-HT-immunoreactive cells, approximately two thirds were immunoreactive for met-enkephalin in both NRP and NRO, whereas in the NRM, one half of the 5-HT cells were immunoreactive for met-enkephalin. These findings clearly indicate an important brain stem and above (rostral) "connection" for serotonin with met-enkephalin in the mammalian central nervous system and pull the "memory molecule" network even tighter.

Passing further down the nervous system and into the periphery, Kanagawa et al.,[80] also in 1987, employed immunohistochemical staining of alternate consecutive sections in the guinea pig superior cervical ganglion to demonstrate the presence of numerous enkephalin-immunoreactive small cells, with intense fluorescence, which also were immunoreactive for 5-HT.

3. GABA

Aronin et al.,[81] in 1986, explored the interrelationship between met-enkephalin and GABA, with respect to their immunoreactive axon terminals in the rat neostriatum and possible synaptic connection with striatonigral neurons. Numerous GABA- and met-enkephalin-immunoreactive boutons were found to form synapses with the cell bodies and dendrites of medium-sized striatonigral neurons. The authors concluded that, within the neostriatum, GABA and met-enkephalin may directly influence caudate output pathways. From the recent demonstrations of $GABA_B$-receptor involvement in long-term potentiation, we find yet another memory molecule connection for met-enkephalin.

4. Dopamine

In 1984 Pasinetti et al.,[82] of Milan, Italy, reported that dopamine enhances met-enkephalin efflux from rat striatal slices. A 1985 publication by Kalivas and Bronson[83] demonstrated that mesolimbic dopamine lesions produce an augmented behavioral response to enkephalin. In a prior section of this chapter, dopamine lesioning, was shown to exert a tonic influence on met-enkephalin gene expression and synthesis.

5. Transmembrane Signaling Molecules

In a 1976 report in *Nature*, Minneman and Iversen[84] reported enkephalin-induced increases of cyclic GMP accumulation in slices of rat neostriatum. Also in 1976, Brand et al.[85] described a met-enkephalin-induced increase in cyclic GMP in neuroblastoma glioma hybrid cells (108 Ce15), which have been shown and known to display many properties that are characteristic of neurons.

With respect to the PI (phosphatidyl inositol) transmembrane signaling pathway and met-enkephalin, it was reported in 1988, by Kley[86] of Germany, that phorbol ester induced a time-dependent increase in cellular proenkephalin mRNA levels and that this increase was further potentiated by low concentrations of the calcium ionophore A23187. In 1986 Ueda et al.,[87] of Kyoto, presented evidence that a met-enkephalin releaser, kyotorphin, enhanced intracellular Ca^{++} in synaptosomes.

IV. VASCULAR SYSTEM

In a 1985 study of cerebral blood flow in miniature swine by the radiolabeled microsphere method, Blum et al.,[88] from the University of Texas (San Antonio), gave central-venous infusions of met-enkephalin into a cannulated jugular vein. Their results revealed that peripheral circulation central intravenous infusion of met-enkephalin into miniature swine significantly increased cerebral blood flow to and into the basal ganglia, the cerebellum, the pons, the inferior and superior parietal cortex, and the frontal cortex. Increases, although not reaching statistical significance, also were encountered in the hippocampus, occipital cortex, and medulla oblongata. Blood flow to the pituitary gland was not altered. It was of interest that those anatomical swine brain areas most responsive to met-enkephalin infusion, with an increase significantly in cerebral blood flow, also were those anatomical brain areas that contain the greatest concentrations of potential met-enkephalin receptors.

In the peripheral circulation, intravascular infusion of met-enkephalin also has been found, in isolated cat hind-limb preparations, to lower perfusion pressures and to cause vasodilatation. In this 1982 study by Moore and Dowling,[89] met-enkephalin infusions were into the femoral artery of the study extremity, with accompanying decreases in arterial pressure. The study animals (cats) in this series of experiments were under phenobarbital general anesthesia.

Intravascular infusion of met-enkephalin also has decreased arterial blood pressure in anesthetized rabbits[90] and dogs.[91] Wightman et al.,[92] in 1987, described decreased vascular resistance in the hindquarters of conscious rabbits after intraarterial infusion of met-enkephalin. A new method permitted simultaneous measuring of blood pressure and blood flow to the study vascular bed. Resistance decreases also were observed in the mesenteric vascular bed in this study. These mesenteric vascular bed changes were coincident with those in the hindquarters. In the anesthetized rabbit study by Rhee et al.,[93] met-enkephalin was shown to reduce the rate of sympathetic nerve discharge, in parallel with the reduction in blood pressure.

Employing a radiolabeled microsphere methodology in rabbits, Eulie et al.,[94] in 1987, investigated the intravascular met-enkephalin-induced decreases in blood pressure and heart rate in these animals. They found that the decrease in blood pressure was due to an increase in blood flow to the skeletal muscular bed. They speculated that the hypotensive effect of intravascular met-enkephalin in anesthetized rabbits may be due in part to its vasodilatory effect in skeletal muscle.

V. IMMUNE RESPONSE SYSTEM

A rather extensive literature has been built up concerning a potent immunostimulant effect of met-enkephalin on multiple levels and on multiple segments of the stimulatory and inflammation-related components of the immune response.

Van Epps and Saland,[95] in 1984 from Albuquerque, NM, described met-enkephalin-induced stimulation of human peripheral blood mononuclear cell chemotaxis. Peak stimulation occurred at met-enkephalin concentrations between 10^{-8} and 10^{-12} *M*. Also in 1984 Foris et al.,[96] in Budapest, Hungary reported met-enkephalin stimulation of antibody-dependent cytotoxicity by rat peritoneal macrophages (thioglycolate elicited). This macrophage functional alteration was associated with an increase in macrophage cyclic GMP (a stimulative transmembrane signaling molecule). In a 1987 report by Martin et al.,[97] from the University of Pennsylvania, macrophage, mast cell, and T-lymphocyte tumor cell lines were found to have high levels of preproenkephalin mRNA. Purified natural and nontumorous macrophages and mast cells also were found to have easily detectable levels of preproenkephalin mRNA.

Met-enkephalin-induced immunostimulation also involves lymphocyte types and subtypes. In an *in vitro* study, Mathews et al.,[98] in 1983, reported that met-enkephalin can enhance the natural cytotoxicity of lymphocytes toward tumor cells. In 1986 Zurawski et al.[99] reported in *Science* that mitogen activation of mouse T helper cells induced abundant cellular preproenkephalin mRNA synthesis. In 1979 Wybran et al.[100] presented suggestive evidence for the presence of met-enkephalin receptors on human blood T lymphocytes. They reported that met-enkephalin increased the ability of human lymphocytes to form rosettes with sheep red blood cells. This response was blocked by naloxone. Carr and Klimpel,[101] in 1986, described a met-enkephalin-induced increase in the generation of cytotoxic T lymphocytes at physiological concentrations (10^{-8} to 10^{-14} *M*).

Systemic administration of met-enkephalin by subcutaneous injection has been described by Plotnikoff et al.[102] as resulting in significant increase in thymic weights and significant decreases in splenic weights, in mice. In 1987 Janovic and Maric[107] presented evidence that systemic administration of met-enkephalin in physiological doses potentiated the development and expression of the inflammatory components of experimental allergic encephalomyelitis (EAE) in rats. Additional *in vivo* immunostimulative effects of met-enkephalin have been described by the Janovic group of Belgrade, Yugoslavia.[103-109] These include low-dose (physiological range) potentiation of specific inflammatory responses to complete Freund's adjuvant, to guinea pig spinal cord injected subcutaneously into the foot pad (rats), to bovine serum albumin, to old tuberculin, to local inflammation, to turpentine, and to adjuvant arthritis.

In 1987 Van Epps and Kutvirt[110] presented an *in vitro* observation that met-enkephalin could induce a concentration-dependent increase in polymorphonuclear neutrophil (PMN) adherence to serum-coated glass. As early events in acute inflammation include the adherence of PMN to capillary endothelial cells prior to migration into tissues, the authors speculated that this met-enkephalin-mediated response might indicate an important met-enkephalin contribution to the inflammatory process.

Also on the met-enkephalin and inflammation point, Stephano and associates have made some important recent observations.[111-114] Studies by this group have shown that a marine bivalve mollusc, *Mytilus edulis*, processes molecules and molecular mechanisms related to those involved in the regulation of vertebrate, mammalian, and human immune response and nervous systems, despite the divergence of man and *Mytilus* some 500 million years ago. The involvement of CD10/NEP-related structures in regulating met-enkephalin-mediated inflammatory responses in man and *Mytilus* was cited in the ''Introduction'' to this chapter. In 1990 Hughes et al. of this group[115] recorded the important observation that *Mytilus* hemocytes can synthesize an immunoreactive interleukin-1 (IL-1). In this study the authors also demonstrated that recombinant human IL-1 can induce the formation of a tumor-necrosis-factor-like substance in *Mytilus*, as well as initiate specific hemocyte conformation changes. The 1991 report of Stephano et al.[111] demonstrated that both the immune and nervous systems of *Mytilus* contain an IL-1-like molecule. They showed that *in vitro* administration of a stable met-enkephalin analogue (DAMA) to *Mytilus* ganglia and hemocytes and to human peripheral blood lymphocytes induces in all the formation of an immunoreactive interleukin-1-like molecule. Immunoreactive IL-1 activity also was identified in *Mytilus* nervous tissue. The stimulative actions of extracted immunoreactive IL-1 on *Mytilus* hemocytes was shown to be antagonized by an IL-1 antibody. The authors speculated from these findings that an interleukin-like molecule must have evolved in evolution earlier than previously thought — i.e., 500 million years ago.

In 1987 Plotnikoff et al.[116] presented evidence that the infusion of met-enkephalin in normal human volunteers and in cancer patients increases CD4/CD8 ratios. In 1988, Faith,[117] of Houston, reported that met-enkephalin significantly inhibited tumor metastasis in mice and significantly enhanced their natural-killer-cell activity. In 1985 Brown and Van Epps[118] described a met-enkephalin-induced increase in the production of gamma interferon.

In the supported-sheep lymphocyte traffic model of the author, infusion of met-enkephalin into the cannulated popliteal node afferent lymphatic produced a prompt, sharp, and prolonged increase in lymphocyte traffic, as monitored by the output of lymphocytes into nodal efferent lymph (chronically cannulated).[119-121] The increase in lymphocyte outputs involved both small recirculating and blast lymphocytes (Figure 1). These findings were

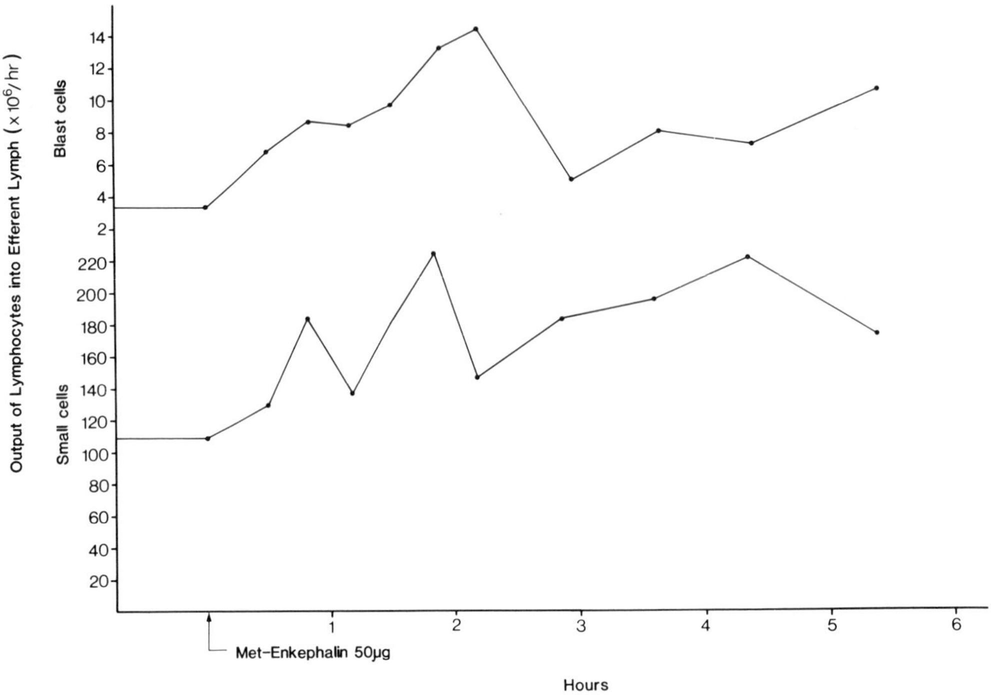

FIGURE 1. Acute infusion of met-enkephalin (50 µg) into a cannulated popliteal lymph node afferent lymphatic of a sheep produces a prompt, sharp, and prolonged increase in the lymphocyte traffic and output of small recirculating lymphocytes into nodal efferent lymph (chronically cannulated). The output of nonrecirculating blast lymphocytes also was increased. The results were interpreted as immunostimulative. (From Moore, T. C., *Immunology*, 52, 511, 1984. With permission.)

interpreted as immunostimulative when compared with other findings in these studies of supported-sheep lymphocyte traffic.[121]

VI. CONCLUSION

The functional reputation of met-enkephalin has long suffered from its discovery and classification with other endogenous ''opioids'' — a name that rather strongly suggests *opium* and its long and well-known depressant effects, from the *Confessions of an English Opium Eater* (1821, by Thomas De Quincey, 1785–1859) to the present. From the data included in this chapter, it is apparent that despite met-enkephalin's inappropriate ''opioid''-type reputation, it is more likely a stimulative molecule in the nervous and immune response systems and a vasodilator in key portions of the vascular system. It also must be considered as, at least, a second-tier ''memory molecule''. Much more work needs to be done in all of these areas.

REFERENCES

1. **Hughes, J., Smith, T. W., Kosterlitz, H. W., Fothergill, L., Morgan, B. A., and Morris, H. R.,** Identification of two related pentapeptides from the brain with potent opiate agonist activity, *Nature,* 258, 577, 1975.
2. **Reiner, A.,** The distribution of proenkephalin-derived peptides in the central nervous system of turtles, *J. Comp. Neurol.,* 259, 65, 1987.
3. **Wollemann, M.,** Recent developments in the research of opioid receptor subtype molecular characterization, *J. Neurochem.,* 54, 1095, 1990.
4. **Venturini, G., Carolei, A., Palladini, G., Margotta, V., and Lauro, M. G.,** Radioimmunological and immunocytochemical demonstration of met-enkaphalin in planaria, *Comp. Biochem. Physiol.,* 74C, 23, 1983.
5. **Zipser, B.,** Identification of specific leech neurons immunoreactive to enkephalin, *Nature,* 283, 857, 1980.
6. **Alumets, J., Hakanson, R., Sundler, F., and Thorell, J.,** Neuronal localisation of immunoreactive enkephalin and beta-endorphin in the earthworm, *Nature,* 279, 805, 1979.
7. **Ford, R., Jackson, D. M., Tetrault, L., Torres, J. C., Assanah, P., Harper, J., Leung, M. K., and Stefano, G. B.,** A behavioral role for enkephalins in regulating locomotor activity in the insect *Leucophaea maderae*: evidence for high affinity kappa-like opioid binding sites, *Comp. Biochem. Physiol.,* 85C, 61, 1986.
8. **Stefano, G. B. and Scharrer, B.,** High affinity binding of an enkephalin analog in the cerebral ganglion of the insect *Leucophaea maderae* (Blattaria), *Brain Res.,* 225, 107, 1981.
9. **Stefano, G. B., Scharrer, B., and Assanah, P.,** Demonstration, characterization and localization of opioid binding sites in the midgut of the insect *Leucophaea maderae, Brain Res.,* 253, 202, 1982.
10. **Duve, H., Thorpe, A., and Scott, A.,** Localization and characterization of opioid-like peptides in the nervous system of the blowfly *Calliphora vomitoria,* in *Comparative Opioid and Related Neuropeptide Mechanisms,* Vol. 1, Stefano, G. B., Ed., CRC Press, Boca Raton, FL, in press.
11. **Remy, C. and Dubois, M. P.,** Localization pour immunofluorescence des peptides analogues a l'-endorphine dans les ganglions infra-oesphagiens du lombricide *Dendrobaena subrubienunda, Eisen. Experimentia* (Basel), 35, 137, 1979.
12. **Nishiitsutsuji-Uwo, J., Endo, Y., Takeda, M., and Saito, H.,** Brain gut peptides in the cockroach with special reference to midgut, in *Comparative Opioid and Related Neuropeptide Mechansisms,* Vol. 1, Stefano, G. B., Ed., CRC Press, Boca Raton, FL, in press.
13. **Coletti-Previero, M.-A., Mattras, H., Zwilling, R., and Previero, A.,** Enkephalin-degrading activity in arthropod hemolymph, *Neuropeptides,* 6, 405, 1985.
14. **Kream, R. M., Zukin, R. S., and Stefano, G. B.,** Demonstration of two classes of opiate binding sites in the nervous tissue of the marine mollusc *Mytilus edulis*: positive homotrophic cooperativity of lower affinity binding sites, *J. Biol. Chem.,* 225, 9218, 1980.
15. **Stefano, G. B. and Martin, R.,** Enkephalin-like immunoreactivity in the pedal ganglion of Mytilus edulis (Bivalvia) and its proximity to dopamine-containing structures, *Cell Tissue Res.,* 230, 147, 1983.
16. **Leung, M. K. and Stefano, G. B.,** Isolation and identification of enkephalin in pedal ganglia of *Mytilus edulis* (Mollusca), *Proc. Natl. Acad. Sci. U.S.A.,* 81, 955, 1984.
17. **Stefano, G. B. and Leung, M. K.,** Presence of met-enkephalin-Arg-6-Phe-7 in molluscan neural tissues, *Brain Res.,* 298, 362, 1984.
18. **Andersen, A. C., L'Hermite, A., Ferrand, R., and Dubois, M. P.,** Immunohistological detection of methionine-enkephalin-like and endorphin-like material in the digestive tract and in the nervous system of the mussel: *Mytilus edulis, L. Gen. Comp. Endocrinol.,* 62, 111, 1986.
19. **Schot, L. P. C., Boer, H. H., Swaab, D. F., and Van Noorden, S.,** Immunocytochemical demonstration of peptidergic neurones in the central nervous system of the pond snail *Lymnae stagnalis* with antisera raised to vertebrates, *Cell Tissue Res.,* 216, 273, 1981.
20. **Takayanagi, H. and Takeda, N.,** Coexistence of FMRFamide, met-enkephalin and serotonin in molluscan neurons, *Comp. Biochem. Physiol.,* 91A, 613, 1988.
21. **Doble, K. E. and Greenberg, M. J.,** The claim rectum is sensitive to FMRFamide, the enkephalins and their common analogs, *Neuropeptides,* 2, 157, 1982.
22. **Martin, R., Forsch, D., Kiehling, C., and Voigt, K. H.,** Molluscan neuropeptide-like and enkephalin-like material coexists in *Octopus* nerves, *Neuropeptides,* 2, 141, 1981.
23. **Georges, D. and Dubois, M. P.,** Methionine-enkephalin-like immunoreactivity in the nervous ganglion and ovary of a protochordate, *Ciona intenstinalis, Cell Tissue Res.,* 236, 165, 1984.
24. **Gold, M. R. and Finger, T. E.,** Localization of enkephalin immunoreactivity in the brain of the lamprey, *Soc. Neurosci. Abs.,* 7, 85, 1982.

25. **Nozaki, M. and Gorbman, A.,** Immunoreactivity for met-enkephalin and substance P in cells of the adenohypophysis of larval and adult sea lampreys, *Petromyzon marinus, Gen. Comp. Endocrinol.,* 57, 172, 1985.
26. **Finger, T. E.,** Enkephalin-like immunoreactivity in the gustatory lobes and visceral nuclei in the brains of goldfish and catfish, *Neuroscience,* 6, 2747, 1981.
27. **Gordon, R. and Levina, S.,** Methionine enkephalin-induced changes in pigmentation of zebrafish (cyprinidae, *brachydanio rerio*) and related species and varieties, measured videodensitometrically, *Gen. Comp. Endocrinol.,* 51, 378, 1983.
28. **Levina, S. E. and Gordon, R.,** Met-enkephalin induced changes in pigmentation of zebrafish *(Brachydanio rerio)* and related species and varieties, in *Proc. Western Regional Conference on Comparative Endocrinology,* 1982, 29.
29. **Levina, S. and Gordon, R.,** Methionine-enkephalin induced changes in pigmentation of zebrafish (cyprinidae, *Brachydanio rerio*) and related species and varieties, measured videodensitometrically. I. Zebrafish, *J. Comp. Endocrinol.,* 51, 370, 1983.
30. **Lindberg, I. and White, L.,** Reptilian enkephalins: implications for the evolution of proenkephalin, *Arch. Biochem. Biophys.,* 245, 1, 1986.
31. **Wolters, J. G., Ten Donkelaar, H. J., and Verhofstad, A. A. J.,** Distribution of some peptides (substance P, [leu]enkephalin, [met]enkephalin) in the brain stem and spinal cord of a lizard, *Varanus exanthematicus, Neuroscience,* 18, 917, 1986.
32. **Reiner, A.,** The distribution of proenkephalin-derived peptides in the central nervous system of turtles, *J. Comp. Neurol.,* 259, 65, 1987.
33. **Du, F. and Dubois, P. M.,** Development and distribution of enkephalin-immunoreactive elements in the chicken spinal cord, *Neuroscience,* 27, 251, 1988.
34. **Margiotta, J. F. and Berg, D. K.,** Enkephalin and substance P modulate synaptic properties of chick ciliary ganglion neurons in cell culture, *Neuroscience,* 18, 175, 1986.
35. **Ryan, S. M., Arnold, A. P., and Elde, R. P.,** Enkephalin-like immunoreactivity in vocal control regions of the zebra finch brain, *Brain Res.,* 229, 236, 1981.
36. **Pickel, V. M., Miller, R., Chan, J., and Sumal, K. K.,** Substance P and enkephalin in transected axons of medulla and spinal cord, *Regul. Pept.,* 6, 121, 1983.
37. **Nahin, R. L. and Micevych, P. E.,** A long ascending pathway of enkephalin-like immunoreactive spinoreticular neurons in the rat, *Neurosci. Lett.,* 65, 271, 1986.
38. **Cesselin, F., Bourgoin, S., Clot, A. M., Hamon, M., and LeBars, D.,** Segmental release of met-enkephalin-like material from the spinal cord of rats, elicited by noxious thermal stimuli, *Brain Res.,* 484, 71, 1989.
39. **Iversen, L. L., Iversen, S. D., Bloom, F. E., Vargo, T., and Guillemin, R.,** Release of enkephalin from rat globus pallidus *in vitro, Nature,* 271, 679, 1978.
40. **Elde, R., Hokfelt, T., Johansson, D., and Terenius, L.,** Immunohistochemical studies using antibodies to leucine enkephalin in initial observations on the nervous system of the rat, *Neuroscience,* 1, 349, 1977.
41. **Simantov, R., Kuhar, M. J., Uhl, G. R., and Snyder, S. H.,** Opioid peptide enkephalin: mapping in rat central nervous system, *Proc. Natl. Acad. Sci. U.S.A.,* 74, 2167, 1977.
42. **Haber, S. N. and Nauta, W. J. H.,** Ramifications of the globus pallidus in the rat as indicated by patterns of immunohistochemistry, *Neuroscience,* 9, 245, 1983.
43. **Cuello, A. C. and Sofroniew, M. V.,** The anatomy of the CNS cholinergic neurons, *Trends Neurosci.,* 7, 74, 1984.
44. **Rye, D. B., Wayner, B. H., Mesulam, M.-M., Mufson, E. J., and Saper, C. B.,** Cortical projections arising from the basal forebrain: a study of cholinergic and noncholinergic components employing combined retrograde tracing and immunohistochemical localization of choline acetyltransferase, *Neuroscience,* 13, 627, 1984.
45. **Beach, T. G., Tago, H., and McGeer, E. G.,** Light microscopic evidence for a substance P-containing innervation of the human nucleus basalis of Meynert, *Brain Res.,* 408, 251, 1987.
46. **Haber, S. N., Broenewegen, E. H., Grove, E. A., and Nauta, W. J. H.,** Efferent connections of the ventral pallidum: evidence for a dual striatopallidofugal pathway, *J. Comp. Neurol.,* 235, 322, 1985.
47. **Martinez-Murillo, R., Blasco, O., Alvarez, F. J.,Villalba, R., Solano, M. L., Montero-Caballero, I., and Rodrigo, J.,** Distribution of enkephalin-immunoreactive nerve fibres and terminals in the region of the nucleus basalis magnocellularis of the rat: a light and electron microscopic study, *J. Neurocytol.,* 17, 361, 1988.
48. **Arluison, M., Vankova, M., Cesselin, F., and Leviel, V.,** Origin of some enkephalin-containing afferents to the ventro-medial region of the globus pallidus in the rat, *Brain Res.,* 25, 25, 1990.
49. **Frey, J. M. and Huffman, R. D.,** Effects of enkephalin and morphine on rat globus pallidus neurons, *Brain Res. Bull.,* 14, 251, 1985.

50. **Kalivas, P. W. and Richardson-Carlson, R.,** Endogenous enkephalin modulation of dopamine neurons in ventral tegmental area, *Am. J. Physiol.*, 251, R243, 1986.
51. **Normand, E., Popovici, T., Onteniente, B., Fellmann, D., Piatier-Tonneau, D., Auffray, C., and Bloch, B.,** Dopaminergic neurons of the substantia nigra modulate preproenkephalin A gene expression in rat striatal neurons, *Brain Res.*, 439, 39, 1988.
52. **Pretel, S. and Ruda, M. A.,** Immunocytochemical analysis of noradrenaline, substance P and enkephalin axonal contacts on serotonin neurons in the caudal raphe nuclei of the cat, *Neurosci. Lett.*, 89, 19, 1988.
53. **Taquet, H., Javoy-Agid, F., Hamon, M., Legrand, J. C., Agid, Y., and Cesselin, F.,** Parkinson's disease affects differentially Met^5- and Leu^5-enkephalin in the human brain, *Brain Res.*, 280, 379, 1983.
54. **Nicoll, R. A., Siggins, G. R., Ling, N., Bloom, F. E., and Guillemin, R.,** Neuronal actions of endorphins and enkephalins among brain regions: a comparative microiontophoretic study, *Proc. Natl. Acad. Sci. U.S.A.*, 74, 2584, 1977.
55. **Lee, H. K., Dunwiddie, T., and Hoffer, B.,** Electrophysiological interactions of enkephalins with neuronal circuitry in the rat hippocampus. II. Effects on interneuron excitability, *Brain Res.*, 184, 331, 1980.
56. **Zieglgansberger, W., French, E. D., Siggins, G. R., and Bloom, F. E.,** Opioid peptides may excite hippocampal pyramidal neurones by inhibiting adjacent inhibitory interneurones, *Science*, 205, 415, 1979.
57. **French, E. D. and Zieglgansberger, W.,** The excitatory response of in vitro hippocampal pyramidal cells to normorphine and methionine-enkephalin may be mediated by different receptor populations, *Exp. Brain Res.*, 48, 238, 1982.
58. **Madison, D. V. and Nicoll, R. A.,** Enkephalin hyperpolarizes interneurones in the rat hippocampus, *J. Physiol. (Lond.)*, 398, 123, 1988.
59. **Andrade, R., Malenta, R. C., and Nicoll, R. A.,** A G-protein couples serotonin and $GABA_B$ receptors to the same channels in hippocampus, *Science*, 234, 1261, 1986.
60. **Haigler, H. J., Cahill, L. F., Crager, M. R., and Charles, C.,** Met-enkephalin, age and anatomy: a microiontophoretic study in the hippocampus, *Life Sci.*, 40, 203, 1987.
61. **Weisner, J. B. and Henriksen, S. J.,** Enkephalin enhances responsiveness to perforant path input while decreasing spontaneous activity in the dentate gyrus, *Neurosci. Lett.*, 74, 95, 1987.
62. **Dunwiddie, T. V. and Su, M.-T.,** Pertussis toxin pretreatment antagonizes the actions of mu- and delta-opiate agonists in hippocampal slices, *Neurosci. Lett.*, 95, 329, 1988.
63. **Kanamatsu, T., Obie, J., Grimes, L., McGinty, J. F., Yoshikawa, K., Sabol, S., and Hong, J.-S.,** Kainic acid alters the metabolism of Met^5-enkephalin and the level of dynorphin A in the rat hippocampus, *J. Neurosci.*, 6, 3094, 1986.
64. **McGinty, J. F., Kanamatsu, T., Obie, J., Dyer, R. S., Mitchell, C. L., and Hong, J.-S.,** Amygdaloid kindling increases enkephalin-like immunoreactivity but decreases dynorphin-A-like immunoreactivity in rat hippocampus, *Neurosci. Lett.*, 71, 31, 1986.
65. **Gall, C.,** Seizures induce dramatic and distinctly different changes in enkephalin, dynorphin, and CCK immunoreactivities in mouse hippocampal mossy fibers, *J. Neurosci.*, 8(6), 1852, 1988.
66. **Wanscher, B., Kragh, J., Barry, D. I., Bolwig, T., and Zimmer, J.,** Increased somatostatin and enkephalin-like immunoreactivity in the rat hippocampus following hippocampal kindling, *Neurosci. Lett.*, 118, 33, 1990.
67. **Tanaka, T., Takeshita, H., Kawahara, R., and Hazama, H.,** Chemical kindling and electrical kindling, *Epilepsy Res.*, 3, 214, 1989.
68. **Tanaka, T., Takeshita, H., Hamasaki, Y., Kawahara, R., and Hazama, H.,** Chemical kindling with met-enkephalin and changes of brain opioid o-receptor sensitivity after electrical kindling, *Epilepsia*, 27, 596, 1986.
69. **Penny, G. R., Afsharpour, S., and Kitai, S. T.,** The glutamate decarboxylase-, leucine enkephalin-, methionine enkephalin- and substance P-immunoreactive neurons in the neostriatum of the rat and cat: evidence for partial population overlap, *Neuroscience*, 17, 1011, 1986.
70. **Maley, B. E., Newton, B. W., Howes, K. A., Herman, L. M., Oloff, C. M., Smith, K. C., and Elde, R. P.,** Immunohistochemical localization of substance P and enkephalin in the nucleus tractus solitarii of the Rhesus monkey, *Macaca mulatta*, *J. Comp. Neurol.*, 260, 483, 1987.
71. **Maley, B., Mullett, T., and Elde, R.,** The nucleus tractus solitarii of the cat: a comparison of golgi impregnated neurons with methionine-enkephalin- and substance P-immunoreactive neurons, *J. Comp. Neurol.*, 217, 405, 1983.
72. **De Lanerolle, N. C. and LaMotte, C. C.,** The human spinal cord: substance P and methionine-enkephalin in immunoreactivity, *J. Neurosci.*, 2, 1369, 1982.
73. **Tashiro, T., Takahashi, O., Satoda, T., Matsushima, R., and Mizuno, N.,** Immunohistochemical demonstration of coexistence of enkephalin- and substance P-like immunoreactivities in axonal components in the lumbar segments of cat spinal cord, *Brain Res.*, 424, 391, 1987.

74. **Senba, E., Yanaihara, C., Yanaihara, N., and Tohyama, M.,** Co-localization of substance P and met-enkephalin-Arg^6-Gly^7-Leu^8 in the intraspinal neurons of the rat, with special reference to the neurons in the substantia gelatinosa, *Brain Res.*, 453, 110, 1988.
75. **Duggan, A. W., Morton, C. R., Zhao, Z. Q., and Hendry, I. A.,** Noxious heating of skin releases immunoreactive substance P in the substantia gelatinosa of the cat: a study with antibody probes, *Brain Res.*, 403, 345, 1987.
76. **Tang, T., Chou, J., Yang, H.-Y. T., and Costa, E.,** Substance P stimulates the release of met^5-enkephalin-ARG^6-PHE^7 and met^5-enkephalin from rat spinal cord, *Neuropharmacology,* 22, 1147, 1983.
77. **Huitinga, I., Van Dijk, C. A., and Groenewegen, H. J.,** Substance P- and enkephalin-containing projections from the interpeduncular nucleus to the dorsal tegmental region in the rat, *Neurosci. Lett.*, 62, 311, 1985.
78. **Leger, L., Charnay, Y., Dubois, P. M., and Jouvet, M.,** Distribution of enkephalin-immunoreactive cell bodies in relation to serotonin-containing neurons in the raphe nuclei of the cat: immunohistochemical evidence for the coexistence of enkephalins and serotonin in certain cells, *Brain Res.*, 362, 63, 1986.
79. **Wiklund, L., Leger, L., and Persson, M.,** Monoamine cell distribution in the cat brain stem, *J. Comp. Neurol.*, 203, 613, 1981.
80. **Kanagawa, Y., Matsuyama, T., Wanaka, A., Yoneda, S., Kimura, K., Kamada, T., Steinbusch, H. W. M., and Tohyama, M.,** Coexistence of enkephalin- and serotonin-like substances in single small intensely fluorescent cells of the guinea pig superior cervical ganglion, *Brain Res.*, 379, 377, 1986.
81. **Aronin, N., Chase, K., and Difiglia, M.,** Glutamic acid decarboxylase and enkephalin immunoreactive axon terminals in the rat neostriatum synapse with striatonigral neurons, *Brain Res.*, 365, 151, 1986.
82. **Pasinetti, G., Goroni, S., DiGiovine, S., Spano, P. F., and Trabucchi, M.,** Dopamine enhances met-enkephaline efflux from striatal slices, *Brain Res.*, 293, 364, 1967.
83. **Kalivas, P. W. and Bronson, M.,** Mesolimbic dopamine lesions produce an augmented behavioral response to enkephalin, *Neuropharmacology,* 24, 931, 1985.
84. **Minneman, K. P. and Iversen, L. L.,** Enkephalin and opiate narcotics increase cyclic GMP accumulation in slices of rat neostriatum, *Nature,* 262, 313, 1976.
85. **Brandt, M., Guillis, R. J., Fischer, K., Buchen, C., Hamprecht, B., Moroder, L., and Wunsch, E.,** Enkephalin regulates the levels of cyclic nucleotides in neuroblastoma × glioma hybrid cells, *Nature,* 262, 311, 1976.
86. **Kley, N.,** Multiple regulation of proenkephalin gene expression by protein kinase C, *J. Biol. Chem.*, 263, 2003, 1988.
87. **Ueda, H., Yoshihara, Y., and Takagi, H.,** A putative met-enkephalin releaser, kyotorphin enhances intracellular Ca^{2+} in the synaptosomes, *Biochem. Biophys. Res. Commun.*, 137, 897, 1986.
88. **Blum, K., Gaskill, H., DeLallo, L., Bridges, A. H., and Hall, W.,** Methionine enkephalin as a possible neuromodulator of regional cerebral blood flow, *Experientia,* 41, 932, 1985.
89. **Moore, R. H., III and Dowling, D. B.,** Effects of enkephalin on perfusion pressure in isolated hindlimb preparations, *Life Sci.*, 31, 1559, 1982.
90. **Eluie, P. J. and Rhee, H. M.,** Reduction by pentolamine of the hypotensive effect of methionine enkephalin in anesthetized rabbits, *Br. J. Pharmacol.*, 83, 783, 1984.
91. **Caffrey, J. L., Woolridge, C. B., and Gaugl, J. F.,** The interaction of endogenous opiates with autonomic circulatory control in the dog, *Circ. Shock,* 17, 233, 1985.
92. **Wrightman, J. M., Schadt, J. C., and Gaddis, R. R.,** Decreased vascular resistance after intra-arterial injection of [met]enkephalin in the hindquarters of conscious rabbits, *J. Pharmacol. Exp. Ther.*, 241, 314, 1987.
93. **Rhee, H. M., Eulie, P. J., and Peterson, D. F.,** Suppression of renal nerve activity by methionine enkephalin in anesthetized rabbits, *J. Pharmacol. Exp. Ther.*, 234, 537, 1985a.
94. **Eulie, P. H., Rhee, H. M., and Laughlin, M. H.,** Effects of [met^5]enkephalin on regional blood flow and vascular resistance in rabbits, *Eur. J. Pharmacol.*, 137, 25, 1987.
95. **Van Epps, D. E. and Salad, L.,** γ-Endorphin and met-enkephalin stimulate human peripheral blood mononuclear cell chemotaxis, *J. Immunol.*, 132, 3046, 1984.
96. **Foris, G., Medgyesi, G. A., Gyimesi, E., and Hauck, M.,** Met-enkephalin induced alterations of macrophage functions, *Mol. Immunol.*, 21, 747, 1984.
97. **Martin, J., Prystowsky, M. B., and Angeletti, R. H.,** Preproenkephalin mRNA in T-cells, macrophages, and mast cells, *J. Neurosci. Res.*, 18, 82, 1987.
98. **Matthews, P. M., Froelich, C. J., Sibbit, W. L., and Bankhurst, A. D.,** Enhancement of natural cytotoxicity by B-endorphin, *J. Immunol.*, 130, 1658, 1983.
99. **Zurawski, Y., Benedik, M., Kamp, B. J., Abrams, J. S., Zurawski, S. M., and Lee, F. D.,** Activation of mouse T helper cells induces abundant preproenkephalin mRNA synthesis, *Science,* 232, 772, 1986.

100. **Wybran, J., Appelboom, T., Famaey, J. P., and Govaerts, A.,** Suggestive evidence for receptors for morphine and methionine-enkephalin on normal human blood T lymphocytes, *J. Immunol.*, 23, 1068, 1979.
101. **Carr, D. J. J. and Klimpel, G. R.,** Enhancement of the generation of cytotoxic T cells by endogenous opiates, *J. Neuroimmunol.*, 12, 75, 1986.
102. **Plotnikoff, N. P., Murgo, A. J., and Faith, R. E.,** Neuroimmunomodulation with enkephalins: effects on thymus and spleen weights in mice, *Clin. Immunol. Immunopathol.*, 32, 52, 1984.
103. **Jankovic, B. D. and Maric, D.,** Enkephalins and autoimmunity: differential effect of methionine-enkephalin on experimental allergic encephalomyelitis in Wistar and Lewis rats, *J. Neurosci. Res.*, 18, 88, 1987.
104. **Jankovic, B. D. and Maric, D.,** Enkephalins modulate *in vivo* immune reactions through delta- and mu-opioid receptors, *Ann. N.Y. Acad. Sci.*, 590, 691, 1988.
105. **Veljic, J., Ranin, J., Maric, D., and Jankovic, B. D.,** Modulation of immunoinflammatory reactions by centrally applied methionine-enkephalin, *Ann. N.Y. Acad. Sci.*, 650, 51, 1992.
106. **Jankovic, B. D.,** Immunorestorative activity of methionine-enkephalin in senescence, *Int. J. Neurosci.*, 51, 185, 1990.
107. **Jankovic, B. D. and Maric, D.,** *In vivo* modulation of the immune system by enkephalins, *Int. J. Neurosci.*, 51, 167, 1990.
108. **Jankovic, B. D. and Maric, D.,** Enkephalins and immunity. I. *In vivo* suppression and potentiation of humoral immune response, *Ann. N.Y. Acad. Sci.*, 496, 115, 1987.
109. **Maric, D. and Jankovic, B. D.,** Enkephalins and immunity. II. *In vivo* modulation of cell-mediated immunity, *Ann. N.Y. Acad. Sci.*, 496, 126, 1987.
110. **Van Epps, D. E. and Kutvirt, S. L.,** Modulation of human neutrophil adherence by γ-endorphin and met-enkephalin, *J. Neuroimmunol.*, 15, 219, 1987.
111. **Stefano, G. B., Smith, E. M., and Hughes, T. K.,** Opioid induction of immunoreactive interleukin-1 in *Mytilus edulis* of human immunocytes: an interleukin-1-like substance in invertebrate neural tissue, *J. Neuroimmunol.*, 32, 29, 1991.
112. **Shipp, M. A., Stefano, G. B., D'Adamio, L., Switzer, S. N., Howard, F. D., Sinisterra, J., Scharrer, B., and Reinherz, E.,** Downregulation of enkephalin-mediated inflammatory responses by CD10/neutral endopeptidase 24, 11, *Nature*, 347, 394, 1990.
113. **Stefano, G. B., Leung, M. K., Zhao, X., and Scharrer, B.,** Evidence for the involvement of opioid neuropeptides in the adherence and migration of immunocompetent invertebrate hemocytes, *Proc. Natl. Acad. Sci. U.S.A.*, 86, 626, 1989a.
114. **Stefano, G. B., Cadet, P., and Scharrer, B.,** Stimulatory effects of opioid neuropeptides on locomotory activity and conformational changes in invertebrate and human immunocytes: evidence for a subtype of delta receptor, *Proc. Natl. Acad. Sci. U.S.A.*, 86, 6307, 1989b.
115. **Hughes, T. K., Jr., Smith, E. M., Cadet, P., Sinisterra, J., Leung, M. K., Shipp, M. A., Scharrer, B., and Stefano, G. B.,** Interaction of immunoactive monokines (IL-1 and TNF) in the bivalve mollusc, *Mytilus edulis, Proc. Natl. Acad. Sci. U.S.A.*, 87, 4426, 1990.
116. **Plotnikoff, N. P., Miller, G. C., Nimeh, N., Faith, R. E., Murgo, A. J., and Wybran, J.,** Enkephalins and T-cell enhancement in normal volunteers and cancer patients, *Ann. N.Y. Acad. Sci.*, 496, 608, 1987.
117. **Faith, R. E.,** Inhibition of pulmonary metastases and enhancement of natural killer cell activity by methionine-enkephalin, *Brain Behav. Immun.*, 2, 114, 1988.
118. **Brown, S. L. and Van Epps, E. E.,** Beta endorphin (BE), met-enkephalin (ME), and corticotrophin (ACTH) modulate the production of gamma interferon (INF) *in vitro, Fed. Proc.*, 44, 949, 1985.
119. **Moore, T. C.,** The effects of inflammatory mediators and anaesthetic agents upon lymphocyte traffic of sheep *in vivo*, Ph.D. thesis, University of Cambridge (Trinity College), England, 1982.
120. **Moore, T. C.,** Modification of lymphocyte traffic by vasoactive neurotransmitter substances, *Immunology*, 52, 511, 1984.
121. **Moore, T. C.,** Mediator and physiological influences on lymphocyte traffic in sheep, in *Immunology of the Sheep*, Morris, B. and Miyasaka, M., Eds., Editiones Roche, Basel, 1985, 327.

Chapter 8

CHOLECYSTOKININ-8 (CCK-8)

CCK-8 (cholecystokinin-8), an eight-amino-acid-residue peptide, is the major cholecystokinin-related neuropeptide in the central nervous system. Cholecystokinin was initially identified in porcine gut in 1968 by Jorpes and Mutt.[1] The chemically identified forms of CCK in the gut were the peptides of 33 (CCK-33) and 39 (CCK-39) residues. It is of interest that synthetic CCK-8 had been synthesized and widely used in biological studies before it became clear that the fragment existed naturally as an endogenous peptide of major functional significance.

The identification of CCK-8 as an endogenous substance was first made in extracts of brain from rats, dogs, and man, by Vanderhaeghen et al.[2] in 1975. They observed that these vertebrate/mammalian brains, including man, contained substantial quantities of a gastrin/CCK-like substance. Their observations were made with a gastrin radioimmunoassay (RIA) with considerable cross reactivity with CCK. The sheep brain sequence of CCK was published in 1978 by Dockray et al.[3] It was found by the use of antibodies against gastrin, due to the considerable cross reactivity between CCK-8 and gastrin and gastrin pentapeptide. CCK-8 cross reacts with some gastrin antisera because a common L-terminal pentapeptide is shared by both molecules. Isolated brain CCK-8 was shown (in 1978 by Dockray, Gregory, et al.[3]) to possess the full chemical and biological properties of synthetic CCK-8. The major active forms of CCK-8 have a sulfated tyrosine at position 7, counting from the C terminus.

Although a rather massive literature has developed concerning CCK and CCK-8 in the central nervous system, its interpretation has been rendered difficult by the cross reactivity of CCK-8 with gastrin pentapeptide and CGRP (calcitonin gene-related peptide). Recent publications from the Karolinska Institute of Stockholm, by Hokfelt and associates,[4-6] have suggested that CCK-8 immunoreactivity in rat primary sensory neurons may actually be CGRP. Hokfelt and associates have pointed out that CGRP and CCK share glycine amino acid residues in the C-terminal region, and they have the same C-terminal amino acid (phenylalanine) that in both molecules is amidated. They further point out that CCK-8 immunoreactivity can be absorbed out with high concentrations of CGRP. Particularly important additional points made by Hokfelt and associates relate to dorsal-root ganglia and spinal cord dorsal horn. They emphasize that while dorsal-root ganglia contain no, or very low, concentrations of CCK as measured by radioimmunoassay, they do contain high concentrations of CGRP immunoreactivity. They make a last point of special significance. It relates to the important point that, although CCK immunoreactivity (as measured by RIA) in the dorsal horn of the spinal cord is not influenced by capsaicin treatment, which is known to cause degeneration of small primary neurons, capsaicin treatment does deplete the levels of CGRP in sensory neurons (measured biochemically), as expected from the well-known and -established actions of capsaicin on small primary neurons (degeneration).

These impressive demonstrations by Hokfelt and associates and the gastrin-gastrin pentapeptide crossreactivity, must put all CCK-CCK-8 observations on "hold" and subject to revision and reevaluation. A major role for CCK-8 as a "long-distance operator" has also been brought into question by the observation, by Seroogy et al.,[7] that few cortical cholecystokinin-immunoreactive neurons have long projections.

The state of CCK-8 in central nervous system function has been brought into important perspective in 1990 publications by S. D. Iversen and associates.[31,32] This group points out that a major problem in assessing CCK-8 function has been the lack of a reliable and specific antagonist that could confidently and reliably identify, without question and without cross

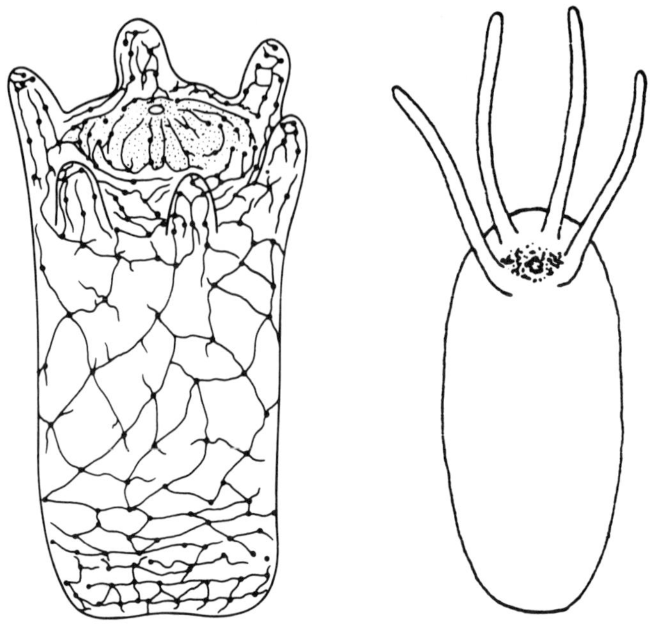

FIGURE 1. Nerve net of *Hydra* (left) and distribution of CCK-8 immunoreactivity in *Hydra* at base of tentacles and at entrance to "oral" and "gut" cavities of *Hydra* (right). (Composite from Encyclopaedia Britannica and Reference 8.)

reactivity, CCK-8-receptor types and subtypes so that their functional roles could be identified. Weak and nonselective antagonists to CCK and CCK-8 merely muddy the picture. Profound interspecies differences in receptor distributions also further muddies the waters. Purification, cloning, and sequencing of the putative receptors (suggested to be CCK-A and CCK-B) would help in restoring confidence in observations in this important area.

The evolutionary trail of CCK-8 or CCK/gastrin is a long one. It has been tracked down the evolutionary tree to cnidarian coelenterata (Hydra)[8] (Figure 1). It has been identified in the neuronal systems of annelids (segmented worms),[9-11] crustacean arthropods,[12] insect arthropods,[13-15] molluscs,[16,17] and primitive vertebrates, including prochordates, cartilaginous fish, bony fish, and amphibians.[18-21]

High levels of CCK-8 have been identified throughout the central nervous system.[22-27] A neurotransmitter role has been suggested for it, yet in the most-studied mammal, the rat, it is not found in long sensory neurons.[28] It has appeared to be an excitatory molecule when applied iontophoretically to nerve cells *in vivo* and *in vitro*.[26,27,29,30]

Recent studies by Iversen and associates,[31-33] at the Neuroscience Research Center in Harlow, England have identified a role of significance for CCK-B receptor blockade in pain control. Devazepide (formerly L-364,718) has been shown to enhance morphine-induced analgesia. A new and more potent CCK-B antagonist (L-365,260) has been shown, by itself, to induce analgesia in the squirrel monkey. These important findings establish, for the first time, that potent, highly selective, and specific blockage of CCK-B receptors may induce analgesia in primates. These observations provide strong support for the 1983 suggestion of Faris et al.[34] that CCK may function as an endogenous modifier of opiate activity.

Vascular effects of CCK and CCK-8 are quite underinvestigated. Hendry et al.,[35] of St. Louis, in 1983, described the presence of CCK-immunoreactive neurons (in the rat and monkey cerebral cortex) that make symmetric synapses and that have intimate associations with blood vessels. The soma and processes of the CCK neurons were found to establish very close nonsynaptic associations with blood vessels, suggesting a possible role for CCK (or perhaps CGRP) in the maintenance and regulation of cerebral blood flow. This observation reminds us of the evolutionary observations regarding CCK-immunoreactive (or perhaps CGRP-immunoreactive) neurons in close contact with neurohemal organs and hemolymph areas in crustacean and insect arthropods.

Solid information relating to CCK-8 on the immune response system front is almost as scarce. McMillen and associates[36] at Yale have shown that CCK-8 may activate human peripheral blood mononuclear cells with an increase in their Ca^{++} levels. They observed that when CCK-8 is added to submitogenic doses of PHA or anti-CD_3 monoclonal antibody, an increase in ^{3}H-thymidine-uptake occurs. The authors speculated that CCK-8 may be a comodulator of CD3-mediated activation of lymphocytes.

In the supported-sheep lymphocyte traffic of the author, single acute infusions of CCK-8 into cannulated popliteal nodal afferent lymphatics, in a limited number of studies, were found to produce delayed (24 h) increases in the outputs of small recirculating lymphocytes into nodal efferent lymph (unpublished data). This response was considered immunostimulative, but considerably more work needs to be done in this important area and with this model. A 1987 report in *Science*, by Flood et al.,[37] identifying CCK as a memory-influencing molecule, is sure to stimulate more research in this area and establishes CCK as at least a putative ''memory molecule'' in addition to its involvement in pain transmission and perception (see above).

REFERENCES

1. **Jorpes, P. E.,** The isolation and chemistry of secretin and cholecystokinin, *Gastroenterostomy,* 55, 157, 1968.
2. **Vanderhaeghen, J. J., Signeau, J. C., and Gepts, W.,** New peptide in the vertebrate CNS reacting with antigastrin antibodies, *Nature,* 257, 604, 1975.
3. **Dockray, G. J., Gregory, R. A., and Hutchinson, J. B.,** Isolation, structure, and biological activity of two cholescystokinin octapeptides from sheep brain, *Nature,* 274, 711, 1978.
4. **Hokfelt, T., Herrera-Marschit, M., Seroogy, K., Ju, G., Staines, W. A., Holets, V., Schalling, M., Ungerstedt, U., Post, C., Rehfeld, J. F., Frey, P., Fischer, J., Dockray, G., Hamaoka, T., Walsh, J. H., and Goldstein, M.,** Immunohistochemical studies on cholecystokinin (CCK)-immunoreactive neurons in the rat using sequence specififc antisera and with special reference to the caudate nucleus and primary sensory neurons, *J. Chem. Neuroanat.,* 1, 11, 1988.
5. **Ju, G., Hokfelt, T., Fischer, J. A., Frey, P., Rehfeld, J. F., and Dockray, G. J.,** Does cholecystokinin-like immunoreactivity in rat primary sensory neurons represent calcitonin gene related peptide?, *Neurosci. Lett.,* 68, 305, 1986.
6. **Ju, G., Hokfelt, T., Brodin, E., Fahrenkrug, J., Fischer, J. A., Frey, P., Elde, R. P., and Brown, J. C.,** Primary sensory neurons of the rat showing calcitonin gene-regulated peptide (CGRP) immunoreactivity and their relation to substance P-, somatostatin-, galanin-, vasoactive intestinal polypeptide- and cholecystokinin-immunoreactive ganglion cells, *Cell Tissue Res.,* 247, 417, 1987a.
7. **Seroogy, K. B., Fallon, J. H., and Leslie, F. M.,** Few cortical cholecystokinin immunoreactive neurons have long projections, *Exp. Brain Res.,* 59, 533, 1985.
8. **Grimmelikhuijzen, C. J. P., Sundler, F. S., and Rehfeld, J. F.,** Gastrin/CCK-like immunoreactivity in the nervous system of Coelenterates, *Histochemistry,* 69, 61, 1980.
9. **Engelhardt, R.-P., Dhainaut-Courtois, N., and Tramu, G.,** Immunohistochemical demonstration of a CCK-like peptide in the nervous system of a marine annelid worm, *Nereis diversicolor* O. F. Muller, *Cell Tissue Res.,* 227, 401, 1982.
10. **Durchon, M.,** Recherches experimentales sur deux aspects de la reporduction chez les Annelides Polychetes: l'epitoquie et al stolonisation, *Ann. Sci. Nat. Zool. Biol. Anim.,* 14, 119, 1952.

11. **Englehardt, R. P., Dhainaut-Courtois, N., and Tramu, C.,** Immunohistochemical demonstration of CCK-like peptide in the nervous system of a marine annelid worm, *Nereis diversicolor* O. F. Muller, *Cell Tissue Res.,* 227, 401, 1982.
12. **Turrigiano, G. G. and Selverston, A. I.,** A cholecystokinin-like hormone activates a feeding-related neural circuit in lobster, *Nature,* 344, 866, 1990.
13. **Nässel, D. R., Ohlsson, L. G., and Cantera, R.,** Metamorphosis of identified neurons innervating thoracic neurohemal organs in the blowfly: transformation of cholecystokininlike immunoreactive neurons, *J. Comp. Neurol.,* 267, 343, 1988.
14. **Lundquist, T. and Nässel, D. R.,** Substance P-, FMRFamide-, and gastrin/cholecystokinin-like immunoreactive neurons in the thoraco-abdominal ganglia of the flies *Drosophila* and *Calliphora, J. Comp. Neurol.,* 294, 161, 1990.
15. **Dockray, G. J., Duve, H., and Thorpe, A.,** Immunochemical characterization of gastrin/cholecystokinin-like peptides in the brain of the blowfly, *Calliphora vomitaria, Gen. Comp. Endocrinol.,* 45, 491, 1981.
16. **Ono, J. K.,** Synaptic connections in the buccal ganglia of *Aplysia* mediated by an identified neuron containing a CCK/gastrin-like peptide co-localized with acetylcholine, *Brain Res.,* 493, 212, 1989.
17. **Bokisch, A. J., Osborne, N. N., and Walker, R. J.,** Actions of cholecystokinin on neurones in the central nervous system of the snail *Helix aspersa, Comp. Biochem. Physiol.,* 75(1), 171, 1983.
18. **Holmgren, S. and Nilsson, S.,** Bombesin-, gastrin, CCK-, 5-hydroxytryptamine-, neurotensin-, somatostatin-, and VIP-like immunoreactivity and catecholamine fluorescence in the gut of the elasmobranch, *Squalus acanthias, Cell Tissue Res.,* 234, 595, 1983.
19. **Sankaran, H., Wong, A., Khan, S. J., Peeke, H. V. S., and Raghupathy, E.,** Bioassayable cholecystokinin in the brain of the goldfish, *Carassius auratus, Neuropeptides,* 9, 103, 1987.
20. **Beinfeld, M. C., Trubatch, J. R., and Brownstein, M. J.,** Cholecystokinin peptides in the brain and pituitary of the bullfrog *Rana catesbiana*: distribution and characterization, *Brain Res.,* 268, 192, 1983.
21. **Thorndyke, M. C.,** Cholecystokinin (CCK)/gastrin-like immunoreactive neurones in the cerebral ganglion of the protochordate ascidians *Styela clava* and *Ascidiella aspersa, Regul. Pept.,* 3, 281, 1982.
22. **Hokfelt, T., Herrera-Marschitz, M., Seroogy, K., Ju, G., Staines, W. A., Holets, V., Schalling, M., Ungerstedt, U., Post, C., Rehfeld, J. F., Frey, P., Fischer, J., Dockray, G., Hamaoka, T., Walsh, J. H., and Goldstein, M.,** Immunohistochemical studies on cholecystokinin (CCK)-immunoreactive neurons in the rat using sequence specific antisera and with special reference to the caudate nucleus and primary sensory neurons, *J. Chem. Neuroanat.,* 1, 11, 1988.
23. **Baile, C. A. and Della-Fera, M. A.,** Central nervous system cholecystokinin and the control of feeding, *Ann. N.Y. Acad. Sci.,* 448, 424, 1985.
24. **Ingram, S. M., Krause, R. G., II, Baldino, F., Jr., Skeen, L. C., and Lewis, M. E.,** Neuronal localization of cholecystokinin mRNA in the rat brain by using *in situ* hybridization histochemistry, *J. Comp. Neurol.,* 287, 260, 1989.
25. **Baile, C. A. and Della-Fera, M. A.,** Central nervous system cholecystokinin and the control of feeding, *Ann. N.Y. Acad. Sci.,* 448, 424, 1985.
26. **Emson, P. C., Lee, C. M., and Rehfeld, J. F.,** Cholecystokinin octapeptide: vesicular localization and calcium dependent release from rat brain *in vitro, Life Sci.,* 26, 2157, 1980.
27. **Dodd, J. and Kelly, J. S.,** The actions of cholecystokinin and related peptides on pyramidal neurons in the mammalian hippocampus, *Brain Res.,* 205, 337, 1981.
28. **Cortes, R., Arvidsson, U., Schalling, M., Ceccatelli, S., and Hokfelt, T.,** *In situ* hybridization studies on mRNAs for cholecystokinin, calcitonin gene-related peptide and choline acetyltransferase in the lower brain stem, spinal cord and dorsal root ganglia of rat and guinea pig with special reference to motoneurons, *J. Chem. Neuroanat.,* 3, 467, 1990.
29. **Yaksh, T. L., Furui, T., Kanawati, I. S., and Go, V. L. W.,** Release of cholecystokinin from rat cerebral cortex *in vivo*: role of GABA and glutamate receptor systems, *Brain Res.,* 406, 207, 1987.
30. **Jeftinija, S., Miletic, V., and Randic, M.,** Cholecystokinin octapeptide excites dorsal horn neurons both *in vivo* and *in vitro, Brain Res.,* 213, 231, 1981.
31. **Dourish, C. T., O'Neill, M. F., Schaffer, L. W., Siegl, P. K. S., and Iversen, S. D.,** The cholecystokinin receptor anatagonist devazepide enhances morphine-induced analgesia but not morphine-induced respiratory depression in the squirrel monkey, *J. Pharmacol. Exp. Ther.,* 255, 1158, 1990.
32. **O'Neill, M. F., Dourish, C. T., Tye, S. J., and Iversen, S. D.,** Blockade of CCK-B receptors by L-365,260 induces analgesia in the squirrel monkey, *Brain Res.,* 534, 287, 1990.
33. **Dourish, C. T., Hawley, D., and Iversen, S. D.,** Enhancement of morphine analgesia and prevention of morphine tolerance in the rat by the cholecystokinin antagonist L-364,718, *Eur. J. Pharmacol.,* 147, 469, 1988.
34. **Faris, P. L., Komisaruk, B. R., Watkins, L. R., and Mayer, D. J.,** Evidence for the neuropeptide cholecystokinin as an antagonist of opiate analgesia, *Science,* 219, 310, 1983.

35. **Hendry, S. H. C., Jones, E. G., and Beinfield, M. C.,** Cholecystokinin-immunoreactive neurons in rat and monkey cerebral cortex make symmetric synapses and have intimate associations with blood vessels, *Proc. Natl. Acad. Sci. U.S.A.,* 80, 2400, 1983.
36. **McMillen, M. A., Ferrara, A., Schaefer, H. C., Goldering, R., Zucker, K. A., and Modlin, I. M.,** Cholecystokinin mediates a calcium signal in human peripheral blood mononuclear cells and is a co-mitogen, *Ann. N.Y. Acad. Sci.,* 594, 399, 1990.
37. **Flood, J. F., Smith, G. E., and Morley, J. E.,** Modulation of memory processing by cholecystokinin: dependence on the vagus nerve, *Science,* 234, 832, 1987.

Chapter 9

BRADYKININ

I. INTRODUCTION

Bradykinin, a nine-amino-acid residue peptide, is one of the most potent and important of the vasoactive neurotransmitter "memory molecules". Its credentials as a major "memory molecule" in the inflammation-heat-pain-and-memory scenario of both the immunological and central nervous systems are comparable to those of substance P. Like other important molecules in the "memory" class, such as serotonin, substance P, and bombesin, it exerts its second-messenger transmembrane-signaling effect through the stimulatory and excitatory cyclic GMP and PI (phosphatidyl inositol) mechanisms and functions as well as nonclassical growth factor with important influences on stimulating protein synthesis and oncogene expression and activity.

Bradykinin holds a central and critical position in biological defense. It is closely involved and interrelated with the actions and effector molecules of five central cascade systems (the blood coagulation-fibrinolysis, the angiotensin-renin, the EDRF/NO/nitric oxide synthase activation, the arachidonic-acid-mobilization, and the complement cascades). Bradykinin is formed by the arginine esterase activity of the bradykinin-forming enzyme, plasma kallikrein. This enzyme acts on an α_2-globulin substrate called kininogen. Plasma kallikrein is formed from prekallikrein, which is also Fletcher Factor of the blood-coagulation cascade. The mechanism of the formation of kallikrein from prekallikrein (Fletcher Factor) is by the direct action of activated Hageman Factor (Factor XII of the blood-coagulation cascade) on prekallikrein. The bradykinin-degrading enzyme (Kininase II) is also the angiotensin-converting enzyme of the angiotenin-renin cascade. Also involved in these events are the coagulation activation of complement by the alternate pathway with C3a- and C5a-induced histamine release and the activation of prostaglandin E_2 formation and thromboxin synthetase of the arachidonic-acid-mobilization cascade. Bradykinin, accordingly, stands at the center of an extremely important interlocking web of vital biological defense.

II. HISTORICAL BACKGROUND

1909. Two French surgeons, Abelous and Bardier, observed a transient fall in blood pressure in man after the intravenous injection of fractions extracted from human urine.

1928. The hypotensive effect of urine was found by Frey and Kraut[1] to arise from a thermolabile nondialyzable substance they termed "kreislauf-hormon" on the assumption that it was a circulating hormone that was formed in some organ other than the kidney and was excreted in the urine.

1930. In searching for the gland of origin for this hypotensive effect, a similar substance was found in the pancreas by Kraut, Frey, and Werle,[2] and it was named "kallikrein" after the Greek word for the pancreas (Kallikreas). The potency of this substance as a vasodilator caused Frey to use it clinically early on in the 1930s in the treatment of patients with arterial occlusion of the legs, but they complained of *pain*.[3]

1936. Kallikrein subsequently was found in other glands, such as the salivary.[3]

1940. Kallikrein was shown, by Werle and Urhaln,[4] to be readily inactivated in blood and tissue and appeared to have little clinical potential as a hypotensive agent.

1948. *Kallidin* is discovered by Werle and Birek.[5] Earlier it had been observed by Werle's group[6] that kallikrein was virtually inactive on the guinea pig isolated intestine, but that the incubation of kallikrein with blood had markedly increased its activity. In the course of pursuing this lead, it was found that kallikrein was an enzyme that reacted with a plasma component to release a small component, a polypeptide, which Werle first called "Darmkontralierender Stoff" and later, thankfully, renamed "Kallidin". The plasma substrate globulin was named "kallidinogen". Their important discovery, interestingly, antedated by 2 years the analogous observation that renin acts in a similar way, and since then the kallikrein-kinin and renin-angiotensin systems have never been too far apart (although not always recognized as such).

1948. Noncardiovascular effects for kallidin soon were found. It was observed by Erspamer[7] to be a potent stimulator of the isolated uterus.

1949. Rocha e Silva et al.[8] observed that trypsin incubated with blood released a substance that contracted guinea pig intestine slowly. Rocha e Silva named this substance "bradykinin", again from the Greek ("bradys" meaning "slow" and "kinein" meaning "to move").

1955. Rocha e Silva[9] also determined bradykinin to be a peptide and, with Andrade,[10] obtained it in pure form. Bradykinin also was obtained in pure form from ox blood by reacting trypsin with globulin, by Elliot et al.[11-13] Elliot and Lewis also isolated the third naturally occurring mammalian kinin, methionyl-lysyl-bradykinin, from ox blood.

1960. An important breakthrough occurred in 1960 when bradykinin was sequenced and synthesized by a group of Swiss chemists, Boissonnas et al.[14] With synthetic bradykinin some of the earlier observations, obtained with crude extracts by Armstrong, could be confirmed, such as its role in pain production.[15] In addition, new biological actions of bradykinin were discovered, such as its important involvement in inflammation.[16-18]

1965. An observation of great importance was made by Ferreira,[19] who identified the existence and presence of a bradykinin-potentiating factor (BPR) in the venom of the Brazilian snake *Bothrops jararaca*.

1967. A major breakthrough occurred in 1967 with the development of sensitive and specific bioassays for bradykinin, by Ferreira and Vane.[20] Employing this sensitive bioassay, Ferreira and Vane[21,22] were able to demonstrate the rapid disappearance of bradykinin from the pulmonary and other vascular beds. The rapid inactivation of bradykinin in blood and tissues seemed to eliminate bradykinin and other kinins from the category or circulating hormones of physiological significance.

1970. In 1970 Rocha e Silva[23] proposed that these substances be considered tissue or "local" hormones, feeling that they satisfied the essential requirements for this group of substances set down in 1930 by Feldberg and Schilf:[24] (1) locally released from diffuse stores in the body, (2) not having specialized glands of secretion, and (3) rapidly inactivated at the site of release. At present, it is appropriate to add rapid local production to Feldberg and Schilf's local release from storage sites, since many of these local hormones (such as bradykinin, the prostaglandins, etc.) do not have storage sites.

1974. The author and associates[25] described an involvement of bradykinin in cellular immunity. An increase in the levels of the bradykinin-forming enzyme, plasma kallikrein, was identified in venous blood draining rejecting and nonimmunosuppressed kidney allografts in rats.

1975. Correa and Graeff of Ribeirao Preto (Sao Paulo) in Brazil[26] reported a role of importance for bradykinin in the central nervous system. With local application of bradykinin, they identified the pars ventralis of the lateral septal area to be the central nervous system site of a hypertensive action of bradykinin. In 1979 Correa with Snyder and associates[27] at Johns Hopkins (Baltimore) employed histochemical techniques to map a bradykinin nervous system in the rat brain: bradykinin-immunoreactive neuronal cell bodies in the hypothalamus, with fibers and varicose projections going to the periacqueductal gray (PAG), the perirhinal and cingulate cortex, the basal ganglia (ventral caudate putamen), and the lateral septal area of the hypothalamus.

1975. The author, Koppelman, and associates[28-32] described an antimalignant-tumor effect of bradykinin. Intralesional injections of bradykinin slowed and stopped the growth of a transplanted SV-40 virus-induced syngeneic fibrosarcoma in hamsters in a manner that was both dose- and injection frequency-dependent and in association with massive small round cell (lymphocyte) infiltration of the injected tumors and with reversal of the profound depressions in tumor and host lymphoid tissue levels of histidine decarboxylase (HDC), the histamine-forming enzyme. In 1976 Koppelmann and the author[33] described bradykinin-induced enhancement *in vitro* of *in vivo*-sensitized hamster spleen cell antitumor cell cytotoxicity.

1977. Fauve and associates,[34-37] at the Pasteur Institute (Paris), made a series of important observations relating to antiinflammatory and antibradykinin products of malignant Lewis carcinomas in rats. The tumor-produced antiinflammatory product was antimacrophage (macrophagotoxic) and prevented macrophage spreading. The compound was of a low molecular weight and appeared structurally quite similar to bradykinin. It also had an antibradykinin effect in that it could prevent bradykinin-induced macrophage spreading if applied to the macrophages before, but not after, bradykinin application. Some of these macrophagotoxic effects could also be mitigated by induced inflammation.

III. STRUCTURE, FORMATION, AND METABOLISM

Bradykinin (BK), a nine-amino-acid residue peptide, is formed by the arginine esterase activity of the bradykinin-forming enzyme kallikrein acting on an α_2-globulin substrate called kininogen. The structure of bradykinin and kallidin (Lys-bradykinin) are shown in Figure 1, along with the major kinin-degrading enzymes and their sites of action on the bradykinin-kallidin molecules, which are identical except for the addition of the Lys amino acid at the N terminus. The only other naturally occurring kinin in mammals is methionyl-kallidin (Met-Lys-bradykinin). In lower animals, particularly insects, longer kinins are present in strings — as many as 18 amino acids, but all with the original nine, as in bradykinin. Some of these insect kinins are more potent as pain and permeability-enhancing substances than bradykinin.

Of the two major types of kallikreins, glandular kallikrein is of a lower molecular weight (27,000 to 43,000), is an acidic glycoprotein, and is formed in the pancreas and glandular organs. It differs from species to species and from plasma kallikrein and uses a low-molecular-weight (LMW) kininogen (48,000 to 70,000) as its substrate. Plasma kallikrein operates in the blood, is the same from species to species, and is of a higher molecular weight than glandular kallikrein (95,000 to 100,000). Plasma kallikrein is formed by the action of Hageman Factor (Factor XII of the blood-coagulation cascade) on prekallikrein (mol wt 130,000), which is formed in the liver and is identical with Fletcher Factor of the blood-coagulation cascade. The substrate for plasma kallikrein is high-molecular-weight (HMW) kininogen (mol wt 100,000 to 200,000). Negatively charged surfaces or particles activate Hageman Factor, which in turn activates prekallikrein.

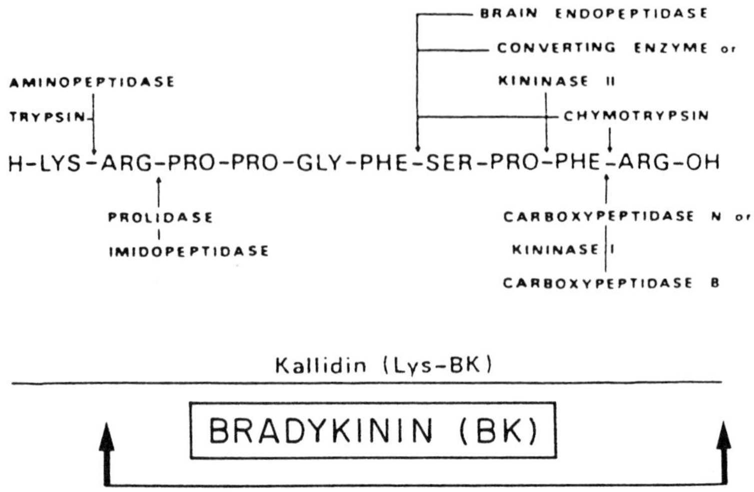

FIGURE 1. Structure, formation, and catabolism of bradykinin (BK) and kallidin (Lys-BK). (From Regoli, D., *Brit. Med. Bull.*, 43, 270, 1987. With permission.)

There are multiple and complex interrelationships between the kallikrein-kinin, coagulation-fibrinolysis, complement renin-angiotensin, and prostaglandin systems (Figures 2 and 3). These triggered enzyme cascades that are of such importance in biologic defense and homeostasis have been well characterized by Colman: "The work of these enzymes illustrates the various control mechanisms possible in proteolysis, including activation of inactive precursors, positive feedback, stochiometric inhibition, multistep amplification, and enzymatic degradation of active products."[38]

As for the degradation of the major kinins, kallidin and bradykinin, of the enzymes active at the amino or N terminus, only aminopeptidase is present in the blood. This enzyme cleaves Lys from Lys-bradykinin and Met-Lys from Met-Lys-bradykinin, but is inactive against bradykinin; the other two enzymes that function at the N terminus (imidopeptidase and prolidase) are located intracellularly and play no significant role in the inactivation of the kinins.

The major enzymes participating in the effective inactivation of the kinins act at the C-terminal end of their molecule and are termed kininase I and kininase II. They are metalloproteins, and their actions may be inhibited by metal-chelating agents. Kininase I (carboxypeptidase *N*) is an exopeptidase that removes the C-terminal Arg from bradykinin and Lys-bradykinin with equal facility, but still leaves behind an active product.[39,40] Oshima et al. have shown kininase I to be a large protein (mol wt 280,000) and different from carboxypeptidase B (mol wt 24,000), a pancreatic enzyme. Kininase I also is active against C3a anaphylotoxin[41] and other peptides containing Lys or Arg at their carboxyl ends.[40,42] Most kininase I is present in the blood.

Kininase II (angiotensin-converting enzyme) is a carboxydipeptidase that splits off the C-terminal dipeptide Phe-Arg and inactivates bradykinin. It also converts angiotension I to angiotensin II (Figure 4). It was originally discovered in horse plasma.[43,44] The conversion of angiotensin I to angiotensin II in plasma was found to be too slow to account for its *in vivo* conversion, and the lung was found to be a major site of conversion by Ng and Vane.[45,46] Most kininase II has been found on the surface of lung endothelial cells and in other vascular

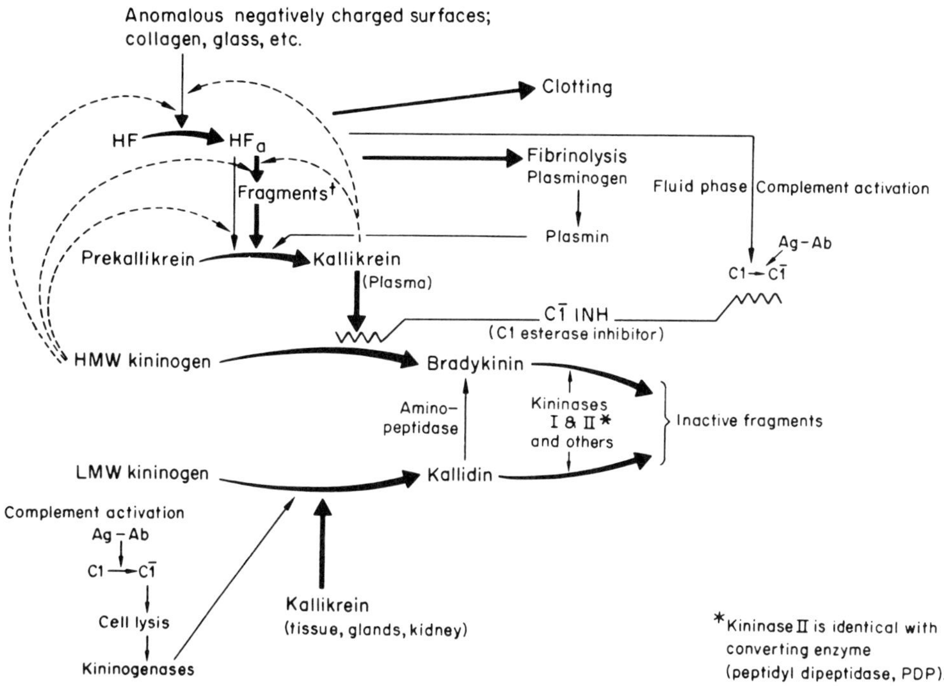

FIGURE 2. Central involvement of bradykinin in multiple and critical defense-system cascades; including the coagulation-fibrinolysis and the complement cascade systems. Hageman Factor (HF) and activated Hageman Factor (HF_a) of the coagulation cascade are shown to play an important role in bradykinin from high-molecular-weight (HMW) kininogen. (From Garrison, J. C., *Goodman and Gillman's The Pharmacological Basis of Therapeutics*, Pergamon Press, New York, 1990, chap. 23. With permission.)

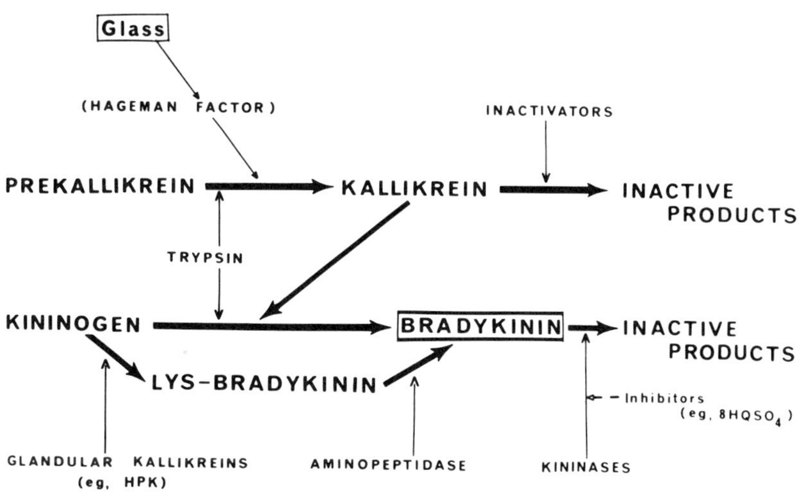

FIGURE 3. Bradykinin at the center of the generalized kallikrein-kinin system of the mammal. (From Dunn, R. S. and Perks, A. B., *Gen. Comp. Endocrinol.*, 26, 165, 1975. With permission.)

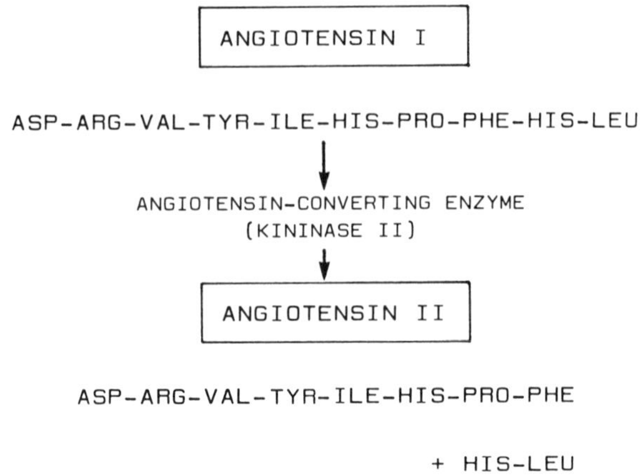

FIGURE 4. The conversion of the ten-amino-acid residue angiotensin I to the nine-amino-acid residue angiotensin II of the angiotensin-renin cascade by angiotensin-converting enzyme which also is kininase II, the bradykinin-degrading/metabolizing enzyme.

beds. It has been well established that angiotensin-I-converting enzyme is the same as kininase II.[47-49] Bradykinin is among the best natural substrates for kininase II, with a Km lower than that of angiotensin I.[50] Kininase is a glycoprotein (mol wt 129,000 to 480,000) with a variable carbohydrate component accounting for the variability of its molecular weight. Bradykinin degradation *in vivo* is quite rapid. With the use of an extracorporeal circulation unit, the half-life of bradykinin was found to be 0.27 min.[51] A comparably short half-life was reported in the cat,[21] employing the blood-bathed organ technique of Vane.[52]

Antikinin activity may be produced by kallikrein inhibitors, which include activated complement-C-1 inactivator, α_2-macroglobulin, antithrombin III (in the presence of heparin), and α_1 antitrypsin.[53] A naturally occurring antikallikrein substance is a protease inhibitor, aprotinin (active substance of Trasylol, Bayer). Aprotinin is obtained from bovine organs and is produced by mast cells. It is specifically a serine esterase inhibitor.[54] An additional role of interest for plasma kallikrein is that is appears to be a physiological activator or prorenin.[55]

IV. RECEPTORS

The short half-life of bradykinin and the complexity and varied nature of its actions on multiple cells, tissues, and organs in multiple species have seriously compromised the identification of bradykinin receptor(s), including purification, cloning, and sequencing. Antagonist peptides have tended to have low affinity for kinin receptors and have short durations of action *in vivo* due to inactivation by peptidases locally or by removal by lung extraction or renal excretion. Lack of specificity also has been a major problem. Some antagonists act as partial agonists, rather than antagonists, in various biological systems, and crossreactivity also has been a major problem. Actions produced by bradykinin and by bradykinin-released mediator molecules, such as histamine, prostaglandin E_2, EDRF (nitric oxide), and others, are difficult to separate and distinguish as purely bradykinin or bradykinin-released substance effects. When the biological response is remote from the kinin-receptor action, the precision of the bioassay may be significantly reduced. Progress in this important but complex field has been reviewed recently by Regoli et al.,[56] Margolius,[57] and Steranka et al.[58]

The multiple actions of bradykinin led Regoli and Barabe,[59] in 1980, to suggest the presence of at least two receptors for bradykinin and to name them B_1 and B_2. Evidence has been presented by multiple investigators for at least two bradykinin receptors. Extensive studies by Regoli et al.[56] have suggested that the contractile effects on the rabbit aorta represent a pure bradykinin-1 (BK_1)-receptor identification system, while contractile effects on the rabbit jugular vein permit characterization of bradykinin-2 (BK_2) receptors. Despite extensive and in-depth studies by the Regoli group and others, much work needs to be done in this critical area.

V. EVOLUTION

> "As the comparative anatomy of former times treated only man and the higher animals, so medicine has hitherto excluded all the pathological phenomena which occur in lower animals. And yet the study of these animals, affording as they do infinitely simpler and move primitive conditions that those in man and vertebrata, really furnishes the key to the comprehension of the complex pathological phenomena which are of special interest in medical science."
>
> Elie Metchnikoff[257]
> Paris, January 15, 1892

Bradykinin has been shown to have two important and interesting evolutionary pathways. One is concerned with external defense, and one with internal defense. The external defense (stings and venoms) is by far the oldest evolution-wise. The internal defense pathway is suprisingly recent and involves only mammals and selected reptiles.

The internal defense pathway is concerned with the blood plasma kallikrein and bradykinin cascade and was the subject of intense and in-depth study in the early and mid-1970s by Erdos and associates,[60,61] Dunn and Perks,[62,63] and Rabito et al.[64] The blood plasma kallikrein-bradykinin cascade enzymes and substrate are shared by all mammals. Birds cannot form bradykinin due to the lack of Hageman Factor (Factor XII) prekallikrein activator. In this regard, it is of considerable interest that all the representatives of two of the four surviving orders of Reptilia studied can make bradykinin and possess the entire complement of enzymes and substrate, as well as the degrading kinin enzyme. These orders are Celonia (turtles) and Crocodilia (crocodiles, alligators, and caimans). Species of turtles studied included *Chelydra serpentina, Pseudemys scripta elegans*, and *Terraphene ornata*. The species of Crocodilia investigated included *Alligator mississippiensis, Crocodylus acutus, Caiman sclerops*, and *Caiman latirostris*. The Crocodilia findings are of particular interest because of the evolutionary position of the Crocodilia as the last living link with the dinosaur-like reptiles of prehistoric times, their comparatively large brains, and their additional evolutionary position as the nearest living relatives of the birds, which evolved from reptilian ancestors, as did man and mammals (but from a different line of reptilian ancestor). Somehow in evolution birds appear to have lost the Hageman Factor and the associated capability of making bradykinin. The presence of Hageman Factor in turtles of the ancient reptilian order of Chelonia stretches back some 200 million years in evolution. By the middle of the Triassic Period (200 million years ago), turtles were both numerous and in possession of essentially all of the basic characteristics and structures of modern-day turtles. Their extraordinary survival, essentially unchanged for so long, is one of the great success stories and "miracles" of evolution.

The lowly snake that was studied *(Natrix rhonbifera)*, of one of the three suborders (Serpentes) of the reptilian order Squamata, was found to possess none of the bradykinin-forming enzymes or substrate.[60] From the bradykinin-forming standpoint, it is comforting to know that we of *Homo sapiens sapiens* are more closely related to the lovable turtle ("And the voice of the turtle is heard in the land," Song of Solomon, *Old Testament*) than to the loathsome snake (loathsome, since that unfortunate affair in the Garden of Eden, also *Old Testament*).

None of the amphibians, bony fish, or cartilaginous fish studied by the various investigators cited earlier were found to possess any of the bradykinin-forming enzymes or substrate, although all were shown to have kininase activity.

The evolutionary trail is much longer in the bradykinin involvement in external defense arena (stings and venoms). Bradykinin's credentials in this critical survival area are quite impressive as well as lengthy. Bradykinin has been identified in the stings of stinging ants of the *Pogonomyrmex* species (harvester ant), the *Solenopsis* species (fire ant), and *Paraponera* species (ponerine ant).[65] Piek et al.,[66,67] in 1989, demonstrated the presence of bradykinin-type kinins in the stings of ants belonging to the subfamilies of Ponerinae *(Paraponera clavata)* and Myrimicinae. The *Paraponera* sting was found to contain a neuroactive kinin.

Bradykinin-type kinins also have been encountered in both social and solitary wasps. Watanabe et al., in 1986,[68] described the presence of Thr^6-bradykinin (BK) and Ala-Arg-Thr^6-bradykinin (BK) in the venom of the social wasp *Polistes rothneyi*. Thr^6-bradykinin is an arthropod and turtle bradykinin that contains the substitution of Thr (threonine) for Ser (serine) at position 6 of the nine-amino-acid residue mammalian bradykinin. It was first identified in nature by Dunn and Perks[63] in 1970 and the red-eared turtle *Pseudemys scripta elegans*. Social wasps sting mainly for defense. Solitary scoliid wasps, on the other hand, sting largely for food.

The nature and actions of bradykinin toxins in the venom of the solitary scoliid wasps is of considerable interest. These solitary wasps generally give one sting for each clearly separate nerve center involved in location, attack, defense, or resistance of the selected prey.[69] Wasps of the family Scoliidae, which prey on the larvae of Coleptera, employ these sting toxins to paralyze and parasitize their intended victims. In 1987 both Piek et al.[70] and Yasuhara et al.[71] described the block of synaptic transmission in the insect central nervous system by toxins from the venom of the scoliid wasp *Megascolia flavifrons*. A mixture of histamine and bradykinin-like active peptides has been described as present in aqueous extracts of venom reservoirs of females of *M. flavifrons*.[71-73] Comparable neurotoxic actions have been identified in the venom of a scoliid wasp of another genus, *Colpa interupta* (F.).[74,75]

In 1990 Piek, Yashura, and associates[76] reported that the venom of *C. interupta* (F.) possessed bradykinin-like activity when tested on a cascade of mammalian smooth-muscle preparations. The venom also was shown to block irreversibly the nicotinic synaptic transmission from the cercal nerve to a giant interneuron in the sixth abdominal ganglion of the cockroach *Periplaneta americana*. In fractionating this venom, only the most active kinin fraction blocked synaptic transmission in the cockroach central nervous system. This most active fraction contained Thr^6BK. The authors then investigated the effects of synthetic Thr^6BK and demonstrated that this synthetic Thr^6BK also could produce irreversible presynaptic activation-induced block of transmission in the *P. americana* cockroach central nervous system.

Bradykinin-type peptides also may be involved in defense/attack activities of venomous amphibians and reptiles. Yasuhara et al.,[77] in 1989, described the isolation of Thr^6BK and three analogues from a methanol extract of the skin of the amphibian frog *Rana rugosa*. Bradykininogen has been encountered in the venoms of an erectile-fanged snake, *Crotalis atrox* (western diamondback rattlesnake), and a true viper, *Vipera berus* (European viper).[78]

VI. NERVOUS SYSTEM

Bradykinin neuronal systems have been identified in the brain.[27] Explorations and investigations in this area are still at the early stages and, from the standpoint of histochemical approaches, are handicapped by the difficulties cited earlier of obtaining reliable antibodies against bradykinin. Some other studies will be presented in the following section on ''Pain''. Studies on canine kidneys provide evidence to suggest that bradykinin acts on postjunctional sites and induces adrenergically induced vasoconstriction by mechanisms that appear dependent upon prostaglandin synthesis.[79]

A. PAIN

As cited in the ''Historical Background'' section, pain was one of the earliest actions of the kallikrein-kinin system to be identified, as Frey tried to use the crude kallikrein's hypotensive effect to alleviate the discomfort of human lower-extremity obliterative arterial disease and arterial insufficiency and got only *pain* for his trouble (the patient's pain). One of the earliest post-Frey identifications of the pain-producing quality of bradykinin came after the sequencing and synthesis of pure bradykinin in 1960. With the real bradykinin at hand, all sorts of studies were possible.

The application of synthetic bradykinin to human blister bases caused pain.[80] In experimental animals, which cannot vocalize or verbalize pain, so-called ''pseudoeffective responses'' are used to indicate painful stimuli. In addition to trying to get away from the source of the pain, these pseudoeffective responses include an increase in heart rate, an increase in blood pressure, and an increase in renal resistance. Pseudoeffective responses have been produced by intra-arterial injections of bradykinin into spleens and other organs of dogs and cats by Ferreira[81] and by Ferreira, Moncada, and Vane.[82,83] Pain also was produced by bradykinin application to the knee joint by Moncada, Ferreira, and Vane.[84] Bradykinin has been reported to stimulate pain receptors and to increase the discharge of afferent sympathetic fibers.[85] This stimulation of cardiac afferent sympathetic nerve fibers may account for some of the pain of cardiac ischemic ''angina''. Bradykinin, a powerful cardiac algesic, has been found to be released by ischemic hearts[86] and is thought to be a major factor in the pain of myocardial ischemia[87] — a factor that may also be potentiated by prostaglandins. In many other tissues, paravascular pain receptors may be sensitized to the algesic effects of bradykinin, by prostaglandin, largely of the E type.[82,88,89] Bradykinin and prostaglandin E were found to be mutually potentiating in their effects on afferent nerve terminals *in vivo* — multifiber strands dissected from saphenous nerves of anesthetized rats.[90]

Chahl,[91] in 1977, also reported that substance P potentiated responses to bradykinin as well as to serotonin and prostaglandin in the dye-leakage response test in rats. In 1986 Shibata et al.,[92] of Japan (Fukuoka and Osaka), made some very important observations regarding the synergistic effects of a combination of bradykinin and substance P on vascular permeability and pain responses in the mouse paw. Their previous studies had demonstrated a major role played by bradykinin in the tissue formalin injection-induced pain response.[93] They cited observations of Bynke et al.[94] that the miosis and aqueous flare produced by bradykinin injection into the vitreous chamber of the rabbit eye could be reduced by substance P antagonists.

The Shibata et al. 1986 study evaluated the pain and vascular permeability responses in mice produced by paw injection of bradykinin, substance P, or a combination of the two substances. Vascular permeability was evaluated by a quantitative dye-leakage test employing pontamine blue. The degree of dye leakage and increase in vascular permeability was found to be greater with the combination of bradykinin and substance P than by a summation of their individual responses. They also demonstrated that use of a substance P antagonist produced a significant reduction in bradykinin-induced vascular permeability increases. Denervation of the sciatic nerve to the study-injection paw also resulted in a significant reduction in bradykinin-induced vascular permeability increases. This denervation observation suggested that the bradykinin response might be sensory nerve mediated. Gamse et al.,[95] in 1980, reported that saphenous nerve division in rats resulted in marked depletion of substance P content of hind-paw skin.

Prostaglandin potentiation of bradykinin-induced hyperalgesia also has received considerable attention.[96-99] In 1988 Taiwo and Levine[96] of San Francisco presented evidence that prostaglandin E_2 (PGE_2) was the principal mediator of bradykinin-induced hyperalgesia in rats. A 1986 study from Japan had suggested PGI_2 as the most potent mediator of bradykin-induced pain responses in dogs.[97] Another 1982 study from Japan, this one employing cats, explored potentiating effects of PGE_2 on bradykinin responses in medial-thalamic nociceptive neurons.[98] In this study, single neurons were recorded from the medial thalamus by stainless steel microelectrodes. Both bradykinin and PGE_2 were administered by arterial infusion. Of the 22 neurons that responded to bradykinin, 13 were potentiated in response by infusion of PGE_2. There was no response of potentiation in the other nine. PGE_2 significantly shortened the mean latency of bradykinin to fire the neuronal activity without changing the duration. Aspirin suppressed the bradykinin-associated increased activity of the medial-thalamic nociceptive neurons, and this aspirin suppression could be overcome by intra-arterial infusion of PGE_2.

In 1984 report from Basel, Switzerland, Schweizer et al.,[99] at Ciba-Geigy described pain-threshold-lowering effects of PGE_2 and PGI_2 on bradykinin-induced pain in the isolated rabbit ear algesia model of Juan and Lembeck.[88,100,101] This potentiation of bradykinin-induced pain responses was found, by Schweizer et al., to be limited to cyclooxygenase-pathway arachidonic acid metabolites. They observed that leukotrienes (LTB_4, LTC_4, and LTD_4) all caused a reversible, concentration-dependent reduction in the pain reflex in this model induced by bradykinin.

A 1974 report from Kiel, Germany, by Mense and Schmidt,[102] contains some especially important information relating to bradykinin as an algesic agent. Their studies involved unmyelinated (group IV) afferent fibers of muscular origin. These unmyelinated fibers with conduction velocity less than 2.5 m/sec constitute more than half the sensory afferents in mammalian nerves and, in cutaneous areas, are considered to play important roles in pain perception, by conducting impulses elicited by noxious stimuli.[103-109] Their study employed chemical stimulation by substances that, at that time (1974), were considered to be pain-producing substances. They utilized bradykinin, histamine, and serotonin. Bradykinin was found to be the most potent of the stimulating substances and the one producing the least degree of desensitization on repeated injections at short intervals.

An important study of bradykinin as a pain mediator was published in 1988 by Steranka et al.[109] In autoradiographic studies, they localized [^{3}H] bradykinin-receptor-binding sites to the pain-reception area of the dorsal root and the spinal cord dorsal horn substantia gelatinosa areas of the guinea pig peripheral and central nervous systems. They also identified [^{3}H] bradykinin labeling over myocardial/coronary visceral afferent nerve fibers. They further reported that bradykinin antagonists could block bradykinin-induced acute vascular pain in the rat and could relieve bradykinin and urate-induced hyperalgesia in the rat paw. A 1989 report from London, by Haley et al.,[110] provided electrophysical evidence of a role for bradykinin as a mediator in prolonged chemical nociception.

In a 1987 report from Tokyo, Iwata et al.[111] described interesting and important findings relating to responses of bradykinin-sensitive tooth-pulp-driven neurons (TPNs) in the cat cerebral cortex. Their study involved properties of single cortical neurons that responded to electrical stimulation of the tooth pulp and to intratooth pulp injections of bradykinin. They recorded the activities of TPNs in both the middle and anterior parts of the coronal gyrus of the cerebral cortex. Bradykinin-sensitive tooth-pulp-driven neurons (BK-TPNs) were found to be located in layer IV of area 36 of the anterior part of the coronal gyrus. It was of considerable interest and importance that these BK-TPNs had a large cutaneous orofacial field and received nociceptive inputs from the facial skin as well as from the tooth pulp. The BK-TPNs also had a higher threshold and longer latency to electrical stimulation than those TPNs that were insensitive to bradykinin. In 1973 and 1979 Inoki et al.[112,113] had demonstrated by radioimmunoassay methodology that bradykinin was released from tooth pulp by noxious stimuli. Narhi,[114] in 1985, had reported that intradental C fibers, but not A fibers, responded to bradykinin.

B. PERIPHERAL NERVOUS SYSTEM

Bradykinin has been shown to produce a long-lasting contraction of the isolated iris of the rabbit eye. A part of this action has been thought to be due to the release of neuronal substance P.[115-117] Multiple substance P and CGRP pools have been identified in sensory nerve fibers in the rabbit iris, and evidence has been presented to identify the rapid-release pool as the bradykinin-sensitive pool.[118]

A 1989 report from Australia describes the participation of a kallikrein-kinin-receptor system in reflexes arising from neural afferents in the dog epicardium.[119]

Bradykinin has been shown to produce depolarization (excitation) and to induce membrane permeability changes in afferent C neurons.[120-123] In 1988 Dray et al.[124] presented evidence that bradykinin-induced stimulation of afferent sensory neuronal fibers is mediated through activation of protein kinase C.

From Heidelberg, Germany in 1988, White and Zimmerman[125] reported that bradykinin-induced release of substance P from nerve endings in the rat cutaneous saphenous nerve neuroma was through a neurochemical action that was independent of electrophysiological excitation, which also has been shown by Otsuka and Konishi,[126] to release neuronal substance P.

C. SPINAL CORD

The autoradiographic identification of the bradykinin-receptor binding of the substantia gelatinosa of the dorsal horn of the guinea pig spinal cord, by Steranka et al.,[109] was cited earlier as the ''Pain'' part of the ''Nervous System'' section of this chapter.

Dray et al.,[127] in 1988, also have described spinal cord depolarization resulting from bradykinin application. In 1990 Dunn and Rang,[128] of the Sandoz Institute (London), observed that application of bradykinin to the spinal cord of the neonatal rat produced depolarizations that could be recorded in either the dorsal or ventral root. Ventral-root responses were inhibited by the substance P antagonist, spantide, whereas those of the dorsal root were not affected. Their findings suggested that bradykinin, by acting through bradykinin receptors, can depolarize the central terminals in the spinal cord of primary afferent nerve fibers.

D. SINGLE CELLS IN CULTURE

In 1984 Reiser et al.,[129] in Wurzburg, Germany, demonstrated bradykinin binding to receptors on neuroblastoma × glioma cells (NG108-15), and Yano et al.,[130] in Japan, described bradykinin-induced enhancement of synaptic transmission when these hybrid cells are cocultured with myotubules. The studies of Yano et al.[130,131] identified PI (phosphatidyl inositol) turnover as the second-messenger root of bradykinin action on these neuronal hybrid-type cells. Reiser et al.[129] attributed bradykinin excitation of these cells to elevation of cellular cyclic GMP levels. Also in 1984 Snider and Richelson[132] observed bradykinin-receptor-mediated cyclic GMP formation in a murine neuroblastoma clone (N1E-115).

Support for both routes of bradykinin action have appeared in subsequent reports. Chuang and Dillon-Carter,[133] of Washington, D.C., in 1988, observed a robust bradykinin-induced PI turnover in hybrid NCB-20 cells. In 1989 Fu et al.,[134] of Japan, described bradykinin-induced production of diacylglycerol (DG) and translocation of protein kinase C (PKC) from cytosol to plasma membrane in these same neuroblastoma-derived hybrid NCB-20 cells. Bradykinin-induced PI turnover also was demonstrated in a novel neuroblastoma × dorsal-root ganglion sensory neuron cell line (F-11) by Francel et al. of the University of Chicago.[135]

In 1989 from London, Burgess et al.,[136] of the Sandoz Institute, reported that bradykinin, which activates polymodal nociceptors, increased levels of cyclic GMP in a capsaicin (pain)-sensitive population of cultured sensory neurons from rat dorsal-root ganglia. This action was found to be from stimulation of guanylate cyclase, with no effect on cyclic AMP. In nonneuronal cells from dorsal-root ganglia, bradykinin increased cyclic AMP, but not cyclic GMP. The bradykinin-induced increase in cyclic GMP in the neurons was blocked completely by removal of extracellular Ca^{++} or by incubation of the cells with calcium channel blockers. Also in 1989 this same group[137] published an interesting study relating to bradykinin and the PI-signaling pathway. They found that bradykinin application to neonatal rat dorsal-root ganglion neurons in culture caused a depolarization (excitation) associated with an inward current and an increase in membrane conductance. They found that direct activation of PKC with phorbol esters mimicked the ability of bradykinin to depolarize these neurons and to increase the rate of ^{45}Ca uptake. Downregulation of PKC blocked both bradykinin- and phorbol-ester-induced ^{45}Ca influx and substantially decreased the proportion of neuronal cells that gave electrophysiological responses to either agent.

Miller and associates,[138,139] at the University of Chicago, in 1988 and 1990, reported their interesting and significant studies relating to bradykinin activation of rat sensory neurons in culture. Higashi et al.,[140] in 1982, had observed that bradykinin powerfully activates sensory neurons. Also in 1982 Yaksh and Hammond[141] presented evidence that this powerful activation of sensory neurons also increased the release of important, powerful, and sensory pain-related neurotransmitters, including 5-HT. These activations are considered of importance in stimulating the transmission of nociceptive information into the central nervous system. The 1990 report of the Miller (University of Chicago) group was concerned with the use of combined patch-clamp-microfluorometric recordings to explore the effects of bradykinin on calcium transients and currents in cultured neonatal rat dorsal-root ganglion neurons. Their experimental approach permitted the use of simultaneous electrophysical-microfluorometric recordings of sensory neurons *in vitro* in culture in order to examine some of the ways in which bradykinin modulates $[Ca^{++}]$ in these cells. Bradykinin was found to depolarize the majority of these neurons (65%). The results of these studies demonstrated that bradykinin regulation of $[Ca^{++}]$ in these neurons depends critically upon the activation state of the cell. The results also suggested mechanisms by which bradykinin may function in the regulation of normal and physiological sensory neuron excitability.

VII. VASCULAR SYSTEM

Vasodilatation and hypotension are major effects of exogenous bradykinin. Following angiotensin-converting-enzyme inhibition by captopril, an increase in bradykinin levels was found, by Mullane and Moncada, to lead to an increase in prostacyclin (prostaglandin I_2) production that enhanced the hypotensive effect of bradykinin.[142,143]

A. CENTRAL NERVOUS SYSTEM EFFECTS

In 1988 Davis and Dostrovsky,[144] of Toronto, described sensory nerve excitation following application of bradykinin to cerebral vessels or intravascular infusion into the sagittal sinus. Bradykinin excited 8 of 15 cells studied. The excited cells generally were those which were spontaneously active and which received nociceptive input from the orofacial region. The sensory neurons of the trigeminal brain stem complex and the thalamus, which were responsive to electrostimulation of the cerebrovasculature, were the sensory neurons studied.

Evidence for bradykinin-involved modulation of cerebral blood flow and cerebral edema was reported by Ellis and associates[145] from the Medical College of Virginia (Richmond) in 1985. In 1987 Ellis et al.,[146] of this Richmond group, reported that topical application of the bradykinin-forming enzyme kallikrein produced pial arteriolar dilatation in the rabbit. This response could be inhibited by a bradykinin antagonist.

In 1987 Rosenblum,[147] also of Richmond, reported that the bradykinin-induced relaxation of pial arterioles in the mouse brain was endothelium dependent. A 1983 report from Munich, Germany, by Whalley et al.,[148] had shown that while bradykinin produced concentration-related relaxations of cat middle cerebral arteries, it was ineffectual in cat basilar arteries. Concentration-related *in vitro* relaxations of the human basilar artery were reported in 1987 from Manchester, England, by Whalley et al.[149]

Unterberg et al.,[150] in 1984 from Munich, published their experiences with the effects of bradykinin on the permeability and diameter of the pial vessels of the cat *in vivo*. They described massive arterial dilatation as occurring immediately after superfusion of bradykinin over an exposed cerebral surface. The maximum increase in arterial diameter of 40% was found at a bradykinin concentration of 4×10^{-5}. Bradykinin superfusion also led to the selective opening of the blood-brain barrier for Na^{+}-fluorescein, but not for FITC-dextran or FITC-albumin.

B. PERIPHERAL EFFECTS

Open arteriovenous shunts[151] and increased blood flow to capillaries[152] by bradykinin has been reported, as well as an increase in vascular permeability.[153] Bradykinin has been observed to produce immediate dilatation in arterioles, with an increase in pressure in capillaries and venules. Endothelial cells contract and pull apart, with partial detachment of intercellular connections. The major work of Majno and associates[154-156] over the years has provided much key information as to the sites and mechanisms of vasoactive substances actions on blood vessels to increase their permeability. His group's initial studies were with "histamine-like" substances (histamine, serotonin, bradykinin) given into subcutaneous tissues, skin, and muscle, with permeability being located histologically by a vascular carbon labeling of blood vessels at leakage points (colloidal carbon particles became deposited against the vascular basement membrane). They found the site of action of bradykinin and these other vasoactive mediators of inflammation to the postcapillary venules, with a pulling apart of endothelial cells in this area to produce gaps through which particulate matter, such as colloidal carbon, could pass to become deposited behind the endothelial cells and against the vascular basement membrane.

In the lung, endothelial leak was produced by bradykinin in bronchial, but not pulmonary, vessels. Epinephrine in the lung did not produce bronchial endothelial gaps, such as bradykinin and the other agents.[157] Bradykinin, on chronic infusion into pulmonary arteries of unanesthetized sheep, has been found to increase fluid and protein flow in chronic lung lymph fistulas in hypoxic lung.[158] In unanesthetized sheep, induced hypercapnic acidosis was found to stimulate *in vivo* production of bradykinin, which, with superimposed hypoxia (which inhibits angiotensin-converting enzyme activity and decreased bradykinin degradation), produced an increase in pulmonary lymph flow and an increase in lung proteins.[159]

Although initial studies by Majno et al. were by injections into tissues, it was also observed later, by Majno with Gabbiani and associates,[160] that intra-arterial injections of the studied vasoactive agents, such as bradykinin, were as effective as tissue injections — endothelial cell gaps and carbon leaks, with the resulting blackening of affected postcapillary venules.

The effect of bradykinin (and histamine) on vascular permeability has been found to be directly on endothelial cells, while that of other permeability agents, such as C5a (des-arg form), leukotriene B4, and formyl tripeptide (FMLP), is polymorph-dependent with rapid involvement of PMNs required.[161] In a 1990 report from the University of California–San Francisco, Frantz et al.[162] presented evidence that bradykinin produces fetal pulmonary vasodilatation by at least two mechanisms: one dependent on PGI_2 production and one independent of PGI_2 production. The PGI_2-independent mechanism appeared to be the predominating one.

In 1986 Gawlowski and Duran[163] described a dose-dependent increase in macromolecular permeability following topical application of bradykinin to the hamster cheek-pouch preparation. Baxter et al.,[164] in 1987, reported studies also employing the hamster cheek-pouch model and with a simple one-dimensional mathematical model that related the number of "leaky" sites in postcapillary venules to the extravasation of macromolecules in terms of effective microvascular permeability. In this study and with this mathematical model, bradykinin topical application produced a fourfold increase in effective microvascular permeability.

In 1986 Mullins[165] of Louisville, KY, studied intra-arterial infusion of bradykinin into the lower extremity of anesthetized dogs. There was an immediate increase in the flow of protein-rich hind-paw lymph. After 210 min of bradykinin infusion, lymph flow was three times greater than baseline, and lymph protein concentration remained doubled.

In 1989 Svensjo and Joyner[166] investigated the effects of continuous and intermittent bradykinin stimulation of hamster cheek-pouch microvessels. They found that intermittent stimulation of microvessels in the cheek bones of hamsters, with bradykinin, produced a maximal response that was not reduced by subsequent 30-min pulses or applications. Reducing the stimulation to 15 min did produce some reduction in the maximum number of venular leaky sites produced by the subsequent and repeated application of bradykinin. Their results indicated that there was only a minor reduction in responsiveness in this hamster cheek-pouch preparation as long as the stimulatory interval was longer than 15 to 30 min. After the termination of a continuous bradykinin stimulation for the development of venular leaking sites, the immediate application of another stimulus leads to the development of another response that is comparable in magnitude to the original response.

Employing Doppler technology, Treede et al.,[167] of Johns Hopkins, in 1990, investigated cutaneous blood flow and vasodilatation changes in the hairy skin of anesthetized monkeys, following intradermal injections of bradykinin. A significant increase in blood flow occurred, with a spreading cutaneous vasodilitation ("flare"-like response), although no actual visible flare developed in this hairy primate with thick and pigmented skin.

In 1990 Feletou and Teisseire,[168] of France, reported that angiotensin-converting-enzyme inhibition (depression of bradykinin degradation) potentiates bradykinin relaxation of isolated porcine resistance arteries, and that the enzyme is present in the vascular walls of these vessels and is not necessarily located in endothelial cells.

Bradykinin antagonism was observed, in 1990 by Burch and DeHaas[169] of Baltimore, to inhibit carrageenin-induced edema formation in rats. In 1981 Amelang et al.,[170] of the University of Houston (Texas), described a response in which combined intra-arterial infusions of bradykinin and histamine resulted in increased lymph flow and protein transport that were greater than encountered with either agent alone.

In considering vascular involvement in bradykinin-induced elevations of metabolism at the local level, most biologic activities, particularly those of a stimulative level, require energy and glucose as a major source of biologic energy production. Local stimulation of energy production is of importance for the meeting of energy needs. Bradykinin-infused intra-arterially in humans increases twofold the blood flow to an extremity. Arterial and venous monitoring during infusion indicated no oxygen consumption difference. The increase in blood flow with bradykinin infusion was associated with increased arteriovenous differences in glucose concentration and an increase in muscle uptake of glucose. These effects were abolished by indomethacin taken orally. It was concluded that bradykinin, in addition to affecting blood flow, also affects glucose metabolism *via* endogenously synthesized prostaglandins.[171] In 1932 Frey et al. observed that kallikrein, in addition to its hemodynamic effects, also lowered blood glucose levels.[172] Thus, it would appear from the metabolic perspective that bradykinin, as a local hormone, does its work in association with local increases in blood flow and glucose entry into restricted and required sites.

VIII. IMMUNE RESPONSE SYSTEM

A. LYMPHOCYTE EFFECTS

1. DNA Synthesis and Mitogenic Effects

Bradykinin *in vitro* has been found, by Perris and Whitfield,[173] to increase the mitogenic activity of rat thymic lymphocytes, as determined by the percentage of cell populations in the metaphase. In addition, bradykinin *in vitro* with rat thymocytes has been found, by Whitfield et al.,[174] to increase twofold the level of uptake of ^{3}H-thymidine in the first hour of incubation. Employing rat lymphocytes from peripheral lymph nodes, the effect of bradykinin *in vitro* on the incorporation or uptake of DNA precursors (^{3}H-thymidine, ^{3}H-uridine, and ^{14}C-leucine) was studied by Stachurska et al.[175] Incorporation of the precursors was markedly enhanced at all concentrations of bradykinin employed.

2. Enhancement of Antitumor Cell Cytolytic Activity of Tumor-Sensitized Spleen Cells

Intralesional injections of bradykinin into transplanted syngeneic SV-40-virus-induced and transplanted fibrosarcomas in hamsters slowed and halted the growth of these tumors. Spleen cells from the approximately 39% of the animals that received tumor cells and did not develop tumors (rejected the cells), and from those whose tumor growth was halted by bradykinin when placed in tissue culture medium with added bradykinin, showed significantly enhanced killing of tumors cells, as determined by the ^{51}Cr-release assay. Specific ^{51}Cr release from target tumor cells produced by adding bradykinin to spleen cells from animals whose tumor growth was halted by bradykinin injections was found, by Koppelmann and the author,[33] to be significantly greater than the effect of adding bradykinin to the spleen cells of normal animals ($p < 0.00025$) or to the spleen cells of animals with noninjected tumors ($p < 0.01$) (Figure 5). Bradykinin added *in vitro* also increased the antitumor cell cytolytic activity of spleen cells from animals that rejected tumor cells (did not develop tumors), in comparison with adding bradykinin to spleen cells from normal animals ($p < 0.025$).

3. Chemotaxis

A 1982 publication from Rostock, Germany, by Paegelow and Lang,[176] described a bradykinin-induced 2.5-fold increase in guinea pig lymphocyte chemotaxis. Bradykinin also increased the chemotactic activities of lymphocytes from rats and cats (modified Boyden chambers).

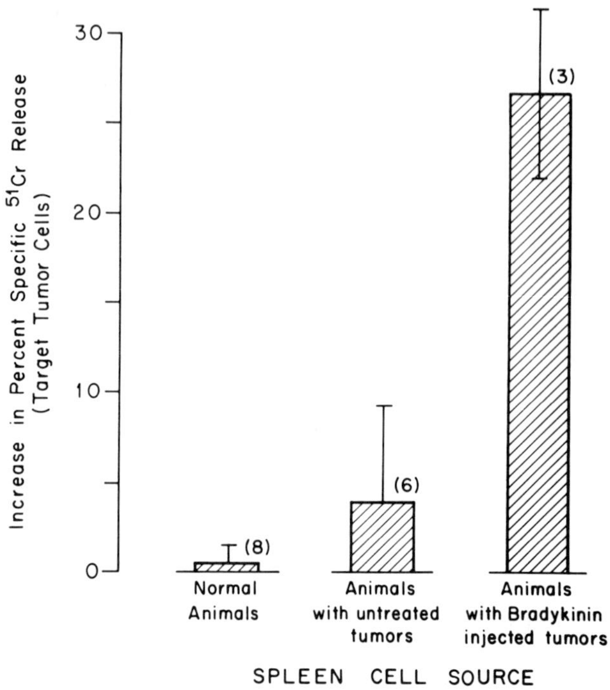

FIGURE 5. Bradykinin-induced enhancement *in vitro* of tumor cytolytic activity of spleen cells of *in vivo* tumor-exposed hamsters. The greatest enhancement of antitumor cytolytic activity (^{51}Cr-release assay) is found in animals whose tumor growth has been halted with intralesional tumor injections with bradykinin.

B. NEUTROPHIL EFFECTS

Nielsen et al.,[177] in 1988, reported from Copenhagen that bradykinin stimulation of human neutrophils (PMNs) resulted in the formation and release of the 5-lipoxygenase products leukotriene B_4 (LTB_4) and 5-hydroxyeicosa-tetraenoic acid (5-HETE) from endogenous arachidonate. The authors interpreted these results as indicating the presence of specific binding receptors for bradykinin on human PMNs, and an additional inflammation-producing mechanism for bradykinin.

C. MACROPHAGE EFFECTS

A 1989 report from Baltimore, by Burch and Tiffany,[178] describes bradykinin-stimulation production of tumor necrosis factor (TNF) and interleukin-1 by murine macrophages. Tumor necrosis factor was the predominant cytokinin resulting from this bradykinin action.

D. ORGAN ALLOGRAFT REJECTION

The acute rejection of an organ allograft, without immunosuppression, in experimental animals, and despite immunosuppression in man, provides the classic example of the most dramatic physiologic response to bradykinin — progressive swelling, edema, pain, and vascular engorgement and disruption, with reddish to blackish discoloration. It was not surprising to the author, from these observations, that elevated plasma levels of kallikrein, the bradykinin-forming enzyme, were found, by the author and associates,[25,179] in venous blood-draining renal allografts in rats and in coronary sinus venous blood draining-rejecting heterotropic heart allografts in rats.

E. CANCER CONTROL

An anti-inflammatory and phagotoxic compound (or compounds) has been found (by Fauve et al.,[180] of the Pasteur Institute of Paris) to be released from malignant cells (teratocarcinoma). Tumor cells producing this substance repulsed macrophages *in vivo* and *in vitro*. This compound was characterized by protease degradation and high-performance gel-filtration chromatography (GFC) to be a peptide of molecular weight between 500 and 2,000 Da and has been called by Fauve et al.[34-37] ''antikinin'' (AK) because of its chemical similarities and biological antagonistic effects to bradykinin. The chromatographic behavior of bradykinin and the anti-inflammatory phagotoxic AK compound were very similar on GFC columns. Both substances were retained far beyond their molecular-weight elution volume, due to cationic charges and probably also due to other absorption effects. The chemical properties also showed many similarities. There was a difference in the sensitivity of bradykinin and the AK compound toward some endoperoxidases. Furthermore, bradykinin was inflammatory, and the other AK compound was anti-inflammatory. During purification by high-resolution GFC, the anti-inflammatory and antimacrophage activities produced by tumor cells could not be dissociated. There also seemed to be some crossreactivity between the AK and the anti-bradykinin (BK) antibodies — good anti-BK antibodies for radioimmunoassay are notoriously difficult to obtain. The AK for these studies was obtained from the supernatant of Lewis lung carcinoma cultures. There was also a clear antagonistic effect of the AK compound and BK on the morphology and spreading characteristics of peritoneal macrophages. Bradykinin caused spreading of macrophages to increase from a normal of 50% to 10%. The AK compound given before BK could prevent the BK-induced increase in macrophage spreading, but not if given after BK.

In the period 1974 to 1978, a series of articles from the author's surgical research laboratory at the Harbor–UCLA Medical Center in Torrance, CA, were published by the author, Koppelmann, and associates,[28-33] relating to the role and involvement of bradykinin and its forming enzyme, kallikrein, in experimental malignant tumor growth, rejection, and control. These studies also included the related role and involvement of histamine and its forming enzyme in these events. The hamster malignant tumors in these studies were syngeneic SV-40-virus-induced fibrosarcomas.

Intralesional injection of bradykinin in doses ranging from 50 to 250 μg slowed and halted the growth of these tumors in a manner that was both dose and injection-frequency associated. Injected tumors were found to have marked infiltration by small lymphocytes (Figure 6). Noninjected tumors had marked depression in the levels of their histamine-forming enzyme, histidine decarboxylase. The histidine decarboxylase levels were most depressed in those malignant tumors that were growing most rapidly. In addition to the bradykinin-injection-induced tumor growth slowing and halting and the marked tumor infiltration with lymphocytes, there was also a marked elevation in tumor and host spleen histidine decarboxylase levels. The associated bradykinin-induced enhancement of antitumor cell cytolytic activity of tumor-sensitized host spleen cells was cited above in detail in the ''Lymphocyte Effects'' section.

The author was quite pleased to see confirmation of the above mid-1970s studies in several mouse tumor study models from Fukuoka, Japan, by Mashiba and Matsunaga[181] in 1985. These authors employed single intralesional injections of 100, 250, 500, and 1,000 μg of bradykinin. Two types of tumors were studied; a homotransplantable sarcoma (sarcoma 180) in ddY mice, and a spontaneous mammary tumor (MM46) of C3H/He mouse origin. The single intralesional injections were at 7 and 8 d after tumor cell injections and when the tumor size had reached about 10 mm. Single injections of 500 μg of bradykinin were found to be effective in inhibiting tumor growth and in decreasing tumor size, as were single injections of 1,000 μg of bradykinin. The bradykinin doses were given in oil or gelatin.

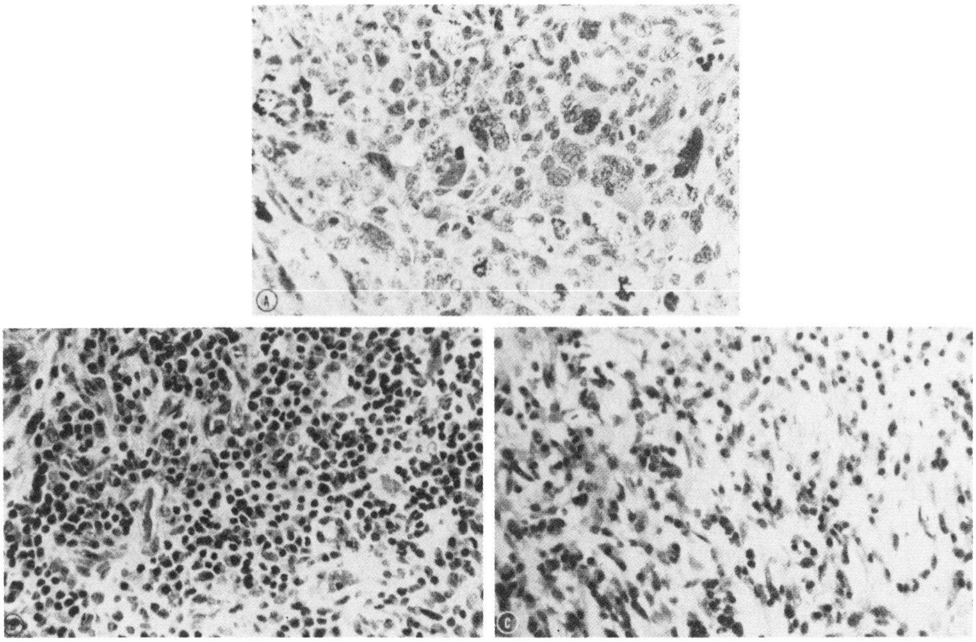

FIGURE 6. Photomicrographs showing (A, top) the essential absence of small mononuclear cell lymphocyte infiltration of saline-injected SV-40-virus-induced and transplanted syngeneic hamster fibrosarcoma; (B, lower left) the massive small mononuclear cell (lymphocyte) infiltration of a daily repeated bradykinin (250 μg)-injected tumor; and (C, lower right) the marked small mononuclear cell (lymphocyte) infiltration of a noninjected malignant fibrosarcomatous tumor of a two-tumor animal in which only one tumor was injected. These findings suggest a systemic as well as a local effect of intralesional tumor injections with bradykinin. (From Koppelmann, L. E., Moore, T. C., Lemmi, C. A. E., and Porter, D. D., *Surgery*, 78, 181, 1975. With permission.)

Control injections were of saline (phosphate-buffered saline, PBS). Bradykinin also was combined with an immunostimulant streptococcal preparation emulsified in oil or gelatin. Tumor-inhibitory effects were not observed with 100 and 250 μg of bradykinin emulsified in oil. It was of particular interest that single intralesional tumor injections of bradykinin would suppress malignant tumor growth for as long as 2 weeks. This finding raises the question of bradykinin-induced long-term potentiation (LTP) of immunological memory and would appear to put into a class with substance P as a mediator of immunological (as well as central nervous system) memory.[182,183]

F. LYMPHOCYTE TRAFFIC EFFECTS

The extremely important and critical studies of both Gowans[184-187] and Ford[188-193] have made major contributions to the identification of the existence, site locations, and significance of the traffic of small recirculating lymphocytes into and through lymph nodes and lymphoid tissues. Gowans identified the postcapillary venule as the anatomical site within the lymph node where the passage of the small recirculating lymphocyte from the nodal arterial blood into the substance of the lymph node occurs. An especially important publication by Schoefl[194] in 1972 established the pathway of these small recirculating lymphocytes into the lymph node to be between lining vascular endothelial cells rather than through the substance of these high endothelial cells (rats) which line the important postcapillary venules. This traffic of lymphocytes is of great importance in both the intiation and the development of the immune response.[187,193]

Major observations made in the early and late 1960s by Majno, Palade, Schoefl, and associates[154-156,194] established the post-capillary venule as the site of action along the vascular tree for the macromolecular vascular leakage caused by bradykinin, histamine, and serotonin. They employed carbon particles as their macromolecular study vehicles and demonstrated by electron microscopic study that endothelial cell contraction was the cause of this macromolecular leakage.

An additional significant contribution to this important area was made in 1979 by Svensjo et al.[195] of Sweden (Uppsala). In a combined morphological and physiological study of bradykinin-induced macromolecular efflux in the everted hamster cheek-pouch-permeability study preparation, these authors demonstrated a prompt, immediate, massive, and dose-related increase of large vascular pores for macromolecular leakage (from 2–3 to 158 leakage sites with 200 ng/ml of bradykinin) and the exclusive occurrence of these leakage sites in the postcapillary venules. These responses were produced by single applications of bradykinin to the cheek-pouch preparation. Continuous 90-min bradykinin superfusions of these pouches (150 ng/ml) increased the number of macromolecular leakage sites from 2 (±1) to 188 (±25), with a peak increase at 20 min after the initiation of the superfusion. No macromolecular leaks were observed in arterioles or capillaries, or in venules greater than 20 μm in diameter. The leakage sites occurred in venules 9 to 16 μm in diameter. Electron photomicrographs showed that the leaks occurred between adjacent endothelial cells, due to separation of loose junctions in the postcapillary venules, which created gaps of approximately 1 μm in diameter — *large enough for a small recirculating lymphocyte to squeeze through*. There was no desensitization or tachyphylaxis to this important bradykinin response. As long as exposure to bradykinin, in single application studies, is less than 5 min and not repeated any more often than every 4 min, repeated applications of a given bradykinin dose, after the return in the number of leakage sites toward the control and pre-application levels, produce a similar increase in the number of macromolecular leakage sites in the postcapillary venule similar to the increase of the initial application.

Extremely important studies in Canberra, Australia, at the Australian National University (John Curtin School) in 1962 and 1963, by Morris and Hall, led to the development of the sheep (large mammal) model for the extended study of the fine and critical points in lymphocyte traffic through the primary peripheral lymph nodes of this smallest large mammal in which these studies practically can be carried out.[196,197]

In 1965 Hall and Morris identified the origin of lymphocytes in the efferent lymph of single lymph nodes as nodal arterial blood (95 to 98% of the cells)[198] and described the immediate effect of antigen administration into the lymph node drainage area on the lymphocyte cell outputs into the study lymph node efferent lymph lymphatic (chronically cannulated).[199] In an important 1967 observation, Hall, Morris, et al.[200] found that with the use of particulate antigen for immunization in the popliteal lymph node drainage of the sheep, an immune response could be limited anatomically to that drainage area so long as nodal efferent lymph could be collected from the efferent lymphatic (chronically cannulated) and diverted from the systemic vasculature and, hence, from the total animal. The especially important 1977 report of Hay and Hobbs[201] established an antigen-induced increase in lymph node blood flow (radioactive microsphere study) as an important contributor to the enhancement of lymphocyte traffic and a stimulation and triggering of the immune response.

The supported-sheep lymphocyte traffic model of the author has permitted short-period (minutes) collection of popliteal lymph node efferent lymph from chronic silastic efferent lymphatic cannulas, without the disruptive effects of unpredictable animal sitting and standing (see Appendix A). Simultaneous popliteal afferent lymphatic cannulation (microsurgery with fiberoptic intense lighting and magnification utilizing human coronary artery bypass technology and instrumentation) has permitted afferent lymphatic cannulation and prompt and

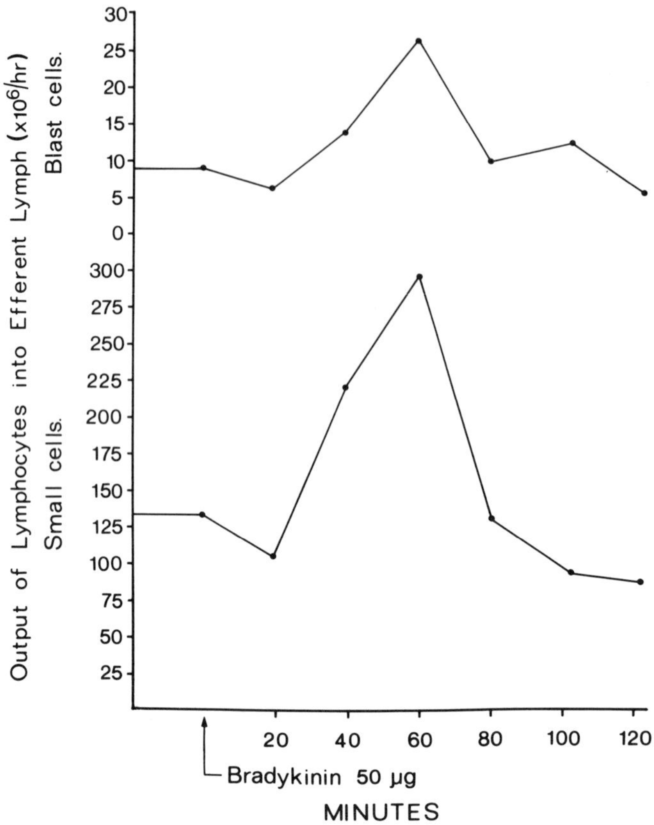

FIGURE 7. The sharp increase of short duration in the output of both blast and recirculating small lymphocytes into a popliteal efferent lymphatic (chronically cannulated) of an awake and nonsedated sheep following acute infusion of bradykinin (50 μg) into a cannulated afferent lymphatic of this popliteal lymph node. A mild dip in the output of both small recirculating and blast lymphocytes occurred in the first 20-min collection postinjection. (From Moore, T. C., *Immunology*, 51, 455, 1984. With permission.)

effective delivery in short infusion time of potent and often very short-lived vasoactive neurotransmitter molecules, such as bradykinin.

Acute infusion of bradykinin (50 μg) into a cannulated popliteal afferent lymphatic of a sheep has been found, by the author and associates[202-204] to increase promptly and sharply the output of both small recirculating and blast lymphocytes into nodal efferent lymph (chronically cannulated). The peak early postacute infusion elevation in the output of both types of lymphocytes was at 60 min postacute infusion (Figure 7). A later (6 h) postinfusion increase in the output of small recirculating lymphocytes also was found (Figure 8). Even during the profound depression in lymphocyte traffic that occurs during general anesthesia, acute infusion of bradykinin into cannulated popliteal lymph node afferent lymphatics has been observed to increase promptly and sharply the output of both small recirculating and blast lymphocytes (Figure 9). This response was targeted toward T cells, as acute bradykinin infusion had little effect on the output into efferent lymph of Ig^+ (B cells) lymphocytes (Figure 10). It was also demonstrated in the author's 1982 and 1984 publications that the bradykinin-induced increase in lymphocyte traffic, time-wise, was on a prostaglandin inhibitory "leash". By giving an infusion of indomethacin into the cannulated popliteal lymph node afferent lymphatic by cannula a short time (1 h) before and immediately prior to bradykinin infusion, the inhibitory effect was blocked and the single infusion of 50 μg of

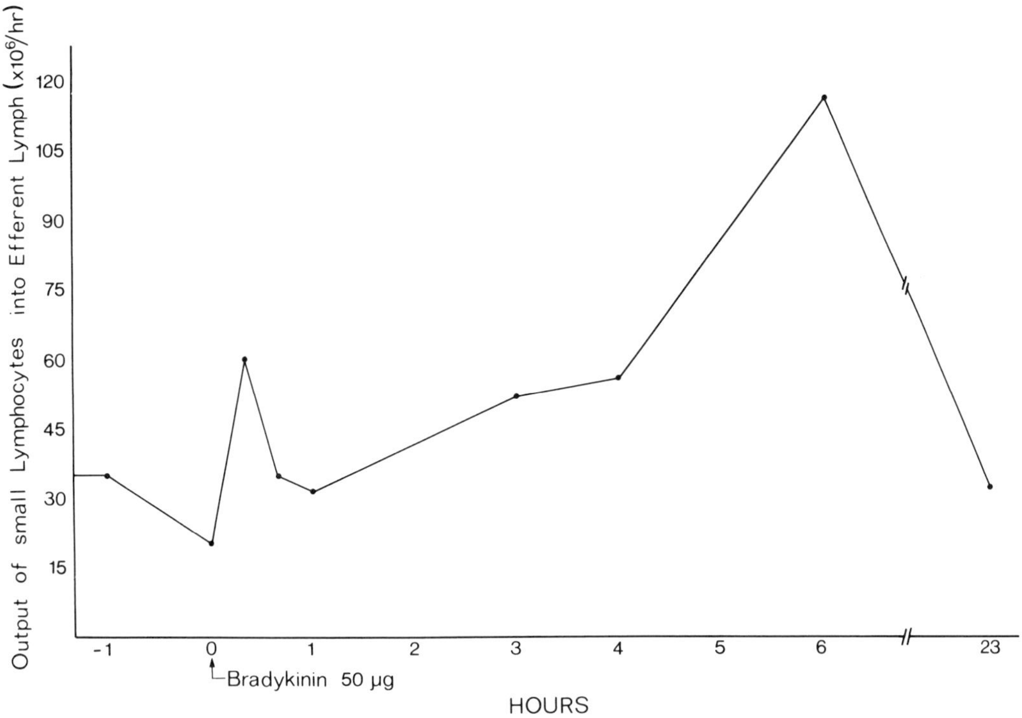

FIGURE 8. Biphasic increase in small-recirculating-lymphocyte outputs into the efferent popliteal lymphatic of a sheep following acute infusion of bradykinin (50 μg) into a cannulated popliteal lymph node afferent lymphatic. The greatest peak in output elevation occurred at 6 h postinfusion, with a return to preinfusion levels at 23 h postinfusion. (Data for graph from studies by Moore, T. C., et al., *IRCS Med. Sci.*, 8, 272, 1980.)

bradykinin produced an increasing elevation in the output and traffic of small recirculating (and blast) lymphocytes, which exceeded 20 and 27 h (Figures 11 to 13). We have here another bradykinin-stimulated LTP relating to the immune response system.

G. IMMUNOLOGICAL SIGNIFICANCE OF BRADYKININ

Bradykinin is an important component of the immune response system as a part of the kallikrein-kinin cascade and amplification system that interrelates closely with the coagulation-fibrinolysis, complement, renin-angiotensin, prostaglandin, and EDRF/NO/nitric oxide and NO synthase systems. Bradykinin is an important stimulator substance. Both its initiation and amplification of major actions (such as pain, vasodilatation, and increased vascular permeability immunostimulation, including LTP) greatly enhance its importance in neurovascular immunology and related events, including immune response and central nervous system "memory". Its involvement in cancer anti-inflammatory events and potentials for use in cancer control give it considerable potential and excitement for further study.

IX. OTHER MOLECULES

A. PROSTAGLANDINS

The increase in the production and the release of prostaglandins is an intriguing action of bradykinin because the prostaglandins intervene both as mediators and potentiators and as modulators and inhibitors of a number of the actions of bradykinin.[205]

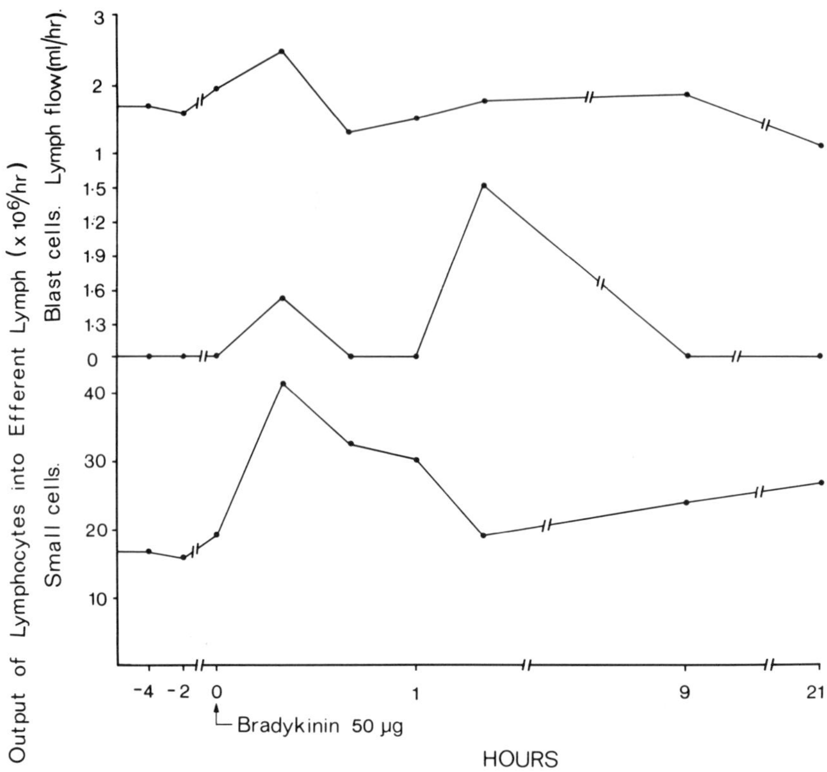

FIGURE 9. Acute infusion of bradykinin (50 μg) into a cannulated sheep popliteal lymph node afferent lymphatic modulates/decreases the profound anesthesia-induced depression of lymphocyte traffic and the output of small recirculating and blast lymphocytes into nodal efferent lymph. (From Moore, T. C., *Am. J. Surg.*, 147, 807, 1984. With permission.)

Bradykinin has been found to stimulate the release of prostaglandins from the kidney.[206,207] It has been found to stimulate the release of prostaglandins from many organs and structures, including the lung,[208] cat spleen,[209] blood vessels,[210-212] dog uterus,[213] and heart.[214,215]

Pulmonary release of prostaglandins from guinea pig lung had been shown, by Piper and Vane,[216] as a consequence of anaphylaxis. *In vivo* release of prostaglandins from rabbit lung by bradykinin also has been shown by Vane and Ferreira.[217] Augmented release of prostaglandin from an organ signifies either enhanced biosynthesis, or diminished degradation, of prostaglandin, since there is no storage of prostaglandin in tissues.[218] Caution also must be exerted in the interpretation of prostaglandin release from organs in *ex vivo* studies, as some prostaglandin may be released from most organs by manipulations required to prepare them for study. Also, studies carried out before the discovery of PGI_2 (prostacyclin) should be reevaluated, since the prostaglandin released may have been prostacyclin. The use of anti-inflammatory nonsteroid drugs (indomethacin, aspirin, meclofenamate) may be employed in evaluating the drug release of prostaglandins and the resulting effects attributable to the released prostaglandins.[219] This is not a simple matter. As kinins may release prostaglandins,[220] prostaglandins may also lead to *in vitro* activation of plasma kallikrein.[221]

Bradykinin has been found to induce a tenfold increase in prostacyclin production from endothelial cells of calf and pig aorta and from human umbilical cord vein.[222] Also, an important mechanism for bradykinin-induced arachidonic acid release from platelets has

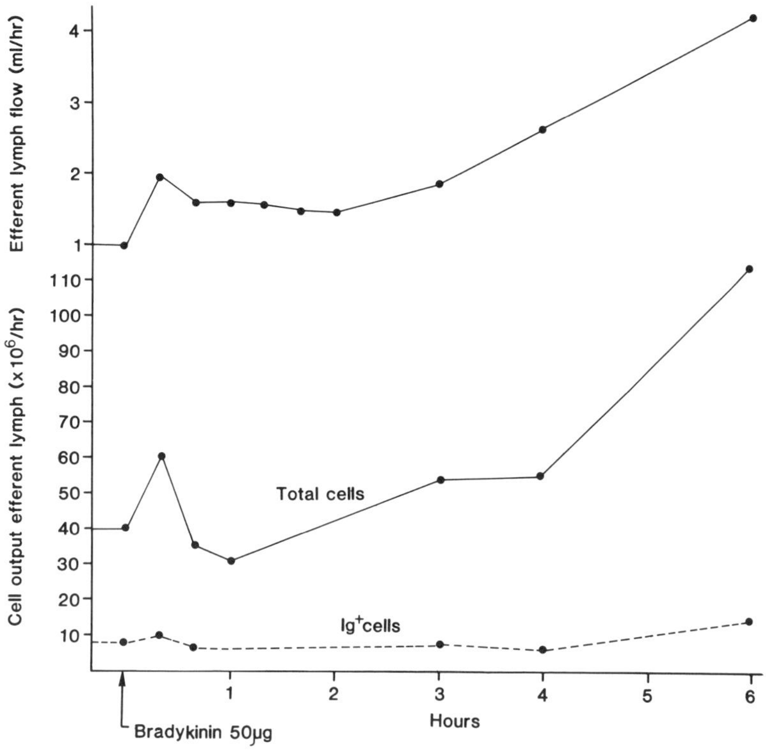

FIGURE 10. Acute infusion of bradykinin-induced increase in lymphocyte traffic is targeted toward T cells, as output of Ig^+ (B cell) small recirculating lymphocytes is little effected. There also is an associated increase in efferent lymph flow. (Data for graft from studies by Moore, T. C., et al., *IRCS Med. Sci.*, 8, 272, 1980).

been described by Bell et al.,[223] whereby bradykinin depletes platelet membrane phosphatidyl inositol, leading to the arachidonic acid release.

There is considerable evidence (cited above) that the bradykinin-stimulated release of prostaglandins may potentiate many of the actions of bradykinin. Of considerable importance is the evidence and observations from Goldberg et al.[224] that prostaglandins may also modulate or inhibit the actions of bradykinin. In a study of the interrelationships of prostaglandins and bradykinin to isolated canine saphenous veins, it was found that endogenous prostaglandins acted to attenuate, rather than to mediate, the venoconstrictor responses to bradykinin. Indomethacin was used to inhibit the synthesis of endogenous prostaglandins. Progressive enhancement of responses to bradykinin was associated with a decreased ability of inhibitors of prostaglandin to enhance the response. When responses (venoconstrictor) to bradykinin were minimal (? prostaglandin depressed), indomethacin enhanced these responses. These are important observations with respect to one of the studies in the author's Ph.D. thesis research of 1980 to 1982,[203] in which bradykinin increased lymphocyte output into efferent lymph, but for a limited time and often with a first 20-min sample decrease in output — events eliminated by prior nodal treatment with indomethacin, which both eliminated the first sample drop in lymphocyte output and extended, without interruption, the bradykinin acceleration in lymphocyte output (? due to elimination of the natural inhibition of this effect by bradykinin-induced production of prostaglandins). Both prostaglandin E_2 and histamine are able to modulate their own stimulating immune responses, triggering actions with respect to cellular responses. No such inhibition at the cellular level appears for bradykinin, and perhaps this effect is taken care of for it by the stimulated production

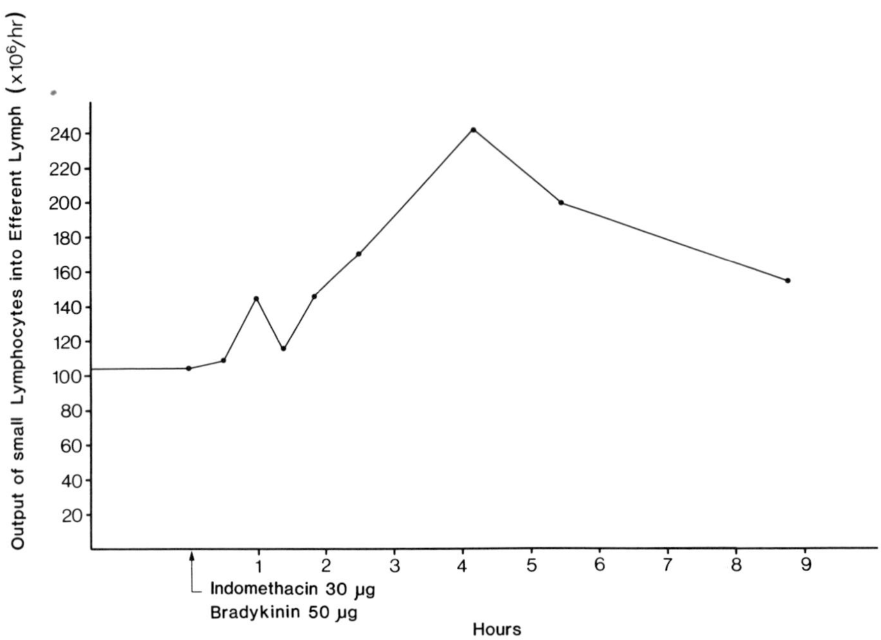

FIGURE 11. Prior infusion of indomethacin into cannulated sheep popliteal lymph node afferent lymphatic prolongs bradykinin infusion-induced increase in small recirculating lymphocyte outputed into efferent lymph up to 9 h postinfusion. (From Moore, T. C., Ph.D. thesis, University of Cambridge, Trinity College, England, 1982. With permission.)

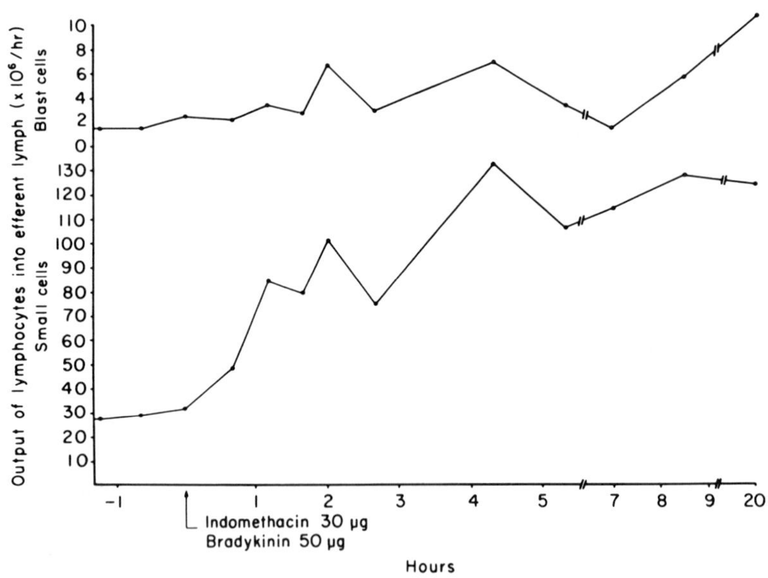

FIGURE 12. Prior infusion of indomethacin into cannulated sheep popliteal lymph node afferent lymphatic greatly prolongs and increases bradykinin-induced increase in lymphocyte traffic and the output of small recirculating and blast lymphocytes into nodal efferent lymph up to 20 h postinfusion of bradykinin (50 μg) as a single infusion. (From Moore, T. C., *Immunology*, 51, 455, 1984. With permission.)

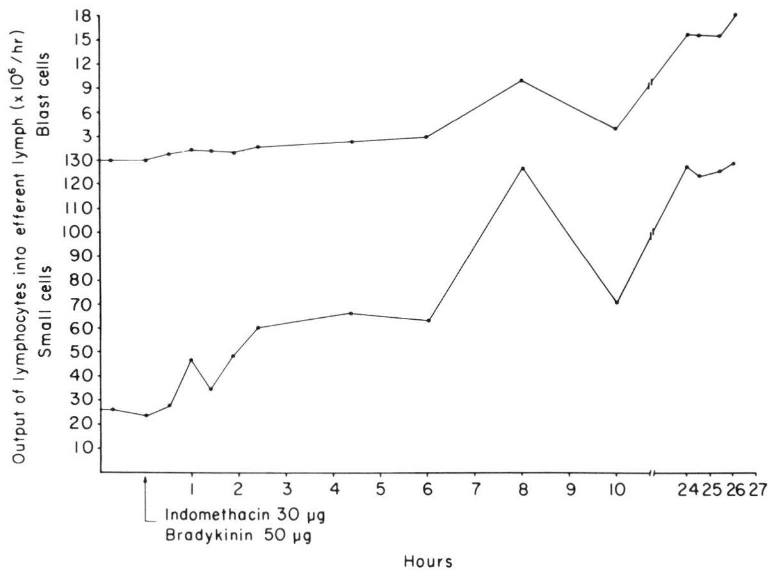

FIGURE 13. Prior infusion of indomethacin into a cannulated sheep popliteal lymph node afferent lymphatic greatly prolongs (more than 27 h) the marked and steady increase in small-recirculating-lymphocyte outputs into the nodal efferent lymphatic (chronically cannulated). This finding suggests that the bradykinin-induced increase in lymphocyte traffic is in a prostaglandin "leash" that is released by the cyclooxygenase-inhibitor indomethacin. It also provides evidence of a potent long-term potentiating effect of bradykinin on this important component of the cellular immune response. (From Moore, T. C., *Immunology*, 51, 455, 1984. With permission.)

of prostaglandins that may modulate, as well as mediate, some of bradykinin's important actions.

In 1982 Furchgott and associates[225] reported that bradykinin-induced relaxation of isolated transverse rings of canine coronary, celiac, superior mesenteric, renal, splenic, pulmonary, gastric, and femoral arteries was endothelium dependent. A similar response was shown for human arteries, but not for those of cats and rabbits where the bradykinin-induced relaxation appears to be prostaglandin-release dependent and may be blocked by cyclooxygenase inhibition. A 1980 report in *Nature*, by Furchgott and Zawadzki,[226] gave birth to this endothelium-dependent phenomenon (EDRF, endothelium-derived relaxing factor). EDRF subsequently was shown by the brilliant work of Moncada and associates of the Wellcome Research Laboratories of Beckenham, Kent (U.K.) to be nitric oxide (NO) and L-arginine-dependent with cyclic GMP and arachidonic acid metabolites involved.[227-236]

In 1985 Heavy et al.,[237] of London, described an aspirin-produced short-lived inhibition of bradykinin-stimulated prostacyclin production in man. In 1986 Clark et al.,[238] of Philadelphia, reported that bradykinin-induced synthesis of prostacyclin in bovine endothelial cells was pertussis toxin sensitive and could be blocked by pertussis toxin/islet-activating protein. Warren et al.,[239] of London, found that intravenous infusion of bradykinin in conscious rabbits stimulated prostaglandin synthesis. In 1989 Burch,[240] of Baltimore, presented evidence of a diacylglycerol (PI pathway) synergy in the bradykinin and thrombin stimulation of prostaglandin synthesis.

Pertussis toxin treatment of fibroblasts, on the other hand, has been found, by Moss et al.[241] in 1988, to enhance sensitivity of these cells to bradykinin-stimulated prostaglandin formation.

B. TRANSMEMBRANE SIGNALING MOLECULES

1. Cyclic GMP Levels of Tissues Elevated *In Vitro* and *In Vivo*

Bradykinin *in vitro* produced a rapid rise in both cAMP and cGMP of guinea pig lung slices. The maximum effect was in 2 min, with a rapid decline to basal levels in 6 to 12 min. The transient nature was presumed to be due to the rapid destruction of bradykinin, since a repeated adding of bradykinin had the same effect and the effect was prolonged by the addition of a kininase II inhibitor (Pro-Glu-Lys-Try-Ala-Pro). Two inhibitors of prostaglandin synthesis (aspirin and indomethacin) completely prevented the rise in cyclic AMP, but had no effect on the elevation in cyclic GMP after adding bradykinin. The cyclic GMP effect was a bradykinin effect, and that on cyclic AMP was found, by Stoner et al.,[242] to be secondary to the bradykinin-induced increase in prostaglandin synthesis.

Bradykinin *in vivo* injected into a rat's paw produced a decrease in tissue cyclic AMP and an increase in cyclic GMP. Carrageenin had a similar effect. The reduction in cyclic AMP was greater with bradykinin, and the effects on cyclic GMP were comparable for the two substances.[243] Earlier studies by these authors had shown acetic-acid-induced pleuritis in rats to be associated with marked reductions in cyclic AMP and increases in cyclic GMP. They also observed that serotonin-, bradykinin-, and histamine-induced bronchospasm had been reduced by cyclic AMP and increased by cyclic GMP.

In 1990 Moncada and associates,[227] at the Wellcome Research Laboratories in Beckenham, Kent, U.K., presented important evidence that the endothelium-dependent vasodilator bradykinin activates the production of cyclic GMP in cultured porcine endothelial cells, via the synthesis of nitric oxide, which in turn stimulates the soluble guanylate cyclase. In 1989 Burgess et al.,[244] of London, had described bradykinin-induced activation of guanylate cyclase and increased cyclic GMP production in rat sensory neurons.

2. Phosphatidyl Inositol (PI) Pathway

In a 1988 report from the National Institutes of Health (N.I.H., Bethesda, MD), Burch et al.[245] reported that phorbol esters and diacylglycerol amplify bradykinin-stimulated prostaglandin in Swiss 3T3 fibroblasts. A 1989 report from Japan, by Fu et al.,[246] described a bradykinin-induced translocation of protein kinase C (PKC) from the cytosol to the membrane in a neuroblastoma cell line.

Also in 1989 Gammon and Morell,[247] of Chapel Hill, NC, described bradykinin-stimulated phosphoinositide hydrolysis and mobilization of arachidonic acid in cultures of rat fetal dorsal root ganglion neurons. In 1986 Lambert et al.,[248] of Duke University (near Chapel Hill), had described bradykinin-induced stimulation of inositol phosphate production in cultured porcine endothelial cells.

C. SUBSTANCE P AND HISTAMINE

Important neuronal and nervous system interrelationships and interactions of bradykinin with substance P have been considered in the "Nervous System" section of this chapter. An important communication from Florence, Italy (in 1988 by Geppetti et al.[249]) described the release by bradykinin of substance P from capsaicin (pain)-sensitive structures in the guinea pig heart. CGRP also was released simultaneously within the substance P.

Several reports have appeared identifying bradykinin-induced release of histamine. In 1986 Ishizaka et al.,[250] of Johns Hopkins reported the occurrence of bradykinin-induced release of histamine and arachidonic acid from mouse mast cells. Bradykinin-stimulated release of histamine from rat and human mast cells was described in 1990 and 1991.[251,252]

Devillier et al.,[253] of Canada (Sherbrooke, Quebec), presented evidence and a speculation regarding a role of the N-terminal arginine in the histamine-releasing activity of bradykinin. They also included substance P in their speculations.

X. HEAT AND INFLAMMATION

A. INFLAMMATION

Bradykinin is an important mediator of many of the important aspects of inflammation — an impressive and, at times, dazzling cascade of events involving interlocking systems, neurological, vascular, humoral, cellular, and biochemical, which are brought into play as a host response to injury or invasion. Localization of the hurt or harm is a first-order priority, and local hormones such as described here provide many aspects of these prompt defense needs. Redness, warmth, swelling, and pain are cardinal attributes of this response, and much has been written about them and the cells and agents that contribute to their development and control.[254-256] The defense contributions of inflammation are not only local, at the site of infection or assault. They may also be systemic. It was observed long ago (Metchnikoff, 1891,[257] 1901[258]) that a local infection substantially increased the general and systemic resistance of the host organism.

The field of inflammation abounds in theories. The recent two-mediator theory of Williams[259] has some points of interest and challenge. This theory relates to the vascular responses of inflammation and considers the chemical mediators of vascular permeability and vascular dilatation separately. In terms of inflammatory edema, prostaglandin is regarded as an important potentiator of histamine- and bradykinin-induced exudation. He suggests that inflammation involves the release (not necessarily concurrently) of mediators important for permeability-increasing ability (not necessarily products of lipoxygenase activity) and vasodilator mediators (especially PGE_2). The magnitude of the exudation in his theory depends on the levels of both types of mediators. This concept, in its simplest form, contends that inflammation involves the separate production of vasodilator mediators (such as prostaglandins) and permeability-increasing mediators (such as histamine and bradykinin). Williams has subsequently found that intradermal injections of archidonate alone resulted in little plasma exudation, but produced an increase in blood flow.[260] Arachidonate also potentiated bradykinin-induced plasma exudation. Both PGE_2 and PGI_2 were potent in increasing blood flow, but little plasma exudation was produced by any of the prostaglandins. Locally injected indomethacin had no effect on basal blood flow and little effect on the degree of exudation produced by bradykinin. Indomethacin did, however, inhibit the vasodilatation and exudation potentiation produced by arachidonate. Both PGE_2 and PGI_2 had similar potency in producing marked potentiation of the plasma exudation that was induced by the intradermal injection of zymosan, which led him to conclude that in the reaction to intradermal injection of zymosan, vasodilatation was the result of the release of arachidonate that was converted to PGE_2 or PGI_2 and that these substances regulated the plasma exudation induced by independently released vascular-permeability-increasing mediators (? histamine, bradykinin, or both). It has also been suggested by others that the vascular permeability produced by bradykinin may be largely due to a venoconstrictor effect of the peptide.[261-263]

Two 1985 publications described the detection of elevated levels of bradykinin in inflammatory exudate produced in tissues by carrageenin and other tissue irritants.[264,265] In the Barlas et al.[264] report, T-kinin (a unique rat kinin) also was found to be elevated.

In 1990 Hargreaves and Costello[266] of the N.I.H. (Bethesda, MD) reported that glucocorticoids could suppress levels of immunoreactive bradykinin in inflamed tissues. They employed microdialysis probes to monitor bradykinin levels.

An especially important publication also appeared in 1990 from Erlangen, Germany, by Kirchhoff, Handwerker, and associates.[267] Their studies examined the hypothesis that the sensitivity of polymodal nociceptors to bradykinin might be increased in inflammed tissue — an important concept of considerable significance. They induced inflammation in the rat dorsal hind-paw skin by subcutaneous injection of blue-stained carrageenin. Three hours after the injection, skin and saphenous nerve were excised, and the chemical sensitivity of mechano-heat-sensitive C- and A-delta fibers were examined *in vitro*, employing repeated superfusions of the receptive fields with 10^{-5} *M* bradykinin. Only units within and at the border of the inflamed area showed signs of sensitization in the form of ongoing activity *and lower thresholds to heat stimuli*. No sensitization to mechanical (von Frey) stimulation occurred. The incidence of bradykinin responsiveness was significantly higher, and the tachyphylaxis to repeated bradykinin application was smaller, inside the inflamed skin than outside or in unconditioned skin. These findings and observations strongly indicated that nociceptive afferents are driven more effectively in inflamed skin, and they represented and suggested an extremely important role of bradykinin in inflammatory pain.

B. HEAT

Bradykinin has long been known to play an important role in the physiological and biochemical responses to thermal heat and thermal burning. Rocha e Silva and Rosenthal,[268] in 1961, reported the release of bradykinin at the skin site of thermal burns. In 1978 Roberts and Wright[269] observed that spinal cord dorsal-horn neurons excited by noxious radiant heat and substance P also may be excited by intra-arterial infusions of bradykinin into the arterial blood supply of the spinal cord.

A 1974 publication by Beck and Handwerker,[270] of Heidelberg, Germany, explored the effects of bradykinin (and serotonin) on various types of cutaneous nerves and made the important "bradykinin and heat" observation that bradykinin injections enhanced spike discharges of unmyelinated nociceptive fibers upon noxious radiation heat stimulation and for a period of 2 to 5 min after a single injection of 5 to 30 μg of bradykinin. This clearly was a long-term potentiation response to heat stimulated by bradykinin. These studies were carried out on the hind legs of cats and following intra-arterial administration of small doses of bradykinin. Spike responses were obtained from single afferent fibers in cutaneous nerves.

In 1977 Kumazawa and Mizumura,[271] of Nagoya, Japan, presented the findings of their studies of thin fiber nerve unitary activities of skeletal muscle (gastroenemius) in the dog stimulation by heat, bradykinin, and other modalities. Bradykinin consistently evoked responses, and it was of particular interest that the responses to thermal heating were consistently obtained in the units that responded to the surface applications of bradykinin. It was of further interest that heating responses were never obtained in units that did not respond to surface chemical stimulation with bradykinin.

Also, in a 1977 report from University College, London, Fitzgerald and Lynn[272] observed that heat-responding units always become more sensitive with repeated heat stimulation, whereas their mechanical sensitivity to stimulation showed no comparable changes when heat stimulation recurred. In 1979 Fitzgerald[273] published an extension of these studies to include the spread of sensitization of polymodal nociceptors in the rabbit from nearby injury and by antidromic nerve stimulation.

Polymodal nociceptors are a group of high-threshold sensory receptors with unmyelinated (C) fibers. Cutaneous polymodal nociceptors have been found to be excited to maximum activity by mechanical and heat stimulation of noxious intensity and by irritant chemicals, such as bradykinin, applied to the skin. A particularly striking property of these units, as cited above, is that severe heating of their receptive fields results in enhanced responsiveness to further heating. This "sensitization" has been observed to be associated with a lowered threshold, a higher frequency of firing at a given temperature, and the development often of background firing.[274-278] These findings raise the question of whether polymodal nociceptors might become sensitized by noxious stimulation outside their receptive fields.

Perl et al.[278] noted in 1974 that extensive skin damage could result in spontaneous firing of polymodal receptors with receptive fields some distance from the site of injury, and

Fitzgerald herself, in 1978,[276] reported that the sensitization of cutaneous nociceptors by spread from a nearby injury could be blocked by local anesthesia.

In Fitzgerald's 1979 reported study,[273] she isolated for study 93 polymodal nociceptor units with unmyelinated axons, from rabbit sural nerves, including some for control data. These study units showed normal sensitization on repeated heating of their receptive fields, measured as a drop in mean heat threshold, as in prior studies. Small injuries were made outside the receptive fields of some polymodal nociceptors. These injuries outside the receptive fields resulted in the development of spontaneous firing and of lowered thresholds to heating of the receptive field. Previous injection of a local anesthetic agent at the site of injury blocked and prevented the spread of the heat sensitization. Antidromic stimulation of the sural nerve proximal to the recording site also produced a heat-sensitization response in the polymodal nociceptors. These observations clearly established the spread of sensitization of polymodal nociceptors from a nearby injury.

In accord with the heat, pain, injury, and memory scenario, it is of particular and especial interest that with the passage of some 12 years from the 1979 Fitzgerald publication, Walters et al.,[279] from Texas, report in the August 16, 1991 issue of *Science* similar neuronal alterations in the marine mollusc *Aplysia* by axonal injury and learning. These findings led the authors to speculate that some learning mechanisms may have evolved from long-lasting adaptive cellular injury in these primitive sensory neurons — and perhaps (the author of this volume also speculates) in considerably less primitive neurons.

Mohan Rao and Bahattacharya,[280] from the Banaras Hindu University in Varanasi, India, in 1988, made the important observation that intracerebroventricularly administered bradykinin in adult rats produced a dose-related increase in rectal temperature. This bradykinin-induced hyperthermia was reduced significantly by pretreatment of the animals with pharmacological agents that selectively reduce rat brain serotonin or prostaglandin activity.

XI. GROWTH FACTOR ACTIONS AND ONCOGENE INFLUENCES

There is increasing evidence that bradykinin, like other stimulatory/excitatory vasoactive neurotransmitter molecules, such as serotonin and bombesin, may also act as a "nonclassical" growth factor and that it may influence, and be influenced by, oncogene expression and activity. Included in these growth factor-like actions are activation of protein formation and alterations of cell structural and biochemical composition. The phenomena of heat sensitization (cited above) and seizure kindling also seem to be associated with similar alterations in protein synthesis and cellular composition.

In 1984 Goldstein and Wall,[281] of Boston, made especially important observations concerning activation of protein formation and cell division by human fetal lung fibroblasts in culture, by bradykinin. They also found that the bradykinin stimulation of protein formation is on a prostaglandin "leash" — much like the lymphocyte-traffic-stimulating effect of bradykinin referred to earlier in this chapter. A 24% increase in protein formation stimulated by bradykinin was increased even further by the presence of indomethacin. The biologically active bradykinin metabolite des-Arg^{8}-bradykinin, formed by cleavage of the C-terminal arginine residue, produced an even greater increase in protein formation (80%) and induced a 61% increase in collagen formation. The action of des-Arg^{9}-bradykinin was not on a prostaglandin "leash", as indomethacin did not alter these levels of protein and collagen formation. In these important studies, bradykinin, and especially its active des-Arg^{9}-bradykinin metabolite, have been shown dramatically to stimulate protein formation and DNA synthesis. In 1986 Marceau and Tremblay,[282] of Quebec, Canada, found a single fibroblast line (rabbit dermis) that responded in a similar way, with a bradykinin-induced increase in

DNA synthesis. This fibroblast line differed from others in this study by doubling faster in culture.

Also in 1984 Straus and Pang,[283] of the University of California – Riverside, reported that bradykinin stimulated [^{3}H]thymidine incorporation and DNA synthesis in resting and serum-deprived NIL8 hamster cells, and stimulated DNA synthesis in IMR-90 human fibroblasts. The human fibroblast cell response occurred only in the presence of indomethacin, which blocked prostaglandin synthesis. Bradykinin-induced [^{3}H]thymidine incorporation by these human fibroblast cells was strongly inhibited by exogenously added PGE_1, PGE_2, PGA_1, PGA_2, PGB_1, and PGB_2.

In 1987 Drinkwater et al.,[284] of Melbourne, Australia, described characterization of the kallikrein (bradykinin- and kallidin-forming enzyme) gene family that encodes epidermal growth factor binding proteins (EGF-BP) type A, type B, and type C. Another member of the kallikrein gene family is the gamma subunit of the 7S nerve growth factor (NGF) complex.

Also in 1987 Parries et al.,[285] of Cornell University (Ithaca, NY) reported highly significant findings demonstrating that expression of a transforming Harvey or Kirsten *ras* gene causes opposing effects in the abilities of platelet-derived growth factor (PDGF) and bradykinin to activate phospholipase-C-mediated phosphoinositide hydrolysis. In NIH 3T3 cells carrying a v-Ha-*ras* gene, the authors found a close correlation between the expression of $p21^{ras}$ and the loss of PDGF-stimulated [^{3}H]$InsP_3$ accumulation. In contrast to this observed *ras*-induced desensitization to $PDGF_1$, *ras*-transformed NIH 3T3 cells exhibited enhanced sensitivity to bradykinin, which was shown to be associated with an elevated level of high-affinity [^{3}H]bradykinin binding. These are observations of great importance and significance, as the family of mammalian H-, K-, and N-*ras* genes encode highly conservative 21-kDa proteins ($p21^{ras}$) that appear intimately to be involved in the control of cellular growth and differentiation. Some mutated *ras* genes, including the v-*ras* genes of the Harvey and Kirsten murine sarcoma viruses, as well as various "activated" *ras* genes found in many human tumor cells, code for altered p21 proteins that are known to be capable of transforming morphologically normal cells into tumorigenic cells.

This same Cornell group of Parries, Racker, and Volonte,[286] in 1988, carried these studies further and expanded them. Intrigued by the numerous examples of significant bradykinin actions in neural cells and stimulated by the 1985 observations of Bar-Sagi and Feramisco[287] that microinjections of *ras* oncogene protein into cells of the P12 rat pheochromocytoma cell line could induce differentiation of these cells (a well-studied model for examining mechanisms of action of NGF, nerve growth factor) in a manner similar to NGF, the authors decided to explore a possible direct link between bradykinin and NGF with respect to the PI cycle of these PC12 cells. They found it demonstrating a bradykinin-induced rapid threefold stimulation of PI turnover in P12 cells cultured with NGF. They also provided evidence that new protein synthesis was required during the incubation to achieve the bradykinin-induced stimulation.

Another 1988 publication (from the Whitehead Institute of Cambridge, MA, by Weinberg and associates[288]) also addressed these same bradykinin-induced and -associated responses impinging on growth factors and oncogenes. They employed normal and transforming H-, K-, and N-*ras* genes transfected into Rat-1 fibroblasts in culture and selected cells that expressed high levels of $p21^{ras}$. Bradykinin was found to have a greatly enhanced affect on PI turnover in these $p21^{ras}$ overexpressor cells. They demonstrated that this effect was due to a 40-fold elevation in the number of bradykinin receptors in these $p21^{ras}$ overexpressor cells.

In 1990 Etscheid et al.,[289] from the University of Chicago, presented evidence for a 16-fold reduction in bradykinin receptors in cultured WI-38 fibroblasts following transformation with the SV-40 virus, an established oncogenic virus.

XII. CONCLUSION

In the "Introduction" part of this chapter, bradykinin was characterized as an immensely important and potent component of the "heat, pain, and memory" scenario of the stimulatory, effector, and memory components of both the immune response and nervous systems, with major and critical involvements in the vascular system as well. Its central position in a number of major biological defense cascade systems and its particularly important growth-factor functions provide additional support for its very major role in biological defense, normal function, and the underpinnings of the molecular basis of learning and memory. Its long, fascinating, and dual role in the evolution of defense and function provide further incentive for continued and expanded investigation of its role in normal physiology and in the molecular biology of normal function and defense. Its established potential for the control of malignant tumor and tissue growth and spread clearly is worth continued in-depth study and serious consideration for pilot clinical use.

REFERENCES

1. **Frey, E. K. and Kraut, H.,** Ein neueskreislaufhormen und seine Wirkung, *Naunyn-Schmiedeberg's Arch. Exp. Pathol. Pharmakol.*, 133, 1, 1928.
2. **Kraut, H., Frey, E. K., and Werle, E.,** Der Nachweiss eines kreislaufhormons in der Pankreasdruse, *Hoppe-Seyler's Z. Physiol. Chem.*, 189, 97, 1930.
3. **Werle, E.,** A short history of the kallikrein-kinin system, in *Fogarty International Center Proceedings, Vol. 27, Chemistry and Biology of the Kallikrein-Kinin System in Health and Disease,* Pisano, J. J. and Austen, K. F., Eds., U.S. Government Printing Office, Washington, D.C., 1976, 1.
4. **Werle, E. and Urhaln, K.,** Uber der Aktivitatzustand des Kallikreins in den Bauchspeicheldruse, *Biochem. Z.*, 304, 387, 1940.
5. **Werle, E. and Birek, U.,** Zurkenntnis des Kallikreins, *Z. Angew. Chem.*, 60A, 53, 1948.
6. **Werle, E., Gotze, W., and Keppler, A.,** Uber die Wirkung des Kallikreins auf den isolierten Darm und uber eine neue darnkontrahierende substanz, *Biochem. Z.*, 289, 217, 1937.
7. **Erspamer, V.,** Active substances in the posterior salivary gland of octopoda, *Acta Pharmacol. Toxicol.*, 4, 213, 1948.
8. **Rocha e Silva, M., Beraldo, W. T., and Rosenfeld, G.,** Bradykinin, a hypotensive and smooth muscle stimulating factor released from plasma globulin by snake venoms and trypsin, *Am. J. Physiol.*, 56, 261, 1949.
9. **Rocha e Silva, M.,** Bradykinin: occurrence and properties, in *Polypeptides which Stimulate Plain Muscle,* Gaddum, J. H., Ed., Livingstone, Edinburgh, 1970, 45–57.
10. **Andrade, S. O. and Rocha e Silva, M.,** Purification of bradykinin by ion exchange chromatography, *Biochem. J.*, 64, 701, 1956.
11. **Elliot, D. F.,** The discovery and characterization of bradykinin, *Handbo. Exp. Pharmacol.*, 25, 7, 1970.
12. **Elliot, D. F. and Lewis, G. P.,** Methionyl-lysyl-bradykinin, a new kinin from ox blood, *Biochem. J.*, 95, 437, 1965.
13. **Elliot, D. F., Lewis, G. P., and Horton, E. W.,** The isolation of bradykinin. A plasma kinin from ox blood, *Biochem. J.*, 74, 15P, 1959.
14. **Boissonnas, R. A., Guttman, S., and Jaquenoud, P. A.,** Synthese de la L-argynyl-L-propyl-glycyl-L-phenalynyl-L-seryl-L-propyl-L-phenylalanyl-L-arginine, un nonapeptide presentant les proprietes de la bradykinine, *Helv. Chim. Acta,* 43, 1349, 1960.
15. **Armstrong, D.,** Pain, *Handbo. Exp. Pharmacol.*, 25, 434, 1970.
16. **Stern, P., Nikulin, A., and Feluga, J.,** The role of histamine and bradykinin in the inflammatory process, *Arch. Int. Pharmacodyn. Ther.*, 140, 528, 1962.
17. **Graham, R. C., Erbert, R. H., Ratnoff, O. D., and Moses, J. M.,** Pathogenesis of inflammation. II. *In vivo* observations of inflammatory effects of activated Hageman factor and bradykinin, *J. Exp. Med.*, 121, 807, 1965.

18. **Lewis, G. P.**, Kinins in inflammation and tissue injury, *Handbo. Exp. Pharmacol.*, 25, 516, 1970.
19. **Ferreira, S. H.**, A bradykinin potentiating factor (BPF) present in the venom of *Bothrops jararaca, Br. J. Pharmacol.*, 24, 163, 1965.
20. **Ferreira, S. H. and Vane, J. R.**, The detection and estimation of bradykinin in the circulating blood, *Br. J. Pharmacol.*, 29, 367, 1967a.
21. **Ferreira, S. H. and Vane, J. R.**, The disappearance of bradykinin and elodoisin in the circulation and vascular beds of the cat, *Br. J. Pharmacol.*, 30, 417, 1967b.
22. **Ferreira, S. H. and Vane, J. R.**, Prostaglandins: their disappearance from and release into the circulation, *Nature*, 216, 868, 1967c.
23. **Rocha e Silva, M.**, *Kinin Hormones*, Charles C Thomas, Springfield, IL, 1970.
24. **Feldberg, W. and Schilf, W.**, *Histamine. Seine Pharmakologie und Bedentung fur die Humoralphysiologie*, Julius Springer, Berlin, 1930.
25. **Moore, T. C., Sinclair, M. C., and Lemmi, C. A. E.**, Elevation in rat plasma kallikrein activity following renal allografting, *Am. J. Surg.*, 127, 289, 1974.
26. **Correa, F. M. A. and Graeff, F. C.**, Central site of the hypertensive action of bradykinin, *J. Pharmacol. Exp. Ther.*, 192, 670, 1975.
27. **Correa, F. M. A., Innis, R. B., Uhl, G. R., and Snyder, S. H.**, Bradykinin-like immunoreactive neuronal systems located in rat brain, *Proc. Natl. Acad. Sci. U.S.A.*, 76, 1489, 1979.
28. **Koppelmann, L. E., Moore, T. C., Lemmi, C. A. E., and Porter, D. D.**, Accumulation of mononuclear cells in tumors with growth slowing and elevation in host splenic histidine decarboxylase activity following repeated tumor injections with bradykinin, *Surgery*, 78, 481, 1975.
29. **Koppelmann, L. E., Moore, T. C., and Porter, D. D.**, Increased plasma kallikrein activity and tumor growth suppression associated with intralesional bradykinin injections in hamsters, *J. Pathol.*, 126, 1, 1978.
30. **Moore, T. C., Koppelmann, L. E., and Lemmi, C. A. E.**, Effect of tumor cell rejection and progressive tumor growth on hamster thymic histidine decarboxylase, *IRCS Med. Sci.*, 5, 493, 1977.
31. **Moore, T. C., Koppelmann, L. E., and Lemmi, C. A. E.**, Decrease in histamine forming enzyme activity of non-metastasizing fibrosarcomas in hamsters with progressive tumor growth, *Ann. Surg.*, 188, 175, 1978a.
32. **Moore, T. C., Koppelmann, L. E., and Lemmi, C. A. E.**, An experimental study of tumor growth and splenic histamine enzymes, *Surg. Gynecol. Obstet.*, 146, 559, 1978b.
33. **Koppelmann, L. E. and Moore, T. C.**, Bradykinin-induced enhancement *in vitro* of sensitized hamster spleen cell anti-tumor cell cytotoxicity, *IRCS Med. Sci.*, 4, 524, 1976.
34. **Fauve, R. M. and Hevin, M. B.**, Inflammation et resistance anti-tumorale. II. Effets antagonistes de la bradykinine et d'une fraction isolee d'un surnageant de culture de cellules malignes sur l'etalement des macrophages, *Ann. Immunol. (Inst. Pasteur)*, 128C, 1079, 1977a.
35. **Fauve, R. M.**, Inflammation and natural immunity, in *Progress in Immunology*, Vol. IV, Fougereau, M. and Daussett, J., Eds., Academic Press, New York, 1980.
36. **Fauve, R. M. and Hevin, M. B.**, Toxic effects of tumour cells on macrophages, in *The Macrophage and Cancer*, James MacBride and Stuart, Edinburgh, 1977b, 264.
37. **Fauve, R. M., Hevin, M. B., and Dauge, M. C.**, Inflammation and activation of macrophages, in *The Macrophage and Cancer*, James MacBride, Edinburgh, 1977, 258.
38. **Colman, R. W.**, The formation of human plasma kinin, *New Engl. J. Med.*, 291, 509, 1974.
39. **Oshima, G., Kato, J., and Erdos, E. G.**, Subunits of human plasma carboxypeptidases N (Kininase I; anaphylaxotonin inactivator), *Biochim. Biophys. Acta*, 365, 344, 1974.
40. **Oshima, G., Kato, J., and Erdos, E. G.**, Plasma carboxypeptidase N, subunits and characteristics, *Arch. Biochem. Biophys.*, 170, 132, 1975.
41. **Bokisch, V. A. and Muller-Eberhard, H. J.**, Anaphylotoxin inactivator of human plasma: its isolation and characterization as a carboxy peptidase, *J. Clin. Invest.*, 49, 2427, 1970.
42. **Erdos, E. G., Nakajima, T., Oshima, G., Gesce, A., and Kato, J.**, Kininases and their interactions with other systems, in *Fogarty International Center Proceedings*, Vol. 27, *Chemistry and Biology of the Kallikrein-Kinin-System in Health and Disease*, Pisano, J. J. and Austen, K. F., Eds., U.S. Government Printing Office, Washington, D.C., 1986, 277.
43. **Skegg, L. T., Kahn, J. R., and Shumway, N. P.**, Preparation and function of the hypertensin-converting enzyme, *J. Exp. Med.*, 103, 295, 1956.
44. **Lentz, K. E., Skeggs, L. T., Woods, K. R., Kahn, J. R., and Shumway, N. P.**, Aminoacid composition of hypertensin II and its biochemical relationship to hypertensin I, *J. Exp. Med.*, 104, 183, 1956.
45. **Ng, K. K. F. and Vane, J. R.**, Conversion of angiotensin I to angiotenin II, *Nature*, 216, 762, 1967.
46. **Ng, K. K. F. and Vane, J. R.**, Fate of angiotensin I in the circulation, *Nature*, 218, 144, 1968.
47. **Cheung, H. S. and Cushman, D. W.**, Inhibition of angiotensin-converting enzyme of rabbit lung by synthetic venom peptides of *Bothrops jararaca, Biochim. Biophys. Acta*, 293, 451, 1973.

48. **Erdos, E. G.,** Conversion of angiotensin I to angiotensin II, *Am. J. Med.,* 60, 749, 1976.
49. **Erdos, E. G.,** Kininases, *Handbo. Exp. Pharmacol.,* 25, 427, 1979.
50. **Erdos, E. G.,** Angiotensin I converting enzyme, *Circ. Res.,* 36, 247, 1975.
51. **McCarthy, D. A., Potter, D. F., and Nicolaides, E. D.,** An *in vivo* estimation of the potencies and half-lives of synthetic bradykinin and kallidin, *J. Pharmacol. Exp. Ther.,* 148, 117, 1965.
52. **Vane, J. R.,** The use of isolated organs for detecting active substances in the circulating blood, *Br. J. Pharmacol.,* 23, 360, 1964.
53. **Fritz, H., Fink, E., and Truscheit, E.,** Kallikrein inhibitors, *Fed. Proc.,* 38, 2753, 1979.
54. **Haberland, G. and McConn, R.,** A rationale for the therapeutic action of aprotinin, *Fed. Proc.,* 38, 2760, 1979.
55. **Rumpf, K. W., Becker, K., Kreusch, U., Schmidt, S., Vetter, R., and Scheler, F.,** Evidence for a role of plasma kallikrein in the activation of prorenin, *Nature,* 283, 482, 1980.
56. **Regoli, D., Rhaleb, N.-E., Dion, S., and Drapeau, G.,** New selective bradykinin receptor antagonists and bradykinin B_2 receptor characterization, *TIPS,* 156, 1990.
57. **Margolius, H. S.,** Tissue kallikreins and kinins: regulation and roles in hypertensive and diabetic diseases, *Ann. Rev. Pharmacol. Toxicol.,* 29, 343, 1989.
58. **Steranka, L. R., Farmer, S. G., and Burch, R. M.,** Antagonists of B_2 bradykinin receptors, *FASEB J.,* 3, 2019, 1989.
59. **Regoli, D. and Barabe, J.,** Pharmacology of bradykinin and related kinins, *Pharmacol. Rev.,* 32, 1, 1980.
60. **Seki, T., Miwa, I., Nakajima, T., and Erdos, E. G.,** Plasma kallikrein-kinin system in nonmammalian blood: evolutionary aspects, *Am. J. Physiol.,* 224, 1425, 1973.
61. **Erdos, E. G., Miwa, I., and Graham, W. J.,** Studies on the evolution of plasma kinins: reptilian and avian blood, *Life Sci.,* 6, 2433, 1967.
62. **Dunn, R. S. and Perks, A. M.,** Comparative studies of plasma kinins: the kallikrein-kinin system in poikilotherm and other vertebrates, *Gen. Comp. Endocrinol.,* 26, 165, 1975.
63. **Dunn, R. S. and Perks, A. M.,** A new plasma kinin in the turtle, *Pseudemys scripta elegans, Experientia,* 26, 1220, 1970.
64. **Rabito, S. F., Binia, A., and Segovia, R.,** Plasma kininogen content of toads, fowl and reptiles, *Comp. Biochem. Physiol.,* 41A, 281, 1972.
65. Encyclopaedia Britannica, Vol. 25, 15th ed., 1985, 907.
66. **Piek, T., Hue, B., Mantel, P., Nakajima, T., Schmidt, J. O., and Yasuhara, T.,** A neuroactive kinin from the venom of the ant *Paraponera clavata,* in *Proc. 30th Dutch Fed. Meet.,* 1989, 289.
67. **Piek, T., Schmidt, J. O., de Jong, J. M., and Mantel, P.,** Kinins in ant venoms — a comparison with venoms of related Hymenoptera, *Comp. Biochem. Physiol.,* 92C, 117, 1989.
68. **Watanabe, M., Yasuhara, T., and Nakajima, T.,** Occurrence of Thr^6-bradykinin and its analogous peptide in the venom of *Polistes rothneyi iwatai,* in *Animal Plant and Microbial Toxins,* Vol. 2, Ohsaka, A., Hayashi, K., and Sawai, Y., Eds., Plenum Press, New York, 1976, 105.
69. **Steiner, A. L.,** Stinging behaviour of solitary wasps, in *Venoms of the Hymenoptera,* Piek, T., Ed., Academic Press, London, 1986, 161.
70. **Piek, T., Hue, B., Mantel, P., Nakajima, T., Schmidt, J. O., and Yasuhara, T.,** A neuroactive kinin from the venom of the ant *Paraponera clavata, Proc. 30th Duth Fed. Meet.,* 1989, 289.
71. **Yashuhara, T., Mantel, P., Nakajima, T., and Piek, T.,** Two kinins isolated from an extract of the venom reservoirs of the solitary wasp *Megascolia flavifrons, Toxicon.,* 25, 527, 1987.
72. **Piek, T., Buitenhuis, A., Simonthomas, R. T., Ufkes, J. G. R., and Mantel, P.,** Smooth muscle contracting factors of the venom of *Megascolia flavifrons, Comp. Biochem. Physiol.,* 75C, 145, 1983.
73. **Piek, T., Mantel, P., and Van Ginkel, C. J. W.,** Megascoliakinin, a bradykinin-like compound in the venom of *Megascolia flavifrons* Fab, *Comp. Biochem. Physiol.,* 78C, 473, 1984.
74. **Piek, T., Hue, B., Mantel, P., Nakajima, T., Pelhate, M., and Yasuhara, T.,** Effects of the venom of *Colpa interrupta* (F.) on vertebrate smooth muscle and synaptic transmission in insect CNS, *Pest. Sci.,* 24, 266, 1988.
75. **Piek, T., Hue, B., Pelhate, M., and Mony, L.,** The venom of the wasp *Campsomeris sexmaculata* (F.) blocks synaptic transmission in insect CNS, *Comp. Biochem. Physiol.,* 87C, 283, 1987.
76. **Piek, T., Hue, B., Mantel, P., Nakajima, T., Pelhate, M., and Yasuhara, T.,** $Threonine^6$-bradykinin in the venom of the wasp *Colpa interrupta* (F.) presynaptically blocks nicotinic synaptic transmission in the insect CNS, *Comp. Biochem. Physiol.,* 96C, 157, 1990.
77. **Yasuhara, T., Ishikawa, O., Nakajima, T., Araki, K., and Tachibana, S.,** The studies on the active peptide on smooth muscle in the skin of *Rana rugosa,* Thr^6-bradykinin and its analogous peptide, Ranakinin-R, *Chem. Pharm. Bull.,* 27, 486, 1979.
78. Encyclopaedia Britannica, Vol. 25, 15th ed., 1985, 909.

79. **Susic, H., Nasjletti, A., and Malik, K. U.**, Bradykinin effects on adrenergic transmission in the canine kidney: relation to prostaglandins, *Am. J. Physiol.*, 241, 146, 1981.
80. **Keele, C. A. and Armstrong, D.**, Substances producing pain and itch, in Philosophical Society Monogr. No. 12, Arnold, London, 1964, 33.
81. **Ferreira, S. H.**, Prostaglandins, aspirin like drugs and analgesia, *Nat. New Biol.*, 240, 200, 1972.
82. **Ferreria, S. H., Moncada, S., and Vane, J. R.**, Prostaglandin and the mechanism of analgesia produced by aspirin-like drug, *Br. J. Pharmacol.*, 49, 86, 1973.
83. **Ferreira, S. H., Moncada, S., and Vane, J. R.**, Some effects of inhibiting endogenous prostaglandin formation on the response of the cat spleen, *Br. J. Pharmacol.*, 47, 48, 1973.
84. **Moncada, S., Ferreira, S. H., and Vane, J. R.**, Inhibition of prostaglandin biosynthesia as the mechanism of analgesia of aspirin-like drugs in the dog knee joint, *Eur. J. Pharmacol.*, 31, 250, 1975.
85. **Uchida, Y. and Murao, S.**, Bradykinin-induced excitation of cardiac sympathetic nerve fibers, *Jpn. Heart J.*, 15, 84, 1974.
86. **Kirmura, E., Hashimoto, F., Furukawa, S., and Hayakawa, H.**, Changes in bradykinin levels in coronary sinus blood after experimental occlusion of a coronary artery, *Am. Heart J.*, 85, 635, 1973.
87. **Staszewska-Barczak, J., Ferreira, S. H., and Vane, J. R.**, An excitatory nociceptive cardiac reflex excited by bradykinin and potentiated by prostaglandins and myocardial ischemia, *Cardiovasc. Res.*, 10, 314, 1976.
88. **Juan, H. and Lembeck, F.**, Action of peptides and other algesic agents on paravascular pain receptors of the isolated perfused rabbit ear, *Naunyn-Schniedeberg's Arch. Pharmacol.*, 283, 151, 1974.
89. **Juan, H. and Seewanns, S.**, Selective reduction by some vasodilators and prostaglandin antagonist SC-19220 of a response to the algesic effect of bradykinin, *Eur. J. Pharmacol.*, 65, 267, 1980.
90. **Chahl, L. A. and Iggo, A.**, The effects of bradykinin and prostaglandin E_1 on rat cutaneous afferent nerve activity, *Br. J. Pharmacol.*, 59, 343, 1977.
91. **Chahl, L. A.**, Interactions of substance P with putative mediators of inflammation and ATP, *Eur. J. Pharmacol.*, 44, 45, 1977.
92. **Shibata, M., Ohkubo, T., Takahashi, H., and Inoki, R.**, Interaction of bradykinin with substance P on vascular permeability and pain response, *Jpn. J. Pharmacol.*, 41, 427, 1986.
93. **Shibata, M.**, Pharmacological study of formalin-induced inflammatory pain response, *Jpn. J. Oral Biol.*, 27, 890, 1985.
94. **Bynke, G., Hakanson, R., Horig, J., and Leander, S.**, Bradykinin contracts the pupillary sphincter and evokes ocular inflammation through release of neuronal substance P, *Eur. J. Pharmacol.*, 91, 469, 1983.
95. **Gamse, P., Holzer, P., and Lembeck, F.**, Decrease of substance P in primary afferent neurones and impairment of neurogenic plasma extravasation by capsaicin, *Br. J. Pharmacol.*, 68, 207, 1980.
96. **Taiwo, Y. O. and Levine, J. D.**, Characterization of the arachidonic acid metabolites mediating bradykinin and noradrenaline hyperalgesia, *Brain Res.*, 458, 402, 1988.
97. **Hori, Y., Katori, M., Harada, Y., Uchida, Y., and Tanaka, K.**, Potentiation of bradykinin-induced nociceptive response by arachidonate metabolites in dogs, *Eur. J. Pharmacol.*, 132, 47, 1986.
98. **Andoh, R., Sakurada, S., Sato, T., Takahashi, N., and Kisara, K.**, Potentiating effects of prostaglandin E_2 on bradykinin and capsaicin responses in medial thalamic nociceptive neurons, *Jpn. J. Pharmacol.*, 32, 81, 1982.
99. **Schweizer, A., Bray, M. A., Brom, R., and Glatt, M.**, Leukotrienes desensitize the nociceptors to the painful stimulus of bradykinin, *Pain*, 20, S249, 1984.
100. **Juan, H.**, Prostaglandins as modulators of pain, *Gen. Pharmacol.*, 9, 403, 1978.
101. **Lembeck, F., Popper, H., and Juan, H.**, Release of prostaglandins by bradykinin as an intrinsic mechanism of its algesic effect, *Naunyn Schmiedeberg's Arch. Pharmacol.*, 294, 69, 1976.
102. **Mense, S. and Schmidt, R. F.**, Activation of group IV afferent units from muscle by algesic agents, *Brain Res.*, 72, 305, 1974.
103. **Bishop, G. H. and Heinbecker, P.**, The afferent functions of non-myelinated or C fibers, *Am. J. Physiol.*, 114, 179, 1935.
104. **Clark, D., Hughes, J., and Gasser, H. S.**, Afferent function in the group of nerve fibers of slowest conduction velocity, *Am. J. Physiol.*, 114, 69, 1935.
105. **Collins, W. F., Nulsen, F. W., and Randt, C. T.**, Relation of peripheral nerve fiber size and sensation in man, *Arch. Neurol. Psychiatry*, 3, 381, 1960.
106. **Reuck De, A. V. S. and Knight, J., Eds.**, *Touch, Heat and Pain, CIBA Foundation Symposium*, Churchill, London, 1966.
107. **Torebjork, H. E. and Hallin, R. G.**, Perceptual changes accompanying controlled preferential blocking of A and C fibre responses in intact human skin nerves, *Exp. Brain Res.*, 16, 321, 1973.
108. **Zotterman, Y.**, Studies in the peripheral nervous mechanism of pain, *Acta Med. Scand.*, 80, 185, 1933.

109. **Steranka, L. R., Manning, D. C., DeHaas, C. J., Ferkany, J. W., and Snyder, S. H.,** Bradykinin as a pain mediator: receptors are localized to sensory neurons, and antagonists have analgesic actions, *Proc. Natl. Acad. Sci. U.S.A.*, 85, 3245, 1988.
110. **Haley, J. E., Dickenson, A. H., and Schachter, M.,** Electrophysiological evidence for a role of bradykinin in chemical nociception in the rat, *Neurosci. Lett.*, 97, 198, 1989.
111. **Iwata, K., Itoga, H., Muramatsu, H., Toda, K., and Sumino, R.,** Responses of bradykinin sensitive tooth-pulp driven neurons in cat cerebral cortex, *Exp. Brain Res.*, 66, 435, 1987.
112. **Inoki, R., Matsumoto, K., Kudo, T., Kotani, Y., and Oka, M.,** Bradykinin as an algesic (pain producing) substance in the pulp, *Naunyn-Schmiedeberg's Arch. Pharmacol.*, 306, 29, 1979.
113. **Inoki, R., Toyama, T., and Yamamoto, I.,** Elaboration of a bradykinin-like substance in dog's canine pulp during electrical stimulation and its inhibition by narcotic and non-narcotic analgesics, *Naunyn-Schmiedeberg's Arch. Pharmacol.*, 279, 333, 1973.
114. **Narhi, M.,** The characteristics of intradental sensory units and their responses to stimulation, *J. Dent. Res.*, 64, 564, 1985.
115. **Bynke, G., Hakanson, R., Horig, J., and Leander, S.,** Bradykinin contracts the pupillary sphincter and evokes ocular inflammation through release of neuronal substance P, *Eur. J. Pharmacol.*, 91, 469, 1983.
116. **Ueda, N., Muramatsu, I., and Fujiwara, M.,** Capsaicin and bradykinin-induced substance P-ergic responses in the iris sphincter muscle of the rabbit, *J. Pharmacol. Exp. Ther.*, 230, 469, 1984.
117. **Wahlestedt, C., Bynke, G., and Hakanson, R.,** Pupillary constriction by bradykinin and capsaicin mode of action, *Eur. J. Pharmacol.*, 106, 577, 1985.
118. **Hakanson, R., Beding, B., Ekman, R., Heilig, M., Wahlestedt, C., and Sundler, F.,** Multiple tachykinin pools in sensory nerve fibers in the rabbit iris, *Neuroscience*, 21, 943, 1987.
119. **Staszewska-Woolley, J. and Woolley, G.,** Participation of the kallikrein-kinin-receptor system in reflexes arising from neural afferents in the dog epicardium, *J. Physiol. (Lond.)*, 419, 33, 1989.
120. **Baccaglini, P. I. and Hogan, P. G.,** Some rat sensory neurones in culture express characteristics of differentiated pain sensory cells, *Proc. Natl. Acad. Sci. U.S.A.*, 80, 594, 1983.
121. **Higashi, H., Veda, N., Gallagher, J. P., and Shinnick-Gallagher, P.,** Chemoreceptors for serotonin (5-HT), acetylcholine (ACh), bradykinin (BK), histamine (H) and γ-aminobutyric acid (GABA) on rabbit visceral afferent neurons, *Brain Res. Bull.*, 8, 23, 1982.
122. **Lindsay, R. M. and Rang, H. P.,** Effects of bradykinin on the membrane currents of rat sensory neurones in tissue culture, *J. Physiol. (Lond.)*, 390, 74p, 1987.
123. **Weinreich, D.,** Bradykinin inhibits a slow spike after hyperpolarization in visceral sensory neurons, *Eur. J. Pharmacol.*, 132, 61, 1986.
124. **Dray, A., Bettaney, J., Forster, P., and Perkins, M. N.,** Bradykinin-induced stimulation of afferent fibres is mediated through protein kinase C, *Neurosci. Lett.*, 91, 301, 1988.
125. **White, D. and Zimmerman, M.,** The bradykinin-induced release of substance P from nerve fibre endings in the rat saphenous nerve neuroma is not related to electrophysiological excitation, *Neurosci. Lett.*, 92, 108, 1988.
126. **Otsuka, M. and Konishi, S.,** Release of substance P-like immunoreactivity from isolated spinal cord of newborn rat, *Nature*, 264, 83, 1976.
127. **Dray, A., Bettaney, J., Forster, P., and Perkins, M. N.,** Activation of a bradykinin receptor in peripheral nerve and spinal cord in the neonatal rat *in vitro*, *Br. J. Pharmacol.*, 95, 1008, 1988.
128. **Dunn, P. M. and Rang, H. P.,** Bradykinin-induced depolarization of primary afferent nerve terminals in the neonatal rat spinal cord *in vitro*, *Br. J. Pharmacol.*, 100, 656, 1990.
129. **Reiser, G., Walter, U., and Hamprecht, B.,** Bradykinin regulates the level of guanosine 3′,5′-cyclic monophosphate (cyclic GMP) in the neural cell lines, *Brain Res.*, 290, 367, 1984.
130. **Yano, K., Higashida, H., Inoue, R., and Nozawa, Y.,** Bradykinin-induced rapid breakdown of phosphatidylinositol 4,5-bisphosphate in neuroblastoma × glioma hybrid NG108-15 cells, *J. Biol. Chem.*, 259, 10201, 1984.
131. **Yano, K., Higashida, H., Hattori, H., and Nozawa, Y.,** Bradykinin-induced transient accumulation of inositol trisphosphate in neuron-like cell line NG108-15 cells, *FEBS Lett.*, 181, 403, 1985.
132. **Snider, R. M. and Richelson, E.,** Bradykinin receptor-mediated cyclic GMP formation in a nerve cell population (murine neuroblastoma clone N1E-115), *J. Neurochem.*, 43, 1749, 1984.
133. **Chuang, D.-M. and Dillon-Carter, O.,** Characterization of bradykinin-induced phosphoinositide turnover in neurohybrid NCB-20 cells, *J. Neurochem.*, 51, 505, 1988.
134. **Fu, T., Okano, Y., Hagiwara, M., Hidaka, H., and Nozawa, Y.,** Bradykinin-induced translocation of protein kinases C in neuroblastoma NCB-20 cell: dependence on 1,2-diacylglycerol content and free calcium, *Biochem. Biophys. Res. Commun.*, 162, 1279, 1989.
135. **Francel, P. C., Miller, R. J., and Dawson, G.,** Modulation of bradykinin-induced inositol trisphosphate release in a novel neuroblastoma × dorsal root ganglion sensory neuron cell line (F-11), *J. Neurochem.*, 48, 1632, 1987.

136. **Burgess, G. M., Mullaney, I., McNeill, M., Coote, P. R., Minhas, A., and Wood, J. N.,** Activation of guanylate cyclase by bradykinin in rat sensory neurones is mediated by calcium influx: possible role of the increase in cyclic GMP, *J. Neurochem.*, 53, 1212, 1989.
137. **Burgess, G. M., Mullaney, I., McNeill, M., Dunn, P. M., and Rang, H. P.,** Second messengers involved in the mechanism of action of bradykinin in sensory neurons in culture, *J. Neurosci.*, 9, 3314, 1989.
138. **Bleakman, D., Thayer, S. A., Glaum, S. R., and Miller, R. J.,** Bradykinin-induced modulation of calcium signals in rat dorsal root ganglion neurons, *in vitro, Mol. Pharmacol.*, 38, 785, 1990.
139. **Thayer, S. A., Perney, T. M., and Miller, R. J.,** Regulation of calcium homeostasis in sensory neurons by bradykinin, *J. Neurosci.*, 8, 4089, 1988.
140. **Higashi, H., Ueda, N., Nishi, S., Gallagher, J. P., and Shinnick-Gallagher, P.,** Chemoreceptors for serotonin (5-HT), acetylcholine (ACh), bradykinin (BK), histamine (H) and γ-aminobutyric acid (GABA) on rabbit visceral afferant neurons, *Brain Res. Bull.*, 8, 23, 1982.
141. **Yaksh, T. L. and Hammond, D. L.,** Peripheral and central substrates involved in the transmission of nociceptive information, *Pain,* 13, 1, 1982.
142. **Mullane, K. M. and Moncado, S.,** Prostacyclin release and the modulation of some vasoactive hormones, *Prostaglandins,* 20, 25, 1980.
143. **Mullane, K. M. and Moncado, S.,** Prostacyclin mediates the potentiated hypotensive effect of bradykinin following captopril treatment, *Eur. J. Pharmacol.*, 66, 355, 1980.
144. **Davis, K. D. and Dostrovsky, J. O.,** Cerebrovascular application of bradykinin excites central sensory neurons, *Brain Res.*, 446, 401, 1988.
145. **Kamitani, T., Little, M. H., and Ellis, E. F.,** Evidence for a possible role of the brain kallikrein-kinin system in the modulation of the cerebral circulation, *Circ. Res.*, 57, 545, 1985.
146. **Ellis, E. F., Heizer, M. L., Hambrecht, G. S., Holt, S. A., Stewart, J. M., and Vavrek, R. J.,** Inhibition of bradykinin- and kallikrein-induced cerebral arteriolar dilation by a specific bradykinin antagonist, *Stroke,* 18, 792, 1987.
147. **Rosenblum, W. I.,** Hydroxyl radical mediates the endothelium-dependent relaxation produced by bradykinin in mouse cerebral arterioles, *Circ. Res.*, 61, 601, 1987.
148. **Whalley, E. T., Wahl, M., and Sampaio, C. A. M.,** Angiotensin-converting enzyme, bradykinin, angiotensin, and cerebral vessel reactivity, *Hypertension,* 5, 34, 1983.
149. **Whalley, E. T., Amure, Y. O., and Lye, R. H.,** Analysis of the mechanism of action of bradykinin on human basilar artery *in vitro, Naunyn-Schmiedeberg's Arch. Pharmacol.*, 335, 433, 1987.
150. **Unterberg, A., Wahl, M., and Baethmann, A.,** Effects of bradykinin on permeability and diameter of pial vessels *in vivo, J. Cerebral Blood Flow Metab.*, 4, 574, 1984.
151. **DePasquale, N. P. and Burch, G. E.,** Digital vascular responses to intra-arterial injections of bradykinin, kallidin, and eledoisin in man, *Circulation,* 34, 211, 1966.
152. **Ludbrook, J., Vincent, A. H., and Walsh, J. A.,** The effect of pyridino/carbamate on the vascular action of bradykinin in the human forearm, *Aust. J. Exp. Biol. Med. Sci.*, 51, 405, 1973.
153. **Haddy, F. J., Emerson, T. E., Scott, J. B., and Dougherty, R. M.,** The effect of kinins on the cardiovascular system, *Handb. Exp. Pharmacol.*, 25, 362, 1979.
154. **Majno, G., Palade, G. E., and Schoefl, G. I.,** Studies on inflammation. II. The site of action of histamine and serotonin along the vascular tree: a topographical study, *J. Biophys. Biochem. Cytol.*, 11, 607, 1961.
155. **Majno, G., Gilmore, V., and Leventhal, M.,** On the mechanism of vascular leakage caused by histamine-type mediators, *Circ. Res.*, 21, 833, 1967.
156. **Majno, G., Shea, S. M., and Leventhal, M.,** Endothelial contraction induced by histamine-type mediators. An electron microscopic study, *J. Cell Biol.*, 42, 647, 1969.
157. **Gabbiani, G., Badonnel, M. C., Gervaboni, C., Portmann, B., and Majno, G.,** Carbon deposition in bronchial and pulmonary vessels in response to vasoactive compounds, *Proc. Soc. Exp. Biol. Med.*, 140, 958, 1972.
158. **Pang, L. M., Stalcup, S. A., O'Brodovich, H., Lipset, J. S., and Mellins, R. B.,** Effects of bradykinin on lung lymph protein and fluid flow, *Fed. Proc.*, 38, 1265, 1979.
159. **O'Brodovich, H. M., Stakup, S. A., Pang, L. M., Lipset, J. S., and Mellins, R. B.,** Bradykinin production and increased endothelial permeability during acute respiratory failure in unanesthetized sheep, *J. Clin. Invest.*, 67, 514, 1981.
160. **Gabbiani, G., Badonnel, M. C., and Majno, G.,** Intra-arterial injections of histamine, serotonin or bradykinin: a topographical study of vascular leakage, *Proc. Soc. Exp. Biol. Med.*, 135, 447, 1970.
161. **Wedmore, C. V. and Williams, T. J.,** Evidence for two types of vascular permeability-increasing mediators: the direct action of histamine and bradykinin, the polymorph-dependent action of C5a, leukotriene B_4, and formyl tripeptide, *Br. J. Pharmacol.*, 73, 209, 1980.
162. **Frantz, E., Soifer, S. J., Clyman, R. I., and Heymann, M. A.,** Bradykinin produces pulmonary vasodilation in fetal lambs: role of prostaglandin production, *J. Appl. Physiol.*, 67, 1512, 1989.

163. **Gawlowski, D. M. and Duran, W. N.,** Dose-related effects of adenosine and bradykinin on microvascular permselectivity to macromolecules in the hamster cheek pouch, *Circ. Res.,* 58, 348, 1986.
164. **Baxter, L. T., Jain, R. K., and Svensjo, E.,** Vascular permeability and interstitial diffusion of macromolecules in the hamster cheek pouch: effects of vasoactive drugs, *Microvasc. Res.,* 34, 336, 1987.
165. **Mullins, R. J.,** Bradykinin causes a prolonged increase in skin microvascular permeability, *J. Surg. Res.,* 40, 540, 1986.
166. **Svensjo, E. and Joyner, W. L.,** The effects of intermittent and continuous stimulation of microvessels in the cheek pouch of hamsters with histamine and bradykinin on the development of venular leaky sites, *Microcirc. Endothel. Lymphat.,* 1, 381, 1984.
167. **Treede, R.-D., Meyer, R. A., Davis, K. D., and Campbell, J. N.,** Intradermal injections of bradykinin or histamine cause a flare-like vasodilatation in monkey. Evidence from laser Doppler studies, *Neurosci. Lett.,* 115, 201, 1990.
168. **Feletou, M. and Teisseire, B.,** Converting enzyme inhibition in isolated porcine resistance artery potentiates bradykinin relaxation, *Eur. J. Pharmacol.,* 190, 159, 1990.
169. **Burch, R. M. and DeHaas, C.,** A bradykinin antagonist inhibits carrageenan edema in rats, *Naunyn-Schmiedeberg's Arch. Pharmacol.,* 342, 189, 1990.
170. **Amelang, E., Prasad, C. M., Raymond, R. M., and Grega, G. J.,** Interactions among inflammatory mediators on edema formation in the canine forelimb, *Circ. Res.,* 49, 298, 1981.
171. **Dietze, G., Wickmayr, M., Mayer, L., Buttger, I., and V. Funcke, H.,** Bradykinin and human forearm metabolism: inhibition of endogenous prostaglandin synthesis, *Hoppe-Seyler's Z. Physiol. Chem.,* 359, 369, 1978.
172. **Frey, E. K., Kraut, H., and Werle, E.,** Uber die blutzuckersenkende Wirkung des Kallikreins (Padutins), *Klin. Wochenschr.,* 11, 846, 1932.
173. **Perris, A. D. and Whitfield, J. F.,** The mitogenic action of bradykinin on thymic lymphocytes and its dependence on calcium, *Proc. Soc. Exp. Biol. Med.,* 130, 1198, 1969.
174. **Whitfield, J. F., MacManus, J. P., and Gillan, D. J.,** Cyclic GMP mediation of bradykinin-induced stimulation of mitotic activity of DNA synthesis in thymocytes, *Proc. Soc. Exp. Biol. Med.,* 133, 1270, 1970.
175. **Stachurska, J., Roszkowski, W., Kopec, M., and Ryzewski, J.,** Effects of bradykinin on rat lymphocytes, *Proc. Soc. Exp. Biol. Med.,* 1532, 408, 1976.
176. **Paegelow, I. and Lange, P.,** Pharmacological studies on lymphocytes. I. Effects of 5-hydroxytryptamine, bradykinin, and lymphokines on the migration of lymphocytes *in vitro, Agents and Actions,* 10, 255, 1982.
177. **Nielsen, O. H., Bukhave, K., Ahnfelt-Ronne, I., and Rask-Madsen, J.,** Source of endogenous arachidonate and 5-lipoxygenase products in human neutrophils stimulated by bradykinin and A23187, *Gut,* 29, 319, 1988.
178. **Burch, R. M. and Tiffany, C. W.,** Bradykinin stimulates the production of tumor necrosis factor and interleukin 1 from macrophages, *Clin. Res.,* 37, 406A, 1989.
179. **Lemmi, C. A. E., Vigran, T. S., Koppelmann, L. E., Yu, B. M., and Moore, T. C.,** Alterations in kinin metabolism of coronary sinus venous blood of heterotopic rat heart allographs and isographs, *Arch. Surg.,* 110, 1070, 1975.
180. **Fauve, R. M., Hevin, G., Jacob, A., Gaillard, J. A., and Jacob, F.,** Anti-inflammatory effects of murine malignant cells, *Proc. Natl. Acad. Sci. U.S.A.,* 71, 4052, 1974.
181. **Mashiba, H. and Matsunaga, K.,** Tumor-inhibitory effect of intralesional injection of bradykinin and immunostimulants in mice, *Cancer Lett.,* 29, 177, 1985.
182. **Moore, T. C., Whitley, G. A., Lami, J. L., and Said, S. I.,** Neuro-vascular immunology: vasoactive neurotransmitters and cellular immunity, *Ann. N.Y. Acad. Sci.,* 594, 403, 1990.
183. **Moore, T. C., Whitley, G. A., Lami, J. L., and Said, S. I.,** Substance P increases and prolongs increased output of T4 (CD4) lymphocytes from lymph nodes of sheep *in vivo:* is it a mediator of immunological memory? *Immunopharmacology,* 20, 207, 1990.
184. **Gowans, J. L.,** The effect of continuous reinfusion of lymph and lymphocytes on the output of lymphocytes from the thoracic duct of unanesthetized rats, *Br. J. Exp. Pathol.,* 38, 67, 1957.
185. **Gowans, J. L.,** The recirculation of lymphocytes from blood to lymph in the rat, *J. Physiol. (Lond.),* 145, 54, 1959.
186. **Gowans, J. L., McGregor, D. D., Cowen, D. M., and Ford, C. E.,** Initiation of immune responses to small lymphocytes, *Nature,* 196, 651, 1962.
187. **Gowans, J. L. and McGregor, D. D.,** Immunological activities of lymphocytes, *Prog. Allergy,* 9, 1, 1965.
188. **Ford, W. L.,** The recruitment of recirculating lymphocytes in the antigenically stimulated spleen, specific and non-specific consequences of initiating a secondary antibody response, *Clin. Exp. Immunol.,* 12, 243, 1972.

189. **Ford, W. L. and Atkins, R. C.,** Specific unresponsiveness of recirculating lymphocytes after exposure to histocompatibility antigen in F1 hybrid rats, *Nature New Biol.*, 234, 178, 1971.
190. **Ford, W. L. and Gowans, J. L.,** The traffic of lymphocytes, *Sem. Haematol.*, 6, 67, 1969.
191. **Ford, W. L. and Simmonds, S. J.,** The tempo of lymphocyte circulation from blood to lymph in the rat, *Cell Tissue Kinet.*, 5, 175, 1972.
192. **Ford, W. L., Smith, M. E., and Andrews, P.,** Possible clues to the mechanism underlying the selective migration of lymphocytes from the blood, in *Cell Recognition,* Curtis, A. G. S., Ed., Cambridge University Press, 1978, 359.
193. **Ford, W. L.,** Lymphocyte migration and immune responses, *Prog. Allergy,* 19, 1, 1975.
194. **Schoefl, G. I.,** The migration of lymphocytes across the vascular endothelium in lymphoid tissues, *J. Exp. Med.*, 136, 568, 1972.
195. **Svensjo, E., Arfors, K.-E., Raymond, R. M., and Grega, G. J.,** Morphological and physiological correlation of bradykinin-induced macromolecular efflux, *Am. J. Physiol.*, 236(4), H605, 1979.
196. **Hall, J. G. and Morris, G.,** The output of cells in lymph from the popliteal nodes of sheep, *Q. J. Exp. Physiol.*, 47, 360, 1962.
197. **Hall, J. G. and Morris, B.,** The lymph-born cells of the immune response, *Q. J. Exp. Physiol.*, 48, 235, 1963.
198. **Hall, J. G. and Morris, B.,** The origin of cells in the efferent lymph from a single lymph node, *J. Exp. Med.*, 121, 901, 1965a.
199. **Hall, J. G. and Morris, B.,** The immediate effect of antigens on the cell output of a lymph node, *Br. J. Exp. Pathol.*, 46, 450, 1965b.
200. **Hall, J. G., Morris, B., Moreno, G. D., and Bessis, M. C.,** The ultra-structure and function of cells in lymph following antigenic stimulation, *J. Exp. Med.*, 125, 91, 1967.
201. **Hay, J. B. and Hobbs, B. B.,** The flow of blood to lymph nodes and its relations to lymphocyte traffic and the immune response, *J. Exp. Med.*, 145, 31, 1977.
202. **Moore, T. C., Hopkins, J., McConnell, I., and Lachmann, P. J.,** Efferent lymph output of lymphocytes following infusions of prostaglandin E_2 and other vasoactive mediators into cannulated afferent popliteal lymphatics in sheep, *IRCS Med. Sci.*, 8, 272, 1980.
203. **Moore, T. C.,** The effects of inflammatory mediators and anaesthetic agents upon lymphocyte traffic of sheep *in vivo,* Ph.D. thesis, University of Cambridge (Trinity College), England, 1982.
204. **Moore, T. C.,** The modulation of prostaglandins of increases in lymphocyte traffic induced by bradykinin, *Immunology,* 51, 455, 1984.
205. **Nasjletti, A., Colina-Chourio, J., and McGiff, J. C.,** Assay of kinins by their effects on canine femoral blood flow, *Proc. Soc. Exp. Biol. Med.*, 150, 493, 1975.
206. **Terragno, N. A., Lonigro, A. J., Malik, K. U., and McGiff, J. C.,** The relationship of renal vasodilator action of bradykinin to the release of prostaglandin E-like substance, *Experentia,* 28, 437, 1972.
207. **McGiff, J. C., Terragno, N. A., Malik, K. U., and Longro, A.,** Release of prostaglandin E-like substance from canine kidneys by bradykinin, *Circ. Res.*, 31, 36, 1972.
208. **Palmer, M. A., Piper, P. J., and Vane, J. R.,** Release of rabbit aorta contracting substance (RCS) and prostaglandins induced by chemical or mechanical stimulation of guinea pig lungs, *Br. J. Pharmacol.*, 49, 226, 1973.
209. **Ferreira, S. H., Moncado, S., and Vane, J. R.,** Some effects of inhibiting endogenous prostaglandin formation on the response of the cat spleen, *Br. J. Pharmacol.*, 47, 48, 1973b.
210. **Damas, J. and Deby, C.,** Liberation de prostaglandines par la bradykinine chez le rat, *C. R. Soc. Biol. Filiales,* 168, 375, 1974.
211. **Terragno, D. A., Crowshaw, K., Terragno, N. A., and McGiff, J. C.,** Prostaglandin synthesis by bovine mesenteric arteries and veins, *Circ. Res.*, 36, 76, 1975.
212. **Nasjletti, A., Colina-Chourio, J., and McGiff, J. C.,** Assay of kinins by their effects on canine femoral blood flow, *Proc. Soc. Exp. Biol. Med.*, 150, 493, 1975.
213. **Terragno, N. A., Terragno, A., Packlezyk, D., and McGiff, J. C.,** Prostaglandins and the regulation of uterine blood flow in pregnancy, *Nature,* 249, 57, 1974.
214. **Needleman, P., Key, S. L., Denny, S. E., Isakson, P. C., and Marshall, G. R.,** The mechanism and modification of bradykinin-induced coronary vasodilatation, *Proc. Natl. Acad. Sci. U.S.A.*, 72, 2060, 1975.
215. **Needleman, P., Marshall, G. R., and Sobel, B. E.,** Hormone interactions in the isolated rabbit heart; synthesis and coronary vasomotor effects of prostaglandins, angiotensin and bradykinin, *Circ. Res.*, 37, 802, 1975.
216. **Piper, P. J. and Vane, J. R.,** Release of additional factors in anaphylaxis and its antagonism by anti-inflammatory drugs, *Nature,* 223, 29, 1969.
217. **Vane, J. R. and Ferreira, S. H.,** Interactions between bradykinin and prostaglandins, in *Fogarty International Center Proceedings*, Vol. 27, *Chemistry and Biology of the Kallikrein-Kinin System in Health and Disease,* Pisano, J. J. and Austen, Eds., U.S. Government Printing Office, Washington, D.C., 1976, 255.

218. **Vane, J. R.**, Inhibition of prostaglandin in synthesis as a mechanism for aspirin-like drugs, *Nature New Biol.*, 231, 232, 1971.
219. **Vane, J. R.**, The mode of action of aspirin-like drugs, *Agents Actions*, 8, 430, 1978.
220. **Terragno, N. A. and Terragno, A.**, Release of vasoactive substances by kinins, *Handbo. Exp. Pharmacol.*, 25, 401, 1979.
221. **Geiger, R. and Whalley, E. T.**, Effects of various prostaglandins on the *in vitro* activation of kallikrein in rat plasma, *Br. J. Pharmacol.*, 73, 224p, 1980.
222. **Hong, S. L.**, Effects of bradykinin and thrombin on prostacyclin synthesis on endothelial cells from calf and pig aorta and human umbilical cord vein, *Thromb. Res.*, 18, 787, 1980.
223. **Bell, R. L., Baenziger, N. L., and Majerus, P. W.**, Bradykinin-stimulated release of arachidonate from phosphatidyl inositol in mouse fibrosarcoma cells, *Prostaglandins*, 20, 269, 1980.
224. **Goldberg, M. R., Chapnick, B. M., Joiner, P. D., Hyman, A. L., and Kadowitz, P. J.**, The influence of prostaglandin synthesis on venoconstrictor responses to bradykinin, *J. Pharmacol. Exp. Ther.*, 198, 357, 1976.
225. **Cherry, P. D., Furchgott, R. F., Zawadzki, J. V., and Jothianandan, D.**, Role of endothelial cells in relaxation of isolated arteries by bradykinin, *Proc. Natl. Acad. Sci. U.S.A.*, 79, 2106, 1982.
226. **Furchgott, R. F. and Zawadzki, J. V.**, The obligatory role of endothelial cells in the relaxation of arterial smooth muscle by acetylcholine, *Nature*, 373, 1980.
227. **Boulanger, C., Schini, V. B., Moncada, S., and Vanhoutte, P. M.**, Stimulation of cyclic GMP production in cultured endothelial cells of the pig by bradykinin, adenosine diphosphate, calcium ionophore A23187 and nitric oxide, *Br. J. Pharmacol.*, 101, 152, 1990.
228. **Knowles, R. G., Palacios, M., Palmer, R. M. J., and Moncada, S.**, Formation of nitric oxide from L-arginine in the central nervous system: a transduction mechanism for stimulation for the soluble guanylate cyclase, *Proc. Natl. Acad. Sci. U.S.A.*, 86, 5159, 1989.
229. **McCall, T. B., Boughton-Smith, N. K., Palmer, R. M. J., Whittle, B. J. R., and Moncada, S.**, Synthesis of nitric oxide from L-arginine by neutrophils, *Biochem. J.*, 261, 293, 1989.
230. **Palmer, R. M. J., Ferrige, A. G., and Moncada, S.**, Nitric oxide release accounts for the biological activity of endothelium-derived relaxing factor, *Nature*, 327, 524, 1987.
231. **Moncada, S., Radomski, M. W., and Palmer, R. M. J.**, Endothelium-derived relaxing factor: identification as nitric oxide and role in the control of vascular tone and platelet function, *Biochem. Pharmacol.*, 37, 2495, 1988.
232. **Gryglewski, R. J., Palmer, R. M. J., and Moncada, S.**, Superoxide anions play a role in the breakdown of EDRF, *Nature*, 320, 454, 1986.
233. **Moncada, S., Palmer, R. M. J., and Gryglewski, R. J.**, Mechanism of action of some inhibitors of endothelium-derived relaxing factor, *Proc. Natl. Acad. Sci. U.S.A.*, 83, 9164, 1986.
234. **Gryglewski, R. J., Moncada, S., and Palmer, R. M. J.**, Bioassay of prostacyclin and endothelium-derived relaxing factor (EDRF) from porcine aortic endothelial cells, *Br. J. Pharmacol.*, 87, 685, 1986.
235. **Palmer, R. M. J., Ashton, D. S., and Moncada, S.**, Vascular endothelial cells synthesize nitric oxide from L-arginine, *Nature*, 333, 664, 1988a.
236. **Palmer, R. M. J., Rees, D. D., Ashton, D. S., and Moncada, S.**, L-arginine is the physiological precursor for the formation of nitric oxide in endothelium-dependent relaxation, *Biochem. Biophys. Res. Commun.*, 153, 1251, 1988.
237. **Heavey, D. J., Barrow, S. E., Hickling, N. E., and Ritter, J. M.**, Aspirin causes short-lived inhibition of bradykinin-stimulated prostacyclin production in man, *Nature*, 318, 186, 1985.
238. **Clark, M. A., Conway, T. M., Bennett, C. F., Crooke, S. T., and Stadel, J. M.**, Islet-activating protein inhibits leukotriene D_4- and leukotriene C_4- but not bradykinin- or calcium ionophore-induced prostacyclin synthesis in bovine endothelial cells, *Proc. Natl. Acad. Sci. U.S.A.*, 83, 7320, 1986.
239. **Warren, J. B., Ritter, J. M., Hickling, N. E., and Barrow, S. E.**, Bradykinin-stimulated prostaglandin synthesis in conscious rabbits, *Br. J. Pharmacol.*, 92, 895, 1987.
240. **Burch, R. M.**, Diacylglycerol in the synergy of bradykinin and thrombin stimulation of prostaglandin synthesis, *Eur. J. Pharmacol.*, 168, 39, 1989.
241. **Moss, J., Hom, B. E., Hewlett, E. L., Tsai, S.-C., Adamik, R., Halpern, J. L., Price, S. R., and Manganiello, V. C.**, Mechanism of enhanced sensitivity to bradykinin in pertussis toxin-treated fibroblasts: toxin increases bradykinin-stimulated prostaglandin formation, *Mol. Pharmacol.*, 34, 279, 1988.
242. **Stoner, J., Manganiello, V. C., and Vaughn, M.**, Effects of bradykinin and indomethacin on cyclic GMP and cyclic AMP in lung slices, *Proc. Natl. Acad. Sci. U.S.A.*, 70, 3830, 1973.
243. **Bertelli, A., Caciagli, F., and Schinetti, M. L.**, Kinins and inflammation: changes induced on the levels of cyclic nucleotides, *Adv. Exp. Med. Biol.*, 70, 97, 1976.
244. **Burgess, G. M., Mullaney, I., McNeill, M., Coote, P. R., Minhas, A., and Wood, J. N.**, Activation of guanylate cyclase by bradykinin in rat sensory neurones is mediated by calcium influx: possible role of the increase in cyclic GMP, *J. Neurochem.*, 53, 1212, 1989.

245. **Burch, R. M., Ma, A. L., and Axelrod, J.,** Phorbol esters and diacylglycerols amplify bradykinin-stimulated prostaglandin synthesis in swiss 3T3 fibroblasts, *J. Biol. Chem.,* 263, 4764, 1988.
246. **Fu, T., Okano, Y., Hagiwara, M., Hidaka, H., and Nozawa, Y.,** Bradykinin-induced translocation of protein kinases C in neuroblastoma NCB-20 cell: dependence on 1,2-diacylglycerol content and free calcium, *Biochem. Biophys. Res. Commun.,* 162, 1279, 1989.
247. **Gammon, C. M., Allen, A. C., and Morell, P.,** Bradykinin stimulates phosphoinositide hydrolysis and mobilization of arachidonic acid in dorsal root ganglion neurons, *J. Neurochem.,* 53, 95, 1989.
248. **Lambert, T. L., Kent, R. S., and Whorton, A. R.,** Bradykinin stimulation of inositol polyphosphate production in porcine aortic endothelial cells, *J. Biol. Chem.,* 261, 15288, 1986.
249. **Geppetti, P., Maggi, C. A., Perretti, F., Frilli, S., and Manzini, S.,** Simultaneous release by bradykinin of substance P- and calcitonin gene-related peptide immunoreactivities from capsaicin-sensitive structures in guinea-pig heart, *Br. J. Pharmacol.,* 94, 288, 1988.
250. **Ishizaka, T., Iwata, M., and Ishizaka, K.,** Release of histamine and arachidonate from mouse mast cells induced by glycosylation-enhancing factor and bradykinin, *J. Immunol.,* 134, 1880, 1985.
251. **Cohan, V. L., MacGlashan, D. W., Jr., Warner, J. A., Lichtenstein, L. M., and Proud, D.,** Mechanisms of mediator release from human skin mast cells upon stimulation by the bradykinin analog, [DArg^0-Hyp^3-DPhe^7] bradykinin, *J. Pharmacol.,* 41, 293, 1991.
252. **Lee, P. Y. and Pearce, F. L.,** Histamine secretion from mast cells stimulated with bradykinin, *Agents Actions,* 30, 67, 1990.
253. **Devillier, P., Drapeau, G., Renoux, M., and Regoli, D.,** Role of the N-terminal arginine in the histamine-releasing activity of substance P, bradykinin and related peptides, *Eur. J. Pharmacol.,* 168, 53, 1989.
254. **Ryan, G. B. and Majno, G.,** Acute inflammation, *Am. J. Pathol.,* 86, 185, 1977.
255. **Vane, J. R. and Ferreira, S. H., Eds.,** *Inflammation,* Springer-Verlag, Berlin, 1978.
256. **Movat, H. Z.,** The kinin system: its relation to blood coagulation, fibrinolysis and the formed elements of the blood, *Rev. Physiol. Biochem. Pharmacol.,* 84, 143, 1978.
257. **Metchnikoff, E.,** *Lectures on the Comparative Pathology of Inflammation: delivered at the Pasteur Institute in 1891* (translated from the French by Starling, F. A. and Starling, E. H.), Dover Publications, New York, 1968.
258. **Metchnikoff, E.,** *L'immunite dans les Maladies Infectieuses,* Masson, Paris, 1901.
259. **Williams, T. J.,** Chemical mediators of vascular reponses in inflammation: a two mediator hypothesis, *Br. J. Pharmacol.,* 61, 447p, 1977.
260. **Williams, T. J.,** Prostaglandin E_2, prostaglandin I_2 and the vascular changes of inflammation, *Br. J. Pharmacol.,* 65, 517, 1979.
261. **Rowley, D. A.,** Venous constriction as the cause of increased vascular permeability produced by 5-hydroxytryptamine, histamine, bradykinin and 48/80 in the rat, *Br. J. Exp. Pathol.,* 45, 56, 1964.
262. **Haddy, F. J.,** The mechanism of bradykinin odema, bradykinin and related kinins, in *Cardiovascular, Biochemical and Neural Actions,* Sicuteri, F., Rocha e Silva, M., and Back, N., Eds., Plenum Press, New York, 1970, 283.
263. **Kline, R. L., Scott, J. B., Haddy, F. J., and Grega, G. J.,** Mechanism of edema formation in canine forelimbs by locally administered bradykinin, *Am. J. Physiol.,* 225, 1051, 1973.
264. **Barlas, A., Sugio, K., and Greenbaum, L. M.,** Release of T-kinin and bradykinin in carrageenin induced inflammation in the rat, *FEBS Lett.,* 190, 268, 1985.
265. **Tissot, M., Regoli, D., and Giroud, J. P.,** Bradykinin levels in inflammatory exudates, *Inflammation,* 9, 419, 1985.
266. **Hargreaves, K. M. and Costello, A.,** Glucocorticoids suppress levels of immunoreactive bradykinin in inflamed tissue as evaluated by microdialysis probes, *Clin. Pharmacol. Ther.,* 48, 168, 1990.
267. **Kirchhoff, C., Jung, S., Reeh, P. W., and Handwerker, H. O.,** Carrageenin inflammation increases bradykinin sensitivity of rat cutaneous nociceptors, *Neurosci. Lett.,* 111, 206, 1990.
268. **Rocha e Silva, M. and Rosenthal, S. R.,** Release of pharmacologically active substances from the rat skin *in vivo* following thermal injury, *J. Pharmacol. Exp. Ther.,* 132, 110, 1961.
269. **Roberts, M. H. T. and Wright, D. M.,** Functional identification of units in the rat dorsal horn responding to a substance P analogue, *J. Physiol. (Lond.),* 281, 33P, 1978.
270. **Beck, P. W. and Handwerker, H. O.,** Bradykinin and serotonin effects on various types of cutaneous nerve fibres, *Pflugers Arch.,* 347, 209, 1974.
271. **Kumazawa, T. and Mizumura, K.,** Thin-fibre receptors responding to mechanical, chemical, and thermal stimulation in the skeletal muscle of the frog, *J. Physiol. (Lond.),* 273, 179, 1977.
272. **Fitzgerald, M. and Lynn, B.,** The sensitization of high threshold mechanoreceptors with myelinated axons by repeated heating, *J. Physiol. (Lond.),* 365, 549, 1977.
273. **Fitzgerald, M.,** The spread of sensitization of polymodal nociceptors in the rabbit from nearby injury and by antidromic nerve stimulation, *J. Physiol. (Lond.),* 297, 207, 1979.

274. **Bessou, P. and Perl, E. R.,** Response of cutaneous sensory units with unmyelinated fibres to noxious stimuli, *J. Neurophysiol.*, 32, 1025, 1969.
275. **Chahl, L. A. and Ladd, R. J.,** Local oedema and general excitation of cutaneous sensory receptors produced by electrical stimulation of the saphenous nerve in the rat, *Pain*, 2, 25, 1976.
276. **Fitzgerald, M.,** The sensitization of cutaneous nociceptors by spread from a nearby injury and its blockade by local anaesthesia, *J. Physiol. (Lond.)*, 278, 44, 1978.
277. **Lynn, B.,** The heat sensitization of polymodal nociceptors in the rabbit and its independence of the local blood flow, *J. Physiol. (Lond.)*, 287, 493, 1979.
278. **Perl, E. R., Kumuzawa, T., Lynn, B., and Kenins, P.,** Sensitization of high threshold receptors with unmyelinated (C) afferent fibres, in *Somatosensory and Visceral Receptor Mechanisms*, Iggo and Ilynsky, Eds., *Prog. Brain Res.*, 43, 263, 1974.
279. **Walters, E. T., Alizadeh, H., and Castro, G. A.,** Similar neuronal alterations induced by axonal injury and learning in *Aplysia*, *Science*, 253, 797, 1991.
280. **Mohan Rao, P. J. R. and Bhattacharya, S. K.,** Hyperthermic effect of centrally administered bradykinin in the rat: role of prostaglandins and serotonin, *Int. J. Hyperther.*, 4, 183, 1988.
281. **Goldstein, R. H. and Wall, M.,** Activation of protein formation and cell division by bradykinin and des-arg^9-bradykinin, *J. Biol. Chem.*, 259, 9263, 1983.
282. **Marceau, F. and Tremblay, B.,** Mitogenic effect of bradykinin and of des-arg^9-bradykinin on cultured fibroblasts, *Life Sci.*, 39, 2351, 1986.
283. **Straus, D. S. and Pang, K. J.,** Effects of bradykinin on DNA synthesis in resting NIL8 hamster cells and human fibroblasts, *Exp. Cell Res.*, 151, 87, 1984.
284. **Drinkwater, C. C., Evans, B. A., and Richards, R. I.,** Mouse glandular kallikrein genes: identification and characterization of the genes encoding the epidermal growth factor binding proteins, *Biochemistry*, 26, 6750, 1987.
285. **Parries, G., Hoebel, R., and Racker, E.,** Opposing effects of a *ras* oncogene on growth factor-stimulated phosphoinositide hydrolysis: desensitization to platelet-derived growth factor and enhanced sensitivity to bradykinin, *Proc. Natl. Acad. Sci. U.S.A.*, 84, 2648, 1987.
286. **Volonte, C., Parries, G. S., and Racker, E.,** Stimulation of inositol incorporation into lipids of PC12 cells by nerve growth and bradykinin, *J. Neurochem.*, 51, 1156, 1988.
287. **Bar-Sagi, D. and Feramisco, J. R.,** Microinjection of the *ras* oncogene protein into PC12 cells induces morphological differentiation, *Cell*, 42, 841, 1985.
288. **Downward, J., De Gunzburg, J., Riehl, R., and Weinberg, R. A.,** $p21^{ras}$-induced responsiveness of phosphatidyl-inositol turnover to bradykinin is a receptor number effect, *Proc. Natl. Acad. Sci. U.S.A.*, 85, 5574, 1988.
289. **Etscheid, B. G., Jamieson, G. A., Jr., Toscas, K., and Villereal, M. L.,** Loss of bradykinin receptors and TPA-stimulated Na^+ influx in SV40-transformed WI-38 cells, *Am. J. Physiol.*, 259, 549, 1990.

Chapter 10

SUBSTANCE P

I. INTRODUCTION

Substance P is the most fascinating of the many important and intriguing vasoactive neurotransmitter and modulator molecules considered in this volume. From current evidence, it clearly is the flagship molecule in the "heat, pain, and memory" memory-molecule scenario.

This is a bit ironic because for most of its 60-year existence as a "known" endogenous substance, it has been a little-noticed "sleeper", Rip Van Winkle-style. This is more than a little surprising when one considers the large volume of important and exciting information presented by von Euler and Gaddum in their 1931 "discovery" publication.

II. HISTORICAL BACKGROUND

1931. von Euler and Gaddum[1] reported the discovery of a novel and extremely potent substance found accidentally in a search for the organ distribution of acetylcholine in the horse. This substance, reduced to a stable powder, was used as a standard and was labeled and called substance P. Its greatest concentrations were shown to be in the brain and the small intestine. In the brain the highest concentration with the greatest bioactivity was demonstrated to be in the critically important basal ganglia. Its blood-pressure-depressing effects are shown to be due to peripheral vasodilatation and with no lung passage destruction of activity. They did not leave much for subsequent investigators to do.

1936. von Euler[2,3] established an identity of substance P as a peptide by showing that it could be inactivated by trypsin preparations. From his studies of 5 years earlier (see above), which identified the high levels of substance P in mammalian (horse) brain, this clearly established substance P as a neuropeptide (the first to be so identified). Also from his 1931 study that also showed a brain differential distribution in substance P concentrations, with the highest in the basal ganglia, substance P, early in its known existence, was established unambiguously as a neuropeptide of immense importance and by its discoverer, all of this long before neuropeptides became famous and popular.

1953. Lembeck,[4] from a study that employed a bioassay methodology for identification of a substance P in spinal cord dorsal roots, with much higher concentrations in dorsal than ventral roots, suggested an amazingly accurate hypothesis that substance P functions as a neurotransmitter released by primary sensory neurons.

1956. von Euler[5] described the presence of substance P in nonmammals. He reported its presence in the brains and intestines of poikolotherm animals, of freshly caught cod (*Gadus calarias* and *G. morrhua*) and dogfish (*Squalus acanthias*). He reported substance P reactivity to be the same for brains and intestines in these lower vertebrates.

1959–1977. Jancso and associates[6-8] carried out a series of important studies into the pain and inflammation-modulating properties of capsaicin, a substance P-releasing and -depleting agent and the pungent principle of the Hungarian red (and hot) pepper. Here we found heat (of a sort) and pain brought together in association with substance P.

1963. von Euler[9] reported finding substance P in subcellular particles (synaptosomes) of peripheral nerves (sediment obtained by high-speed centrifugation of press juice from bovine vagus and splenic nerves). This also was the first identification of substance P in the nervous system nerve supply to a major immune response system organ (spleen) — a major contribution to neurovascular immunology and the nervous-to-immune-systems "connection".

1970. Chang and Leeman[10] provided a major breakthrough in reporting the chemical characterization and amino acid sequence of this 11-amino-acid residue peptide. They purified substance P from extracts of bovine hypothalamus, employing the powerful sialogogic activity of substance P as a bioassay.

1974. Otsuka and Konishi,[11-15] by radioimmunoassay, established the presence of substance P in high concentration in primary sensory neurons. By immunohistochemistry, they demonstrated substance P-immunoreactivity in the cell bodies of sensory neurons in sensory ganglia and in the terminals of these neurons peripherally and centrally in the spinal cord and brain stem. They also describe substance P-induced depolarization/excitation of spinal neurons.

1976. Henry,[16] in iontophoretic studies, provided the first physiological evidence that substance P is selectively involved in nociception. He also made the particularly important observation that all dorsal-horn neurons excited by the local iontophoretic application of substance P were also activated by noxious radiant heat stimulation of the cutaneous receptive field. With these observations, Henry brought heat into the "heat, pain, and memory" scenario for substance P as a major memory molecule.

1977. Jessell and Iversen[17] focused a riveting spotlight on substance P and opened a veritable Pandora's box in their report in *Nature* that morphine blocks the release of substance P in the mammalian nervous system. This report of immense importance sent thousands of researchers scrambling to rewrite their research grants.

1980. Hanley, Iversen, and associates[18-21] described the presence of high-affinity binding sites for [^{3}H]substance P in membrane preparations from rat brain. These binding sites were shown to possess a structural specificity similar to that encountered in other studies of the effects of substance P and related peptides on central neurons. They also made the particularly important observation that substance P stimulates an accelerated turnover in membrane phosphatidyl inositides (PI). This was the first demonstration of substance P-induced activation and stimulation of neuronal membrane PI cycle transmembrane signaling — an extremely important observation in light of subsequent findings related to substance P functions, including memory.

1982. The author[22] described (in his University of Cambridge Ph.D. thesis) the *in vivo* immunostimulative effect of substance P in increasing the traffic of lymphocytes through primary peripheral lymph nodes of sheep. The substance P used in these studies was provided and given on the recommendation of the author's college advisor on the neuroscience and neurochemistry aspects of his thesis research, Dr. Leslie L. Iversen, FRS. In this communication the author also suggested a physiological and molecular commonality between immunological and central nervous system memories.

1982. Crystal and Davies,[23] in a study of brains of patients dying of Alzheimer's disease, presented raw data from control human brains, indicating that tissue levels of substance P were 3.7 times higher in the hippocampus of those nondiseased and control brains than in six other major cortical areas of the brains studied. In Alzheimer's disease brains, the substance P levels of the hippocampus were reduced to less than 50% of the levels in control/"normal" brains.

1984. The author[24] published in *Immunology* a report of the Cambridge studies, including the substance P-induced immunostimulative effect on lymphocyte traffic *in vivo* in sheep and again suggests the existence of a "brain-immune response" axis with interplay of effector and modulator molecules of these two systems, nervous and immune response.

1984. Payan et al.[25,26] described the presence of high-affinity receptors for substance P on human lymphocytes.

1987. Duggan et al.,[27] in a study with antibody probes, demonstrated that the application of skin and subcutaneous tissue levels of tissue-destructive noxious thermal heat (50 to 52°C) to peripheral areas of anesthetized cats produces releases of substance P from axon terminals in the substantia gelatinosa of lamina I of the dorsal horn of the spinal cord of these cats. This release was not inhibited by the anesthesia employed during these studies. These findings provided a further tightening of the heat and pain component links of the "heat, pain, and memory" scenario chain.

1989. The author et al.[28] reported marked prolongations (in excess of 6 d) in the lymphocyte-traffic- and lymphocyte-output-stimulating effects of single acute substance P infusions (50 μg) into cannulated afferent lymphatics of sheep popliteal lymph nodes. These greatly prolonged immunostimulative effects suggest an immune response system long-term potentiation (LTP) effect similar to the LTP phenomenon observed in the mammalian hippocampus following electrical or chemical stimulation. The central nervous system LTP has long been considered a model for memory and learning. Perhaps it is so for the immune response system as well.

1990. The author et al.[29] reported that substance P-stimulated marked and prolonged increases in lymphocyte output from primary peripheral lymph nodes of sheep *in vivo* are targeted to the all-important CD4 subset of T lymphocytes and are associated with altered, earlier, and accelerated antibody output, with immunization occurring 6 d after a single infusion of substance P (50 μg) into a cannulated afferent lymphatic. It is suggested that substance P may be a mediator of immunological memory and of value when an increased output and traffic of CD4 T lymphocyte is desired. It was again suggested that substance P may be involved in the molecular alterations that may occur in the establishment of both types of memory.

1991. Kowall et al.,[30] in the August 15, 1991 issue of the *Proceedings of the National Academy of Science U.S.A.*, reported that substance P injected into the brains or systemic circulation of rats blocked β-amyloid-induced experimental Alzheimer's disease in these animals — a truly remarkable observation of great significance and importance.

III. DISCOVERY AND EARLY STUDY

In 1930 a 25-year-old Ulf von Euler arrived in London from Stockholm to do a brief 6-month stint in the National Institute for Medical Research laboratory of Henry Dale in the London suburb of Hampstead. As the young and comparatively inexperienced son of a world-famous biochemist (who, only 6 months before, had received a Nobel Prize for his work on enzymes in the fermentation of sugar), this clearly was intended to be a short-term introductory and overview exposure to British superscience, then world preeminent in both the biological and physical sciences.

On arrival, Ulf was assigned a young "local", J. H. Gaddum, to work with in some routine laboratory "mop-up" experiments relating to acetylcholine, discovered 16 years earlier by Dale (a discovery that ultimately, and 6 years after, also fetched him a Nobel Prize). A short time before Ulf's arrival, Dale and Dudley[30] published a report in the *Journal of Physiology* identifying the presence of acetylcholine in the fresh spleens of oxen and horses. This young team of von Euler and Gaddum was asked to sort out the distribution of this substance in other organs — a seemingly routine task.

They started with alcoholic extracts that were made from various tissues obtained from horses by mincing the fresh tissue in five times its weight of alcohol. After removing the alcohol under reduced pressure, the extracts were tested on the longitudinal muscle of the rabbit's isolated intestine. This tissue appeared to be quite suitable, since it contracts in very low concentrations of acetylcholine and is very insensitive to histamine.

At this point, routine and expected observations ceased, as some tissues, particularly those from brain and intestine, produced contractions that came on more slowly than those due to acetylcholine or choline and were unaffected by doses of atropine that abolished the effects of these substances. It was clear to these young scientists that the novel and unexpected effect was due to some substance other than a choline ester or histamine.

The unexpected stimulant substance that attracted their attention was relatively stable to heating in acid solution, but was easily destroyed by brief boiling in alkalai. The alkalai instability differentiated this novel substance from adenylic acid or adenine.

In *in vivo* studies, extracts from almost any organ, when injected intravenously into a rabbit under ether, produced a fall in blood pressure that was unaffected by atropine. This phenomenon, first described in 1900 by Osborne and Vincent,[1] had been recognized as due neither to histamine, which does not produce a fall in blood pressure in the etherized rabbit, nor to choline or choline esters, which produce no effect after atropine.

The young investigators then undertook to purify their novel substance. They selected an alcohol suspension of minced horse small intestine muscle. Addition of a bit of sulfuric acid increased the yield and made filtration easier. The suspension obtained was filtered after an hour of standing with stirring at room temperature, and placed in a cold room overnight. The next morning, alcohol was removed under reduced pressure and fats were removed with ether. These measures left a watery extract that had high activity in the blood-pressure-depressing and intestine-stimulating bioassays. This watery extract was found to be very stable when made just acid to congo-red. If the watery extract was concentrated, made alkaline with sodium carbonate, and dried by rubbing-up with anhydrous sodium sulfate, as much as 90% of the activity could be extracted with absolute alcohol. The addition of sulfuric acid to the water extract solution in absolute alcohol produced a precipitate that retained its activity with rabbit blood-pressure-reducing activity in doses as low as 0.01 mg.

The precipitate contained about 65% ash and was found to be stable when stored as a powder. The powder standards of this substance were labeled and called "P" (hence substance P).

In physiological studies of the active principle of the powder P, water solutions (clear) were found to exert their effect on the blood pressure of atropinized rabbits due to peripheral dilatation. The blood-pressure-depressing effect was produced equally well by direct injections into the aorta or the superior vena cava (indwelling tubes). A shorter latent period was experienced with injections into the aorta.

These intrepid investigators then determined the distribution of activity in different organs. High and essentially equal concentrations of activity were identified in the brain and small intestines. Activity as milligrams of "P" per gram of tissue of brain and small intestine ranged from 10 to 20 times that of the next highest organ (stomach), with a rapid falloff thereafter. They were especially interested in the unexpectedly high activity of "P" in extracts from the brain. In Figure 6 of their publication in the *Journal of Physiology* (72, 74, 1931), they demonstrated that extracts from the basal ganglia have much greater activity in the two bioassay systems than those from the cerebral hemisphere (Figure 1). This is an especially important and significant observation with respect to recent findings

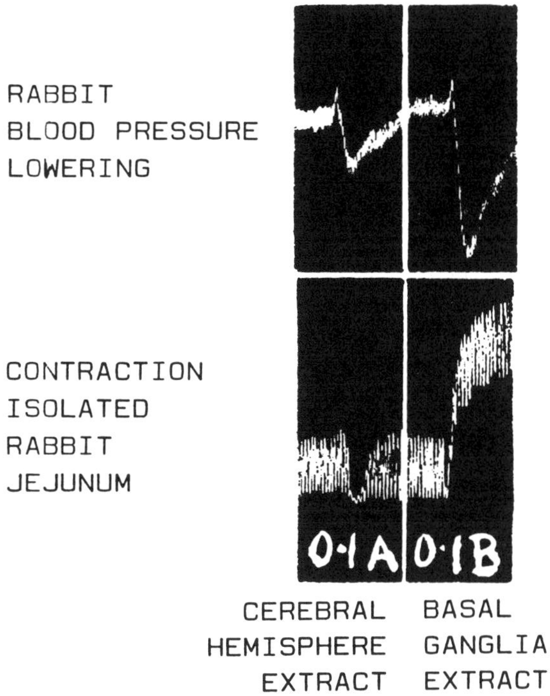

FIGURE 1. Figure 6 (with new labeling by the author of this book) from 1931 von Euler and Gaddum substance P discovery paper demonstrating much higher substance P levels in basal ganglia area of the brain than in the cerebral hemispheres.

relating to substance P contribution to, and involvement in, memory, immunological as well as central nervous system, and in the search for the molecular mechanisms of learning and memory.

From a perspective of 60 years later, with the exceptions of the pain-related findings of Lembeck, Jessell, and Iversen and the more recent memory-related observations, these two young investigators just about said it all. Even the name they gave their powdery standards, substance P, has resisted the erosions of time.

Five years later, von Euler[2,3] established substance P as a peptide and a neuropeptide by demonstrating that it could be inactivated by trypsin preparations.

IV. STRUCTURE, RECEPTORS, AND ANTAGONISTS

The amino acid sequence of the 11-amino-acid residue substance P was established by Chang and Leeman[10] in 1970. Important pioneering work by the remarkable and ubiquitous Vittorio Erspamer,[31] of Rome, Italy, has established substance P as a member of a family of closely related peptides that have been termined tachykinins. The tachykinin family of peptides share a common C-terminal sequence (Phe-X-Gly-Leu-Met NH_2). Many of the nonmammalian tachykinins have been derived from amphibian skin. One of the most potent, eledoisin, was discovered in the salivary gland tissues of the octopus, a molluscan cephalopod. Amino acid sequences of the mammalian substance-P tachykinin has been compared with the sequences of nonmammalian tachykinins, by Erspamer (Figure 2). In his 1981 publications, Erspamer[31] suggested that substance P might not be the only tachykinin in mammals. A year earlier he had speculated that there might be more than one substance P/

1981 ERSPAMER TABULATION OF TACHYKININ PEPTIDE FAMILY

Molluscan tachykinins	
Pyr-Pro-Ser-Lys-Asp-Ala-Phe-Ile-Gly-Leu-Met-NH_2	Eledoisin
Amphibian tachykinins	
Pyr-Ala-Asp-Pro-Asn-Lys-Phe-Tyr-Gly-Leu-Met-NH_2	Physalaemin
Pyr-Ala-Asp-Pro-Lys-Thr-Phe-Tyr-Gly-Leu-Met-NH_2	Lys^5,Thr^4-Physalaemin
Pyr-----Asn-Pro-Asn-Arg-Phe-Ile-Gly-Leu-Met-NH_2	Uperolein
Asp-Val-Pro-Lys-Ser-Asp-Gln-Phe-Val-Gly-Leu-Met-NH_2	Phyllomedusin
Asp-Glu-Pro-Lys-Pro-Asp-Gln-Phe-Val-Gly-Leu-Met-NH_2	Glu^2,Pro^5-Kasinin*
Asp-Pro-Pro-Asp-Pro-Asp-Arg-Phe-Tyr-Gly-Met-Met-NH_2	Hylambatin*
Mammalian tachykinins	
Arg-Pro-Lys-Pro-Gln-Gln-Phe-Phe-Gly-Leu-Met-NH_2	Substance P

* Yasuhara et al., in preparation.

FIGURE 2. 1981 Erspamer[31] tabulation of tachykinin peptide family.

NONMAMMALIAN TACHYKININS

Species	Peptide	Sequence
Mollusca	Eledoisin	pGlu-Pro-Ser-Lys-Asp-Ala-Phe-Ile-Gly-Leu-Met-NH_2
Fishes	Scyliorhinin I	Ala-Lys-Phe-Asp-Lys-Phe-Tyr-Gly-Leu-Met-NH_2
	Scyliorhinin II	Ser-Pro-Ser-Asn-Ser-Lys-Cys-Pro-Asp-Gly-Pro-Asp-Cys-Phe-Val-Gly-Leu-Met-NH_2
Amphibia	Physalaemin	pGlu-Ala-Asp-Pro-Asn-Lys-Phe-Tyr-Gly-Leu-Met-NH_2
	Lys-Thr-Physalaemin	pGlu-Ala-Asp-Pro-Lys-Thr-Phe-Tyr-Gly-Leu-Met-NH_2
	Uperolein	pGlu-Pro-Asp-Pro-Lys-Ala-Phe-Tyr-Gly-Leu-Met-NH_2
	Phyllomedusin	pGlu-Asn-Pro-Lys-Arg-Phe-Ile-Gly-Leu-Met-NH_2
	Kassinin	Asp-Val-Pro-Lys-Ser-Asp-Gln-Phe-Val-Gly-Leu-Met-NH_2
	Glu-Pro-Kassinin	Asp-Glu-Pro-Lys-Pro-Asp-Gln-Phe-Val-Gly-Leu-Met-NH_2
	Entero-Kassinin	Asp-Glu-Pro-Asn-Ser-Asp-Gln-Phe-Ile-Gly-Leu-Met-NH_2
	Hylambatin	Asp-Pro-Pro-Asp-Pro-Asp-Arg-Phe-Tyr-Gly-Met-Met-NH_2
	Entero-Hylambatin	Asp-Pro-Pro-Asn-Pro-Asp-Arg-Phe-Tyr-Gly-Met-Met-NH_2

FIGURE 3. 1988 tabulation of nonmammalian tachykinins. (After Negri, L. and Melchiorri, R., *Regul. Peptides,* 22, 13, 1988.)

tachykinin receptor in mammals. Both speculation/prophecies turned out to be remarkably accurate.

Since Erspamer's 1981 report,[31] the number of identified nonmammalian tachykinins has expanded from 8 to 12, and mammalian tachykinins from 1 (substance P) to 4, with the addition of neurokinins A and B (NK-A and NK-B) and neuropeptide K (NP-K). Nonmammalian tachykinins have been encountered in species ranging from cnidarian *Hydra*, to molluscs, to fishes, to amphibians, to reptiles, and to birds (Figure 3). These mammalian tachykinins share a similar C-terminal amino acid sequence, but differ widely in the amino acid compositions of their N-terminal portions (Figure 4). NP-K is an N-terminally extended form of NK-A.

Three mammalian tachykinin receptor types have generally been accepted. The NK-1 receptors, assayed in rat salivary gland, is substance P "preferring", the NK-2 type, assayed in hamster urinary bladder, is NK-A preferring, and NK-3, assayed in rat cerebral cortex, is NK-B preferring. The McKnight, Iversen, et al. group,[21,32-35] in addition to the above studies, presented evidence in 1988 for the existence of an NK-4 type of mammalian

MAMMALIAN AND NONMAMMALIAN TACHYKININS

Substance P	Arg-Pro-Lys-Pro-Gln-Gln-Phe-Phe-Gly-Leu-Met-NH_2
Physalaemin	pGlu-Ala-Asp-Pro-Asn-Lys-Phe-Tyr-Gly-Leu-Met-NH_2
Eledoisin	pGlu-Pro-Ser-Lys-Asp-Ala-Phe-Ile-Gly-Leu-Met-NH_2
Kassinin	Asp-Val-Pro-Lys-Ser-Asp-Gln-Phe-Val-Gly-Leu-Met-NH_2
Neurokinin A	His-Lys-Thr-Asp-Ser-Phe-Val-Gly-Leu-Met-NH_2
Neurokinin B	Asp-Met-His-Asp-Phe-Phe-Val-Gly-Leu-Met-NH_2

FIGURE 4. 1990 comparison of mammalian tachykinins (substance P and neurokinins A and B) with non-mammalian tachykinins. (After Iversen, L. L. et al.[21])

tachykinin receptor. This is a receptor that mediates smooth-muscle contraction in the guinea pig trachea. In this preparation the rank order of potencies is similar to that seen at the NK-2 receptor.

Progression in the development of antagonist drugs for the various mammalian tachykinin receptors has been disappointingly slow and has obscured understanding of multiple and important functions of substance P and those of related mammalian tachykinins. The 1990 development of Spantide II as an antinociceptive substance P antagonist with no apparent neurotoxicity or motor impairment may hold some promise in this difficult area of slow progress.[36]

The biological half-life of substance P, like bradykinin, is short. It is rapidly inactivated by proteolytic degradation in the brain, in peripheral tissue, and in blood.[20,37]

V. EVOLUTION

> "It might therefore be of interest to look into the evolutionary history of the neurotransmitters. In primitive animals septal synapses and axoaxonic or axo-dendritic contacts serve as a simple mechanism for stimulus transfer, complemented by more sophisticated processes in more developed species. Amino acids such as GABA, aspartate, glutamate, and others were available at an early stage and still appear to serve as transmitters. From amino acids the step was not far to peptides, which are now very much the center of interest, offering at the same time an almost inexhaustable source of compounds. From amino acids a number of amines would be readily formed by decarboxylation, further specialized by additional transformations at various stages. In this context it is of interest to note that among the catecholamines, dopamine was discovered much later than the more specialized substances and was probably the primary one during evolution, still serving as an important transmitter at various evolutionary stages."
>
> U. S. von Euler, 1982[3]

A. INTRODUCTION

Like met-enkephalin, substance P appears to be a highly conserved defense- and survival-related vasoactive neuropeptide. Whereas Stephano and associates have tracked met-enkephalin back more than 500 million years to the bivalve marine mollusc mussel *Mytilus edulis*, substance P has been tracted back even further to the cnidarian coelenterate *Hydra*, one of the most primitive of multicellular animals to possess a nervous system (Figure 5). Of the other major vasoactive neurotransmitter (and memory) molecules, only 5-hydroxytryptamine (serotonin), bombesin, neurotensin, and CCK-8 have been tracked back this far.

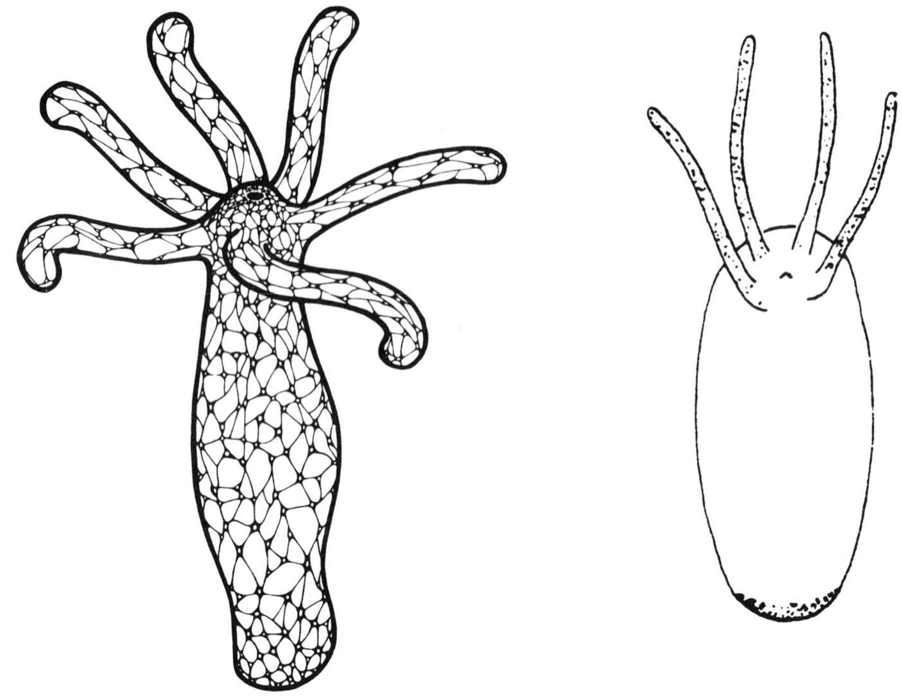

FIGURE 5. Nerve net of the cnidarian coelenterate *Hydra* (a) (After Buchsbaum, R. et al., *Animals without Backbones,* 3rd ed., University of Chicago Press, Chicago, 1982, 100.) and locations of *Hydra* nerve cells containing substance P (b) (After Grimmelikhuijzen et al., *Histochemistry*, 71, 325, 1981.)

B. CNIDARIAN COELENTERATA

In 1981 Grimmellikhuijzen, Emson, and associates[38] reported the identification of substance P-like immunoreactivity in the cnidarian *Hydra*. As shown in Figure 5, the nervous system of *Hydra*, an animal with only two body layers (entoderm and ectoderm), is simply a "nerve net". The cnidarian coelenterates were the first animals to possess a nervous system of any type, and the simple and primitive "nerve net" was it.

C. PLATYHELMINTHES/FLATWORMS

A 1986 publication by Gustafsson et al.[39] presented evidence for the existence of substance P-like immunoreactivity in the nervous system and neurons of a platyhelminthes/flatworm, the tapeworm, *Diphylobothrium dendriticum*.

D. ANNELIDA

A 1980 report by Aros et al.[40] established the presence of substance P in the central nervous system of the slightly more advanced Annelid segmented worm *Lumbricus terrestis*.

E. ARTHROPODA

1. Chelicerata

An extensive literature has developed regarding the existence, location, and presumed functions of substance P in arthropods. In the most primitive of the arthropod assemblages, the Chelicerate (horseshoe crab, spiders, scorpions, mites, and ticks) substance P has been identified, by Chamberlain and Engbretson[41] in 1982, by means of immunoreactivity to be present in lateral eye and the protocerebrum of the horseshoe crab, *Limulus polyphemus*.

In this very careful and thorough study from Syracuse University, Chamberlain and Engbretson[41] showed that small-diameter substance P-immunoreactive fibers innervated the corneal epidermis of the *Limulus* lateral eye and appeared to be a part of a generalized epidermal innervation. The optic medulla also was found to contain neurons with substance P-immunoreactive perinuclear caps in the ganglion cell layer and immunoreactive fibers of undetermined origin in the posterolateral regions of the neuropile. The central body was observed to contain many substance P-immunoreactive fibers in its neuropile, some or all of which appeared to arise from neurons adjacent to the central body anteromedial tips. Substance P-immunoreactive neurons within the curve of the central body were found to give rise to processes that were shown to join the central neuropile of the protocerebrum. Substance P-immunoreactive innervation of the corpora pedunculata included processes that terminated among Kenyon cell somata and processes in the peduncular neuropile. All of these corpora pedunculata substance P-immunoreactive processes appeared to arise from a bilateral cluster of neurons along the midline. Strong substance P immunoreactivity was encountered in and near the central body of the *Limulus* protocerebrum (Figure 6). Of particular interest with respect to the vascular ''connection'' in this primitive Chelicerata arthropod *Limulus* was the discovery of substance P-immunoreactive nerve fibers in the vascular glia. Small beaded substance P-immunoreactive fibers were stained in the ganglion cell layer of the medulla. These fibers were found to belong to a population of neural fibers that coursed among the vascular glial cells that filled the space between the neural structures of the protocerebrum and its tough vascular sheath. In the medullar ganglion cell layer, the vascular glia form a superstructure that holds and supports the neural somata. As Fahrenbach,[42] in ultrastructural studies of this anatomical area of *Limulus*, found only one type of nerve fibers within the vascular glia, Chamberlain and Engbretson speculated from their studies that this type of fiber, termed ''neurosecretory'' by Fahrenbach, was a substance P-containing fiber. Beyond this neurovascular ''connection'', substance P clearly is of great importance in the nervous system of *Limulus*.

Extended studies by Mancillas and Selverson,[43] in 1984, confirmed the presence of substance P in the lateral eye of *Limulus* and carried out physiological and anatomical investigations that demonstrated a role of substance P in the lateral eye in the modulation of photosensitivity.

2. Crustacea

Several reports also have appeared relative to the presence of substance P in the crustacean assemblage of Arthropoda. In 1981 Mancillas, Bloom, and associates[44] described in *Nature* the identification and localization of substance P in the retina and eye stalk of the lobster. In 1988 Goldberg et al.[45] reported finding substance P immunoreactivity in the somatogastric nervous system of the crab *Cancer borealis* and the lobsters *Panulirus interruptus* and *Homarus americanus*.

3. Insecta

In the highest of the arthopodan assemblages, Insecta, substance P has been identified in the central nervous system of the cockroach *Periplaneta americana*[46] and in the cerebral ganglion of locusts.[47] In 1980 and 1983 El-Sahly et al.[48,49] described the presence and location of substance P immunoreactivity in the nervous system of the dipteran hoverfly *Eristalis den* and in the brain corpora cardiaca and corpora allata of the lepidopteran moth *Manduca sexta*.

In 1990 Nässel and associates,[50] of Stockholm and Uppsala, described in great detail the locations and projections of substance P-like immunoreactive neurons, projections, and pathways in the much-studied fruitfly *Drosophila melanogaster*. In the brain, ten substance

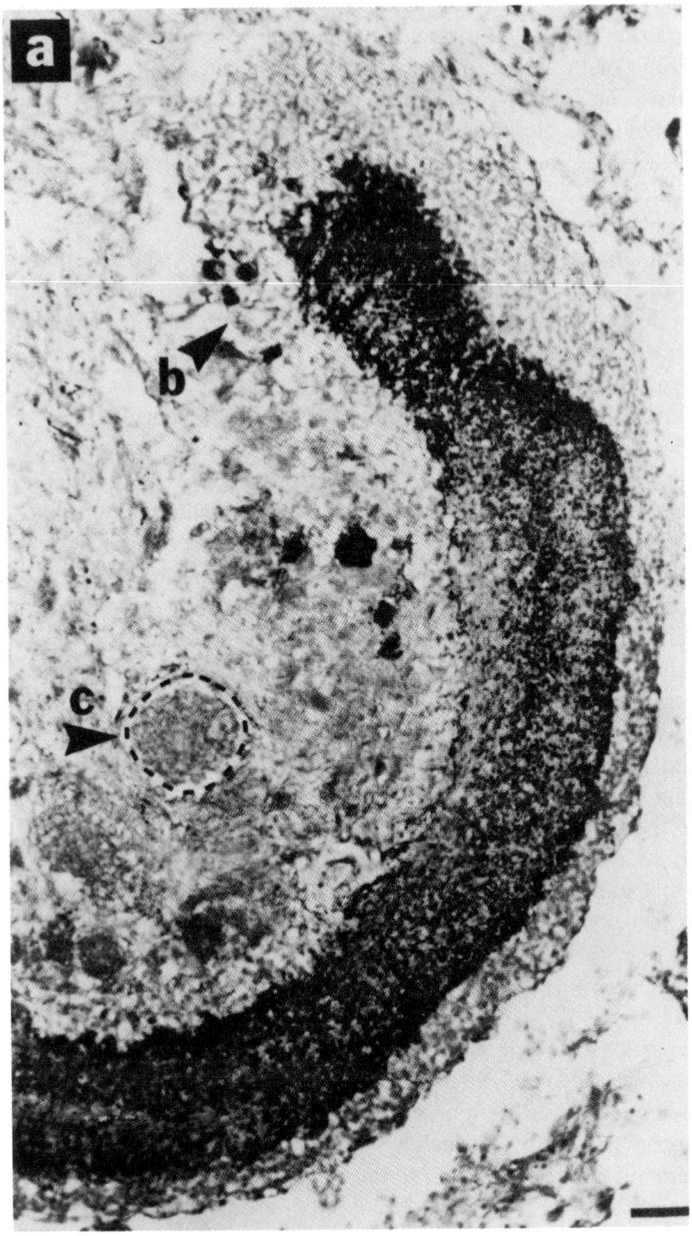

FIGURE 6. Substance P-like immunoreactivity (black) in and near the protocerebrum of the chelicerate arthropod horseshoe crab *Limulus polyphemus*. (From Chamberlain, S. C. and Engbretson, G. A., *J. Comp. Neurol.*, 208, 304, 1982. With permission).

P-like immunoreactive neurons were identified individually. Two of these substance P neurons were found to establish extensive bilateral connections with dorsal and ventral protocerebral neuropile. Two other neurons possessed cell bodies that were located ventrally in the subesophageal ganglion and that sent neuronal processes throughout the subesophageal ganglion neuropile. Of the ten substance P-like immunoreactive neurons encountered in the thoracoabdominal ganglia, eight had large cell bodies that were located ventrally in the thoracic ganglia, and two possessed small cell bodies that were found posteriorly in the abdominal ganglia. Thoracic substance P-like immunoreactive neurons that could be resolved in detail were discovered to form processes in the dorsal thoracic and abdominal neuropile, as well as processes that coursed through the cervical connective and into the subesophageal ganglia. They also sent processes running into the dorsal neural sheath of the thoracic ganglia, which formed an extensive network of varicose terminals over the thoracic ganglia. The authors interpreted their findings as suggesting that a substance P-like vasoactive neuropeptide may be released into the circulation from terminals in the neural sheath, which approximates the circulatory system of *Drosophila*. They also speculated that a substance P-like neuropeptide might be released and function in neurotransmission, neuromodulation, or both, in interneuronal-type activity in the *Drosophila* brain. This is a particularly important major and in-depth study relating to the phylogenetic development of substance P in both the nervous and vascular systems of a much-studied insect of immense importance in genetics, *Drosophila melanogaster*.

F. MOLLUSCA

In Mollusca, substance P immunoreactivity has been identified in the nervous system of the pond snail *Lymnaea stagnalis*.[51]

Osborne and Dockray,[52,53] of Oxford and Liverpool, U.K., made the important observation in 1982 that substance P and serotonin were colocalized in specific neurons of the pulmonate gastropod snail *Helix aspersa*. These are especially important findings because they join in the same neuronal cell and cells four major "memory molecules" (serotonin, substance P, bombesin, and CCK-8) whose function in memory processes has been established in two types of animal species whose ancestors diverged more than 500 million years ago — the molluscans (*Helix* and *Aplysia*) and mammals (man and rodents) — and whose presence (all four) has been identified in the nervous systems of the most primitive of multicellular cnidarian coelenterate animals (*Hydra* and *Renilla*).

G. NONMAMMALIAN VERTEBRATES

1. Bony Fish/Osteichthyes

The habenula-interpeduncular system of the brain stem has been regarded as a useful example of a phylogenetically preserved portion of the vertebrate brain. The habenula-interpeduncular system is a major connecting route between the forebrain and limbic centers in the midbrain.[54] In the goldfish, the habenular nuclei are epithalamus paired structures that are connected to the unpaired interpeduncular nucleus of the tegmentum mesencephali. Mammalian studies have suggested that habenular neurons may utilize substance P. Substance P has been shown to be present in the habenula-interpeduncular system of the amphibian frog *Rana catesbiana* and the amphibian newt *Triturus cristatus*.[55,56]

Stimulated by these observations, Villani et al.,[57] of the University of Bologna in Italy, initiated a series of studies of substance P immunoreactivity in the habenula-interpeduncular nervous system of the brain stem of the common goldfish *Carassius auratus*. Their studies involved identification of substance P-like immunoreactivity in this system under normal conditions and after habenular ablation. In their 1991 publication,[57] they reported that, under normal conditions, intense substance P-like immunoreactivity was observed in the neuropilar structure of the interpeduncular nucleus, with immunoreactivity concentrated in the ventral zone of the nucleus. As anatomical data have suggested a large afference of the interpeduncular nucleus is derived from the habenula, the authors speculated that substance P might be involved in the habenula-interpeduncular "connection" in goldfish. To confirm this speculation, they carried out surgical ablation of the habenular nuclei and demonstrated a severe loss of substance P-like immunoreactivity in the interpeduncular nucleus. The authors considered that their findings were in substantial agreement with studies in the rainbow trout[58] and the lizard,[59] where no substance P-like immunoreactive cell bodies were found in the habenular nuclei, but where substance P-like immunoreactive neuronal fibers and projections were found to be passing through these nuclei.

In a 1986 report, Ekstrom and Korf[58] demonstrated the presence of substance P-like immunoreactivity in the photosensory pineal organ of the rainbow trout *Salmo gairdneri* Richardson (Teleosti).

2. Amphibia

In 1989 Schmidt et al.,[60] of Bremen, Germany, in a study of retinofugal projection sites of the visual systems of the salamander urodele species *Salamandra salamandra* and *Batrachoseps attenuatus*, identified the presence of substance P-like immunoreactivity in four tectal lamina, two of them in the stratum opticum. In 1980 Gaudino and Fasolo[61] reported the identification of substance P-related peptides in the hypothalamus of amphibians.

3. Reptilia

An evolutionary study of immense "tie-in" importance and significance was published by Reiner and associates[62] in 1983. Although the main thrust of their investigation was directed toward the substance P-containing striatotegmental path in reptiles, it also studied parallel organization and distribution of serotoninergic and dopaminergic fibers and cells. Their study considered two divergent species, the chelonian turtle *Chrysemys scripta* and the crocodilian *Caiman crocodilus*. As pointed out in the prior chapter on bradykinin, crocodiles and birds share one path of reptilian evolutionary development; the lowly turtle and man appear to share another. This certainly was shown to be so with respect to bradykinin and plasma kallikrein metabolic studies. Accordingly, emphasis here will be placed on the substance P findings relating to our closer "relative", the turtle.

In the 1983 study by Reiner and associates,[62] substance P immunoreactivity was found to be distributed selectively throughout the central nervous system of the turtle *Chrysemys scripta*. Within the telencephalon, the largest collection of substance P-immunoreactive neuronal cell bodies and fibers was encountered in the basal ganglia (paleostriatal complex) (Figure 7). In the midbrain the largest collection of substance P-immunoreactive fibers and terminals was found in the ventrolateral tegmentum (Figure 8). A dense accumulation of substance P-containing fibers was present in the ventral paleostriatum (Figure 9). Substance P-containing fibers also were observed in the olfactory tubercle, the primordium hippocampi, the lateral septal nucleus, the cellular layer of the dorsal cortex, and the molecular and cellular layers of the medial and dorsomedial cortices. Many substance P-immunoreactive neuronal cell bodies were identified in the primodium hippocampi, the lateral septal nucleus, and the medial and dorsomedial cortices.

Substance P-like immunoreactivity within the ventral paleostriatum extended as far caudally as the rostral diencephalon. At this point it became continuous with a substance P-containing fiber bundle that was coursing within, and ventral to, the ventral peduncle of the forebrain bundle. These fibers were found to be continuous with several fields of substance P-containing fibers and terminals in the midbrain tegmentum. It was of particular interest that the substance-P fiber plexuses in the midbrain tegmentum overlapped tegmental cell fields known to contain dopaminergic cell bodies. Tyrosine hydroxylase immunohistochemistry was employed to identify the dopaminergic neuronal cell bodies and fiber projections. Three dopaminergic cell groups were differentiated, and substance P fibers were shown to be present within all of them. The most intense plexus of substance P-immunoreactive fibers was found to invest a large region of the ventral portion of the midbrain tegmentum (VTA).

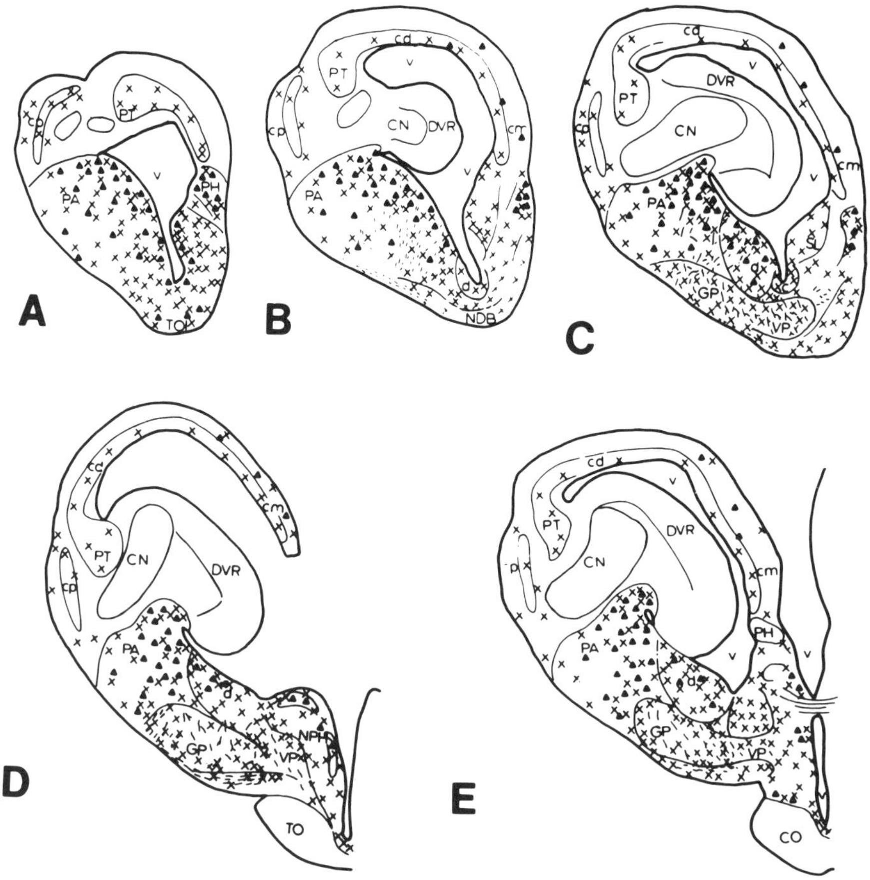

FIGURE 7. Substance P immunoreactivity in the forebrain of a direct terrestrial vertebrate ancestor of mammals and man, the turtle *Chrysemys scripta*. This illustration presents a series of schematic drawings in the transverse plane of the turtle telencephalon from the level of the olfactory tubercle (A) to the rostral optic tract (E). Substance P-containing neuronal cell bodies are presented as filled triangles. Many of these are shown in the paleostriatum augmentation (PA) and the primordium hippocampi (PH) in transverse section A. These many small substance P-immunoreactive neuronal cell bodies are found, in these areas, in a rich substance P-containing neuropile. Substance P-containing fine fibers and terminals are presented as X's, and coarse, thick substance P-containing fibers are shown as broken lines. Additional key areas are the globus pallidus (GP) and the ventral paleostriatum (VP). (From Brauth, S. E., et al., *J. Comp. Neurol.*, 219, 305, 1983. With permission.)

Substance P-containing fibers and terminals also were shown to be present in the superficial and deep layers of the tectum as well as within the periventricular cell layer of the central gray of the midbrain, in the interpeduncular nucleus, and in the lateral reticular formation. Numerous substance P-immunoreactive cell bodies were identified within the stratum griseum profundum of the tectum at all depths of the layer. Serotoninergic innervation of these areas of the *Chrysemys* central nervous system also was extensive. Many 5-HT-immunoreactive neuronal cell bodies and fibers were observed throughout the tegmental areas. Dense accumulations of 5-HT-immunoreactive fibers were present at all levels of the ventral tegmentum and extended caudally to the level of the interpeduncular nucleus. Many 5-HT-immunoreactive 5-HT cell bodies were identified in the caudal tegmental levels and were found to be present within all of the three discrete dopaminergic cell fields.

From the above it would appear that substance P is strategically located to participate in, and influence, a large number of central nervous system actions and activities of

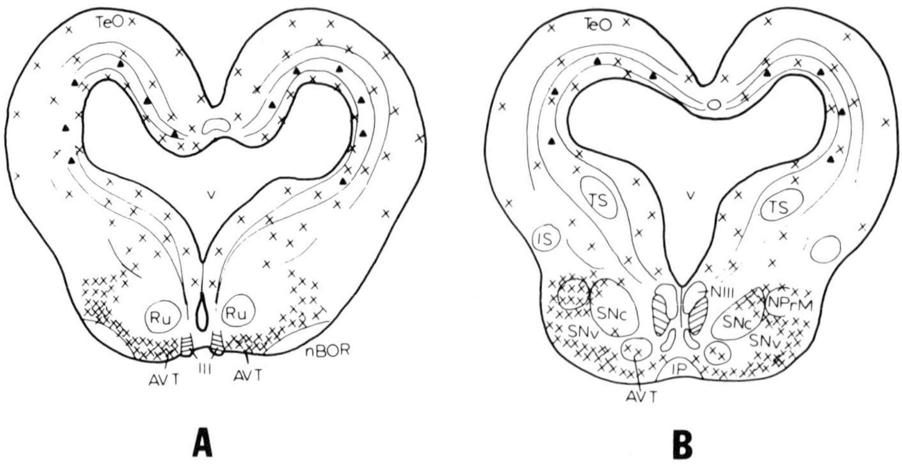

FIGURE 8. Substance P neurons and fibers in the midbrain of the turtle *Chrysemys scripta* (presented as in Figure 7). At the level of the emergence of the oculomotor nerve (A), a rich plexus of substance P-containing fibers is shown to invest the rostral area ventralis of Tsai (AVT). More caudally (B) and at the level of the interpeduncular nucleus (IP), a heavy substance P fiber plexus is shown to invest the region of the substantia nigra (SN). The heaviest and most intense substance P-immunoreactive staining in this area was found in a plexus ventral to the substantia nigra pars compacta (SNc). This substance P-containing fiber plexus was found to invest a large region of the ventral portion of the midbrain tegmentum. (From Brauth, S. E., et al., *J. Comp. Neurol.*, 219, 305, 1983. With permission.)

importance. The close association and interrelationships of these substance P cells and fibers with those of the serotoninergic and dopaminergic neuronal systems greatly enhance these potentials and opportunities. These arrangements may have contributed to the survival, essentially unchanged, of the turtle for more than 250 million years — a testimonial and impressive ''advertisement'' for evolutionary conservatism.

The authors of this outstanding publication include in their report a most valuable schematic drawing (Figure 10) to aid the reader in visualizing the distributions and anatomical connections, approximations, and interrelationships of substance P, dopamine, and serotonin in the reptilian, avian, and mammalian mesencephalons. This illustration facilitates an enhanced appreciation of the many striking parallels that exist with respect to neurotransmitter organization of the striatotegmental pathways in representative members of all three of the classes of terrestrial amniote vertebrates. It also clearly identifies our closer evolutionary relationship as *Homo sapiens sapiens* mammals to the friendly and benign turtle than to the somewhat less friendly and benign crocodile and vulture.

The authors conclude that features of basal ganglion organization, such as the descending substance P striatotegmental pathways and the ascending dopaminergic tegmentostriate pathways, appear to have evolved at an early point in amniote evolution (certainly the ancestral stem reptiles, if not earlier) and have been retained in extant amniote species, including mammals and man.

In additional substance P reptile studies, substance P immunoreactivity has been demonstrated in the parietal eye system of the lizard *Uta stansburiana*[59] and in the brain and spinal cord of the lizard *Varanus exanthematicus*.[63]

4. Birds/Aves

In birds, substance P immunoreactivity has been demonstrated in the spinal cord of the domestic fowl *Gallus domesticus*, where fibers and varicosities were most dense in lamina

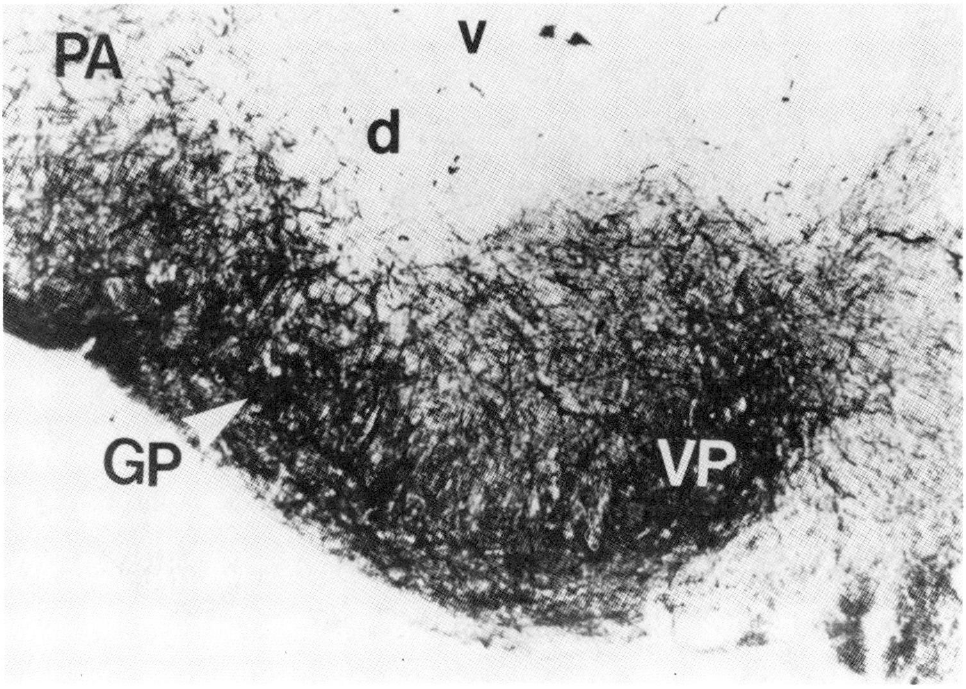

FIGURE 9. Photomicrograph demonstrating dense substance P-fiber immunoreactivity (black) in the globus pallidus (GP) and the ventral paleostriatum (VP) at the level of section E in Figure 7. The reader may recall that the basal ganglia (including the globus pallidus) was the mammalian brain area found to contain the largest concentration of substance P in the 1931 ''discovery'' publication of von Euler and Gaddum. It is of interest that this ''connection'' has persisted over 200 million years of evolution. (From Brauth, S. E., et al., *J. Comp. Neurol.*, 219, 305, 1983. With permission.)

I and lamina II of the dorsal horn.[64] It also has been shown to be present in an important subpopulation of displaced ganglion cells (DGCs) in the retina of the pigeon *Columba livia*.[65]

VI. NERVOUS SYSTEM

A. PAIN

Physiological evidence that substance P might be involved in nociception was derived from iontophoretic studies carried out by Henry[16] and reported by him in 1976. In these studies, all dorsal-horn neurons that were excited by the local iontophoretic application of substance P also were activated by noxious radiant-heat stimulation of the cutaneous receptive field of these dorsal-horn neurons.

In 1977 Jessell and Iversen[17] made their dramatic observation that morphine could inhibit stimulus-induced release of substance P, a putative pain transmitter, from central nervous system tissues of the rat. They employed spinal trigeminal nerve nuclei that had been dissected from freshly chilled 0.8-mm-thick coronal sections of the rat brain stem, by means of a steriomicroscope. They also found that the morphine-induced inhibition of release of substance P from these trigeminal nuclei could be blocked by naloxone, the morphine antagonist.

A 1980 report by Henry,[66,67] with Yashpal and Wright, extended the heat, pain, and substance P ''connection'' by giving substance P intrathecally to rats. They observed that this route of substance P administration reduced the reaction time for a tail-flick response

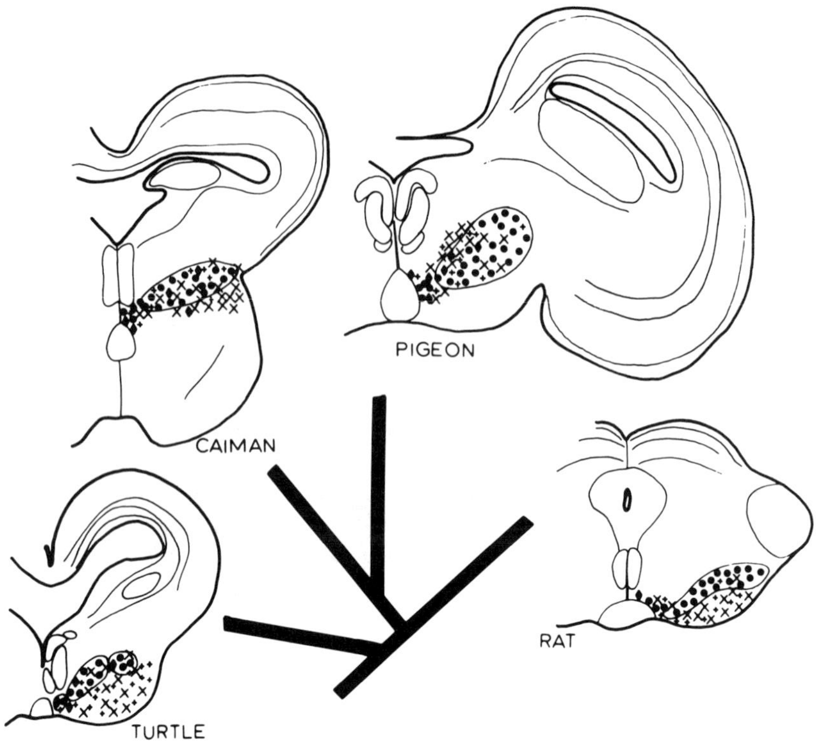

FIGURE 10. Schematic drawing showing the distribution of substance P fibers (X), dopaminergic cell bodies (filled circles), serotonin fibers (pluses), and serotonin cell bodies (filled diamonds), at the level of the midbrain tegmentum of the chelonian reptile (turtle), the crocodilian reptile (caiman), a bird (pigeon), and a mammal (rat). From the location, concentration, and anatomical interrelation of these three major vasoactive neurotransmitter and "memory" molecules, it is comforting to know that we of *Homo sapiens sapiens* share a common evolutionary descent with the friendly and lovable turtle rather than with the somewhat less friendly and less lovable crocodile. "And the voice of the turtle is heard in the land" — Song of Solomon. (From Brauth, S. E., et al., *J. Comp. Neurol.*, 219, 305, 1983. With permission. And from the Bible, hopefully with permission.)

to a peripherally applied noxious radiant-heat stimulus. This effect was observed as early as 30 sec after administration and reached a maximum response at 60 sec with the response persisting for as long as 10 min. Higher doses of substance P produced behavioral responses that consisted of immediate vocalization, restlessness, and scratching on the inside of the container. These responses were followed by teeth chattering and a stiff tail. These animals clearly were unhappy, substance-P induced. They also were learning, for up to 3 min after administering the substance P (preadministration tail-flick no-substance P controls had already been carried out), merely touching the tail to place it over the noxious heat source, with no noxious stimulus whatsoever, provoked a brisk withdrawal of the tail, and vocalization. Additional studies by Cridland and Henry,[68] in 1986, suggested that the receptor involved in the facilitation of the noxious heat-induced tail-flick reflex was a substance P receptor rather than a receptor for one of the other endogenous kinins. In 1988 Cridland and Henry[69] reported that the tail-flick reflex was inhibited by a substance P antagonist.

The 1987 report of Duggan et al.,[27] from Canberra, Australia, was of particular importance because it demonstrated for the first time a direct cause and effect between tissue-destructive peripheral noxious heat and central nervous system release of substance P in established sensory pain impulse reception areas. They employed the innovative antibody microprobe technique of Duggan and Hendry.[70] Barbiturate-anesthetized spinal cats were employed in this study. The noxious thermal stimulus used was hind-paw immersion in water heated to 50 and 52°C for 10 min. Significant amounts of endogenous substance P were detected as released in the substantia gelatinosa area of lamina I of the dorsal horn of these animals, and the release was further increased by elevating the temperature of the immersion of water to 52°C. These results made it quite clear that substance P is released centrally in the nervous system when cutaneous nociceptors are excited by heat. It also is of particular importance that this release was found to occur despite the presence of general anesthesia during the course of these studies.

Kellstein et al.,[71] of the Medical College of Virginia, presented evidence in 1990 that substance P selectively modulates c fiber-evoked discharges of dorsal-horn nociceptive neurons, and provided evidence that substance P functions as a neuromodulator of primary afferent nociception and that the effects of substance P are selective to nociceptive transmission by c fibers. Also in 1990 Wiesenfeld-Hallin, Folkers, and associates[36] described the specific antagonist effect of intrathecal Spantide II (a new substance P antagonist) on substance P- and c fiber-conditioning stimulation-induced facilitation of the nociceptive flexor reflex in the rat. They also relate that Spantide II administration is not associated with signs of the motor impairment or neurotoxicity that plagued Spantide I use.

A 1991 publication from Canberra, Australia, by Morton, Hutchison, and Hendry,[72] is of particular importance as it addresses the perplexing problem of the functional interrelationship of substance P and morphine, including opiate dependence and withdrawal. Their studies were carried out *in vivo* and utilized the antibody microprobe technique of Duggan and Hendry cited earlier in this section. They employed anesthetized spinalized cats that had been rendered morphine dependent by a treatment schedule (twice the daily intraperitoneal injections of morphine hydrochloride in doses ranging from 2 to 20 mg/kg for 3.5 d) shown in previous studies[73] to induce morphine tolerance and dependence in feline lumbar dorsal-horn neurons. Prior studies involving morphine-naive cats had shown that excitation of these study neurons by impulses in unmyelinated primary afferents could be reduced by intravenous administration of morphine, with no effect by naloxone. In morphine-pretreated cats, the response was altered, with no effect by morphine and excitaion by naloxone microiontophoretically applied into the substantia gelatinosa region of the dorsal horn. These findings were interpreted by the authors and their associates as indicating opiate tolerance and dependence.

In the 1991 report[72] the authors (Morton et al.) measured the release of CGRP as well as that of substance P. Both noxious cutaneous thermal stimulation and high-intensity electrical stimulation of a hind-limb nerve (tibial nerve in this study) stimulated substance P and CGRP release in the region of the substantia gelatinosa. Intravenous administration of naloxone to these morphine-dependent cats increased the nociceptive excitation of lumbar dorsal-horn neurons, but it did not alter the stimulation-evoked release of substance P or CGRP in the superficial gray matter dorsal to these neurons. Furthermore, the release of both of these neuropeptides did not differ significantly from that observed under similar experimental conditions in morphine-naive cats. The authors interpreted the results of these studies as indicating that alterations in the release of substance P or CGRP from the central terminals of nociceptive primary afferents does not occur during states of opiate dependence and withdrawal and, accordingly, does not contribute to the classical signs of these phenomena in morphine-dependent animals.

B. PERIPHERAL NERVOUS SYSTEM

The development and availability of radioimmunoassay and immunohistochemical methodology has made possible the identification of substance P in the peripheral as well as central nervous systems. Substance P has been found to be present in high concentration in primary sensory neurons, including the cell bodies in sensory ganglia and the terminals of these neurons both centrally in the spinal cord, brain stem sensory nuclei, and basal ganglia, and peripherally in the distal ends of sensory fibers. Hokfelt et al.,[74] in 1975, localized substance P in the peripheral nervous system to small, unmyelinated fibers within sensory nerves and identified the presence of substance P in the peripheral as well as central terminals of these fibers.

Substance P has been estimated to be present in from 10 to 20% of the total population, in spinal sensory ganglia, and to be the major nociceptive neurotransmitter in the small c fibers. Substance P-staining fibers also have been observed to be abundant in visceral afferent nerves. Gamse, Lembeck, and Cuello,[75] in 1979, reported that approximately 10% of all vagus nerve fibers contained substance P. In 1980 Katz and Karten[76] estimated that 30 to 60% of vagus nerve fibers in the rabbit jugular and nodose ganglia were substance P containing. Substance P also has been identified in the trigeminal nerve and in nerves that supply the pars flaccida of the tympanic membrane.[77,78]

Stimulation of sensory nerves has been shown to result in substance P release from peripheral terminals. In the 1985 reported study of White and Helme,[79] from Melbourne, Australia, electrical stimulation of the sciatic nerve in rats has been shown to produce a release of immunoreactive substance P in blister fluid in an established blister model that permits evaluation of substance P release from cutaneous nerve terminals. Neonatal pretreatment of study rats with capsaicin eliminated this response.

C. SPINAL CORD

Lembeck, in 1953,[4] described a differential distribution between the spinal cord and spinal roots for substance P, and a higher concentration in the dorsal horn than the ventral horn. These important findings led him to suggest a role as a primary afferent neurotransmitter for substance P.

Lembeck's suggestion received significant support in the findings of Otsuka and associates,[11-15] who demonstrated the high concentrations of substance P in dorsal roots and showed that it could be released *in vitro* from spinal cord preparations by electrical stimulation of dorsal roots. The studies of this important group also identified potent excitatory effects of substance P on spinal cord neurons.

The high concentration of substance P in the substantia gelatinosa superficial area of lamina I of the dorsal horn of the spinal cord and the direct release of substance P by noxious thermal heat in the peripheral cutaneous area, demonstrated by Duggan et al.,[27] was commented on in the ''Pain'' section of this chapter.

By the use of immunocytochemistry, degenerative lesions, transections, and retrograde transport studies, alone or in combination, the sources of substance P in the spinal cord have been suggested to be from the dorsal-root ganglia,[74,80-84] from the reticular area of the medulla oblongata,[85-87] and from intrinsic neurons within the dorsal horn of the spinal cord.[88,89]

In a 1983 communication from the University of Chicago and Cornell University Medical College, Pickel, Miller, and associates[90] employed light and electron microscopic immunocytochemical identification of substance P in the rostral and caudal ends of transected axons of the rat medulla (caudal medulla) and spinal cord (C2–C3 levels), employing unilateral knife cuts. Their findings with descending pathways suggested that they had interrupted long projecting substance P-containing neuronal axons in descending bulbospinal pathways, which appeared to arise from the venterolateral reticular area of the brain stem, with some axons also arising from the raphe nuclei. Thus, with different methodology, the authors confirmed the descending substance P bulbospinal pathways. With respect to ascending substance P-containing neuronal pathways, three previously undescribed ascending substance P-containing neuronal pathways were identified. These were found in the dorsal columns and the dorsolateral faciculis, and as a tract below the central canal. The ultimate paths and functions of these second-order neurons that are substance P containing were not addressed in these studies, but clearly are of considerable interest and likely of major significance.

A 1991 report by Salter and Henry,[91] of McGill University, is of considerable interest. They tested the effects of substance P, neurokinin A, and physalaemin on functionally identified dorsal-horn neurons of the cat *in vivo*. The cats were anesthetized with sodium pentobarbital or were anemically decerebrated. Extracellular single-unit recordings were made in the lumbar spinal cord, and the three tachykinins were applied by iontophoresis. Each of the neurons studied was classified as wide dynamic range (WDR), nonnociceptive, nociceptic specific, or propioceptive. The response to tachykinin application for each neuron was recorded and determined. Each of the three tachykinins produced a characteristic excitatory response that was delayed in onset, slow in developing, and prolonged (? LTP). Physalaemin had a preferential excitatory effect on nociceptive neurons, as did substance P. On comparing the proportions of nociceptive neurons excited by the three neuropeptides, this type was not found to be differentially sensitive to any of the three. On the other hand, the nonnociceptive neurons were preferentially excited by neurokinin A and physalaemin, with none of these neurons being excited by substance P.

With respect to the endogenous mammalian tachykinins (physalaemin is amphibian derived), the authors interpreted their findings as identifying a differential excitatory effect of substance P on nociceptive neurons, and as supporting a proposed role for substance P in the transmission specifically of nociceptive inputs at the first afferent synapse, with a distinct functional role for neurokinin A differing from that for substance P. A similar preferential excitatory effect on nociceptive vs. nonnociceptive neurons had been shown in earlier reports by Henry.[16,92]

D. BRAIN STEM

The very high substance P concentration in the basal ganglia area of the brain stem was demonstrated clearly in the 1931 discovery publication of von Euler and Gaddum.[1] This was in comparison with other tissues and organs of the whole body, as well as with other segments of the central nervous system.

A flurry of mapping and tracking studies in the late 1970s and early 1980s, by Hokfelt, Ljungdahl, Cuello, Kanazawa, Jessell, Emson, Costa, Paxinos, and others, provided a mountain of invaluable data concerning the central and peripheral substance P-ergic neuronal system, including its massive size and wide distribution. The monumental 1978 publication of Ljungdahl, Hokfelt, and Nilsson[89] stands as a classic of in-depth scholarly investigation and documentation. Their study identified as many as 30 groups of substance P-immunoreactive cell bodies and an extremely wide distribution of neuronal fiber projections, peripheral as well as central.

In the forebrain (telencephalon and diencephalon), the heaviest network of substance P-immunoreactive cell bodies and projections was encountered in the medial amygdaloid nucleus. Microknife-cut findings suggested that the majority of the fibers were intrinsic to the amygdaloid nucleus. Substance P-immunoreactive fibers in the frontal cortex and septum appeared to originate from extrinsic sources, with those projecting to cortical areas seeming to arise from regions caudal to the mesencephalon.

The principal forebrain substance P nervous system projections are from the caudate-putamen neostriatum in the basal ganglia area to the substantia nigra dopamine cell body concentration area where multiple and repetitive synapses of striatonigral substance P-containing fibers onto dopaminergic neurons are encountered.

There are two other major forebrain substance P nervous system projection pathways in the forebrain. One is from the amygdala to the stria terminalis and onto the hypothalamus, and the other is from the medial habenula to the lateral habenula and on to the ventral tegmental area (VTA), which surrounds the nucleus interpeduncularis (the habenula-interpeduncular pathway). The VTA shares with the substantia nigra top billing as a major homesite for dopaminergic neuronal cell bodies. Dopaminergic projections from the substantia nigra and the VTA have major ascending projections of immense functional significance.

Hindbrain projections of the substance P nervous system have a major projection from the raphe nuclei, which are home base for most, if not all, neuronal cell bodies of the serotoninergic nervous system.

Central and ascending fibers and ganglia of the substance P nervous system, from the periphery via cranial and spinal sensory nerves, provide essential sensory input for the establishment of learning and memory at the central and molecular level.

The great mappers and trackers of the late 1970s left little for posterity to do with respect to the substance P nervous system.

In 1984 Hancock,[93] of the University of Texas (Galveston), employing a two-color immunoperoxidase staining methodology, identified punctate substance P-immunoreactive processes on neuronal cells containing both substance P and 5-HT. These cells were found in both the raphe nuclei and the VTA region of the medulla of the hamsters employed in this study. The author considered that this close association with accompanying substance P release in the brain stem might affect bulbospinal 5-HT/SP (substance P) pathways which could, in turn, influence sympathetic preganglionic neurons.

Also in 1984 Ronnekliev et al.[94] described the presence and distribution of substance P-immunoreactive neurons in and about the hypothalamus and pituitary of the Rhesus monkey. They discovered substance P-immunoreactive neuronal cell bodies in the arcuate nucleus and in the median eminence and the periventricular area of the dorsal tuberal region. A rich network of substance P-containing neuronal fibers was identified in the arcuate region. The neuronal fiber stain for substance P immunoreactivity was especially dense in the external line of the median eminence and in the external layer of the infundibular stalk. Immunoreactive fibers also were observed in the internal layer of the pituitary stalk, and in the neural lobe of the pituitary gland, outside the hypothalamus, a dense network of substance P-immunoreactive fibers was identified in the globus pallidus.

Li et al.,[95] from the People's Republic of China, in 1990, made some important observations concerning serotonin- and substance P-containing neurons in the midbrain periaqueductal gray (PAG) and nucleus raphe dorsalis (NRD), which send projection fibers to the central amygdaloid nucleus of the rat. Fibers were sent to the central amygdaloid nucleus bilaterally, with an ipsilateral dominance. The PAG neurons sending these ascending fibers to the amygdala were found to be located largely in the venterolateral PAG subdivision and in the ventral parts of the medial PAG subdivision at the middle and caudal levels of the PAG. The authors employed a double-labeling method that combined retrograde horseradish peroxidase (HRP) tracing and immunocytochemical methodologies to obtain these important observations. They emphasized the importance of the amygdala complex in the limbic system and its involvement in the endogenous control of pain (and memory).[96-98]

Also in 1990 the same group from China,[99] and with the same methodology, reported that substance P-immunoreactive neurons identified in the medial and commissural subnuclei of the nucleus tractus solitarii (NTS) of the rat send axons to the nucleus accumbens. This finding also has significance in the heat, pain, and memory scenario, as the NTS is a major site of termination of visceral afferent sensory and pain-impulse transport fibers. Like the amygdala, the nucleus accumbens also is an important structure in the limbic system.

A 1991 report by West and Michael,[100] from Atlanta, explored the effects of substance P injections into the ventral tegmentum on unit activity in mesolimbic terminal regions. These studies were of particular importance because of the tachykinin neuropeptides; substance P is known to have a discrete distribution and discrete actions in the mammalian central nervous system. In the rat brain, large concentrations of substance P have been identified in the VTA, in close proximity to the dopaminergic cell bodies of origin of the mesolimbic, mesocortical, and nigrostriatal systems.[89,101] Several lines of evidence have suggested that substance P might have an influence on dopaminergic neurons in this area. These include studies that have shown that microinjections of substance P into the VTA produces an increase in spontaneous motor activity, rearing, and exploration.[102-108] It has been suggested that these behavioral alterations are mediated through the dopaminergic mesolimbic system that projects from the VTA to the nucleus accumbens and to the olfactory tubercle.[109] Support for this speculation can be found in experimental studies that have demonstrated the substance P-induced behavioral activations can be blocked by lesioning of dopaminergic terminals in the nucleus accumbens with 6-hydroxydopamine[108] or by the use of dopamine antagonists.[107] Additional support is derived from studies showing that injection of substance P or a stable analogue into the VTA increases dopamine metabolism[104,110,111] and that iontophoretic application of substance P activates dopaminergic neurons in the substantia nigra in electrophysiological studies.[112,113] These multiple observations strongly suggest that substance P activation of dopaminergic neurons in the VTA may elevate the levels of dopamine-mediated neurotransmission throughout the mesolimbic system.

In their 1991 reported study, West and Michael[100] recorded firing rates of neurons in areas receiving projections from the mesolimbic dopamine system during injections of substance P (3 μg in 0.5 μl saline) into the VTA of rats anesthetized with chloral hydrate. Significant alterations in firing rates were found in 84% of the units recorded, in the nucleus accumbens and the olfactory tubercle. Alterations were mostly decreases in the nucleus accumbens (21 of 25 units recorded) and mostly increases in the olfactory tubercle (13 of 18 units recorded). Neither saline injections into the VTA or substance P injections 2 mm dorsal to the VTA had any effect. A dopamine antagonist (haloperidol) blocked the effects. The authors interpreted their results as indicating that activation of dopaminergic neurons by substance P injections into the VTA can produce changes in the activity of neurons that receive dopaminergic neuronal projections, such as the nucleus accumbens and the olfactory tubercle.

E. BRAIN

In a 1974 study carried out shortly after the commercial availability of synthetic substance P, Phillis and Limacher, of the University of Saskatchewan (Canada), investigated the potency of substance P as an exciting agent on cerebral cortical Betz cells of rats in comparison with other potent excitatory neuropeptides including bradykinin and bombesin.[114] Of the endogenous peptides, the highest level of excitation was achieved by substance P both at spontaneously active unidentified cells (91%) and on Betz cells (94%). Whereas the amphibian peptide physalaemin equaled the excitability rate of substance P on unidentified spontaneously active cortical neuronal cells (91%), its excitability rate on Betz cells fell to 71%. Bradykinin's excitability rate for spontaneously active neuronal cortical cells was 78%, and for Betz cells it rose to 97%. Bombesin excited 60% of the spontaneously active cells. Both substance P and bradykinin were unique in that their excitation of neuronal cells often continued for minutes after the termination of topical application (? LTP).

The bulk of studies relating to substance P neuronal cell bodies, fibers, and terminals in mammals have been carried out on rats. From these studies Cuello et al.,[115] in 1982, presented a valuable schematic drawing of the substance P nervous system of the rat brain (Figure 11).

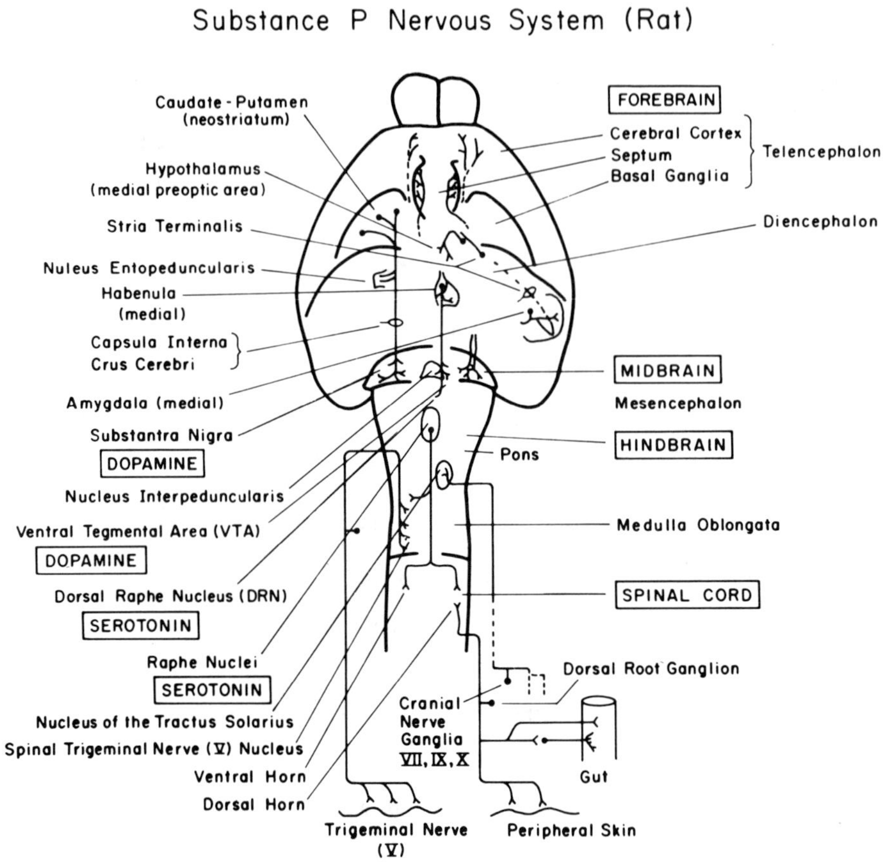

FIGURE 11. Substance P nervous system of the mammalian rat brain. (After Cuello, A. C. et al.[115])

A 1985 report from the Karolinska Institute in Stockholm, by Lindefors et al.,[116] described the regional rat brain differences in the distribution of substance P and 20 other mammalian tachykinins: neurokinin A (NK-A) and neurokinin B (NK-B). NK-A and NK-B had similar tissue distributions and concentrations and differed significantly from those of substance P. The ratio of tissue concentrations of NK-A and NK-B (antisera-determined), as compared with those of substance P, were higher in the frontal cortex and hippocampus and lower in the pons/medulla oblongata than in the other regions studied.

Cuello, Mai, and associates,[117] from Oxford and Dusseldorf in 1986, published an interesting study of substance P location and pathways in the human brain, employing not only fresh but also paraffin-embedded sections from decades ago. They not only demonstrated that archival material could be used for substance P immunoreactivity study, but they also extended prior observations. Novel findings included the identification of substance P-immunoreactive neuronal cell bodies in most cortical areas and distribution of substance P-containing elements in the tuberal region, the corpus striatum, and the substantia nigra, where a novel relationship to blood vessels was encountered. With respect to the somatosensory areas, they concluded that the relationship that exists between substance P and areas that receive direct or indirect input from primary sensory afferents is substantially more stable across species boundaries than is the relationship between substance P and the cortex, amygdaloid complex, and the interpeduncular nuclei. The presence of substance P neuronal cell bodies in most cortical areas in humans was a change from the rat, where none were found, and comparable to other primates, including the baboon and marmoset. They emphasized that studies carried out on rodents cannot be extrapolated directly to man. In addition to the cerebral cortex substance P-immunoreactive cell bodies cited above, a dense plexus of substance P-immunoreactive terminals identified in both monkeys and man are not found in rodents.[118]

Studies by this group relating to the hippocampus and the cortical areas are of particular interest. They identified three discrete layers of substance P immunoreactivity in the hippocampus. They found intense staining, with some columnar appearance, in the second layer of hippocampal and entorhinal cortex. In these areas, terminal branching was demonstrated in the proximity of substance P-immunoreactive pyramidal cells as well as within substance P-immunoreactive clusters of stellate cells of the entorhinal cortex (nose-brain). A diversity of the frequency of substance P-immunoreactive neuronal cell bodies in the cortex, in terms of location, concentration, and variability of perikaryal and dendritic morphology, was observed. Substance P-immunoreactive neurons located in the cortical white matter were spindle-shaped bipolar cells with their dendrites oriented along the fiber pathways. Immunoreactive small pyramidal cells were encountered more often in the deep cortical layers, whereas substance P-immunoreactive varicosities were present in all layers, but predominantly in layer 3 and in the vicinity of positive cell bodies. These findings suggest a capability of the substance P nervous system in the cortical areas for significant and likely important functional specificity and unique and discrete activity. They also would seem to indicate that despite the phylogenetic antiquity of substance P as a major neuropeptide, there is still space relating to this important transmitter and ''memory molecule'' for a divergence between ''mice and men''.

Two particularly valuable reports appeared in 1987 concerning substance P in the hippocampus of humans. The report of Del Fiacco et al.,[119] from the University of Cagliari (Italy), presented important information concerning the location of substance P immunoreactivity in both the newborn and the adult. They reviewed the comparatively limited, yet significant, literature relating to substance P in the human central nervous system,[120-125] as well as that concerned with alterations of substance P in discrete brain nuclei in human neurological disorders, including Alzheimer's disease and Huntington's chorea.[126,127] In their study, Del Fiacco et al.[119] found substance P immunoreactivity to be abundant in neuronal structures of the human hippocampal formation and the parahippocampal gyrus. They used an indirect immunofluorescence technique. In the newborn, they found a large number of substance P-containing cell bodies (perikarya) in the hippocampus. The largest number of cell bodies were identified in the stratum oriens of the hippocampus and were especially numerous in the important (learning and memory) CA1 (Cornu Ammonis 1) area. Many of these cell bodies were spindle shaped in appearance, and their proximal branches could often be traced for considerable lengths parallel to the fibers of the alveus layer of the hippocampus. In addition, irregularly shaped triangular and multipolar substance P-immunoreactive neuronal cell bodies also could be identified. Some of these neurons and projections seemed to merge in the stratum pyramidale layer of the hippocampus, with scattered substance P-immunoreactive cell bodies being encountered in the dorsal CA1, CA2, and CA3 fields. Sector CA4 was found to contain numerous substance P-immunoreactive cell bodies that had a wide variety of morphological characteristics, including oval, pear-shaped, fusiform, triangular, and multipolar. Nerve fibers and terminals increased in numbers from the deeper to the more superficial layers of the hippocampus. Short neuronal-fiber tracts of intense substance P immunoreactivity were abundant in the CA4 sector. The strata radiata, lacunosum, and moleculare displayed a punctiform substance P immunoreactivity of progressively increasing density, and horizontal sections revealed the presence of strongly fluorescent beaded axons running in the molecular layer, parallel to the long axis of the hippocampus.

In the human adult hippocampus, Del Faccio et al.[119] found a comparative decrease in the number of substance P-immunoreactive cell bodies in the stratum oriens, with the largest number, as in the newborn hippocampus, being located in the CA1 sector. Here most of the cell bodies were spindle shaped, with their long axis parallel to the fibers of the alveus layer. The hilus of the fascia dentata retained a large number of substance P-immunoreactive cell bodies of a variety of sizes and shapes.

Whereas the distribution of substance P-immunoreactive neuronal fibers and terminals in the hippocampal alveus, stratum oriens, and strata radiatum, lacunosum, and moleculare layers of the adult were similar to that encountered in the newborn hippocampus, a striking difference was encountered in the adult distribution pattern in the stratum pyramidale and in the fascia dentata. While substance P-immunoreactive neuronal terminals are comparatively scarce in the CA1 sector, there is an abrupt and marked increase in these terminals in the CA2 sector where a dense network of intense substance P-immunoreactive fibers with large varicosities surrounds the neuronal cell bodies of this sector. The CA3 and CA4 sectors of this stratum pyramidale layer show a similar pattern of intense substance P-fiber immunoreactivity.

In the fascia dentata of the adult human hippocampus, neuronal terminal substance P immunoreactivity is densely packed in the superficial three fourths of the stratum moleculare. In addition, a rich plexus of substance P-containing nerve fibers and terminals was observed to border the molecular margin of the stratum granulare.

The authors interpreted the findings of these studies, including the demonstration of numerous substance P-containing neuronal cell bodies in the stratum oriens and in the polymorphic layer of the fascia dentata, to indicate that in man a conspicuous source of substance P immunoreactivity and connectivity is intrinsic to the hippocampal formation.

These unique distinctive patterns of substance P immunoreactivity in neuronal cell bodies, fibers, and terminals in the human newborn and adult are likely to be of considerable importance and significance with respect to the critical involvement of the hippocampus in learning and memory and the cellular and molecular plasticity alterations that occur with growth and development.

These findings are in accordance with ontogenic studies in the rat[128,129] in which substance P-immunoreactive cell bodies are more numerous in the newborn than in the adult, whereas substance P-immunoreactive nerve fibers and terminals are more densely distributed in specific areas, as well as being more precisely organized, in the adult.

The other 1987 publication (by Sakamoto et al.[130] from Japan, the U.S., and France) is also of great importance and significance as well as being beautifully illustrated. The authors have provided an unusually informative and helpful schematic drawing of the main anatomical features of the hippocampal formation, as gleaned from major works by Brodal, Ramon y Cajal, Carpenter and Sutin, and Lorente de No[131-134] (Figure 12). They have, to great instructional effect, employed this same schematic representation to present the distribution of substance P-immunoreactive neuronal cell bodies, fibers, and terminals, as encountered in the hippocampi of brains of newborn humans dying in the early (17, 18, and 20) days of life from sudden-infant-death syndrome (SIDS) (Figure 13). For any residual skeptics, they also have provided an impressive photomicrograph from the same orientation as the two schematic drawings (Figure 14).

The investigations of Sakamoto et al.[130] were carried out employing both monoclonal and polyclonal (serum) antibodies (against substance P) developed at Osaka University which had been shown in prior studies to give similar staining results. For staining they use a modified peroxidase antiperoxidase (PAP) staining methodology.

Numerous substance P-immunoreactive neuronal cell bodies were identified in the stratum oriens throughout the entire rostrocaudal extent of the hippocampus and forming a

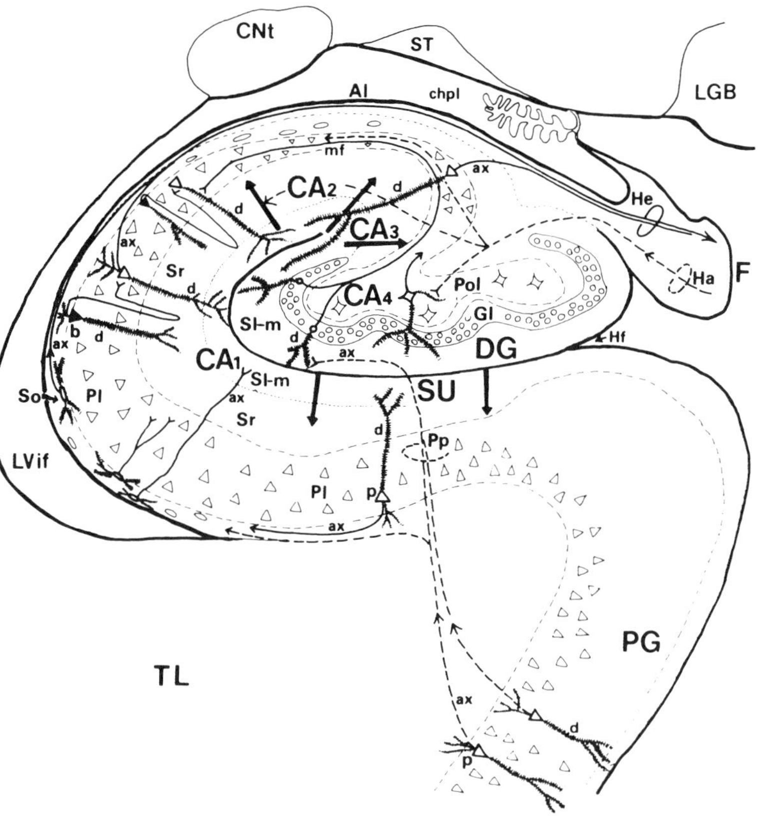

FIGURE 12. Schematic transverse section drawing illustrating the main anatomical features of the human hippocampal formation. The sequential arrangement of the dentate gyrus (DG) and the Cornu Ammonis areas CA4, CA3, CA2, and CA1 of the hippocampus proper are shown, as are the subiculum (SU) and the parahippocampal gyrus (PG) of the entorhinal area. Axons (ax), dendrites (d), mossy fibers (mf), and pyramidal cells (p) are shown, as well as the perforant path (Pp) from the entorhinal area and afferent fibers to the hippocampus (Ha) and efferent fibers from the hippocampus (He). The choroid plexus (chpl) and the inferior horn of the lateral ventrical (LVif) are shown surrounding a significant portion of the hippocampus. Hippocampal layers from the outside in are alveus (Al), the stratum oriens (So), the pyramidal layer (Pl), the stratum radiatum (Sr), and the stratum lacunosum moleculare (Sl-m). In the CA4 area a granular layer (Gl) and a polymorphic layer (Pol) also are found. The fimbria of the fornix (F) constitutes a major access site. This excellent schematic drawing by the authors was based on work by Brodal,[5] Ramon y Cajal,[6] Carpenter and Satin,[7] and Lorente de No.[8] (From Sakamoto, N., et al., *Neuroscience*, 22, 801, 1987. With permission.)

cellular band that extended into the subiculum. Most of the substance P-immunoreactive cell bodies in the stratum oriens were long, elliptical, and bipolar and were lined up with their long axes parallel to the course of the alveus layer. Substance P-immunoreactive staining in this area ranged from moderate to strong. A few substance P-immunoreactive bipolar and multipolar neurons, with moderate or strong staining, were identified in the superficial part of the pyramidal layer of CA2, CA3, and CA4, with numerous immunoreactive cell bodies being found in the polymorphic layer and in the hilus of the dentate gyrus. It was of interest that other neuronal cell bodies (in the polymorphic layer) that did not have cytoplasmic staining did have their surface membranes covered with substance P-immunoreactive puncta. Substance P-immunoreactive punctate profiles also were encountered throughout the pyramidal layer in the subiculum and CA1, CA2, and CA3 regions, as well as in the granular layer of the dentate gyrus. The density of substance P-immunoreactive axons was higher in the granular than in the pyramidal layer. Numerous substance P-immunoreactive fibers,

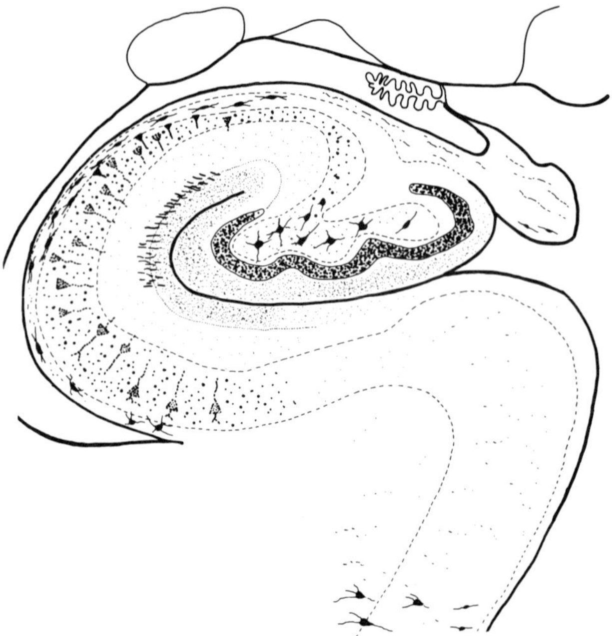

FIGURE 13. Substance P-immunoreactive neuronal cell bodies and fibers in the human hippocampal formation (same transverse plane as shown in Figure 12). Substance P-immunoreactive cell bodies (perikarya) are shown in solid black. The dotted pyramidal cells have substance P terminals on their surface. (From Sakamoto, N., et al., *Neuroscience*, 22, 801, 1987. With permission.)

predominantly very fine fibers, were shown to be present in the stratum lacunosum moleculare of the hippocampus and of the dentate gyrus. A few thick varicose immunoreactive fibers also were found in this layer. Deep in this stratum lacunosum moleculare layer, and adjacent to the stratum radiatum, numerous thick substance P-immunoreactive varicose fibers also were found. In addition, substance P-immunoreactive neuronal cell bodies and scattered varicose fibers were shown to be present throughout the length of the indusium griseum, with shapes similar to those found in the stratum oriens of the hippocampus proper.

Particularly important observations from this impressive study included the identification of substance P-immunoreactive cell bodies in the stratum oriens and polymorphic layers of the human hippocampus, as well as the findings of some neuronal cell bodies in the most superficial part of the pyramidal layer, which appeared to correspond to hippocampal basket cells. The intercommunications of stratum oriens cells were extensive and included axons from outer-zone cells to the molecular layer and axons from the inner zone to the alveus layer, ramifying within the stratum oriens or passing into the pyramidal layer. Immunoreactive hippocampal basket cells, which are similar to pyramidal cells, were found largely near the border between the pyramidal layer and the stratum oriens. Their axons, interestingly, did not enter the alveus layer; instead they looped back through the statum radiatum to form dense substance P-immunoreactive plexes about pyramidal cell bodies. It was of further interest that substance P-immunoreactive neuronal cell bodies in the dentate gyrus do not give rise to fibers that pass beyond the hippocampus — an extensive, tight, and intensive internal organization that possesses the potential for considerable integrative activity with respect to input sensory information.

An intriguing additional observation was the discovery of very large substance P-immunoreactive boutons on hippocampal neuronal cell bodies, which were quite distinctive

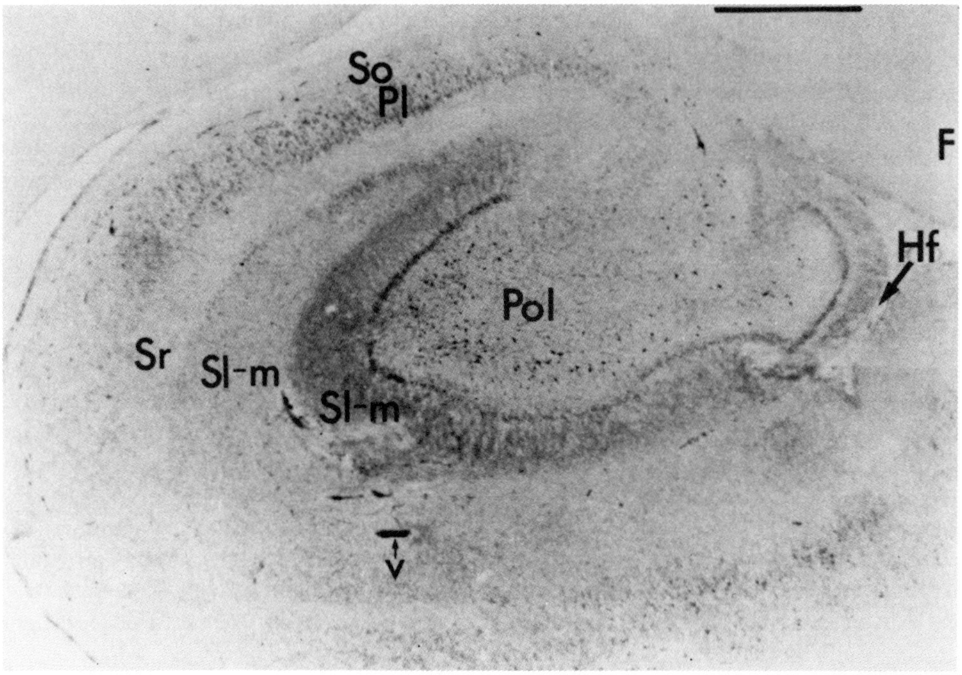

FIGURE 14. Photomicrograph showing substance P immunoreactivity in the human hippocampus in the same orientation as shown in Figures 12 and 13. Intensely stained substance P-immunoreactive cell bodies are shown in the polymorphic layer (Pol) and in the stratum oriens (So). Substance P-immunoreactive terminal boutons are rich in both the pyramidal (Pl) and granular layers. Substance P immunoreactivity was found to be particularly intense in axon networks in that part of the stratum lacunosum moleculare immediately adjacent to the granular layer. (From Sakamoto, N., et al., *Neuroscience*, 22, 801, 1987. With permission.)

from substance P-immunoreactive staining elsewhere in the brain. The distribution of substance P-immunoreactive boutons related to pyramidal neuronal cell bodies and was found by the authors to change with age. Whereas in infants these boutons are located largely on neuronal cell bodies, in adults they are found to have migrated from the cell bodies to the neuropile where they are likely on axons and likely contacting unstained dendrites. The authors interpreted these important and intriguing observations as suggesting significant synaptic remodeling during maturation.

The unique and highly significant findings of these two 1987 reports provide the author of this volume with an irresistable temptation to speculate, from the perspective of the ''heat, pain, and memory'' scenario, that human growth and development expose this unique and remarkable substance P nervous system, and related nervous systems, with continuing and accelerating exposures to sensory inputs, including intellectual, which may play a basic and fundamental role in the increasing capacity for learning and memory at the cellular and molecular plasticities levels.

The issue of ontogenic plasticity of substance P-immunoreactive structures in the primate brain was explored in a 1990 report (From Kyoto, Japan, by Yamashita et al.,[135]) of various cerebrocortical areas of the brains of macaque monkeys from embryonic day 120 to adult life. They found that the distribution and developmental ontogeny of substance P immunoreactivities differed among various cortical areas studied. In association areas (areas FD, PE, TA, and TE) large numbers of substance P-immunoreactive cell bodies were present at embryonic day 140 (E140), and at birth this number decreased progressively until postnatal day 60 (P60), at which time the adult number of substance P-immunoreactive cell bodies was reached. In cortical areas OC and FA (von Bonin and Bailey), on the other hand, the authors found high densities of substance P-immunoreactive nerve fibers and terminals, despite the low numbers of substance P-immunoreactive cell bodies.

If substance P is so important in growth, development, learning, and memory, it must also play a role in the important area of sex. This potentially serious gap in information was filled recently by three important and provocative reports, one of them in 1991.[136-138] The 1989 publication of Dornan and Malsburg[136] described effects of intracerebral injections of substance P on the copulatory sexual behavior of sexually experienced adult male rats, which presumably had some degree of "memory" for this sort of thing. Bilateral injection of substance P into the medial preoptic-anterior hypothalamic area produced marked changes in several components of male copulatory behavior, with latencies (shortened) being the most affected. Significant shortening of the interval to initiate copulation also was observed. These responses were not produced by CCK injections, and with substance P injections were eliminated by prior castration of the study rats. Injection of an undiluted substance P antiserum into the medial preoptic-anterior-hypothalamic area of these rats produced a dramatic impairment of male copulatory behavior, with significantly lengthened mount, intromission, and ejaculation latencies. The authors interpreted this important information as providing the first support for a role of substance P in the neural regulation of male rat sexual behavior and as suggesting that a substance P sexually dimorphic pathway forms part of the neural circuitry important for the expression of male rat sexual behavior.

In 1991 Neal and Newman,[137] from the University of Michigan, described identification of a substance P neuronal projection from the medial nucleus of the amygdala and the bed nucleus of the stria terminalis to the medial preoptic area of the male Syrian hamster brain and speculated that these projections might play a role in the modulation of reproductive behavior (sex) in this species.

An important 1991 report from Lausanne and Zurich, Switzerland, by Barakat-Walter et al.,[139] addressed the ontologic plasticity area relating to substance P expression in neuronal cells at various stages in embryonal development. Cultured chick dorsal-root ganglion (DRG) cells were studied at E6 and E10 embryonic days (before and after establishment of functional connections with central or peripheral targets). Expression of both substance P and preprotachykinin mRNA were investigated. DRG cell cultures studied at E6 showed a significant hybridization signal for preprotachykinin mRNA and substance P-like immunoreactivity in virtually all of the neurons studied (98%). By day E10 (after establishment of functional connections), only 60% of neurons showed substance P immunoreactivity. The addition of extracts from spinal cord, brain, skin, or muscle to DRG cells cultured at day E6 reduced by 20% the percentage of neurons that expressed preprotachykinin mRNA and substance P immunoreactivity. The authors interpreted their findings as suggesting that factors issued from target tissues inhibit substance P expression by a subset of primary sensory neurons and that these target factor(s) act on the transcriptional control of the preprotachykinin gene.

VII. VASCULAR SYSTEM

The principal systemic vascular effects of substance P were presented clearly in the 1931 "discovery" publication of von Euler and Gaddum.[1] The depression in rabbit systemic arterial blood pressure was both recognized and correctly identified as being due to peripheral vasodilatation. It also was demonstrated that the effect could be produced equally well by direct injection of substance P into either the aorta or the superior vena cava (indwelling cannulas). The vascular effect was so well characterized by the authors that they employed this effect as one of the two major responses to measure and quantitate the substance P concentrations in various body tissues, including various parts of the central nervous system.

The involvement of substance P in "neurogenic inflammation" and its effects on local blood flow and vascular permeability will be considered in more detail in the portion of this chapter dealing with inflammation.

A 1988 report from Madrid, by Catalan et al.,[140] explored substance P-stimulated responses in protein kinase C (PKC) translocation in bovine brain microvessels. The authors found that 85% of the total PKC in untreated brain microvessels was located in the cytosolic fraction. Substance P caused a shift (translocation) in PKC activity to the membrane fraction, with an accompanying loss of activity in the cytosolic fraction. This was shown to be a dose-dependent response. The authors interpreted their findings as suggesting that substance P might be involved in the regulation of processes underlying protein phosphorylation in and at the blood-brain barrier.

In 1990 Matis et al.,[141] of the University of Pennsylvania, observed that substance P induces the expression of an endothelial-leukocyte adhesion molecule (ELAM-1) by microvascular endothelium. The authors exposed organ cultures of neonatal human foreskins for 45 min to substance P. Dermal mast cells were uniformly degranulated by substance P, as judged by ultrastructural analysis. After subsequent culture in medium alone for 6 h, superficial venules of explants exposed to substance P showed evidence of ELAM-1 induction, as documented histochemically employing the H4/18 monoclonal antibody. Preincubation of explants with the mast cell degranulation inhibitor, cromolyn sodium, blocked the ability of substance P to induce ELAM-1. The authors interpreted these results as suggesting that substance P released endogenously by dermal fibers on physiologic or electrical stimulation may be important in the regulation of endothelial-leukocyte interactions *in vivo*.

Also in 1990 Galligan et al.,[142] from Portland, OR, presented evidence that substance P mediates neurogenic vasodilatation in extrinsically denervated guinea pig submucosal arterioles. The authors employed a new optical television video measuring method of determining vascular (arteriole) diameter, developed recently by Neild.[143,144] The guinea pig submucosal plexus (small intestine) was studied *in vitro*. Extrinsic denervation of this plexus results in complete loss of immunofluorescence for catecholamines, neuropeptide Y (NPY), and calcitonin gene-related peptide (CGRP). Application of exogenous substance P to this preparation produced vasodilatation of all submucosal arterioles. This occurred in both normal and extrinsically denervated arterioles. A strong correlation was demonstrated between the finding of noncholinergic vasodilatation in response to ganglionic stimulation and the presence of substance P-immunoreactive fibers running from the stimulated ganglion to the observed arterioles. This correlation did not exist for VIP (vasoactive intestinal peptide). The authors interpreted these findings as indicating that intrinsic intestinal substance P-containing nerve fibers remain to supply submucosal arterioles after sympathetic efferent and sensory afferent extrinsic nerve fibers are removed.

From Florence, Italy, a 1990 report by Ziche et al.[145] described substance P-stimulated neovascularization *in vivo* and substance P-stimulated proliferation of cultured endothelial cells *in vitro* — a mitogenic effect. The effect of substance P on the growth of capillary vessels *in vivo* (neovascularization) employed slow-release preparations of substance P implanted into the avascular cornea of rabbits, with daily vessel growth monitoring through a slit-lamp stereo microscope. Substance P was observed to produce marked neovascularization (capillary vessel growth). A selective NK-1-receptor agonist produced similar results. Addition of substance P to serum-free cultured endothelial cells (bovine adrenal and human umbilical cord veins) increase proliferation *in vitro* of both cell lines in a concentration-dependent manner. From these observations the authors suggested an activity of substance P in trophic actions relating to the vascular system. A mitogenic and "growth factor" effect also is quite apparent.

In an interesting and important 1991 report from Waterloo (Canada, not Belgium), Globus et al.[146] presented evidence supporting a mitogenic role for substance P in amphibian (newt) forelimb regeneration and supporting an involvement of the inositol phospholipid signaling pathway in this process. The essential trophic role of nerves in this limb regeneration process is the promotion of mitosis in the blastema cells of the regenerate. The authors showed the presence of substance P at the site of action in the regenerate by immunohistochemistry and by radioimmunoassay. They also identified a diminution of substance P immunoreactivity in the regenerate following denervation of the limb. In addition, substance P was shown to be extremely mitogenic at very low concentrations in cultured limb regenerate blastema cells. The mitogenic influence of intact nerves to a regenerate was suppressed in a dose-dependent manner when sensory ganglia (cocultured transfilter to the blastema) were treated with increasing concentrations of antiserum to substance P in the culture medium. The authors demonstrated substance P-induced stimulation of the incorporation of [^{3}H]-thymidine into the DNA of regenerate blastemal mesenchyme cells and a rapid dose-dependent reduction of inositol phospholipids and an accumulation of radiolabeled inositol phosphates (increased PI turnover) in the newt forelimb blastema, in response to stimulation by substance P.

Substance P was reported (in *Nature* in 1985 by Nilsson, A. M. von Euler, and Dalsgaard[147]) to exert a mitogenic effect on mammalian connective tissue cells. In 1986 Salo and Baguna[148] reported a substance P mitogenic effect in regenerating planarians. These important studies would appear to propel substance P into the "nonclassical" (? "neoclassical") growth factor category along with serotonin, bradykinin, and bombesin — also fellow "memory molecules".

In 1990 a particularly important article (from University College, London, by Ravelic et al.[149]) appeared in *Circulation Research*. The authors studied substance P release from the endothelium of normal and capsaicin-treated rat hind-limb vasculature by increased blood flow *in vivo*. They showed that rat hind-limb vasculature releases substance P, when subjected to rapid increases in blood flow through this vascular bed. This rapid-blood-flow-induced substance P release also was shown to occur during high blood flow after the rats had been treated with capsaicin and when loss of substance P-containing nerves had been verified by immunohistochemistry. Air treatment to remove endothelial cells from arteries, but not arterioles or capillaries, in this rat hindlimb preparation, eliminated this high-blood-flow-induced substance P release. These significant observations indicated that substance P release arose from arterial endothelial cells rather than from perivascular nerves. The autoradiographic demonstration of receptors for substance P on the endothelium of dog carotid and renal arteries,[150,151] and the ultrastructural localization of substance P in the endothelial cells of rat coronary, femoral, and mesenteric arteries,[152,153] give added significance to the findings of Ravelic et al., as does the stimulated release of substance P from rat coronary artery endothelial cells.[153]

VIII. IMMUNE RESPONSE SYSTEM

In 1982 the author[22] published one of the earliest, if not *the* earliest, descriptions of a substance P-induced *in vivo* immunostimulatory response. Single acute infusions of a small amount of substance P (50 μg) into a cannulated afferent lymphatic of a popliteal lymph node of a sheep resulted in a prompt, marked, and prolonged increase in the output of small recirculating and blast lymphocytes into the efferent lymph of the efferent lymphatic, which was chronically cannulated, as was the afferent lymphatic. These studies were carried out in awake, supported, nonsedated, and nonmedicated adult sheep. This report (Ph.D. thesis) concluded with a suggestion that there was likely a commonality in mechanisms at the molecular level in immune response system and central nervous system "memories" (a concept and consideration persisting to and included in this book). Additional reports of these and related studies appeared in 1984 and 1985.[24,154]

In 1983 Payan et al.,[155] of San Francisco, described a substance P-induced stimulation of lymphocyte proliferation *in vitro* in the absence of other stimuli and followed this important observation with a number of publications identifying the presence and number of substance P receptors on a variety of lymphocyte types, including cell lines and blast forms.[156-158] Hartung et al.,[159] in 1986, described activation of macrophages by substance P to be associated with the induction of an oxidative burst and thromboxane release.

In 1985 Ruff, Wahl, and Pert,[160] from the National Institute of Mental Health (Bethesda, MD) reported substance P stimulation of human monocyte chemotaxis to be mediated by a substance P receptor. In another study in 1986, Wiederman, Stertl, and Pert[161] reported substance P binding to intact lymphatic tissue. This was an autoradiographic visualization study that employed slide-mounted tissue sections of rat spleen. Substance P was found to bind rapidly to an apparently single class of noninterreacting high-affinity site. The ligand selectivity pattern suggested to the authors that the substance P binding sites in this lymphoid system tissue (spleen) are similar to substance P receptors encountered in other tissues, including the brain, T lymphocytes, and macrophages. Substance P receptors were found to be highly concentrated in the antigen-trapping spleen marginal zone, with low densities being found in the red pulp. A 1987 report by this group[162] also identifies the presence of substance P binding to a substance P receptor in bronchus-associated lymphoid tissues in the rat.

Geppetti et al.,[163] from Rome and Florence, observed in 1987 that the thymus of the rat and hamster, unlike that of the mouse and guinea pig, contained measurable amounts of substance P-like immunoreactivity. In an ontogenetic study they found that levels of substance P-like immunoreactivity increased from the time of birth up to 60 d of age and decreased thereafter. Capsaicin treatment depleted thymic substance P-like immunoreactivity content in both newborn and adult rats. Rats treated with capsaicin as newborns, but not as adults, were found to have lower thymus weights as compared with controls. Rats treated with capsaicin as adults experienced partial time-dependent recovery of thymic substance P-like immunoreactivity content. The authors considered their findings to demonstrate the existence of capsaicin-sensitive structures containing substance P in the rat thymus — a major immune response system organ. Possible functions of these structures, including trophic actions, were considered and in relation to activities influenced by, and involved in, both afferent and efferent activities of sensory (as opposed to autonomic) neurons.

Wagner et al.,[164] of Munich, also in 1987, described the substance P-induced increase in interferon-gamma production by cultured human peripheral blood mononuclear cells stimulated by pokeweed mitogen (PWM). The increase was up to 18% and was statistically significant. The substance P-induced increase of rat interferon-gamma production of PHA-stimulated cells was not statistically significant.

Eglezos, Helme, and associates,[165] at Monash University in Melbourne, Australia, in 1987, reported an especially important *in vivo* finding relating to a substance P modulation of the primary antibody response to *in vivo* immunization in rats. In their study they used 6- to 12-week-old rats that had been treated neonatally with capsaicin. They examined the primary antibody response of popliteal lymph-node lymphocytes to antigenic stimulation *in vivo* with sheep red blood cells (SRBC). They found that the number of lymphocytes secreting antigen-specific immunoglobulins was decreased in the capsaicin-treated rats by more than 85%. This reduced response was reversed by subcutaneous infusion of substance P at the immunization site immediately following antigenic stimulation with SRBC. The authors concluded that the capsaicin-immunosuppressive effect was mediated by an associated depletion of substance P in the regional, popliteal, lymph-node reaction area, as this was reversed by the local-immunization-site administration of substance P to the presumably substance P-depleted regional lymph node. In another 1987 publication, this same Australian group of Helme et al.[166] related that neonatal capsaicin treatment of rats reduced SRBC immunization antibody production by more than 80% in both the direct and indirect blood-agar plaque assays. They again reported reversal of this reduction by substance P infusions over a 4-h period at the injection site immediately after the immunization. They demonstrated that this response had a substance-P threshold at approximately substance $P_{1\text{-}11}$ of 1.0×10^{-5} *M*. Substance $P_{1\text{-}7}$ was without effect, whereas substance $P_{5\text{-}11}$ at 1.0×10^{-5} *M* had the same reversal effect as substance $P_{1\text{-}11}$. These findings suggested to the authors that the reversal effect was located in the carboxy-terminal segment of the substance P molecule and that the observed reduced antibody formation response of neonatal capsaicin-treated animals to SRBC immunization antigenic stimulus was due to an effect of capsaicin on the substance P-containing primary afferent nerves, rather than due to a toxic effect of capsaicin on the immune response system.

Fink, Weihe, and associates, of the Johannes Gutenberg University (Department of Anatomy) in Mainz, Germany, in 1987,[167] made some important observations regarding substance P innervation of guinea pig and other mammalian lymph nodes. They showed that substance P-containing nerve fibers supplied afferent and efferent lymphatic vessels and virtually all of the hilar, capsular, cortical, and medullary blood vessels of the lymph nodes. In addition, numerous substance P-containing nerve fibers were seen in the lymph-node medullary cords where they appeared to form close contacts with reticulum cells and lymphocytes. They concluded that substance P innervation of lymph nodes was dual — sensory as well as autonomic.

A 1988 publication, by Popper et al.,[168] of UCLA and Harvard, was of particular interest and significance. These authors tracked the localization of substance P-containing sensory fibers and receptor binding sites for substance P (and other sensory neuropeptides) in canine mesenteric lymph nodes. Substance P-receptor binding sites were localized to arterioles and venules in T-cell regions and in 25 to 30% of germinal centers. The association of substance P and T cell regions in regional lymph nodes is of special importance with respect to the author's *in vivo* findings (to be related shortly).

In another 1988 report, Weihe, Fink, and Moller[169] also addressed that important issue concerning substance P innervation of primary peripheral lymph nodes (substance P in nerves supplying the immune system). They related that in all species studied, both the thymus gland and the lymph nodes received innervation by vascular and nonvascular fibers that contained substance P (and CGRP) immunoreactivity. Nonvascular substance P-immunoreactive nerve fibers of lymph nodes were identified in medullary cords, in the deep cortex, and in the internodular regions of the outer cortex where they formed close contacts with immune cells and in particular with T cell lymphocytes. Nonvascular substance P-immunoreactive nerve fibers in the thymic medulla outnumbered those in the thymic cortex and, as in the lymph nodes, also were in close contact with T lymphocytes. From the demonstrated presence of substance P in primary afferents, the authors concluded that tachykinin innervation of lymphoid tissues is mainly, if not exclusively, of sensory origin. On the sensory side, this anatomical arrangement would appear to provide a pipeline and circuit for immunological "memory".

An additional important 1988 publication by Fink and Weihe[170] provided further details of their important studies of the innervation of immune response system lymphoid tissues — a critical nervous system-immune response system "connection". They studied lymph nodes from both somatic and visceral regions from a variety of mammals, including guinea pig, rat, cat, pig, mouse, and human. Lymph nodes studied included cervical, mesenteric, axillary, and inguinal. They observed that substance P (and CGRP) immunoreactivity had the widest distribution and was present in the perivascular plexus (media-adventitia border), with paravascular and many nonvascular fibers traveling in close contact with lymphoid cells. Substance P immunoreactivity was found in numerous paravascular fibers. It was of particular interest that substance P-immunoreactive nonvascular sensory fibers in the lymph-node substance supplied both afferent and efferent lymph vessels. In addition, numerous substance P-immunoreactive vascular and nonvascular sensory nerve fibers were identified in lymph-node capsular regions — "early warning" centers for afferent lymph coming to the node, as will be discussed (with illustrations/figures) later in this section.

Fink and Weihe also observed that neuropeptide Y/dopamine β-hydroxylase (DBH and presumably noradrenergic) immunoreactivity prevailed in the perivascular plexus (media-adventia border) and only rarely branched off into lymphoid parenchyma. Alternate staining of adjacent sections demonstrated that substance P/CGRP-immunoreactive fibers were different from NPY/DBH-immunoreactive fibers. The authors concluded that NPY/DBH innervation of lymph nodes was sympathetic noradrenergic, while nerves immunoreactive for coexisting substance P and CGRP were of sensory origin and that these distinct sensory and autonomic peptidergic neuronal pathways linking the nervous system with lymph nodes likely play different roles in bidirectional neuroimmunomodulation. These beautiful and scholarly studies go to the anatomical heart of neurovascular immunology and suggest a multiplicity of neurovascular immune response functional potentials and investigational directions.

In 1989 Schumann and Gardner[171] of Stanford described substance P-induced modulation of the voltage-gated K^+-conductance in T lymphocytes. They employed whole-cell recordings from Jurkat E6-1 human T lymphocytes. Their interesting and important finding was that substance P reduces K^+ conductance in their model by decreasing current peak amplitude and by accelerating the rate at which current inactivates. The authors pointed out that their observations of a substance P-induced fall in human T lymphocyte membrane voltage-gated K^+ conductance was "reminiscent" of a 1985 report in *Nature*, by Stanfield, Nakajima, and Yamaguchi,[172] showing that substance P raises neuronal-membrane excitability by reducing inward rectification, and a 1988 report, by Nakajima, Nakajima, and Inoue,[173] that substance P inhibits potassium channels in brain neurons.

Jonakait and Schotland[174] of Rutgers (New Jersey), in 1990, described a most interesting nervous-system/immune-response-system interaction. They reported that substance P increases dramatically in cultures of rat superior cervical ganglion (SCG) nerve cells in culture grown in the presence of conditioned medium from Con A-stimulated splenocytes. The response was mimicked by growing SCG explants in the presence of human recombinant interleukin-1, but not human recombinant interleukin-2. Nerve growth factor (NGF) was not involved in mediating this effect, as antibodies to NGF included in the culture failed to alter the lymphokine (IL-1)-induced increase in substance P. In addition, the effect appeared to be somewhat substance P specific, as the levels of norepinephrine, serotonin, acetylcholine, and their forming enzymes were not similarly affected. The authors found that spinal cord dorsal-root ganglia responded with only modest increases in substance P. They speculated that the actions of lymphokines in stimulating substance P formation and release might be a ganglion-specific action that might contribute to the promotion of recovery following injury.

An additional and quite major lymph-node-innervation report of significance appeared from Heidelberg, Germany, by Kurkowski et al.[175] These authors studied substance P-immunoreactive nerve fibers in tracheobronchial lymph nodes of the guinea pig, by both light and electron microscopy. These lymph nodes are of critical defense importance in providing lower-respiratory-tract protections at the immune response level to a variety of hazardous exposures: noninfectious as well as infectious. Substance P immunoreactivity was identified in perivascular nerves and in neuronal axons of the medullary lymphatic tissue. Ultrastructurally, all substance P-immunoreactive axons in the medullary lymphatic tissue were unmyelinated (? nociceptive type), and all displayed vesicle-containing varicosities. Retrograde neuronal tracing combined with immunohistochemistry demonstrated a sensory origin of these substance P-immunoreactive unmyelinated nerve fibers from dorsal-root ganglia. These tracing and immunohistochemistry studies also revealed an additional neuronal population from these tracheobronchial lymph nodes and lymphoid tissues that contained no substance P immunoreactivity. The authors interpreted these findings as providing evidence for several types of sensory nerve fibers innervating lymph nodes.

The authors of this particularly important publication made a number of key observations in both their light and electron microscopic studies. In addition to antibodies against substance P, antibodies against CGRP, CCK, and dynorphin also were obtained and used.

In the light microscopic investigations, substance P-immunoreactive nerve fibers were found to be entering the lymph-node hilus both via large blood vessels and as part of separate nerve bundles with no vascular connection. Substance P-immunoreactive nerve fibers were observed along hilar, medullary, and paracortical blood vessels and also were encountered in nonvascular areas scattered in the medullary and paracortical lymphoid tissue without proximity to blood vessels.

Of particular importance was the identification of substance P-immunoreactive nerve fibers in the lymph-node capsule as well as traversing the especially important subcapsular lymph sinus collecting area for afferent lymph from the drainage area of the lymph node — an important early warning site for cells and substances entering the lymph node from its peripheral tissue drainage ''water-shed'' (lymph-shed).

The authors in their light microscopic studies investigated several hundreds of substance P-immunoreactive neuronal axons, and all, without exception, had CGRP colocalization. Some substance P-immunoreactive nerve fibers also contained dynorphin immunoreactivity. These dynorphin-also nerve fibers were found exclusively in the lymphatic tissue of the hilus and medulla. Even in these regions the combined substance P- and dynorphin-immunoreactive fibers comprised less than 50% of the substance P-immunoreactive neuronal fiber population. In the same regions, substance P-immunoreactive neuron fibers also were shown to colocalize with CCK and in approximately equal numbers to the dynorphin colocalization numbers. Furthermore, CCK-immunolabeled varicose fibers, which lacked substance P immunoreactivity, were identified in a vascular area surrounding small hilar arteries.

Additional information of considerable importance also was obtained in the ultrastructural electron microscopy studies by Kurkowski et al. Unmyelinated axon bundles, apart from nonreactive fibers, were found to contain substance P-immunoreactive axons. These axons, at times, were marginally opposed to the Schwann cytoplasm. Numerous small clear and a few large dense-core substance P-immunoreactive vesicles were found to be visible in varicose fiber profiles. The substance P-immunoreactive fiber bundles often were seen to abut to the walls of small arterial or venous vessels. Substance P-immunoreactive neuronal fibers abutting the walls of small arterial or venous vessels in the substance of the lymph nodes, at times, were seen to be separated from vascular myocytes only by a basal lamina (Figure 15). In this figure substance P-immunoreactive neuronal fibers are stained black (immunogold). Substance P-immunoreactive neuronal fibers in the lymphatic tissue were observed to have the same ultrastructural appearance as the substance P-immunoreactive vascular axons. These substance P-immunoreactive axons in the lymphatic tissue were seen in close vicinity to reticular cells (Figure 16) and to lymphocytes and plasma cells (Figure 17).

Injection of fluorogold into the right major paratracheal lymph node labeled both small (<20 μm) and large (approx. 40 μm) neuronal cell bodies in cervical dorsal-root ganglia

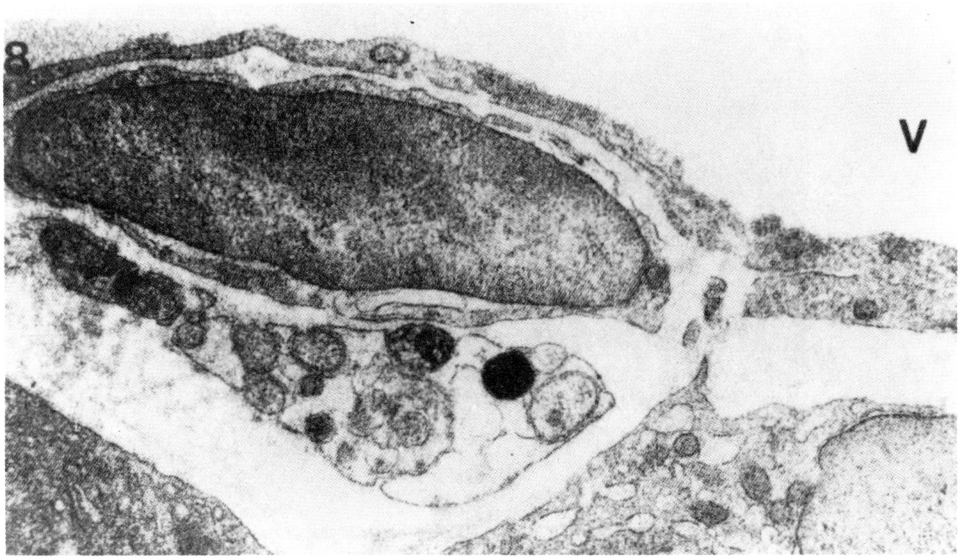

FIGURE 15. Electron photomicrograph identifying an unmyelinated nerve fiber at the wall of a small blood vessel (v) of a guinea pig tracheobronchial lymph node. This unmyelinated nerve fiber contains both substance P-immunoreactive axons (black) and nonreactive axons. This photomicrograph provides an excellent demonstration of a close and important Neurovascular "connection" in this major and central immune response organ. (From Kurkowski, R., et al., *Peptides*, 11, 13, 1990).

of segmental levels 3 to 8. Fluorogold labeling was found to occur in a decreasing number of neuronal cell bodies when checking the ganglia in a caudal direction. Approximately two times as many labeled neuronal cell bodies were seen and identified in the C3 ganglion as in the C8 ganglion. Double immunofluorescence staining demonstrated substance P and CGRP immunoreactivity in some of the fluorogold-labeled small neuronal cell bodies.

In the above studies it was of interest that the substance P (SP)- and CGRP (SP/CGRP)-immunoreactive neuronal fibers differed from the SP/(GRP/DYN)- and the SP/CGRP/CCK-immunoreactive neuronal axons that were restricted to the lymphatic tissue and did not accompany blood vessels. The authors considered this finding as supporting previous suggestions,[176] from silver impregnation and electron microscopy, that the nerve fibers that supply the vascular tree may differ from the nerve fibers that supply the lymphocytic elements. However, the authors, using electron microscopy, were unable to demonstrate any ultrastructural differences between substance P-immunoreactive vascular and nonvascular axons. They speculated that the earlier observed differences, or possible differences, might have been due to electron microscopically visualized nonsubstance P-immunoreactive neuronal fibers. In addition, the authors pointed out observations by others in which sympathetic nervous system axons that display catecholamine fluorescence[170] or immunoreactivity to dopamine-β-hydroxylase[170] are restricted to the vascular tree and do not traverse the lymphatic tissue independent of blood vessels.

The authors also emphasized that their retrograde tracing experiments clearly established the presence of sensory innervation of lymph nodes by SP/CGRP-immunoreactive neuronal fibers. This had been hypothesized earlier by Fink and Weihe.[167,169,170] The authors further emphasized that since a great proportion of SP/CGRP-immunoreactive axons within the lymph node were obviously related to blood vessels and formed varicose, vesicle-containing swellings at their blood vessel locations, at least a dual vascular innervation of lymph nodes, sympathetic and sensory, had to be assumed.

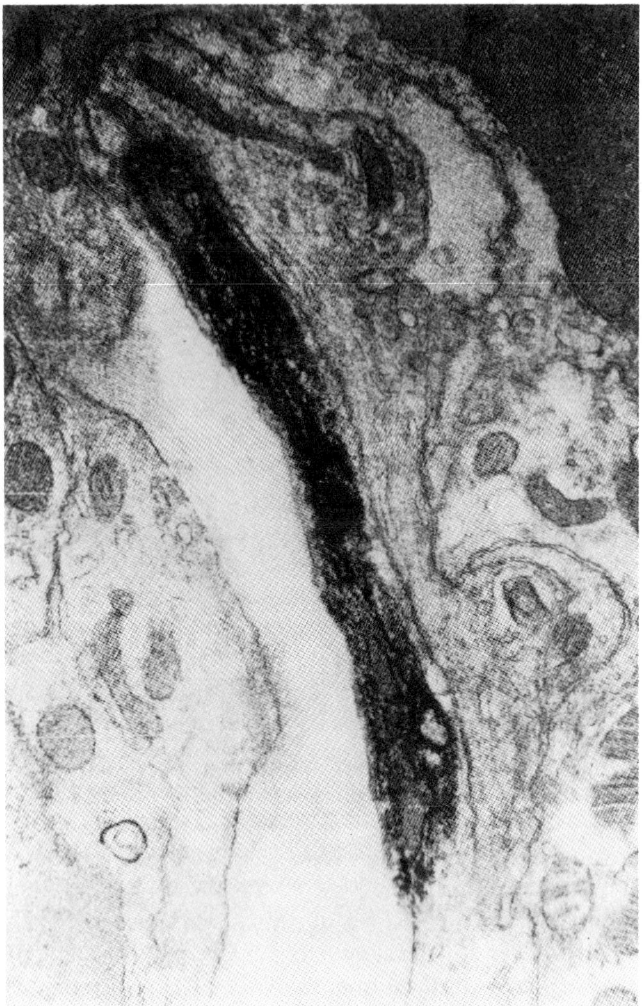

FIGURE 16. Electron photomicrograph showing a longitudinal profile (black) of a substance P-immunoreactive axon in a guinea pig tracheobronchial lymph node lymphatic reticulum. This immunoreactive axon is partially devoid of Schwann cell covering. (From Kurkowski, R., et al., *Peptides*, 11, 13, 1990. With permission.)

The authors summed up their especially important and significant findings as being strongly indicative of at least three different populations of sensory neurons innervating tracheobronchial lymph nodes of the guinea pig, based on their neurochemistry and distribution pattern: (1) SP/CGRP-immunoreactive neurons projecting both to blood vessels and to lymphatic tissue, (2) SP/CGRP/DYN- and/or SP/CGRP/CCK-immunoreactive neurons that project exclusively to lymphatic tissue, and (3) non-substance P-immunoreactive neurons projecting to lymphatic tissue (and possibly to blood vessels as well). The origin of the CCK-only immunoreactivity seen at hilar vessels could not be determined.

The excellent anatomical studies of Fink, Weihe, and associates[167,169,170] and of Kurkowski, Kummer, and Heym,[175] from the ancient and neighboring universities of Heidelberg and Mainz, have filled in an important and critical gap in our understanding of nervous system control and influence over the immune response system at the core and heart of the immune response system, the regional lymph node. Knowledge of its substance P, pepti-

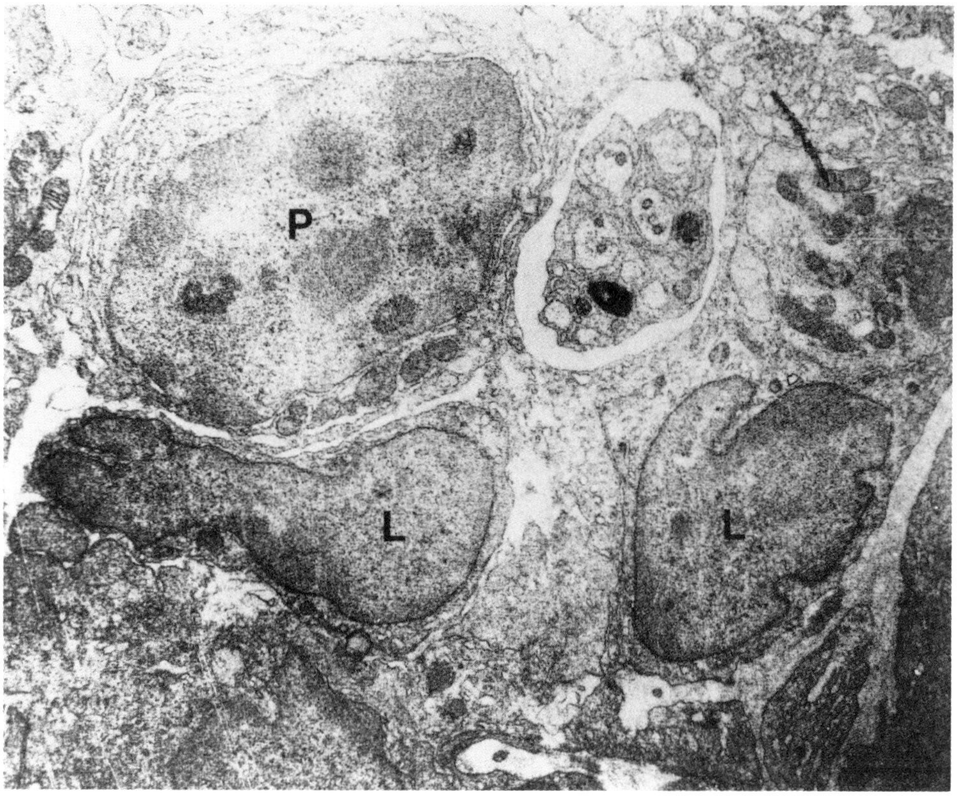

FIGURE 17. Electron photomicrograph demonstrating an unmyelinated nerve fiber in a guinea pig tracheobronchial lymph node with several nonreactive and two substance P-immunoreactive axons (black). This unmyelinated and substance P-immunoreactive axon-containing nerve fiber is in close proximity to a plasma cell (P) and to two lymphocytes (L). (From Kurkowski, R., et al., *Peptides*, 11, 13, 1990. With permission.)

dergic, and sensory innervation at both the light and electron microscopic ultrastructural levels is of the utmost importance. With their work a key missing link has been supplied and a major advance has been achieved with respect to the neurovascular component of the neuroimmune response axis.

The brilliant Heidelberg and Mainz findings also suggest the value of a return to the classical lymph-node anatomical and ultrastructural studies of Anderson and Anderson,[177-179] which were published in the mid- and late 1970s. In 1975 Anderson and Anderson[177] made critically valuable additions to our knowledge of mammalian lymph-node vascular anatomy and of the neurovascular connection. They observed that a distinct vasculature supplied each lobule of lymphatic tissue and that these units contained numerous arteriovenous communications and many venous sphincters immediately beyond the specialized high-endothelial-cell areas of the postcapillary venules (Figure 18). The postcapillary venule is the anatomical site shown by the work of Gowans[180,181] to be the entry point of small recirculating lymphocytes into the lymph-node substance from the blood vascular system. In the 1975 studies by Anderson and Anderson,[177] both the arteriovenous communications and the venous sphincters were innervated by unmyelinated nerve fibers. In the arteriovenous communications, focal discontinuities were seen at sites where unmyelinated nerve fibers passed through the adventitia of the vessels and approximated the outer borders of smooth-muscle cells. Varicosities frequently were observed in these exposed nerve segments. These arteriovenous

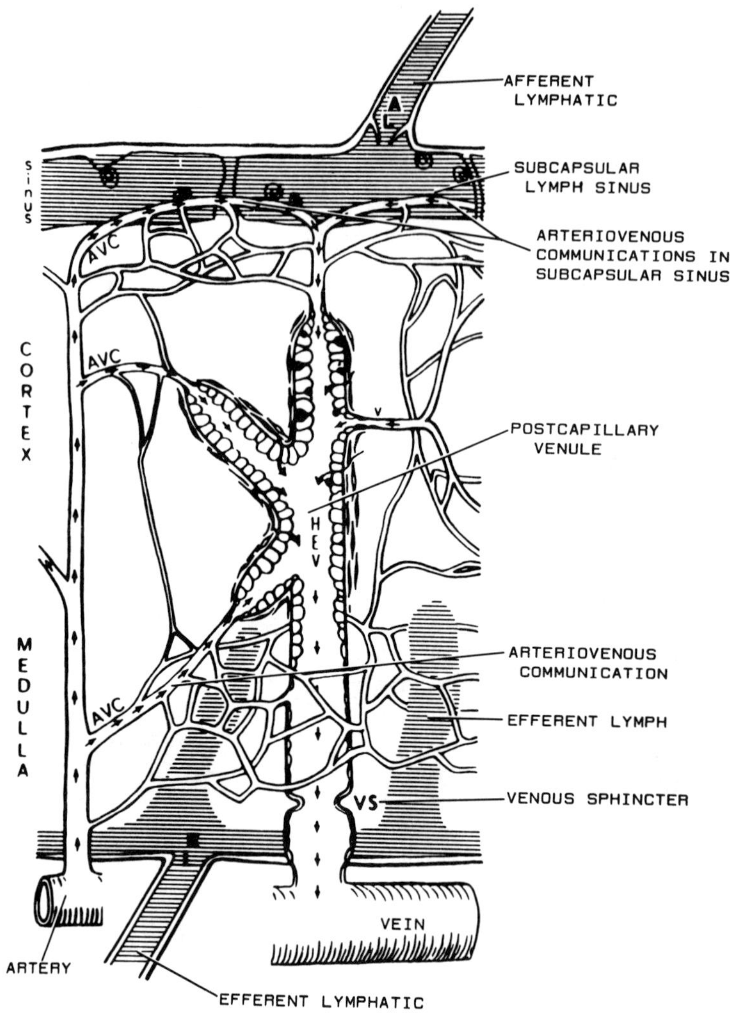

FIGURE 18. Schematic drawing showing rat lymph node neurovascular ''connections'' (after Anderson and Anderson[177]). Numerous unmyelinated nerve-fiber varicosity approximations to critical lymph node arteriovenous communications and venous sphincters controlling postcapillary venule high endothelial areas were shown (by Anderson and Anderson) in these critical small recirculation areas. They also were shown to be present in the arteriovenous communications that projected into the subcapsular lymph sinus that received afferent lymph directly from peripheral tissues and hence constituted an important first and early warning system with respect to peripheral tissue responses to defense-triggering assaults.

communications were found throughout the diffuse cortical and corticomedullary areas of the lymph nodes. They also formed prominent arcades immediately beneath the subcapsular lymph sinus into which lymph from the afferent lymphatics was collected. At times these arteriovenous communications and their unmyelinated neuronal fiber connections appeared to extend into these important afferent lymph-collecting subcapsular lymph sinuses. This arrangement constituted an important anatomical relationship for prompt and early detection of drainage-area biochemical alterations with the capability of effective and synchronously coordinated responses within the lymph node and other portions of the lymphoid immune resposne system, by means of its neurovascular connections.

In 1976 Anderson and Anderson[183] reported additional studies carried out in this lymph-node anatomy area by scanning and transmission electron microscopy combined with perfusion techniques. Their findings have provided especially useful information regarding lymphocyte traffic from the blood vascular system in postcapillary venule lymph-node areas and into the lymphoid substance of the lymph nodes. In the rats studied, the postcapillary venules in the recirculating small lymphocyte passage area are lined with high endothelial cells and are called high endothelial venules (HEV). In their studies Anderson and Anderson found that blood lymphocytes selectively adhere to HEV surfaces by means of microvilli that adhere to shallow pits on the luminal surface of the high endothelial cells and that the intracellular contact points resist hydrodynamic and osmotic sheering forces. Following this attachment, the lymphocytes enter apical clefts between endothelial cells where they assume a motile configuration that is associated with the loss of microvilli, the formation of irregular surface folds, and the development of a cytoplasmic polarity (which is oriented in the direction of movement). They observed that 92% of the lymphocytes that exhibited motile configurations were oriented away from the venular lumen. Adhering to adjacent endothelial cells by macular and villous contacts, the lymphocytes pass through extracellular spaces in the HEV and then cross successive laminations in the perivascular sheath to enter the interstitial spaces of the lymph node (Figure 19). Transmission electron microscopic studies by the authors showed that the HEV venular wall and its surrounding sheath (composed of three to five layers of overlapping reticular cell plates and connective tissue) were particularly well designed to permit unidirectional flow in one direction (node into HEV lumen) and unidirectional recirculating lymphocyte flow or passage in the opposite direction (HEV lumen into the node) up a chemotactic gradient and with minimal vascular leakage about the migrating lymphocyte.[179]

The venous sphincters beyond the high-endothelial-cell areas of the postcapillary venules also were found, by Anderson and Anderson,[177] to have unmyelinated nerves that were accompanied by Schwann processes and that passed through the vascular adventitia and approximated the smooth bundles of these venous sphincters.

These anatomical arrangements would appear to offer multiple opportunities for neurovascular sensory detections and appropriate motor and axon reflex responses with respect to the traffic of cells and substances into and through lymphoid reacting centers. This traffic has been shown (by the work of Gowans,[182] Ford,[183] and others) to be of central and major importance in the initiation and development of the immune response. The recent contributions of the Heidelberg and Mainz groups with respect to the substance P, peptidergic, and sensory innervation of regional lymph nodes certainly puts the icing on this important cake.

In 1989 the author and associates[128] reported additional substance P and lymphocyte traffic studies. In multiple studies involving acute infusions of substance P (50 μg) into cannulated popliteal lymph-node afferent lymphatics in the author's supported-sheep lymphocyte traffic study model, marked and unusually prolonged increases in lymphocyte traffic were produced, as monitored by the output of lymphocytes into nodal efferent lymph in nodal efferent lymphatics, chronically cannulated. This increase in lymphocyte output included both small recirculating and blast lymphocytes (Figure 20). It was of particular interest that a single acute infusion could produce such a prolonged stimulation and associated increase in lymphocyte traffic — indeed a long-term potentiation (LTP). This effect also was demonstrated in an acute substance P infusion study in which the profound and prolonged (48 h) anesthesia-associated depression of lymphocyte traffic was overcome and for an extended period (Figure 21).

A further extension of these studies by the author and associates was published in 1990.[29] The greatly prolonged substance P-induced increase in lymphocyte traffic and lymphocyte output was found to be targeted to the T4 (CD4) subset of T lymphocytes (Figures 22 and 23). A brief, first-15-min lymph collection postinfusion depression in lymphocyte output

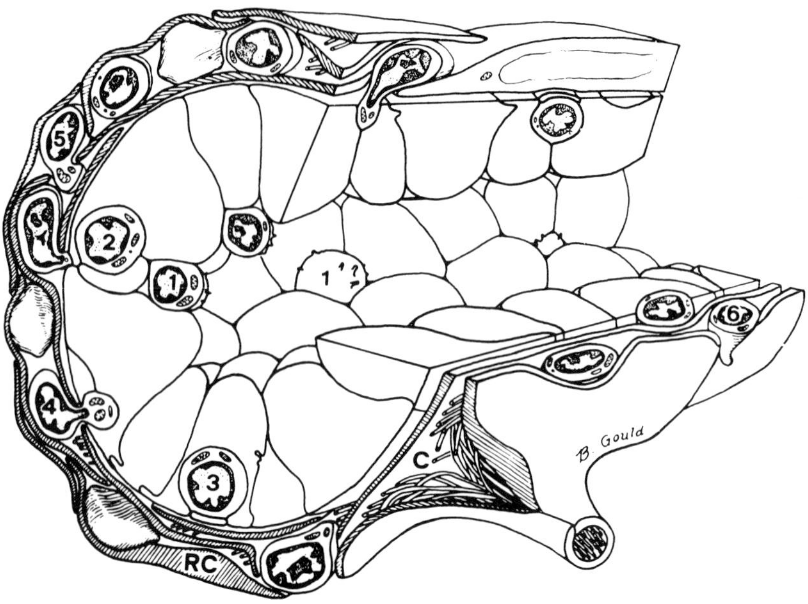

FIGURE 19. Steps and stages by which a lymphocyte passes from the blood in the high endothelial postcapillary venule of the rat to pass through the venule wall between endothelial cells to enter the substance of the lymph node. (From Anderson, A. O. and Anderson, N. D., *Immunology*, 31, 731, 1976. With permission.)

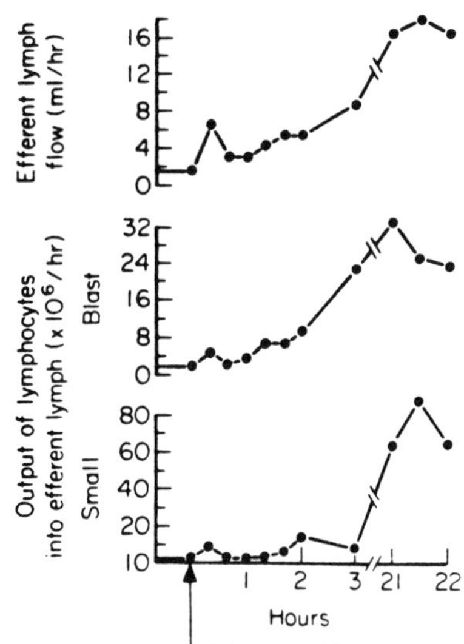

FIGURE 20. Prolonged elevations in the output of both small recirculating and blast lymphocytes into the efferent lymph of a sheep popliteal lymph node following a single acute infusion of substance P (50 μg) into a cannulated afferent lymphatic of this lymph node. (From Moore, T. C., et al., *Immunology*, 67, 109, 1989. With permission.)

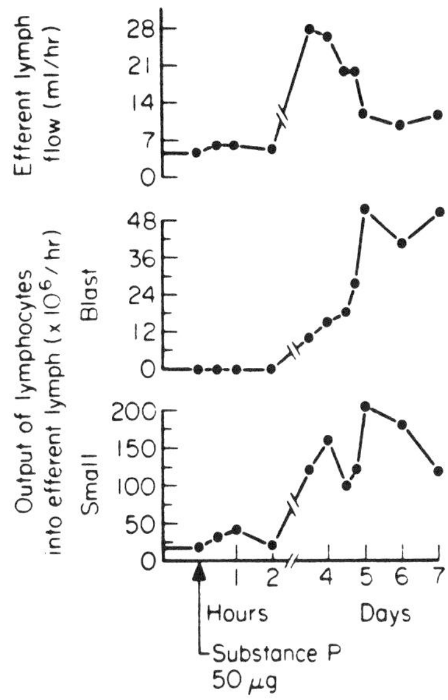

FIGURE 21. Shows the remarkable capability of a single acute infusion of substance P (50 μg) into a cannulated sheep popliteal lymph afferent lymphatic to overcome a profound and prolonged (48 h) anesthesia-induced depression of lymphocyte traffic and an associated profound depression in the output of lymphocytes into nodal efferent lymph. (From Moore, T. C., et al., *Immunology*, 67, 109, 1989. With permission.)

was found to be associated with a marked and equally brief increase in the efferent lymph levels of thromboxane B_2 (a cyclooxygenase-pathway arachidonic acid metabolite). A remarkable and accelerated antibody response (Figure 24) when lymph-node drainage-area immunization was carried out 6 d following a single acute substance P (50 μg) infusion into the cannulated nodal afferent lymphatic suggested a functional immunological memory response and led the author to suggest that substance P might be a mediator of immunological memory and that immunological memory and nervous system memory might have a common molecular basis involving substance P, as well as other "memory molecules" considered in this volume.

IX. OTHER MOLECULES

A. DOPAMINE

There are multiple and extensive anatomical relationships and interrelationships between the substance P nervous system and the equally massive, widespread, and extremely important dopaminergic nervous system.[89,184] The precise functional roles of most of these connections remain to be established. Some individual studies may give a glimpse of some of the interrelated actions of these two major molecules. Bannon et al.[185] of Yale, in 1986, reported that the dopamine antagonist haloperidol decreases substance P and preprotachykinin mRNA levels in rat striatonigral neurons. A 1984 report from France by Oblin et al.[186] described studies in which repeated treatments with drugs that impaired dopaminergic transmission also produced a consistent reduction in substance P immunoreactivity in the rat substantia nigra. In 1990 Tamiya et al.,[187] of Japan, employed an immunoelectron microscopic mirror technique to show that substance P afferents have synaptic contacts with dopaminergic in the important VTA (ventral tegmental area) of the dopaminergic rostral projection pathways.

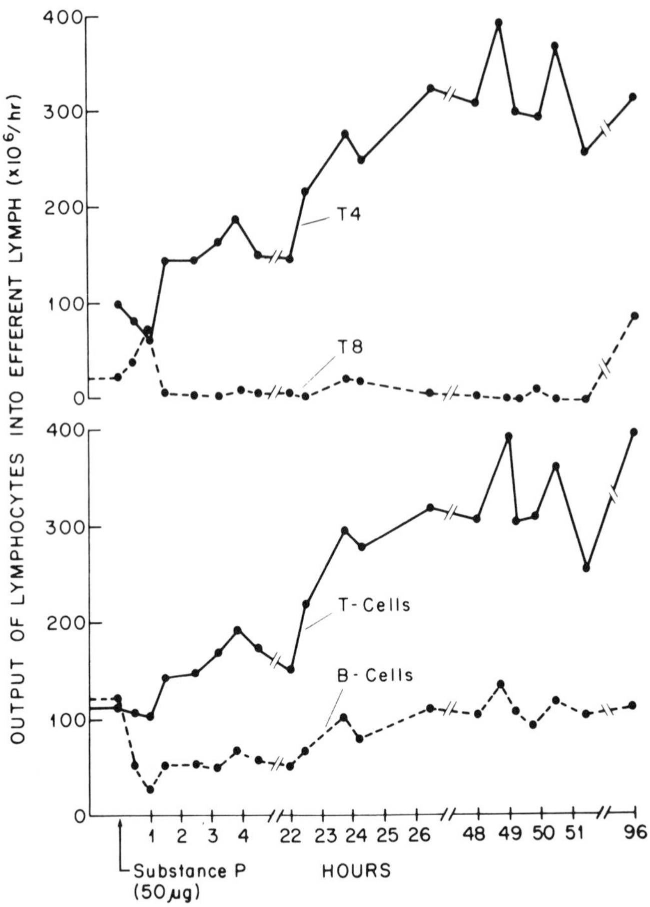

FIGURE 22. The prompt, marked, prolonged, and T4 (CD4)-lymphocyte-targeted increase in small-circulating-lymphocyte outputs into nodal efferent lymph of a sheep popliteal lymph node following a single acute infusion of substance P (50 μg) into a cannulated afferent lymphatic of this lymph node. (From Moore, T. C., et al., *Immunopharmacology*, 20, 207, 1990. With permission.)

Also in 1990, Bolam and Smith[188] of Oxford, in a double and triple immunocytochemical study, showed that many substance P- and GABA-forming enzyme boutons were associated with dopaminergic cells in the substantia nigra. In an electron microscopic component of this study it was demonstrated that the perikarya and dendrites of the dopaminergic neurons received symmetrical synaptic input from terminals that displayed immunoreactivity for substance P or glutamate decarboxylase (the GABA-forming enzyme). A small portion of the substance P-positive boutons were seen to form asymmetrical synapses. Some of the substance P-immunoreactive boutons that were found to be in contact with dopaminergic neurons in the substantia nigra also displayed GABA immunoreactivity.

Reid, Hokfelt, Folkers, and associates[189] at the Karolinska Institute in Stockholm in 1990, described a substance P-induced increase in ipsilateral striatal dopamine levels by ipsilateral injection of substance P into the substantia nigra, pars reticulata. This response was prevented by the coadministration of substance P and Spantide II, a new and nontoxic substance P antagonist.

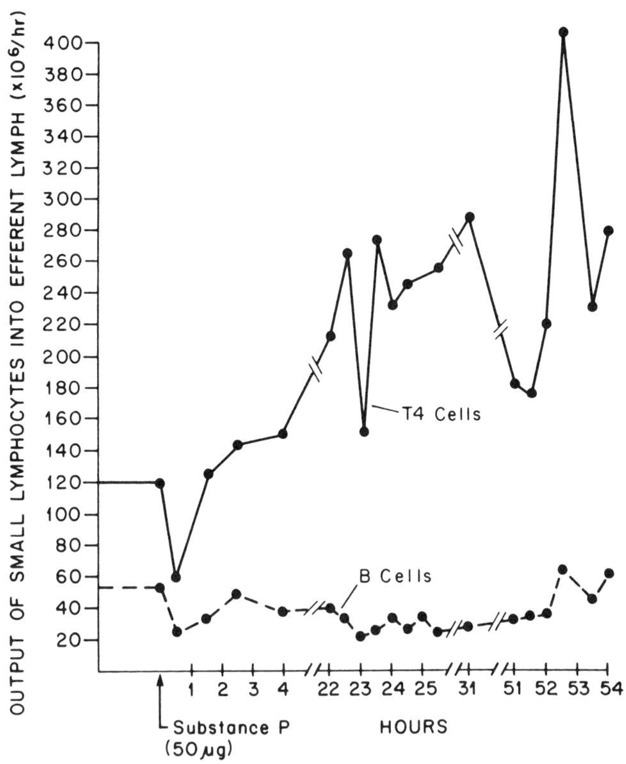

FIGURE 23. Marked and prolonged T4 (CD4) small-recirculating-lymphocyte-targeted increase in lymphocyte traffic into sheep-popliteal lymph-node efferent lymph following a single acute infusion of substance P (50 µg) into a cannulated afferent lymphatic of this lymph node. (From Moore, T. C., et al., *Immunopharmacology*, 20, 207, 1990. With permission.)

B. SEROTONIN

The anatomical mapping studies cited above for dopamine and substance P have also shown extensive connections with the serotoninergic nervous system, both at raphe nuclei cell body locations and in projections throughout the central nervous system.

The colocalization of 5-HT and substance P in descending bulbospinal neurons with cell bodies in the medullary raphe nuclei was described in 1978 by Hokfelt et al.[190] and Chan-Palay et al.[191] In an immunoelectron microscopy study, Pelletier et al.[192] demonstrated 5-HT to coexist with substance P within large dense-cored vesicles in the spinal cord. In 1988 Fried et al.[193] reported findings that suggested a differential storage of substance P and 5-HT within synaptosomes isolated from the ventral spinal cord, with 5-HT stored in both small and large vesicles and substance P stored only in large vesicles. These observations suggested and provided a basis for differential release of substance P and 5-HT from the same presynaptic neuronal terminal. A 1989 reported study by this same group[194] explored the role of substance P and other tachykinins on the potassium-evoked release of 5-HT from superfused tissue slices of rat ventral spinal cord where most 5-HT and substance P are stored in the same nerve endings. They found that substance P, but not NKA or NKB, potentiates the evoked release of 5-HT in a dose-dependent manner. In an additional differential storage study, this group also included CGRP and studied both dorsal and ventral horns of the spinal cord. They found that in the dorsal spinal cord, substance P and CGRP had almost identical distributions (large dense-cored vesicles) on differential and gradient centrifugation in linear sucrose gradients following lysis of a synaptosomal fraction. In the

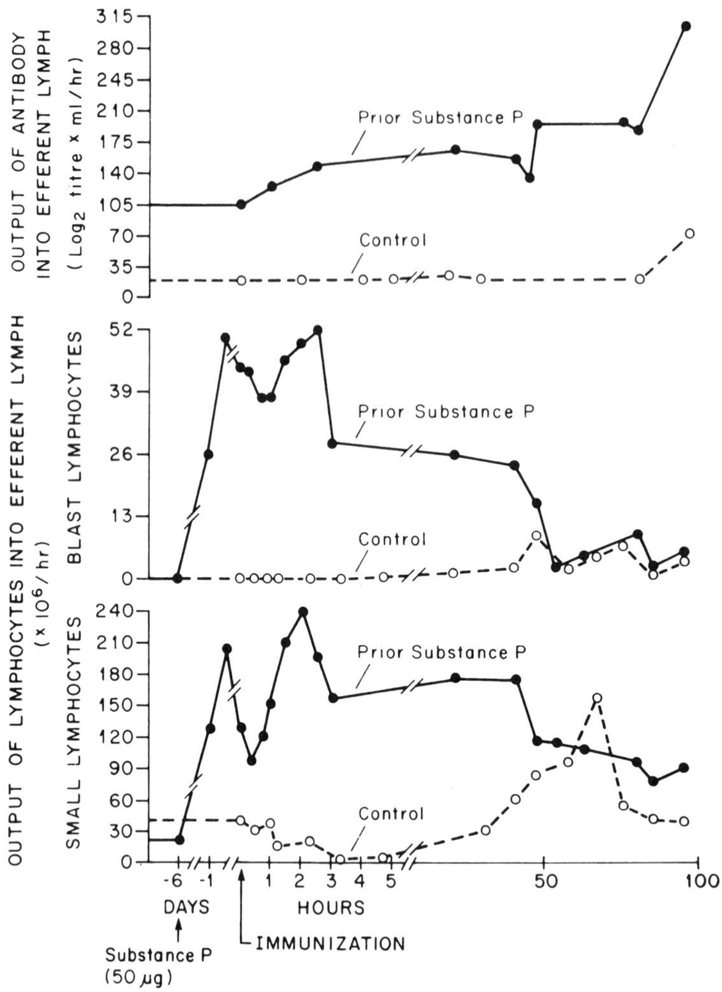

FIGURE 24. The marked substance P-induced alterations in lymphocyte traffic and associated antibody response to immunization in popliteal lymph-node drainage area (killed *Salmonella muenchen* bacteria) carried out 6 d following a single acute infusion of substance P into a cannulated popliteal lymph-node afferent lymphatic. The customary "shutdown"/recruitment period of diminished output of small recirculating lymphocytes into nodal efferent lymph was eliminated and there was a prompt and accelerated increase in the output into nodal efferent lymph of antibodies to the immunizing *S. muenchen* antigen. (From Moore, T. C., et al., *Immunopharmacology*, 20, 207, 1990.)

ventral spinal cord, substance P and 5-HT had different distribution patterns, with 5-HT mainly in small vesicles, substance P in large vesicles, and both inside synaptosomes.

In 1983 Agnati et al.[195] described studies on the functional role of the coexistence of 5-HT and substance P. They showed that substance P reduces the affinity and increases the density of [^{3}H]-5-HT binding sites.

In a study of substance P and 5-HT interactions in a peripheral setting in 1990, Khalil and Helme,[196] from Australia, used the blister model of inflammation to explore substance P and 5-HT interactions. No effect was observed in simultaneous perfusion of both mediators. However, sequential perfusion changes were observed in plasma extravasation and vasodilatation, with 5-HT extending these responses to substance P. This effect did not occur in capsaicin-pretreated rats. The authors interpreted these findings as providing evidence, for the first time *in vivo*, that serotonin can modulate an inflammatory response to substance P by way of a mechanism that involves capsaicin-sensitive sensory fibers.

Also in 1990, Meller et al.,[197] of the University of Iowa, reported that neonatal capsaicin treatment abolished the nociceptic responses to intravenous 5-HT in the rat, a vagal-afferent-mediated effect. The authors concluded that these results demonstrated that the 5-HT-induced, vagally mediated inhibition of the nociceptive tail-flick reflex and accompanying pseudo-affective responses are mediated by capsaicin-sensitive afferents. This finding strongly suggested to the authors that the afferent fibers subserving the nociceptive response are predominantly c fiber vagal afferents and that these afferents are different from those vagal afferents that mediate the cardiovascular responses to intravenous 5-HT.

C. GLUTAMATE

De Biasi and Rustioni[198] reported in 1988, from the University of North Carolina, that approximately 70% of the neurons in the rat dorsal-root ganglia (DRG), by microscopic immunocytochemistry, are immunoreative to an antiserum for glutamate conjugated to hemocyanin. In addition, most of the substance P-immunoreactive DRG are also immunopositive for the glutamate conjugate. The smaller of the glutamate-immunoreactive neurons are the ones that are most likely immunoreactive also for substance P. By employing a postembedding immunogold method and electron microscopy, the authors were able to show that glutamate and substance P exist in primary afferent terminals in the superficial area of the dorsal horn of the spinal cord in the rat. Substance P was shown to be present only in the dark type of glutamate-labeled scalloped terminals. This type of terminal had many agranular vesicles of different sizes and a few large granular vesicles. These were considered to be the endings most likely of small unmyelinated primary afferent fibers. From these findings the authors proposed that glutamate is the neurotransmitter in primary afferents mediating input from different peripheral receptor classes, including nociceptors, and that effects of glutamate and substance P on spinal dorsal-horn neurons may result from the corelease of these two mediators from the same dorsal-root afferent terminal.

An especially important report concerning actions and activities of sensory neurotransmitters in the spinal cord dorsal horn also appeared in 1988 in *Nature*. This highly significant study (from Harvard and Columbia by Womack, MacDermott, and Jessell[199]) was directed toward the activities of primary sensory afferent terminals in the dorsal horn of the mammalian spinal cord, which release excitatory amino acids, such as glutamate and aspartate, and peptide transmitters, such as substance P. In order to determine the mechanisms by which substance P contributes to afferent sensory synaptic neurotransmission, the authors monitored the levels of intracellular calcium in single isolated rat dorsal-horn neurons. They found that substance P produced a prolonged elevation in calcium concentration by mobilizing its release from intracellular stores. The authors concluded that this elevation likely contributes to the long-term changes (? LTP) in the excitable properties of dorsal-horn neurons, which has been shown[200] to occur following afferent sensory fiber stimulation. One of the authors had reported earlier (also in *Nature*) that L-glutamate elevates intracellular calcium in substance P-sensitive dorsal-horn neurons by increasing calcium influx.[201] The authors interpreted these important and exciting findings as providing a direct demonstration of intracellular calcium changes in response to neuropeptides, such as substance P, in mammalian central neurons and as indicating that there is a convergent regulation of intracellular calcium in dorsal-horn neurons by two different classes of sensory neurotransmitters (substance P and glutamate) that are coreleased from the same afferent sensory terminals.

An important observation relating to substance P and glutamate interaction was reported from Iowa State University by Kangrga and Randic[202] in 1990. They demonstrated with a rat spinal cord slice/dorsal-root ganglion preparation that substance P can release endogenous glutamate and aspartate. They showed that high-density repetitive stimulation of a dorsal root produced a Ca^{++}-dependent increase in the basal release of both glutamate and aspartate. Low concentrations of substance P caused a selective increase in the basal release of glutamate, while a higher concentration increased the basal release of both glutamate and asparate. The authors concluded that the enhancement of the basal and the dorsal-root stimulation release of glutamate and aspartate by substance P may have important physiological implications in the area of strengthening synaptic connections in the spinal dorsal horn. In 1990 Randic et al.[203] reported that substance P, when applied simultaneously with or prior to glutamate, caused a potentiation in the glutamate current in 65% of the freshly dissociated rat dorsal-horn neuronal cells tested. Substance P enhancement of glutamate-receptor-activated conductance is an important observation. Also in 1990, this group[204,205] reported that the application of substance P (2 to 5 $\times$ 10^{-7} *M*) to the rat spinal dorsal horn *in vitro* produced a significant increase in the concentration of glutamate in the spinal-slice perfusate (a higher concentration was required to induce aspartate release). Neonatal capsaicin treatment blocked substance P-induced increase in glutamate release in this preparation, but not that of CGRP-induced increase in glutamate release.

Because excitatory amino acid (EAA) neurotransmission has been implicated in long-term synaptic events, Somers and Beckstead,[206] of the Medical University of South Carolina, decided to conduct a study to ascertain whether or not the EAA-containing corticostriatal projection might influence substance P biosynthesis in neurons of the rat's basal ganglia — site of the major and highest substance P concentration, as determined by von Euler and Gaddum[1] in 1931. Their report, published in 1990, described their findings which included a reduction in caudatoputamen content of EAA by frontal cortical ablation or by chronic intracerebroventricular infusion of methionine sulfoximine (MS). After 7 d, synaptosomal levels of aspartate and glutamate in the caudate putamen (striatum) were reduced by 15 and 24% by ablation and by 61 and 48% by intraventricular MS infusion. With both treatments, substance P in the substantia nigra was increased by 38%. These treatments decreased preprotachykinin (PPT) mRNA in the rostral striatum (caudate putamen) by only 17%. Thus, caudate-putamen decreases in EAA and substance P following these "treatments" were comparable. These findings indicated that levels of substance P in the substantia nigra and levels of PPT mRNA in the caudate-putamen striatum depend on corticostriatal axons that use the excitatory amino acid transmitters aspartate and glutamate. The authors interpreted their findings as suggesting that a large reduction in EAA neurotransmission can influence steady-state levels of substance P in striatal neurons.

D. PROSTAGLANDINS

In 1983 Hartung and Toyka,[207] of Dusseldorf, Germany, described substance P-induced activation of macrophages accompanied by an oxidative burst and thromboxane release. In 1986 these authors with Wolters[159] reported substance P bindings to guinea pig macrophages and related responses, as cited earlier in the "Immune Response" section of this chapter. In 1988 this group[208] demonstrated that cultured astrocytes from the cortices of neonatal Lewis rats could be stimulated by substance P to produce both prostaglandin E_2 and thromboxane B_2.

In 1991 Marriott et al.,[209] from London, described some interesting studies relating to substance P-induced release of prostaglandins from astrocytes, with special attention being given to regional specialization of this release and to correlation with phosphoinositol (PI) metabolism. This group had noted earlier (Cholewinski et al.[210]) that astrocytes from the spinal cord were unique in their response to substance P. The authors found that the addition of substance P to astrocytes cultured from the neonatal rat spinal cord stimulated a time- and concentration-dependent increase in the accumulation of PI and in the release of prostaglandin (PG) D_2 and PGE_2. Both the basal and stimulated levels of PI, PGD_2, and PGE_2 were reduced to similar levels by indomethacin. In interesting contrast, astrocytes cultured from the cerebral cortex and the cerebellum were observed to have no substance P-stimulated increase in PI accumulation or in prostaglandins release. Release of the two prostaglandins was stimulated by the calcium ionophore A23187, and PI accumulation and PG release were evoked from cortical astrocytes that were incubated in the presence of serum. The authors concluded that substance P-stimulated PI accumulation and PG release from cultured rat neonatal astrocytes are regionally specialized in favor of cells derived from the spinal cord.

E. HISTAMINE

A component of the substance P involvement in neurogenic inflammation and the axon reflex has been suggested to be due to, and related to, a substance P-induced release of histamine, and the mast cell has been incriminated as a likely source for this release.

Injection of substance P into human skin or the rat hind limb has been found to release histamine into the circulation.[211,212] Histamine-containing mast cells are located in large numbers around blood vessels and have been shown to form neuroeffector junctions.[213-217]

Antidromic stimulation of sensory nerves has been found to produce mast-cell degranulation in tissue,[218] and the resulting vasodilatation (flare) has been found to be blocked in appearance by H_1-histamine-receptor antihistamines. Rat mast cells have been observed to respond directly to substance P application by releasing histamine.[219,220] In addition, pretreatment of human subjects with histamine-H_1-receptor antagonists has prevented the vasodilatation flare produced by intradermal substance P injection.[221-223]

A recent (1990) report from Japan by Tani et al.[224] describes a histamine-induced release of substance P. The authors studied substance P immunoreactivity in both the neuronal soma and the terminal fibers of nasal mucosa projection areas in the trigeminal ganglion after short-term application of histamine to the nasal mucosa of the guinea pig. They found that substance P immunoreactivity was decreased in the soma of the trigeminal-nasal-mucosa-projection neurons, but not in the terminals and that the maximum decrease was at 1 to 3 h after histamine application.

F. CGRP

Most substance P-containing sensory neurons also contain CGRP.[225] The precise role of CGRP in this colocalization remains somewhat of a mystery. A 1988 report from Harrow, England, by Brain and Williams,[226] suggests that substance P and CGRP may exert a mutually modulatory effect. They report that injection of substance P with CGRP into human skin converts a long-lasting vasodilatation (expected to be produced by CGRP) into a transient response. They attribute this interesting finding to the action of proteases from mast cells stimulated by substance P.

X. HEAT AND INFLAMMATION

A. INFLAMMATION

The concept of a neurogenic component to inflammation existed long before the discovery of substance P. A substantial amount of evidence has been accumulated over the years to indicate an important involvement of the sensory nervous system in inflammation as a defense-type response. The presence of vasodilator fibers in peripheral nerves was established before an involvement of these fibers in inflammation was recognized. Goltz,[227] in 1874, identified the presence of vasodilator fibers in the sciatic nerve, and in 1876 Stricker[228] located the vasodilator fibers in the dorsal root. In 1901 Bayliss[229] established that antidromic (retrograde, against-the-grain) stimulation of dorsal-root fibers produced vasodilation in the skin by mechanisms that did not involve the autonomic nervous system.

Bruce, in 1910 and 1913,[230,231] and Breslauer,[232] in 1919, reported that transection of the sensory nerve prevented the hyperemia and edmea produced by mustard oil application to skins and conjunctiva if the nerve had degenerated.

The concept of the "axon reflex" was devised to explain how sensory and motor inflammation could be transmitted in opposite directions in the same nerve fiber and how peripheral terminals of "afferent" sensory fibers might receive activating impulses from more distant and more central terminals and segments of the same nerve. An important advance in this area was achieved in 1930 by Hinsey and Gasser,[233] who identified the component of the dorsal-root-mediating vasodilatation on antidromic stimulation of sensory nerves on the very small unmyelinated c fibers.

The studies of Jancso and associates,[6-8,234-238] from the Medical University Szeged (Hungary) in the 1950s and 1960s, provided additional information of high significance. They established that antidromic stimulation of peripheral sensory nerves produced increased vascular permeability as well as vasodilatation, and introduced capsaicin into the study of these responses, both in initiating and in blocking neurogenic inflammation and its component parts. Capsaicin is a fascinating neurotoxic agent that is the pungent and "hot" component of the Hungarian red pepper and brings a "heat" beyond the thermal into this interesting panorama of inflammation and the combined vascular and nervous systems contribution to this defense-type response to tissue injury and "foreign" nonself invasion. In 1977 Jancso et al.[8] presented evidence that the systemic administration of capsaicin to newborn and adult rats produced long-lasting and even permanent damage to the chemosensitive unmyelinated c fibers. The permanent damage produced by neonatal administration provided a powerful new tool and model for the investigation of the neurogenic component of inflammation.

In their 1967 paper, Jancso et al.[7] also made the interesting observation — relating to an interaction between the sensation of pain and inflammation — that in certain species (frog, pigeon, or hen) capsaicin does not cause pain, nor does it induce inflammation.

The 1975 studies of Hokfelt et al.,[74] of Sweden and Stockholm, brought substance P very much into this field and picture by their identification of the presence and anatomical localizations of substance P in the small unmyelinated c fibers within afferent sensory nerves as the vasoactive neurotransmitter peptide in the peripheral as well as the central ending terminals of these nerves.

An extensive literature has been built up concerning the production of inflammation and components of the inflammatory response by the application of substance P by a variety of anatomical routes. Some interesting recent reports have further defined and explored these responses.

In 1988 Lotz, Vaughn, and Carson,[239] from the Research Institute of the Scripps Clinic in La Jolla, CA, reported that substance P could induce the release of interleukin-1 (IL-1), tumor necrosis factor α, and interleukin-6 from human blood monocytes. Also in 1988 Serra et al.,[240] from the University of Verona (Italy), described events associated with substance P activation of human neutrophils. These responses included induction of a respiratory burst, as measured by oxygen consumption and hydrogen peroxide production as well as activation of PI turnover and secretion of specific and azurophilic granules (cytochalasin-B dependent).

In 1989 Albert et al.,[241] of Freiburg (Max Planck) and Munich (University) in Germany, described a new approach and model for the study of a role for substance P in inflammation relating to local neural mechanisms. As a nonimmunologic mast cell stimulation model, they utilized staphylococcal enterotoxin-B (SEB)-induced immediate-type skin reactions in unsensitized monkeys. They obtained evidence suggesting that substance P plays a predominant role in mediating intradermal SEB challenge in unsensitized monkeys. With a rabbit substance P antiserum directed against the C-terminal region of substance P, a concentration-dependent inhibition of SEB-induced skin reactivity was shown. An antiserum directed against the mast-cell-activating N-terminal segment of substance P produced total blockade of the reaction. Inhibition of the reaction also was achieved with capsaicin pretreatment of the skin.

Also in 1989, Khalil and Helme,[242] of Australia, reported rat hind-footpad-blister model studies indicating that the plasma extravasation response to substance P consists of an early substance P-mediated response that is independent of other mediators, and a late response that involves interactions between substance P, mast cell mediators, and prostaglandins. An early histamine-independent response also was found to be produced by a C-terminally directed analogue of substance P, SP_{5-11}. The late histamine-dependent response was found to be mediated by an N-terminally directed analogue, SP_{1-7}. The authors speculated that a possible sequence of events might occur during an inflammatory response to substance P, which would involve an inflammatory response to substance P that would involve independent actions of its C and N terminals.

A 1990 publication from CIBA-GEIGY (Summit, NJ), by O'Byrne et al.,[243] presented data indicating that both substance P and prostaglandin-E_2 levels in joint fluid of rabbits are elevated following intra-articular injection of recombinant human interleukin-1. Also on the arthritis front, Duggan and associates,[244] now in Edinburgh, U.K., utilized antibody probes to investigate substance P levels in the spinal cord of anesthetized spinal cats to study the development of kaolin and carrageenin joint-injection-induced acute arthritis. Spinal substance P release was found after joint stimulation in seven of ten cats studied, and the release was predominantly — and after a delay of several hours — in the superficial dorsal horn, the dorsal column of the spinal cord, and at the dorsal surface of the spinal cord.

A human substance P involvement in disease came from Ankara, Turkey, by Kurkcuoglu and Cakar[245] in 1990 and was concerned with increased substance P immunoreactivity in the active edges of psoriatic plaques. By direct immunocytochemical study of biopsy material, they identified increased numbers of substance P-containing cutaneous nerves in the active edges of psoriatic plaques and identified pronounced substance P-immunoreactive nerves in the papillary dermis, as compared to normal skin. They considered these findings to be of significance, as it has been suggested that the initial event in the development of a psoriatic plaque is an increased blood flow in the papillary dermis.

An additional blister model study of interest was published in 1990 by Khalil and Helme.[246] They found that substance P-induced plasma extravasation and vasodilatation, when perfused over the blister base, were inhibited significantly by both the mu-opioid-receptor agonist (DAGO) and the delta-receptor agonist (DSLET) in a naloxone-reversible manner. The DSLET inhibitory action was 100 times more potent than that of DAGO. Neonatal capsaicin treatment inhibited the inhibitor actions produced by these opioid-receptor agonists.

B. HEAT

The first physiological evidence that substance P was selectively involved in nociception was provided in 1976 by Henry at McGill University, Montreal,[16] in studies involving the use of thermal-heat noxious stimulation of an extremity. The effect of substance P on functionally identified units in the cat spinal cord was studied. All dorsal-horn neurons excited by the local iontophoretic application of substance P were also activated and excited by noxious radiant-heat stimulation in the receptive field. In 1980 Yashpal, Wright, and Henry[66] reported that substance P and a partial homologue reduced the thermal-heat nociceptive tail-flex response latency when the substance P was given intrathecally to the rat. In this study the noxious radiant-heat stimulus with intrathecal substance P not only produced a decrease in reaction time, but, even more significantly according to the authors, produced, with higher substance P doses, immediate behavioral responses that consisted of an immediate vocalization, restlessness, and scratching on the inside of the container, followed at times by teeth chattering and a stiff tail for 1 to 2 min. It was of particular interest that, for up to 3 min after substance P intrathecal administration, mere touching of the tail to place it over the noxious thermal-heat source produced a brisk withdrawal of the tail and vocalization by these rats that had had multiple pre-substance P administration noxious thermal-heat applications to their tail. This observations would appear to suggest a "heat, pain, and memory" scenario type of memory response.

In 1979 Hayes and Tyers[247] had reported that intrathecal administration of substance P to the rat spinal cord had produced a short latency reduction in reaction time to the hind-paw lick in the hot-plate test.

The immensely important 1987 report of Duggan et al.[27] established directly that the application of destructive levels of thermal heat (50 to 52°C) to a peripheral area (hind paw) in the cat under anesthesia produced a consistent release of substance P, as detected by antibody probes from axon terminals in the substantia gelatinosa of the most superficial layer (lamina I) in the dorsal horn of the spinal cord. This study clearly identified a direct thermal-heat-stimulus-induced release of substance P in a critical thermal-pain reception area of the central nervous system.

XI. MEMORY AND BEHAVIOR

A. MEMORY

1. Rodent Memory

A considerable amount of work has been carried out in the past 15 years relating to positive and negative effects of various peptides and neuropeptides on learning and memory. Much of this work has been carried out in negatively reinforced learning situations where peptide-induced arousal may complicate and confuse the findings.

Nonetheless, a body of work does suggest that substance P, either on peripheral or central nervous system administration, may affect conditioned avoidance behavior in rodents.[248,249] Central administration effects have appeared to depend on the sites of injection. The findings of Huston and Staubli,[250-253] of the University of Dusseldorf, indicate that substance P injections into the vicinity of the septum or medial forebrain bundle facilitate retention of avoidance learning, whereas injection into the substantia nigra or the amygdala impairs this retention.

In 1984 Gaffori et al.[254] found that injection of the N-terminal fragment of substance P into the nucleus accumbens enhanced the retention of a passive avoidance habit. In this regard, it is of interest that a 1990 report from the Huston group at Dusseldorf also finds that substance P enhancement of inhibitory avoidance learning is mediated by its N-terminal sequence.[255,256] The authors used the intraperitoneal injection route in this study.

Schlesinger et al.,[257] at the University of Colorado in 1983, found that peripheral posttrial administration of substance P significantly enhanced the retention of a single-trial passive avoidance habit and that substance P-treated animals were more resistant to extinction after being trained on an active avoidance task than were vehicle-treated control mice. In these studies the effects were dose dependent and were found only if substance P was administered immediately after the training. These authors also reported in 1983 that substance P reversed or ameliorated the amnestic effects of electroconvulsive shock and cyclohexamine-induced retrograde amnesia.[258] Peripheral administration of substance P was by subcutaneous injection.

In 1986 Schlesinger et al.[259] made some very important and significant observations concerning substance P enhancement of memory in hunger-motivated learning tasks that did not involve the complicating arousal factor. They studied a food-deprived, heterogeneous strain of mice that were trained on two different discrimination tasks for food reinforcement. In one experiment animals were trained to make spacial discriminations in a T maze, and in the other experiment they were trained to make visual discriminations in this maze. Immediately after training they were given subcutaneous injections of substance P (1 ng/g) or vehicle. After 24 h, reversal habit acquisition was diminished due to presumed persistence of the substance P-induced greater strength of the original habit, a strengthening of the original habit not found in the vehicle-treated animals. In the visual-discrimination-trained animal group, substance P treatment significantly enhanced task performance retention. In the visual-training group task retraining in the same task was carried out at 1, 2, 3, and 7 d after the original training, and at each interval the substance P-treated mice retained the task better than control animals. The authors interpreted these important findings as indicating that substance P facilitated and enhanced the remembrance of the original task in treated animals, as compared with vehicle-control-treated animals. It was considered to be of considerable importance that substance P improved the retention of a visual-discrimination habit for at least 1 week after the original learning. The authors stressed the point that the observed substance P enhancement of memory in an appetitively motivated learning task gave them more confidence that this neuropeptide affects learning and memory performance in ways that are not mediated by changes in arousal.

In 1988 the Colorado group of Schlesinger and associates[260] described studies designed to explore interactions between substance P and nigrostriatal 5-HT. They utilized avoidance training in these studies. They employed both neurochemical lesioning and alterations in 5-HT levels and receptors. They found that 5,7-dihydroxytryptamine lesions of the substantia nigra, but not the caudate, attenuated the retention-enhancing effects of posttraining substance P injection. Attenuation produced by substantia nigra lesioning was reversed by posttraining substance P injection. They also found that passive avoidance training and posttraining substance P injections did not alter substantia nigra 5-HT levels, but did increase 5-HT_1-receptor density. They speculated regarding substance P and 5-HT relationships in the learning and memory processes.

Also in 1988 Nagel and Huston[261] described enhanced inhibitory avoidance learning produced by posttrial injections of substance P into the basal forebrain.

The collective findings of these multiple studies of the influence of posttrial-injected substance P (subcutaneously, intraperitoneally, or intracranially) on the performance of learning tasks strongly suggest that substance P interacts with variables and molecular mechanisms involved in memory formation and learning.

2. Human Memory

The 1982 report of Crystal and Davies[23] in the *Journal of Neurochemistry*, concerning a substance P deficiency in the cerebral cortices of Alzheimer's disease patients, was a major contribution both to an understanding of an important component of Alzheimer's disease and to a further delineation of substance P at the human level in central nervous system functioning at the molecular level and in human memory. The paper by Crystal and Davies, despite its importance and publication in a prestigious journal, was strangely both doubted and ignored for a considerable period of time.[262,263]

In looking at the raw data in the results section of the Crystal and Davies publication, the author of this book[29] calculated from their numbers for control — non-Alzheimer's disease brains — that tissue levels of substance P were 3.7 times higher in the hippocampus than in six other major cortical areas of the brain studied by them. In the Alzheimer's disease brains, substance P levels of the hippocampus were less than 50% of the levels of substance P in the hippocampus of control/"normal" brains.

The observations by Crystal and Davies were finally confirmed in a 1987 publication by Beal and Mazurek,[264] from Harvard and the neurology service at the Massachusetts

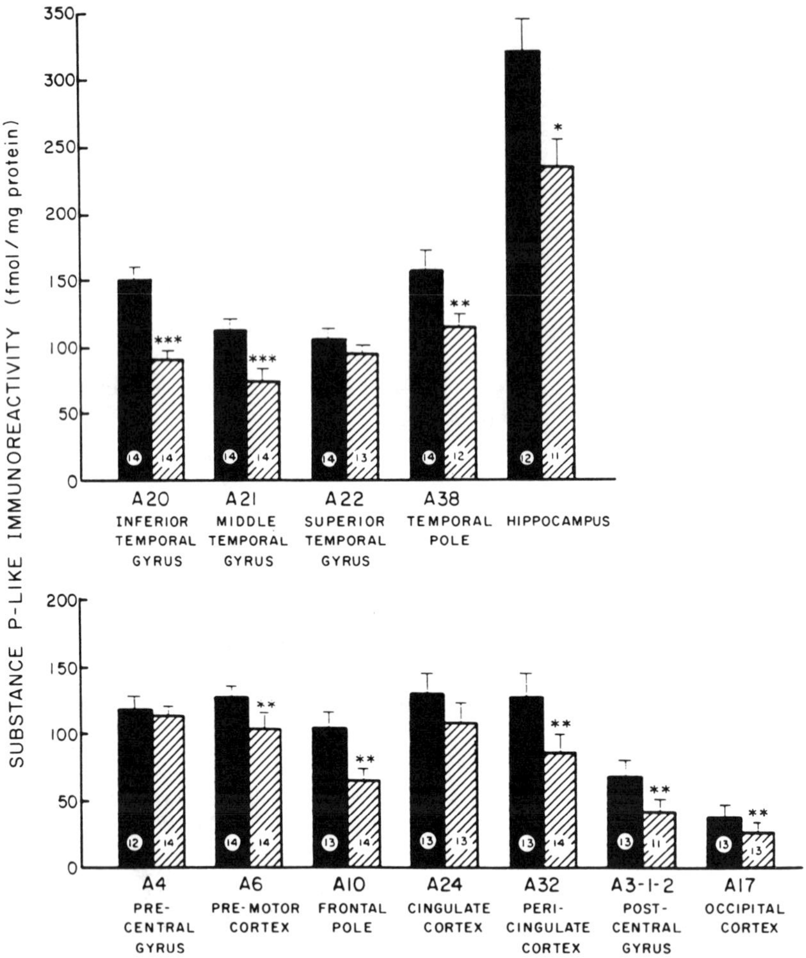

FIGURE 25. Graph prepared from data in 1987 publication of Beal and Mazurek,[264] comparing substance P levels in hippocampus and other brain areas in control (black) and Alzheimer's disease human brains (hatched).

General Hospital (Figure 25), and in a 1991 publication by Quigley and Kowall,[265] from the same service, hospital, and university. The Beal and Mazurek report also confirms the high hippocampus levels of substance P in comparison with other brain cortical areas contained in the 1982 Crystal and Davies study. Although the hippocampus substance P-level reduction in Alzheimer's brains in percent is only 22% as compared with more than 50% in the Crystal and Davies study, in absolute amounts of substance P (fmol/mg protein) the Alzheimer's disease brain hippocampus substance P-level fall is approximately two times as great as the next two highest falls (85 fmol/mg protein vs. approx. 45 fmol/mg protein, Figure 26).

In a 1990 report by Sakurada et al.,[266] from Japan and Sweden, a 35% reduction in substance P in the hippocampus was found if cases of Alzheimer's and/or multi-infarct dementia were combined and compared with an equally limited number of controls.

The 1991 Quigley and Kowall[265] study was an immunohistochemical one, and its principal finding was a significant depletion of substance P immunoreactivity of cortical perikarya (neuronal cell bodies).

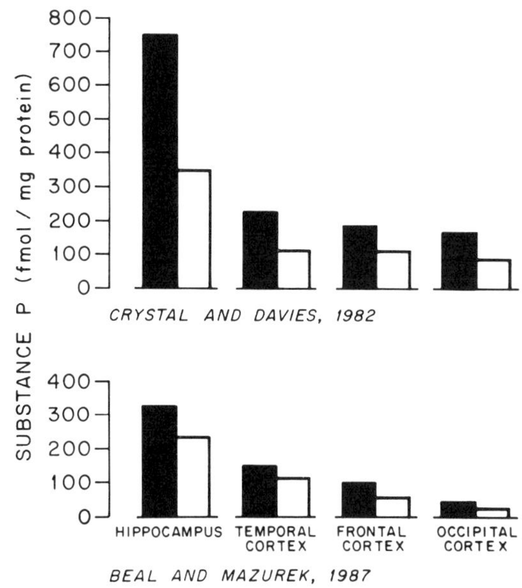

FIGURE 26. Graph prepared from raw data of Crystal and Davies[23] and Beal and Mazurek[264] papers concerned with substance P levels in hippocampus and brain areas of control (black) and Alzheimer's disease brains (white).

An especially important report by Kowall, Yanker, and associates,[30] of Boston and Harvard (MGH and Children's), appeared in the August 15, 1991 issue of *Proceedings of the National Academy of Science, U.S.A.*, describing the blocking of experimental Alzheimer's disease (produced in rats by β-amyloid) by substance P. Substance P was effective both on coadministration into the rat cerebral cortex with β-amyloid and by systemic administration. Substance P prevented both the β-amyloid-induced neuronal loss and expression of characteristic Alz-50 protein antigen appearance in cortical neurons.

3. Immunological Memory

The 1989 and 1990 reports by the author and associates[28,29] have been considered in the "Immune Response System" section of this chapter. The remarkable prolongation of the lymphocyte-output-stimulating response and the targeting of this response to the all-important CD4 subset of T cells represents a long-term-potentiation type of response. The stimulation and accelerating affect on antibody production as long as 6 d following substance P infusion into the afferent lymphatics of the study popliteal lymph node draining the immunization area suggests a memory type of response and prompted the suggestion that substance P might be a mediator of immunological memory.

B. BEHAVIOR

Intracranial and intrathecal administration of substance P has been found to elicit a number of fairly dramatic behavioral responses. Some have been thought to be mediated by activation of intracranial dopaminergic pathways.

Iversen and associates, first at the University of Cambridge and later at the Neuroscience Research Centre (Merck, Sharp and Dohme Research Laboratories, Harlow, Essex, U.K.), have carried out extensive investigations in this area.[267-275] Infusions of substance P through permanent indwelling cannulae positioned in the VTA of the A10 dopaminergic mesolimbic area of rats was found to produce a stimulation of locomotor activity (photo cell monitoring). Behavioral analysis of this stimulatory response was obtained in an open-field apparatus that permitted quantification of exploratory as well as locomotive responses. These studies identified increased activity in both the peripheral and central areas of the field, producing a distribution of behavior that was considered to reflect high levels of patterned exploration. Selective destruction of the mesolimbic dopamine innervation with 6-hydroxydopamine or dopamine antagonist drug treatment (haloperidol) prevented these substance P-infusion-induced responses.

This group also explored infusion of substance P into the substantia nigra of rats. Bilateral infusion at first produced only a weak stimulation of locomotor behavior, but on repeated applications, an intense full-body grooming, as a dramatic stereotyped behavior, occurred. This also may be associated as a repetitious species-specific behavior, with sniffing, licking, and gnawing. Interestingly, unilateral injection of substance P into the substantia nigra was found to produce rotational behavior contralateral to the side of injection, a response considered to be due to nigrostriatal-tract-activation induction of asymmetry in the striatum, resulting in rotation. A 1988 report by this group described a NK-3 tachykinin agonist (senktide) induction of a yawning syndrome in the rat.[276]

1987 and 1990 reports by Papir-Kricheli et al.,[277,278] from Jerusalem, explored behavior effects of receptor-specific substance P agonists. They found that selective activation of NK-1 (SP-P) receptors with the synthetic agonist senktide evoked compulsive scratching, licking, and biting responses without alterations in sensory responsiveness. Activation of NK-3 (SP-N) receptors with senktide produced thermal and mechanical hypoalgesia and transient flaccid paralysis, and at smaller doses evoked ''wet-dog shakes''. Some of these senktide responses may have been due to neurotoxic properties of this drug.

XII. CONCLUSION

Substance P is a most remarkable molecule. It was identified by its discoverer, U.S. von Euler, not only as a neuropeptide, but also as a very important neuropeptide with heavy concentration in the all-important and function-critical basal ganglia area of the mammalian (horse) brain.[1-3] All of this was achieved decades before the term neuropeptide was even invented and before Canadian neurosurgeons demonstrated, in the 1950s, the importance of the human basal ganglia and hippocampus in the laying down of memory, as the results of bilateral hippocampus resection to control intractable epilepsy became apparent.

Substance P clearly is the keystone molecule in the ''heat, pain, and memory'' scenario. A major piece of this puzzle was put in place by Duggan and associates[27] of Australia, who demonstrated that noxious thermal stimuli to an anesthetized cat's lower extremity could produce substance P release in the spinal cord dorsal-horn pain-reception critical substantia gelatinosa area. Earlier work (by Jancso and co-workers[6-8] of Hungary) with capsaicin provided strong evidence for a role of importance for substance P in pain transmission and in neurogenic inflammation. Substance P also holds impeccable evolutionary credentials. It has been tracked down the phylogenetic tree to the earliest multicellular animal and to the most primitive animal to possess a nervous system, the cnidarian coelenterate *Hydra*.[38]

Roles of importance for substance P in both central nervous system memory and in immunological memory also have been established. In the early and mid-1980s, the Schlesinger group[257-259] of Colorado provided impressive evidence for a significant role for substance P in mammalian (mice) memory, and in the same period Crystal and Davies[23] of New York described serious depletion in human hippocampal substance P levels in patients dying of Alzheimer's disease. The author's 1990 report[29] in *Immunopharmacology*, identifying a role of importance for substance P in immunological memory, and the 1991 report of the Boston/Harvard group[30] in the *Proceedings of the National Acadmy of Science, U.S.A.*, describing the prevention of experimental Alzheimer's disease in rats by substance P, have added further fuel to the substance P and memory ''fire''.

As the premier neuropeptide, substance P will continue to serve as a ''motherlode'' for those seeking to discover research nuggets in the important fields of nervous-system science, vascular-system science, immune response-system science, and neurovascular immunology.

REFERENCES

1. **von Euler, U. S. and Gaddum, J. H.,** An unidentified depressor substance in certain tissue extracts, *J. Physiol. (Lond.),* 72, 74, 1931.
2. **von Euler, U. S.,** Untersuchungen uber Sustanz P, die atropinfeste darmerregende und gefasserweiternde Substanz aus Darm und Hirn, *Naunyn-Schmiedeberg's Arch. Pharmacol.,* 181, 181, 1936.
3. **von Euler, U. S.,** Chemical neurotransmission — yesterday and today, Ciba Foundation 33rd Annual Lecture, Cavendish Centre, London, 1982, 3.
4. **Lembeck, F.,** Zur Frage der zentralen Ubertragung afferenter Impulse. III. Mitteilung das Vorkommem und die Bedeutung der Substanz P in den dorsalen Wurzein des Ruckenmarks, *Naunyn-Schmiedeberg's Arch. Pharmacol.,* 219, 197, 1953.
5. **von Euler, U. S. and Ostlund, E.,** Occurrence of a substance P-like polypeptide in fish intestine and brain, *Br. J. Pharmacol.,* 11, 323, 1956.
6. **Jancso, N. and Jancso-Gabor, A.,** Dauerausschaltung der chemischen Schmerzempfindlichkeit durch Capsaicin, *Naunyn-Schmiedeberg's Arch. Pharmacol.,* 236, 142, 1959.
7. **Jancso, N., Jancso-Gabor, A., and Szolcsanyi, J.,** Direct evidence for neurogenic inflammation and its prevention by denervation and by pretreatment with capsaicin, *Br. J. Pharmacol.,* 31, 138, 1967.
8. **Jancso, N., Kiraly, E., and Jancso-Gabor, A.,** Pharmacologically induced selective degeneration of chemosensitive primary sensory neurones, *Nature,* 270, 741, 1977.
9. **von Euler, U. S.,** Substance P in subcellular particles in peripheral nerves, *Ann. N.Y. Acad. Sci.,* 104, 449, 1963.
10. **Chang, M. M. and Leeman, S. E.,** Isolation of a sialogogic peptide from bovine hypothalamic tissue and its characterization as substance P, *J. Biol. Chem.,* 245, 4784, 1970.
11. **Otsuka, M. and Konishi, S.,** Electrophysiology of mammalian spinal cord *in vitro, Nature,* 252, 733, 1974.
12. **Konishi, S. and Otsuka, M.,** Excitatory action of hypothalamic substance P on spinal motoneurones of newborn rats, *Nature,* 252, 734, 1974.
13. **Konishi, S. and Otsuka, M.,** The effects of substance P and other peptides on spinal neurons of the frog, *Brain Res.,* 65, 397, 1974.
14. **Takahashi, T., Konishi, S., Powell, D., Leeman, S. E., and Otsuka, M.,** Identification of the mononeuron-depolarizing peptide in bovine dorsal root as hypothalamic substance P, *Brain Res.,* 73, 59, 1974.
15. **Otsuka, M., Konishi, S., Yanagisawa, M., Tsunoo, A., and Akagi, H.,** Role of substance P as a sensory transmitter in spinal cord and sympathetic ganglia, in *Substance P in the Nervous System,* CIBA Foundation Symposium, Pitman, London, 1991, 13.
16. **Henry, J. L.,** Effects of substance P on functionally identified units in cat spinal cord, *Brain Res.,* 114, 439, 1976.
17. **Jessell, T. M. and Iversen, L. L.,** Opiate analgesics inhibit substance P release from rat trigeminal nucleus, *Nature,* 268, 549, 1977.
18. **Hanley, M. R., Sandberg, B. E. B., Lee, C. M., Iversen, L. L., Brundish, D. E., and Wade, R.,** Specific binding of ^{3}H-substance P to rat brain membranes, *Nature,* 286, 810, 1980.
19. **Hanley, M. R., Lee, C. M., Jones, L. M., and Michell, R. H.,** Similar effects of substance P and related peptides on salivation and on phosphatidylinositol turnover in rat salivary glands, *Mol. Pharmacol.,* 18, 78, 1980.
20. **Iversen, L. L.,** Substance P, *Br. Med. Bull.,* 38, 277, 1982.
21. **Iversen, L. L., McKnight, A. T., Foster, A. C., Young, S. C., and Williams, B. J.,** *Pharmacology of the Tachykinin System,* Munksgaard, Copenhagen, 1990, 363.

22. **Moore, T. C.,** The effects of inflammatory mediators and anaesthetic agents upon lymphocyte traffic of sheep *in vivo*, Ph.D. thesis, University of Cambridge (Trinity College), England, 1982.
23. **Crystal, H. A. and Davies, P.,** Cortical substance P-like immunoreactivity in cases Alzheimer's disease and in senile dementia of the Alzheimer's type, *J. Neurochem.*, 38, 1781, 1982.
24. **Moore, T. C.,** Modification of lymphocyte traffic by vasoactive neurotransmitter substances, *Immunology,* 52, 511, 1984.
25. **Payan, D. G., Brewster, D. R., and Goetzl, E. J.,** Sterospecific receptors for substance P on cultured human IM-9 lymphoblasts, *J. Immunol.*, 133, 3260, 1984.
26. **Payan, D. G., Brewster, D. R., Missirian-Bastian, A., and Goetzl, E. J.,** Substance P recognition by a subset of human T lymphocytes, *J. Clin. Invest.*, 74, 1532, 1984.
27. **Duggan, A. W., Morton, C. R., Zhao, Z. Q., and Hendry, I. A.,** Noxious heating of skin releases immunoreactive substance P in the substantia gelatinosa of the cat: a study with antibody probes, *Brain Res.*, 403, 345, 1987.
28. **Moore, T. C., Lami, J. L., and Spruck, C. H.,** Substance P increases lymphocyte traffic and lymph flow through peripheral lymph nodes of sheep, *Immunology,* 67, 109, 1989.
29. **Moore, T. C., Whitley, G. A., Lami, J. L., and Said, S. I.,** Substance P increases and prolongs increased output of T4 (CD4) lymphocytes from lymph nodes of sheep *in vivo*: is it a mediator of immunological memory?, *Immunopharmacology,* 20, 207, 1990.
30. **Kowall, N. W., Beal, M. F., Busciglio, J., Duffy, L. K., and Yankner, B. A.,** An *in vivo* model for the neurodegenerative effects of B amyloid and protection by substance P, *Proc. Natl. Acad. Sci. U.S.A.*, 88, 7247, 1991.
31. **Erspamer, V.,** The tachykinin peptide family, *Trends Neurosci.*, 4, 267, 1981.
32. **Iversen, L. L., Watling, K. J., McKnight, A. T., Williams, B. J., and Lee, C. M.,** Multiple receptors for substance P and related tachykinins, in *Proc. 4th SCI-RSC Medicinal Chemistry Symposium* (Topics in Medicinal Chemistry), Special Publication No. 65, Leeming, P., Ed., Royal Society of Chemistry, London, 1988, 1.
33. **McKnight, A. T., Maguire, J. J., Elliott, N. J., Fletcher, A. E., Foster, A. C., Tridgett, R., Williams, B. J., and Iversen, L. L.,** Pharmacological specificity as antagonists at tachykinin receptors of novel synthetic cyclic peptides, *Br. J. Pharmacol.*, in press.
34. **McKnight, A. T., Maguire, J. J., Varney, M. A., Williams, B. J., and Iversen, L. L.,** Characterisation of tachykinin receptors using selectivity of agonists and antagonists: evidence for an NK-4 type, *Regul. Pept.*, 22, 126, 1988.
35. **McKnight, A. T., Maguire, J. J., Williams, B. J., Foster, A. C., Tridgett, R., and Iversen, L. L.,** Pharmacological specificity of synthetic peptides as antagonists at tachykinin receptors, *Regul. Pept.*, 22, 127, 1988.
36. **Wiesenfeld-Hallin, Z., Xu, X.-J., Kristensson, K., Hakanson, R., Feng, D.-M., and Folkers, K.,** Antinociceptive and substance P antagonistic effects of intrathecally injected Spantide II in rat: no signs of motor impairment or neurotoxicity, *Regul. Pept.*, 29, 1, 1990.
37. **Berger, H., Fechner, K., Albrecht, E., Niedrich, H.,** Substance P: *in vitro* inactivation by rat brain fractions and human plasma, *Biol. Pharmacol.*, 28, 3173, 1979.
38. **Grimmelikhuijzen, C. J. P., Balfe, A., Emson, P. C., Powell, D., and Sundler, F.,** Substance P-like immunoreactivity in the nervous system of hydra, *Histochemistry,* 71, 325, 1981.
39. **Gustafsson, M. K. S., Lehtonen, M. A. E., and Sundler, F.,** Immunocytochemical evidence for the presence of mammalian neurohormonal peptides in neurons of the tapeworm, *Diphyllbothrium dendriticum, Cell Tissue Res.*, 243, 41, 1986.
40. **Aros, B., Wenger, T., Vigh, B., and Vigh-Techman, I.,** Immunohistochemical localization of substance P and ACTH-like activity in the central nervous system of the earthworm Lumbricus terrestris, *Acta Histochem.*, 66, 262, 1980.
41. **Chamberlain, S. C. and Engbretson, G. A.,** Neuropeptide immunoreactivity in *Limulus*. I. Substance P-like immunoreactivity in the lateral eye and protocerebrum, *J. Comp. Neurol.*, 208, 304, 1982.
42. **Fahrenbach, W. H.,** The brain of the horseshoe crab (Limulus polyphemus). I. Neuroglia, *Tissue Cell,* 8, 395, 1976.
43. **Mancillas, J. R. and Selverston, A. I.,** Neuropeptide modulation of photo-sensitivity. II. Physiological and anatomical effects of substance P on the lateral eye of *Limulus, J. Neurosci.*, 4, 847, 1984.
44. **Mancillas, J. R., McGinty, J. F., Selverston, A. I., Karten, H., and Bloom, F. E.,** Immunocytochemical localization of enkephalin and substance P in retina and eyestalk neurones of lobster, *Nature,* 293, 576, 1981.
45. **Goldberg, D., Nusbaum, M. P., and Marder, E.,** Substance P-like immunoreactivity in the stomatogastric nervous system of the crab *Cancer borealis* and the lobsters *Panulirus interruptus* and *Homarus americanus, Cell Tissue Res.*, 252, 515, 1988.

46. **Verhaert, P. and De Loof, A.,** Substance P-like immunoreactivity in the central nervous system of the blattarian insect *Periplaneta americana L.*, revealed by a monoclonal antibody, *Histochemistry,* 83, 501, 1985.
47. **Benedeczky, I., Kiss, J. Z., and Somogyi, P.,** Light and electron microscopic localization of substance P-like immunoreactivity in the cerebral ganglion of locusts with a monoclonal antibody, *Histochemistry,* 75, 123, 1982.
48. **El-Sahly, M., Abou-El-Ela, R., Falkmer, S., Grimelius, L., and Wilander, E.,** Immunohistochemical evidence of gastroenteropancreatic neurohormonal peptides of vertebrate type in the nervous system of a dipteran insect, the hoverfly, *Eristalis aenus, Regul. Pept.,* 1, 187, 1980.
49. **El-Sahly, M., Falkmer, S., Kramer, K. J., and Speirs, R. D.,** Immunohistochemical investigations of neuropeptides in brain corpora cardiaca and corpora allata of an adult lepidopteran insect, *Manduca sexta* (L.), *Cell Tissue Res.,* 232, 295, 1983.
50. **Nässel, D. R., Lundquist, T., Hoog, A., and Grimelius, L.,** Substance P-like immunoreactive neurons in the nervous system of *Drosophila, Brain Res.,* 507, 225, 1990.
51. **Schot, L. P. C., Boer, H. H., Swaab, D. F., and Van Noorden, S.,** Immunocytochemical demonstration of peptidergic neurons in the central nervous system of the pond snail *Lymnaea stagnalis* with antisera to biologically active peptides of vertebrates, *Cell Tissue Res.,* 216, 273, 1981.
52. **Osborne, N. N. and Dockray, G. J.,** Bombesin-like immunoreactivity in specific neurons of the snail *Helix aspersa* and an example of coexistence of substance P and serotonin in an invertebrate neuron, *Neurochem. Int.,* 4, 175, 1982.
53. **Osborne, N. N., Cuello, A. C., and Dockray, G. J.,** Substance P and cholecystokinin-like peptides in *Helix* neurons and cholecystokinin and serotonin in a giant neuron, *Science,* 216, 409, 1982.
54. **Contestabile, A., Villani, L., Fasolo, A., Franzoni, M. F., Gribaudo, L., Oktedalen, O., and Fonnum, F.,** Topography of cholinergic and substance P pathways in the habenulo-interpeduncular system of the rat. An immunocytochemical and microchemical approach, *Neuroscience,* 21, 253, 1987.
55. **Kemali, M. and Guglielmotti, V.,** The distribution of substance P in the habenulo-interpeduncular system of the frog shown by an immunohistochemical method, *Arch. Ital. Biol.,* 122, 269, 1984.
56. **Taban, C. H. and Cathieni, M.,** Distribution of substance P-like immunoreactivity in the brain of the newt *(Triturus cristatus), J. Comp. Neurol.,* 216, 453, 1983.
57. **Villani, L., Guarnieri, T., Salsi, U., and Bollini, D.,** Substance P in the habenulo-interpeduncular system of the goldfish, *Brain Res. Bull.,* 26, 225, 1991.
58. **Ekstrom, P. and Korf, H.-W.,** Substance P-like immunoreactive neurons in the photosensory pineal organ of the rainbow trout, *Salmo gairdneri* Richardson (Teleostei), *Cell Tissue Res.,* 246, 359, 1986.
59. **Engbretson, G. A., Brecha, N., and Rainer, A.,** Substance P-like immunoreactivity in the parietal eye visual system of the lizard *Uta stansburiani, Cell Tissue Res.,* 227, 543, 1982.
60. **Schmidt, A., Roth, G., and Ernst, M.,** Distribution of substance P-like, leucine-enkephalin-like, and bombesine-like immunoreactivity and acetylcholinesterase activity in the visual system of salamanders, *J. Comp. Neurol.,* 288, 123, 1989.
61. **Gaudino, G. and Fasolo, A.,** Substance P-related peptides in the hypothalamus of amphibia, *Cell Tissue Res.,* 211, 241, 1980.
62. **Brauth, S. E., Reiner, A., Kitt, C. A., and Karten, H. J.,** The substance P-containing striatotegmental path in reptiles: an immunohistochemical study, *J. Comp. Neurol.,* 219, 305, 1983.
63. **Wolters, J. G., Donkelaar, H. J., and Verhofstad, A. A. J.,** Distribution of some peptides (substance P, [leu]enkephalin, [met]enkephalin) in the brain stem and spinal cord of a lizard, *Varanus exanthematicus, Neuroscience,* 18, 917, 1986.
64. **Lavalley, A. L. and Ho, R. H.,** Substance P, somatostatin, and methionine enkephalin immunoreactive elements in the spinal cord of the domestic fowl, *Gallus domesticus, J. Comp. Neurol.,* 213, 406, 1982.
65. **Britto, L. R. G. and Hamassaki, D. E.,** A subpopulation of displaced ganglion cells of the pigeon retina exhibits substance P-like immunoreactivity, *Brain Res.,* 546, 61, 1991.
66. **Yashpal, K., Wright, D. M., and Henry, J. L.,** Substance P analogue reduces tail flick latency when given intrathecally to the rat, *Neurosci. Abstr.,* 6, 490, 1980.
67. **Henry, J. L.,** Relation of Substance P to Pain Transmission: Neurophysiological Evidence, *Ciba Foundation Symposium 91, Pitman, London,* 1982, 206.
68. **Cridland, R. A. and Henry, J. L.,** Comparison of the effects of substance P, neurokinin A, physalaemin and eledoisin in facilitating a nociceptive reflex in the rat, *Brain Res.,* 381, 93, 1986.
69. **Cridland, R. A. and Henry, J. L.,** Facilitation of the tail-flick reflex by noxious cutaneous stimulation in the rat: antagonism by a substance P analogue, *Brain Res.,* 462, 15, 1988.
70. **Duggan, A. W. and Hendry, I. A.,** Laminar localization of the sites of release of immunoreactive substance P in the dorsal horn with antibody-coated microelectrodes, *Neurosci. Lett.,* 68, 134, 1986.

71. **Kellstein, D. E., Price, D. D., Hayes, R. L., and Mayer, D. J.,** Evidence that substance P selectively modulates C-fiber-evoked discharges of dorsal horn nociceptive neurons, *Brain Res.*, 526, 291, 1990.
72. **Morton, C. R., Hutchison, W. D., and Hendry, I. A.,** Intraspinal release of substance P and calcitonin gene-related peptide during opiate dependence and withdrawal, *Neuroscience*, 43, 593, 1991.
73. **Johnson, S. M. and Duggan, A. W.,** Tolerance and dependence of dorsal horn neurones of the cat: the role of the opiate receptors of the substantia gelatinosa, *Neuropharmacology*, 20, 1033, 1981.
74. **Hokfelt, T., Kellerth, J. O., Milsson, G., and Pernaw, B.,** Experimental immunohistochemical studies on the localization and distribution of substance P in cat primary sensory neurones, *Brain Res.*, 100, 235, 1975.
75. **Gamse, R., Lembeck, F., and Cuello, A. C.,** Substance P in the vagus nerve, *Naunyn-Schmiedeberg's Arch. Pharmacol.*, 306, 37, 1979.
76. **Katz, D. M. and Karten, H. J.,** Substance P in the vagal sensory ganglia: localization in cell bodies and pericellular arborizations, *J. Comp. Neurol.*, 193, 549, 1980.
77. **Cuello, A. C., Del Fiacco, M., and Paxinos, G.,** The central and peripheral ends of the substance P-containing sensory neurones in the rat trigeminal system, *Brain Res.*, 152, 499, 1978.
78. **Widemar, L., Hellstrom, S., Schultzberg, M., and Stenfors, L.-E.,** Autonomic innervation of the tympanic membrane, *Acta Otolaryngol., (Stockh.)*, 100, 58, 1985.
79. **White, D. M. and Helme, R. D.,** Release of substance P from peripheral nerve terminals following electrical stimulation of the sciatic nerve, *Brain Res.*, 336, 27, 1985.
80. **Barbut, D., Polak, J. M., and Wall, P. D.,** Substance P in spinal cord dorsal horn decreases following peripheral nerve injury, *Brain Res.*, 205, 289, 1981.
81. **Chan-Palay, V. and Palay, S. L.,** Immunocytochemical identification of substance P cells and their processes in rat sensory ganglia and their terminals in the spinal cord: light microscopic studies, *Proc. Natl. Acad. Sci. U.S.A.*, 74, 3597, 1977.
82. **Hokfelt, T., Kellerth, J. O., Nilsson, G., and Pernow, B.,** Substance P: localization in the central nervous system and in some primary sensory neurons, *Science*, 190, 889, 1975.
83. **Jessell, T., Tsunoo, A., Kamazawa, I., and Otsuka, M.,** Substance P: depletion in dorsal horn of rat spinal cord after section of peripheral processes of primary sensory neurons, *Brain Res.*, 168, 247, 1979.
84. **Takahashi, T. and Otsuka, M.,** Regional distribution of substance P in the spinal cord and nerve roots of the cat and the effect of dorsal root section, *Brain Res.*, 87, 1, 1975.
85. **Bowker, R. M., Steinbusch, H. W. M., and Coulter, J. D.,** Serotonergic and peptidergic projections to the spinal cord demonstrated by a combined retrograde HRP histochemical and immunocytochemical staining method, *Brain Res.*, 211, 412, 1981.
86. **Jessell, T. M., Iversen, L. L., and Cuello, A. C.,** Capsaicin-induced depletion of substance P from primary sensory neurons, *Brain Res.*, 152, 183, 1978.
87. **Naftchi, N. E., Abrahams, S. T., St. Paul, H. M., Lowman, E. W., and Schlosser, W.,** Localization and changes of substance P in spinal cord of paraplegic cats, *Brain Res.*, 153, 507, 1978.
88. **Barber, R. P., Vaugh, J. E., Slemmon, J. R., Salvaterra, P. M., Roberts, E., and Leeman, S. E.,** The origin, distribution and synaptic relationships of substance P axons in rat spinal cord, *J. Comp. Neurol.*, 184, 331, 1979.
89. **Ljungdahl, A., Hokfelt, T., and Nilsson, G.,** Distribution of substance P-like immunoreactivity in the central nervous system of the rat. I. Cell bodies and nerve terminals, *Neuroscience*, 3, 861, 1978.
90. **Pickel, V. M., Miller, R., Chan, J., and Sumal, K. K.,** Substance P and enkephalin in transected axons of medulla and spinal cord, *Regul. Pept.*, 6, 121, 1983.
91. **Salter, M. W. and Henry, J. L.,** Responses of functionally identified neurones in the dorsal horn of the cat spinal cord to substance P, neurokinin A and physalaemin, *Neuroscience*, 43, 601, 1991.
92. **Henry, J. L., Sessle, B. J., Lucier, G. E., and Hu, J. W.,** Effects of substance P on nociceptive and non-nociceptive trigeminal brain stem neurons, *Pain*, 8, 33, 1980.
93. **Hancock, M. B.,** Substance P-immunoreactive processes on 5HT/SP-immunoreactive medullary cells, *Brain Res. Bull.*, 13, 559, 1984.
94. **Ronnekliev, O. K., Kelly, M. J., and Eskay, R. L.,** Distribution of immunoreactive substance P neurons in the hypothalamus and pituitary of the Rhesus monkey, *J. Comp. Neurol.*, 224, 51, 1984.
95. **Li, Y.-Q., Jia, H.-G., Rao, Z.-R., and Shi, J.-W.,** Serotonin-, substance P- or leucine-enkephalin-containing neurons in the midbrain periaqueductal gray and nucleus raphe dorsalis send projection fibers to the central amygdaloid nucleus in the rat, *Neurosci. Lett.*, 120, 124, 1990.
96. **Rodgers, R. J.,** Influence of intra-amygdaloid opiate injections on shock thresholds, tail-flick latencies and open field behavior in the rats, *Brain Res.*, 153, 211, 1979.
97. **Xu, D. Y., Zhou, Z. F., and Han, J. S.,** Amygdaloid serotonin and endogenous opioid substances (OLS) are important for mediating electroacupuncture analgesia and morphine analgesia in the rabbit, *Acta Physiol. Sinica*, 37, 162, 1985.

98. **Yu, L. C. and Han, J. S.,** The antinociceptive related neural pathway from midbrain periaqueductal gray to amygdaloid nucleus, *Chinese Sci. Bull.*, 33, 68, 1988.
99. **Li, Y.-Q., Rao, Z.-R., and Shi, J.-W.,** Substance P-like immunoreactive neurons in the nucleus tractus solitarri of the rat send their axons to the nucleus accumbens, *Neurosci. Lett.*, 120, 194, 1990.
100. **West, C. H. K. and Michael, R. P.,** Substance P injections into the ventral tegmentum affect unit activity in mesolimbic terminal regions, *Brain Res. Bull.*, 26, 229, 1991.
101. **Brownstein, M. J., Mroz, E. A., Kizer, S., Palkovits, M., and Leeman, S. E.,** Regional distribution of substance P in the brain of the rat, *Brain Res.*, 116, 299, 1976.
102. **Deutch, A. Y., Maggio, J. E., Bannon, M. J., Kalivas, P. K., Tam, S.-Y., Goldstein, M., and Roth, R. H.,** Substance K and substance P differentially modulate mesolimbic and mesocortical systems, *Peptides,* 6, 113, 1985.
103. **Eison, A. S., Eison, M. S., and Iversen, S. D.,** The behavioral effects of a novel substance P analogue following infusion into the ventral tegmental area or substantia nigra of rat brain, *Brain Res.*, 238, 137, 1982.
104. **Kalivas, P. W.,** Interactions between neuropeptides and dopamine neurons in the ventromedial mesencephalon, *Neurosci. Biobehav. Rev.*, 9, 573, 1985.
105. **Kalivas, P. W., Deutch, A. Y., Maggio, J. E., Mantyh, P. W., and Roth, R. H.,** Substance K and substance P in the ventral tegmental area, *Neurosci. Lett.*, 57, 241, 1985.
106. **Kelley, A. E., Cador, M., and Stinus, L.,** Behavioral analysis of the effect of substance P injected into the ventral mesencephalon on investigatory and spontaneous motor behavior in the rat, *Psychopharmacology (Berlin),* 85, 37, 1985.
107. **Kelley, A. E., Stinus, L., and Iversen, S. D.,** Behavioral activation induced in the rat by substance P infusion into ventral tegmental area: implication of dopaminergic A10 neurones, *Neurosci. Lett.*, 11, 335, 1979.
108. **Stinus, L., Kelley, A. E., and Iversen, S. D.,** Increased spontaneous activity following substance P infusion into dopaminergic areas, *Nature,* 276, 626, 1978.
109. **Fink, J. S. and Smith, G. P.,** Mesolimbicocortical dopamine terminal fields are necessary for normal locomotor and investigatory exploration in rats, *Brain Res.*, 199, 359, 1980.
110. **Cador, M., Rivet, J.-M., Kelley, A. E., LeMoal, M., and Stinus, L.,** Substance P, neurotensin and enkephalin injections into the ventral tegmental area: comparative study on dopamine turnover in several forebrain structures, *Brain Res.*, 486, 357, 1989.
111. **Elliott, P. J., Alpert, J., Bannon, M. J., and Iversen, S. D.,** Selective activation of mesolimbic and mesocortical dopamine metabolism in the rat brain by infusion of a stable substance P analogue into the ventral tegmental area, *Brain Res.*, 363, 145, 1986.
112. **Collingridge, G. L. and Davies, J.,** Actions of substance P and opiates in the rat substantia nigra, *Neuropharmacology,* 21, 715, 1982.
113. **Davies, J. and Dray, A.,** Substance P in the substantia nigra, *Brain Res.*, 107, 623, 1976.
114. **Phillis, J. W. and Limacher, J. J.,** Substance P excitation of cerebral cortical Betz cells, *Brain Res.*, 69, 158, 1974.
115. **Cuello, A. C., Priestly, J. V., and Mathews, M. R.,** Localization of substance P in neuronal pathways, in *Substance P in the Nervous System,* Ciba Symposium 91, Pittman, London, 1982, 55–79.
116. **Lindefors, N., Brodin, E., Theodorsson-Norheim, E., and Ungerstedt, U.,** Regional distribution and *in vivo* release of tachykinin-like immunoreactivities in rat brain: evidence for regional differences in relative proportions of tachykinins, *Regul. Pept.*, 10, 217, 1985.
117. **Mai, J. K., Stephens, P. H., and Cuello, A. C.,** Substance P in the human brain, *Science,* 17, 709, 1986.
118. **Beach, T. G. and McGeer, E. G.,** Neocortical substance P neurons in the baboon: an immunohistochemical finding, *Neurosci. Lett.*, 44, 265, 1983.
119. **Del Fiacco, M., Levanti, M. C., Dessi, M. L., and Zucca, G.,** The human hippocampal formation and parahippocampal gyrus: localization of substance P-like immunoreactivity in newborn and adult post-mortem tissue, *Neuroscience,* 21, 141, 1987.
120. **Charnay, Y., Paulin, C., Chayvialle, J. A., and Dubois, P. M.,** Distribution of substance P-like immunoreactivity in the spinal cord and dorsal root ganglia of the human foetus and infant, *Neuroscience,* 10, 41, 1983.
121. **Cooper, P. E., Fernstrom, M. H., Rorstad, O. P., Leeman, S. E., and Martin, J. B.,** The regional distribution of somatostatin, substance P and neurotensin in human brain, *Brain Res.*, 218, 219, 1981.
122. **Cuello, A. C., Polak, J. M., and Pearse, A. G. E.,** Substance P: a naturally occurring transmitter in human spinal cord, *Lancet,* II, 1054, 1976.
123. **de Lanerolle, N. C. and La Motte, C. C.,** The human spinal cord: substance P and methionine-enkephalin immunoreactivity, *J. Neurosci.*, 2, 1369, 1982.

124. **La Motte, C. C. and de Lanerolle, N. C.,** Human spinal neurons: innervation by both substance P and enkephalin, *Neuroscience,* 6, 713, 1981.
125. **Langevin, H. and Emson, P. C.,** Distribution of substance P, somatostatin, and neurotensin in the human hypothalamus, *Brain Res.,* 246, 65, 1982.
126. **Emson, P. C., Arregui, A., Clement-Jones, V., Sandberg, B. E. B., and Rossor, M.,** Regional distribution of methionine-enkephalin and substance P-like immunoreactivity in normal brain and in Huntington's disease, *Brain Res.,* 119, 147, 1980.
127. **Kanazawa, I., Bird, E., O'Connell, R., and Powell, D.,** Evidence for a decrease in substance P content of substantia nigra in Huntington's chorea, *Brain Res.,* 120, 387, 1977.
128. **Inagaki, S., Sakanaka, M., Shiosaka, S., Senba, E., Takatsuki, K., Takagi, H., Kawai, Y., Minagawa, H., and Tohyama, M.,** Ontogeny of substance P-containing neuron system of the rat: immunohistochemical analysis. I. Forebrain and upper brain stem, *Neuroscience,* 7, 251, 1982.
129. **Sakanaka, M., Inagaki, S., Shiosaka, S., Senba, E., Takagi, H., Takatsuki, K., Kawai, Y., Iida, H., Hara, Y., and Tohyama, M.,** Ontogeny of substance P-containing neuron system of the rat: immunohistochemical analysis. II. Lower brainstem, *Neuroscience,* 7, 1097, 1982.
130. **Sakamoto, N., Michel, J.-P., Kopp, N., Tohyama, M., and Pearson, J.,** Substance P- and enkephalin-immunoreactive neurons in the hippocampus and related areas of the human infant brain, *Neuroscience,* 22, 801, 1987.
131. **Brodal, A.,** *Neurological Anatomy,* Oxford University Press, Oxford, England, 1981.
132. **Ramon y Cajal, S.,** *Histologie du Systeme Nerveux de l'homme et des Vertebres,* 2, 762, Norbert Maloine, Paris, 1909/1911.
133. **Carpenter, M. D. and Sutin, J.,** *Human Neuroanatomy,* Williams & Wilkins, Baltimore, 1983.
134. **Lorente de No, R.,** Studies on the structure of the cerebral cortex. II. Continuation of the study of the ammonic system, *J. Psychol. Neurol.* (Lpz.), 46, 113, 1934.
135. **Yamashita, A., Shimizu, K., and Hayashi, M.,** Ontogeny of substance P-immunoreactive structures in the primate cerebral neocortex, *Dev. Brain Res.,* 57, 197, 1990.
136. **Dornan, W. A. and Malsbury, C. W.,** Peptidergic control of male rat sexual behavior: the effects of intracerebral injections of substance P and cholecystokinin, *Physiol. Behav.,* 46, 547, 1989.
137. **Neal, C. R., Jr. and Newman, S. W.,** Prodynorphin- and substance P-containing neurons project to the medial preoptic area in the male Syrian hamster brain, *Brain Res.,* 546, 119, 1991.
138. **Swann, J. M. and Newman, S. W.,** Effect of castration and testosterone treatment on substance P levels within the vomeronasal pathway of the male golden hamster, *Soc. Neurosci. Abstr.,* 13, 1576, 1987.
139. **Barakat-Walter, I., Affolter, H. U., and Droz, B.,** Expression of substance P and preprotachykinin mRNA by primary sensory neurons in culture: regulation by factors present in peripheral and central target tissues, *Mol. Brain Res.,* 10, 107, 1991.
140. **Catalan, R. E., Martinez, A. M., Aragones, M. D., and Fernandez, I.,** Substance P stimulates translocation of protein kinase C in brain microvessels, *Biochem. Biophys. Res. Commun.,* 164, 595, 1989.
141. **Matis, W. L., Lavker, R. M., and Murphy, G. F.,** Substance P induces the expression of an endothelial-leukocyte adhesion molecule by microvascular endothelium, *J. Invest. Dermatol.,* 94, 492, 1990.
142. **Galligan, J. J., Jiang, M.-M., Shen, K.-Z., and Surprenant, A.,** Substance P mediates neurogenic vasodilatation in extrinsically denervated guinea-pig submucosal arterioles, *J. Physiol. (Lond.),* 420, 267, 1990.
143. **Neild, T. O.,** Measurement of arteriole diameter changes by analysis of television images, *Blood Vessels,* 26, 48, 1989.
144. **Neild, T. O., Shen, K.-Z., and Surprenant, A.,** Vasodilatation of arterioles by acetylcholine released from single neurones in the guinea-pig submucosal plexus, *J. Physiol. (Lond.),* 420, 247, 1990.
145. **Ziche, M., Morbidelli, L., Pacini, M., Geppetti, P., Alessandri, G., and Maggi, C. A.,** Substance P stimulates neovascularization *in vivo* and proliferation of cultured endothelial cells, *Microvasc. Res.,* 40, 264, 1990.
146. **Globus, M., Smith, M. J., and Swani, V.-G.,** Evidence supporting a mitogenic role for substance P in amphibian limb regeneration; involvement of the inositol phospholipid signaling pathway, *Ann. N.Y. Acad. Sci.,* 632, 396, 1991.
147. **Nilsson, J., von Euler, A. M., and Dalsgaard, C.-J.,** Stimulation of connective tissue growth by substance P and substance K, *Nature,* 315, 61, 1985.
148. **Salo, E. and Baguna, J.,** Stimulation of cellular proliferation and differentiation in the intact and regenerating planarian Dugesia(G) tigrina by the neuropeptide substance P, *J. Exp. Zool.,* 237, 129, 1986.
149. **Ralevic, V., Milner, P., Hudlicka, O., Kristek, F., and Burnstock, G.,** Substance P is released from the endothelium of normal and capsaicin-treated rat hind-limb vasculature, *in vivo,* by increased flow, *Circ. Res.,* 66, 1178, 1990.

150. **Stephenson, J. A., Burcher, E., and Summers, R. J.,** Autoradiographic demonstration of endothelium-dependent ^{125}I-Bolton-Hunter substance P binding to dog carotid artery, *Eur. J. Pharmacol.*, 124, 377, 1986.
151. **Stephenson, J. A. and Summers, R. J.,** Autoradiographic analysis of receptors on vascular endothelium, *Eur. J. Pharmacol.*, 134, 35, 1987.
152. **Loesch, A. and Burnstock, G.,** Ultrastructural localisation of serotonin and substance P in vascular endothelial cells of rat femoral and mesenteric arteries, *Anat. Embryol. (Berlin)*, 178, 137, 1988.
153. **Milner, P., Ralevic, V., Hopwood, A. M., Feher, E., Lincoln, J., Kirkpatrick, K. A., and Burnstock, G.,** Ultrastructural localisation of substance P and choline acetyltransferase in endothelial cells of rat coronary artery and release of substance P and acetylcholine during hypoxia, *Experientia*, 45, 121, 1989.
154. **Moore, T. C.,** Mediator and physiological influences on lymphocyte traffic in sheep, in *Immunology of the Sheep*, Morris, B. and Miyasaka, M., Eds., Editiones Roche, Basel, 1985, 327–354.
155. **Payan, D. G., Brewster, D. R., and Goetzl, E. J.,** Specific stimulation of human T lymphocytes by substance P, *J. Immunol.*, 131, 1613, 1983.
156. **Organist, M. L., Harvey, J., McGillis, J. P., Mitsuhashi, M., Melera, P., and Payan, D. G.,** Characterization of a monoclonal antibody against the lymphoblast substance P receptor, *J. Immunol.*, 139, 3050, 1987.
157. **Payan, D. G., Brewster, D. R., Missirian-Bastian, A., and Goetzel, E. J.,** Substance P recognition by a subset of human T lymphocytes, *J. Clin. Invest.*, 14, 1532, 1984.
158. **McGillis, J. P., Organist, M. L., and Payan, D. G.,** Immunoaffinity purification of membrane protein constituents of the IM-9 lymphoblast SP receptor, *Anal. Biochem.*, 164, 502, 1987.
159. **Hartung, H. P., Wolters, K., and Toyka, K. V.,** Substance P: binding properties and studies on cellular responses in guinea pig macrophages, *J. Immunol.*, 136, 3856, 1986.
160. **Ruff, M. R., Wahl, S. M., and Pert, C. B.,** Substance P receptor-mediated chemotaxis of human monocytes, *Peptides*, 6(2), 107, 1985.
161. **Wiederman, C. J., Sertl, K., and Pert, C. B.,** Substance P receptors in rat spleen: characterization and autoradiographic distribution, *Blood*, 68, 1398, 1986.
162. **Wiedermann, C. J.,** Substance P receptors in bronchus-associated lymphoid tissue of rat, *Ann. N.Y. Acad. Sci.*, 496, 205, 1987.
163. **Geppetti, P., Maggi, C. A., Zecchi-Orlandini, S., Santicioli, P., Meli, A., Frilli, S., Spillantini, M. G., and Amenta, F.,** Substance P-like immunoreactivity in capsaicin-sensitive structures of the rat thymus, *Regul. Pept.*, 18, 321, 1987.
164. **Wagner, F., Fink, R., Hart, R., and Dancygier, H.,** Substance P enhances interferon-γ production by human peripheral blood mononuclear cells, *Regul. Pept.*, 19, 355, 1987.
165. **Eglezos, A., Helme, R. D., Dandie, G. W., Andrews, P. V., and Boyd, R. L.,** Substance P-mediated modulation of the primary antibody response, *Adv. Exp. Med. Biol.*, 237, 499, 1987.
166. **Helme, R. D., Eglezos, A., Dandie G. W., Andrews, P. V., and Boyd, R. L.,** The effect of substance P on the regional lymph node antibody response to antigenic stimulation in capsaicin-pretreated rats, *J. Immunol.*, 139, 3470, 1987.
167. **Fink, T., Nohr, D., Leibold, A., and Weihe, E.,** Peptidergic innervation of guinea-pig lymph nodes: chemoanatomical link for sensory and autonomic neuroimmunomodulation?, *Neuroscience*, 22(Suppl.), 5814, 1987.
168. **Popper, P., Mantyh, C. R., Vigna, S. R., Maggio, J. E., and Mantyh, P. W.,** The localization of sensory nerve fibers and receptor binding sites for sensory neuropeptides in canine mesenteric lymph nodes, *Peptides*, 9, 257, 1988.
169. **Weihe, E., Fink, T., and Moller, S.,** Substance P in nerves supplying the immune system: a messenger role of tachykinin in sensory neuroimmunomodulation? *Regul. Pept.*, 22, 186, 1988.
170. **Fink, T. and Weihe, E.,** Multiple neuropeptides in nerves supplying mammalian lymph nodes: messenger candidates for sensory and autonomic neuroimmunomodulation? *Neurosci. Lett.*, 90, 39, 1988.
171. **Schumann, M. A. and Gardner, P.,** Modulation of membrane K+ conductance in T-lymphocytes by substance P via a GTP-binding protein, *J. Membr. Biol.*, 111, 133, 1989.
172. **Stanfield, P. R., Nakajima, Y., and Yamaguchi, K.,** Substance P raises neuronal membrane excitability by reducing inward rectification, *Nature*, 315, 498, 1985.
173. **Nakajima, Y., Nakajima, S., and Inoue, M.,** Pertussis toxin-insensitive G protein mediates substance P-induced inhibition of potassium channels in brain neurons, *Proc. Natl. Acad. Sci. U.S.A.*, 85, 3643, 1988.
174. **Jonakait, G. M. and Schotland, S.,** Conditioned medium from activated splenocytes increases substance P in sympathetic ganglia, *J. Neurosci. Res.*, 26, 24, 1990.
175. **Kurkowski, R., Kummer, W., and Heym, C.,** Substance P-immunoreactive nerve fibers in tracheobronchial lymph nodes of the guinea pig: origin, ultrastructure and coexistence with other peptides, *Peptides*, 11, 13, 1990.

176. **Novotny, G. E. and Kliche, K. O.,** Innervation of lymph nodes: a combined silver impregnation and electron-microscopic study, *Acta Anat.*, 127, 243, 1986.
177. **Anderson, A. O. and Anderson, N. D.,** Studies on the structure and permeability of the microvasculature in the normal rat lymph nodes, *Am. J. Pathol.*, 80, 387, 1975.
178. **Anderson, A. O. and Anderson, N. D.,** Lymphocyte migration from high endothelial venules in rat lymph nodes, *Immunology,* 31, 731, 1976.
179. **Anderson, A. O., Anderson, N. D., Hoffman, R. D., and White, J. D.,** Cytoskeletal control of lymphocyte recirculation, *Fed. Proc.*, 37, 591, 1978.
180. **Gowans, J. L.,** The effect of continuous reinfusion of lymph and lymphocytes on the output of lymphocytes from the thoracic duct of unanaesthetized rats, *Br. J. Exp. Pathol.*, 38, 67, 1957.
181. **Gowans, J. L.,** The recirculation of lymphocytes from blood to lymph in the rat, *J. Physiol. (Lond.),* 145, 54, 1959.
182. **Gowans, J. L. and McGregor, D. D.,** Immunological activities of lymphocytes, *Prog. Allergy,* 9, 1, 1965.
183. **Ford, W. L.,** Lymphocyte migration and immune responses, *Prog. Allergy,* 19, 1, 1975.
184. **Lungdahl, A., Hokfelt, T., Nilsson, G., and Goldstein, M.,** Distribution of substance P-like immunoreactivity in the central nervous system of the rat. II. Light microscopic localization in relation to catecholamine-containing neurons, *Neuroscience,* 3, 945, 1978.
185. **Bannon, M. J., Lee, J.-M., Giraud, P., Young, A., Affolter, H.-U., and Bonner, T. I.,** Dopamine antagonist haloperidol decreases substance P, substance K, and preprotachykinin mRNAs in rat striatonigral neurons, *J. Biol. Chem.*, 261, 6640, 1986.
186. **Oblin, A., Zivkovic, B., and Bartholini, G.,** Involvement of the D-2 dopamine receptor in the neuroleptic-induced decrease in nigral substance P, *Eur. J. Pharmacol.*, 105, 175, 1984.
187. **Tamiya, R., Hanada, M., Kawai, Y., Inagaki, S., and Takagi, H.,** Substance P afferents have synaptic contacts with dopaminergic neurons in the ventral tegmental area of the rat, *Neurosci. Lett.*, 110, 11, 1990.
188. **Bolam, J. P. and Smith, Y.,** The GABA and substance P input to dopaminergic neurones in the substantia nigra of the rat, *Brain Res.*, 529, 57, 1990.
189. **Reid, M. S., Hokfelt, T., Herrera-Marschitz, M., Hakanson, R., Feng, D. M., Folkers, K., Goldstein, M., and Ungerstedt, U.,** Intranigral substance P stimulation of striatal dopamine release is inhibited by Spantide II: a new tachykinin antagonist without apparent neurotoxicity, *Brain Res.*, 532, 175, 1990.
190. **Hokfelt, T., Ljungdahl, A., Steinbusch, H., Verhofstad, A., Nilson, G., Brodin, E., Pernow, B., and Goldstein, M.,** Immunohistochemical evidence of substance P-like immunoreactivity in some 5-hydroxytryptamine containing neurons in the rat central nervous system, *Neuroscience,* 3, 517, 1978.
191. **Chan-Palay, V., Jonsson, G., and Palay, S. L.,** Serotonin and substance P coexist in neurons of the rat's central nervous system, *Proc. Natl. Acad. Sci. U.S.A.*, 75, 1582, 1978.
192. **Pelletier, G., Steinbusch, H. W. M., and Verhofstad, A. A. J.,** Immunoreactive substance P and serotonin present in the same dense-core vesicles, *Nature,* 293, 71, 1981.
193. **Fried, G., Franck, J., and Brodin, E.,** Differential distribution of 5-hydroxytryptamine and substance P in synaptosomal vesicles of rat ventral spinal cord, *Neurosci. Lett.*, 91, 315, 1988.
194. **Fried, G., Franck, J., Brodin, E., Born, W., Fischer, J. A., Hiort, W., and Hokfelt, T.,** Evidence for differential storage of calcitonin gene-related peptide, substance P and serotonin in synaptosomal vesicles of rat spinal cord, *Brain Res.*, 499, 315, 1989.
195. **Agnati, L. F., Fuxe, K., Benefenati, F., and Zini, I.,** On the functional role of coexistence of 5-HT and substance P in bulbospinal 5-HT neurons. Substance P reduces affinity and increases density of ^{3}H-5-HT binding sites, *Acta Physiol. Scand.*, 117, 299, 1983.
196. **Khalil, Z. and Helme, R. D.,** Serotonin modulates substance P-induced plasma extravasation and vasodilatation in rat skin by an action through capsaicin-sensitive primary afferent nerves, *Brain Res.*, 527, 292, 1990.
197. **Meller, S. T., Lewis, S. J., Ness, T. J., Brody, M. J., and Gebhart, G. F.,** Neonatal capsaicin treatment abolishes the nociceptive responses to intravenous 5-HT in the rat, *Brain Res.*, 542, 212, 1990.
198. **De Biasi, S. and Rustioni, A.,** Glutamate and substance P coexist in primary afferent terminals in the superficial laminae of spinal cord, *Proc. Natl. Acad. Sci. U.S.A.*, 85, 7820, 1988.
199. **Womack, M. D., MacDermott, A. B., and Jessell, T. M.,** Sensory transmitters regulate intracellular calcium in dorsal horn neurons, *Nature,* 334, 351, 1988.
200. **Woolf, C. and Wiesenfeld-Hallin, A.,** Substance P and calcitonin gene-related peptide synergistically modulate the gain of nociceptive flexor withdrawal reflex in the rat, *Neurosci. Lett.*, 60, 226, 1986.
201. **MacDermott, A. B., Mayer, M. L., Westbrook, G. L., Smith, S. J., and Barker, J. L.,** NMDA-receptor activation increases cytoplasmic concentration in cultured spinal cord neurons, *Nature,* 321, 519, 1986.
202. **Kangrga, I. and Randic, M.,** Tachykinins and calcitonin gene-related peptide enhance release of endogenous glutamate and aspartate from the rat spinal dorsal horn slice, *J. Neurosci.*, 10, 2026, 1990.

203. **Randic, M., Hecimovic, H., and Ryu, P. D.,** Substance P modulates glutamate-induced currents in acutely isolated rat spinal dorsal horn neurones, *Neurosci. Lett.*, 117, 74, 1990.
204. **Kangrga, I., Larew, J. S. A., and Randic, M.,** The effects of substance P and calcitonin gene-related peptide on the efflux of endogenous glutamate and aspartate from the rat spinal dorsal horn *in vitro, Neurosci. Lett.*, 108, 155, 1990.
205. **Kangrga, I. and Randic, M.,** Tachykinins and calcitonin gene-related peptide enhance release of endogenous glutamate and aspartate from the rat spinal dorsal horn slice, *J. Neurosci.*, 10, 2026, 1990.
206. **Somer, D. L. and Beckstead, R. M.,** Striatal preprotachykinin and preproenkephalin mRNA levels and the levels of nigral substance P and pallidal Met^5-enkephalin depend on corticostriatal axons that use the excitatory amino acid neurotransmitters aspartate and glutamate: quantitative radioimmunocytochemical and in situ hybridization evidence, *Mol. Brain Res.*, 8, 143, 1990.
207. **Hartung, H. P. and Toyka, K. V.,** Activation of macrophages by substance P: induction of oxidative burst and thromboxane release, *Eur. J. Pharmacol.*, 89, 301, 1983.
208. **Hartung, H. P., Heininger, K., Schafer, B., and Toyka, K. V.,** Substance P and astrocytes: stimulation of the cyclooxygenase pathway of arachidonic acid metabolism, *FASEB J.*, 2, 48, 1988.
209. **Marriott, D. R., Wilkin, G. P., and Wood, J. N.,** Substance P-induced release of prostaglandins from astrocytes: regional specialisation and correlation with phosphoinositol metabolism, *J. Neurochem.*, 56, 259, 1991.
210. **Cholewinski, A. J., Hanley, M. R., and Wilkin, G. P.,** A phosphoinositide response in astrocytes: evidence for regional heterogeneity, *Neurochem. Res.*, 13, 389, 1988.
211. **Skofitsch, G., Donnerer, J., Petronijevic, S., Saria, A., and Lembeck, F.,** Release of histamine by neuropeptides from the perfused rat hindquarter, *Naunyn Schmiedeberg's Arch. Pharmacol.*, 322, 153, 1983.
212. **Heavey, D. J., Fuller, R. W., Barnes, P. J., Ind, P. W., Brown, M. J., and Dollery, C. T.,** Release of histamine by intradermal injection of substance P in man, in *Proc. 9th Int. Congress Pharmacology,* 1984, 1067.
213. **Stach, W.,** Morphologisch Beziehungen zwischen Mastzellen und vegetativer Endformation, *Z. Mikrosk. Anat. Forsch.*, 67, 257, 1961.
214. **Heine, H. and Forster, F. J.,** Zur morphologie der Beziehungen zwischen Mast- und Muskelzellen bei Saugetieren, *Acta. Anat.*, 89, 387, 1974.
215. **Newson, B., Dahlstrom, A., Enerback, L., and Ahlman, H.,** Suggestive evidence for a direct innervation of mucosal mast cells. An electromicroscopic study, *Neuroscience,* 10, 565, 1983.
216. **Weisner-Henzel, L., Schulz, B., Vakilzadeh, F., and Czarnetzki, B. M.,** Electron microscopial evidence for a direct contact between nerve fibres and mast cells, *Acta Dermatovener (Stockholm),* 61, 465, 1981.
217. **Skofitsch, G., Savitt, J. M., and Jacobowitz, D. M.,** Suggestive evidence for a functional unit between mast cells and substance P fibres in the rat diaphragm and mesentery, *Histochemistry,* 2, 5, 1985.
218. **Kiernan, J. A.,** The involvement of mast cells in vasodilatation due to axon reflexes in injured skin, *Q. J. Exp. Physiol.*, 57, 311, 1972.
219. **Johnson, A. R. and Erdos, E. G.,** Release of histamine from mast cells by vasoactive peptides, *Proc. Soc. Exp. Biol. Med.*, 142, 1253, 1973.
220. **Fewtrell, C. M. S., Foreman, J. C., Jordan, C. C., Oehme, P., Renner, H., and Stewart, J. M.,** The effects of substance P on histamine and 5-hydroxytryptamine release in rat, *J. Physiol. (Lond.),* 330, 393, 1982.
221. **Foreman, J. C., Jordan, C. C., Oehme, P., and Renner, H.,** Structure-activity relationships for some substance P-related peptides that cause wheal and flare reactions in human skin, *J. Physiol. (Lond.),* 335, 449, 1983.
222. **Hagermark, O., Hokfelt, T., and Pernow, B.,** Flare and itch induced by substance P in human skin, *J. Invest. Dermatol.*, 71, 233, 1978.
223. **Coutts, A. A., Jorizzio, J. L., Greaves, M. W., and Burnstock, G.,** Mechanism of vasodilatation due to substance P in humans, *Br. J. Dermatol.*, 105, 354, 1981.
224. **Tani, E., Senba, E., Kokumai, S., Masuyama, K., Ishikawa, T., and Tohyama, M.,** Histamine application to the nasal mucosa induces release of calcitonin gene-related peptide and substance P from peripheral terminals of trigeminal ganglion: a morphological study in the guinea pig, *Neurosci. Lett.*, 112, 1, 1990.
225. **Costa, M.,** Tachykinins in the sensory and peripheral nervous system, *Regul. Pept.*, 22, 3, 1988.
226. **Brain, S. D. and Williams, T. J.,** Substance P regulates the vasodilator activity of calcitonin gene-related peptide, *Nature,* 335, 73, 1988.
227. **Goltz, F.,** Uber Gefasserweiternde Nerven, *Pfluger's Arch. Gen. Physiol.*, 9, 174, 1874.
228. **Stricker, S.,** Untersuchungen Uber die Gefassnerven-Wurzeln des Ischiadicus, *Sitzb Kaiserl Akad Wissensch Wien* LXXIV Ab III, 3, 173, 1876.

229. **Bayliss, W. M.**, On the origin from the spinal cord of the vasodilator fibres of the hind limb and on the nature of these fibres, *J. Physiol. (Lond.)*, 26, 173, 1901.
230. **Bruce, A. N.**, Uber die Beziehung der sensiblen Nervenendigungen zum Entzundungsvorgang, *Arch. Exp. Pathol. Pharmacol.*, 63, 424, 1910.
231. **Bruce, A. N.**, Vaso-dilator axon reflexes, *Q. J. Exp. Physiol.*, 6, 339, 1913.
232. **Breslauer, F.**, Die Pathogenese der trophischen Gewebsschaden nach der Nervenverletzung, *Dt. Z. Chir.*, 150, 50, 1919.
233. **Hinsey, J. C. and Gasser, H. S.**, The component of the dorsal root mediating vasodilatation and the Sherrington contracture, *Am. J. Physiol.*, 92, 679, 1930.
234. **Jancso, N.**, Role of the nerve terminals in the mechanism of inflammatory reactions, *Bull. Millard Filmore Hosp. (Buffalo)*, 7, 53, 1960.
235. **Jancso, N.**, Inflammation and the inflammatory mechanisms, *J. Pharm. Pharmacol.*, 13, 577, 1961.
236. **Jancso, N., Jancso-Gabor, A., and Takats, I.**, Pain and inflammation induced by nicotine, acetylcholine and structurally related compounds and their prevention by desensitizing agents, *Acta Physiol. Hung.*, 19, 113, 1961.
237. **Jancso, N. and Jancso-Gabor, A.**, Neurogenic inflammatory responses and their pharmacological analysis, in *2nd Int. Pharmacol. Meeting, Prague, Biochemical Pharmacology,* Conference issue, Abstract No. 605, Pergamon Press, Oxford, England, 1963.
238. **Jancso, N.**, Neurogenic inflammatory responses, *Acta Physiol. Hung.*, 24 (Suppl.), 3, 1964.
239. **Lotz, M., Vaughan, J. H., and Carson, D. A.**, Effect of neuropeptides on production of inflammatory cytokines by human monocytes, *Science,* 241, 1218, 1988.
240. **Serra, M. C., Bazzoni, F., Bianca, V. D., Greskowiak, M., and Rossi, F.**, Activation of human neutrophils by substance P, *J. Immunol.*, 141, 2118, 1988.
241. **Alber, G., Scheuber, P. H., Reck, B., Sailer-Kramer, B., Hartmann, A., and Hammer, D. K.**, Role of substance P in immediate-type skin reactions induced by staphylococcal enterotoxin B in unsensitized monkeys, *J. Allergy Clin. Immunol.*, 84, 880, 1989.
242. **Khalil, Z. and Helme, R. D.**, Sequence of events in substance P-mediated plasma extravasation in rat skin, *Brain Res.*, 500, 256, 1989.
243. **O'Byrne, E. M., Blancuzzi, V. J., Wilson, D. E., Wong, M., Peppard, J., Simke, J. P., and Jeng, A. Y.**, Increased intra-articular substance P and prostaglandin E_2 following injection of interleukin-1 in rabbits, *Int. J. Tissue React.*, 12, 11, 1990.
244. **Schaible, H.-G., Jarrott, B., Hope, P. J., and Duggan, A. W.**, Release of immunoreactive substance P in the spinal cord during development of acute arthritis in the knee joint of the cat: a study with antibody microprobes, *Brain Res.*, 529, 214, 1990.
245. **Kurkcuoglu, N. and Cakar, N.**, Substance-P immunoreactivity in active edges of psoriatic plaques, *Clin. Exp. Dermatol.*, 1(15), 467, 1990.
246. **Khalil, Z. and Helme, R. D.**, Modulation of a peripheral inflammatory response to substance P by locally administered opioid receptor agonists, *Neuropeptides,* 17, 45, 1990.
247. **Hayes, A. G. and Tyers, M. B.**, Effects of intrathecal and intracerebroventricular injections of substance P on nociception in the rat and mouse, *Br. J. Pharmacol.*, 66, 488, 1979.
248. **Hecht, K., Oehme, P., Wachtel, J., Pawlownaljowschina, I., Poppei, M., and Airapetjanze, M.**, Lernen und Gedachtnis von Ratten, *Pharmazie,* 38, 208, 1983.
249. **Wetzel, W. and Matthies, H.**, Effect of substance P on the retention of a brightness discrimination task in rats, *Acta Biol. Med. Germanica,* 41, 647, 1982.
250. **Huston, J. and Staubli, U.**, Post-trial injection of substance P into lateral hypothalamus and amygdala, respectively, facilitates and impairs learning, *Behav. Neural Biol.*, 27, 244, 1979.
251. **Huston, J. and Staubli, U.**, Retrograde amnesia produced by post-trial injection of substance P into substantia nigra, *Brain Res.*, 159, 468, 1978.
252. **Staubli, U. and Huston, J.**, Differential effects on learning by ventromedial vs. lateral hypothalamic injection of substance P, *Pharmacol Biochem. Behav.*, 10, 783, 1979.
253. **Staubli, U. and Huston, J.**, Facilitation of learning by post-trial injection of substance P into the medial septal nucleus, *Behav. Brain Res.*, 1, 245, 1980.
254. **Gaffori, O., Stewart, J., and deWied, D.**, Influence of substance P and its fragments on passive avoidance behavior, *Experientia,* 14, 89, 1984.
255. **Oitzl, M.-S., Hasenohrl, R. U., and Huston, J. P.**, Reinforcing effects of peripherally administered substance P and its C-terminal sequence $pGlu^6$-SP6-11 in the rat, *Psychopharmacology,* 100, 308, 1990.
256. **Hasenohrl, R. U., Gerhardt, P., and Huston, J. P.**, Substance P enhancement of inhibitory avoidance learning: mediation by the N-terminal sequence, *Peptides,* 11, 163, 1990.
257. **Schlesinger, K., Lipsitz, D., Peck, P., Pelleymounter, M., Stewart, J., and Chase, T.**, Substance P enhancement of passive and active avoidance conditioning in mice, *Pharmacol. Biochem. Behav.*, 19, 655, 1983.

258. **Schlesinger, K., Lipsitz, D., Peck, P., Pelleymounter, M., Stewart, J., and Chase, T.,** Substance P reversal of electroconvulsive shock and cycloheximide-induced retrograde amnesia, *Behav. and Neural Biol.,* 39, 30, 1983.
259. **Schlesinger, K., Pelleymounter, M. A., Van De Kamp, J., Bader, D. L., Stewart, J. M., and Chase, T. N.,** Substance P facilitation of memory: effects in an appetitively motivated learning task, *Behav. Neural Biol.,* 45, 230, 1986.
260. **Pelleymounter, M. A., Schlesinger, K., Wehner, J., Hall, M. E., and Stewart, J. M.,** Nigral 5-HT and substance P-induced enhancement of passive avoidance retention, *Behav. Brain Res.,* 29, 159, 1988.
261. **Nagel, J. A. and Huston, J. P.,** Enhanced inhibitory avoidance learning produced by post-trial injections of substance P into the basal forebrain, *Behav. Neural Biol.,* 49, 374, 1988.
262. **Ferrier, I. N., Cross, A. J., Johnson, J. A., et al.,** Neuropeptides in Alzheimer type dementia, *J. Neurol. Sci.,* 62, 159, 1983.
263. **Yates, C. M., Harmr, A. J., Rosie, R., et al.,** Thyrotropin-releasing hormone, luteinizing hormone releasing hormone and substance P immunoreactivity in postmortem brain from cases of Alzheimer-type dementia and Down's Syndrome, *Brain Res.,* 258, 45, 1983.
264. **Beal, M. F. and Mazurek, M. F.,** Substance P-like immunoreactivity is reduced in Alzheimer's disease cerebral cortex, *Neurology,* 37, 1205, 1987.
265. **Quigley, B. J., Jr. and Kowall, N. W.,** Substance P-like immunoreactive neurons are depleted in Alzheimer's disease cerebral cortex, *Neuroscience,* 41, 41, 1991.
266. **Sakurada, T., Alufuzoff, I., Winblad, B., and Nordberg, A.,** Substance P-like immunoreactivity, choline acetyltransferase activity and cholinergic muscarinic receptors in Alzheimer's disease and multi-infarct dementia, *Brain Res.,* 521, 329, 1990.
267. **Eison, A. S., Iversen, S. D., Sandberg, E. B., Watson, S. P., Hanley, M. R., and Iversen, L. L.,** A novel substance P analogue, DiMe-C7: evidence for stability in the rat brain and prolonged control actions, *Science,* 215, 188, 1982.
268. **Iversen, S. D.,** Behavioral effects of substance P through dopaminergic pathways in the brain. Substance P in the nervous system, Ciba Foundation Symposium 91, Pitman, London, 1982, 307.
269. **Iversen, S. D.,** Brain dopamine systems and behaviour, in *Handbook of Psychopharmacology,* Vol. 8, Iversen, S. D. et al., Eds., Plenum Press, New York, 1977, 333.
270. **Iversen, S. D.,** Striatal functions and sterotyped behaviour, in *Psychobiology of the Striatum,* Cools, A., et al., Eds., North-Holland, Amsterdam, 1977, 99.
271. **Iversen, S. D.,** Behaviour after neostriatal lesions in animals, in *The Neostriatum,* Divac, I. and Oberg, G., Eds., Pergamon Press, Oxford, 1979, 195.
272. **Iversen, S. D.,** Brain chemistry and behaviour, *Psychol. Med.,* 10, 527, 1980.
273. **Kelley, A. E. and Iversen, S. D.,** Behavioural response to bilateral injection of substance P into the substantia nigra of the rat, *Brain Res.,* 158, 474, 1978.
274. **Stoessl, A. J., Dourish, C. T., and Iversen, S. D.,** The NK-3 tachykinin receptor agonist senktide elicits 5-HT-mediated behaviour following central or peripheral administration in mice and rats, *Br. J. Pharmacol.,* 94, 285, 1988.
275. **Stoessl, A. J., Dourish, C. T., Young, S., Williams, B. T., Iversen, S. D., and Iversen, L. L.,** Senktide, a selective neurokinin B-like agonist, elicits serotonin-mediated behaviour following intracisternal administration in the mouse, *Neurosci. Lett.,* 80, 321, 1987.
276. **Stoessl, A. J., Dourish, C. T., and Iversen, S. D.,** Behavioural evidence for CNS tachykinin-cholinergic interactions, *Regul. Peptides,* 22, 178, 1988.
277. **Papir-Kricheli, D., Gilon, C., Chorev, M., Selinger, Z., and Devor, M.,** Neurochemical mediators of the behavioural effects of receptor-selective substance P agonists administered intrathecally in the rat, *Neuropharmacology,* 29, 1055, 1990.
278. **Papir-Kricheli, D., Frey, J., Laufer, R., Gilon, C., Chorev, M., Selinger, Z., and Devor, M.,** Behavioral effects of receptor-specific substance P agonists, *Pain,* 31, 263, 1987.

Chapter 11

NEUROTENSIN

I. INTRODUCTION

Neurotensin was isolated in 1973, by Carraway and Leeman,[1] from a bovine hypothalamic fraction in the process of attempts to purify substance P. It was a bioactive principle that eluted before substance P activity on ion-exchange chromatography. Because of its vasoactive properties and its location in neural tissue, it was named "neurotensin". It contains 13 amino acid residues (Glu-Leu-Tyr-Glu-Asn-Lys-Pro-Arg-Arg-Pro-Tyr-Ile-Leu). The neurotensin amino acid sequence does not appear within that of other neuroendocrine or neurovascular substances or "families". Only the amphibian peptide zenopsin (isolated from skin extracts of *Xenopus laevis*, the African clawed frog) seems to resemble neurotensin in structure. The five-amino-acid sequence at the C-terminal end of xenopsin has been reported by Said[2] to be quite similar to that of neurotensin. The phylogenetic antiquity of neurotensin has been established by its identification in the primitive nervous system nerve net of the cnidarian coelenterate *Hydra*, by Grimmelikhuijzen et al.[3] (Figure 1).

Neurotensin produces a number of major systemic and localized actions, such as hypothermia, hypotension, vasodilatation, increased permeability, and hyperglycemia. The effects on peripheral administration appear directed at glucose metabolism. There are high-affinity binding sites for neurotensin on brain membranes, synaptosomes, and mast cells.

Vascular and immune-response-system-related actions of neurotensin include high-affinity binding of neurotensin to mast cells in the periphery, chemotactic action on human neutrophils, and capillary permeability increases in rats. Quirion et al.,[4] in 1980, suggested mast cell degranulation and histamine liberation to be responsible for the systemic hypotensive effects of neurotensin. They found that the hypotensive effect of neurotensin could be blocked by antihistamines, by prior depletion of mast cell monoamines by compound 48/80, or by use of disodium cromoglycate to prevent mast cell release of histamine. The vascular effects are complex.[5] Neurotensin-induced increase in capillary permeability, observed by Chahl[6] in 1979, also is likely associated with the histamine release from mast cells response.

The hypothalamus, thalamus, cortex, and pituitary intracranially, and the small intestine of rat and calf, have been found to be rich in neurotensin. The greatest portion of the body's neurotensin is located in the gastrointestinal tract in neuroendocrine cells of the gut mucosa, called N cells. These cells predominate in the gut mucosal villi. There is a paucity of neurotensin-containing neurons in the periphery. Neurotensin-containing nerve fibers are found in the dorsal-horn substantia gelatinosa layer of the spinal cord. Seybold and Elde,[7] in 1980, demonstrated that section of the dorsal root did not alter this distribution of neurotensin-containing nerve fibers in the spinal cord and concluded that neurotensin is not contained in primary afferent nerve fibers arriving at this pain-sensitive area of the spinal cord.

A neuroinhibitory effect of neurotensin was observed by Young et al.[8] in 1977. They found that neurotensin inhibits the spontaneous and glutamate-evoked firing of locus coeruleus neurons. In addition, Clindeschmidt and McGuffin,[9] in 1977, reported that neurotensin decreased reactivity to painful stimuli when it was injected into the cisterna magna.

Prostaglandin system influences on neurotensin action may be identified by inhibitors of PG synthesis (indomethacin, aspirin, mefenamic acid) that markedly inhibit the dose-dependent contraction of isolated rat portal vein produced by neurotensin.[10] The authors

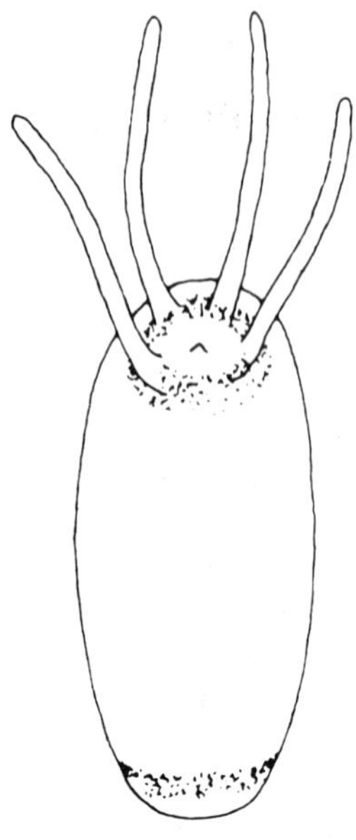

FIGURE 1. Distribution of neurotensin in the cnidarian coelenterate *Hydra*. (After Grimmelikhuijzen et al.[3])

concluded that receptors on rat portal vein for the neurotensin response appeared to depend on a functioning prostaglandin biosynthetic pathway.

Neurotensin has been reported (by Nemeroff et al.,[11] in 1977) to have a neuroleptic-like action in producing hypothermia and in potentiating the action of barbiturates and ethanol in sedation and hypothermia. They also demonstrated that neurotensin antagonizes amphetamine-induced increases in locomotor activity. McCarthy et al.,[12] in 1979, found that microiontophoretic application of neurotensin decreased the activity of dopamine-sensitive neurons in the nucleus accumbens. Further information regarding a dopamine-related activity of neurotensin was published by Haubrich et al.[13] in 1982. They showed that neurotensin inhibits the synthesis of dopamine within nerve terminals and is able to reduce dopamine content in the striatum.

Nemeroff and associates,[14] at Duke University in 1986, reported an important study relating to neurotensin actions involving the dopaminergic systems of the central nervous system. They observed that central administration of neurotensin by injection intracisternally suppressed locomotor activity induced by *d*-amphetamine, but not that induced by scopolamine or caffeine. They interpreted their findings as indicating that neurotensin modulates the activity of the mesolimbic (but not the nigrostriatal) dopaminergic system and that neurotensin does not simply produce a nonspecific impairment of locomotor activity.

On the pain front, an interesting report appeared in 1987 from the University of California–San Francisco, by Fang et al.,[15] describing a dose-dependent antinociceptive effect of neurotensin microinjected into the rostroventromedial medulla of the rat. They cited earlier behavioral studies in which injection of neurotensin into the central nervous system inhibited mouse and rat hot-plate and tail-flick responses.[16-18] In their 1987 study, microinjection of neurotensin into the rostroventromedial medulla produced dose-dependent inhibition of the tail-flick reflex response to noxious thermal heat (42 to 45°C).

Another 1987 study from the University of Dusseldorf (Germany), by Mai et al.,[19] addressed the anatomical location of neurotensin in the human brain. A widespread, but uneven, distribution of neurotensin-immunoreactive cells and processes was found throughout the central nervous system, with prominence in the medial hypothalamus and various regions of the limbic system, including the amygdaloid body, the septal area, the bed nucleus of the stria terminalis, and the piriform cortex. This human distribution of neurotensin-immunoreactive cell bodies and processes corresponded well with that reported for rodents. By comparison mapping, they demonstrated an anatomical relation between almost all the catecholaminergic cell clusters and the neurotensin-containing fibers.

A 1989 report from the Karolinska Institute of Stockholm, by von Euler et al.,[20] described reduction in the binding of dopamine D-2 receptors in the rat forebrain by intraventricular injection of neurotensin. Significant reduction in binding occurred in both the striatum and in the nucleus accumbens. This study demonstrated an *in vivo* effect of neurotensin on the binding of D-2 receptors. Neurotensin mRNA in the rat forebrain, with unexpected abundance in the hippocampus and subiculum, was reported in 1989 by Alexander et al.[21]

A 1990 publication from Paris, by Masuo et al.,[22] described the effects of unilateral lesion of the rat substantia nigra with 6-hydroxydopamine on the endogenous contents of neurotensin and its binding-site densities in the striatum and the substantia nigra. They observed dramatic increases in neurotensin contents in both the ipsi- and contralateral striata, with no changes in nigral neurotensin levels on either side. Neurotensin binding sites decreased in the ipsilateral striatum and substantia nigra, a response that reflected the destruction of dopamine neuronal elements in these areas. The authors regarded these findings as suggesting strongly a dopaminergic control of striatal neurotensin systems and an indication that unilateral loss of this control may lead to major bilateral alterations in neurotensin levels.

Another 1990 report from Washington State University, by Kalivas and associates,[23] provides biochemical evidence that neurotensin may bind directly to dopamine and that this action may be at least partly responsible for observed antagonism by neurotensin of dopamine-induced locomotor activity. This finding suggests that dopamine-antagonist-like effects of neurotensin may not be receptor mediated. These observations are of importance, as nearly 100% of the neurotensin-containing neurons in the A10 dopamine region of the ventromedial mesencephalon of the rat are colocalized with dopamine,[24,25] and these neurons project to the medial nucleus accumbens and the medial prefrontal cortex.[26,27]

Although all of the above information places neurotensin in the inhibitory and activity-depressant column, little is known of its activity with respect to the immune response system. The acute infusion by the author[28-30] into cannulated afferent lymphatics of sheep popliteal lymph nodes (in the author's supported-sheep lymphocyte traffic study model) produces prompt, sharp, and sustained falls in the traffic of small recirculating lymphocytes (as monitored by the output into nodal efferent lymph and lymphatic of these small recirculating lymphocytes) (Figure 2). The output of nonrecirculating blast lymphocytes also was reduced sharply. The reduction persisted up to 6 h. The sheep in these studies were neither anesthetized nor sedated. These findings, in comparison with other inhibitory vasoactive neurotransmitter molecules, were interpreted as immunosuppressive/immunomodulatory (Figure 3).

II. CONCLUSION

The neuroleptic, sedation-potentiating, hypothermia-inducing, antinociceptive, and lymphocyte-traffic-depressing effects of neurotensin rather clearly place this neuropeptide molecule in the inhibitory and modulatory category. The phylogenetic antiquity of this molecule, and its extensive distribution throughout the central nervous system, suggest a role of importance for this molecule. Continued in-depth study of the role and function of neurotensin in the nervous, vascular, and immune response systems clearly is in order.

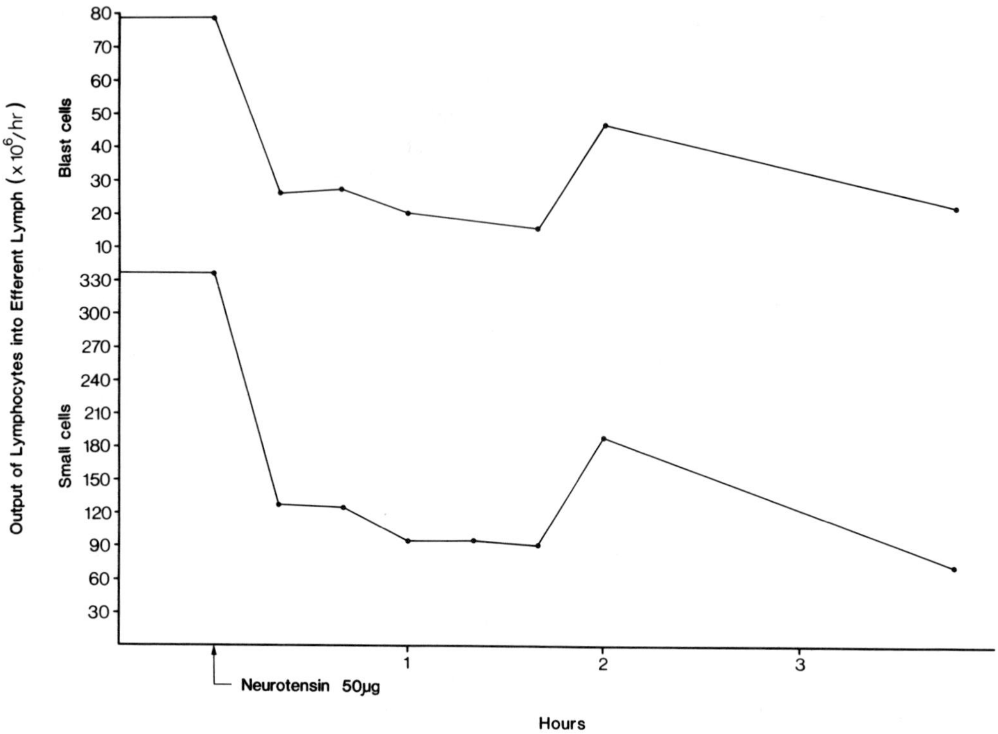

FIGURE 2. Acute infusion of neurotensin (50 μg) into cannulated afferent lymphatics of sheep popliteal lymph nodes produces a prompt, sharp, and marked decrease in the output of both small recirculating and nonrecirculating blast lymphocytes into the efferent lymphatic of the study popliteal lymph node, chronically cannulated. This depression in lymphocyte traffic was interpreted as immunomodulatory/suppressive. (From Moore, T. C., Ph.D. thesis, University of Cambridge, 1982. With permission.)

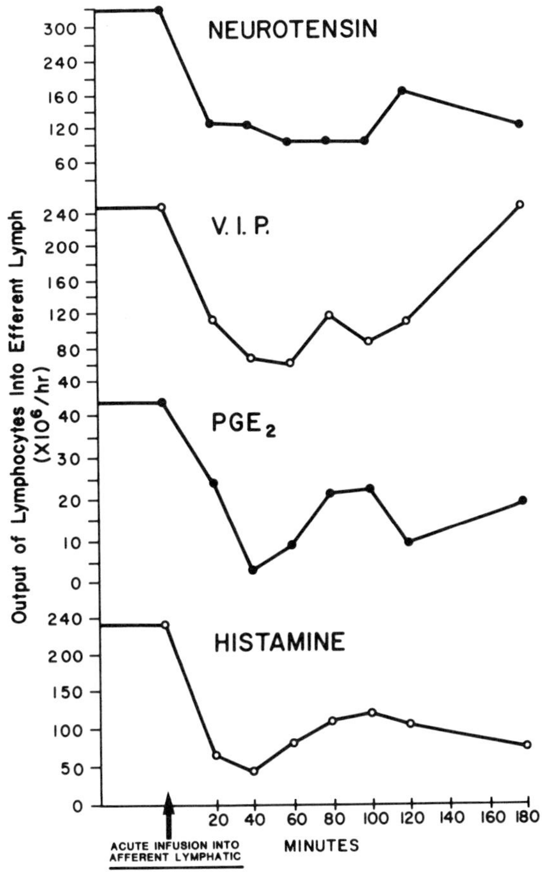

FIGURE 3. Neurotensin-induced depressions in lymphocyte traffic/outputs compared with other molecules such as vasoactive intestinal peptide (VIP). The histamine depressant effect was based on a limited number of studies of short duration in hours and requires considerable additional study. (From Moore, T. C., in *Immunology of the Sheep*, Editiones Roche, Basel, 1985. With permission.)

REFERENCES

1. **Carraway, R. and Leeman, S. E.,** The isolation of a new hypotensive peptide, neurotensin, from bovine hypothalami, *J. Biol. Chem.*, 248, 6854, 1973.
2. **Said, S. I.,** Peptides common to the nervous system and the gastro-intestinal tract, in *Frontiers in Neuroendocrinology*, Vol. 6, Martini, L., and Ganong, W. F., Eds., Raven Press, New York, 1980b, 293.
3. **Grimmelikhuijzen, C. J. P., Balfe, A., Emson, P. C., Powell, D., and Sundler, F.,** Substance P-like immunoreactivity in the nervous system of *Hydra, Histochemistry*, 71, 325, 1981.
4. **Quirion, R., Rioux, F., Regoli, D., and St. Pierre, S.,** Compound 48/80 inhibits neurotensin-induced hypotension in rats, *Life Sci.*, 27, 1889, 1980a.
5. **Bissette, G., Manberg, P., Nemeroff, C. B., and Prange, A. J., Jr.,** Neurotensin, a biologically active peptide, *Life Sci.*, 23, 2173, 1978.
6. **Chahl, L. A.,** The effect of putatative peptide neurotransmitters on cutaneous vascular permeability in the rat, *Naunyn Schmiedeberg's Arch. Pharmacol.*, 309, 159, 1979.
7. **Seybold, V. and Elde, R.,** Immunohistochemical studies of peptidergic neurons in the dorsal horn of the spinal cord, *J. Histochem. Cytochem.*, 28, 367, 1980.

8. **Young, W. S., III, Uhl, G. R., and Kuhar, M. J.,** Iontophoresis of neurotensin in the area of the locus coereleus, *Brain Res.,* 150, 431, 1978.
9. **Clineschmidt, B. V. and McGuffin, J. C.,** Neurotensin administered intracisternally inhibits responsiveness of mice to noxious stimuli, *Eur. J. Pharmacol.,* 46, 395, 1977.
10. **Rioux, F., Quirion, R., Leblanc, M. A., Regoli, D., and St. Pierre, S.,** Possible interactions between neurotensin and prostaglandins in the isolated rat portal vein, *Life Sci.,* 27, 259, 1980.
11. **Nemeroff, C. B., Bissette, G., Prange, A. J., Loosen, P. T., Barlow, T. S., and Lipton, M. A.,** Neurotensin: central nervous system effects of a hypothalamic peptide, *Brain Res.,* 128, 485, 1977.
12. **McCarthy, P. A., Walker, R. J., Yajima, H., Kitagawa, K., and Woodruff, G. N.,** The action of neurotensin on neurones in the nucleus accumbens and cerebellum of the rat, *Gen. Pharmacol.,* 10, 331, 1979.
13. **Haubrich, D. R., Martin, G. E., Pflueger, A. B., and Williams, M.,** Neurotensin effects on brain dopaminergic systems, *Brain Res.,* 231, 216, 1982.
14. **Skoog, K. M., Cain, S. T., and Nemeroff, C. B.,** Centrally administered neurotensin suppresses locomotor hyperactivity induced by *d*-amphetamine but not by scopolamine or caffeine, *Neuropharmacology,* 25, 777, 1986.
15. **Fang, F. G., Moreau, J.-L., and Fields, H. L.,** Dose-dependent antinociceptive action of neurotensin microinjection into the rostroventromedial medulla of the rat, *Brain Res.,* 420, 171, 1987.
16. **Clineschmidt, B. V., Martin, G. E., and Veber, D. F.,** Antinocisponsive effects of neurotensin and neurotensin-related peptides, *Ann. N.Y. Acad. Sci.,* 400, 283, 1982.
17. **Clineschmidt, B. V., McGuffin, J. C., and Bunting, P. B.,** Neurotensin: antinocisponsive action in rodents, *Eur. J. Pharmacol.,* 54, 129, 1979.
18. **Martin, G. E., Naruse, T., and Papp, N. L.,** Antinociceptive and hypothermic actions of neurotensin administered centrally in the rat, *Neuropeptides,* 1, 447, 1981.
19. **Mai, J. K., Triepel, J., and Metz, J.,** Neurotensin in the human brain, *Neuroscience,* 22, 499, 1987.
20. **von Euler, G., Meister, B., Hokfelt, T., and Fuxe, K.,** Intraventricular injection of neurotensin reduces the binding of dopamine D-2 receptors in the rat forebrain, *Acta Physiol. Scand.,* 137, 309, 1989.
21. **Alexander, M. J., Miller, M. A., Dorsa, D. M., Bullock, B. P., Melloni, R. H., Jr., Dobner, P. R., and Leeman, S. E.,** Distribution of neurotensin/neuromedin N mRNA in rat forebrain: unexpected abundance in hippocampus and subiculum, *Proc. Natl. Acad. Sci. U.S.A.,* 86, 5202, 1989.
22. **Masuo, Y., Pelaprat, D., Montagne, M.-N., Scherman, D., and Rostene, W.,** Regulation of neurotensin-containing neurons in the rat striatum and substantia nigra. Effects of unilateral nigral lesion with 6-hydroxydopamine on neurotensin content and its binding site density, *Brain Res.,* 510, 203, 1990.
23. **Adachi, D. K., Kalivas, P. W., and Schenk, J. O.,** Neurotensin binding to dopamine, *J. Neurochem.,* 54, 1321, 1990.
24. **Hokfelt, T., Everitt, B. J., Theodorsson-Norheim, E., and Golstein, M.,** Occurrence of neurotensin-like immunoreactivity in subpopulations of hypothalamic mesencephalic, and medullary catecholamine neurons, *J. Comp. Neurol.,* 222, 543, 1984.
25. **Seroogy, K. B., Mehta, A., and Fallon, J. H.,** Neurotensin and cholecystokinin coexistence within neurons of the ventral mesencephalon: projections to forebrain, *Exp. Brain Res.,* 68, 277, 1987.
26. **Kalivas, P. W. and Miller, J. S.,** Neurotensin neurons in the ventral tegmental area project to the medial nucleus accumbens, *Brain Res.,* 300, 157, 1984.
27. **Studler, J. M., Kitabgi, P., Tramu, G., Herve, D., Glowinski, J., and Tassin, J. P.,** Extensive co-localization of neurotensin with dopamine in rat meso-cortico-frontal dopaminergic neurons, *Neuropeptides,* 11, 95, 1988.
28. **Moore, T. C.,** The effects of inflammatory mediators and anaesthetic agents upon lymphocyte traffic of sheep in vivo, Ph.D. thesis, University of Cambridge (Trinity College), England, 1982.
29. **Moore, T. C.,** Modification of lymphocyte traffic by vasoactive neurotransmitter substances, *Immunology,* 52, 511, 1984.
30. **Moore, T. C.,** Mediator and physiological influences on lymphocyte traffic in sheep, in *Immunology of the Sheep,* Morris, B. and Miyasaka, M., Eds., Editiones Roche, Basel, 1985, 327.

Chapter 12

BOMBESIN

I. INTRODUCTION

1970 was a vintage year for the discovery of bombesin and bombesin-like peptides. In this year, Erspamer and associates,[1] in Rome, Italy, discovered, isolated, and characterized bombesin, a 14-amino-acid-residue peptide, and alytesin from extracts of the skin of two European discoglossid frogs, *Bombina bombina* and *Alytes obstetricans*. These discoveries resulted from a long-term project by the Erspamer group for the screening of amphibian skin for bioactive peptides. This highly productive and creative project already had led to the discovery, in 1961, of eledosin and, later, of psysalaemin, phyllokinin, caerulin, phyllocaerulin, and phyllomedusin.

A few months earlier, Nakajima et al.[2] reported the discovery of an 11-amino-acid-residue vasoactive peptide in the skin of the American frog *Rana pipiens*. This blood-pressure-raising peptide was named ranatensin. On further study, this blood-pressure-altering peptide was found by this group[3] to have differential blood-pressure effects depending on the species of animal used in testing. Hypertensive effects were observed in dogs, rabbits, rats, and guinea pigs. Cats were unaffected, while frank lowering of blood pressure (hypotension) was observed in primates (monkeys). Early studies of structure and pharmacological actions demonstrated clearly that ranatensin and bombesin belong to the same vasoactive peptide family and initiated an intense search for the complete spectrum of biological activity of these peptides, as well as for other similar amphibian peptides and related peptide counterparts in mammalian tissues, especially brain and gut.

A truly extraordinary worldwide search for bombesin-like peptides in amphibian skin was initiated and carried on by the Italian group of Erspamer and the Japanese group of Nakajima. These two groups, working both independently and in collaboration, explored amphibian skin peptides from approximately 500 species worldwide. These massive efforts proved quite productive, as 13 different bombesin-like peptides were identified, characterized, and sequenced. They were divided into three subfamilies: three bombesins, seven litorin/ranatensins, and three phyllolitorins.

The concurrent search for mammalian counterparts of these unusually potent amphibian skin peptides (? venoms) proved to be equally productive, as these and other groups identified a series of mammalian counterparts in brain, gut, and other tissues. The Erspamer group of Italy had an impressive track record of isolating potent vasoactive neurotransmitter peptide substances from amphibian skin, which had important mammalian, primate, and human counterparts. These included psysalaemin (substance P), caerulein (gastrin/CCK), and bradykinin and phyllokinin (bradykinins). As cited earlier in this book, all three of these amphibian-mammalian peptides have been tracked back far beyond amphibians and vertebrates phylogenetically. Substance P has been identified in the cnidarian coelenterate *Hydra*, a member of the most primitive multicellular animal phyla with nervous systems (nerve "nets").

An important advance was achieved by McDonald, Bloom, Mutt, and associates,[4] who isolated, in 1978, a pig gastric fundus peptide that caused the release of gastrin. These investigators characterized this peptide and named it "gastrin-releasing peptide" (GRP). This 27-amino-acid-residue peptide clearly was a mammalian counterpart and member of the bombesin family of vasoactive neuropeptides, with gut as well as other tissue locations, as the carboxyl-terminal amino acid sequences of bombesin and GRP were identical in nine of ten amino acid residues and identical in eight of eight carboxyl-terminal residues.

In 1983 and 1984 Minamino et al.,[5,6] described the identification of two mammalian bombesin-like peptides found in porcine spinal cords: neuromedin B and neuromedin C. Neuromedin C, in amino acid sequence, is structurally identical to the carboxyl-terminal decapeptide fragment of GRP (GRP_{18-27}) and, by a 1987 nomenclature agreement, will be designated as GRP_{18-27}. Neuromedin B is a 32-amino-acid-residue peptide (NMB_{1-32}) that has an active decapeptide fragment, NMB_{23-32}, which has a relationship similar to that of GPP_{1-27} and GRP_{18-27}. Human and porcine NMB_{1-32} have identical NMB_{23-32} sequences at the decapeptide level, but differ at positions 15 and 21. From structural determinations, Minamino et al. suggest that there are two families of endogenous mammalian bombesin-like peptides in the NMC-GRP family and the NMB family.

II. EVOLUTION

In 1988 Renda et al.[7] described studies of bombesin-like immunoreactivity in the annelid roundworm *Lumbricus terrestris* (earthworm). Their studies were carried out on the heads (first six segments) of several species of earthworms. Bombesin-immunoreactive elements were found to be quite numerous in a narrow zone of the head that corresponded to the 3rd and 4th segments where both the so-called head (cerebral or supraesophageal ganglion) and the subesophageal ganglion with corresponding connectives are located. Bombesin-immunoreactive neuronal elements were discovered inside the cerebral ganglion, in neurons of varying size. The number of these neurons was greatest in the middle part of the ganglion. Bombesin immunoreactivity in this rather primitive animal was located largely in the neuronal cell body cytoplasma and was rarely encountered in the neuronal processes and fibers. Fibers, when seen, were most often identified in the most anterior parts of the cerebral ganglion. It was of particular interest that bombesin-immunoreactive epithelial elements were seen in the cuticular monolayer that covered the surface of these animals. The cuticular distribution appeared relatively standard and characteristic. In the more anterior sections, bombesin-immunoreactive epithelial elements were located on each dorsal half of the head cuticala where they formed a symmetrical cluster of 15 to 20 adjacent columnar cells. More caudally, the ventral numbers of immunoreactivity increased and converged in a regular, continuous semicircle at the end of the 4th segment. The authors interpreted these findings as suggesting that the bombesin-immunoreactive elements might be involved in some specific functions such as orientation, choice of food, locomotion, and other regulatory actions.

A 1982 report by Fritsch et al.[8] described the presence of bombesin immunoreactivity in the central nervous system and in the longitudinal branchial bars of the phlebobranchiate ascidian *Ciona intestinalis*. In 1988 Pestarino,[9] of Genoa, Italy, identified bombesin-immune reactivity in the stolido-branchiate ascidan *Styela plicata*. Bombesin-like immunoreactivity was shown in nerve fibers in the cerebral ganglion. In addition, a few small pseudounipolar bombesin-like immunoreactive neuronal cell bodies were identified in the cerebral ganglion. Endocrine cells of the "open" type with immunoreactivity were present in, and lining, the inner wall of the branchial sac and in the stomach. The combined nervous and endocrine systems location of bombesin-like immunoreactivity in the acidian suggested to the authors a regulatory role of importance for bombesin in the primitive tunicate chordate.

Another 1988 publication from Genoa, by Tagliaferro et al.,[10] presented interesting data on bombesin-like immunoreactivity in several lower vertebrates. Their studies included two cartilaginous fishes (*Scyliorhinus canicula* and *Squatina aculeata*), one lungfish *(Protopterus annectens)*, and two amphibian urodeles (*Ambystoma tigrinum* and *Hydromantes italicus*). Bombesin-like immunoreactivity was identified in both endocrine cells and nerve fibers along the gastrointestinal tract of all the species studied. In the cartilaginous fish studied, bombesin-like immunoreactivity was localized in the pyloric stomach and in the spiral valve.

In the stomachless lungfish, immunoreactivity for bombesin is found only in the spiral valve. Bombesin-like peptides and 5-HT seemed to be colocalized in some of the intestinal endocrine cells and nerve fibers of the cartilaginous fish, especially in the walls of blood vessels. The results of these studies suggested a tendency for bombesin-like-immunoreactivity cells to be concentrated in the stomach, for immunoreactive fibers to be increased, and for the number of bombesin-like-immunoreactive cells to be reduced during lower vertebrate evolution.

Another Italian publication of interest, by Zaccone et al.[11] from Messina, reported the finding of GRP/bombesin-immunoreactive sacciform gland cells in the epidermis of the clingfish *Lepadogaster candollei*. These sacciform cells of fish epidermis are unicellular serous glands that constitute an important category of secretory elements of fish integument. The secretory contents of these cells include protein, enzymes, and serotonin. The same sacciform gland epidermal cells are immunostained with antibody to both GRP and bombesin. Staining was positive for secretions of a large sacciform gland cells of the dorsal, lateral, and ventral portions of the skin epithelium. This report was the first for bombesin/GRP immunoreactivity for exocrine glands of the fish epidermis. These findings were of particular interest, as the authors pointed out, because one important function of these exocrine glands and peptides, including the identified bombesin, could be a defensive response to predators, since these sacciform glands may function in the synthesis and secretion of toxic principles, crinotoxins. The authors speculated that the sacciform gland cells of the teleost epidermis may combine the characteristics of the granular glands of amphibian skin.

III. NERVOUS SYSTEM

Costa and associates[12] from Australia have explored the presence and possible roles of bombesin-like peptides in enteric neurons. They and others have found bombesin/GRP-like immunoreactivity to be located in nerve fibers throughout the mammalian gastrointestinal tract in a number of species. In the stomach, bombesin immunoreactivity is encountered in cell bodies of the gastric myenteric ganglia and in nerve fibers in these ganglia. They also are found in nerve fibers in the muscle coats and in the mucosa of both the antral and nonantral areas. In the small intestine, bombesin-immunoreactive cell bodies also are present in myenteric ganglia and in the nerve fibers in both the myenteric and submucous plexes.

In the spinal cord, bombesin immunoreactivity was highest in the dorsal-horn area in a study reported in 1988 by Go and Yaksh.[13] The reactivity was greatest in the sacral spinal cord dorsal horn as compared with more rostral segments. Reactivity also was identified in dorsal-root ganglion cells, with the possibility, expressed by the authors, that bombesin immunoreactivity might also be present in primary afferent fibers to the spinal cord.

In a 1988 investigation by Panula et al.[14] of Finland, bombesin immunoreactivity in the spinal cord of the rat in the dorsal horn was observed to be most concentrated in the superficial laminae (substantia gelatinosa) of the dorsal horn in the critical nociceptive sensory input area. In this area the immunoreactivity was not only concentrated, but highly concentrated, and with a distribution that resembled that of substance P. In the studies by Panula et al., capsaicin released immunoreactivity for substance P, but not for bombesin, from slices of rat spinal cord. The authors interpreted this finding as suggesting that substance P and bombesin are present in different spinal cord nerve terminals.

In a 1985 study by Decker et al.,[15] neonatal capsaicin treatment reduced both substance P and bombesin/GRP immunoreactivity in terminals in the superficial laminae of the spinal cord. By radioimmunoassay study these authors also found high concentrations of bombesin-like immunoreactivity in the hypothalamus, thalamus, medulla, and spinal cord. While neonatal capsaicin treatment reduced bombesin-like immunoreactivity in the spinal cord, it did not reduce bombesin-like immunoreactivity in the thalamus. The authors interpreted these findings as indicating that the source of bombesin-like immunoreactivity in the spinal cord likely is of sensory origin and from capsaicin-sensitive dorsal-root ganglion cells.

Unilateral rhizotomy abolished virtually all bombesin/GRP-immunoreactive terminals in the ipsilateral dorsal horn without effecting that on the contralateral side and with little effect on the immunostaining. Immunoreactive cell bodies in bombesin/GRP studies also were shown to be present in lamina X near the central canal.

In the Panula et al.[14] studies and those of others, the small ganglion cells in the spinal sensory ganglia and in the trigeminal ganglion were found to contain both substance P-immunoreactive and bombesin/GRP-immunoreactive sites. Peripheral arborizations of bombesin/GRP-immunoreactive sensory ganglion cells were observed (adrenal medulla, middle ear, lung) by the Panula group. In addition, Fux et al.,[16] of Stockholm, described bombesin/GRP immunoreactivity and substance P immunoreactivity in the same cells of spinal ganglia.

Panula et al.[14] also explored the localization of bombesin/GRP in the brain stem and brain of the rat, in their excellent and comprehensive studies. Bombesin/GRP-immunoreactive neurons were identified mainly in the hypothalamus and the pons/medulla, where the trigeminal complex and the nuclei involved in autonomic function regulation were shown to have bombesin/GRP immunoreactivity. In these studies, intraventricular colchicine treatment to inhibit axon transport generally was required to identify bombesin/GRP immunoreactivity in cell bodies. Hypothalamic locations stained included the preoptic area, the suprachiasmic nucleus, and the ventral and medial parvocellular part of the paraventricular nucleus.

In the pons/medulla, bombesin/GRP-immunoreactive neuronal cell bodies were found in the laterodorsal nucleus, the nucleus of the solitary tract, and some other medullary nuclei. A high density of bombesin-GRP-immunoreactive neuronal fibers and terminals was encountered in the parabrachial nucleus, the nucleus of the solitary tract, and the trigeminal complex, including the substantia gelatinosa of the trigeminal nucleus.

Of the number of locations of bombesin/GRP immunoreactivity in the central nervous system, the most important, in the judgment of Panula et al.,[14] are the hypothalamus, the medullary nuclei involved in the regulation of autonomic function, and the sensory systems.

A 1991 publication by Brown and associates,[17] of the University of California–San Diego, continued important work of this group in exploring significant involvements of bombesin in the functioning of the autonomic nervous system and in the phenomenon of "satiety" and associated food-intake stimulation or inhibition.[18-21] Their 1991 study[17] was carried out to explore the possibility that bombesin-like-immunoreactive neuronal cell bodies in the hypothalamus were the source of bombesin-like immunoreactivity in nerve terminals within the dorsal vagal complex. They utilized a combined retrograde-tracing and immunohistochemical methodology. After retrograde-tracer injections were made into the dorsal vagal complex (an area of a dense plexus of bombesin-immunoreactive terminals),[22] they found cells in hypothalamus that were labeled with both retrograde tracer and bombesin immunoreactivity and were located in the parvocellular portion of the paraventricular nucleus from which some 30% of the bombesin fiber projections were to the dorsal vagal complex. These results suggested to the authors that the paraventricular nucleus of the hypothalamus is a major source of bombesin terminals in the dorsal vagal complex and that this pathway may mediate some of the autonomic nervous system changes that result from bombesin injection into the central nervous system (identified in previous studies by the authors). Some of these changes result from bombesin-induced elevations in plasma levels of epinephrine and glucagon, with resulting hyperglycemia, suppression of feeding, etc. Their studies indicate that bombesin cells in other hypothalamic nuclei, the bed nucleus of the striaterminalis (NTS), and the amygdala do not project to the dorsal vagal complex. The findings of this group of authors suggest that the central effects of bombesin probably are mediated by the autonomic, rather than neuroendocrine, system. These results of the Brown group also are of interest with respect to recent (1989) observations of Ishikawa and Tache[23] that infusion of bombesin directly into the dorsal vagal complex inhibits vagally stimulated and mediated gastric acid secretion.

IV. VASCULAR SYSTEM

The principal vascular system effect of bombesin (determined by Erspamer[1]) in primates (monkeys) is a lowering of the systemic arterial blood pressure (hypotension). This is similar to the vascular system effect of ranatensin in the Geller et al.[3] early studies in primates (monkeys). In nonprimate mammals (dogs, cats, rabbits, and rats), bombesin was found by Erspamer to increase blood pressure. Ranatensin, in the Geller et al. studies, produced a rise in the blood pressure of all nonprimate mammals except the cat, which experienced no blood pressure changes. The blood pressure differentials observed in primate and nonprimate mammals might be traceable to different levels of bombesin-induced epinephrine and norepinephrine released into the systemic circulation of nonprimate mammals as compared with primate mammals (see above work by Brown et al.).

V. IMMUNE RESPONSE SYSTEM

In 1982 (and subsequently) the author[24-26] described a potent *in vivo* immunostimulative effect of bombesin on acute infusion (50 μg) into a cannulated afferent lymphatic of the popliteal lymph node of the sheep. There was a prompt, sharp, and prolonged increase in lymphocyte traffic through this completely intact and otherwise unmollested primary peripheral lymph node, as mirrored in the output of both small recirculating and blast lymphocytes into the efferent lymph, with the efferent lymphatic chronically cannulated (Figure 1). The level of this traffic, in the studies of Gowans, Ford, Morris, Hall, Hay, and others, has been shown to play a major role in the initiation and development of the immune response *in vivo* (where it counts).[27-33]

The bombesin effect was evaluated in five separate experiments and consisted of a single acute (1 to 2 min) infusion of bombesin (50 μg) into the afferent lymphatic. Multiple 20-min preinfusion efferent lymph collections were made prior to bombesin infusion, and collections were made every 20 min postinfusion for 6 h. The postinfusion output of small recirculating lymphocytes into efferent lymph increased in all five experiments. The increases ranged from 146 to 486% of preinfusion values (196, 156, 183, 195, and 486%). The output of blast lymphocytes into efferent lymph also was increased in all five experiments, with increases ranging from 129 to 567% (129, 173, 242, 260, and 567%). In three of the five experiments the peak increase in blast-lymphocyte output occurred before the peak increase in small-recirculating-lymphocyte output. The blast lymphocytes do not recirculate, and their place of origin is the study popliteal lymph node. Efferent lymph flows increased postbombesin infusion in four of the five experiments. The bombesin-induced increases in lymph outputs into efferent lymphatics ranged from 115 to 166% of preinfusion levels (115, 145, 163, and 166%).

The author also published a part of the above data in *Immunology* in 1984,[25] and in November of 1984 presented the bombesin-immunostimulative experiences in comparison with other immunostimulative substances (and "memory" molecules) at an international symposium at the Basel Institute for Immunology in Switzerland[26] (Figure 2).

Ruff, Pert, and associates,[34] at the National Institute of Mental Health (Bethesda, MD) in 1985, reported an important immunostimulative effect of bombesin. Utilizing a Boyden chamber chemotaxis assay method, they were able to demonstrate that human buffy-coat

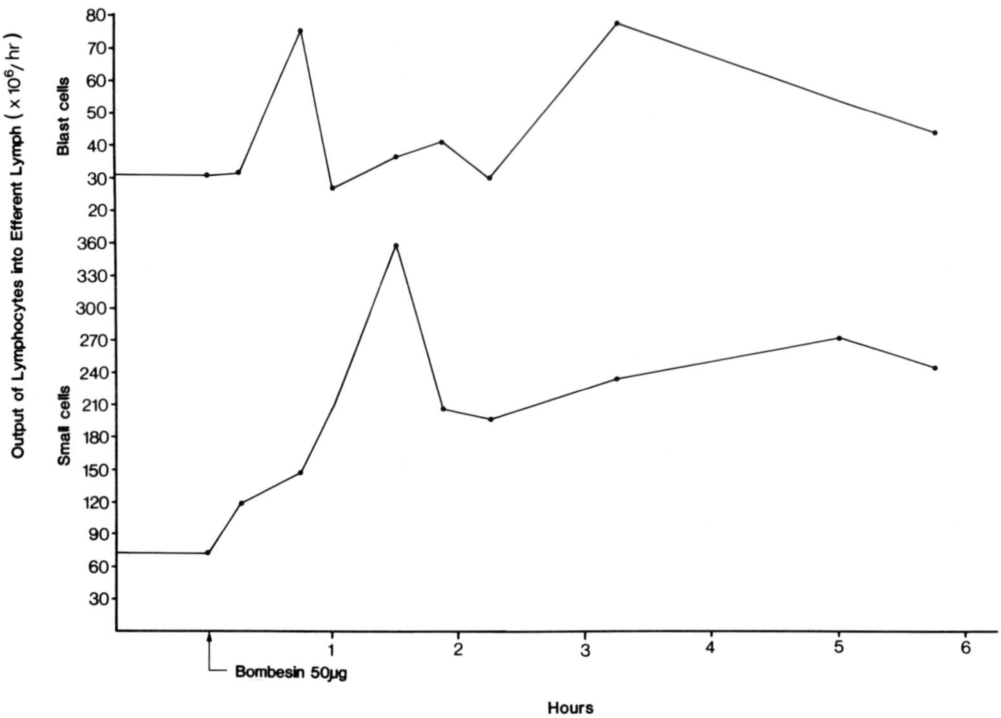

FIGURE 1. Acute infusion of bombesin (50 μg) into a cannulated popliteal lymph-node afferent lymphatic of a sheep produces a prompt, sharp, and prolonged elevation in the output of small recirculating lymphocytes into nodal efferent lymph (chronically cannulated). The output of blast lymphocytes also was elevated. (From Moore, T. C., *Immunology,* 52, 511, 1984. With permission.)

monocytes respond chemotactically to bombesin. GRP also was shown to be active in this assay, but with slightly less potency than bombesin.

In 1986 Wiedermann with Ruff, Pert, and associates,[35] of Bethesda, MD and Washington, D.C., described the presence of bombesin in adherent human peripheral blood mononuclear cells and in human and guinea pig alveolar macrophages. They employed radioimmunoassay, immunohistochemistry, and high-pressure liquid chromatography with the use of monospecific antibodies.

In 1988 Wiedermann and Pert of Bethesda, with Bowden and Adamson of Winnipeg, Canada,[36] found enhanced secretion of bombesin by macrophages that had been exposed *in vivo* to silica. Bowden and Adamson had carried out extensive prior studies into foreign-particle-induced lung fibrosis, including the actions of silica.[37-39] This study involved the use of fibrogenic particles (silica) and nonfibrogenic inert particles (carbon) instilled into the lungs of rats by intratracheal injection. Alveolar macrophages obtained at 6 weeks post-silica instillation and at a time fibrosis (from prior studies) had been shown to begin were found, after 4 h of culturing, to secrete 50% more bombesin than control values for untreated lungs. Bombesin production from carbon-treated lungs was at control (untreated) levels.

In the June 1991 issue of *Immunology*, De La Fuente et al. of Madrid[40] presented some interesting and important data concerning modulation of phagocyte function of murine (mouse) peritoneal macrophages by bombesin and related neuropeptides (GRP and neuromedin C [$GRP_{18\text{-}27}$]). These neuropeptides were shown *in vitro* to stimulate all steps of the phagocytic process: adherence to substrate, chemotaxis, ingestion of cells *(Candida albicans)* and inert particles (latex beads), and production of superoxide ion. A dose-response relationship was

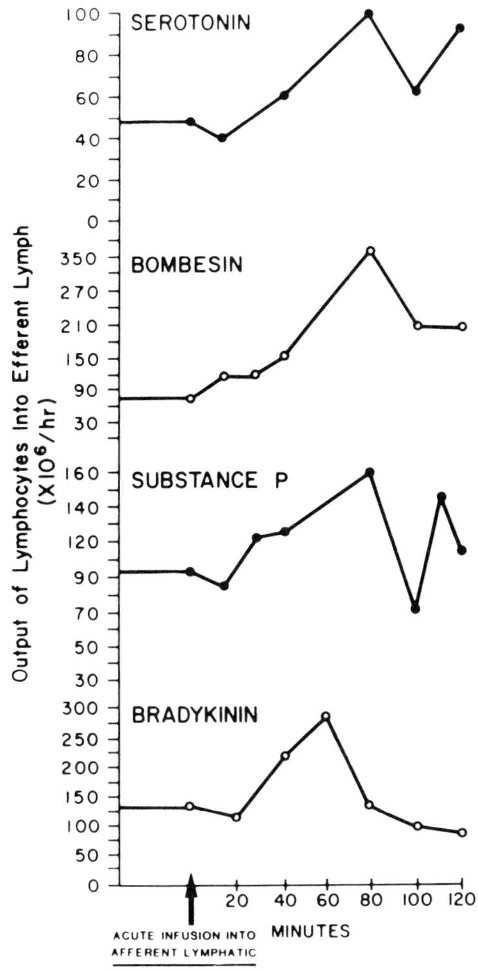

FIGURE 2. The lymphocyte traffic- and output-enhancing effect of bradykinin in the supported-sheep model is comparable to that of other stimulatory vasoactive neurotransmitter and ''memory molecules''. (From Moore, T. C., in *Immunology of the Sheep,* Editiones Roche, Basel, 1985. With permission.)

observed, with maximum stimulation of phagocytosis between 10^{-12} and 10^{-9} *M*. A significant increase in PI turnover was noted at 60 sec, and a significant decrease of cyclic AMP was found at 30 sec.

VI. HEAT, PAIN, AND MEMORY

A. HEAT

In 1977 Brown, Rivier, and Vale[20] published in *Science* the first report of a central nervous system action of bombesin. This report was concerned with a potent thermoregulatory effect by bombesin in the rat, by central administration (into the cysternum magnum). The initial resposne was to decrease core temperature of cold-exposed rats. Additional studies suggested that the central nervous system effect of bombesin on body temperature varied in parallel with ambient temperature (? induced poikilothermia, like a toad). Heat-stressed animals became hyperthermic. These observations raised the question of bombesin in the systemic fever components of infection and inflammation and in associated alterations in immune response system function.

A 1990 report by Stump et al.[41] from Colorado State University (Ft. Collins), again addressed the intriguing and elusive role of intracranial bombesin in the regulation of core body temperature. In an effort to dissociate general activity effects from thermoregulatory effects, they employed a horizontal temperature gradient. They found that temperature selection of cooler ambient temperatures was not consistently associated with decreased body temperatures, and considered that this finding calls the set-point interpretation into question. Avery et al.[42] (1981) observed that central administration of bombesin increased responses of experimental animals to escape heat in a warm environment. Their studies involved injection of bombesin into lateral cerebral ventricles. This Avery and Hawkins group,[43] in 1983, also found that intracranial administration of bradykinin produced an increase in responses to radiant heat in a cold environment (heat-reinforcement tasks). *Perhaps these experimental animals, when in a cold environment, had their memories stimulated by bombesin with respect to remembrances of the pleasures of a bit warmer situation and environment.*

The 1990 study by Stump, McCoy, and Avery[41] demonstrated that bombesin acts centrally in the rat to produce behavioral selection of cooler ambient temperatures.

Lin and Lin,[44] of Taipei, Taiwan, in 1986, reported that microiontophoretic application of bombesin into the preoptic anterior hypothalamic area resulted in inhibition of the majority of cold-responsive neurons (62.5%) and excitation of 50% of the warm-responsive neurons, as recorded in the preoptic anterior hypothalamic area, and concluded that bombesin in this area of the hypothalamus excites warm-responsive neurons and inhibits cold-resposnive neurons.

B. PAIN

In 1988 Cowan,[45] of Philadelphia, considered the behavioral effects of intracranial administration of bombesin to be possibly pain related and involved with altered skin sensations. The excessive grooming response within a minute or so of intracerebroventricular administration of bombesin in rats was first described for bombesin in 1977 by Brown et al.[20] and also identified by this group in 1980,[21] also for GRP, and characterized as a stereotyped scratching of the head with the forepaw and the hindpaw. The studies of Cowan and associates[46,47] have shown that bombesin acts at both spinal and supraspinal levels in inducing the washing/scratching response in mice and rats, as it is equipotent by intracerebroventicular or intrathecal routes of administration. These studies also have shown that guinea pigs, rabbits, and Rhesus monkeys respond to intracranial administration of bombesin by displaying a range of behavioral responses that are suggestive of altered skin sensation. They also make the especially important observation and point that goldfish, frogs, chickens, and pigeons are quite unaffected (behavior-wise) by bombesin administration centrally or peripherally. The earlier report in the "Evolution" section of this chapter that bombesin is present in an antipredator defense crinotoxin area of fish skin is of interest here. It may be that fish and amphibians that use bombesin with other toxins and venoms and stings may themselves be immune or resistant to their untoward effects, whereas mammals that have no such skin-involved protective devices are not immune or resistant. Gmerek and Cowan,[48] in 1983, described the bombesin-induced grooming as such a robust behavior that neither adrenalectomy nor hypophysectomy significantly affected it.

The 1988 publication of Cowan[45] explored the relative grooming-stimulative effects in rats of bombesin, $GRP_{18\text{-}27}$, and neuromedia B on intracranial administration. They employed stereotoxic placement of polyethylene cannulas in a lateral cerebral ventricle of each rat. They found that both bombesin and $GRP_{18\text{-}27}$ produced a dose-related increase in grooming, with the components of grooming (washing, scratching, licking) being qualitatively similar for both agents. Neuromedin B had no significant effect on grooming. Bombesin was approximately 30 times more potent as a grooming inducer than was GRP_{18-27}, with respect to the maximum number of grooming episode scores, although the grooming syndromes, as cited above, were qualitatively similar. Neuromedin B was almost inactive. Bombesin also had a longer duration of action than either GRP_{18-27} or neuromedin B (? LTP).

An important potential involvement of bombesin in sensory pain transmission is suggested by its location in the superficial areas of mammalian spinal cord dorsal horns and its depletion from sensory-pain-important areas by capsaicin treatment.

C. MEMORY

In 1988 Flood and Morley,[49] of the University of California–Los Angeles, described important memory-processing effects of bombesin and GRP in mice. They had previously shown that feeding mice immediately following training enhances memory retention[50] and that one of the gastrointestinal hormones released during a meal, preprocholecystokinin (CCK), also enhances memory retention after peripheral administration. These CCK studies involved modulated retention for foot-shock avoidance training.[50-56] The CCK improved retention in both a time- and a dose-dependent manner. The ability of CCK to enhance retention was blocked by vagotomy.[57] The 1988 report[49] demonstrated that bombesin, GRP, and GRP_{14-27} all, after peripheral administration, enhanced memory retention. The effects of both bombesin and GRP were time dependent, and both reversed amnesia induced by scopolamine (an anticholinergic). With intracerebroventicular administration, higher doses were required to produce an effect on memory retention. This finding suggested to the authors that the memory-enhancing effect was mediated predominantly through a peripheral mechanism (? by sensory neuronal input). Doses of these neuropeptides that enhanced memory retention after peripheral administration failed to increase serum glucose levels. This finding suggested that glucose modulation was not the mechanism by which bombesin and GRP modulate and enhance memory processing. Vagotomy inhibited the memory-enhancing effects of both bombesin and GRP. This particularly important observation suggests that these neuropeptides elicit their memory-enhancement effects by stimulating ascending vagal pathways (? sensory afferent connection to higher up — like maybe the hippocampus).

In 1981 Dodd and Kelly,[58] from the MRC Neurochemical Pharmacology Unit and the Department of Pharmacology at the University of Cambridge (U.K.), made important observations regarding excitatory effects of bombesin and sulfated CCK-8 on the memory-important CA1 (Cornu Ammonis 1) pyramidal cells of the mammalian hippocampus. These investigators selected this site as one considered to be innervated by a peptidergic pathway. Intracellular recordings from pyramidal cells of the CA1 region of the hippocampal slice preparation were used to ascertain changes in excitability and associated changes in membrane potential and resistance produced by pressure application of the bombesin and CCK-8 into the cell body layer from an independently mounted multibarrelled micropipet. The action of bombesin was excitatory and was accompanied by a clear depolarization. The latency of the response to bombesin was markedly larger than that to CCK-8 on other cells and to later glutamate application to the same cell. In all studies a comparison was made of the actions of bombesin and CCK-8 with those of glutamate, which was released by iontophoresis from an adjacent barrel of the same multibarrelled pipet.

VII. GROWTH FACTOR ACTIONS AND ONCOGENE INFLUENCES

The 1981 report in *Science* by Moody et al.[59] describing the release in high densities of bombesin-like peptides from human small-cell lung cancer (SCLC) cells was a major stimulus to bombesin research, especially in the growth factor and oncogene arenas.

Growth factors, classical and nonclassical, have been found to have a variety of important actions in both normal and pathological biological events, including embryogenesis, growth and development, cancer, the nervous and vascular systems, and the immune response system. Serotonin, bradykinin, and bombesin (with established vasoactive neurotransmitter credentials) have recently been recognized as a novel, nonclassical group of growth factors.

The plot becomes even more interesting, as comparatively recent (1983 to 1985) observations have identified important links between growth factors and their receptors and oncogene products.

The 1983 report by Rozengurt and Sinnett-Smith,[60] from the Imperial Cancer Research Fund, Lincoln's Inn Fields in London, that bombesin could stimulate DNA synthesis and cell division in cultures of Swiss 3T3 fibroblast cells was of major importance. Prior experience with this interesting cell line by the Rozengurt group gave added significance to these observations. This group has effectively used this fibroblast cell line in culture as a model system to explore mechanisms of action of growth factors. Swiss 3T3 fibroblast cells cease to proliferate and become arrested in the G_0/G_1 phase of the cell cycle when they use up the nutrient growth-promoting activity of the medium. Cells thus rendered quiescent can be made to reinitiate DNA synthesis and cell division by adding fresh and nutrient-complete serum to the culture medium. The interesting aspect of this scenario is that, in serum-free medium, the addition of defined growth-promoting agents (growth factors) will also start these cells synthesizing DNA and dividing.

Studies carried out by Rozengurt[61] and published in *Science* in 1986 have shown that an important aspect of the action of these growth-promoting factors is the existence, generally, of potent, specific, and required synergistic interactions. This highly significant finding indicated that extracellular growth-promoting factors likely bind to different receptors and, as a result, generate multiple complementary intracellular signals that act and interact synergistically to initiate a proliferative response.

In 1985 Zachary and Rozengurt[62] presented evidence that other members of the bombesin vasoactive neurotransmitter family, such as GRP, also bind with high affinity to Swiss 3T3 fibroblast cells and stimulate mitogenesis in a manner mediated by a single class of high-affinity receptors. They also demonstrated that this receptor is distinct from those of other mitogens for these Swiss 3T3 fibroblast cells.

Also in 1985, Cuttitta, Moody, and associates,[63] in Washington, D.C. and Bethesda, MD, published data in *Nature* indicating that bombesin-like peptides can function as autocrine (self-stimulating) growth factors in human SCLC and that these effects can be blocked *in vitro* and *in vivo* by antibodies directed against the C-terminal region of the bombesin/GRP molecule.

A 1986 publication by Letterio et al.,[64] from the University of California–San Francisco, described a potent stimulatory effect of bombesin on Swiss 3T3 fibroblast cells in nanomolar concentrations that increased levels of *c-myc* mRNA and stimulated DNA synthesis. Incubation of the 3T3 cells for 2 h with pertussis toxin completely blocked the bombesin stimulatory effects on *c-myc* mRNA and DNA synthesis, suggesting to the authors a pertussis-toxin-sensitive pathway for the stimulative actions of bombesin.

Rozengurt and associates,[65] in a 1987 report, further explored this interesting finding. They first identified major bombesin and GRP actions on Swiss 3T3 fibroblast cells. These included binding of the peptides to high-affinity receptors on the 3T3 cells, stimulation of PI turnover, mobilization of Ca^{++} from intracellular stores, activation of protein kinase C, inhibition of epidermal growth factor binding, and induction of the cellular oncogenes *c-fos* and *c-myc*. This 1987 report confirmed the effect of pertussis toxin in blocking mitogenesis at an early stage, but they also demonstrated that pertussis toxin did not alter the rapid turnover of PI, Ca^{++} mobilization, or protein kinase C activation. They concluded that inhibition of bombesin-induced mitogenesis by pertussis toxin can be dissociated from actions of pertussis toxin on the phospholipase C signaling pathway.

In a 1988 publication, Rozengurt[66] provided a series of important observations concerning the unique actions of this remarkable vasoactive neuropeptide (bombesin) on these remarkable cells (Swiss 3T3 fibroblasts). The occasion for this report was an international symposium on bombesin held in October 1987 in Rome in honor of Professor Vittorio Erspamer of Rome and its University. Professor Erspamer's many outstanding contributions over several years to the study of neuropeptides and their submammalian counterparts included the discovery of bombesin.

The Rozengurt studies showed that bombesin, at namomolar concentrations, markedly enhanced the ability of fresh serum to stimulate DNA synthesis in confluent and quiescent cultures of Swiss 3T3 fibroblast cells. Even in the presence of a low concentration of serum (3.5%), bombesin stimulates proliferation of these cells. In serum-free medium, bombesin induced DNA synthesis in the absence of any added growth-promoting agents, with a half-maximum effect at a concentration of 1 n*M*. In this respect, bombesin resembles platelet-derived growth factor (PDGF) and PDGF-like fibroblast-derived growth factor (FDGF) in being able to act as a sole mitogen for Swiss 3T3 cells, in contrast to other modulators of cell growth (growth factors) that are only active in synergistic combinations in this cell line.

This report also identified DNA synthesis by bombesin to be potentiated markedly by a variety of other extracellular factors, including insulin, PDGF, and FDGF. These mitogens were found to increase both the maximum response elicited by bombesin and to decrease the bombesin concentration required to produce a half-maximum response — a decrease from 1 to 0.3 n*M*. It was also established that endogenous mammalian analogues of bombesin, including GRP, act as functional agonists of bombesin in the stimulation of DNA synthesis.

In a study of specific bombesin and bombesin-like peptide receptors on 3T3 cells, Rozengurt used radioiodinated GRP (^{125}I-GRP). He found that ^{125}I-GRP bound to intact, quiescent Swiss 3T3 cells in a specific, saturable, and reversible manner. Scatchard analysis indicated the presence of a single class of high-affinity binding sites. The number of binding sites per cell was approximately 1.25×10^5.

VIII. CONCLUSION

Bombesin, a 2-decade-old gift to biomedical science from Erspamer and *Bombina bombina*, has proved to be a most fascinating molecule with many surprises. Mammalian bombesin-like peptides have been found to include gastrin-releasing peptide and to have multiple stimulatory and excitatory effects involving the nervous, vascular, and immune response systems. It also appears to function as a nonclassical growth factor, with related oncogene effects as well. Bombesin also appears to qualify for inclusion in the "heat, pain, and memory" scenario, with increasing credentials as a "memory molecule". The discovery of the release of high densities of bombesin-like peptides from human SCLC cells, by Moody and associates, was a major bombshell and thrust bombesin and bombesin-like peptides very much into the clinical arena and triggered intense research efforts in this area. Bombesin and bombesin-like peptides may be expected to remain "hot" research items for many years to come.

REFERENCES

1. **Erspamer, V.,** Discovery, isolation, and characterization of bombesin-like peptides, *Ann. N.Y. Acad. Sci.*, 547, 3, 1988.
2. **Nakajima, T., Tanimura, T., and Pisano, J. J.,** Isolation and structure of a new vasoactive peptide, *Fed. Proc.*, 29, 282, 1970.
3. **Geller, R. G., Govier, W. C., Pisano, J. J., Tanimura, T., and Van Clineschmidt, B.,** The action of ranatensin, a new peptide from amphibian skin, on the blood pressure of experimental animals, *Br. J. Pharmacol.*, 40, 605, 1970.
4. **McDonald, T. J., Jornvall, H., Nilsson, G., Vagne, M., Ghatel, M., Bloom, S. R., and Mutt, V.,** Characterization of a gastrin releasing peptide from procine non-antral gastric tissue, *Biochem. Biophys. Res. Commun.*, 90, 227, 1979.
5. **Minamino, N., Kangawa, K., and Matsuo, H.,** Neuromedin B: a novel bombesin-like peptide identified in porcine spinal cord, *Biochem. Biophys. Res. Commun.*, 114, 541, 1983.
6. **Minamino, N., Kangawa, K., and Matsuo, H.,** Neuromedin B and neuromedin C[a]. Two mammalian bombesin-like peptides identified in porcine spinal cord and brain, *Ann. N.Y. Acad. Sci.*, 547, 373, 1988.
7. **Renda, T., D'Este, L., and Campo, S.,** Bombesin-like immunoreactivity in nervous and epithelial elements of the earthworm (*Lumbricus terrestris* L.), *Ann. N.Y. Acad. Sci.*, 547, 466, 1988.
8. **Fritsch, H. A. R., Van Noorden, S., and Pearse, A. G. E.,** Gastrointestinal and neurohormonal peptides in the alimentary tract and cerebral complex of *Ciona intestinalis* (Ascidiaceae), *Cell Tissue Res.*, 223, 369, 1982.
9. **Pestarino, M.,** Occurrence of bombesin-like peptide immunoreactivity in a Tunicate, the Ascidian *Styela plicata, Ann. N.Y. Acad. Sci.*, 547, 461, 1988.
10. **Tagliaferro, G., Bonini, E., Faraldim, G., Farina, L., and Rossi, G.,** Distribution and ontogeny of VIP-like immunoreactivity in the gastro-entero-pancreatic system of a cartilaginous fish *Scyliorhinus stellaris, Cell Tissue Res.*, 253, 23, 1988.
11. **Zaccone, G., Fasulo, S., Licata, A., Ainis, L., and Lo Cascio, P.,** GRP/bombesin-immunoreactive sacciform gland cells in the epidermis of the clingfish, *Lepadogaster candollei, Ann. N.Y. Acad. Sci.*, 547, 464, 1988.
12. **Furness, J. B., Miller, A. S., and Costa, M.,** The presence and possible roles of bombesin-like peptides in enteric neurons, *Ann. N.Y. Acad. Sci.*, 547, 76, 1988.
13. **Go, V. L. W. and Yaksh, T. L.,** Quantification of bombesin-like peptides in mammalian spinal cord, *Ann. N.Y. Acad. Sci.*, 547, 70, 1988.
14. **Panula, P., Nieminen, O., Falkenberg, M., and Auvinen, S.,** Localization and development of bombesin/GRP-like immunoreactivity in the rat central nervous system, *Ann. N.Y. Acad. Sci.*, 547, 54, 1988.
15. **Decker, M. W., Towle, An. C., Bissette, G., Mueller, R. A., Lauder, J. M., and Nemeroff, C. B.,** Bombesin-like immunoreactivity in the central nervous system of capsaicin-treated rats: a radioimmunoassay and immunohistochemical study, *Brain Res.*, 342, 1, 1985.
16. **Fuxe, K., Agnati, L. F., McDonald, T., Locatelli, V., Hokfelt, T., Dalsgaard, C. J., Battistini, N., Yanaihara, N., Mutl, V., and Cuello, A. C.,** Immunohistochemical indications of gastrin releasing peptide-bombesin-like immunoreactivity in the nervous system of the rat. Codistribution with substance P-like immunoreactive nerve terminal systems and coexistence with substance P-like immunoreactivity in dorsal root ganglion cell bodies, *Neurosci. Lett.*, 37, 17, 1983.
17. **Costello, J. F., Brown, M. R., and Gray, T. S.,** Bombesin immunoreactive neurons in the hypothalamic paraventricular nucleus innervate in the dorsal vagal complex in the rat, *Brain Res.*, 542, 77, 1991.
18. **Brown, M., Allen, R., Villarreal, J., Rivier, J., and Vale, W.,** Bombesin-like activity: radioimmunologic assessment in biological tissues, *Life Sci.*, 23, 2721, 1978.
19. **Brown, M. R., Rivier, J., and Vale, W. W.,** Bombesin affects the central nervous system to produce hyperglycemia in rats, *Life Sci.*, 21, 1729, 1977.
20. **Brown, M., Rivier, J., and Vale, W.,** Bombesin: potent effects on thermoregulation in the rat, *Science*, 196, 988, 1977.
21. **Brown, M., Marki, W., and Rivier, J.,** Is gastrin releasing peptide mammalian bombesin? *Life Sci.*, 27, 125, 1980.
22. **Panula, P., Yand, H.-Y. T., and Costa, E.,** Neuronal location of the bombesin-like immunoreactivity in the central nervous system of the rat, *Regul. Pept.*, 4, 275, 1982.
23. **Ishikawa, T. and Tache, Y.,** Bombesin microinjected into the dorsal vagal complex inhibits vagally stimulated gastric acid secretion in the rat, *Regul. Pept.*, 24, 187, 1989.
24. **Moore, T. C.,** The effects of inflammatory mediators and anaesthetic agents upon lymphocyte traffic of sheep in vivo, Ph.D. thesis, University of Cambridge (Trinity College), England, 1982.

25. **Moore, T. C.,** Modification of lymphocyte traffic by vasoactive neurotransmitter substances, *Immunology,* 52, 511, 1984.
26. **Moore, T. C.,** Mediator and physiological influences on lymphocyte traffic in sheep, in *Immunology of the Sheep,* Morris, B. and Miyasaka, Eds., Editiones Roche, Basel, 1985, 327.
27. **Gowans, J. C.,** The recirculation of lymphocytes from blood to lymph in the rat, *J. Physiol. (Lond.),* 145, 54, 1959.
28. **Ford, W. L., Smith, M. E., and Andrews, P.,** Possible clues to the mechanism underlying the selective migration of lymphocytes from the blood, in *Cell Recognition,* Curtis, A. G. S., Ed., Cambridge University Press, England, 1978, 359.
29. **Hall, J. G. and Morris, B.,** The output of cells in lymph from the popliteal nodes of sheep, *Q. J. Exp. Physiol.,* 47, 360, 1962.
30. **Hall, J. G. and Morris, B.,** The lymph-born cells of the immune response, *Q. J. Exp. Physiol.,* 48, 235, 1963.
31. **Hall, J. G. and Morris, B.,** The origin of cells in the efferent lymph from a single lymph node, *J. Exp. Med.,* 121, 901, 1965a.
32. **Hall, J. G. and Morris, B.,** The immediate effect of antigens on the cell output of a lymph node, *Br. J. Exp. Pathol.,* 46, 450, 1965b.
33. **Hay, J. B. and Hobbs, B. B.,** The flow of blood to lymph nodes and its relations to lymphocyte traffic and the immune response, *J. Exp. Med.,* 145, 31, 1977.
34. **Ruff, M., Schiffmann, E., Terranova, V., and Pert, C. B.,** Neuropeptides are chemoattractants for human tumor cells and monocytes: a possible mechanism for metastasis, *Clin. Immunol. Immunopathol.,* 37, 387, 1985.
35. **Wiedermann, C. J., Goldman, M. E., Plutchok, J. J., Sertl, K., Kaliner, M. A., Johnston-Early, A., Cohen, M. H., Ruff, M. R., and Pert, C. B.,** Bombesin in human and guinea pig alveolar macrophages, *J. Immunol.,* 137, 3928, 1986.
36. **Wiedermann, C. J., Adamson, I. Y. R., Pert, C. B., and Bowden, D. H.,** Enhanced secretion of immunoreactive bombesin by alveolar macrophages exposed to silica, *J. Leukoc. Biol.,* 43, 99, 1988.
37. **Bowden, D. H. and Adamson, I. Y. R.,** Adaptive responses of the pulmonary macrophagic system to carbon. I. Kinetic studies, *Lab. Invest.,* 38, 422, 1978.
38. **Bowden, D. H. and Adamson, I. Y. R.,** The role of cell injury and the continuing inflammatory response in the generation of silicone pulmonary fibrosis, *J. Pathol.,* 144, 149, 1984.
39. **Bowden, D. H. and Adamson, I. Y. R.,** Bronchial and alveolar lesions in the pathogenesis of crocidolite induced pulmonary fibrosis in mice, *J. Pathol.,* 147, 257, 1985.
40. **De La Fuente, M., Del Rio, M., Ferrandez, M. D., and Hernanz, A.,** Modulation of phagocytic function in murine peritoneal macrophages by bombesin, gastrin-releasing peptide and neuromedin C, *Immunology,* 73, 205, 1991.
41. **Stump, B. S., McCoy, J. G., and Avery, D. D.,** The effects of intraventricular injections of bombesin on temperature selection in the rat, *Brain Res. Bull.,* 25, 173, 1990.
42. **Avery, D., Hawkins, M. F., and Wunder, B. A.,** The effects of injections of bombesin into lateral cerebral ventricles on behavioral thermoregulation, *Neuropharmacology,* 20, 23, 1981.
43. **Hawkins, M. F. and Avery, D. D.,** Effects of centrally-administered bombesin and adrenalectomy on behavioral thermoregulation and locomotor activity, *Neuropharmacology,* 22(11), 1249, 1983.
44. **Lin, K. S. and Lin, M. T.,** Effects of bombesin on thermoregulatory responses and hypothalamic neuronal activities in the rat, *Regul. Int. Comp. Physiol.,* 20, 303, 1986.
45. **Cowan, A.,** Behavioral effects of bombesin, *Ann. N.Y. Acad. Sci.,* 547, 204, 1988.
46. **Gmerek, D. E., Cowan, A., and Vaught, J. L.,** Intrathecal bombesin in rats: effects on behaviour and gastrointestinal transit, *Eur. J. Pharmacol.,* 94, 141, 1983.
47. **Cowan, A., Khunawat, P., Zu Zhu, X., and Gmerek, D. E.,** Effects of bombesin on behavior, *Life Sci.,* 37, 135, 1985.
48. **Gmerek, D. E. and Cowan, A.,** Studies on bombesin-induced grooming in rats, *Peptides,* 4, 907, 1983.
49. **Flood, J. F. and Morley, J. E.,** Effects of bombesin and gastrin-releasing peptide on memory processing, *Brain Res.,* 460, 314, 1988.
50. **Flood, J. F., Smith, G. E., and Morley, J. E.,** Modulation of memory processing by cholecystokinin: dependence on the vagus nerve, *Science,* 234, 832, 1987.
51. **Cohen, S. L., Knight, M., Tamminga, C. A., and Chase, I. N.,** Cholecystokinin-octapeptide effects on conditioned-avoidance behavior, stereotypy and catalepsy, *Eur. J. Pharmacol.,* 83, 213, 1982.
52. **Deupree, D. and Hsiao, S.,** Cholecystokinin octapeptide, proglumide, and passive avoidance in rats, *Peptides,* 8, 25, 1987.

53. **Fekete, M., Szabo, A., Balazs, M., Penke, B., and Telgedy, G.,** Effects of intraventricular administration of cholecystokinin octapeptide sulfate ester and unsulfated cholecystokinin octapeptide on active avoidance and conditioned feeding behavior of rats, *Acta Physiol. Hung.*, 58, 39, 1981.
54. **Fekete, M., Kadas, T., Penke, B., and Telgedy, G.,** Modulation of passive avoidance behavior by cholecystokinin octapeptide in the rat, *Neuropeptides,* 1, 301, 1981.
55. **Flood, J. F., Hernandez, E. N., and Morley, J. E.,** Modulation of memory processing by neuropeptide Y., *Brain Res.,* 421, 280, 1987.
56. **Katsuura, G. and Itoh, S.,** Prevention of experimental amnesia by peripherally administered cholecystokinin octapeptide in the rat, *Drug Dev. Res.,* 7, 269, 1986.
57. **Smith, G. P., Jerome, C., Cushin, B. J., Etermo, K., and Sumansky, J.,** Abdominal vagotomy blocks the satiety effect of cholecystokinin in the rat, *Science,* 212, 1036, 1982.
58. **Dodd, J. and Kelly, J. S.,** The actions of cholecystokinin and related peptides on ramidal neurones of the mammalian hippocampus, *Brain Res.,* 205, 337, 1981.
59. **Moody, T. W., Pert, C. B., Gazdar, A. F., Carney, D. N., and Minna, J. D.,** High levels of intracellular bombesin characterize human small-cell lung carcinoma, *Science,* 214, 1246, 1981.
60. **Rozengurt, E. and Sinnett-Smith, J.,** Bombesin stimulation of DNA synthesis and cell division in cultures of Swiss 3T3 cells, *Proc. Natl. Acad. Sci. U.S.A.,* 80, 2936, 1983.
61. **Rozengurt, E.,** Early signals in the mitogenic response, *Science,* 234, 161, 1986.
62. **Zachary, I. and Rozengurt, E.,** High-affinity receptors for peptides of the bombesin family in Swiss 3T3 cells, *Proc. Natl. Acad. Sci. U.S.A.,* 82, 7616, 1985.
63. **Cuttitta, F., Carney, D. N., Mulshine, J., Moody, T. W., Fedorko, J., Fischler, A., and Minna, J. D.,** Bombesin-like peptides can function as autocrine growth factors in human small cell lung cancer, *Nature,* 316, 823, 1985.
64. **Letterio, J. J., Coughlin, S. R., and Williams, L. T.,** Pertussis toxin-sensitve pathway in the stimulation of *c-myc* expression and DNA synthesis by bombesin, *Science,* 1117, 1986.
65. **Zachary, I., Millar, J., Nanberg, E., Higgins, T., and Rozengurt, E.,** Inhibition of bombesin-induced mitogenesis by pertussis toxin: dissociation from phospholipase C pathway, *Biochem. Biophys. Res. Commun.,* 146, 456, 1987.
66. **Rozengurt, E.,** Bombesin-induction of cell proliferation in 3T3 cells, *Ann. N.Y. Acad. Sci.,* 547, 277, 1988.

Chapter 13

VASOACTIVE INTESTINAL PEPTIDE (VIP)

I. INTRODUCTION

Vasoactive intestinal peptide (VIP) is a 28-amino-acid-residue vasoactive neurotransmitter with potent and important modulatory actions on the nervous, vascular, and immune response systems. It was isolated from porcine duodenum and characterized by Said and Mutt at the Karolinska Institute in Stockholm, Sweden, in 1970 to 1971.[1-3] VIP contains 28 amino acids and belongs to a "family" of compounds that includes GIP (gastric inhibitory peptide), glucagon, and secretin. Amino acid sequences are similar within the VIP family of mammalian brain-gut, vaso- and neuroactive peptides. VIP is an important component of the nervous systems of lower animals, including the earthworm.[4] The amino acid sequences of human, rat, cow, and pig VIP are identical and differ at only three amino acid locations in chicken VIP and at five locations in elasmobranch *Scyliorhinus canicula* (dogfish) VIP.[5] They all share the same ten NH2-terminal-half amino acid sequences.

VIP is a potent vasodilator and secretogogue, as well as a muscle-relaxing agent. It is produced by nonbeta cells of the pancreatic endocrine isles of Langerhan. VIP concentrations are higher in the colon, ileum, and jejunum than higher up in the alimentary tract. It has been found by Said and Mutt[6] to produce smooth-muscle relaxation in the lung.

Its wide distribution in and involvement in the central and peripheral nervous systems were first recognized in 1976 by Said and Rosenberg.[7] In 1977 Giachetti, Said, and associates[8] reported identification of VIP in the brain and its localization in and release from isolated nerve terminals. Convincing evidence was presented by Said et al. in 1980 to support a neurotransmitter role for VIP.[9] VIP was shown by Giachetti and Said[10] to accumulate in constricted sciatic nerves in a manner suggesting fast antegrade axonal flow. Subcellular fractionation of VIP-rich areas of the brain show VIP to be concentrated in synaptosomal fractions in the company of better-established neurotransmitters (noradrenalin and dopamine).[11] VIP has a discrete regional distribution in the brain, being found in the cerebral cortex, hypothalamus, amygdala, hippocampus, and corpus striatum. VIP also has a wide distribution in the peripheral nervous system. It appears to play an important part in intrinsic small intestine smooth-muscle function and coordination (present in both Meissner's and Auerbach's plexuses). It has been suggested, by Said and associates[12] from experimental data, to mediate noradrenergic relaxation of tracheal smooth muscle. Dense plexes of neurons containing VIP have been found in discrete areas of the central nervous system and in peripheral organs (gastrointestinal, genitourinary, pancreas). VIP is concentrated in nerve endings where it can be released by high K^+-ion concentration in a Ca^{++}-ion-dependent manner. VIP also can be released by electrical stimulation of nerves, such as the vagus. It is found in fine terminal varicosities of neurons as well as in cell bodies.[9,13,14] The greatest concentration of VIP is within the nervous system. Despite its ability to influence numerous body functions, its short biologic half-life (1 to 2 min) makes it unlikely to function as a circulating hormone. As with other agents discussed earlier, it would appear to fit more closely the local hormone picture, in addition to its apparent neurotransmitter role. Like many of these agents, it has a long evolutionary history, being found, among other places, in the nervous structures of the earthworm.[4]

VIP has been observed to modulate acetylcholine release and to elevate intestinal smooth-muscle cyclic AMP levels.

Prostaglandin system influences on VIP actions, which may be seen following inhibition of PG synthesis (indomethacin), were found to inhibit VIP-stimulated intestinal fluid secretion — a potent action of VIP. Isolated bowel loops were used in rats. VIP produced a highly significant increase in intralaminal-fluid accumulation. This response was completely inhibited by indomethacin.[15] Watery diarrhea is a serious problem in humans with pancreatic or neuroendocrine tumors that secrete excessive amounts of VIP (VIPomas).

II. NERVOUS SYSTEM

Major advances in the study of VIP in the nervous system, as well as in other areas, were presented at a recent international VIP conference organized and cochaired by Said and Mutt, codiscoverers and early investigators of VIP. These important findings and observations were published in a 1988 volume (No. 527) of the *Annals of the New York Academy of the Sciences*. This volume is strongly recommended to the reader.

A report in this 1988 volume, by Moody et al.,[16] describes different central nervous system receptors for VIP and secretin, despite their close sequence similarity. The greatest density of secretin receptors was in the cerebellum. Density of VIP receptors was high in the cerebral cortex, hippocampus, striatum, and thalamus, but not in the cerebellum, medulla, or pons. VIP also was considerably more potent than secretin in stimulating cyclic AMP production by rat cortical brain slices.

Also in this 1988 volume, Besson[17] presents detailed data concerning VIP receptors in the brain and pituitary of the rat. This report was an extension of prior *in vitro* VIP-receptor localization studies by this author and associates.[18,19] VIP-receptor bindings were high in the external plexiform layer of the olfactory bulb; moderate to low in the cerebral cortex, basal forebrain, and basal ganglia; and high in the suprachiasmic nucleus of the hypothalamus, amygdala, thalamus, dentate gyrus of the hippocampal formation, and the superior colliculus of the brain stem. These findings were in general agreement with the immunohistochemical localizations of VIP neurons in the central nervous system.

The 1988 volume also contained important communications from Morrison[20] and from Magistretti and associates[21] relating to especially important actions and anatomical organization of VIP-containing neurons and their projections in the neocortex, and the close interrelationship of these neurons and projections with those of the noradrenergic neuronal processes projecting to the cerebral cortex from the locus ceruleus. The VIP neurons and projections are found to be arranged radially, with the neurons being bipolar and the projections largely local. These neurons and projections interact with projections from the noradrenergic system that project tangentially/horizontally throughout the neocortex. The morphological characteristics and distribution of the VIP-containing neuronal cell bodies in the neocortex are well suited for a role in local and radial regulation of cortical activity. While the anatomical arrangement of noradrenergic projection fibers suggest a global regulatory or activity-influencing role, those of VIP suggest positioning for more local regulatory and activity-influencing actions. Both VIP and norepinephrine increase glycogenolysis, but by independent mechanisms.[22,23]

Magistretti and Schorderet,[24,25] in 1984 and 1985, reported that both VIP and norepinephrine independently increase cyclic AMP synthesis, when applied separately to the rodent cerebral cortex, and, when applied together, act synergistically to increase cyclic AMP in this brain area. They also observed that histamine, as well as norepinephrine, potentiates increases in cyclic AMP elicited by VIP in the mouse cerebral cortical slices. Magistretti et al.[21] also have observed that the α_1-adrenergic potentiation by norepinephrine of increases in cyclic AMP activity of cortical brain slices by VIP was antagonized by indomethacin, suggesting a prostaglandin involvement in this action.

In 1985 Ferron, Siggins, and Bloom[26] made the important observation that VIP acts synergistically with norepinephrine in depressing the spontaneous-discharge rate in cerebral cortical neurons. These synergetic effects of VIP and norepinephrine in both stimulating cyclic AMP formation and depressing the firing rate of cortical neurons suggest that VIP- and norepinephrine-containing terminals converge on the same target cells.

Morrison[20] suggested that both of these neurotransmitter and modulator systems may act on glial cells and cells of the vasculature as well as on neuronal cells in the neocortex and that the actions may be by both synaptic and nonsynaptic release. He considers the synaptic-neuronal mode to predominate in the intermediate laminar layers of the cortex, with the nonsynaptic glial and vascular regulating occurring largely in layer I.

VIP also has been found to be a major neuropeptide in the periphery as well as in the brain. VIP-containing neuronal fibers are widely distributed in the periphery and particularly in the cardiovascular system and in the respiratory, gastrointestinal, and genitourinary tracts, where they innervate both smooth muscle and glands.[27] VIP also has been identified in nerve fibers in the dorsal horn of the spinal cord.[28,29] Peripheral nervous systems involvement in VIP immunoreactivity includes the autonomic nervous system. This is addressed in the 1988 volume by Costa et al.,[30] who describe the extensive involvement of VIP immunoreactivity in the autonomic nervous system, including the colocalization with multiple transmitters and the difficulties this presents in sorting out the individual actions of any single transmitter.

A 1989 publication by Schaad, Schorderet, and Magistretti,[31] of Geneva and Lausanne, Switzerland, extends prior studies and describes VIP-stimulated accumulation of cyclic AMP in mouse cerebral cortical slices to be potentiated by noradrenaline, histamine, adenosine, baclofen (a $GABA_B$-receptor agonist), phorbol esters, and ouabain. The potentiation with histamine and adenosine, as with norepinephrine, was found to be inhibited by indomethacin. It was not inhibited by protein kinase C inhibition.

Also in 1989 Wang and Aghajanian[32] reported that VIP directly excites locus ceruleus neurons by inducing a largely Na^+-dependent inward current, and this response is mediated through a pertussis-toxin-insensitive G protein.

Brenneman et al.,[33] in 1988, described in *Nature* an important neuroprotective effect of VIP and core pentapeptides of VIP and peptide T. In 1990 Pincus et al.,[34] of Cornell University Medical Center (New York City), examined a role for cyclic AMP in the regulation of neuroblast mitosis and survival by VIP. They found that VIP increased the cyclic AMP content of cultured rat sympathetic neuroblasts and that cyclic AMP stimulated DNA synthesis and survival in neuroblast cultures, replicating the effects of VIP. The authors concluded that the VIP-cyclic AMP-signaling pathway may be involved in the regulation of neuronal development.

Magistretti, Palacios, and associates,[35] of Lausanne and Basel, in 1990, described a phylogenetic autoradiographic analysis of the distribution of VIP binding sites in the vertebrate central nervous system. This study included fish, frog, snake, pigeon, rat, mouse, guinea pig, cat, and monkey brains. Their findings demonstrated that VIP recognition sites appear early in vertebrate evolution. Although VIP recognition sites were not found in teleost bony-fish brains, they were present in anurian amphibian brains. In all of the above species studied, VIP recognition sites were more localized in forebrain areas. These included the cerebral cortex, the hippocampal formation, basal ganglia, the amygdala, some thalamic and hypothalamic nuclei, and the olfactory bulb. The highest densities of VIP binding sites were found in rodent and pigeon brains. Cat and monkey brains were characterized by high levels of nonspecific binding. A good conservation in the distribution of VIP binding sites was found; and in all species studied, with the exception of the snake, the cerebral cortex was densely labeled.

In 1991 Brenneman and associates,[36] at the National Institute of Mental Health (Bethesda, MD) reported that spontaneous electrical activity regulates VIP expression in dissociated spinal cord cell cultures. Dissociated spinal-cord/dorsal-root ganglia (SC-DRG) cells in culture from embryonic mice were employed in these studies. This study was designed to investigate activity-dependent expression of VIP in these cells, as expressed in VIP mRNA with and without electrical blockade by tetrodotoxin (TTX). Application of 8-bromo-cyclic AMP, NMDA, substance P, muscimol, calcium ionophore (A23187), and VIP to electrically active cultures resulted in a two- to threefold increase in VIP mRNA, while phorbol ester (PMA) and 8-bromo-cyclic GMP had no effect. In electrically inactive cultures, a three- to fourfold increase in VIP mRNA was produced by phorbol ester, cyclic AMP, and VIP, while the other molecules had no effect. The authors interpreted these findings as indicating that regulation of VIP gene expression in embryonic spinal cord neurons shows a temporal sensitivity to TTX-induced electrical blockade and that this response may be mediated by multiple neurotransmitter inputs that converge on cyclic AMP- and calcium-related processes in an activity-dependent manner. They concluded that electrical activity during development not only regulates neuronal development, but also controls the phenotypic expression of VIP. It is suggested that this activity dependence may be comparable to functionally dependent plasticity in adult neurons.

Also in 1991 Lamperti et al.,[37] from Boston and Worcester, described characterization of the gene and messages for VIP in rat and mouse. They found that all the detectable mRNA from this gene contains both VIP and PHI (peptide histidine isoleucine) coding sequences.

III. VASCULAR SYSTEM

The 1988 volume also contained a valuable discussion by Edvinsson et al.[38] of the role of VIP (and PHI) in the cerebral vascular circulation. Well-developed plexuses of VIP-containing nerve fibers have been shown to be present at the adventitia-medial border in the walls of cerebral cortical pial arteries and in the major arteries to be found at the base of the brain.[39-42] Vessels arising from the rostral part of the circle of Willis (anterior and middle cerebral arteries) tend to have more VIP-immunoreactive fibers than those from the caudal portions of the circle of Willis (anterior and middle cerebral arteries).

Studies by Said and associates, and others, showed that VIP exerts a vasodilatory effect on cerebral vessels both *in vitro*[40,42-45] and *in situ*.[46-48] Said and associates[48-50] and McCulloch and Edvinsson[47] also reported VIP-induced increases in cerebral circulation.

In 1991 Hokfelt and associates,[51] at the Karolinska Institute in Stockholm, described VIP-immunoreactive neuronal systems in the basal hypothalamus of the rat, with reference to the portal (intracranial) vasculature. They employed a combined immunohistochemical and *in situ* hybridization methology. Their studies were to explore the classical 1955 concept of Harris,[52] which holds that the secretion of hormones from the anterior pituitary is controlled by factors that are produced in the brain, released in the median eminence, and transported by the portal venous vascular system to the anterior pituitary gland. The findings of the 1991 Hokfelt studies demonstrated that VIP is present in at least three systems of direct neuroendocrine importance. These systems are in the nerves that are related to the blood vessels in the median eminence and that presumably are involved in the control of blood through the portal system, and in the parvocellular, paraventricular neurons, and are presumably involved in stress-induced prolactin release and in magnocellular neurons after certain experimental manipulations, including hypophysectomy.

IV. IMMUNE RESPONSE SYSTEM

Immune-response-system-related actions of VIP include VIP binding to blood mononuclear cells, leading to a production of an increase in their cyclic AMP levels. This finding was reported in 1981 by Guerrero et al.[53] The binding was time, temperature, and pH dependent and was reversible, saturatable, and specific. There appeared to be two classes of binding sites. At one site, VIP was displaceable by secretin, with a 400-fold lower affinity. Secretin is a member of the VIP "family", along with glucagon and GIP. The mononuclear cells were from human peripheral blood (60% T cells, 25% B cells, and 15% monocytes). VIP was found to be a potent and efficient stimulator of blood mononuclear cell cyclic AMP.

A 1979 publication by O'Dorisio et al.[54] reported the identification of VIP in leukocytes and lymphocytes, normal and leukemic — a major contribution to identifying an involvement of importance for VIP in the immune response.

VIP also is concentrated in mast cells, where it appears to be associated with histamine and is released by histamine releasers. This important observation was reported in *Nature* in 1978 by Said and associates.[55] Mast cells in this study were obtained from rat lung and peritoneal cavity. Also in 1978, Said and associates described the finding of VIP in rabbit platelets.[56]

O'Dorisio and associates have shown that there are receptors for VIP on both neurons and lymphocytes, and VIP acts on both cell types by the cyclic AMP second-messenger mechanism.[57] They also have demonstrated that VIP has an inhibitory effect on lymphocyte transformation by mitogens.[58] Ottaway,[59] in 1984, reported that exposure of T cells to VIP produced a dose-dependent decrease of VIP receptors on these cells and affected their migration *in vivo* in such a way that T cells with reduced VIP receptors localized less well after intravenous transfer in Peyers patches and mesenteric lymph nodes than did untreated cells with greater numbers of VIP receptors.

In the supported-sheep lymphocyte traffic model, the author, in 1982[60] and again in 1984[61] and 1985,[62] reported that a single acute infusion of VIP (50 μg) into a cannulated afferent lymphatic of a popliteal lymph node produces a prompt, sharp, and often prolonged depression in the output of small recirculating lymphocytes into popliteal lymph node efferent lymph and lymphatics, chronically cannulated in awake and nonsedated sheep (Figure 1). In 1988 the author with Spruck and Said (discoverer of VIP with Mutt) reported additional and extended studies of this VIP-induced depression of lymphocyte traffic.[63] The depression in the output of small recirculating lymphocytes was found to be prolonged beyond 20 h when preinfusion levels in efferent lymph of both recirculating and blast lymphocytes were at relatively normal levels, indicating a minimum of nodal drainage area in inflammation (Figures 2 and 3). No consistent alteration in blast-lymphocyte outputs was encountered in these circumstances. When blast-lymphocyte outputs were somewhat elevated preinfusion, suggesting some nodal drainage area inflammation, the output of nonrecirculating blast lymphocytes also was promptly and sharply depressed (Figure 4).

Also in 1988 the author with Spruck and Said[64] published a report in *Immunopharmacology* relating to *in vivo* depression of lymphocyte traffic in sheep by VIP and HIV (AIDS)-related peptides. The lymphocyte-traffic-depressing effects of peptide TDNYT, a core pentapeptide of VIP (VIP[7–11]) was found to have lymphocyte-traffic-depressing effects similar to the whole VIP molecule (VIP[1–28]) (Figures 5 and 6). The lymphocyte-traffic-depressing effects of peptide TTNYT were similar to both peptide TDNYT and VIP (Figure 7). This finding was of particular interest with respect to the variance of only one amino acid residue between TTNYT and peptide TDNYT (VIP[7–11]) and the nature of TTNYT as a core pentapeptide of peptide T (peptide T[4–8]), a putative anti-HIV/AIDS virus drug. This finding also was of interest with respect to the exciting findings published in *Nature* in 1988, by Brenneman et al.,[33] describing a protective effect of VIP, TDNYT, and TTNYT, as well as peptide T, against neuronal damage from culture-added HIV virus envelope gp120 molecule. The lymphocyte traffic effects of peptide TTNYT and VIP were found to be

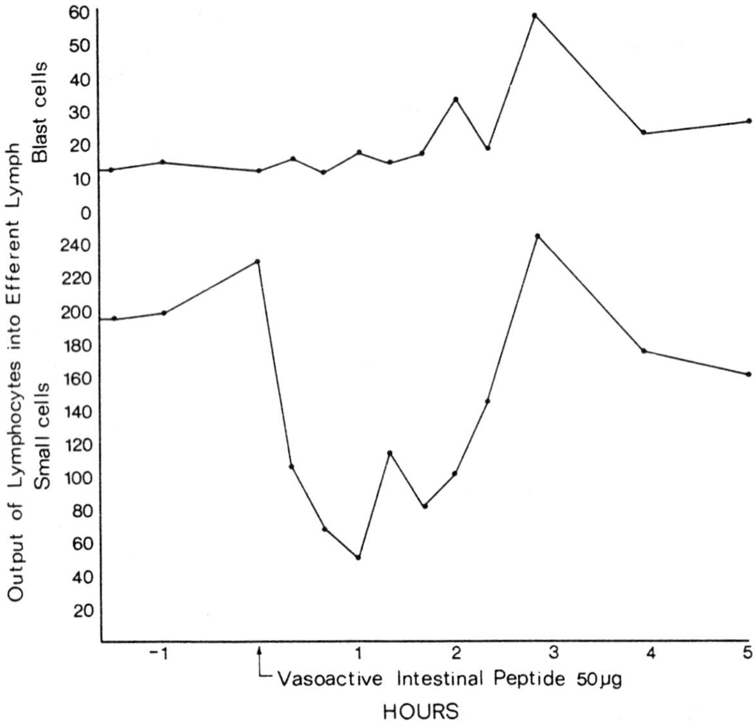

FIGURE 1. Lymphocyte traffic effects of an acute infusion of 50 µg of vasoactive intestinal peptide (VIP) into a cannulated afferent lymphatic of a popliteal lymph node of a sheep. There was a prompt and sharp depression in the output of small recirculating lymphocytes into the efferent lymphatic and lymph (chronically cannulated). The output of nonrecirculating blast lymphocytes was not effected in this study. (From Moore, T. C., *Immunology,* 52, 511, 1984. With permission.)

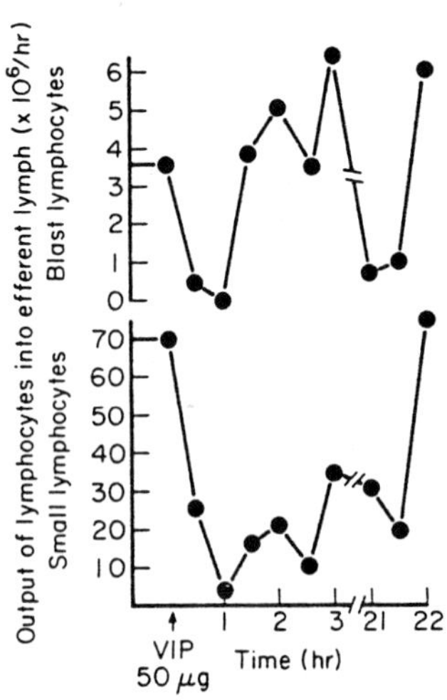

FIGURE 2. A prompt and sharp depression in traffic and output of small recirculating lymphocytes following acute infusion of VIP (50 µg) into a cannulated popliteal lymph-node afferent lymphatic. The depression in small recirculating lymphocytes in this study persisted for more than 20 h. Blast-lymphocyte output was depressed only in the initial postinfusion determinations in the early postinfusion period. (From Moore, T. C., Spruck, C. H., and Said, S. I., *Immunology,* 64, 475, 1988. With permission.)

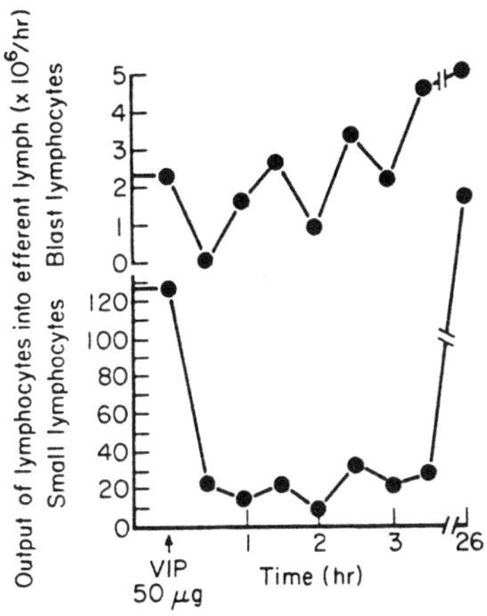

FIGURE 3. A prompt and persistent depression in the output of small recirculating lymphocytes into efferent lymph of a sheep popliteal lymph node following acute infusion of VIP (50 μg) into a cannulated afferent lymphatic of this lymph node. Blast-lymphocyte outputs fluctuated with an upward trend. (From Moore, T. C., Spruck, C. H., and Said, S. I., *Immunology,* 64, 475, 1988. With permission.)

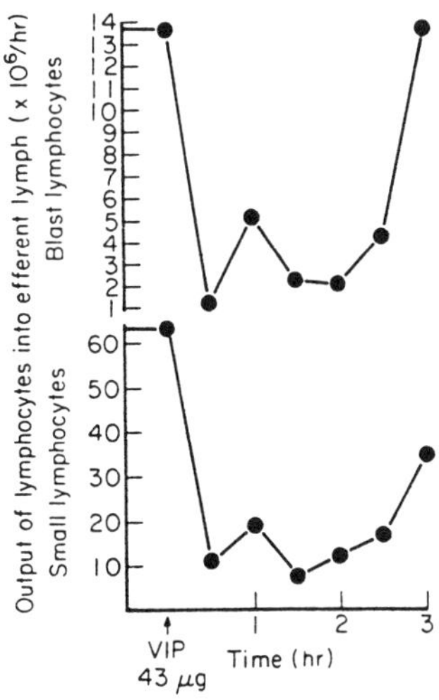

FIGURE 4. A prompt and sharp depression in the output into efferent lymph of both small recirculating and blast lymphocytes after acute infusion of VIP (50 μg) into a cannulated popliteal afferent lymphatic. In this study the preinfusion outputs of blast lymphocytes was quite high in comparison to the outputs of blast lymphocytes shown in Figures 2 and 3. (From Moore, T. C., Spruck, C. H., and Said, S. I., *Immunology,* 64, 475, 1988. With permission.)

comparable with respect both to small recirculating lymphocytes and to nonrecirculating blast lymphocytes that originate in the study lymph node (Figure 8).

In 1990 the author and associates[65] presented evidence for a differential and comparable effect of VIP and peptide TTNYT on the traffic of T4 (CD4), T8 (CD8), and B lymphocytes in this sheep model (Figure 9).

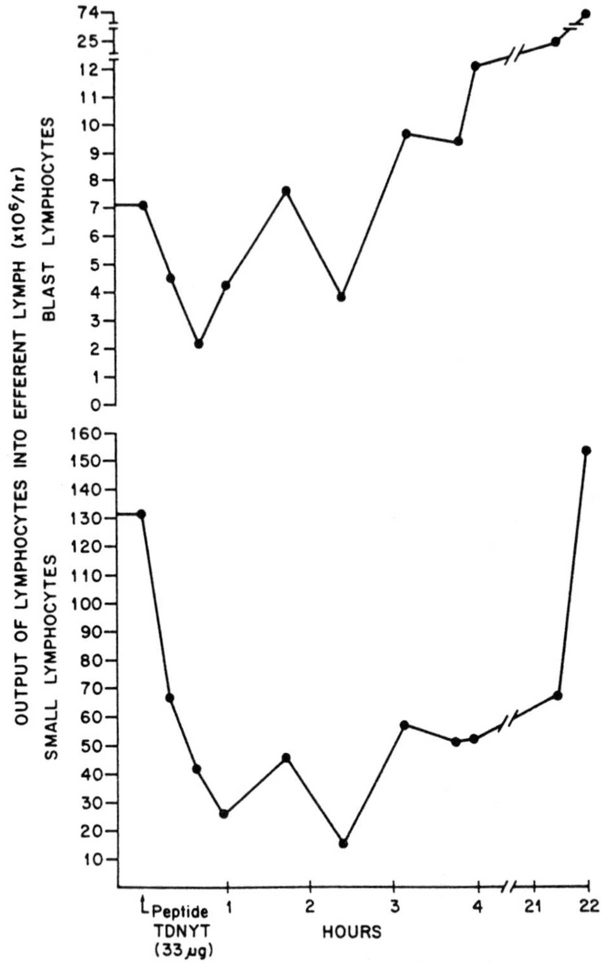

FIGURE 5. The lymphocyte-output-depressing effects of an acute popliteal lymph-node afferent lymphatic infusion of 33 μg of peptide TDNYT, the core pentapeptide of VIP (VIP[7–11]). The output-depressing effect on small recirculating lymphocytes was prolonged more than 20 h, while that on blast lymphocytes was variable and with an upward trend, somewhat like that shown in Figure 2.

V. CONCLUSION

In the 2 decades that have passed since the discovery of vasoactive intestinal peptide (VIP) by Said and Mutt, much has been learned about the multiple and important actions of this major vasoactive neuropeptide molecule in the nervous, vascular, and immune response systems, in addition to the gut, the source of its initial discovery, and characterization. Like many of the important vasoactive neurotransmitter and neuromodulator molecules considered in this book, VIP also has impressive phylogenetic credentials. As the text and references section of this chapter will clearly show, the multiple and extensive contributions to the study of this important molecule by its codiscoverer, S. I. Said, have been truly monumental and are continuing.

The high density of VIP receptors in the cerebral cortex, hippocampus, striatum, and thalamus, and its close interrelationship with the noradrenergic nervous system in these and other areas, and their mutual and cooperative elevation of cyclic AMP levels in brain neuronal

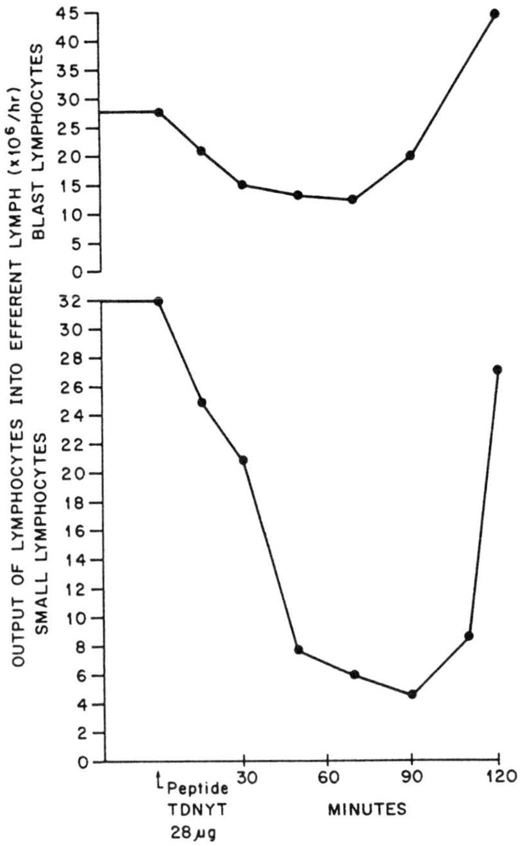

FIGURE 6. The lymphocyte-traffic- and output-depressing effects of an acute infusion of 28 μg of peptide TDNYT (VIP[7–11]) into a cannulated popliteal lymph-node afferent lymphatic. (From Moore, T. C., Spruck, C. H., and Said, S. I., *Immunopharmacology,* 16, 181, 1988. With permission.)

cells are of rather obvious, but as yet undetermined, significance with respect to normal and abnormal function. The demonstration by Brenneman and associates of a neuroprotective effect of VIP against damage by the HIV/AIDS virus envelope gp120 is of considerable basic clinical importance and, with other studies by this group, suggests that the HIV/AIDS virus may employ VIP receptors on neuronal cells and CD4 lymphocytes for entry into these cells.

Professor O'Dorisio, of The Ohio State University, has exerted a pioneering and leadership role in the study of the involvement of VIP in the immune response system. Her identification of an effect of VIP in elevating lymphocyte cyclic AMP levels through VIP receptors on these cells was a major discovery and helped place VIP in the immunoinhibitory class of molecules. *In vivo* studies by the author and by the author, Said, and associates identified a lymphocyte-traffic-depressing effect of VIP in sheep, with differential effects on small recirculating lymphocytes of the CD4 subtype of T cells.

The 1991 report by Hokfelt and associates provided clear evidence for a very major role for VIP in neuroendocrine regulation.

Much remains to be learned about the multiple and complex actions of VIP and its interactions with other important molecules and their receptors, including norepinephrine and the HIV/AIDS virus gp120 envelope.

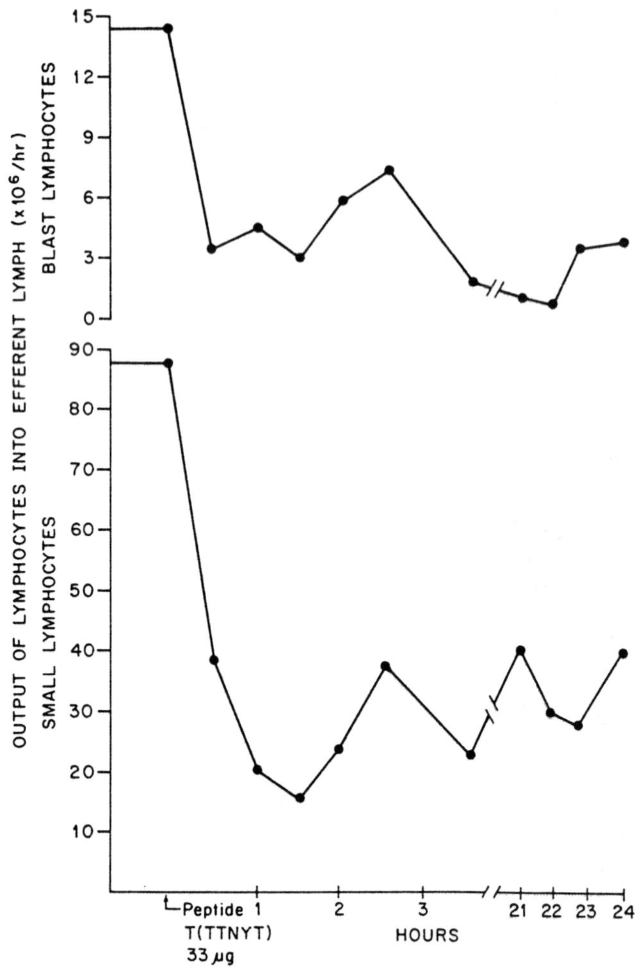

FIGURE 7. Graph demonstrating the lymphocyte-output-depressing effects of peptide TTNYT, a core pentapeptide of Peptide T, an HIV(AIDS)-related peptide from the gp120 external envelope of the HIV(AIDS) virus. This pentapeptide differs from peptide TDNYT (VIP[7–11]) by only one amino acid residue. (From Moore, T. C., Spruck, C. H., and Said, S. I., *Immunopharmacology*, 16, 181, 1988. With permission.)

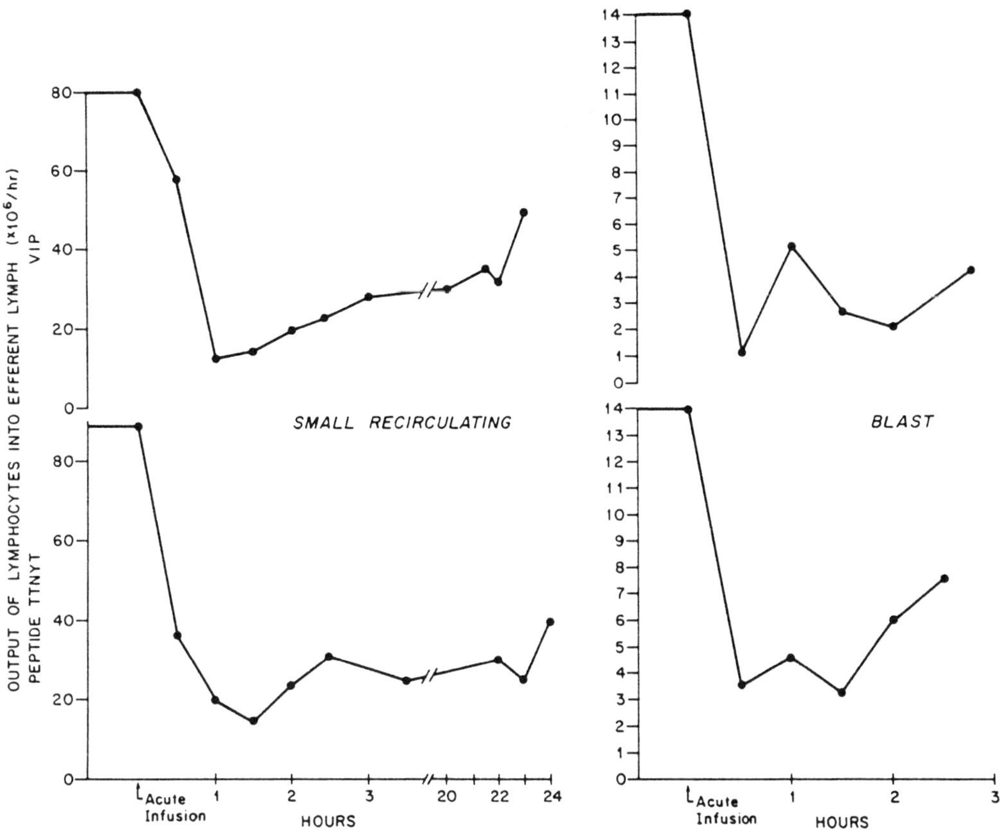

FIGURE 8. In studies of comparable preinfusion popliteal efferent lymph outputs of small recirculating and blast lymphocytes, acute infusions of peptide TTNYT (33 μg) and 50 μg of VIP (VIP[1–28]) into cannulated popliteal node afferent lymphatics were found to produce comparable effects (depressions) on the outputs of both small recirculating and blast lymphocytes. Comparable protective effects by VIP and peptide TTNYT on cultured neurons exposed to HIV gp120 have been reported by Brenneman, D. E., et al.[33] (From Moore, T. C., Spruck, C. H., and Said, S. I., *Immunopharmacology,* 16, 181, 1988. With permission.)

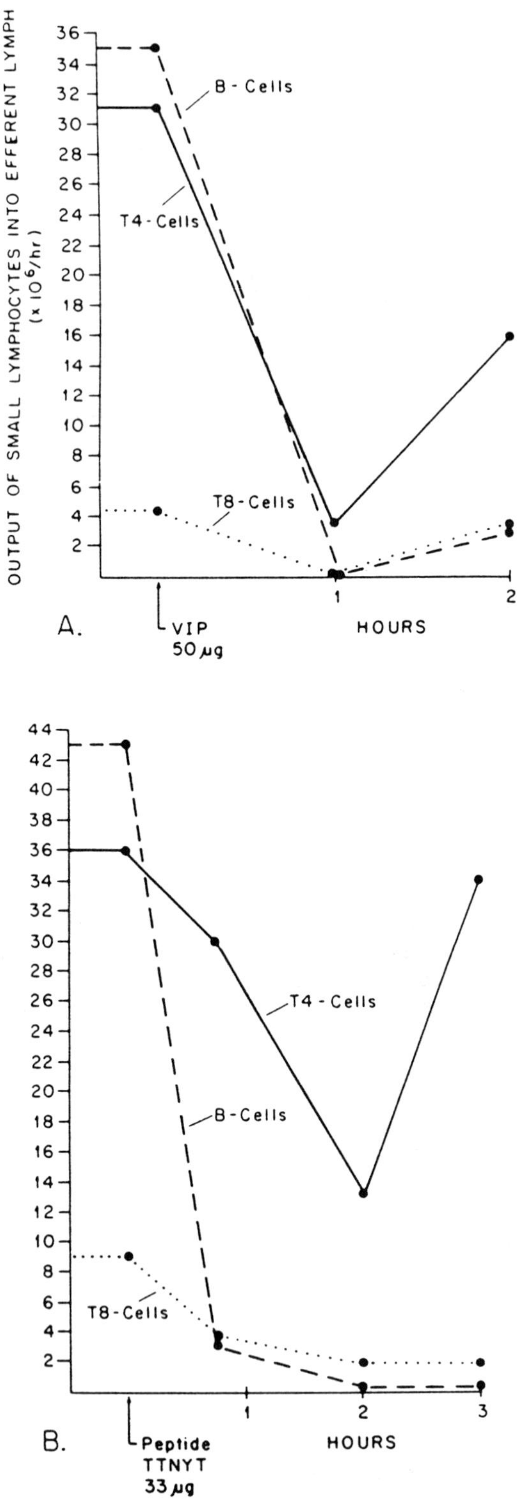

FIGURE 9. Acute infusions of VIP (VIP[1–28]) and peptide TTNYT (Peptide T[4–8]) into cannulated sheep popliteal lymph-node afferent lymphatics produced comparable effects on the outputs of T4 (CD4), T8 (CD8), and B lymphocytes into nodal efferent lymph, chronically cannulated, in awake, nonsedated, and supported sheep. (From Moore, T. C., et al., *Ann. N.Y. Acad. Sci.*, 594, 403, 1990. With permission.)

REFERENCES

1. **Said, S. I. and Mutt, V.**, Potent peripheral and splancnic vasodilator peptide from normal gut, *Nature,* 225, 863, 1970a.
2. **Said, S. I. and Mutt, V.**, Polypeptide with broad biological activity: isolation from gut, *Science,* 169, 1217, 1970b.
3. **Said, S. I. and Mutt, V.**, Isolation from porcine intestinal wall of a vasoactive octacosa peptide related to secretion and gastrin, *Eur. J. Pharmacol.,* 28, 199, 1972.
4. **Sundler, F., Hakanson, R., Alumets, J., and Walles, B.**, Neuronal localization of pancreatic peptide (PP) and vasoactive intestinal peptide (VIP) immunoreactivity in the earthworm *(Lumbricus terrestis), Brain Res. Bull.,* 2, 61, 1977.
5. **Dimaline, R., Young, J., Thwaites, D. T., Lee, C. M., and Thorndyke, M. C.**, Amino acid sequence of a biologically active vasoactive intestinal peptide from the elasmobranch *Scyliorhinus canicula, Ann. N.Y. Acad. Sci.,* 527, 621, 1988.
6. **Said, S. I. and Mutt, V.**, Relationship of spasmogenic and smooth muscle relaxant peptides from normal lung to other vasoactive compound, *Nature,* 265, 84, 1977.
7. **Said, S. I. and Rosenberg, R. N.**, Vasoactive intestinal peptide: abundant immunoreactivity in neural cell lines and nervous tissue, *Science,* 192, 907, 1976.
8. **Giachetti, A., Said, S. I., Reynolds, R. C., and Koniges, F. C.**, Vasoactive intestinal peptide in brain. Localization in and release from isolated nerve terminals, *Proc. Natl. Acad. Sci. U.S.A.,* 74, 3424, 1977.
9. **Said, S. I., Giachetti, A., and Nicosia, S.**, VIP: possible functions as a neural peptide, in *Neural Peptides and Neuronal Communication,* Costa, E. and Trabucchi, M., Eds., Raven Press, New York, 1980, 75.
10. **Giachetti, A. and Said, S. I.**, Axonal transport of vasoactive intestinal peptide in sciatic nerve, *Nature,* 281, 574.
11. **Giachetti, A., Said, S. I., Reynolds, R. C., and Koniges, F. C.**, Vasoactive intestinal peptide in brain. Localization in and release from isolated nerve terminals, *Proc. Natl. Acad. Sci. U.S.A.,* 74, 3424, 1977.
12. **Matsuzaki, Y., Hamasaki, Y., and Said, S.** Vasoactive intestinal peptide: a possible transmitter of non-adrenergic relaxation of guinea pig airways, *Science,* 210, 1252, 1980.
13. **Said, S. I.**, Vasoactive intestinal peptide (VIP): isolation, distribution, biological action, structure-function relationships and possible functions, in *Gastrointestinal Hormones,* Glass, G. B. J., Eds., Raven Press, New York, 1980, 246.
14. **Said, S. I.**, Peptides common to the nervous system and the gastrointestinal tract, in *Frontiers in Neuroendocrinology,* Vol. 6, Martini, L. and Ganong, W. F., Eds., Raven Press, New York, 1980, 293.
15. **Albuquerque, R. H., Owens, C. W. I., and Bloom, S. R.**, A study of vasoactive intestinal peptide (VIP) stimulated intestinal fluid secretion in rat and its inhibition by indomethacin, *Experentia,* 35, 1496, 1979.
16. **Moody, T. W., Getz, R., Fremeau, R. T., Jr., and Shaffer, M. M.**, Vasoactive intestinal peptide and secretin receptors in rat brain, *Ann. N.Y. Acad. Sci.,* 527, 657, 1988.
17. **Besson, J.**, Distribution and pharmacology of vasoactive intestinal peptide receptors in the brain and pituitary, *Ann. N.Y. Acad. Sci.,* 527, 204, 1988.
18. **Besson, J., Dussaillant, M., Marie, J. C., Rostene, W., and Rosselin, G.**, *In vitro* autoradiographic localization of vasoactive intestinal peptide (VIP) binding sites in the rat central nervous system, *Peptides,* 5, 339, 1984.
19. **Besson, J., Sarrieau, A., Vial, M., Marie, J. C., Rosselin, G., and Rostene, W.**, Characterization and autoradiographic distribution of vasoactive intestinal peptide binding sites in the rat central nervous system, *Brain Res.,* 398, 329, 1986.
20. **Morrison, J. H.**, Functional implications of the radial organization of VIP-containing neurons in the neocortex, *Ann. N.Y. Acad. Sci.,* 527, 130, 1988.
21. **Magistretti, P. J., Dietl, M. M., Hof, P. R., Martin, J.-L., Palacios, J. M., Schaad, N., and Schorderet, M.**, Vasoactive intestinal peptide as a mediator of intercellular communication in the cerebral cortex, *Ann. N.Y. Acad. Sci.,* 527, 110, 1988.
22. **Magistretti, P. J., Morrison, J. H., Shoemaker, W. J., Sapin, V., and Bloom, F. E.**, Vasoactive intestinal polypeptide induces glycogenolysis in mouse cortical slices: a possible regulatory mechanism for the local control of energy metabolism, *Proc. Natl. Acad. Sci. U.S.A.,* 78, 6535, 1981.
23. **Magistretti, P. J., Morrison, J. H., Shoemaker, W. J., and Bloom, F. E.**, Effect of 6-hydroxydopamine lesions on norepinephrine-induced [^{3}H]glycogene hydrolysis in mouse cortical slices, *Brain Res.,* 261, 159, 1983.
24. **Magistretti, P. J. and Schorderet, M.**, VIP and noradrenaline act synergistically to increase cyclic AMP in cerebral cortex, *Nature,* 308, 280, 1984.
25. **Magistretti, P. J. and Schorderet, M.**, Norepinephrine and histamine potentiate the increases in cAMP elicited by vasoactive intestinal polypeptide in mouse cerebral cortical slices: mediation by α_1-adrenergic and H1-histaminergic receptors, *J. Neurosci.,* 5, 362, 1985.

26. **Ferron, A., Siggins, G. R., and Bloom, F. E.,** Vasoactive intestinal polypeptide acts synergistically with norepinephrine to depress spontaneous discharge rate in cerebral cortical neurons, *Proc. Natl. Acad. Sci. U.S.A.,* 82, 8810, 1985.
27. **Sundler, F., Ekblad, E., Grunditz, T., Hakanson, R., and Uddman, R.,** Vasoactive intestinal peptide in the peripheral nervous system, *Ann. N.Y. Acad. Sci.,* 527, 143, 1988.
28. **Emson, P. C., Gilbert, R. F. T., Loren, I., Fahrenkrug, J., Sundler, F., and Schaffalitzky de Muckadell, O. B.,** Development of vasoactive intestinal peptide (VIP) containing neurons in the rat brain, *Brain Res.,* 177, 437, 1979.
29. **Gibson, S. J., Polak, J. M., Bloom, S. R., and Wall, P. D.,** The distribution of nine peptides in rat spinal cord with special emphasis on the substantia gelatinosa and on the area around the central canal (Lamina X), *J. Comp. Neurol.,* 201, 65, 1981.
30. **Costa, M., Furness, J. B., Gibbins, I. L., Morris, J. L., Bornstein, J. C., Llewellyn-Smith, I. J., and Murphy, R.,** Colocalization of VIP with other neuropeptides and neurotransmitters in the autonomic nervous system, *Ann. N.Y. Acad. Sci.,* 527, 103, 1988.
31. **Schaad, N. C., Schorderet, M., and Magistretti, P. J.,** Accumulation of cyclic AMP elicited by vasoactive intestinal peptide is potentiated by noradrenaline, histamine, adenosine, baclofen, phorbol esters, and ouabain in mouse cerebral cortical slices: studies on the role of arachidonic acid metabolites and protein kinase C, *J. Neurochem.,* 53, 1941, 1989.
32. **Wang, Y.-Y. and Aghajanian, G. K.,** Excitation of locus coeruleus neurons by vasoactive intestinal peptide: evidence for a G-protein-mediated inward current, *Brain Res.,* 500, 107, 1989.
33. **Brenneman, D. E., Westbrook, G. L., Fitzgerald, S. P., Ennist, D. L., Elkins, K. L., Ruff, M. R., and Pert, C. B.,** Neuronal cell killing by the envelope protein of HIV and its prevention by vasoactive intestinal peptide, *Nature,* 335, 639, 1988.
34. **Pincus, D. W., DiCicco-Bloom, E. M., and Black, I. B.,** Vasoactive intestinal peptide regulation of neuroblast mitosis and survival: role of cAMP, *Brain Res.,* 514, 355, 1990.
35. **Dietl, M. M., Hof, P. R., Martin, J.-L., Magistretti, P. J., and Palacios, J. M.,** Autoradiographic analysis of the distribution of vasoactive intestinal peptide binding sites in the vertebrate central nervous system: a phylogenetic study, *Brain Res.,* 520, 14, 1990.
36. **Agoston, D. V., Eiden, L. E., Brenneman, D. E., and Gozes, I.,** Spontaneous electrical activity regulates vasoactive intestinal peptide expression in dissociated spinal cord cell cultures, *Mol. Brain Res.,* 10, 235, 1991.
37. **Lamperti, E. D., Rosen, K. M., and Villa-Komaroff, L.,** Characterization of the gene and messages for vasoactive intestinal polypeptide (VIP) in rat and mouse, *Mol. Brain Res.,* 9, 217, 1991.
38. **Edvinsson, L.,** Role of vasoactive intestinal peptide and peptide histidine isoleucine in the cerebral circulation, *Ann. N.Y. Acad. Sci.,* 527, 378, 1988.
39. **Edvinsson, L., Fahrenkrug, J., Hanko, J., Owman, C., Sundler, F., and Uddman, R.,** VIP (vasoactive intestinal polypeptide)-containing nerves of intracranial arteries in mammals, *Cell. Tissue Res.,* 208, 135, 1980.
40. **Edvinsson, L. and McCulloch, J.,** Distribution and vasomotor effects of peptide HI (PHI) in feline cerebral blood vessels *in vitro* and *in situ, Regul. Pept.,* 10, 345, 1985.
41. **Gibbins, I. L., Brayden, J. E., and Bevan, J. A.,** Perivascular nerves with immunoreactivity to vasoactive intestinal polypeptide in cephalic arteries of the cat: distribution, possible origins and functional implications, *Neuroscience,* 13, 1327, 1984.
42. **Larsson, L.-I., Edvinsson, L., Fahrenkrug, J., Hakanson, R., Owman, C., Schaffalitzky de Muckadell, O., and Sundler, F.,** Immunohistochemical localization of vasodilatory peptide (VIP) in cerebrovascular nerves, *Brain Res.,* 113, 400, 1976.
43. **Duckles, S. P. and Said, S. I.,** Vasoactive intestinal polypeptide as a neurotransmitter in the cerebral circulation, *Eur. J. Pharmacol.,* 78, 371, 1982.
44. **Lee, T. J.-F., Saito, A., and Berezin, I.,** Vasoactive intestinal polypeptide-like substance: the potential transmitter for cerebral vasodilation, *Science,* 224, 898, 1984.
45. **Bevan, J. A., Buga, G. M., Moskowitz, M. A., and Said, S. I.,** *In vitro* evidence that vasoactive intestinal peptide is a transmitter of neuro-vasodilation in the head of the cat, *Neuroscience,* 19, 597, 1986.
46. **Edvinsson, L., McCulloch, J., and Uddman, R.,** Feline cerebral veins and arteries: comparison of autonomic innervation and vasomotor responses, *J. Physiol. (Lond.),* 325, 161, 1982.
47. **McCulloch, J. and Edvinsson, L.,** Cerebral circulatory and metabolic effects of vasoactive intestinal polypeptide, *Am. J. Physiol.,* 238, H449, 1980.
48. **Wei, E. P., Kontos, H. A., and Said, S. I.,** Mechanism of action of vasoactive intestinal polypeptide on cerebral arterioles, *Am. J. Physiol.,* 239, H765, 1980.
49. **Heistad, D. D., Marcus, M. L., Said, S. I., and Gross, P. M.,** Effect of acetylcholine and vasoactive intestinal peptide on cerebral blood flow, *Am. J. Physiol.,* 239, H73, 1980.
50. **Wilson, D. A., O'Neill, J. T., Said, S. I., and Traystman, R. J.,** Vasoactive intestinal polypeptide and the canine cerebral circulation, *Circ. Res.,* 48, 138, 1981.

51. **Ceccatelli, S., Fahrenkrug, J., Villar, M. J., and Hokfelt, T.,** Vasoactive intestinal polypeptide/peptide histidine isoleucine immunoreactive neuron systems in the basal hypothalamus of the rat with special reference to the portal vasculature: an immunohistochemical and *in situ* hybridization study, *Neuroscience,* 43, 483, 1991.
52. **Harris, G. W.,** *Neural Control of the Pituitary Gland,* Edward Arnold, London, 1955.
53. **Guerrero, J. M., Prieto, J. C., Elorza, F. L., Ramirez, R., and Goberna, R.,** Interaction of vasoactive intestinal peptide with human blood mononuclear cells, *Mol. Cell. Endocrinol.,* 21, 151, 1981.
54. **O'Dorisio, M. S., Cataland, S., and O'Dorisio, T. M.,** Demonstration of vasoactive intestinal peptide (VIP) in normal and leukemic leukocytes, *Endocrinology,* 61, 153, 1979.
55. **Cutz, E., Chan, W., Track, N. S., Goth, A., and Said, S. I.,** Release of vasoactive intestinal peptide in mast cells by histamine liberators, *Nature,* 275, 661, 1978.
56. **Giachetti, A., Goth, A., and Said, S. I.,** Vasoactive intestinal peptide (VIP) in rabbit platelets and rat mast cells, *Fed. Proc.,* 37, 1129, 1978.
57. **O'Dorisio, M. S., Wood, C. L., and O'Dorisio, T. M.,** Vasoactive intestinal peptide and neuropeptide modulation of the immune response, *J. Immunol.,* 135, 792, 1985.
58. **Wenger, G. D. and O'Dorisio, M. S.,** Vasoactive intestinal peptide: messenger in a neuroimmune axis, *Ann. N.Y. Acad. Sci.,* 594, 104, 1990.
59. **Ottaway, C. A.,** *In vitro* alteration of receptors for vasoactive intestinal peptide changes the *in vivo* localization of mouse T cells, *J. Exp. Med.,* 160, 1054, 1984.
60. **Moore, T. C.,** The effects of inflammatory mediators and anaesthetic agents upon lymphocyte traffic of sheep *in vivo,* Ph.D. thesis, University of Cambridge (Trinity College), England, 1982.
61. **Moore, T. C.,** Modification of lymphocyte traffic by vasoactive neurotransmitter substances, *Immunology,* 52, 511, 1984.
62. **Moore, T. C.,** Mediator and physiological influences on lymphocyte traffic in sheep, in *Immunology of the Sheep,* Morris, B. and Miyasaka, M., Eds., Editiones Roche, Basel, 1985, 327.
63. **Moore, T. C., Spruck, C. H., and Said, S. I.,** Depression of lymphocyte traffic in sheep by vasoactive intestinal peptide (VIP), *Immunology,* 64, 475, 1988.
64. **Moore, T. C., Spruck, C. H., and Said, S. I.,** *In vivo* depression of lymphocyte traffic in sheep by VIP and HIV (AIDS)-related peptides, *Immunopharmacology,* 16, 181, 1988.
65. **Moore, T. C., Whitley, G. A., Lami, J. L., and Said, S. I.,** Neuro-vascular immunology: vasoactive neurotransmitters and cellular immunity, *Ann. N.Y. Acad. Sci.,* 594, 403, 1990.

Chapter 14

CALCITONIN-GENE-RELATED PEPTIDE (CGRP)

Calcitonin-gene-related peptide (CGRP) is a 37-amino-acid vasoactive neurotransmitter molecule that frequently is colocalized in neurons and projections with substance P. It is an important vasoactive neurotransmitter peptide with stimulatory and vasodilatory properties. Its precursor is encoded by alternatively spliced messenger RNA from the primary RNA transcript of the calcitonin gene.

CGRP has been reported to be a transmitter in both the peripheral and central nervous systems, with colocalization with substance P and tachykinins in peripheral sensory neurons,[1] in primary afferent fibers in the dorsal column of the spinal cord,[2] and in the limbic system of the brain.[3] It has been reported to have an excitatory effect on peripheral ganglion cells.[4] Findings following intrathecal administration suggest CGRP may act as an excitatory agent in nociceptive pathways.[5] It has been reported to stimulate PI turnover.[6]

CGRP also has potent effects on the vascular system. Intradermal injection of CGRP, in femtomole doses, has been reported to produce microvascular dilation, with increased blood flow and persistent local reddening. A relaxant effect on strips of rat aorta *in vitro* was found to be an endothelial-cell-dependent mechanism.[7] An interesting interrelationship exists with substance P relating to the vascular system. The extensive colocalization of substance P and CGRP in the peripheral nervous system, including lymph node innervation, suggests important regulatory and modulatory interreactions. A 1988 report by Brain and Williams[8] provided evidence that injection of CGRP with substance P into human skin converts the long-lasting vasodilatation induced by CGRP into a transient response.

A particularly important 1986 report from London and Stockholm, by Woolf and Wiesenfeld-Hallin,[9] provided a significant contribution to CGRP and substance P interrelationships and colocalization in peripheral sensory nerves. In a study of intrathecally administered substance P and CGRP (as well as L-glutamate and L-aspartate) on the excitability of rat flexor alpha-motor neurons activated monosynaptically by Ia afferents and polysynaptically by high-mechanothreshold cutaneous afferents, they found that at doses that did not modify the monosynaptic reflex, substance P, CGRP, and some factor released from sacral c fibers increased the excitability of the nociceptive reflex for prolonged periods, in a multiplicative fashion. The excitatory amino acids had no such action. The authors suggested that one role for c-afferent neuropeptides, such as CGRP and substance P, could be to function as long-acting gain modulation/enhancement of nociceptive inputs into the spinal cord. A 1988 study by Cridland and Henry[5] identified a role for CGRP in nociception.

With respect to the immune response system, CGRP has been observed to have relaxant effects on both vascular and extravascular capsular smooth muscle of the isolated blood-perfused spleen of anesthetized dogs.[10] Its concentration has been found to be increased in the dorsal-root ganglia of adjuvant arthritic rats.[11]

REFERENCES

1. **Yamano, M., Hillyard, C. J., Girgis, S., MacIntire, I., Emson, P. C., and Tohyama, M.,** Presence of substance P-like immunoreactive neurone system from the parabrachial area to the central amygdaloid nucleus of the rat with reference to coexistence with calcitonin-gene related peptide, *Brain Res.*, 451, 179, 1988.
2. **Tamatani, M., Senba, E., and Tohyama, M.,** Calcitonin gene-related peptide and substance P containing primary afferent fibers in the dorsal column of the rat, *Brain Res.*, 495, 122, 1989.
3. **Skofitsch, G. and Jacobwitz, D. M.,** Calcitonin gene-related peptide coexists with substance P in capsaicin sensitive neurons and sensory ganglia of the rat, *Peptides,* 6, 747, 1985.
4. **Dun, N. J. and Mo, N.,** Calcitonin gene-related peptide evokes distinct types of excitatory response in guinea pig coeliac ganglion cells, *Brain Res.*, 476, 256, 1989.
5. **Cridland, R. A. and Henry, J. L.,** Effects of intrathecal administration of neuropeptides on a spinal nociceptive reflex in the rat: VIP, galanin, CGRP, TRH, somatostatin and angiotensin II, *Neuropeptides,* 11, 23, 1988.
6. **Laufer, R. and Changeux, J.-P.,** Calcitonin gene-related peptide and cyclic AMP stimulate phosphoinositide turnover in skeletal muscle cells. Interaction between two second messenger systems, *J. Biol. Chem.*, 264, 2683, 1989.
7. **Brain, S. D., Williams, T. J., Tippins, J. R., Morris, H. R., and MacIntyre, I.,** Calcitonin gene-related peptide is a potent vasodilator, *Nature,* 313, 54, 1985.
8. **Brain, S. D. and Williams, T. J.,** Substance P regulates the vasodilator activity of calcitonin gene-related peptide, *Nature,* 335, 73, 1988.
9. **Woolf, C. and Wiesenfeld-Hallin, Z.,** Substance P and calcitonin gene-related peptide synergistically modulate the gain of the nociceptive flexor withdrawal reflex in the rat, *Neurosci. Lett.*, 66, 226, 1986.
10. **Withrington, P. G.,** The relaxant properties of human calcitonin gene-related peptide on vascular and extravascular capsular smooth muscle of the isolated blood-perfused spleen of the anaesthetized dog, *Br. J. Pharmacol.*, 96, 823, 1989.
11. **Kuraishi, Y., Nanayama, T., Ohno, H., Fugii, N., Otaka, A., Yajima, H., and Satoh, M.,** Calcitonin gene related peptide increases in the dorsal root ganglia of adjuvant arthritic rats, *Peptides,* 10, 447, 1989.

Chapter 15

SHORT-CHAIN LIPIDS

I. INTRODUCTION

Short (20-carbon)-chain lipids, known collectively as eicosanoids, are oxidation products of arachidonic acid, a polyunsaturated fatty acid which is a basic constituent of cells and their lipid membranes. Arachidonic acid is an essential fatty acid that may be formed from linoleic acid by the action of desaturase enzymes, or ingested directly. Arachidonic acid does not exist freely within cells. It is present in bound form as an integral part of phospholipids, triglycerides, and cholesterol esters. Phospholipids are the principal source of arachidonic acid, with the three major phospholipids binding arachidonic acid being phosphatidyl inositol, phosphatidyl choline, and phosphatidyl ethanolamine. These phospholipids are bound together in a bilayer to form the membranes of cells. Release of arachidonic acid from phospholipid stores requires hydrolysis of the ester linkage that joins arachidonic acid to the glycerol backbone of the phospholipid. Phospholipase A_2 is a major hydrolyzing enzyme in the eicosanoid production mechanism and constitutes, by its release and supply of free, unbound arachidonic acid, the rate-limiting step in the eicosanoid synthetic pathway. Phospholipase A_2 is a calcium-dependent enzyme, and any stimulus (electrical or chemical) that leads to free calcium fluxes across the cell membrane will activate this enzyme and lead to eicosanoid production.

The two major pathways of eicosanoid production are the cyclooxygenase and the lipoxygenase (Figure 1). The prostaglandins and thromboxanes are the principal products of the cyclooxygenase-pathway activation, including action of isomerases and synthetases. The lipoxygenase pathway leads to the formation of leukotrienes, lipoxins, and related compounds. In addition, epoxyeicosatrienic acid (EET) may be formed by the monooxygenase activity of cytochrome P-450. Arachidonic acid also is susceptible to autooxidation, with the resultant production of various hydroperoxy acids (HPETEs). These various pathways and sequential product formations constitute the arachidonic acid cascade (Figures 2 and 3). Of the major lipoxygenase enzymes (12-, 15-, and 5-lipoxygenase), 5-lipoxygenase is one of the most important in defense-related activities and acts on free arachidonic acid by way of 5-HPETE formation to produce the basic leukotriene (LTA_4) that can function as a substrate for enzymes producing the potent LTB_4 or the LTC_4-LTD_4-LTE_4 cascade of cysteinyl leukotrienes. These three lipoxygenases have been purified to homogeneity, and the 5- and 15-lipoxygenase cDNAs have been cloned recently.[1-4] The 5- and 15-lipoxygenases, interestingly, have many homologous sequences as well as regions homologous to plant lipoxygenases, such as soybean lipoxygenase.

The eicosanoids are extraordinarily versatile in their multiple biological activities. They are formed by essentially every tissue in the body and are extensively distributed throughout the animal kingdom. The actions of the various eicosanoids may vary from organ to organ and may differ substantially from species to species and within a single species and organ, with respect both to organ function and stages of organ development.

These remarkable arachidonic acid cascade molecules play important roles in the vascular system (blood clotting, inflammation, control of vascular tone), in the nervous system (including sleep producing PGD_2 and wake stimulation by PGE_2), and in the immune response system (both stimulating and inhibiting).

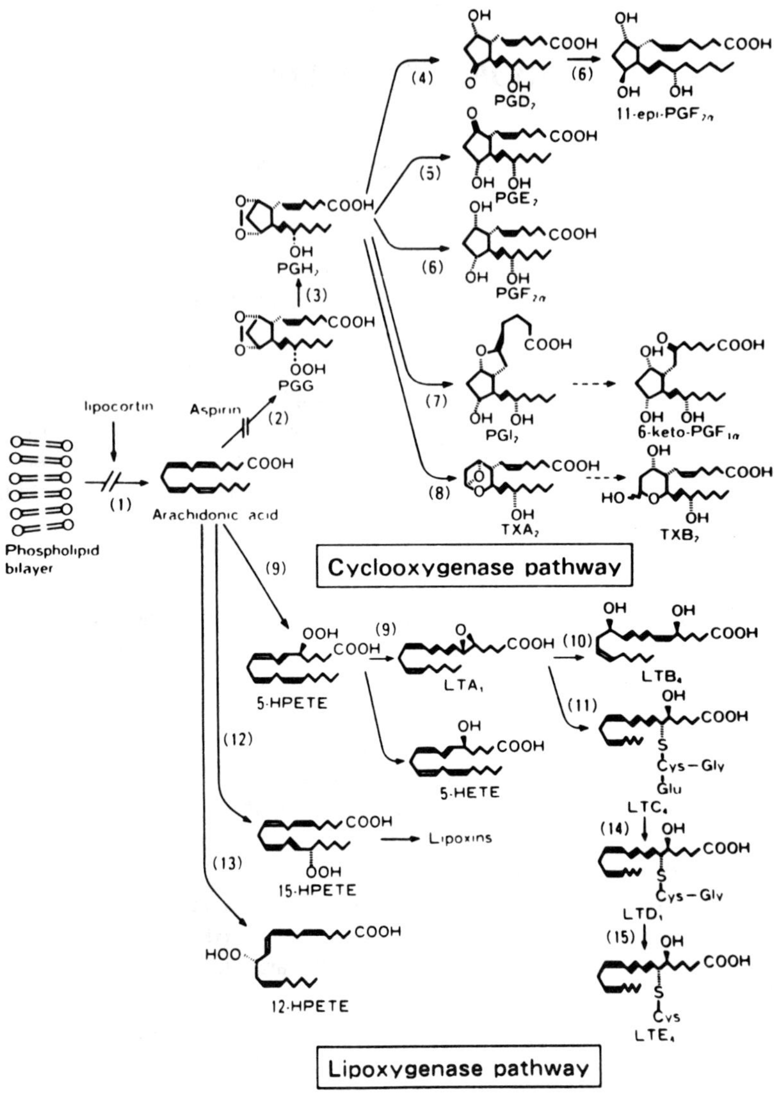

FIGURE 1. Short-chain (20-carbon) lipid metabolites of arachidonic acid by cyclooxygenase and lipoxygenase pathways with typical hair-pin molecular configurations. (From Shimizu, T. and Wolfe, L. S., *J. Neurochem.*, 55, 1, 1990. With permission.)

II. HISTORICAL BACKGROUND

1930. Two American gynecological surgeons, Kurzrok and Leib,[5] observed that strips of human uterus relax or contract when exposed to human semen.

1933–1936. Goldblatt[6,7] in England and von Euler[8-11] in Sweden independently observed vasodepressor and smooth-muscle-contracting activity in seminal fluid and from accessory genital glands. von Euler identified the active material as a lipid soluble acid and named it "prostaglandin".

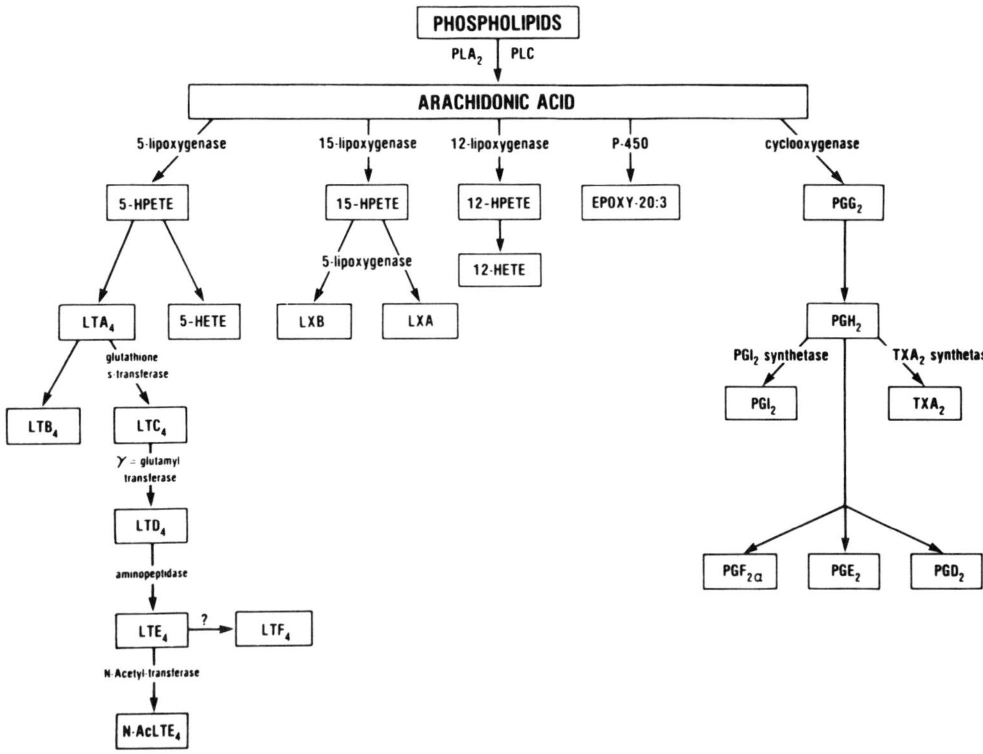

FIGURE 2. Arachidonic acid cascade illustrating the major pathways, including cyclooxygenase and lipoxygenase. (From Fuerstein, G. and Hallenbeck, J. M., *FASEB J.*, 1, 186, 1987. With permission.)

1960. Bergstrom and Sjovall,[12] in Sweden, isolated prostaglandin F from sheep prostate glands.

1962–1963. The chemical structures of prostaglandins E and F were characterized by Bergstrom, Ryhase, Samuelsson, and Sjovall.[13-15]

1964. Biosynthesis of prostaglandin E_2 (PGE_2) from arachidonic acid was achieved using homogenates of sheep seminal vesicle, by Bergstrom, Danielson, and Samuelsson[16] and by Van Dorp et al.,[17] and the flood gates were opened on prostaglandin research.

1968. Williams et al.[18] observed increased levels of prostaglandin in malignant tissue (medullary carcinoma of the thyroid).

1970. Milton and Wendlandt[19,20] of London identified a major role of PGE in systemic hyperthermia. In their initial study, they extracted a prostaglandin-like substance from the cerebrospinal fluid of a cat with an elevated body temperature. The volume was too small to assay. They got around this rather formidable obstacle by injecting a series of available prostaglandins into the third ventricle of conscious cats and observed the effects on rectal temperature. Very small doses of PGE_1 produced high elevations in temperature (as high as 3°C), with an immediate rise postinjection, accompanied by violent shivering and piloerection.

1971. Vane[21] described inhibition of prostaglandin synthesis as a mechanism of action of aspirin-like drugs, in a major *Nature* report, and opened an entirely new field of study and use of nonsteroid anti-inflammatory drugs.

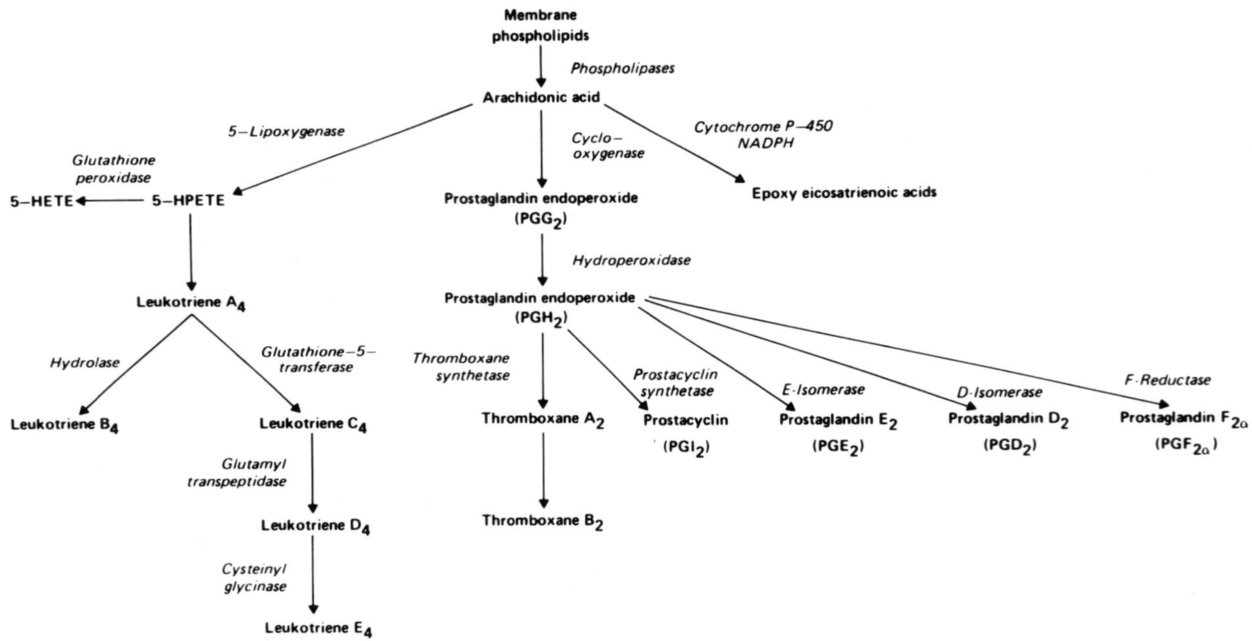

FIGURE 3. Arachidonic acid metabolic pathways with enzymatic components. (From Williams, K. I. and Higgs, G. A., *J. Pathol.*, 156, 101, 1988. With permission.)

1973. Two unstable cyclic endoperoxides, prostaglandin G_2 and prostaglandin H_2, were isolated and identified by Hamberg and Samuelsson[22] and by Nugtern and Hazelhof.[23]

1974. Hamberg and Samuelsson[24,25] discovered lipoxygenase enzymes that catalyze the oxygenation of arachidonic acid and other unsaturated fatty acids to the corresponding hydroperoxides. In contrast to the widely distributed cyclooxygenase, the lipoxygenases, in early studies, appeared to be restricted in location to white blood cells, platelets, and lung.

1974. An involvement of prostaglandins of the E series in cellular immunity *in vivo* was first observed. The author and Jaffe[26] reported (in *Transplantation*) the elevation of prostaglandin E in association with organ (heart) allografting in the rat. Prostaglandin E levels were elevated in both the heart allograft and in host lymphoid tissues (spleen and thymus). In a related study prostaglandin E levels also were found to be elevated in coronary sinus venous blood coming from the nonimmunosuppressed and rejecting allografted heart.[27] The allografting was heterotopic into the host abdominal aorta.

1975. The thromboxanes were discovered. The structure of thromboxane A_2 (TXA_2) and that of its metabolite thromboxane B_2 (TXB_2) were identified by Hamberg, Svensson, and Samuelsson.[28]

1976. The discovery of prostacyclin (PGI_2) was reported in *Nature* by Moncada et al.[29]

1977. Abdel-Halim, Hamberg, and associates[30] identified prostaglandin D_2 (PGD_2) as major prostaglandin in homogenates of rat brain.

1979. The leukotrienes were discovered by Samuelsson and associates.[31-37] This group of compounds from the lipoxygenase pathway of arachidonic acid metabolism was identified and characterized structurally by the Samuelsson group. The name "leukotriene" was suggested by Samuelsson because these compounds were derived from leukocytes and had a conjugated triene structure. The leukotrienes were shown to be noncyclized 20-carbon carboxylic acids with one or two oxygen substituents and three conjugated double bonds. Attack by lipoxygenase at position C5 on the arachidonic acid molecule inserts molecular oxygen to form an unstable 5-hydroperoxy acid intermediate (5-HPETE) that may be converted into another unstable intermediate, leukotriene A_4 (LTA_4), by the loss of a molecule of H_2O. LTA_4, an unstable epoxide with the characteristic structure of all leukotrienes, is the entrance compound to the leukotrienes. In their early studies, the Samuelsson group showed the leukotrienes to be potent substances in inflammatory reactions.

1979. Hayaishi and associates[38] of Japan reported PGD_2-synthetase (isomerase) levels to be highest in the brain, spinal cord, and intestine of the rat, with highest brain levels in the hypothalamus and thalamus. They suggested PGD_2 to be an important neuromodulator.

1980. Dy et al.[39] of Paris reported *in vivo* allografting-stimulated increase in PGE_2 production in both primary and secondary mixed lymphocyte cultures (MLCs).

1981. Mertin and Stackpoole,[40] in *Nature,* made an especially important observation that establishes the obligatory requirement of PGE_2 for the initiation of a cell-mediated immune response. They reported that anti-PGE_2 antiserum suppressed the development of experimental allergic encephalomyelitis (EAE) in rats and prevented the development of graft-vs.-host (GVH) and host-vs.-graft (HVG) responses in mice. In the GVH and HVG studies, where the kinetics could be evaluated, the anti-PGE_2 antiserum had to be given in the early hours (preferably at 0 hour) to suppress the development of these cell-mediated immune responses. Their findings suggested strongly that the effects of the anti-PGE_2 serum were on the induction and on the early phases of the cell-mediated immune response. These *in vivo* observations were of immense importance in understanding the role of PGE_2 in the cellular immune response.

1982. Hayaishi and associates[41,81] described induction of sleep by microinjection of PGD_2 into the preoptic area of conscious rats.

III. EVOLUTION

A. INVERTEBRATES

In 1984 Rich et al.[42] reported that leukotriene B_4 (LTB_4), a 5-lipoxygenase product of arachidonic acid oxidation, causes aggregation of dissociated cells of the Porifera marine sponge *Microciona prolifera.* Prior studies had shown this calcium-dependent aggregation also to be caused by a species-specific aggregation factor (MAF) and by calcium ionophores. No other lipoxygenase products of arachidonic acid were found to produce this aggregation.

The first report of the presence of eicosanoids in invertebrates appeared in 1969, by Weinheimer and Straggins.[43] In an in-depth study of the chemistry of cnidarian coelenterates, these authors discovered two eicosanoids, 15-epi-PGA_2 and its acetate methyl ester, in the gorgonian coral *Plexaura homomalla.* They were present in very large quantities. In 1985 Clare, Crisp, and associates,[44] from North Wales (Menai Bridge), reported (in the *Proceedings of the Royal Society*) that this eicosanoid substance in a fraction of the dried cortex of the same gorgonian coral, *P. homomalla,* also caused the eggs of the arthropod crustacean barnacle *Balanus balanoides* to hatch. They presented several lines of evidence to indicate that this hatching substance is in a prostaglandin. This evidence included the demonstration that aspirin prevented the formation of this hatching substance. It also is of interest that this barnacle hatching substance is nonspecific in that it may be produced by a variety of thoracian barnacles.

In 1974 Destephano et al.[45] described the presence, synthesis, and physiological function of eicosanoids in an invertebrate (in an arthropod insect). They identified the synthesis of prostaglandin by reproductive tissue of the male house cricket *Acheta domesticus.* This prostaglandin played a role in the egg-laying behavior of this house cricket.

In 1981 Loher et al.[46] from the University of California–Berkeley, detailed (in the *Proceedings of the National Academy of Science, U.S.A.*) their in-depth and important studies of the role of prostaglandins in the egg-laying of the Australian field cricket *Teleogryllus commodus.* Insemination of the female of this cricket terminates premating behavior that consists of strong circadian-controlled locomotion and attraction to the calling male. A successful mating also had been found by Loher to provoke increased ovoposition. As egg laying in this cricket does not appear to be released by the mechanical stimulation of mating or by the transfer of spermatophores from testectomized males, these authors, in an earlier report, had proposed the existence of a chemical "mating factor" that originates in the male's testis and is transferred during mating to the female (via a spermatophore) where it causes massive egg release. In their 1981 publication, the authors described the presence of PGE_2 in the spermathecae of mated females and its virtual absence from those of virgins and from male spermatophores. They also identified the presence of arachidonic acid, a PGE_2 precursor, in the spermathecae of virgin females. Substantial prostaglandin biosynthetic also was found in the spermatophores and spermathecae of mated crickets. The authors also showed that only nanogram quantities of PGE_2 are necessary to release oviposition, when injected into the oviduct. This is a rather remarkable resemblance of mammalian mating in which, during copulation, prostaglandins are transferred in seminal fluid to females where they elicit a number of physiological effects.

Further studies by this Berkeley group demonstrated an increase in spermathecal arachidonic acid immediately after mating, which could be accounted for by the amounts of arachidonic acid found in spermatophores.[47] Additional studies by this group tracked the pathway of arachidonic acid from a male to a female spermatheca and from there into the hemolymph (a vascular ''connection'').[47-53] Following injection of radioactive arachidonic acid into newly mated males, the labeled arachidonic acid was found not only in the testes and the accessory glands, but also in the phospholipids of the daily produced spermatophore. Following mating, radioactivity was identified in the female's spermatheca and later in her circulating hemolymph. By direct measurement, PGE_2 concentrations also were shown to increase in female hemolymph after mating.[49]

In molluscs, prostaglandins have been reported to influence spawning in abalone,[54] egg production by snails,[55] and ion fluxes with inhibition of sodium transport in freshwater mussels.[56] In a 1977 publication in *Science,* Morse et al.,[54] from the University of California–Santa Barbara, described hydrogen-peroxide-induced activation of prostaglandin endoperoxide synthetase (cyclooxygenase) as an inducer of spawning by gravid male and female red abalone, *Haliotis rufescens.* This response was blocked by exposure of the animals to aspirin. They also found abalone eggs to be a rich source of prostaglandin endoperoxide synthetase. Spawn liberated from abalones of either sex can trigger ''responsive'' spawning by animals of the opposite sex. Cyclooxygenase activation, as by hydrogen peroxide, also has been observed (by the authors) with crude enzyme from rabbit kidney and prostate and from the Pacific gorgonian soft coral *Lophogorgia chilensis.* These substances and extracts influenced only the cyclooxygenase pathway of arachidonic acid oxidation. They had no effect on lipoxygenase-pathway activation. They also found hydrogen peroxide activation of cyclooxygenase to induce synchronous spawning in female and male mussels, *Mytilus californianus.*

In 1979 Graves and Dietz[56] reported that PGE_2 inhibited sodium transfer in freshwater mussels, and in 1986 Kunigelis and Saleudin[55] described the involvement of prostaglandin in egg production in the freshwater molluscan gastropod *Helisoma.* A marine opisthobranch mollusc, *Tethys fimbria,* also has provided the first example of *in vivo* storage of prostaglandins. In a 1990 publication, Cimino et al.,[57] of Naples, Italy, described *in vivo* conversion of prostaglandins into corresponding lactones and the accumulation of PGE_2-1 and PGE_3-1 15-lactones in the appendages. Detachment of these appendages from the molested mollusc produced *in vivo* conversion of PGE_2 and PGE_3 lactones back to PGE_2 and PGE_3. These results provide the first example of an *in vivo* mechanism by means of which prostaglandins can be stored and, when needed, released. This clearly is an important defense-related response and likely involves ''memory'' at some level.

In 1987 Piomelli, Kandel, and associates,[58] of Columbia University, described (in *Nature*) the actions of lipoxygenase metabolites of arachidonic acid as second messengers for presynaptic inhibition of neurotransmission by sensory cells of the gastropod marine mollusc *Aplysia californica.* These inhibitory actions on identified *Aplysia* neurons were similar to those produced by histamine and by the endogenous inhibitory tetrapeptide, FMRF amide. The authors reported arachidonic acid to be a prominent component of *Aplysia* neuronal phospholipids: approximately 10% of the total fatty acid content. They also described metabolism of arachidonic acid by *Aplysia* neurons through the lipoxygenase 12- and 5-lipoxygenase pathways, as well as through the cyclooxygenase pathway.

The authors expanded these studies in major 1988 and 1989 publications.[59,60] Their studies suggested that lipoxygenase metabolites are involved in the inhibitory actions of FMRF amine and that the active ''second messenger'' derives from a hydroperoxy acid, possibly 12-HPETE, rather than from a hydroxy acid such as the leukotrienes. An additional active metabolite of 12-HPETE was described in 1988.[59] This metabolite, 12-KETE, was formed from endogenous lipid stores in *Aplysia* L14 and L10 cells (of the *Aplysia* histaminergic nervous system) by the application of histamine or 12-KETE to these cells and elicited changes in membrane potentials similar to those evoked by histamine. An additional active metabolite of 12-HPETE, 6-HEPETE, was identified by these authors in 1989.[60] This metabolite was found to elicit the late component (hyperpolarization/inhibition) of the dual-action response in *Aplysia* L14 cells. 12-HPETE had been found to mimic the total dual-action response of *Aplysia* L14 ink-motor neuronal cells to histamine and to stimulation of the L32 nerve cell circuit of the *Aplysia* abdominal ganglion. L32 cells in *Aplysia* are considered putative histaminergic neuronal cells.

In eichinoderms, prostaglandins have been observed to influence egg maturation in starfish[61] and to prevent polyspermia in sea urchins.[62] Shuel et al.,[62] in 1984, reported that cyclooxygenase-inhibiting drugs promoted polyspermic fertilization in sea urchins.

B. NONMAMMALIAN VERTEBRATES

One-, two-, and three-series prostaglandins have one, two, and three double bonds and are produced by cyclooxygenase activity on fatty acid substrates. Monoenoic (single double-bond) compounds are produced from eicosatrienoic acid substrate (ETA), dienoic (double double-bond) compounds of the two-series are produced from arachidonic acid, and trienoic (triple double-bond) compounds are produced from eicosa pentaenoic acids (EPA). All three substrates are found in elasmobranch,[63] teleost fish,[64-66] and amphibian tissues.[67-71] Significant amounts of arachidonic acid and EPA substrates have been found in tissues of these lower vertebrates, whereas levels of ETA were considerably lower. Environmental temperature also has been found to influence prostaglandin production and function. Both arachidonic acid and EPA have been found to be increased in association with acclimation to cold temperature in poikilotherms.[72,73]

In addition to the prostaglandin effects on ovulation and spawning observed in invertebrates, prostaglandins exert important effects on the cardiovascular systems of these lower vertebrates (elasmobranches, teleost fish, and amphibians). In the European eel, PGE_1, PGE_2, and PGA_2 decreased aortic blood pressure.[74] PGE_2 depressed aortic blood pressure in the rainbow trout, *Salmo gairdneri,* with an associated antidiuretic effect and hypotension. Bradycardia is followed by tachycardia and hyperventilation in the carp, *Cyprinus carpio* L.

Piomelli et al.,[75] in 1985, reported that elasmobranches differed from teleost bony fish in responses of gill vasculature to prostacyclin (PGI_2) and suggested a divergence in vascular actions of prostacyclin during vertebrate evolution. The responses of the elasmobranches (*Scyliorhinus stellaris* and *Torpedo marmorata*) were of transient vasoconstriction followed by dilation, whereas those of the teleost fish (*Anguilla anguilla, Conger conger, Scoroaena porcus,* and *Solea solea*) were of increased gill vascular resistence.

Prostaglandins were found to have potent cardiovascular effects in amphibians, with the PGEs, PGAs, and PGIs being hypotensive, and the PGFs and PGDs being hypertensive (bullfrogs).[67,76]

Prostaglandins also have been found to play important roles in gestation maintenance and in the induction of birth in reptiles.[77] Guillette et al.,[78] in 1988, described the synthesis of prostaglandins by the female reproductive tract of the viviparous lizard *Sceloporus jarrovi.* This group also found exogenous indomethacin to inhibit luteolysis and to delay parturition in this lizard.[79]

IV. PROSTAGLANDINS AND THROMBOXANE

A. NERVOUS SYSTEM

In the central nervous system, prostaglandins, particularly PGD_2 and PGE_2, have been shown (largely in recent years) to have three major functions. These are (1) sleep-wake regulation, (2) fever-hypothermia regulation, and (3) involvement in LTP/memory. In the peripheral nervous system, a major prostaglandin function, particularly for PGE_2 and PGI_2, is in pain transmission in sensory nerves, especially in areas of inflammation where the prostaglandins function in concert with bradykinin, histamine, substance P, and other vasoactive neurotransmitter molecules.

The extraordinary productive and important studies of Hayaishi and associates (in Kyoto and Osaka, Japan), during the past decade have dominated the studies of the roles of prostaglandins, especially PGD_2 and PGE_2, in the mammalian brain, particularly with respect to sleep-wake regulation and temperature regulation. The sleep-wake regulation studies were initiated by a serendipitous finding of somnolence during a study of PGD_2 production of hypothermia on injection into the preoptic area of the rat brain.[80] The preoptic-area microinjection produced a fall in the colonic temperature. During these experiments the investigators also observed that the PGD_2 injections into this area appeared to induce sleep. In a specific study to explore this provocative finding, both PGD_2 and control saline were microinjected into the preoptic area.[81] While control rats kept awake, the PGD_2-injected rats experienced markedly induced slow-wave sleep. Within 1 min after the microinjection, PGD_2 caused a relaxed body extension of the rats, with their heads lowered and eyes closed (asleep). The effect of PGD_2 on inducing slow-wave sleep was sustained throughout the experimental session of 200 min. The sleep patterns of the rats returned to the control level 220 to 350 min after the PGD_2 injection. When PGE_2 was injected into this area (preoptic), the rats kept awake with curled posture until their colonic temperatures reached the maximum elevation. The sleep-inducing effects of PGD_2 were not observed when it was microinjected into the posterior hypothalamus.

Studies by Hayaishi with Matsumura and associates, in the 1989 to 1991 period, clearly established a role for PGE_2 in wakefulness and dissociated the wakefulness response from the hyperthermia/fever response.[82-85] They also dissociated the hyperthermia and wakefulness responses anatomically, with hyperthermia/fever being in the preoptic area and wakefulness in the posterior hypothalamus. A brilliant summary to date of this extremely important work was published by Osamu Hayaishi in the August 1991 issue of *FASEB Journal* and is strongly recommended to the reader.

The important involvement of PGE_2 in hyperthermia/fever production was established more than 20 years ago (1970) by Milton and Wendlandt[19,20] and was cited in some detail in the "Historical Background" section of this chapter. This important observation has been confirmed numerous times over the years, including the significant observations of the Hayaishi group, cited above. PGE_2 has been identified as the mediator of fever induced by a wide variety of pyrogenic agents, including amphotericin, endotoxin, tumor neurosis factor, and interleukin-1.[86-88] In 1988 from the University of Wisconsin–Madison, Gollman and Rudy,[89] in a study of comparative pyrogenic effects of endogenous prostanoids, concluded that only E-series prostaglandins were sufficiently potent to merit serious consideration as mediators of pyrexia.

Arachidonic acid and its metabolites have been suggested to be involved in the establishment and maintenance of the LTP/memory model phenomenon. In 1989 Bliss and associates,[90] of Mill Hill (London), reported in *Nature* that arachidonic acid induces a long-term activity-dependent enhancement of synaptic transmission in the hippocampus, whereas the induction of LTP (long-term potentiation) is considered to be triggered by the postsynaptic entry of Ca^{++} through the channel associated with the NMDA glutamate receptor; its maintenance is considered, at least in part and by some, to be mediated by presynaptic mechanisms. In order to explain how postsynaptic happenings can generate an increase in presynaptic neurotransmitter release, the authors postulated the existence of a "retrograde messenger" that can carry information from the postsynaptic side of the synapse to the recently active presynaptic terminals. The authors proposed arachidonic acid or one of its lipoxygenase metabolites (why not cyclooxygenase?) as candidates for this retrograde messenger role. In their 1989 *Nature* publication, they described studies in which weak activation of the perforant path to the hippocampus in the presence of arachidonic acid leads to a slow-onset persistent increase in synaptic efficacy both *in vivo* and *in vitro*. The activity-dependent potentiation produced was observed to be accompanied by an increase in glutamate release. This response was found to be nonadditive with tetanus-induced LTP. The authors interpreted these findings as indicating a role for arachidonic acid as a retrograde messenger in the later, but not the initial, stages of LTP. While arachidonic acid alone did not produce long-term after-effects, it did potentiate the long-term effects of subthresholds of weak high-frequency trains of electrical stimulus of the perforant path *in vivo*. Of particular interest in these studies were the findings that these effects of arachidonic acid were not blocked by lipoxygenase-pathway inhibition (NDGA administration), nor by blockade of the NMDA glutamate receptor (AP5 administration). Interpreted *positively* by the author of this book, these findings strongly suggest that the arachidonic acid effect *is mediated by a non-NMDA receptor and by a cyclooxygenase metabolite of arachidonic acid*. The most likely cyclooxygenase product, from the presence of its synthetic enzymes and receptors in the hippocampus and limbic areas and from its stimulatory nature, is PGE_2. Direct study of the involvement of both PGE_2 and PGD_2 in the stimulation and inhibition of LTP would be of particular interest. From the clearly established involvement of PGE_2 in nervous system generation of heat/fever (see above) and pain (see below), it would appear to pass the ''heat, pain, and memory'' scenario test for consideration as a ''memory molecule''. Furthermore, memories are better established when awake (PGE_2) than when asleep (PGD_2).

Prostaglandins, particularly PGE_2 and prostacyclin (PGI_2), long have been associated with pain transmission and enhancement. Ferreira et al.,[91] in 1978, identified the hyperalgesic effects of PGI_2 and PGE_2. Lembeck, Popper, and Juan,[92] in 1973, described release of prostaglandin by bradykinin as an intrinsic mechanism of its algesic effect. In 1981 Juan[93] identified a dependence of histamine-evoked pain on prostaglandins release. In this study, Juan, of Graz, Austria, showed in a rabbit pain model that arachidonic-acid-liberating phospholipase A_2 exogenously applied enhanced the algesic effects of both histamine and bradykinin, and this enhancement was blocked by indomethacin. The algesic effect of histamine also was enhanced by direct application of prostacyclin.

B. VASCULAR SYSTEM

Vasodilator and hypotensive effects of prostaglandins were among the earliest identified physiological effects of prostaglandins. In 1974, reports by Kilbom and Wennmalm,[94] Messina et al.,[95] and Herbanczynska-Cedro et al.[96] established the importance of prostaglandins in the reactive hyperemia response following arterial occlusion in man and in experimental animals. Additional publications have confirmed and extended these studies.[97,98]

In 1985 two important studies of *in vitro* evaluation of blood-flow-velocity/wall-shear-stress appeared to indicate prompt and sharp elevations of endothelial cell prostacyclin (PGI_2) production. A study from New York Hospital/Cornell[99] employed confluent monolayers of bovine aortic endothelium cells placed in a lucite flow chamber. Step increases in shear stress caused rapid rises in prostacyclin production. Once maximum production was reached, additional PGI_2 could be produced by stimulation of the endothelial monolayer with arachidonic acid or the calcium ionophore A23187. In another 1985 report by Frangos et al.,[100] from Houston (Rice and Baylor), primary human endothelial cell cultures were exposed to steady-flow or pulsatile shear stress. With the onset of steady-flow shear stress, there was a sudden increase in prostacyclin production which decreased to the prestress state in several minutes. Pulsatile shear stress increased prostacyclin production to two times the steady-flow shear-stress level and to 16 times that of human endothelial cells in stationary/non-stressed culture. Cyclooxygenase-pathway inhibition by ibuprofen prevented prostacyclin production by both steady-flow and pulsatile shear stress. These findings indicate that the nature of blood flow may have an important and perhaps controlling effect on the function and metabolism of vascular endothelial cells. An earlier study had demonstrated that flow shear stress may also lead to histamine production.

In an *in vivo* study of prostaglandin mediation of blood-flow-velocity-induced arteriolar vasodilation, Koller and Kaley,[101] of New York Medical College, in 1990, explored the ''flow-dependent'' vasodilator response in rats, employing a parallel occlusion method where temporary occlusion of an arteriole increased red blood cell velocity in proximal parallel arteriolar branches. They found that increases in flow velocity consistently were followed by proportional delayed (6 to 15 sec) increases in arteriolar diameter. Cyclooxygenase inhibition, but not EDRF-synthesis inhibition, blocked this velocity-increase-induced dilator response completely. The authors interpreted these results as indicating that the arterioles are sensitive to increases in blood-flow velocity (wall shear stress) and that in response to this shear stress, they release prostaglandins that produce vasodilatation. The authors also commented that this phenomenon in the skeletal muscle microcirculation suggested a new regulatory mechanism *in vivo* that, by modulation of vascular resistence in the microvascular network, has the role of normalizing wall shear stress and providing for substantial increase in tissue blood flow — prostaglandin mediated.

Publications in recent years have described prostaglandin and prostacyclin mediation of vasodilation by dopamine (humans),[102] histamine (pigs),[103] and bradykinin (dogs).[104,105] In a study of the effect of application of histamine to the brain surface of newborn pigs on pial arteriole dilatation, as viewed through a closed cranial window, Mirro et al.,[103] of the University of Tennessee–Memphis, observed that histamine produced a concentration-dependent vasodilatation that was accompanied by significant increases in cerebrospinal-fluid levels of PGE_2 and 6-keto-PGF_{12} (the stable prostacyclin metabolite). Levels of TXB_2, PGD_2, and PGF_2 were not altered. Pretreatment with indomethacin blocked the histamine-induced cerebral vasodilatation.

Mullane and Moncada,[105] of London and Beckenham, reported in 1980 that prostacyclin mediates the potentiated hypotensive effect of bradykinin after captopril (an inhibitor of antiotensin-converting enzyme, the bradykinin-degrading enzyme). These studies were carried out in normal dogs, and the hypotensive effects of captopril and bradykinin were reduced significantly by indomethacin. These findings suggested that the acute hypotensive effect of captopril in the dog might result from the local release of prostacyclin stimulated by increased plasma bradykinin levels. Reinvestigation of this blockade of the hypotensive effect of captopril by indomethacin in anesthetized dogs was carried out and reported in 1990 by Ferreira[104] (of Ribeirao Preto, Brazil) and associates. They employed both the bradykinin blood-bathed cascade bioassay technique of Vane (tonus of cat terminal ileum as assay tissue) and blockade of the response of the assay tissues with a specific bradykinin-receptor antagonist (L-349b). Treatment of the animals with indomethacin blocked or reversed the hypotensive effect of captopril without affecting the increased level of bradykinin in arterial blood (cascade bioassay determined). Captopril also potentiated the hypotension induced by intra-arterial administration of bradykinin, and indomethacin reduced this effect. The addition of captopril or similar-acting enalpril to the cascade superfusing blood in an extracorporeal circuit caused long-lasting increases in bradykinin levels (bioassay determined). The results of these studies were interpreted by the authors as supporting the hypothesis that immediate hypotension induced by captopril involves a prostaglandin-dependent component that possibly results from increased bradykinin levels generated in the vicinity of the captopril action.

Downey et al.,[106] from the University of Colorado–Denver, in 1988, described enhancement of pulmonary inflammation by PGE_2, with evidence for a vasodilator effect. The inflammation was induced in the rabbit-lung-inflammation model by the instillation of fragments of the fifth component of complement. The extent of the localized area of acute inflammation was quantified by the total number of neutrophils and protein recoverable by bronchoalveolar lavage (BAL) and by serial scintography employing labeled neutropils. The intrabronchial administration of PGE_2 produced a twofold increase in the accumulation of neutrophils on scintography, a fivefold increase in BAL neutrophils, and a threefold increase in protein. These effects were attenuated by a vasoconstrictor (antiotensin II) and mimicked by a vasodilator (nitroprusside). These findings suggested that the effects were due to vasodilatation per se. Radioactive microsphere study showed a local blood-flow decrease of 50% produced by the C5 complement fragments, which could be prevented by concomitant administration of PGE_2 or nitroprusside. The authors concluded that the vasodilatation induced by PGE_2 was associated with the enhancement of acute pulmonary inflammation in this rabbit model and was largely responsible for the proinflammatory effect of PGE_2, which they also could produce by intravenous administration of PGE_2 in this model.

C. IMMUNE RESPONSE SYSTEM

Prostaglandin E_2 (PGE_2) is the major immune-response-system-involved and immunostimulative arachidonic acid oxidation molecule by either pathway: cyclooxygenase or lipoxygenase. There is substantial support for this from *in vivo* studies of endogenous PGE_2 activity, despite a large *ex vivo/in vitro* literature that suggests an immunomodulatory/immunosuppressive role for PGE_2.[107-119] In an *ex vivo/in vitro* study published by Scodras et al.[120] in 1990 (in *Cellular Immunology*), it was stated that "PGE_2 is recognized as a pansuppressor molecule which suppresses the activation of all cells of the immune response."

Ex vivo/in vitro studies of PGE_2 and other vasoactive neurotransmitter molecules have largely been carried out on immune-response-related cells in tissue culture and deprived of the blood and lymph vascular and nervous system connections that are so essential to their *in vivo* function and activities. *Ex vivo* studies of these sorts run a significant risk of being inconclusive, contradictory (especially to *in vivo* observations), misleading, and plain wrong and artifactual. The "bottom line" must remain *in vivo* in evaluating the molecular, biochemical, and physiological significance and function of these potent vasoactive neurotransmitter and neuromodulator molecules of the neurovascular immunity branch of the neuroimmune axis.

The 1974 report by the author and Jaffe[26] (in *Transplantation*) of an involvement of PGE_2 in organ (heart) allograft rejection in rats was one of the earliest, if not *the* earliest, description of a role of increased PGE_2 production and release in a cellular immune response. Following heterotopic (into abdominal aorta) heart allografting between inbred strains of rats, marked increases (as much as 33-fold) in PGE_2 levels were identified in heart allografts as well as in host splenic and thymic lymphoid tissue. The peak increase in PGE_2 levels in both heart allograft and host lymphoid tissue occurred at 7 d post-heart allografting. An increase in PGE_2 levels in heart allograft coronary venous sinus (venous drainage of the heart) also was identified and with a peak elevation at 7 d postallografting.[27]

In an interesting 1980 report by Dy et al.[39] of Paris, in a combined *in vivo* and *in vitro* study, a marked increase in PGE_2 production and release was identified during secondary mixed lymphocyte cultures (MLC) between donor and recipient spleen cells post-skin allografting in mice. The difference between allografted and control animals was found to begin at 6 d postgrafting and reached a maximum at 12 d. It remained elevated until 20 d postgrafting, at which time a decrease began. Macrophages were identified by the authors to be the cells responsible for the increase in PGE_2 production in this response.

An especially important report concerning the role of PGE_2 in the cellular immune reaction *in vivo* appeared in *Nature* in 1981. This publication by Mertin and Stackpoole[40] of London (Harrow, Middlesex, U.K.) clearly and unambiguously established the absolute and obligatory requirement of PGE_2 to establish a cellular immune response *in vivo*. Antiserum raised against PGE_2 significantly suppressed the development of the cellular immune responses of experimental allergic encephalomyelitis (EAE) in rats and prevented the development of HVG and GVH responses in mice.

In the Mertin and Stackpoole *Nature* study, EAE was induced in Lewis rats. Treatment with anti-PGE_2 serum was given subcutaneously beginning in the day of antigen innoculation (day 0) and was continued up to 21 to 23 d post-antigen innoculation. This treatment protocol clearly suppressed the development of EAE. Not only were there significantly fewer animals developing overt ("clinical") signs of EAE, but also the onset of the disease was delayed in those animals in which signs of the disease appeared.

As the use of EAE in rats was found by the authors to be too complicated to permit dose-response studies, as well as studies relating the time of anti-PGE_2-serum administration to its inhibitory action on the development *in vivo* of the cellular immune reaction, they elected to employ mice and the HVG and GVH reactions *in vivo*. They observed that anti-PGE_2-antiserum treatment significantly suppressed both HVG and GVH reactions in these animals. Subcutaneous injections of 30 μl of anti-PGE_2 antiserum per day from day 0 to day 2 after challenge of CBA mice with A strain spleen cells almost completely inhibited the HVG reaction. The authors also observed that the inhibitory effect of a single dose of anti-PGE_2 antiserum, when given immediately after the injection of donor cells, was dose dependent in both HVG and GVH studies. Studies of the effect of the time of administration indicated that the anti-PGE_2 antiserum was most effective during the first hours after the challenge of the animals with donor cells. This finding was most pronounced in the HVG reactions where the inhibitory effect of a single dose of anti-PGE_2 antiserum (120 μl) given at time 0 was much greater than the effect when the antiserum was given 6 h after the injection of donor cells. In GVH studies it was of interest that an inhibitory effect was still found when the anti-PGE_2 antiserum was given as late as 24 h after donor-cell injection. Nonetheless, in these studies the level of inhibition was greatest in the first hours after the inoculation of parental-strain donor cells.

A 1984 report by Mertin and Stackpoole (in *Transplantation,* with Shumway[121]) extended the studies of the 1981 *Nature* report. They showed that the anti-PGE_2 responses with respect to the suppression of the HVG and GVH responses in mice could be duplicated by antiserum raised against PGE_1. They also demonstrated that the anti-PGE_1 inhibitions could be blocked by "add back" experiments in which the animals were treated with PGE_1. They also showed that exogenous PGE_1 in low and "physiological" concentrations enhanced the HVG and GVH cellular immune reactions, but in high, "pharmacological", and toxic doses suppressed these cellular immune responses and snuffed out the lives of these experimental animals. The doses of exogenous PGE_1 exceeding those required for full restoration of the HVG and GVH responses proved to be both toxic and suppressive. The toxic responses to "pharmacological" (i.e., high) doses of PGE_1 included dull and rough appearance of the fur, hunching of the back, apathy, and death. In this regard, a 1977 comment by Pelus and Strausser[122] in a "minireview" in *Life Sciences* about prostaglandins and the immune response is very much in order: "Although Pgs (prostaglandins) are normally found in tissues in nanogram quantities or less, many experiments which demonstrated Pg-mediated immune cell inhibition employ quantities at least one thousand times larger. In such experiments, the results obviously reflect pharmacological rather than physiological effects."

1979 and 1980 publications by Johnston, Hay, and Movat,[123,124] of Toronto, presented important *in vivo* data from sheep studies regarding the effect of various inflammatory stimuli on prostaglandin production, as monitored in afferent and efferent lymphatics. Carrageenin, delayed hypersensitivity, and heat-killed *Escherichia coli* were used. Levels of both PGE and PGF were increased promptly in the acute inflammation studies, with a predominance of PGE. With delayed hypersensitivity (PPD), prostaglandin levels increased early in the response (4 to 6 h), with PPD they increased at 10 to 20 h, and after allogenic lymphocyte injection they increased at 127 h. Outputs were measured at 2-h intervals. In their 1980 report the authors used radioactive arachidonic acid and radioactive PGE_2 and PGF_2, in association with subcutaneous drainage-area injection of *E. coli*. PGE_2 levels were several times higher in afferent lymph coming from the acute inflammation drainage sites, and PGF_2 appeared to be metabolized, with metabolites being detected. This was not so for PGE_2. The highest levels of the prostaglandins were in the first 10 h, with a return to normal by 24 h.

In a 1989 report, the author and associates[125] described a prompt (1-h) and marked increase in the output of both PGE_2 and thromboxone B_2 (TXB_2) into the popliteal lymph node efferent lymph of sheep following the injection of formalin-killed *Salmonella muenchen* bacteria into the nodal lymph drainage area. The peak elevations in PGE_2 levels occurred at 2 to 4 h after injection, with prompt returns toward normal and preinjection levels (Figures 4 and 5). Similar increases in TXB_2 levels were encountered with an earlier (15 min) increase in levels (Figure 6).

Direct injections of PGE_2 into cannulated afferent lymphatics of sheep popliteal lymph nodes by the author et al.[126-128] have decreased lymphocyte traffic, as mirrored in the output of small recirculating lymphocytes into nodal efferent lymph (chronically cannulated). These were early and short-term observations. Clearly, longer-term observations (in excess of 48 h) and the influence in antibody production will be needed to evaluate the significance of these findings as well as the use of various doses of PGE_2. The 50 μg used in acute infusions may be at "pharmacological" rather than physiological levels, and these matters also need to be sorted out.

A 1989 *in vivo* report from Paris by Tissot et al.[129] is of interest and importance. The authors relate that during the development of an acute inflammatory reaction in the pleural cavity of the cat, macrophages at a distant site (peritoneal cavity) displayed increased capacity to release prostanoids (prostaglandins, prostacyclin, and thromboxane). The *in vivo* levels of prostanoids in the peritoneal cavity reached a maximum 24 h after the induction of pleurisy, regardless of the induction agent used.

Early studies in the 1977 to 1982 period (by Bennett and associates[130-133] of London) had suggested a prognostic role of importance for PGE_2 levels in human breast cancers, with elevated levels being associated with a poor prognosis and increased risk for metastasis. These observations have not been confirmed by others[134-138] and have been on the whole, retracted in a 1989 article by the Bennett group.[139] Watson et al.,[138] in a 1987 publication from the Clinical Surgery Department of the Royal Infirmary of Edinburgh, Scotland, reported an in-depth study of 78 women with early breast cancer and no evidence of metastatic disease, which concluded that there was no significant relationship between the breast cancer tissue levels of prostaglandins and estrogen receptors, tumor size, presence of lymph-node metastasis, or disease-free interval following primary tumor treatment. They concluded that it was unlikely that the levels of the prostaglandins studied (PGE_2 and PGF_2) within breast carcinomas play a fundamental role in the prognosis of the disease.

An interesting twist concerning the role of prostaglandins, other than PGE_2, in cellular immune responses was published in 1979 in *Nature*[140] and the *Proceedings of the National Academy of the Sciences, U.S.A.*[141] by Fitzpatrick and Stringfellow of the Upjohn Company of Kalamazoo, MI. Studying a moderate and a highly metastatic B16 melanoma cell line ($B16F_1$ as moderate and $B16F_{10}$ as highly metastatic), the authors discovered that production

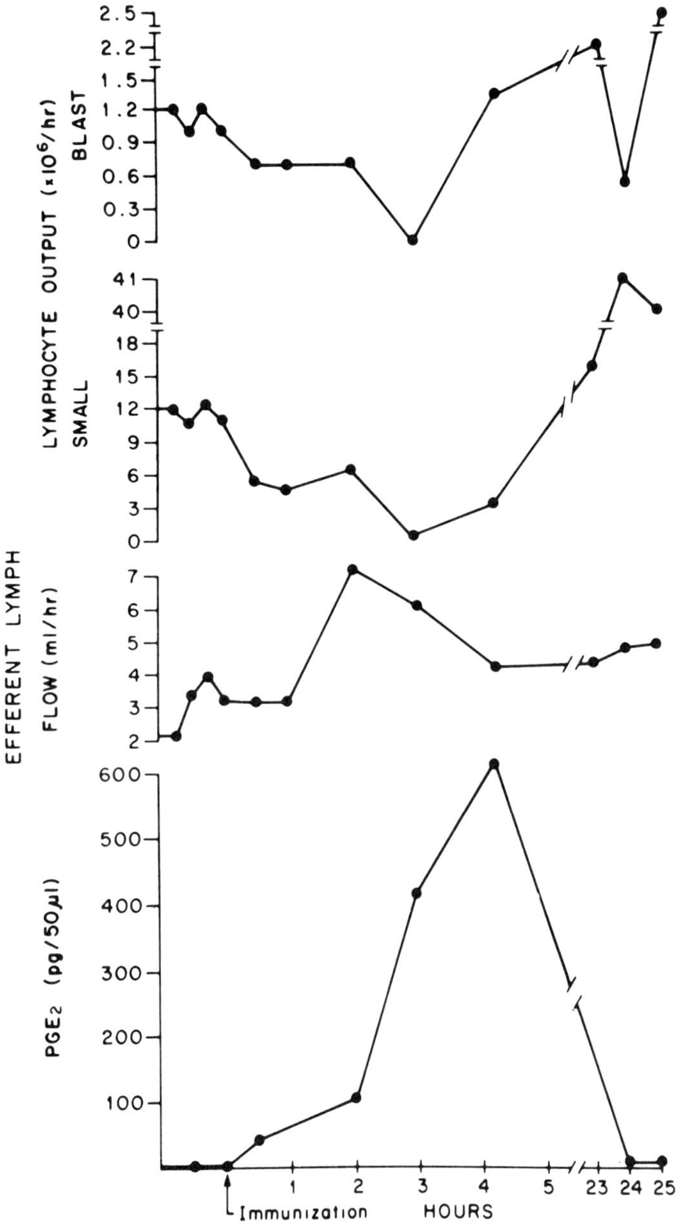

FIGURE 4. The prompt and sharp onset of lymphocyte output "shutdown"/recruitment decrease in lymphocyte traffic and output into popliteal nodal efferent lymph of sheep within minutes after nodal drainage-area immunization with killed *Salmonella muenchen* bacteria. This "shutdown"/recruitment is accompanied by an equally prompt and massive increase in prostaglandin E_2 (PGE_2) levels in nodal efferent lymph. There is an associated increase in efferent lymph flow. (From Moore, T. C. et al., *Immunopharmacology,* 17, 73, 1989. With permission.)

of PGD_2 by these cell lines *in vitro* was inversely related to their metastatic potential, with the highly metastatic line ($B16F_{10}$) producing less PDG_2 than the moderately metastatic cell line. Among the interpretations of this finding, one must consider that PGD_2 is immunosuppressive and anti-inflammatory (opposite of PGE_2) and that the tumor line producing less of the spread-inhibitory and anti-inflammatory PGD_2 is more likely to spread and

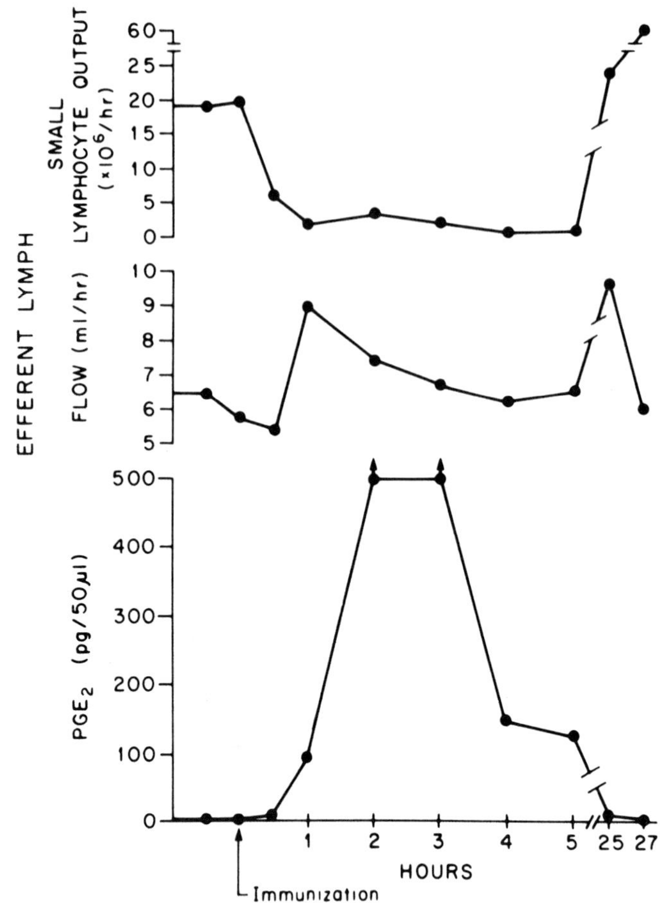

FIGURE 5. The postimmunization sheep popliteal lymph node efferent lymph lymphocyte output "shutdown"/recruitment was associated with a massive and prompt early increase in efferent lymph PGE_2 levels and an equally early and prompt increase in efferent lymph flow. (From Moore, T. C. et al., *Immunopharmacology,* 17, 73, 1989. With permission.)

metastasize with less of an anti-inflammatory antimetastatic "shield" to overcome. In their *Nature* report, the authors found that indomethacin increased the metastatic rate of the high-PGD_2-producing and less-metastatic $B16F_1$ cells *in vivo,* whereas exogenous reconstitution of their PGD_2 levels reversed the indomethacin-induced increase in their metastatic rate *in vivo.*

As PGE_2 and PGD_2, positional isomers (Figure 7), have opposite effects in the central nervous system, with PGE_2 being the fever and wakefulness molecule, and PGD_2 the sleep and hypothermia molecule, perhaps these two molecules have opposite functions in the immune response system, with PGE_2 being immunostimulative and proinflammation *in vivo* and PGD_2 being the reverse.

Of the thromboxanes, only thromboxane A_2 (TXA_2) is biologically active. It is extremely short-lived, difficult to study, and largely involved in clot-associated vasospasm for control of blood loss locally. It appears to have little involvement in nervous system function, vascular system function beyond local vasoconstriction, or immune response system function.

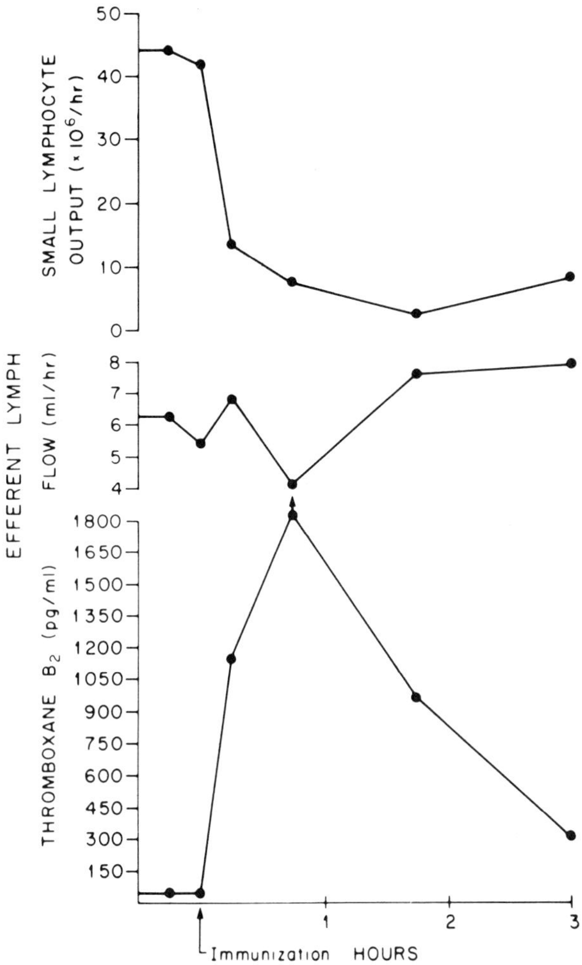

FIGURE 6. Postimmunization sheep popliteal node efferent lymph output "shutdown"/recruitment was associated with markedly increased efferent lymph levels of thromboxane B_2 and an increase in efferent lymph flow. It was of interest that the peak elevation in efferent lymph levels of thromboxane B_2 was associated with a brief depression in efferent lymph flow. (From Moore, T. C. et al., *Immmunopharmacology,* 17, 73, 1989. With permission.)

V. LEUKOTRIENES, LIPOXINS, HETES, AND OTHERS

Leukotrienes, lipoxins, and other noncyclooxygenase products of the oxidative metabolism of arachidonic acid are comparative newcomers to the defense and inflammation armamentarium. Their formation and structure have been considered in the "Introduction" section of this chapter. Their roles in normal physiology, including involvement in the normal functioning *in vivo* of the nervous, vascular, and immune response systems has remained obscure.

The cysteinyl leukotrienes (LTC_4, LTD_4, and LTE_4) provide an established molecular existence for the long-recognized slow-reacting substance of anaphylaxis (SRS-A) and appear to have their primary effect on smooth-muscle cells and other cells with contractile capability. These molecules are potent inducers of airway obstruction by spasmotic activation of airway smooth muscle and have been implicated as causative factors in human asthma,[142] and

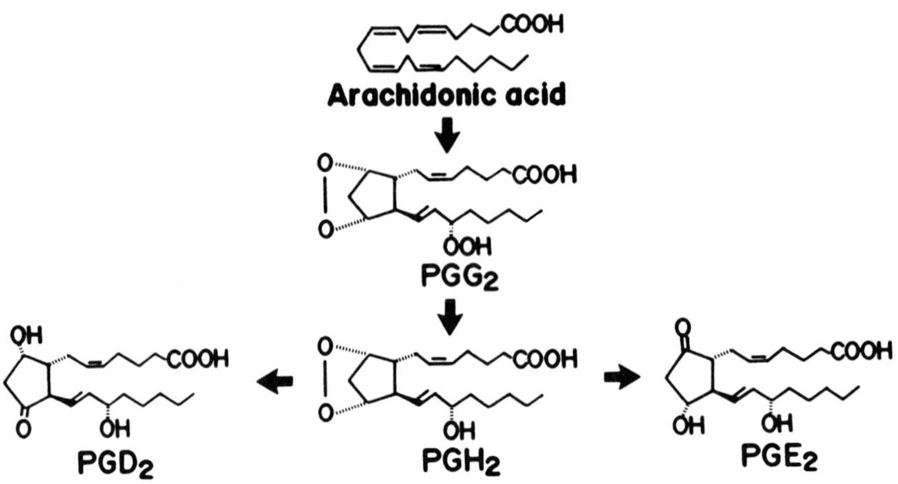

FIGURE 7. PGD_2 and PGE_2 as positional isomers. (After Hayaishi, D., *FASEB J.*, 5, 2579, 1991.)

Samuelsson and associates have described an *in vivo* study of the effects of LKC_4 and LKD_4 on topical application to the vascular network of the hamster cheek pouch. Intense arteriolar construction was produced. It was short lived and was followed consistently by a marked and dose-dependent extravasation of macromolecules from postcapillary venules.

The noncysteinyl leukotriene LTB_4 is involved in acute inflammatory reactions and is leukocyte produced and leukocyte activating, with influences on leukocyte chemotaxis. Both the cysteinyl and noncysteinyl leukotrienes are products of the 5-lipoxygenase enzyme that plays a central role in the biosynthesis of both leukotrienes and lipoxins.[143] The 15-lipoxygenase enzyme also is involved in synthesis of the lipoxins (lipoxin A and lipoxin B). Of these lipoxins (lipoxogenase-interaction products), lipoxin A (LXA) has been suggested to be a stimulator of superoxide-ion generation, chemotaxis, and spasmogenesis. LXA also was shown to activate protein kinase C (PKC) by the Samuelsson group of Stockholm,[144] discoverers of both the leukotrienes and the lipoxins. This group also reported that both LXA and LXB may block human NK cell activity.

Because of the high potency, short ''lives'', and local actions of these lipoxygenase enzymes metabolites of arachidonic acid, the lack of knowledge of the molecular makeup of their forming enzymes and tissue and cellular receptors, and the absence of specific and long-acting agonists and antagonists, much remains to be learned about the normal and physiological functioning of these intriguing and potent defense-related molecules of the arachidonic acid cascade.

IV. CONCLUSION

The prostaglandins, thromboxane, leukotrienes, lipoxins, etc. are short-chain (20-carbon) molecules derived from the mobilization and metabolism of arachidonic acid, with multiple, varied, and often conflicting actions of immense biological significance. The discovery of prostaglandin and its identification as a lipid by von Euler in the 1930s opened a Pandora's box of biomedical research. Its vascular effect as a vasodilator/blood-pressure-depressor was the first physiological action encountered, which led to its discovery in prostatic seminal fluid and its logical naming as ''prostaglandin'' by von Euler.

These major short-chain lipid molecules, called eicosanoids, play major roles in the function of the nervous, vascular, and immune response systems and in neurovascular immunology. The 1974 publication by the author and Jaffe in *Transplantation* was the first identification of the involvement of a prostaglandin and eicosanoid, prostaglandin E_2 (PGE_2), in a cellular immune reaction — the rejection of a rat organ (heart) allograft. The 1981 report of Mertin and Stackpoole in *Nature* clearly and unambiguously established the obligatory requirement for PGE_2 to mount a cellular immune response *in vivo*. Support for this early involvement of PGE_2 in an *in vivo* immune response was presented in the 1989 report in *Immunopharmacology* by the author, Said, and associates.

The immensely important and exciting studies of Hayaishi and associates of Kyoto and Osaka, Japan, concerning the conflicting roles of PGE_2 and PGD_2 (positional isomers) in wakefulness and sleep regulation in the mammalian brain raise many potentials for additional study, including a possible immunomodulatory/suppressant role for PGD_2 in the periphery and in the cellular immune response (author's suggestion).

The eicosanoids surely will remain an immensely rewarding area for biomedical research for many years to come.

REFERENCES

1. **Balcarek, J. M., Theisen, T. W., Cook, M. N., Varrichio, A., Hwang, S.-M., Strohsacker, M. W., and Crooke, S. T.,** Isolation and characterization of a cDNA clone encoding rat 5-lipoxygenase, *J. Biol. Chem.*, 263, 13937, 1988.
2. **Dixon, R. A. F., Jones, R. E., Diehl, R. E., Bennett, C. D., Kargman, S., and Rouzer, C. A.,** Cloning of the cDNA for human 5-lipoxygenase, *Proc. Natl. Acad. Sci. U.S.A.*, 85, 416, 1988.
3. **Matsumoto, T., Funk, C., Radmark, O., Hoog, J.-O., Jornvall, H., and Samuelsson, B.,** Molecular cloning and amino acid sequence of human 5-lipoxygenase, *Proc. Natl. Acad. Sci. U.S.A.*, 85, 26, 1988.
4. **Sigal, E., Craik, C. S., Highland, E., Grunberger, D., Costello, L. L., Dixon, R. A. F., and Nadel, J. A.,** Molecular cloning and primary structure of human 15-lipoxygenase, *Biochim. Biophys. Res. Commun.*, 157, 457, 1988.
5. **Kurzrok, R. and Lieb, C. C.,** Biochemical studies of human semen. II. The action of semen on the human uterus, *Proc. Soc. Exp. Biol. Med.*, 28, 268, 1930.
6. **Goldblatt, M. W.,** A depressor substance in seminal fluid, *J. Soc. Chem. Ind. (Lond.)*, 52, 1056, 1933.
7. **Goldblatt, M. W.,** Properties of human seminal fluid, *J. Physiol. (Lond.)*, 84, 208, 1935.
8. **von Euler, U. S.,** Zur Kenntnis der pharmakalogishen Wirkungen von Nativsekreten und Extrakten mannlicher accessorischer Geschlechtsdrusen, *Naunyn Schmeidebergs Arch. Exp. Pathol. Pharmakol.*, 175, 78, 1934.
9. **von Euler, U. S.,** Uber die spezifische blutdrucksende Substanz des menschlichen Prostata-und Samenblasensekretes, *Klin. Wochenschr.*, 14, 1182, 1935.
10. **von Euler, U. S.,** On the specific vasodilating and plain muscle stimulating substance from accessory genital glands in man and certain animals (prostaglandin and vestiglandin), *J. Physiol. (Lond.)*, 88, 213, 1936.
11. **von Euler, U. S.,** Some aspects of the actions of prostaglandins. The First Heymans Memorial Lecture, *Arch. Int. Pharmacodyn. Ther.*, 202(suppl.), 295, 1973.
12. **Bergstrom, S. and Sjovall, J.,** The isolation of prostaglandin F from sheep prostate glands, *Acta Chem. Scand.*, 14, 1693, 1960.
13. **Bergstrom, S., Ryhage, R., Samuelsson, B., and Sjovall, J.,** The structure of prostaglandins E, F, and F_2, *Acta Chem. Scand.*, 16, 501, 1962.

14. **Bergstrom, S., Ryhage, R., Samuelsson, B., and Sjovall, J.,** Prostaglandins and related factors. 15. The structures of prostaglandin E_1, F_{1x} and F_{1B}, *J. Biol. Chem.*, 238, 3555, 1963.
15. **Samuelsson, B.,** Prostaglandins and related factors. 17. The structure of prostaglandin E_1, *J. Am. Chem. Soc.*, 85, 1878, 1963.
16. **Bergstrom, S., Danielson, H., and Samuelsson, B.,** The enzymatic formation of prostaglandin E_2 from arachidonic acid, prostaglandins and related factors, *Biochim. Biophys. Acta*, 90, 207, 1964.
17. **Van Dorp, D. A., Beerthuis, R. K., Nugteren, D. H., and Vonheman, H.,** The biosynthesis of prostaglandins, *Biochim. Biophys. Acta*, 90, 204, 1964.
18. **Williams, E. D., Karim, S. M. M., and Sandler, M.,** Prostaglandin secretion by medullary carcinoma of the thyroid. A possible cause of associated diarrhea, *Lancet*, 1, 22, 1968.
19. **Milton, A. S.,** Thermoregulatory actions of eicosanoids in the central nervous system with particular regard to the pathogenesis of fever, *Ann. N.Y. Acad. Sci.*, 559, 392, 1989.
20. **Milton, A. S. and Wendlandt, S.,** A possible role for prostaglandin E_1 as a modulator for temperature regulation in the central nervous system of the cat, *J. Physiol. (Lond.)*, 207, 76p, 1970.
21. **Vane, J. R.,** Inhibition of prostaglandin in synthesis as a mechanism for aspirin-like drugs, *Nature New Biol.*, 231, 232, 1971.
22. **Hamberg, M. and Samuelsson, B.,** Detection of an endoperoxide intermediate in prostaglandin biosynthesis, *Proc. Natl. Acad. Sci. U.S.A.*, 70, 899, 1973.
23. **Nugteren, D. H. and Hazelhof, E.,** Isolation and properties of intermediates in prostaglandin biosynthesis, *Biochim. Biophys. Acta*, 326, 448, 1973.
24. **Hamberg, M. and Samuelsson, B.,** Prostaglandin endoperoxides. Novel transformation of arachidonic acid in human platelets, *Proc. Natl. Acad. Sci. U.S.A.*, 71, 3400, 1974a.
25. **Hamberg, M. and Samuelsson, B.,** Prostaglandin endoperoxides. VII. Novel transformation of arachidonic acid in guinea pig lung, *Biochem. Biophys. Res. Commun.*, 61, 942, 1974b.
26. **Moore, T. C. and Jaffe, B. M.,** Prostaglandin E levels of heterotopic rat heart allografts and of host lymphoid tissues at intervals post-grafting, *Transplantation*, 18, 383, 1974.
27. **Jaffe, B. M. and Moore, T. C.,** Elevated prostaglandin E activity of rat heart allograft coronary sinus venous blood, *IRCS Med. Sci.*, 2, 1417, 1974.
28. **Hamberg, M., Svensson, J., and Samuelsson, B.,** Thromboxane: a new group of biologically active compounds derived from prostaglandin endoperoxides, *Proc. Natl. Acad. Sci. U.S.A.*, 72, 2994, 1975.
29. **Moncada, S., Gryglewski, R., Bunting, S., and Vane, J. R.,** An enzyme isolated from arteries transforms prostaglandin endoperoxides to an unstable substance that inhibits platelet aggregation, *Nature*, 263, 663, 1976.
30. **Abdel-Halim, M. S., Hamberg, M., Sjoquist, B., and Anggard, E.,** Identification of prostaglandin D_2 as a major prostaglandin in homogenates of rat brain, *Prostaglandins*, 14, 633, 1977.
31. **Samuelsson, B., Borgeat, P., Hammarstrom, S., and Murphy, R. C.,** Introduction of a nomenclature: leukotrienes, *Prostaglandins*, 17, 785, 1979.
32. **Borgeat, P. and Samuelsson, B.,** Arachidonic acid metabolism in polymorphonuclear leukocytes: effects of ionophone A 23187, *Proc. Natl. Acad. Sci. U.S.A.*, 76, 2148, 1979a.
33. **Borgeat, P. and Samuelsson, B.,** Metabolism of arachidonic acid in polymorphonuclear leukocytes. Structural analysis of novel hydroxylated compounds, *J. Biol. Chem.*, 254, 7865, 1979b.
34. **Murphy, R. C., Hammarstrom, S., and Samuelsson, B.,** Leukotriene C: a slow reacting substance from murine mastocytoma cells, *Proc. Natl. Acad. Sci. U.S.A.*, 76, 4275, 1979.
35. **Corey, E. J., Clark, D. A., Goto, G., Marfat, A., Mioskioski, C., Samuelsson, B., and Hammarstrom, S.,** Stereospecific total synthesis of a slow reacting substance of anaphylaxis, leukotriene C-1, *J. Am. Chem. Soc.*, 102, 1436, 1980.
36. **Dahlen, S.-E., Bjork, J., Hedqvist, P., Arfors, K.-E., Hammarstrom, S., Lindgren, J.-A., and Samuelsson, B.,** Leukotrienes promote plasma leakage and leukocyte adhesion in post-capillary venules: *in vivo* effects with reference to the acute inflammatory response, *Proc. Natl. Acad. Sci. U.S.A.*, 78, 3887, 1981.
37. **Palmblad, J., Malmsten, C. L., Uden, A.-M., Radmark, O., Engstedt, L., and Samuelsson, B.,** Leukotriene B4 is a potent and stereospecific stimulator of neutrophil chemotaxis and adherence, *Blood*, 58, 658, 1981.
38. **Shimizu, T., Mizuno, N., Amano, T., and Hayaishi, O.,** Prostaglandin D_2, a neuromodulator, *Proc. Natl. Acad. Sci. U.S.A.*, 76, 6231, 1979.
39. **Dy, M., Astoin, M., Rigaud, M., and Hamburger, J.,** Prostaglandin (PG) release in the mixed lymphocyte culture; the effect of presensitization by a skin allograft; nature of the PG-producing cell, *Eur. J. Immunol.*, 10, 121, 1980.
40. **Mertin, J. and Stackpoole, A.,** Anti-PGE antibodies inhibit *in vivo* development of cell-mediated immunity, *Nature*, 294, 456, 1981a.

41. **Hayaishi, O.,** Sleep-wake regulation by prostaglandins D_2 and E_2, *J. Biol. Chem.*, 263, 14593, 1988.
42. **Rich, A. M., Weissmann, G., Anderson, C., Vosshall, L., Haines, K. A., Humphreys, T., and Dunham, P.,** Calcium dependent aggregation of marine sponge cells is provoked by leukotriene B_4 and inhibited by inhibitors of arachidonic acid oxidation, *Biochem. Biophys. Res. Commun.*, 121, 863, 1984.
43. **Weinheimer, A. J. and Spraggins, R. L.,** The occurrence of two new prostaglandin derivatives (15)epi-PGA2 and its acetate methyl ester in the gorgonian *Plexaura homomalla:* chemistry of coelenterates XV, *Tetrahedron Lett.*, 59, 5185, 1969.
44. **Clare, A. S., Walker, G., Holland, D. L., and Crisp, D. J.,** The hatching substance of the barnacle, *Balanus balanoides* (L.), *Proc. R. Soc. Lond. B*, 224, 131, 1985.
45. **Destephano, D. E., Brady, U. E., and Lovins, R. E.,** Synthesis of prostaglandin by reproductive tissue of the male house cricket, *Acheta domesticus, Prostaglandins,* 6, 71, 1974.
46. **Loher, W., Ganjian, I., Kubo, I., Stanley-Samuelson, D., and Tobe, S. S.,** Prostaglandins: their role in egg-laying of the cricket *Teleogryllus commodus, Proc. Natl. Acad. Sci. U.S.A.*, 78, 7835, 1981.
47. **Stanley-Samuelson, D. W. and Loher, W. J.,** Arachidonic acid and other long-chain polyunsaturated fatty acids in spermatophores and spermathecae of *Teleogryllus commodus:* significance in prostaglandin-mediated reproductive behavior, *Insect Physiol.*, 29, 41, 1983.
48. **Stanley-Samuelson, D. W. and Loher, W.,** Evolutionary aspects of prostaglandins and other eicosanoids in invertebrates, *Prog. Comp. Endocrinol.*, 614, 1990.
49. **Ai, N., Komatsu, S., Kubo, I., and Loher, W.,** Manipulation of prostaglandin-mediated oviposition after mating in *Teleogryllus commodus, Int. J. Invert. Repr. Dev.*, 10, 33, 1986.
50. **Stanley-Samuelson, D. W.,** Physiological roles of prostaglandins and other eicosanoids in invertebrates, *Biol. Bull.*, 173, 92, 1987.
51. **Stanley-Samuelson, D. W. and Loher, W.,** Prostaglandins in insect reproduction, *Ann. Entomol. Soc. Am.*, 79, 841, 1986.
52. **Stanley-Samuelson, D. W., Peloquin, J. J., and Loher, W.,** Egg-laying in response to prostaglandin injections in *Teleogryllus commodus, Physiol. Entomol.*, 11, 213, 1986.
53. **Stanley-Samuelson, D. W., Jurenka, R. A., Blomquist, G. J., and Loher, W.,** Sexual transfer of prostaglandin precursor in the field cricket, *Teleogryllus commodus, Physiol. Entomol.*, 12, 347, 1987.
54. **Morse, D. E., Duncan, H., Hooker, N., and Morse, A.,** Hydrogen peroxide induces spawning in mollusks, with activation of prostaglandin endoperoxide synthetase, *Science,* 196, 298, 1977.
55. **Kunigelis, S. C. and Saleuddin, A. S. M.,** Reproduction in the freshwater gastropod *Helisoma:* involvement of prostaglandin in egg production, *Int. J. Invertebr. Repr. Dev.*, 10, 159, 1986.
56. **Graves, S. Y. and Pietz, T. H.,** Prostaglandin E_2 inhibition of sodium transport in the freshwater mussel, *J. Exp. Biol.*, 210, 195, 1979.
57. **Cimino, G., Crispino, A., Di Marzo, V., Sodano, G., Spinella, A., and Villani, G.,** A marine mollusc provides the first example of *in vivo* storage of prostaglandins: prostaglandin-1,15-lactones, *Experientia,* 47, 56, 1990.
58. **Piomelli, D., Volterra, A., Dale, N., Siegelbaum, S. A., Kandel, E. R., Schwartz, J. H., and Belardetti, F.,** Lipoxygenase metabolites of arachidonic acid as second messengers for presynaptic inhibition of *Aplysia* sensory cells, *Nature,* 328, 38, 1987.
59. **Piomelli, D., Feinmark, S. J., Shapiro, E., and Schwartz, J. H.,** Formation and biological activity of 12-ketoeicosatetraenoic acid in the nervous system of *Aplysia, J. Biol. Chem.*, 263, 16591, 1988.
60. **Piomelli, D., Shapiro, E., Zipkin, R., Schwartz, J. H., and Feinmark, S. J.,** Formation and action of 8-hydroxy-11,12-epoxy-5,9,14-icosatrienoic acid in *Aplysia:* a possible second messenger in neurons, *Proc. Natl. Acad. Sci. U.S.A.*, 86, 1721, 1989.
61. **Meijer, L. P., Maclouf, J., and Bryant, R. W.,** Contrasting effects of fatty acids on oocyte maturation in several starfish species, *Prostaglandins Leukot. Med.*, 23, 179, 1986.
62. **Schuel, H., Traeger, E., Schuel, R., and Boldt, J.,** Anti-inflammatory drugs promote polyspermic fertilization in sea urchins, *Gamete Res.*, 10, 9, 1984.
63. **Ogata, H. and Nomura, T.,** Isolation and identification of prostaglandin E_2 from the gastrointestinal tract of shark *Triakis scillia, Biochim. Biophys. Acta,* 388, 84, 1975.
64. **Anderson, A. A., Fletcher, T. C., and Smith, G. M.,** Prostaglandin biosynthesis in the skin of the plaice, *Pleuronectes olatessa, Comp. Biochem. Physiol.*, 70C, 195, 1981.
65. **Herman, C. A., Zimmerman, P. R., and Doolittle, K. D.,** Prostaglandin synthesis in goldfish heart, *Carassius auratus, Gen. Comp. Endocrinol.*, 54, 478, 1984.
66. **Gibson, R. A., Kneebone, R., and Kneebone, G. M.,** Comparative levels of arachidonic acid and eicosapentaenoic acid in Malaysian fish, *Comp. Biochem. Physiol.*, 78C, 325, 1984.
67. **Herman, C. A.,** Prostaglandins in lower vertebrates, *Prog. Comp. Endocrinol.*, 608, 1990.
68. **Eiceman, G. A., Guavao, V., Doolittle, K. D., and Herman, C. A.,** Determination of prostaglandin precursors in frog tissue using selected ion monitoring in gas chromatographic-mass spectrometric analysis, *J. Chromatogr.*, 236, 97, 1982.

69. **Herman, C. A., Shinholser, R. L., and Lujan, M. D.,** Comparative effects of prostaglandin E_2 and prostaglandin E_3 on water flow and cyclic AMP in the urinary bladder of the frog, *Rana pipiens, Prostaglandins,* 29, 629, 1985.
70. **Herman, C. A., Robleto, D. O., Mata, P. L., and Lujan, M. D.,** Attenuated cardiovascular effects of prostaglandin I_2 and prostaglandin F_{2a} in cold acclimated American bullfrogs, *Rana catesbeiana, J. Exp. Zool.,* 238, 167, 1986.
71. **Herman, C. A., Hamberg, M., and Granstrom, E.,** Quantitative determination of prostaglandins E1, E2, and E3 in frog tissue, *J. Chromatogr.,* 394, 353, 1987.
72. **Baranska, J. and Woldawer, P.,** Influence of temperature on the composition of fatty acids and on lipogenesis in frog tissues, *Comp. Biochem. Physiol.,* 28, 553, 1969.
73. **Sellner, P. A. and Hazel, J. R.,** Desaturation and elongation of the unsaturated fatty acids from thermally acclimated rainbow trout *Salmo gairdneri, Arch. Biochem. Biophys.,* 213, 58, 1982.
74. **Wales, N. A. M. and Gaunt, T.,** Hemodynamic, renal, and steroidogenic actions of prostaglandins E1, E2, A2 and F2a in European eels, *Gen. Comp. Endocrinol.,* 62, 327, 1986.
75. **Piomelli, D., Pinto, A., and Tota, B.,** Divergence in vascular actions of prostacyclin during vertebrate evolution, *J. Exp. Zool.,* 233, 127, 1985.
76. **Herman, C. A., McCloskey, M. M., and Doolittle, K. D.,** Prostaglandins of the one, two, and three series affect blood pressure in the American bullfrog, *Rana catesbeiana, Gen. Comp. Endocrinol.,* 48, 491, 1982.
77. **Guillette, L. J., Jr.,** Prostaglandins and reproduction in reptiles, *Prog. Comp. Endocrinol.,* 603, 1990.
78. **Guillette, L. J., Jr., Herman, C. A., and Dickey, D. A.,** Synthesis of prostaglandins by tissues of the viviparous lizard *Sceloporus jarrovi, J. Herpetol.,* 22, 180, 1988.
79. **Guillette, L. J., Jr.,** The evolution of viviparity in fishes, amphibians and reptiles: an endocrine approach, in *Hormones and Reproduction in Fishes, Amphibians and Reptiles,* Norris, D. O. and Jones, R. E., Eds., Plenum Press, New York, 1987, 523.
80. **Ueno, R., Narumiya, S., Ogorochi, T., Nakayama, T., Ishikawa, Y., and Hayaishi, O.,** Role of prostaglandin D_2 in the hypothermia of rats caused by bacterial lipopolysaccharide, *Proc. Natl. Acad. Sci. U.S.A.,* 79, 6093, 1982.
81. **Ueno, R., Ishikawa, Y., Nakayama, T., and Hayaishi, O.,** Prostaglandin D_2 induces sleep when microinjected into the preoptic area of conscious rats, *Biochem. Biophys. Res. Commun.,* 109, 576, 1982.
82. **Matsumura, H., Honda, K., Goh, Y., Ueno, R., Sakai, T., Inoue, S., and Hayaishi, O.,** Awakening effect of prostaglandin E_2 in freely moving rats, *Brain Res.,* 481, 242, 1989.
83. **Nishino, S., Mignot, E., Fruhstorfer, B., Dement, W. C., and Hayaishi, O.,** Prostaglandin E_2 and its methyl ester reduce cataplexy in canine narcolepsy, *Proc. Natl. Acad. Sci. U.S.A.,* 86, 2483, 1989.
84. **Matsumura, H., Honda, K., Choi, W. S., Inoue, S., Sakai, T., and Hayaishi, O.,** Evidence that brain prostaglandin E_2 is involved in physiological sleep-wake regulation in rats, *Proc. Natl. Acad. Sci. U.S.A.,* 86, 5666, 1989.
85. **Hiroshima, O., Hayashi, H., Ito, S., and Hayaishi, O.,** Molecular mechanism of sleep-wake regulation: roles of prostaglandins D2 and E2, *FASEB J.,* 5, 2575, 1991.
86. **Gigliotti, F., Shenep, J. L., Lott, L., and Thornton, D.,** Induction of prostaglandin synthesis as the mechanism responsible for the chills and fever produced by infusing Amphotericin B., *J. Infect. Dis.,* 156, 784, 1987.
87. **Coceani, F., Lees, J., and Bishai, I.,** Further evidence implicating prostaglandin E_2 in the genesis of pyrogen fever, *Am. J. Physiol.,* 254, R46, 1988.
88. **Nakamura, H., Seto, Y., Motoyoshi, S., Kadokawa, T., and Sunahara, N.,** Recombinant human tumor necrosis factor causes long-lasting and prostaglandin-mediated fever, with little tolerance in rabbits, *J. Pharmacol. Exp. Ther.,* 245, 336, 1988.
89. **Gollman, H. M. and Rudy, T. A.,** Comparative pyrogenic potency of endogenous prostanoids and of prostanoid-mimetics injected into the anterior hypothalamic/preoptic region of the cat, *Brain Res.,* 449, 281, 1988.
90. **Williams, J. H., Errington, M. L., Lynch, M. A., and Bliss, T. V. P.,** Arachidonic acid induces a long-term activity dependent enhancement of synaptic transmission in the hippocampus, *Nature,* 341, 739, 1989.
91. **Ferreira, S. H., Nakamura, M., and Salete de Abren Castro, M.,** The hyperalgesic effects of prostacyclin and prostaglandin E_2, *Prostaglandins,* 16, 31, 1978.
92. **Lembeck, R., Popper, H., and Juan, H.,** Release of prostaglandins by bradykinin as an intrinsic mechanism of its algesic effect, *Naunyn Schmiedebergs Arch. Pharmacol.,* 294, 69, 1976.
93. **Juan, H.,** Dependence of histamine-evoked nociception on prostaglandin release, *Agents Actions,* 11, 706, 1981.
94. **Kilbom, A. and Wennmalm, A.,** Endogenous prostaglandins as local regulators of blood flow in man: effect of indomethacin on reactive and functional hyperaemia, *J. Physiol. (Lond.),* 257, 109, 1976.

95. **Messina, E. J., Weiner, R., and Kaley, G.,** Arteriolar reactive hyperaemia: modification by indomethacin, *Physiologist,* 17, A287, 1974.
96. **Herbaczynska-Cedro, K., Staszewska-Barczak, K., and Janczewska, H.,** The release of prostaglandin-like substances during reactive and functional hyperaemia in the hind leg of the dog, *Polish J. Pharmacol. Pharm.,* 26, 167, 1974.
97. **Carlsson, I. and Wennmalm, A.,** Effect of different prostaglandin synthesis inhibitors on post-occlusive blood flow in human forearm, *Prostaglandins,* 26, 241, 1983.
98. **Nowak, J. and Wennmalm, A.,** A study on the role of endogenous prostaglandins in the development of exercise-induced and post-occlusive hyperaemia in human limbs, *Acta Physiol. Scand.,* 106, 365, 1979.
99. **Grabowski, E. F., Jaffe, E. A., and Weksler, B. B.,** Prostacyclin production by cultured endothelial cell monolayers exposed to step increases in shear stress, *Circulation,* 66, II53, 1984.
100. **Frangos, J. A., Eskin, S. G., McIntire, L. V., and Ives, C. L.,** Flow effects on prostacyclin production by cultured human endothelial cells, *Science,* 227, 1477, 1985.
101. **Koller, A. and Kaley, G.,** Prostaglandins mediate arteriolar dilation to increased blood flow velocity in skeletal muscle microcirculation, *Circ. Res.,* 67, 529, 1990.
102. **Horton, R., Nadler, J., Manogian, C., and Lee, F.,** Prostacyclin is a key mediator of the vasodilator action of dopamine in humans, *Adv. Prostaglandin Thromboxane Leukotriene Res.,* 17, 753, 1987.
103. **Mirro, R., Busija, D. W., Armstead, W. M., and Leffler, C. W.,** Histamine dilates pial arterioles of newborn pigs through prostanoid production, *Am. J. Physiol.,* 254, H-1023, 1988.
104. **Pontieri, V., Lopes, O. U., and Ferreira, S. H.,** Hypotensive effect of captopril, *Hypertension,* 15, 55, 1990.
105. **Mullane, K. M. and Moncada, S.,** Prostacyclin mediates the potentiated hypotensive effect of bradykinin following captopril treatment, *Eur. J. Pharmacol.,* 66, 355, 1980.
106. **Downey, G. P., Gumbay, R. S., Doherty, D. E., Labrecque, J. F., Henson, J. E., Henson, P. M., and Worthen, G. S.,** Enhancement of pulmonary inflammation by PGE_2: evidence for a vasodilator effect, *J. Appl. Physiol.,* 64(2), 728, 1988.
107. **Smith, J. W., Steiner, A. L., and Parker, C. W.,** Human lymphocyte metabolism. Effects of cyclic and noncyclic nucleotides on stimulation by phytohemagglutinin, *J. Clin. Invest.* 50, 442, 1971.
108. **Goodwin, J. C., Bankhurst, A. D., and Messner, R. P.,** Suppression of human T-cell mitogenesis by prostaglandin; existence of a prostaglandin-producing suppressor cell, *J. Exp. Med.,* 146, 1719, 1977.
109. **Goodwin, J. S., Messner, R. P., and Peake, J. T.,** Prostaglandin suppression of mitogen-stimulated lymphocytes *in vitro*. Changes with mitogen dose and preincubations, *J. Clin. Invest.,* 62, 753, 1978.
110. **Leung, K. H. and Koren, H. S.,** Regulation of human natural killing. II. Protective effect of interferon on NK cells from suppression by PGE_2, *J. Immunol.,* 129, 1742, 1982.
111. **Honda, M. and Steinberg, A. D.,** Effects of prostaglandin E_2 on responses of T-cell subsets to mitogen and autologous non-T-cell stimulation, *J. Immunol. Immunopathol.,* 33, 111, 1984.
112. **Chouaib, S., Chatenoud, L., Klatzmann, D., and Fradelizi, D.,** The mechanisms of inhibition of human IL 2 production. II. PGE_2 induction of suppressor T lymphocytes, *J. Immunol.,* 132, 1851, 1984.
113. **Tripp, C. S., Wyche, A., Unanue, E. R., and Needleman, P.,** The functional significance of the regulation of macrophage Ia expression by endogenous arachidonate metabolites *in vitro, J. Immunol.,* 137, 3915, 1986.
114. **Kunkel, S. L., Wiggins, R. C., Chensue, S. W., and Larrick, J.,** Regulation of macrophage tumor necrosis factor production by prostaglandin E_2, *Biochem. Biophys. Res. Commun.,* 137, 404, 1986.
115. **Choualb, S., Robb, R. J., Welte, K., and Dupont, B.,** Analysis of prostaglandin E_2 effect on T-lymphocyte activation, abrogation of prostaglandin E_2 inhibitory effect by the tumor promotor 12.0 tetradecanoyl phorbol-13 acetate, *J. Clin. Invest.,* 80, 333, 1987.
116. **Rutherford, R. B.,** C3-mediated release of prostaglandin from human monocytes: evidence for independent regulation of thromboxane and prostaglandin synthesis, *J. Leukocyte Biol.,* 42, 524, 1987.
117. **Stephan, R. N., Conrad, P. J., Saizawa, M., Dean, R. E., and Chaudry, I. H.,** Prostaglandin E_2 depresses antigen-presenting cell function of peritoneal macrophages, *J. Surg. Res.,* 44, 733, 1988.
118. **Otterness, I. G., Bliven, M. L., Eskra, J. D., Reinke, M., and Hanson, D. C.,** The pharmacologic regulation of interleukin-1 production: the role of prostaglandins, *Cell. Immunol.,* 114, 385, 1988.
119. **Pollak, R., Dumble, L. J., Wiederkehr, J. C., Maddux, M. S., and Moran, M.,** The immunosuppressive properties of new oral prostaglandin E_1 analogs, *Transplantation,* 50, 834, 1990.
120. **Scodras, J. M., Parhar, R. S., Kennedy, T. G., and Lala, P. K.,** Prostaglandin-mediated inactivation of natural killer cells in the murine decidua, *Cell. Immunol.,* 127, 352, 1990.
121. **Mertin, J., Stackpoole, A., and Shumway, S. J.,** Prostaglandins and cell-mediated immunity, *Transplantation,* 37, 396, 1984.
122. **Pelus, L. M. and Strausser, H. R,** Prostaglandins and the immune response, *Life Sci.,* 20, 903, 1977.

123. **Johnston, M. G., Hay, J. B., and Movat, H. Z.,** The distribution of prostaglandins in afferent and efferent lymph from inflammatory sites, *Am. J. Pathol.,* 99, 695, 1980.
124. **Johnston, M. G., Hay, J. B., and Movat, H. Z.** Kinetics of prostaglandin production in various inflammatory lesions, measured in draining lymph, *Am. J. Pathol.,* 95, 225, 1979.
125. **Moore, T. C., Spruck, C. H., Lami, J. L., and Said, S. I.,** Prompt elevations of PGE_2 and thromboxane A_2 metabolites in peripheral node efferent lymph of sheep following drainage area immunization, *Immunopharmacology,* 17, 73, 1989.
126. **Moore, T. C., Hopkins, J., McConnell, I., and Lachmann, P. J.,** Efferent lymph output of lymphocytes following infusions of prostaglandin E_2 and other vasoactive mediators into cannulated afferent popliteal lymphatics in sheep, *IRCS Med. Sci.,* 8, 272, 1980.
127. **Moore, T. C.,** The effects of inflammatory mediators and anaesthetic agents upon lymphocyte traffic of sheep *in vivo,* Ph.D. thesis, University of Cambridge (Trinity College), England, 1982.
128. **Moore, T. C. and Lachmann, P. J.,** Chronic depression of lymphocyte traffic by longterm administration of prostaglandins to peripheral lymph nodes of sheep, *Surgery,* 95, 45, 1984.
129. **Tissot, M., Strzalko, S., Thuret, A., and Giroud, J. P.,** Prostanoid release by macrophages at a distance from an inflammatory site, *Br. J. Exp. Pathol.,* 70, 525, 1989.
130. **Bennett, A.,** Prostaglandins: relationships to breast cancer and its spread, in *Endocrine Relationships in Breast Cancer,* Stoll, Ed., W. Heinemann, London, 1982, 156.
131. **Bennett, A., Charlier, E. M., McDonald, A. M., Simpson, J. S., Stamford, I. F., and Zebro, T.,** Prostaglandins and breast cancer, *Lancet,* ii, 624, 1977.
132. **Bennett, A., Berstock, D. A., Raja, B., and Stamford, I. F.,** Survival time after surgery is inversely related to the amounts of prostaglandins extracted from human breast cancers, *Br. J. Pharmacol.,* 66, 451, 1979.
133. **Wilson, A. J., Baum, M., Bennett, A., Griffiths, K., Nicholson, R. I., and Stamford, I. F.,** Lymph node status prostaglandin and oestrogen receptors are independent variables in human primary breast cancer, *Clin. Oncol.,* 6, 379, 1980.
134. **Campbell, F. C., Hayes, D. F., Evans, D. F. et al.,** Prostaglandin E_2 synthesis by tumour epithelial cells and estrogen receptor status of primary breast cancer, *Langenbecks Arch. Chir.,* 357, 209, 1982.
135. **Fulton, A., Roi, L., Howard, L., Russo, J., Brooks, S., and Brennan, M. J.,** Tumour-associated prostaglandins in patients with primary breast cancer: relationship to clinical parameters, *Breast Cancer Res. Treat.,* 2, 331, 1982.
136. **Karmall, R. A., Welt, S., Thaler, H. T., and Lefevre, F.,** Prostaglandins in breast cancer: relationship to disease stage and hormone status, *Br. J. Cancer,* 48, 689, 1983.
137. **Rolland, P. H., Martin, P. M., Jaquemier, J., Rolland, A. M., and Toga, M.,** Prostaglandins in human breast cancer. Evidence suggesting that an elevated prostaglandin production is a marker of high metastatic potential for neoplastic cell, *J. Natl. Cancer Inst.,* 64, 1061, 1980.
138. **Watson, D. M. A., Kelly, R. W., and Miller, W. R.,** Prostaglandins and prognosis in human breast cancer, *Br. J. Cancer,* 56, 367, 1987.
139. **Bennett, A., Stamford, I. F., Berstock, D. A., Dische, F., Singh, L., and A'Hern, R. P.,** Breast cancer, prostaglandins and patient survival, *Br. J. Cancer,* 59, 268, 1989.
140. **Stringfellow, D. A. and Fitzpatrick, F. A.,** Prostaglandin D_2 controls pulmonary metastatsis of malignant melanoma cells, *Nature,* 282, 76, 1979.
141. **Fitzpatrick, F. A. and Stringfellow, D. A.,** Prostaglandin D_2 formation by malignant melanoma cells correlates inversely with cellular metastatic potential, *Proc. Natl. Acad. Sci. U.S.A.,* 76, 1765, 1979.
142. **Samuelsson, B., Dahlen, S.-E., Lindgren, J. A., Rouzer, C. A., and Serhan, C. N.,** Leukotrienes and lipoxins: structures, biosynthesis, and biological effects, *Science,* 237, 1171, 1987.
143. **Ramstedt, U., Ng, J., Wigzell, H., Serhan, C. N., and Samuelsson, B.,** Action of novel eicosanoids lipoxin A and B on human natural killer cell cytotoxicity: effects on intracellular cAMP and target cell binding, *J. Immunol.,* 135, 3434, 1985.
144. **Hansson, A., Serhan, C. N., Ingelman-Sundberg, M., and Samuelsson, B.,** Activation of protein kinase C by lipoxin A and other eicosanoids. Intracellular action of oxygenation products of arachidonic acid, *Biochem. Biophys. Res. Commun.,* 134, 1215, 1985.

Chapter 16

TRANSMEMBRANE SIGNALING MOLECULES

I. STIMULATORY

A. PI

These transmembrane signaling molecules function in a remarkably similar way both in inhibitory/modulatory and stimulatory/excitatory effects on both neuronal and immune response cells. Stimulatory/excitatory effects are mediated by the PI (phosphatidylinositol) dual second-messenger system that involves IP_3 (inositol triphosphide) and DG (diacylglycerol). IP_3 acts by an elevation in intracellular Ca^{++}, and DG leads to protein kinase C (PKC) activation.[1-3]

Analogues of these dual messengers, calcium ionophore A-23187 for IP_3 and phorbol ester for DG, when utilized together, have been found *in vitro* to duplicate or simulate an antigen- or mitogen-induced activation of lymphocytes.[4,5] *In vivo* their combined use in the author's supported-sheep lymphocyte traffic model has produced responses that closely resemble the lymphocyte traffic responses to lymph node drainage immunization with particulate antigen, including prompt ''shutdown''/recruitment decreases in lymphocyte outputs into efferent lymph, with associated increases in efferent lymph flow, with later (20 h) marked increases in lymphocyte outputs in the efferent lymph of the stimulated popliteal lymph node (Figures 1 and 2).[6]

B. CYCLIC GMP

Cyclic GMP is a second transmembrane stimulatory molecule, while cyclic AMP with adenosine is an inhibulatory/modulatory molecule. In the supported-sheep lymphocyte traffic model, acute infusion of a lipophilic analogue of cyclic GMP into a cannulated popliteal lymph node afferent lymphatic has been found to produce a prompt and sharp increase in the output of small recirculating lymphocytes into study node efferent lymph (Figure 3).[7]

II. INHIBITORY

A. CYCLIC AMP

In both the nervous system and the immune response system, activation of the cyclic AMP transmembrane signaling mechanism has had an inhibitory and modulatory effect. This has been discussed in detail in chapters relating to individual molecules.

Acute infusion of a lipophilic analogue of cyclic AMP into a cannulated afferent lymphatic in the supported-sheep lymphocyte traffic model has produced a prompt, sharp, and prolonged depression in lymphocyte traffic and in the output of small recirculating lymphocytes into efferent lymph (Figure 4).[7]

B. ADENOSINE

Adenosine, like cyclic AMP, functions largely as an inhibitory molecule. The existence and roles of A_1 and A_2 receptors for adenosine are not entirely resolved. In 1990 Phillis,[8] of Wayne State University (Detroit), described a potent depressant action on cortical neuronal firing of a select adenosine A_2 receptor and interpreted this finding as suggesting that the A_2 receptor is involved in the modulation/decreasing of cerebrocortical neuronal firing by adenosine.

Adenosine also has been reported to have an antinociceptive effect on both central and peripheral administration.[9,10] Intrathecal administration of adenosine analogues also has been

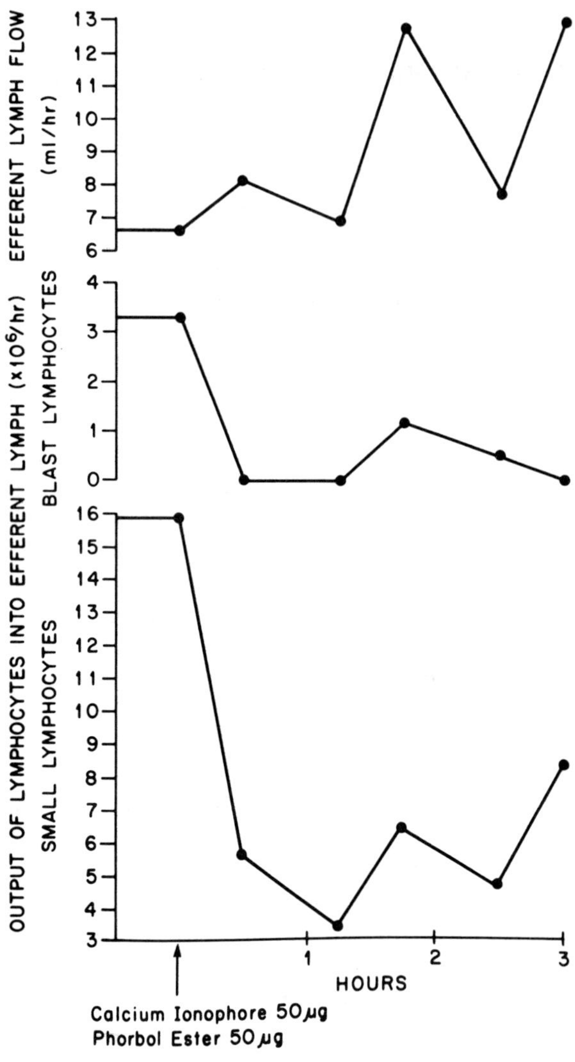

FIGURE 1. Activation of the stimulatory phosphoinositol (PI) transmembrane signaling pathway by injection of pathway analogues, phorbol ester and calcium ionophore, into the drainage area of a supported-sheep popliteal lymph node whose efferent lymphatic has been chronically cannulated, has mimicked the drainage-area "shutdown"/recruitment response, with increased efferent lymph flow produced by drainage-area immunization with killed *Salmonella muenchen* bacteria, as described and illustrated (Figures 4 and 5) in Chapter 15. The phorbol ester used was phorbol 12-myristate-13 acetate (PMA), and the calcium ionophore was A23187 (both from Sigma Chemical Co., St. Louis, MO). (From Moore, T. C., et al., *Immunopharmacology,* 21, 1, 1991.)

found to produce antinociception.[11-15] In 1990 Aran and Proudfit,[16] of the University of Illinois–Chicago, described a synergistic effect of adrenergic and noradrenergic agonists in modifying/decreasing nociceptic transmission in the spinal cord of rats. They concluded that the purinergic component of this interaction likely is mediated, at least in part, by adenosine A_2 receptors.

A 1990 publication also identifies an anti-inflammatory action of adenosine-receptor agonists. Schreier et al.,[17] of Parke-Davis in Ann Arbor, described anti-inflammatory actions of adenosine agonist on both the exudative and cellular phases of carrageenn-induced pleural inflammation in rats.

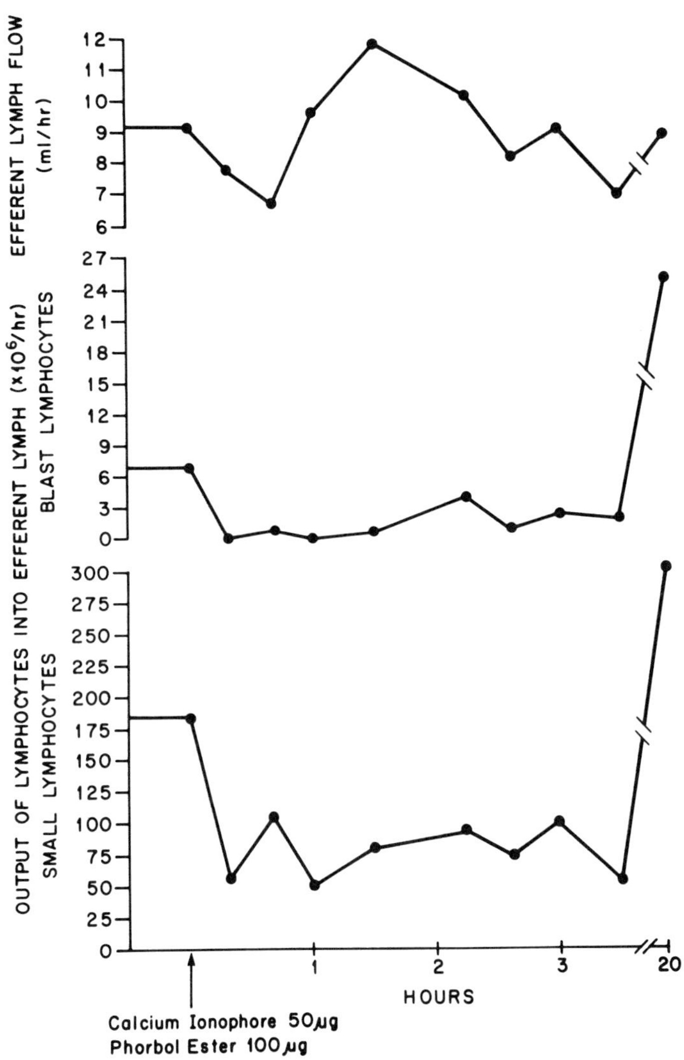

FIGURE 2. Phorbol ester and calcium ionophore activation of sheep popliteal lymph node drainage-area PI transmembrane signaling pathway, with prompt decreases in lymphocyte outputs into nodal efferent lymph, with associated increases in efferent lymph flow and later (20 h) with marked increases in the outputs of both small recirculating and blast lymphocytes, as encountered with drainage-area immunization in this supported-sheep lymphocyte traffic model of the author. (From Moore, T. C., et al., *Immunopharmacology,* 21, 1, 1991.

A report by Barraco and Stephano[18] in 1990 established a long evolutionary role for adenosine in the animal central nervous system, as adenosine was shown by them to inhibit neurotransmitter releases from the pedal ganglia of the marine bivalve mollusc *Mytilus edulis*. The greatest release inhibition involved serotonin and dopamine.

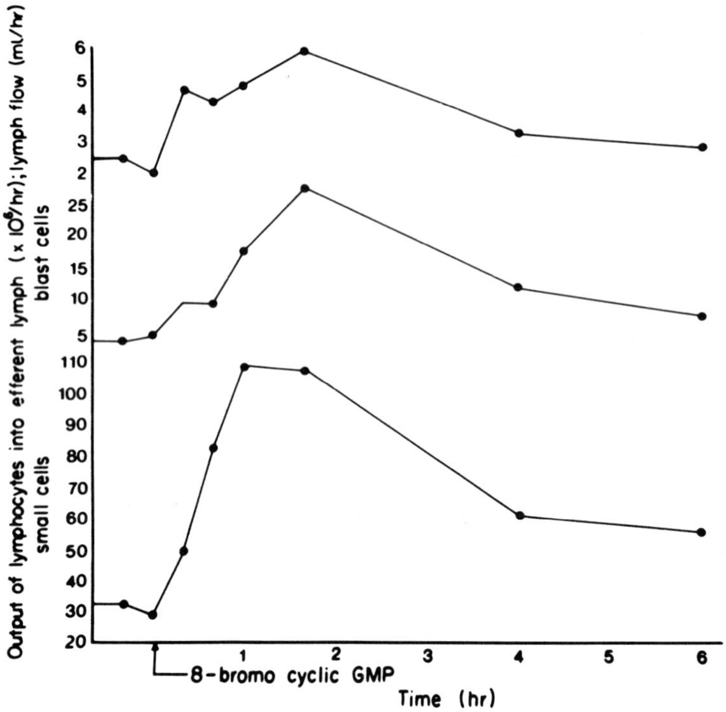

FIGURE 3. Acute infusion of a lipophilic analogue of cyclic GMP (8-bromo cyclic GMP, Sigma Chemical Co., St. Louis, MO) into a cannulated sheep popliteal lymph node afferent lymphatic produces a prompt, sharp, and prolonged increase in the output of small recirculating and blast lymphocytes into nodal efferent lymph (chronically cannulated) and with an accompanying increase in efferent lymph flow. (From Moore, T. C. and Lachmann, P. J., *Immunology,* 47, 423, 1982.)

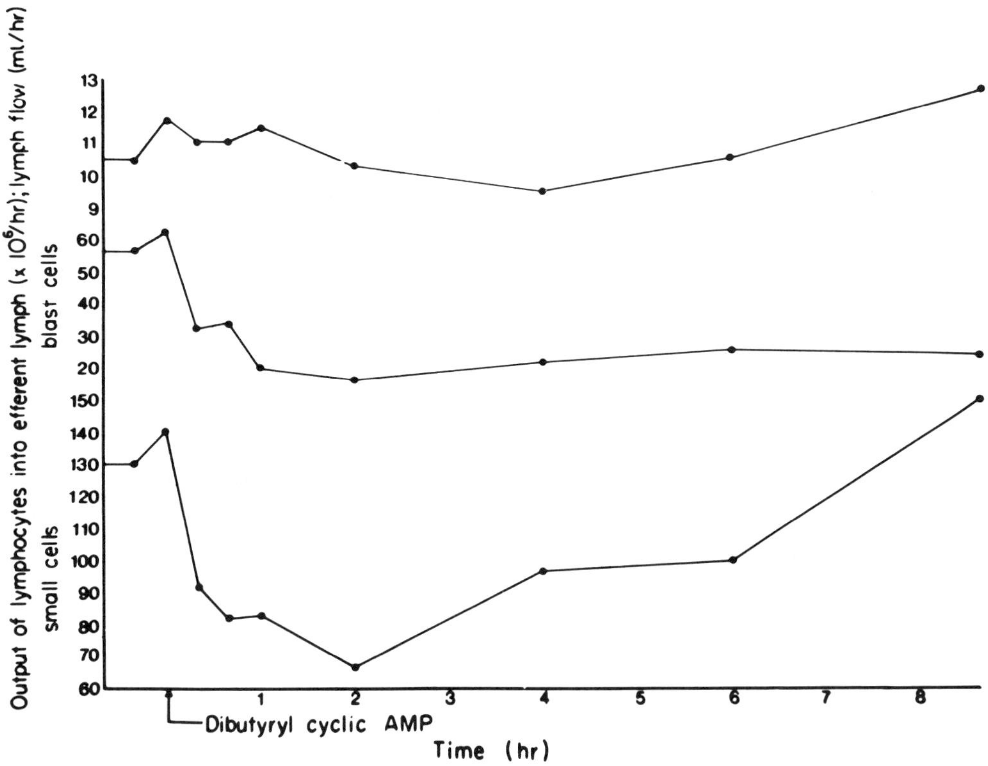

FIGURE 4. Acute infusion of a lipophilic analogue of cyclic AMP (dibutyryl cyclic AMP, Sigma Chemical Co., St. Louis, MO) into a cannulated sheep popliteal lymph node afferent lymphatic produces a prompt, sharp, and prolonged decrease in the output of small recirculating and blast lymphocytes into nodal efferent lymph, chronically cannulated. (From Moore, T. C. and Lachmann, P. J., *Immunology,* 47, 423, 1982.)

REFERENCES

1. **Berridge, M. J., Dawson, R. M., Downes, C. P., Helsop, J. P., and Irvine, R. F.,** Changes in the levels of inositol phosphates after agonist-dependent hydrolysis of membrane phosphoinositides, *Biochem. J.,* 212, 473, 1983.
2. **Berridge, M. J. and Irvine, R. F.,** Inositol triphosphate, a novel second messenger in cellular signal transduction, *Nature,* 312, 315, 1984.
3. **Nishizuka, Y.,** The role of protein kinase C in cell surface signal transduction and tumour production, *Nature,* 308, 693, 1984.
4. **Klaus, G. G. B., O'Garra, A., Bijsterbosch, M. K., and Holman, M.,** Activation and proliferation signals in mouse B cells. VIII. Introduction of DNA synthesis in B cells by a combination of calcium ionophores and phorbol myristate acetate, *Eur. J. Immunol.,* 16, 92, 1986.
5. **Klaus, G. G. B.,** Transmembrane signalling by B lymphocyte receptors, *Immunol. Suppl.,* 1, 35, 1988.
6. **Moore, T. C., Lami, J. L., and Said, S. I.,** *In vivo* influences of phorbol ester and calcium ionophore on lymphocyte traffic, lymph flow and efferent lymph levels of thromboxane B2 in sheep, *Immunopharmacology,* 21, 1, 1991.
7. **Moore, T. C. and Lachmann, P. J.,** Cyclic AMP reduces and cyclic GMP increases the traffic of lymphocytes through peripheral lymph nodes of sheep *in vivo, Immunology,* 47, 423, 1982.
8. **Phillis, J. W.,** The selective adenosine A_2 receptor agonist, CGS 21680, is a potent depressant of cerebral cortical neuronal activity, *Brain Res.,* 509, 328, 1990.

9. **Holmgren, M., Hednar, J., Mellstrand, T., Nordberg, G., and Hedner, T.,** Characterization of the antinociceptive effects of some adenosine analogues in the rat, *Naunyn Schmiedebergs Arch. Pharmacol.*, 334, 290, 1986.
10. **Yarbrough, G. G. and McGuffin-Clineschmidt, J. C.,** *In vivo* behavioral assessment of central nervous system purinergic receptors, *Eur. J. Pharmacol.*, 76, 137, 1981.
11. **DeLander, G. E. and Hopkins, C. J.,** Interdependence of spinal adenosinergic, serotonergic and noradrenergic systems mediating antinociception, *Neuropharmacology,* 26, 1791, 1987.
12. **DeLander, G. E. and Hopkins, C. J.,** Involvement of A_2 adenosine receptors in spinal mechanisms of antinociception, *Eur. J. Pharmacol.*, 139, 215, 1987.
13. **Holmgren, M., Hednar, J., Mellstrand, T., Nordberg, G., and Hedner, T.,** Characterization of the antinociceptive effects of some adenosine analogues in the rat, *Naunyn Schmiedebergs Arch. Pharmacol.*, 334, 290, 1986.
14. **Post, C.,** Antinociceptive effects in mice after intrathecal injection of 5'-*N*-ethylcarboxamide adenosine, *Neurosci. Lett.*, 51, 325, 1984.
15. **Sawynok, J., Sweeney, M. I., and White, T. D.,** Classification of adenosine receptors mediating antinociception in the rat spinal cord, *Br. J. Pharmacol.*, 88, 923, 1986.
16. **Aran, S. and Proudfit, H. K.,** Antinociception produced by interactions between intrathecally administered adenosine agonists and norepinephrine, *Brain Res.*, 513, 255, 1990.
17. **Schrier, D. J., Lesch, M. E., Wright, C. D., and Gilbertsen, R. B.,** The antiinflammatory effects of adenosine receptor agonists on the carrageenin-induced pleural inflammatory response in rats, *J. Immunol.*, 145, 1874, 1990.
18. **Barraco, R. A. and Stefano, G. B.,** Pharmacological evidence for the modulation of monoamine release by adenosine in the invertebrate nervous system, *J. Neurochem.*, 54, 2002, 1990.

Chapter 17

CASCADE MOLECULES

I. INTRODUCTION

> "Some explosive devices contain different kinds of powder so the first effect is followed by a second or by a cascade of effects. One could perhaps think of several transmitters acting like that."
>
> U. S. von Euler[1]

As our knowledge of the cellular immune response and related neuronal and vascular components increases, the great importance and interrelationship of major cascade molecules in contributing to and enhancing chemical plasticity in both the nervous and immune response systems becomes increasingly apparent. These major cascade molecules are closely associated and interrelated with the vascular system. These cascades include the coagulation-fibrinolysis cascade, the kinins, EDRF (endothelium-derived relaxing factor)/nitric oxide/NO, the complement cascade, the arachidonic acid cascade, and the angiotensin-renin cascade.

II. COAGULATION — FIBRINOLYSIS

The remarkable phenomena of coagulation and fibrinolysis permit a physiological and biochemical walking of a fine line between exsanguination (leading to death) and vascular and lymphatic clot plugging (leading to loss of limb or organ, with associated threat to life). The extraordinarily effective functioning of this system truly qualifies for phenomenon classification. The coagulation-fibrinolysis cascade is the fundamental core cascade setting in motion the other and interrelated cascades that are so critical and vital to the biological survival of mammals. The author, as a clinical, vascular, and transplant surgeon, stands in a respectful awe of the extraordinary functioning of this system.

III. COMPLEMENT

Complement was discovered some 100 years ago as a blood vascular system activity in cell-free blood serum that could kill certain microbes. Activation of the complement system by both the classical and alternate pathways (Figure 1) represents an interlocking and major defense cascade with renewal and major defense molecule spin-offs, including the potent anaphylatoxins $C3_a$ and $C5_a$. The releasing of major vasoactive neurotransmitter and modulator molecules, such as histamine and other key molecules in the inflammatory response, is involved in the complement contribution to basic biological defense. Additional molecules such as those relating to chemotaxis, immune adherence, and phagocytosis (Figure 2) and to cytolytic membrane damage, and histamine also are involved (Figure 3).

Like the other cascade systems, the complement system involves a multicomponent enzyme cascade in which each activated component releases a small and active peptide or peptides and, in the process of the molecular split, activates a subsequent enzyme-acting larger component, and on down the cascade. Complement receptors CR1 and CR3 have been reported to be responsible for the phagocytic and adhesive properties of neutrophils, while the C5a receptor has been identified as mediating the proinflammatory and chemotactic actions of the complement anaphylotoxin C5a. The C5a receptor also stimulates chemotaxis,

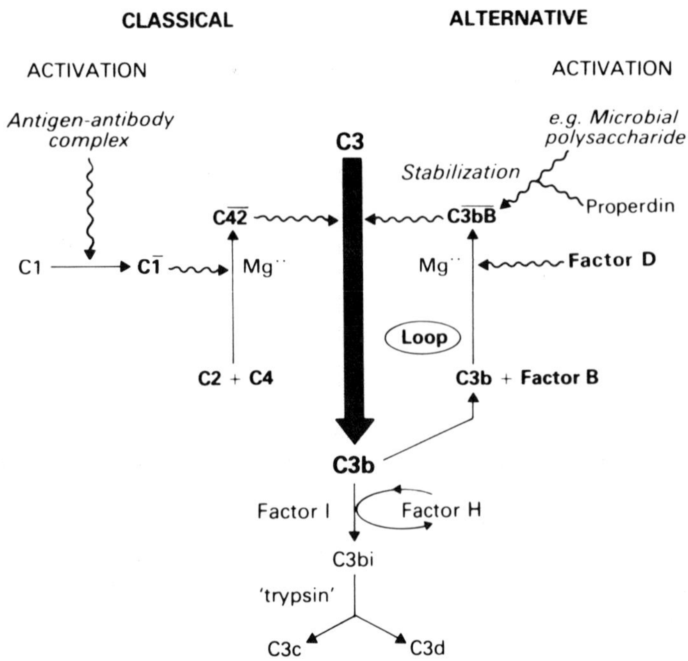

FIGURE 1. Complement cascade through the critical and central C3 stage illustrating the two major pathways — classical and alternative — and their levels and molecules of involvement. The central event in complement activation is the cleavage of C3. The classical pathway was discovered first and is antibody dependent. The alternate pathway was discovered later, is antibody independent, and is considered to be the phylogenetically older of the two pathways, existing prior to the evolutionary development of the lymphoid immune response system in vertebrates. (From Roit, I., *Essential Immunology,* 5th ed., Blackwell Scientific, Oxford, England, 1984. With permission.)

granule enzyme release, and superoxide anion production, as well as stimulates upregulation of the expression and activity of the adhesion molecule MAC-1 and the CR1 receptor and a decrease in the cell-surface glycoprotein 100 MEL-14 on neutrophils.[2-9]

In 1991 Gerard and Gerard,[10] of the Harvard Medical School (Boston), described (in *Nature*) the cloning of the chemotactic receptor for human C5a anaphylotoxin. COS-7 cells transfected with this receptor clone cDNA were capable of responding to C5a with an increase in the labeling of phosphatidylinositols when cultures were prelabeled with $^{32}P_i$. When compared with other receptors coupled to GTP-binding proteins, it was of particular interest that a 25% degree of sequence identity of this C5a receptor with the human receptor for substance K was discovered.[11]

In the author's supported-sheep lymphocyte traffic model, it was of interest that two different means of complement activation have resulted in different lymphocyte traffic responses.[12] Drainage-area injections of complement-activating substances, inulin and cobra venom factor (CVF), produced divergent actions on the pattern of lymphocyte traffic. Inulin produced a prompt and sharp depression in lymphocyte traffic, as monitored by the output of small recirculating lymphocytes into study node efferent lymph, chronically cannulated, as well as in the volume of efferent lymph output (Figure 4). The depression in small lymphocyte traffic was from 143 $\times$ 10^6 cells per hour (45% of preinjection levels) and from 79 $\times$ 10^6 cells per hour (14% of preinjection levels). Blast lymphocytes in both studies were reduced to 0.

Cobra-venom-factor injection into nodal drainage areas, in three separate studies, showed consistent, sharp, and prompt elevations in the output of both small recirculating and blast

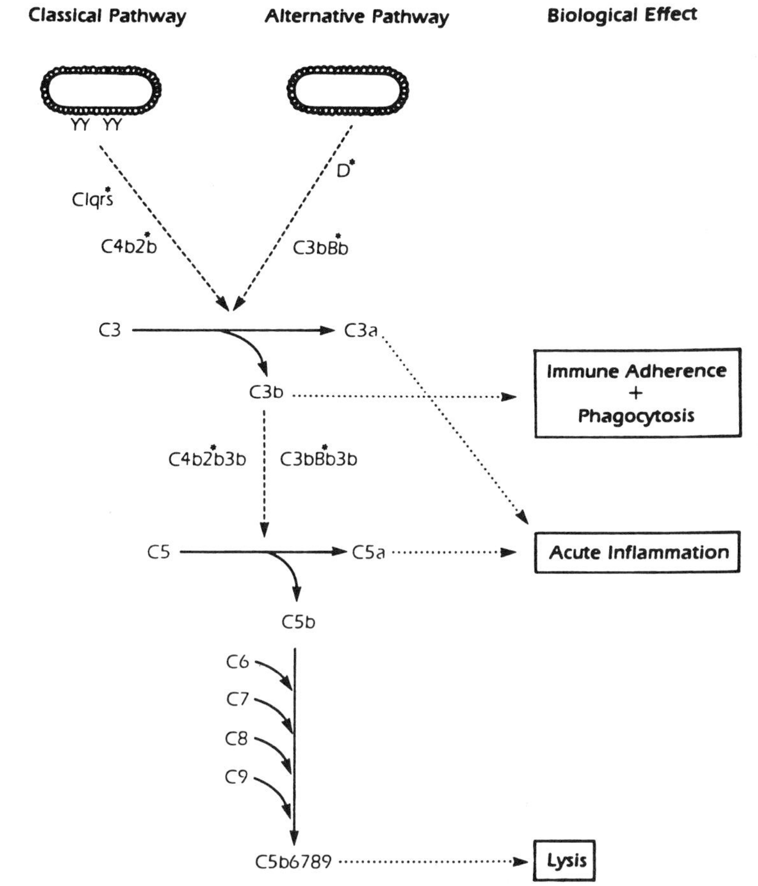

FIGURE 2. Complement cascade through the entire nine components, with identification of biological effects. (From Jose, P. J., *Br. Med. Bull.*, 43, 336, 1987. With permission.)

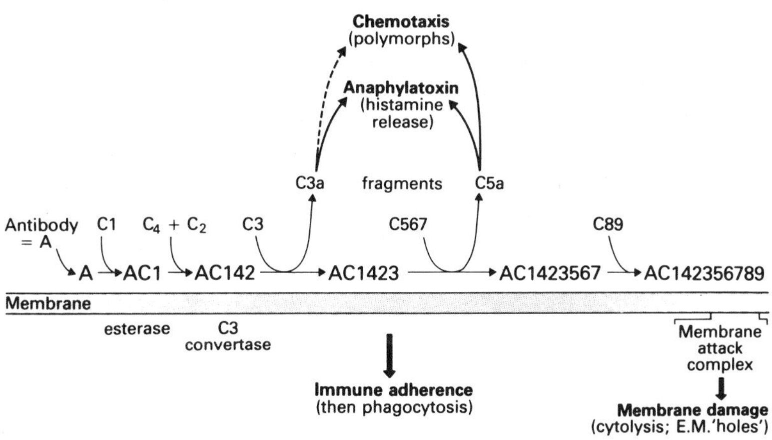

FIGURE 3. Complement cascade with reference to actions at the anatomical level of the cell membrane and including biological effects such as chemotaxis and cell membrane attack and damage and the release of important and potent soluble mediators, such as the major vasoactive neurotransmitter histamine. (From Roit, I., *Essential Immunology,* 5th ed., Blackwell Scientific, Oxford, England, 1984. With permission.)

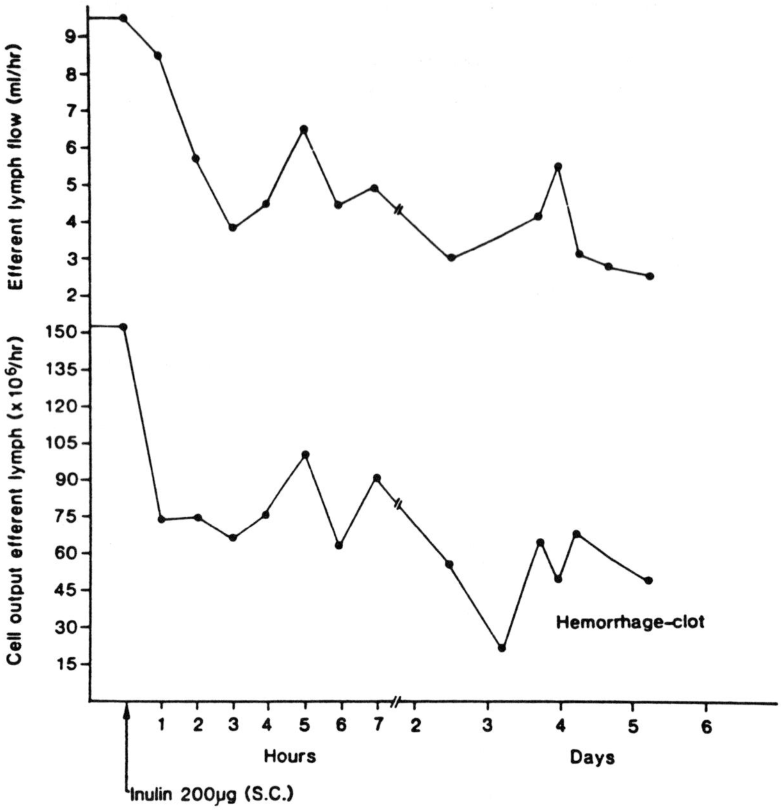

FIGURE 4. The prompt, sharp, and progressive decrease in the output into sheep peripheral (prefemoral), lymph node efferent lymph of small recirculating lymphocytes following injection of inulin (200 μg) subcutaneously into the lymph node lymphatic drainage area. There was an accompanying decrease in efferent lymph flow. (From Moore, T. C., *IRCS Med. Sci.*, 12, 681, 1984. With permission.)

lymphocytes into efferent lymph (Figure 5). Small recirculating lymphocyte outputs increased to 242, 500, and 324% above preinjection levels. In addition, there were marked increases in blast-lymphocyte outputs. These were from 0.02×10^6 to 16.2×10^6 cells per hour, from 1.3×10^6 to 26.4×10^6 cells per hour, and from 12.6×10^6 to 19.8×10^6 cells per hour. The responses to both inulin and CVF were both prompt and prolonged as well.

These differences may be related to differences in the mechanisms of action of these two molecules on the alternate pathway of complement activation. CVF, with Factor B of the alternate pathway, produces a C3 convertase (CVF-B6) that cannot be inactivated by the control proteins, Factor H and Factor I. Inulin, on the other hand, exerts its effect on the alternate pathway by a different method. Inulin binds C3b so that it is protected from activation, and in this circumstance, complement is activated by the C3b feedback cycle.

IV. KININS

The kinin cascade, in which bradykinin is the most potent product, is closely intertwined with the coagulation cascade as well as with the other cascades. These interrelationships are presented in Figures 2 and 3 in Chapter 9, relating to bradykinin and discussed in detail in that chapter.

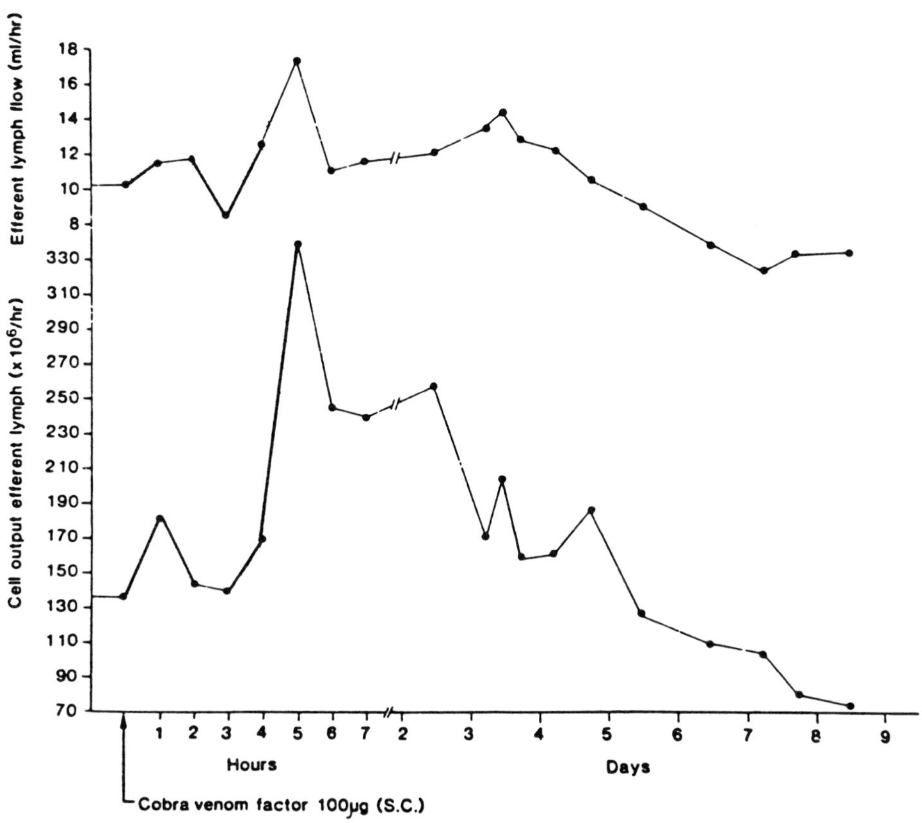

FIGURE 5. The prompt, sharp, and prolonged increase in the output of small recirculating lymphocytes into sheep peripheral lymph node efferent lymph following subcutaneous injection of cobra venom factor (100 μg) into the lymphatic drainage area of the node. Efferent lymph flow also was increased. (From Moore, T. C., *IRCS Med. Sci.*, 12, 681, 1984. With permission.)

V. EDRF/NITRIC OXIDE/NO

A. GENERAL AND VASCULAR

In a major 1980 report in *Nature,* Furchgott and Zawadzki[13] described the obligatory role of endothelial cells in the relaxation of arterial smooth muscle by acetylcholine. Their studies were carried out on the descending thoracic aortas of white male New Zealand rabbits. They showed conclusively that relaxation of isolated preparations of these vessels by acetylcholine required the presence of endothelial cells. The relaxing substance produced was given the name endothelium-derived relaxing factor (EDRF). It was soon established that acetylcholine was by no means the only agent to produce endothelium-dependent relaxation of isolated arteries. Also in 1980 Furchgott and associates[14] described endothelium-dependent relaxation of rabbit aorta by the calcium ionophore A23187. In short order virtually all of the major endogenous vasoactive neurotransmitter and immune response stimulatory molecules (and "memory molecules" or "memory-molecule candidates") were shown to produce endothelium-dependent vascular relaxation through the EDRF mechanism.

In 1981 and 1982 the Furchgott group[15,17] and Chand and Altura[16] described EDRF activation by bradykinin. Also in 1981, the Furchgott group[18] added substance P to the "list" and, in 1983, they[19] added CCK-8 and kassinin as well as substance P. 1983 was the year for both histamine[20,21] and serotonin[22,23] to join the "club". In 1985 Brain et al.[24] added CGRP.

These very exciting findings triggered an intense search for the molecular characterization of EDRF. In 1983 Rapoport and Murad[25] made the significant observation that agonist-induced endothelium-dependent relaxations were mediated through cyclic GMP. Also in 1983 this group with Draznin[26] reported in *Nature* that these relaxations were mediated through a cyclic-GMP-dependent protein phosphorylation.

The grand prize for the molecular characterization of EDRF went to Moncada and associates at the Wellcome Research Laboratories in Beckenham, Kent, U.K. In a series of brilliant investigative maneuvers, they leap-frogged the pack in hot pursuit and identified EDRF as nitric oxide (NO).

In 1986 Moncada with Palmer and Gryglewski,[27-29] in a series of studies to explore the mechanisms of action of some inhibitors of EDRF, discovered that superoxide anion is involved in the breakdown of EDRF, which is quite unstable, with a half-life of only a few seconds (less than 5). They demonstrated that the breakdown of EDRF could be blocked by the use of superoxide dismutase (SOD) and Ca^{++}. This remarkable breakthrough was achieved by the piggy-back combination of microcarrier bead endothelial cell cultures in column and the superfusion cascade bioassay methodology of Vane, which had been used 10 years earlier (1976), by Moncada with Vane, to discover prostacyclin (PGI_2). This approach eliminated the short half-life problem and permitted the accumulation of an adequate amount of EDRF for direct study and molecular characterization. In 1987 the Moncada group,[30] employing chemiluminescence methodology, reported in *Nature* that NO release accounts for the biological activity of EDRF, and in 1988 they reported,[31] also in *Nature,* that vascular endothelial cells synthesize NO from L-arginine. The extraordinary productivity of this remarkable investigative group has continued to enrich the exploding literature relating to this remarkable molecule and its extensive and diverse biological activities.[30-46]

B. IMMUNE RESPONSE SYSTEM

In a 1980 stimulatory study of cyclic GMP metabolism in macrophages, Bromberg and Pick,[47] of Tel Aviv, reported a regulation of macrophage cyclic GMP levels by calcium, and stimulation of cyclic GMP synthesis in macrophages by NO-generating agents. This important observation relating to the immune response system was made in the same year as the discovery of EDRF and 7 years before the identification of EDRF as NO by Moncada and associates. Macrophage metabolism of L-arginine and NO in recent years has experienced a rebirth of enthusiasm. In 1987 Hibbs et al.,[48] at the University of Utah–Salt Lake City, reported that L-arginine is required for the expression of the activated-macrophage-effector mechanism causing selective metabolic inhibition in tumor target cells. In 1988 Hibbs et al.[49] published evidence that NO is an activated-macrophage-effector molecule. In 1989 Stuehr and Nathan[50] reported in the *Journal of Experimental Medicine* that NO was a macrophage product for cytostasis and respiratory inhibition in tumor target cells. Earlier (1985 and 1987) studies by Stuehr with Marletta[51,52] explored mammalian (mouse) macrophage nitrate biosynthesis in response to LPS, BCG, and gamma interferon. In 1990 Mulsch and associates[53] of Freiburg, Germany, described L-arginine-dependent NO formation by bone-marrow-derived macrophages that had been stimulated with LPS and bacterial lipopeptide.

Additional metabolic pathways in macrophage synthesis of NO were explored by Stuehr et al.[54] in 1990. They identified FAD (flavin dinucleotide) and GSH (glutathione) as two additional cofactors for full NO synthase activity, in addition to NADPH and BH_4 (tetrahydrobiopterin). Their findings suggested that an NADPH-utilizing FAD flavoprotein might participate in the NO-synthetic reaction by NO synthase. An additional 1990 publication by the Stuehr (Cornell/New York) group[55] suggested that there are at least two forms of NO synthase from metabolic and inhibitor data: an inducible, tetrahydrobiopterin- and flavin-dependent activity exemplified by the macrophage enzyme, and a constitutive, Ca^{++}-dependent activity exemplified by the endothelial cell enzyme.

In 1990 Kirk et al.,[56] of Johns Hopkins, identified a cloned murine T lymphocyte to synthesize NO, in addition to other immune-response-system-related cells, such as macrophages and neutrophils. These investigators employed cytokine (IL-2) stimulation of the T lymphocytes to generate NO synthesis.

In a series of three major 1990 publications with relevance to NO synthesis and the immune response system, the Moncada group[42,43,45] described the inhibition of endotoxin-induced NO synthesis by immunosuppressive corticosteroids. In a December 1990 publication in the *Proceedings of the National Academy of Science, U.S.A.*, Moncada with Palmer and Radomski[45] made the very important observation that glucocorticoids inhibit the expression of an inducible, but not the constitutive, nitric oxide synthase in vascular endothelial cells. They reported that vascular endothelial cells (porcine aortic endothelial cells) contain both a constitutive NO synthase that is Ca^{++}-dependent and an inducible Ca^{++}-independent NO synthase that is distinct from the constitutive enzyme and that may be stimulated by activation with LPS and gamma interferon. The generation of the inducible NO synthase enzyme and NO production had a lag period of 2 h, with a maximum NO production between 6 and 12 h, with NO production maintained through the duration of the experiment, 48 h (? LTP). The expression of the vascular-endothelium-inducible NO synthase was protein synthesis dependent, blocked by cyclohexamide. The long duration (? LTP) and the protein synthesis requirement stimulate thoughts and speculations with respect to memory/laying-down phenomenon and kindling metabolic responses in the CNS.

In September of 1991, Stuehr et al.[57] of Cornell University in New York reported (in *Proceedings of the National Academy of Science, U.S.A.*) the purification and characterization of the cytokine-induced macrophage NO synthase (NOSmac). The NOS was induced in a mouse macrophage cell line by LPS and gamma interferon and was found to be a flavoprotein that contained not only the flavin FAD, but also the flavin FMN (flavin mononucleotide). It was considered unusual by the authors that induced macrophage NO synthase would contain both FAD and FMN. The only other known mammalian enzyme containing both enzymes is NADPH-cytochrome P-450 reductase. With respect to points of difference between induced (macrophage) and constitutive (brain) NO syntheses, the induced NOS are somewhat heterogeneous in size and smaller than constitutive NOS, appear to be catalytically active as a dimer rather than as a monomer, and do not require added calcium ions or calmodulin for activity. Also in 1991, Yi et al. from Kyoto University (Japan) cloned NOS from rat neutrophils[58] and rat macrophages.[59] The neutrophil NOS appeared to be a calmodulin-independent enzyme.

Also in 1991, Steuhr et al.[60] described the use of the flavoprotein inhibitor, diphenyleneiodonium, as an inhibitor of both macrophages and endothelial cell NOS and suggested this molecule and its analogues as a new class of NOS inhibitors.

C. NERVOUS SYSTEM

A new and fascinating role for EDRF/NO was described by Garthwaite et al.[61] of Liverpool, U.K., in 1988 in *Nature*. They identified the occurrence of EDRF release on activation of NMDA-glutamate receptors in the mammalian (rat) brain and suggested a role for EDRF/NO as an intracellular messenger in the brain. They proposed that the NMDA-receptor activation by glutamate, which they showed to release EDRF, is responsible for the associated brain (cerebellum) increases in cyclic GMP. They speculated that EDRF/NO might be involved in the hippocampal LTP/memory phenomenon.

In 1989 Moncada with Knowles, Palacios, and Palmer[32] described the production of NO from L-arginine in the central nervous system, in a major publication in *Proceedings of the National Academy of Science, U.S.A.* They proposed this as a transducer mechanism for the stimulation of soluble guanalate cyclase. They also demonstrated the presence of the NO-synthesizing enzyme, NO synthase, in adult rat brain. In 1990 this group[62] described the use of L-NMMA (*N*-monomethyl-arginine), an analogue and competetive inhibitor of L-arginine, to block the formation of NO in the brain by NO synthase activity, as L-arginine is the substrate of the NO synthase enzyme. They also reported that the NO synthase in adult rat forebrain has a Ki for *N*-NARG of 0.4 μM. *N*-NARG is another NO synthase inhibitor that has been found to be a more potent inhibitor (than L-NMMA) of the NMDA-induced formation of cyclic GMP in slices of immature rat cerebellum.

Also in 1989, Bredt and Snyder[63] of Johns Hopkins reported studies confirming that NO mediates glutamate-linked enhancement of cyclic GMP levels in the cerebellum. They demonstrated that the NOS inhibitor L-NMMA blocks the enhancement of cyclic GMP levels by glutamate and NMDA and that this blockade could be reversed by added arginine. In 1990 Bredt and Snyder described the isolation and purification of NOS as a calmodulin-requiring enzyme[64] and used this purified NOS enzyme to prepare affinity-purified NOS antibodies for the anatomical location of NOS in the rat brain.[65] Dramatic variations in NOS immunoreactivity were found to occur throughout the rat brain, with the densest staining being found in the cerebellar molecular and granule cell layers. Abundant staining also was identified in the dentate gyrus of the hippocampus (a memory and LTP area), the bed nucleus of the striaterminalis, the islands of Callejae, and the horizontal limb of the diagonal band of Broca, as well as in the superior and inferior colliculi. They found somewhat lesser staining in the cerebral cortex in the superficial layers. In the June 27, 1991 issue of *Nature* this highly productive group[66] dropped another bombshell with their report of the cloning and expression of NOS. They found NOS to be a protein of 1,429 amino acids, with an approximate molecular mass of 160 K. Cloning of a cDNA for brain NOS revealed recognition sites for calmodulin, NADPH, and the flavins FAD and FMN (flavin mononucleotide), as well as phosphorylation sites. By comparing the amino acid sequence of NOS with that of all known sequenced protein in GeneBank (FASTA program), they discovered that the closest homology was to rat cytochrome P-450 reductase. These two proteins were similar at the C-terminal half of NOS where over 641 amino acids had a 36% identity and a 58% close homology to cytochrome P-450 reductase. The N-terminal half of NOS had no significant homology with any known protein. The number of recognition sites identified suggest to the authors that brain NOS is regulated by many different factors. Also in 1991 this Johns Hopkins group[67] reported that NO mediates glutamate-induced neurotoxicity in primary cortical cultures of neurons from fetal rats. Neurons were obtained from the hippocampus, the cortex, and the caudate putamen. The neurotoxicity of glutamate was prevented selectively by inhibitors of NOS.

In 1990 Wood et al.[68] of G. D. Searle (St. Louis) reported that inhibition of NOS blocks NMDA-, quisqualate-, and kainate-dependent increases of cerebellar cyclic GMP *in vivo*. This report is of particular significance, as the first *in vivo* demonstration that NO is an important messenger molecule in the brain (cerebellum), with actions mediating the actions of all of the major glutamate-receptor agonists.

In 1991 two important publications came from the University of Liverpool group of Garthwaite and associates. They described NMDA-receptor activation in the memory-important hippocampus of the rat to be through the L-arginine/nitric-acid pathway induction of cyclic GMP.[69] They also reported an in-depth study of excitatory amino acid receptors coupled to the NO/cyclic GMP pathway in the rat cerebellum during development and including important changes from immaturity to maturity (adult rats).[70] It was of particular interest that the absolute cyclic GMP response to kainate was maximally effective in both adult and immature tissues, remaining relatively constant throughout development. NMDA produced very large accumulations of cyclic GMP in neuronal slices from 8-d-old animals, being about two times those produced by kainate. With NMDA a striking reduction in cyclic

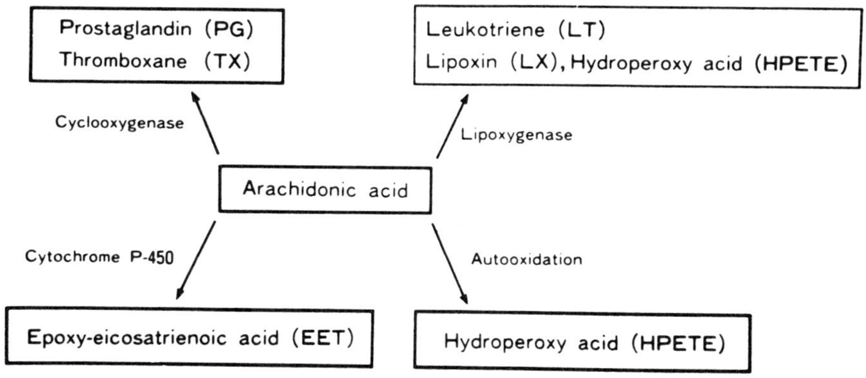

FIGURE 6. Principal pathways of the arachidonic acid cascade. (From Shimizu, T. and Wolfe, L. S., *J. Neurochem.*, 55, 1, 1990. With permission.)

GMP production occurred during maturation, particularly between 8 and 14 d. The adult response to NMDA, with respect to cyclic GMP generation, was only 10% that of kainate. The effect of AMPA became larger during development. Being 20% of the response to kainate at 8 d, it rose to 70% of the kainate response in adult rats. Quisqualate, while starting at a lower response base, behaved in a manner similar to AMPA.

VI. ARACHIDONIC ACID

The fascinating cascades deriving from the multiple pathways of metabolizing mobilized arachidonic acid provide considerable molecular capabilities for multiple levels of modulation of both a stimulatory and inhibitory nature (Figure 6). These are dealt with in some detail in Chapter 15, but this chapter would be incomplete without inclusion of this important and interrelated cascade.

REFERENCES

1. **von Euler, U. S.,** Substance P in the nervous system, Ciba Foundation Symposium No. 92, Pitman, London, 1982, 124.
2. **Jose, P. J.,** Complement-derived peptide mediators of inflammation, *Br. Med. Bull.*, 43, 336, 1987.
3. **Ahearn, J. M. and Fearon, D. T.,** Structure and function of complement receptors, CR1 (CD35) and CR21 (CD21), *Adv. Immunol.*, 46, 183, 1989.
4. **Kishimoto, T. K., Larson, R. S., Corbi, A. S., Dustin, L. M., Staunton, D. E., and Springer, T. A.,** The leukocyte integrins, *Adv. Immunol.*, 46, 149, 1989.
5. **Johnson, R. J. and Chenoweth, D. E.,** Labeling the granulocyte C5a receptor with a unique photoreactive probe, *J. Biol. Chem.*, 260, 7161, 1985.
6. **Rollins, T. E. and Springer, M. S.,** Identification of the polymorphonuclear leukocyte C5a receptor, *J. Biol. Chem.*, 260, 7157, 1985.
7. **Gerard, N. P. and Gerard, C.,** Construction and expression of a novel recombinant anaphylatoxin, C5a-N19, as a probe for the human C5a receptor, *Biochemistry*, 29, 9274, 1990.
8. **Hugli, T. E. and Meuller-Eberhard, H. J.,** Anaphylatoxins: C3a and C5a, *Adv. Immunol.*, 26, 1, 1978.
9. **Kishimoto, T. K., Jutila, M. A., Berg, A. L., and Butcher, E. C.,** Neutrophil Mac-1 and MEL-14 adhesion proteins inversely regulated by chemotactic factors, *Science*, 245, 1238, 1989.

10. **Gerard, N. P. and Gerard, C.,** The chemotactic receptor for human C5a anaphylatoxin, *Nature,* 349, 614, 1991.
11. **Gerard, N. P., Eddy, R., Shows, T. B., and Gerard, C.,** The human neurokinin A (substance K) receptor, *J. Biol. Chem.,* 265, 20455, 1990.
12. **Moore, T. C.,** Drainage area complement activation by injections of inulin and cobra venom factor alters regional lymph node lymphocyte traffic, *IRCS Med. Sci.,* 12, 681, 1984.
13. **Furchgott, R. F. and Zawadzki, J. V.,** The obligatory role of endothelial cells in the relaxation of arterial smooth muscle by acetylecholine, *Nature,* 288, 373, 1980.
14. **Zawadzki, J. V., Cherry, P. D., and Furchgott, R. F.,** Comparison of endothelium-dependent relaxation of rabbit aorta by A23187 and by acetylcholine, *Pharmacologist,* 22 (Abstr.), 271, 1980.
15. **Cherry, P. D., Furchgott, R. F., and Zawadzki, J. V.,** The indirect nature of bradykinin relaxation of isolated arteries: endothelial dependent and independent components, *Fed. Proc.,* 40 (Abstr.), 689, 1981.
16. **Chand, N. and Altura, B. M.,** Acetylcholine and bradykinin relax intrapulmonary arteries by acting on endothelial cells: role in lung vascular diseases, *Science,* 213, 1376, 1981.
17. **Cherry, P. D., Furchgott, R. F., Zawadzki, J. D., and Jothianandan, D.,** The role of endothelial cells in the relaxation of isolated arteries by bradykinin, *Proc. Natl. Acad. Sci. U.S.A.,* 79, 2105, 1982.
18. **Zawadzki, J. V., Furchgott, R. F., and Cherry, P. D.,** The obligatory role of endothelial cells in the relaxation of arterial smooth muscle by substance P, *Fed. Proc.,* 40 (Abstr.), 689, 1981.
19. **Zawadzki, J. V., Furchgott, R. F., and Cherry, P. D.,** Endothelium-dependent relaxation of arteries by substance P, kassinin and octa-cholecystokinin, *Fed. Proc.,* 42 (Abstr.), 619, 1983.
20. **Van de Voorde, J. and Leusen, I.,** Role of endothelium in the vasodilator responses of rat thoracic aorta to histamine, *Eur. J. Pharmacol.,* 87, 113, 1983.
21. **Satoh, H. and Inui, J.,** Endothelial cell-dependent relaxation and contraction induced by histamine in the isolated guinea-pig pulmonary artery, *Eur. J. Pharmacol.,* 97, 321, 1984.
22. **Cocks, T. M. and Angus, J. A.,** Endothelium dependent relaxation of coronary arteries by noradrenaline and serotonin, *Nature,* 305, 627, 1983.
23. **Cohen, R. A., Shepherd, J. T., and Vanhoutte, P. M.,** 5-Hydroxytryptamine can mediate endothelium-dependent relaxation of coronary arteries, *Am. J. Physiol.,* 245, 1077, 1983.
24. **Brain, S. D., Williams, T. J., Tippins, J. R., Morris, H. R., and MacIntyre, I.,** Calcitonin gene-related peptide is a potent vasodilator, *Nature,* 313, 54, 1985.
25. **Rapoport, R. M. and Murad, F.,** Agonist-induced endothelium-dependent relaxations in rat thoracic aorta may be mediated through cGMP, *Circ. Res.,* 52, 352, 1983.
26. **Rapoport, R. M., Draznin, M. B., and Murad, F.,** Endothelium-dependent relaxation in rat aorta may be mediated through cyclic GMP-dependent protein phosphorylation, *Nature,* 306, 174, 1983.
27. **Moncada, S., Palmer, R. M. J., and Gryglewski, R. J.,** Mechanism of action of some inhibitors of endothelium-derived relaxing factor, *Proc. Natl. Acad. Sci. U.S.A.,* 83, 9164, 1986.
28. **Gryglewski, P. J., Palmer, R. M., and Moncada, S. A.,** Superoxide anion is involved in the breakdown of endothelium-derived relaxing factor, *Nature,* 320, 454, 1986.
29. **Gryglewski, R. J., Moncada, S., and Palmer, R. M. J.,** Bioassay of prostacyclin and endothelium-derived relaxing factor (EDRF) from porcine aortic endothelial cells, *Br. J. Pharmacol.,* 87, 685, 1986.
30. **Palmer, R. M. J., Ferrigo, A. G., and Moncada, S.,** Nitric oxide release accounts for the biological activity of endothelium-derived relaxing factor, *Nature,* 327, 524, 1987.
31. **Palmer, R. M. J., Ashton, D. S., and Moncada, S.,** Vascular endothelial cells synthesize nitric oxide from L-arginine, *Nature,* 333, 664, 1988.
32. **Knowles, R. G., Palacios, M., Palmer, R. M. J., and Moncada, S.,** Formation of nitric oxide from L-arginine in the central nervous system: a transducer mechanism for stimulation of the soluble guanylate cyclase, *Proc. Natl. Acad. Sci. U.S.A.,* 86, 5159, 1989.
33. **Palmer, R. M. J., Rees, D. D., Ashton, D. S., and Moncada, S.,** L-Arginine is the physiological precursor for the formation of nitric oxide in endothelium-dependent relaxation, *Biochem. Biophys. Res. Commun.,* 153, 1251, 1988.
34. **Moncada, S., Palmer, R. M. J., and Higgs, E. A.,** The discovery of nitric oxide as the endogenous nitrovasodilator, *Hypertension,* 12, 365, 1988.
35. **Palacios, M., Knowles, R. G., Palmer, R. M. J., and Moncada, S.,** Nitric oxide from L-arginine stimulates the soluble guanylate cyclase in adrenal glands, *Biochem. Biophys. Res. Commun.,* 165, 802, 1989.
36. **Knowles, R. G., Salter, M., Brooks, S. L., and Moncada, S.,** Anti-inflammatory glucocorticoids inhibit the induction by endotoxin of nitric oxide synthase in the lung, liver and aorta of the rat, *Biochem. Biophys. Res. Commun.,* 172, 1042, 1990.
37. **Radomski, M. W., Palmer, R. M. J., and Moncada, S.,** Endogenous nitric oxide inhibits human platelet adhesion to vascular endothelium, *Lancet,* 2, 1057, 1987.

38. **Radomski, M. W., Palmer, R. M. J., and Moncada, S.,** Comparative pharmacology of endothelium-derived relaxing factor, nitric oxide and prostacyclin in platelets, *Br. J. Pharmacol.*, 92, 181, 1987.
39. **Palmer, R. M. J. and Moncada, S.,** A novel citruline-forming enzyme implicated in the formation of nitric oxide by vascular endothelial cells, *Biochem. Biophys. Res. Commun.*, 158, 348, 1989.
40. **Garthwaite, J., Garthwaite, G., Palmer, R. M. J., and Moncada, S.,** NMDA receptor activation induces nitric oxide synthesis from arginine in rat brain slices, *Eur. J. Pharmacol.*, 172, 413, 1989.
41. **Knowles, R. G., Palacios, M., Palmer, R. M. J., and Moncada, S.,** Kinetic characteristics of nitric oxide synthase from rat brain, *Biochem. J.*, 269, 207, 1990.
42. **Rees, D. D., Cellek, S., Palmer, R. M. J., and Moncada, S.,** Dexamethasone prevents the induction by endotoxin of a nitric oxide synthase and the associated effects on vascular tone: an insight into endotoxin shock, *Biochem. Biophys. Res. Commun.*, 173, 541, 1990.
43. **Di Rosa, M., Radomski, M., Carnuccio, R., and Moncada, S.,** Glucocorticoids inhibit the induction of nitric oxide synthase in macrophages, *Biochem. Biophys. Res. Commun.*, 172, 1246, 1990.
44. **Radomski, M. W., Palmer, R. M. J., and Moncada, S.,** An L-arginine/nitric oxide pathway present in human platelets regulates aggregation, *Proc. Natl. Acad. Sci. U.S.A.*, 87, 5193, 1990.
45. **Radomski, M. W., Palmer, R. M. J., and Moncada, S.,** Glucocorticoids inhibit the expression of an inducible, but not the constitutive, nitric oxide synthase in vascular endothelial cells, *Proc. Natl. Acad. Sci. U.S.A.*, 87, 10043, 1990.
46. **Moncada, S., Rees, D. D., Schulz, R., and Palmer, R. M. J.,** Development and mechanism of a specific supersensitivity to nitrovasodilators after inhibition of vascular nitric oxide synthesis *in vivo, Proc. Natl. Acad. Sci. U.S.A.*, 88, 2166, 1991.
47. **Bromberg, Y. and Pick, E.,** Cyclic GMP metabolism in macrophages. I. Regulation of cyclic GMP levels by calcium and stimulation of cyclic GMP synthesis by NO-generating agents, *Cell. Immunol.*, 52, 73, 1980.
48. **Hibbs, J. B., Jr., Vavrin, Z., and Taintor, R. R.,** L-Arginine is required for expression of the activated macrophage effector mechanism causing selective metabolic inhibition in target cells, *J. Immunol.*, 138, 550, 1987.
49. **Hibbs, J. B., Taintor, R. R., Vavrin, Z., and Rachlin, E. M.,** Nitric oxide: a cytotoxic activated macrophage effector molecule, *Biochem. Biophys. Res. Commun.*, 157, 87, 1988.
50. **Stuehr, D. J. and Nathan, C. F.,** Nitric oxide. A macrophage product for cytostasis and respiratory inhibition in tumour target cells, *J. Exp. Med.*, 169, 1543, 1989.
51. **Stuehr, D. J. and Marletta, M. A.,** Mammalian nitrate biosynthesis: mouse macrophages produce nitrite and nitrate in response to *Escherichia coli* lipopolysaccharide, *Proc. Natl. Acad. Sci. U.S.A.*, 82, 7738, 1985.
52. **Stuehr, D. J. and Marletta, M. A.,** Induction of nitrite/nitrate synthesis in murine macrophages by BCG infection, lymphokines, or interferon-γ, *J. Immunol.*, 139, 518, 1987.
53. **Hauschildt, S., Bassenge, E., Bessler, W., Busse, R., and Mulsch, A.,** L-Arginine-dependent nitric oxide formation and nitrite release in bone marrow-derived macrophages stimulated with bacterial lipopeptide and lipopolysaccharide, *Immunology,* 70, 332, 1990.
54. **Stuehr, D. J., Kwon, N. S., and Nathan, C. F.,** FAD and GSH participate in macrophage synthesis of nitric oxide, *Biochem. Biophys. Res. Commun.*, 168, 558, 1990.
55. **Gross, S. S., Stuehr, D. J., Aisaka, K., Jaffe, E. A., Levi, R., and Griffith, O. W.,** Macrophage and endothelial cell nitric oxide synthesis: cell-type selective inhibition by N^G-aminoarginine, N^G-nitroarginine and N^G-methylarginine, *Biochem. Biophys. Res. Commun.*, 170, 96, 1990.
56. **Kirk, S. J., Regan, M. C., and Barbul, A.,** Cloned murine T lymphocytes synthesize a molecule with the biological characteristics of nitric oxide, *Biochem. Biophys. Res. Commun.*, 173, 660, 1990.
57. **Stuehr, D. J., Cho, H. J., Kwon, N. S., Weise, M. F., and Nathan, C. F.,** Purification and characterization of the cytokine-induced macrophage nitric oxide synthase: an FAD- and FMN-containing flavoprotein, *Proc. Natl. Acad. Sci. U.S.A.*, 88, 7773, 1991.
58. **Yui, Y., Hattori, R., Kosuga, K., Elizawa, H., Hiki, K., Ohkawa, S., Ohnishi, K., Terao, S., and Kawai, C.,** Calmodulin-independent nitric oxide synthase from rat polymorphonuclear neutrophils, *J. Biol. Chem.*, 266, 3369, 1991.
59. **Yui, Y., Hattori, R., Kosuga, K., Eizawa, H., Hiki, K., and Kawai, C.,** Purification of nitric oxide synthase from rat macrophages, *J. Biol. Chem.*, 266, 12544, 1991.
60. **Stuehr, D. J., Fasehun, O. A., Kwon, N. S., Gross, S. S., Gonzalez, J. A., Levi, R., and Nathan, C. F.,** Inhibition of macrophage and endothelial cell nitric oxide synthase by diphenyleneiodonium and its analogs, *FASEB J.*, 5, 98, 1991.
61. **Garthwaite, J., Charles, S. L., and Chess-Williams, R.,** Endothelium-derived relaxing factor release on activation of NMDA receptors suggests role as intercellular messenger in the brain, *Nature,* 336, 385, 1988.

62. **Knowles, R. G., Palacios, M., Palmer, R. M. J., and Moncada, S.,** Kinetic characterisitcs of nitric oxide synthase from rat brain, *Biochem. J.*, 269, 207, 1990.
63. **Bredt, D. S. and Snyder, S. H.,** Nitric oxide mediates glutamate-linked enhancement of cGMP levels in the cerebellum, *Proc. Natl. Acad. Sci. U.S.A.*, 86, 9030, 1989.
64. **Bredt, D. S. and Snyder, S. H.,** Isolation of nitric oxide synthase, a calmodulin-requiring enzyme, *Proc. Natl. Acad. Sci. U.S.A.*, 87, 682, 1990.
65. **Bredt, D. S., Hwang, P. M., and Snyder, S. H.,** Localization of nitric oxide synthase indicating a neural role for nitric oxide, *Nature*, 347, 768, 1990.
66. **Bredt, D. S., Hwang, P. M., Glatt, C. E., Lowenstein, C., Reed, R. R., and Snyder, S. H.,** Cloned and expressed nitric oxide synthase structurally resembles cytochrome P-450 reductase, *Nature*, 351, 714, 1991.
67. **Dawson, V. L., Dawson, T. M., London, E. D., Bredt, D. S., and Snyder, S. H.,** Nitric oxide mediates glutamate neurotoxicity in primary cortical cultures, *Proc. Natl. Acad. Sci. U.S.A.*, 88, 6368, 1991.
68. **Wood, P. L., Emmett, M. R., Rao, T. S., Cler, J., Mick, S., and Iyengar, S.,** Inhibition of nitric oxide synthase blocks *N*-methyl-D-aspartate-, quisqualate-, kainate-, harmaline-, and pentylenetetraxole-dependent increases in cerebellar cyclic GMP *in vivo*, *J. Neurochem.*, 55, 346, 1990.
69. **East, S. J. and Garthwaite, J.,** NMDA receptor activation in rat hippocampus induces cyclic GMP formation through the L-arginine-nitric oxide pathway, *Neurosci. Lett.*, 123, 17, 1991.
70. **Southam, E., East, S. J., and Garthwaite, J.,** Excitatory amino acid receptors coupled to the nitric oxide/cyclic GMP pathway in rat cerebellum during development, *J. Neurochem.*, 56, 2072, 1991.

Chapter 18

ANCILLARY MOLECULES

I. INTRODUCTION

The principal ancillary molecules are immune response cell generated or related, as contrasted with the coagulation-fibrinolysis and the vascular-system-generated cascade molecules considered in Chapter 17. The majority of these immune-response-related molecules, lymphokines, and cytokines, are lymphocyte and macrophage produced. These molecules also are largely stimulatory and defense related.

Growth hormones and oncogenes fill out this category of ancillary molecules. In addition to the "classical" growth factors (EGF, PDGF, FGF, and TGF_1) which act through receptor tyrosine kinases,[1] a new class of nonclassical growth factors has been identified and includes bombesin, bradykinin, and serotonin: molecules that have cell growth and proliferation capabilities[2] in addition to their many other actions. These nonclassical vasoactive neurotransmitter growth factors act through G-protein-coupled receptors, and their mitogen signaling pathways have been found to be pertussis-toxin sensitive. They also have been found to be involved in the activation and expression of oncogenes (*c-fos, c-jun,* and *c-myc*). These extremely important responses have been dealt with in detail in the chapters relating to these molecules.

II. LYMPHOKINES AND CYTOKINES

Lymphokines and cytokines have a number of names, which include interleukins. The original concept was that each interleukin, either lymphokine or cytokine, had a single biological function. That has long gone by the board. As the number of interleukins has proliferated, so has the number of functions for each interleukin. As knowledge of these molecules has increased, the breadth of the spectrum of their physiological and biochemical actions and interactions has expanded greatly and has "metastasized" to include the growth factors and oncogenes as well as in a seamless web of ancillary and related defense molecules, including "memory molecules".

An interrelationship between a vasoactive neurotransmitter molecule (histamine) and a lymphokine (lymphotoxin) in cellular immunity was reported in 1974 by the author and Granger.[3,4] An accumulation of lymphotoxin-secreting lymphoid cells in rat heart allografts was found with an increase both in the size and number of lymphoid cells in the heart allografts and in the level of lymphotoxin (LT) excretion per cell, with a maximum at 10 d postallografting[4] (Figure 1). A sharp increase in lymphocyte accumulation in allografts and in the LT excretion per cell occurred at 6 d postallografting. In this study, lymphocyte-rich preparations were prepared from minced autologous and allografted heart of heterotopic allografted rats at intervals of 3 to 19 d after grafting. The increase in cell accumulation in numbers and in LT excretion per cell closely correlated with the increase in histamine-forming HDC (histidine decarboxylase) activity across this same histocompatability barrier in another organ (kidney) allografting study (see Figure 20 in Chapter 4).

In an *in vivo* study of the action of LT on histamine metabolism in the rat spleen, the author, with Granger and Lemmi,[3] found that direct (laporotomy) intrasplenic injection of Lewis rat supernatants (Con A and PHA stimulated) with established LT activity resulted in an increase in histamine liberation and, at 5 d after direct intrasplenic LT injection, a marked and significant ($p < 0.002$) increase in splenic HDC enzyme activity.

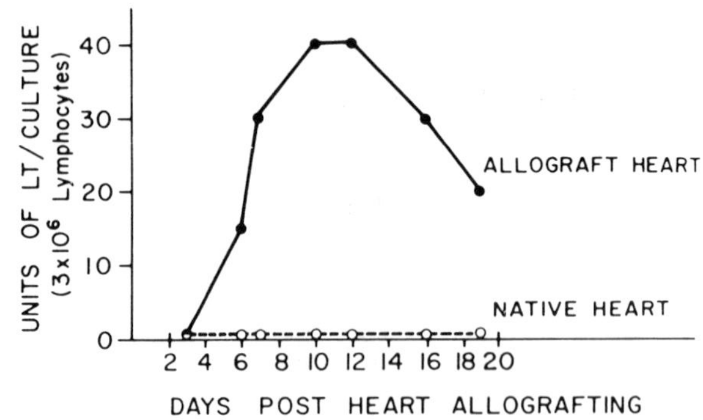

FIGURE 1. The marked and progressive increase in the lymphotoxin (LT) levels of lymphocytes (per cell) extracted at intervals from allografted rat hearts at intervals postallografting in comparison with the ''normal''/ native hearts of the allografted animals. (From Granger, G. A. and Moore, T. C., *Surgery,* 76, 542, 1974. With permission.)

These studies represent one of the earliest, if not *the* earliest, demonstrations of a direct *in vivo* involvement of a lymphokine (lymphotoxin) in a cellular immune reaction (involving organ allografting) and in an *in vivo* demonstration of a lymphokine (lymphotoxin) effect as increasing a lymphoid organ (spleen) level of a vasoactive-neurotransmitter (histamine)-forming enzyme HDC.

III. GROWTH FACTORS

One of the most important recent developments in the biological sciences relates to the recognition and identification of the extraordinary capacity of soluble mediators in serum and culture media to influence cell growth, proliferation, and differentiation. Evidence for the involvement of growth-type factors in cell proliferation derived from the early days of tissue culture when it was recognized that supplementation with serum was required for cell growth and proliferation.

Nerve growth factor (NGF) was the first of these growth factors to be identified and characterized as a growth factor. It was followed by epidermal growth factor (EGF), platelet-derived growth factor (PDGF), fibroblast growth factors (FGFs), and transforming growth factors (TGFs). Some growth factors also are classified as interleukins, such as T-cell growth factor (IL-2) and the colony-stimulating factors (CSFs).

Thrombin, of the blood-clotting process, doubles as a cascade molecule and as a growth factor. It functions as a potent mitogen for fibroblasts, smooth muscle, and endothelial cells.[5-7] Vasoactive neurotransmitter molecules, such as bradykinin, serotonin, substance P, and bombesin, are being recognized increasingly also as a new class of growth factors that act through G-protein-coupled receptors.[2,8,9] These new or ''neoclassical'' growth factors differ from the older (actually not too old) ''classical'' growth factors that act through tyrosine kinases.[1]

A 1990 report by Chiou et al.[10] described an obligatory requirement for polypeptide growth factors PDGF or FGF for interleukin-1-mediated prostaglandin-E_2 production by fibroblasts. Costimulation with one of these growth factors was required. From these observations the authors speculated that PDGF and FGF may play additional roles in connective tissues, including an indirect role in inflammatory processes.

It is of further interest and of relevance to this book that Berk and Alexander[11] recently (1989) described vasoactive effects of growth factors.

IV. ONCOGENES

The tyrosine kinase "connection" also is found in the oncogene arena as well as in growth factor actions. The transforming gene of the Rous sarcoma virus, v-*src,* encodes an enzyme that phosphorylates tyrosine residues on proteins, a tyrosine kinase ($pp60^{v\text{-}src}$), the first identified retroviral tyrosine-specific protein kinase.

The c-*src* gene also is notable as the first protooncogene to be identified in the human genome. Bishop, Varmus, and associates, at the University of California–San Francisco, employed a viral v-*src* probe derived from the Rous sarcoma virus (RSV) to identify an analogous gene in human DNA and earned themselves a Nobel Prize in the process.

As the vasoactive neurotransmitter and modulator molecules discussed earlier in this book, oncogenes also have a long evolutionary history. d-*src,* a gene homologous to vertebrate c-*src,* has been cloned from *Drosophila melanogaster.* Drosophila and vertebrate c-*src* proteins has been discovered to be approximately 40% identical. Nematodes and slime molds have been found to possess similar genes. In *Drosophila,* three classes of *src* transcripts are independently regulated during embryonic and pupal development periods, with maximal *src* expression correlating with differentiation rather than proliferation.

The identification of interrelationships of oncogenes and growth factors (first reported in 1983 by Doolittle et al.[12] of La Jolla, Pasadena, Bethesda, and Boston and by Waterfield et al.[13] of London, Uppsala, and St. Louis) was an especially important advance in achieving a better understanding of these two important and intertwined fields. These two groups of investigators independently demonstrated from the sequence analysis of the human PDGF gene that the human PDGF gene shared sequences with the *sis* oncogene, the transforming gene of the simian sarcoma virus (SSV). Antisera to a synthetic peptide of the *sis* viral oncogene recognizes human PGDF. In addition, anti-PDGF antibodies have been observed to inhibit acute transformation by SSV, and anti-PDGF antibodies have been found to block the growth of some SSV-transformed cell lines.

EGF was initially recognized for its interesting ability to stimulate precocious eyelid opening and tooth eruption in newborn mice. In 1984 Downward, with Waterfield and associates,[14] dropped another bombshell in describing in *Nature* a close similarity of EGF receptor and V-*erb*-B oncogene sequences from avian erythroblastosis virus (AEV). Each of six peptides derived from the human EGF receptor very closely matched a part of the deduced sequence of the V-*erb*-B-transforming protein of AEV. In all, the peptides contained 83 amino acid residues, 73 of which were shared with V-*erb*-B.

In 1985 Sherr et al.[15] reported that the c-*fms* protooncogene product is related to the receptor for mononuclear phagocyte growth factor, CSF-1. The feline c-*fms* protooncogene is a 170-kDa glycoprotein with tyrosine kinase activity and is expressed on mature cat macrophages from peritoneal inflammatory exudates and spleen. The receptor for the murine colony-stimulating factor (CSF-1) also is restricted to cells of the mononuclear phagocytic lineage and also is a 165-kDa glycoprotein with an associated tyrosine kinase. The authors found that rabbit antisera to a recombinant v-*fms*-coded polypeptide precipitated the feline c-*fms* product and was also found to cross-react with a 165-kDa glycoprotein from mouse macrophages. This putative product of the murine c-*fms* gene was observed to exhibit tyrosine kinase activity in immune complexes and to bind specifically murine CSF-1 and, in the presence of the growth factor, was phosphorylated on tyrosine in membrane preparations. The authors concluded that the murine c-*fms* protooncogene product and the CSF-1 receptor are related, and possibly identical, molecules.

In 1991 Trouche et al.[16] of France published evidence that c-*fos* transcriptional activation by interleukin-2 in mouse CTL-L2 cells (a murine T-cell lymphocyte line) is mediated through two distinct signal pathways converging on the same enhancer element. These distinct signal-transduction pathways have the same DNA target on the c-*fos* promoter, the serum response element.

In 1987 Bishop[17] observed that "oncogenes impinge on the chain of command that directs the response to polypeptide growth factors." It is likely that this includes "nonclassical" as well as "classical" growth factors.

REFERENCES

1. **Yarden, Y. and Ullrich, A..,** Growth factor receptor tyrosine kinases, *Ann. Rev. Biochem.*, 57, 443, 1988.
2. **Seuwen, K. and Pouyssegur, J.,** Serotonin as a growth factor, *Biochem. Pharmacol.*, 39, 985, 1990.
3. **Moore, T. C., Granger, G. A., and Lemmi, C. A. E.,** *In vivo* elevation in rat splenic histidine decarboxylase activity by intra-splenic injection of Con A and Con A activated lymphokines, *IRCS Med. Sci.*, 2, 1458, 1974.
4. **Granger, G. A. and Moore, T. C.,** Accumulation of lymphotoxin secreting lymphoid cells in rat heart allografts, *Surgery,* 76, 542, 1974.
5. **Chen, L. B. and Buchanan, J. M.,** Mitogenic activity of blood components. I. Thrombin and prothrombin, *Proc. Natl. Acad. Sci. U.S.A.*, 72, 131, 1975.
6. **Perez-Rodriguez, R., Franchi, A., and Pouyssegur, J.,** Growth factor requirements of Chinese hamster lung fibroblasts in serum free media: high mitogenic action of thrombin, *Cell Biol. Int. Rep.*, 5, 347, 1981.
7. **Seuwen, K., Chambard, J. C., L'Allemain, G., Magnaldo, I., Paris, S., and Pouyssegur, J.,** Thrombin as a growth factor: mechanism of signal transduction. In *Blood Cells and Arteries in Hypertension and Atherosclerosis,* Meyer, P. and Marche, P., Eds., Raven Press, New York, 1989, 217.
8. **Zachary, I., Penella, J. W., and Rozengurt, E.,** A role for neuropeptides in the control of cell proliferation, *Dev. Biol.*, 124, 295, 1987.
9. **Vicentini, L. M. and Villereal, M. L.,** Serum, bradykinin and vasopressin stimulate release of inositol phosphates from human fibroblasts, *Biochem. Biophys. Res. Commun.*, 123, 663, 1984.
10. **Chiou, W. J., Bonin, P. D., and Singh, J. P.,** Obligatory action of polypeptide growth factors for the IL-1-mediated prostaglandin E_2 production in fibroblasts. Potential role of growth factors in modulation of tissue response to IL-1, *J. Immunol.*, 145, 2155, 1990.
11. **Berk, B. C. and Alexander, R. W.,** Vasoactive effects of growth factors, *Biochem. Pharmacol.*, 38, 219, 1989.
12. **Doolittle, R. F., Hunkapiller, M. W., Devare, S. G., Robbins, K. C., Aaronson, S. A., and Antoniades, H. N.,** Simian sarcoma virus onc gene, v-sis, is derived from the gene (or genes) encoding a platelet-derived growth factor, *Science,* 221, 275, 1983.
13. **Waterfield, M. D., Scrace, G. T., Whittle, N., Stroobant, P., Johnsson, A., Wasteson, A., Westermark, G., Heldin, C.-H., Huang, J. S., and Deuel, T. F.,** Platelet-derived growth factor is structurally related to the putative transforming protein $p28^{sis}$ of simian sarcoma virus, *Nature,* 304, 35, 1983.
14. **Downward, J., Yarden, Y., Mayes, E., Scrace, G., Totty, N., Stockwell, P., Ullrich, A., Schlessinger, J., and Waterfield, M. D.,** Close similarity of epidermal growth factor receptor and v-erb-B oncogene protein sequences, *Nature,* 307, 521, 1984.
15. **Sherr, C. J., Rettenmier, C. W., Sacca, R., Roussel, M. F., Look, A. T., and Stanley, E. R.,** The c-fms proto-oncogene product is related to the receptor for the mononuclear phagocyte growth factor, CSF-1, *Cell,* 41, 665, 1985.
16. **Trouche, D., Robin, P., Robillard, O., Sassone-Corsi, P., and Harel-Bellan, A.,** c-fos transcriptional activation by IL-2 in mouse CTL-L cells is mediated through two distinct transductional pathways converging on the same enhancer element, *J. Immunol.*, 147, 2398, 1991.
17. **Bishop, J. M.,** Trends in oncogenes, in *Oncogenes and Growth Factors,* Bradshaw, R. A. and Prentis, S., Eds., Elsevier, Amsterdam, 1987, 1.

Chapter 19

HEAT, PAIN, AND MEMORY

I. INTRODUCTION

''In never-yielding felters linked of adamant to bind this recreant to the high-ridged rocks. For this is he who stole the flame of All-Working Fire, thine own bright flower, and gave to mortal man.''

Prometheus Bound
Aeschylus, 525–456 B.C.

''Then Prometheus gave wisdom, which is strength —
Prometheus saw, and waked the legioned hopes
which sleep within folded Elysian flowers,
And love he sent to bind
The disunited tendrils of that vine
which bears the wine of life, the human heart;
And he tamed Fire which, like some beast of prey,
Most terrible, but lovely, played beneath
The frown of man;
He gave man speech, and speech created thought,
Such, Prometheus gave to Man, for which he hangs
Withering in destined Pain.''

Prometheus Unbound, Act II, Scene IV
Percy Bysshe Shelley, 1792–1822

Thermal heat is one of the most potent of stimuli to pain and resultant memory. From Prometheus to the present, wisdom, hope, love, fire, heat, thought, speech, pain, and human civilization have been closely and intricately associated, intertwined, and interrelated. The ancient Greek name ''Prometheus'' also meant ''forethought'', or ''think ahead'', or just ''think''. In the ancient Greek drama, *Prometheus Bound,* Aeschylus credits Prometheus with bringing not only fire and the means of survival to man, but civilization and the arts and sciences as well — including the related essentials of learning and memory. For this he was chained by Zeus to a rock in the Caucasus mountains where he suffered the indignity, as well as the discomfort and pain, of having his immortal and regenerative liver chewed out each night by hungry vultures.

The ''heat, pain, and memory'' scenario has been referred to earlier in this book and has received considerable attention in the chapters on bradykinin and substance P (major thermal pain impulse vasoactive neurotransmitters and major ''memory'' molecules).

II. HEAT

Aversion to excessive heat is not unique to man, other primates, mammals, and vertebrates. Even the lowly snail may respond promptly and vigorously to thermal pain. In an interesting and pertinent 1983 publication in *Science,* Kavaliers et al.[1] of London (Ontario) reported that the gastropod mollusc terrestrial snail *Cepaea nemoralis,* when placed on a 40°C hot plate, lifted the anterior portion of its foot. This was a reproducible and stereotypical response to heat stress and discomfort and occurred as an elevation of the anterior portion of the extended foot. The latency of this response was observed to be influenced by morphine and by naloxone in a dose-dependent and time-dependent manner. Morphine was found to increase the time taken to respond, whereas naloxone decreased it. In addition, naloxone abolished the effect of morphine. The authors interpreted these valuable findings as indicating that opiate systems have a functional role in snail thermal behaviors and that this role is comparable in many respects to mammalian thermal behaviors as they relate to pain impulses and their blockade and unblockade.

The biological phenomenon of inflammation is a major defense mechanism of man and mammals, and it calls on multiple vasoactive neurotransmitter molecules considered in this book (histamine, bradykinin, serotonin, substance P, prostaglandins, etc.). The word itself strongly suggests fire and heat (set on fire, inflame). Three of the four major components of biological inflammation (*calor, rubor, dolor*) are clearly heat and pain related, while the *tumor* (swelling) is a direct consequence of these and the involved vasoactive stimulative and excitatory molecules cited above. Elevated local and systemic temperature (fever) are important defense and survival components of this critical response to foreign, nonself invasion.

Fever has even been observed to be an important survival factor in nonmammalian vertebrates. In 1975 and in 1977 reports appeared in *Science,* by Kluger et al.,[2] and in *Nature*, by Cavert and Reynolds,[3] indicating fever-induced increased survival of the reptilian lizard *Dipsosaurus dorsalis* and the teleost goldfish *Carassius auratus* during infection with the Gram-negative bacterium *Aeromonas hydrophilia.* This bacterium is pathogenic to bony fish, amphibians, and reptiles and produces a life-threatening hemorrhagic septicemia.

It is important to make a distinction between physiological and "beneficial" fever and life-threatening hyperpyrexia and malignant hyperthermia. In 1971 Nahas et al.[4] of New York and Paris reported that the rate of locomotion of human polymorphonuclear leukocytes (PMNs) was markedly enhanced by temperatures between 32 and 42°C by a direct measurement technique. The maximum increase, ninefold, was at 42°C. In 1982 Duff and Durum[5] of Yale observed that the T-cell proliferation response to IL-1 and IL-2 was greatly increased at 39°C as compared with 37°C. From these findings they questioned the indiscriminate use of antipyretic agents as potentially harmful. Also in 1982, Yerushalmi et al.[6] of Israel and France reported that combined therapy employing either exogenous interferon or interferon inducers together with local hyperthermia exerted a greater antitumor effect in mice bearing transplantable Lewis lung carcinoma than did either treatment alone.

In 1983 Jampel, Duff, Gershon, and associates[7] at Yale described explorations of the effect of temperature on the *in vitro* immune response of unprimed murine spleen cells to sheep red blood cells (SRBC). They found that hyperthermia (39.5°C) potently augments the plaque-forming cell response. The temperature-sensitive events occurred early in the culture period and appeared targeted to T-helper cells and during the priming stage for T-helper cells. A 1980 report from Lucknow, India, by Gupta and Agarwal,[8] also found an *in vitro* temperature-induced enhancement of human T-cell responses to PHA (phytohemagglutinin). They observed that T cells from peripheral blood and tonsils and thymocytes had a significantly higher proliferative response to PHA at 40°C than at 37°C. Non-T cells and B cells did not demonstrate any temperature-induced change in proliferation response.

In 1984 Dinarello et al.[9] published evidence that recombinant human interferon produced fever not by endotoxin contamination but by an intrinsic pyrogenic effect of inducing prostaglandin E_2 (PGE_2) in the hypothalamus. This finding is of particular interest with respect to recent (1991) observations of Hayaishi,[10] of Osaka, Japan, regarding roles of PGE_2 and PGD_2 in the molecular mechanisms of sleep-wake regulation. The wakefulness effect of PGE_2 is located in the posterior hypothalamus and could also be dissociated from a pyrogenic effect of PGE_2 in the preoptic area. These studies clearly demonstrated that the awaking effect of PGE_2 was independent of the pyrogenic effect. Of further interest in this important area (''hot topic'') was the 1978 finding of Heron and Berg[11] (of Aarhus, Denmark, reported in *Nature*) that the actions of interferon are potentiated at elevated temperatures. The effects of interferon on lymphocyte killer activity was assayed by a one-way mixed lymphocyte reaction (MLR) and a cell-mediated lympholysis (CML) reaction and temperature between 35 and 41°C were tested. Maximum enhancement of activity was generated at 38 and 39°C, with a falloff at 40°C and higher temperatures. These investigators also demonstrated that an antiviral property of interferon also was temperature dependent, with greater activity at higher temperatures.

Mention also should be made of the so-called ''heat-shock proteins'' (HSPs). This is a rather complex and somewhat confusing field with a rather misleading name, as these induced proteins are by no means exclusively induced by heat, and shock, in the traditional clinical sense, is in no way involved. This field was born in 1962 by the observation of Ritossa[12] that chromosome puffing in the fly *Drosophila melanogaster* occurred after exposure to heat. This field has been reviewed quite well recently (1988) by Pulla,[13] of Geneva, Switzerland, in *Immunology Today*. These proteins and their genes have been found in virtually every living cell tested (from tomatoes to yeast, to bacteria, to flies, and to humans). These proteins (''in search of functions'') have been classed according to their apparent mass with three main families, 20 kDa, 70 kDa, and 90 kDa, with the 70-kDa family being the most conserved and best characterized. In addition to heat, these proteins may also be induced by heavy metals, inhibition of energy metabolism, amino acid analogues, serum, ethanol, oxygen free radicals, ischemia and reperfusion injury, wounding and tissue injury, inflammation, mitogens, phorbol esters, lymphokines, and immunization (just about everything).[13-17] It would be of interest to know if these multistimulated and elusive molecules have any involvement in memory: immunological as well as central nervous system.

A report of major importance in this area of heat responses was published in 1976 by Perl et al.[18] of Chapel Hill, NC. They cited the 1920s work of Lord Adrian of Cambridge and Zotterman[19,20] relating to fatigue as a feature of somatic sense-organ behavior resulting from repeated activation by natural stimuli. This fatigue was manifested by a depression of response that varied in magnitude and duration according to the type of receptor and the nature of the stimulus. The studies of Perl et al. were with the polymodal nociceptors that differed in showing an enhanced, rather than depressed, response after activation. They characterized this enhanced responsiveness on activation as *sensitization*. These polymodal nociceptors are one of the most common types of cutaneous sense organs with unmyelinated (C) fibers. They characteristically have elevated and high thresholds for all forms of natural stimulation, in comparison with other sensory elements.

In the 1976 study by Perl et al.,[18] the activity of primary sensory units with unmyelinated fibers was recorded from fine bundles dissected from the central end of cut cutaneous nerves of anesthetized cats, monkeys, and rabbits. To achieve the desired preparation, splitting of nerve filaments was carried out until the activity of a single fiber conducting at C-fiber velocity (less than 1.1 m/sec) was of relatively large amplitude and of sufficiently distinctive shape and size to be unequivocally identified in the direct-writing oscillograph recording.

Polymodal receptors have been found to give consistent responses to noxious heat applied to uninjured skin surface. The phenomenon of heat-induced increased responsiveness on repeated applications (sensitization) was shown by the authors in studies of primate skin (Figure 1). In these studies identical heating cycles were repeated every 200 msec, and the temperature application was increased from a starting temperature of 30°C, by 5° steps, to a maximum of 52°C, after which the temperature was rapidly lowered to the starting level of 30°C until the next sequence 200 msec later. The results shown in Figure 1 clearly show a progressive increase in the total number and maximum frequency of discharges evoked by the third and fifth cycles of stimulation (B and C) as compared to those elicited by the

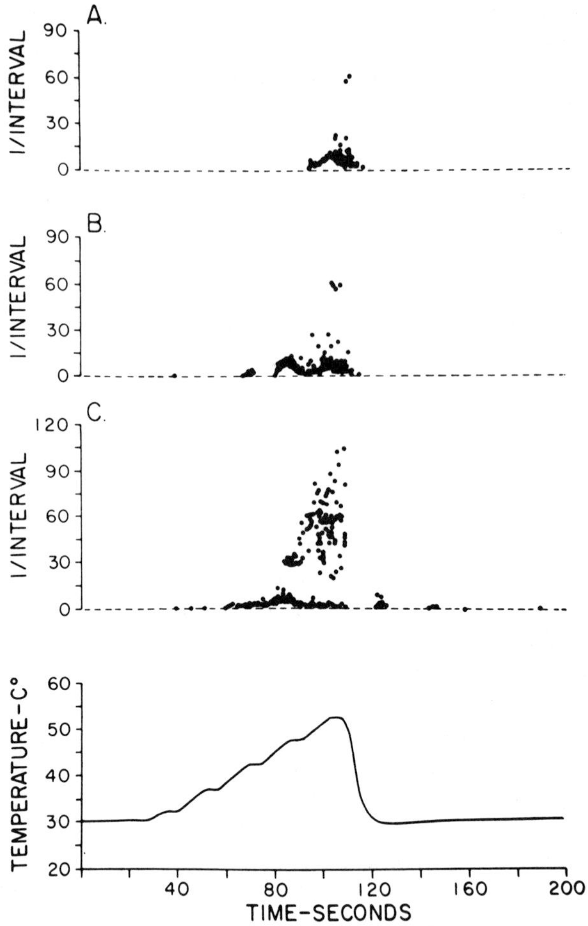

FIGURE 1. Electrical discharges evoked by primate skin polymodal receptors in response to repeated applications of increasing levels of noxious heat to uninjured skin surface. Identical heating cycles were repeated every 200 msec, and the temperature application was increased from a starting temperature of 30°C by 5° steps to a maximum of 52°C, after which the temperature was rapidly returned to the starting level of 30°C until the next sequence 200 msec later. The results shown in this figure identify a progressive increase in the number and frequency of electrical discharges evoked by the third and fifth cycles of stimulation (B and C), as compared with those evoked in the initial sequence shown as A. In B and C, a progressive lowering of the threshold may be noted, with the discharge appearing at lower temperatures. The after-discharge shown in C represents the development of important background activity. (From Perl, E. R., et al., *Prog. Brain Res.*, 43, 263, 1976. With permission.)

initial sequence (A). A careful examination of this figure also shows a progressive lowering of the threshold with the discharge appearing at lower temperatures. Equally close attention to this figure also should identify an after-discharged that represents the development of background activity. This can best be seen (in C of this figure) as impulses that appear after the end of the heating sequence. In this regard, it should be noted that polymodal receptors, with receptive fields in unstimulated and uninjured skin, do not ordinarily exhibit background activity as demonstrated in this figure. After noxious stimulation, however, most polymodal nociceptive units do develop a persisting and ongoing discharge. Once this heat-induced increased responsiveness has appeared in these nociceptive polymodal units, it may last for quite long periods; in excess of 3 h has been observed (? long-term potentiation in the periphery as a result of repeated thermal pain stimulations). Perl et al. also demonstrated

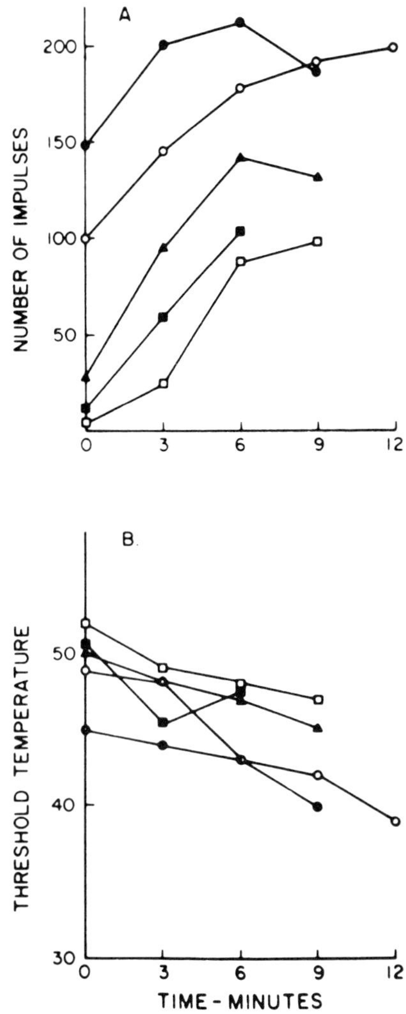

FIGURE 2. In this study successive identical heating cycles were repeated every 3 min for up to 12 min to uninjured primate skin, and electrical discharges from polymodal nociceptive receptors were recorded from five animals. With the passage of time the number of electrical impulses generated increased, while the thresholds for electrical-impulse generation decreased. In two of the five animals (40%), the threshold for electrical-impulse generation on repeated heat stimulations moved from the noxious level (over 45°C) to a level within the physiological range (under 40°C). (From Perl, E. R., et al., *Prog. Brain Res.*, 43, 263, 1976. With permission.)

this response in a study of five separate primate polymodal nociceptive receptors (Figure 2). In this study successive identical heating cycles were repeated every 3 min for up to 12 min. Heightened responsiveness to heat stimulation appeared within 30 sec after heat excitation. In some studies, as in Figure 2, the threshold on repeated heat stimulations moved from the noxious level (over 45°C) to a level within physiological range (under 40°C). These authors also observed that while the process leading to increased responsiveness of polymodal nociceptors is local, the process may spread to adjacent and nearby regions. They speculated as well that the spread to nearby areas might be due to the release of a diffusable substance (? substance P, bradykinin, serotonin, or all three, and/or others as well). Perl et al. used Sherrington's definition of "noxious" to mean "tissue damaging".

These findings of Perl et al. are presented in detail by the author of this book because of their considerable importance in demonstrating the presence of a thermal-pain-induced

LTP-type response in the peripheral nervous system and the obvious question this raises regarding a similar response or a propagation of this type of response in the central nervous system (spinal cord, brain, and hippocampus), a possible involvement in memory (immunological as well as central nervous system), and a possible involvement of vasoactive neurotransmitter "memory" and thermal pain transmission molecules, such as substance P, bradykinin, serotonin, bombesin, CGRP, CCK-8, etc., in all of this.

The author and associates[21-23] (including Storm) have carried out a series of studies in the supported-sheep lymphocyte traffic model to explore influences of burning heat and of radio-frequency-wave magnetic-loop-induced deep-tissue hyperthermia on the traffic of lymphocytes through primary peripheral popliteal or prefemoral lymph nodes of sheep *in vivo*. Local magnetic-loop induced hyperthermia was produced in the nodal drainage area, over the study lymph node, and selectively over and in the sheep's brain (local brain "fever").

In these studies localized deep hyperthermia was produced by the Magnetrode radio-frequency magnetic-loop induction apparatus. This apparatus was developed at UCLA in a joint effort by Professor F. K. Storm of the Department of Surgery (Division of Surgical Oncology) and by Professors W. H. Harrison and R. S. Elliott of the School of Engineering and Applied Science, and it is manufactured by Henry Medical Electronics, Inc., Los Angeles, CA.

The Magnetrode applicators are self-resonant, noncontact circular structures that are composed of a single-turn coil of a rolled conducting sheet that overlaps itself in a nonconducting manner with built-in impedance-matching circuitry. The element parameters have been selected to produce very large circulating currents in the structure. These currents in circular electrodes create a strong electromagnetic field into which the body part to be heated is placed. Because the body or body segment is nonmagnetic, interaction is solely with the electric field, which consists of concentric circular flux lines. This establishes an E-field, which is considered to be parallel to tissue-layer boundaries and provides effective localized deep hyperthermia without surface-tissue injury. Details concerning design, engineering, and heat distribution of this apparatus have been reported elsewhere.[24-26] During hyperthermia studies, deep tissue core temperatures were monitored by a temperature-recording miniprobe (IT-18 tissue-implantable thermocouple, Sensortek, Clifton, NJ) and recorded on a battery-operated temperature monitor (Thermalert model TH-8, Bailey Instruments, Saddle Brook, NJ). The levels of deep hyperthermia employed in the sheep studies ranged from 40 to 45°C.

The increase in the output of small recirculating lymphocytes into study lymph node efferent lymph (chronically cannulated) when the deep hyperthermia is localized over the study lymph node is shown in Figure 3.[22] The graph in this figure shows a prompt and sharp increase in the output of small recirculating lymphocytes, which is closely correlated with the core deep tissue temperature recorded in the area of the lymph node. This prompt response is of interest with respect to the studies of Perl et al.[18] cited earlier in this section, in which heightened responsiveness to heat stimuli appeared within 30 sec after heat excitation. The remarkable temperature-sensitive nature of the hyperthermia-associated increase in the output of small recirculating lymphocytes into study-node efferent lymph with hyperthermia over the study lymph node is shown in Figure 4 in which a temporary power turnoff in the hyperthermia machine resulted in a prompt and sharp decrease in lymphocyte output into study-node efferent lymph. This decrease was equally promptly altered by a return of machine power and associated deep tissue core temperature. Figure 5 shows the lymphocyte traffic and output response when the localized deep hyperthermia was in the study-node lymph drainage area and not over the node itself. The increase in lymphocyte output was prompt and sharp. The increase in study-lymph-node efferent lymph flow was slower and to a lesser degree than the increase in lymphocyte output. After termination of hyperthermia, lymph flow experienced a further increase, with no increase in small recirculating lymphocyte output. The efferent lymph flow during and after hyperthermia in the nodal lymph drainage

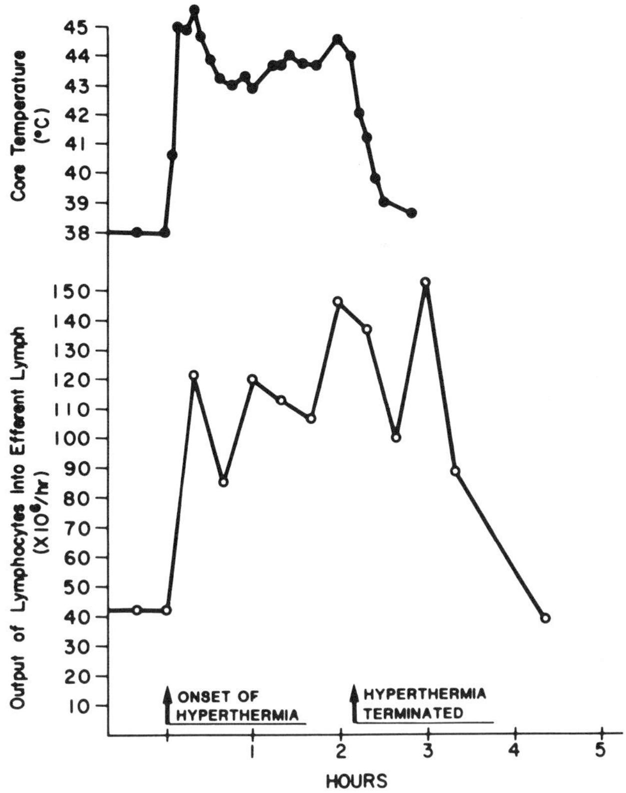

FIGURE 3. The prompt and sharp rise in the output of small recirculating lymphocytes into the efferent lymph of a sheep popliteal lymph node associated with localized deep hyperthermia over the study lymph node. The increase in lymphocyte output was correlated closely with the increase in core temperature in the area of the study lymph node. (From Moore, T. C., et al., *J. Surg. Oncol.*, 35, 75, 1987. With permission.)

area was variable. Figure 6 shows an increase in lymphocyte output with little change in efferent lymph flow during localized nodal lymph drainage area deep hyperthermia, with a decrease in efferent lymph flow at termination of deep hyperthermia. In other studies, volume of efferent lymph flow and small recirculating lymphocyte outputs have been shown to be independent responses in this supported-sheep lymphocyte traffic model.[27] Figure 7 presents findings with lymph-drainage-area hyperthermia when blast lymphocytes were present in study-node efferent lymph prior to the induction of deep hyperthermia. Here a sharp and marked increase in the output into study-lymph-node efferent lymph of blast lymphocytes occurs, as well as an increase in the output of small recirculating lymphocytes. In this study an increase in efferent lymph flow also occurred during the period of deep tissue hyperthermia. All of these studies were carried out in awake and unsedated sheep.

The effects of localized magnetic-loop-induced sheep brain deep hyperthermia on lymphocyte traffic and outputs from peripheral lymph nodes (popliteal, prefemoral, and prescapular) were evaluated in five studies.[21] Sharp and prompt increases in the output of small recirculating lymphocyte outputs occurred in all studies during the periods of localized brain hyperthermia/"fever". Representative findings are presented in Figures 8 and 9. The increases in lymphocyte outputs above prehyperthermia baselines ranged from 280 to 530% of baseline levels, with a mean increase of 398%. The author of this book is personally indebted to the late Professor Bede Morris of Canberra, Australia for personally cannulating the sheep prescapular lymph node used in this study. They were cannulated as a technique

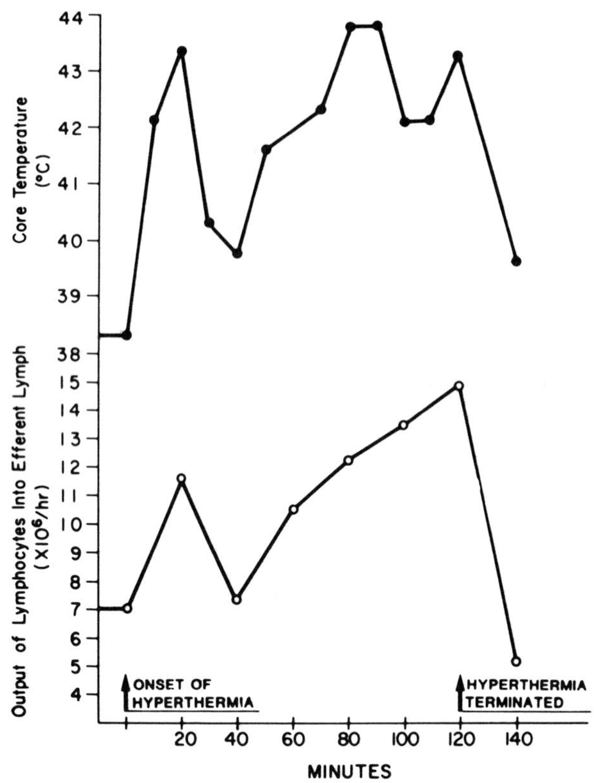

FIGURE 4. The temperature-sensitive nature of the deep hyperthermia-associated increase in the output of small recirculating lymphocytes into popliteal efferent lymph of a sheep with deep hyperthermia over the study lymph node. A temporary power turnoff in the deep hyperthermia machine resulted in a prompt decrease in both core temperature in the area of the study lymph node, and in the output of lymphocytes. Both changes were altered promptly by a return of machine power. (From Moore, T. C., et al., *J. Surg. Oncol.*, 35, 75, 1987. With permission.)

demonstration during a visit by Professor Morris to southern California as a Harbor–UCLA Department of Surgery Visiting Professor.

In a completed but unpublished study[23] the effects of sheep popliteal lymph-node lymph-drainage-area electrocautery dissection, as compared with no heat, cold-knife soft-tissue dissection, in the drainage area in 23 popliteal lymph nodes were compared with respect to their respective modulating effects on the profound and prolonged (up to 48 h) depressions in lymphocyte traffic and outputs into study-node efferent lymph. The electrocautery-dissection-associated small-recirculating-lymphocyte outputs into study-node efferent lymph were significantly higher than those in control cold-knife-dissection animals at both 24- and 48-h periods after general anesthesia (ketamine and xylazine) and operation ($p = 0.0005$ and 0.0001). After 40 h, there was a marked rebound substantially above control cold-knife and normal levels of efferent lymphocyte outputs. Representative findings are presented in Figures 10 and 11. The marked and prolonged elevations in small-recirculating-lymphocyte outputs, in association with study-node lymph-drainage-area cautery dissection, raise questions concerning the LTP and possible heat/burn-associated mediator molecules, such as substance P and bradykinin, in these findings.

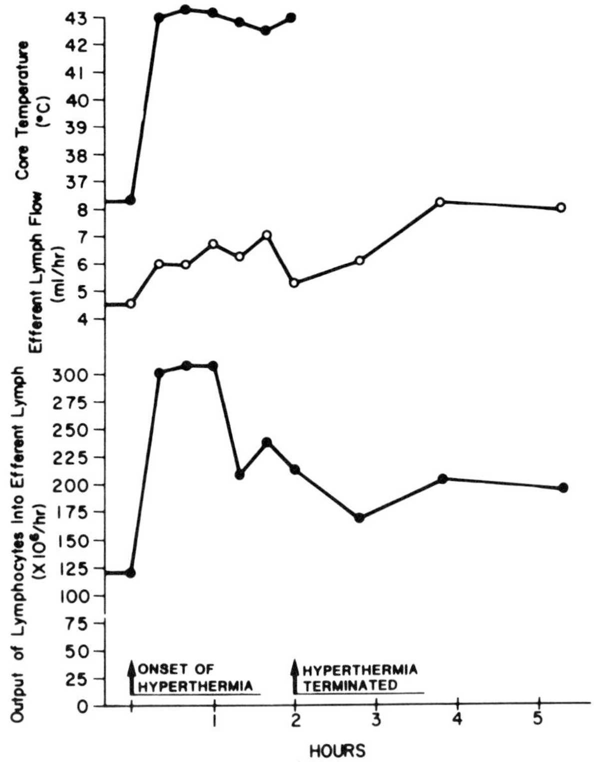

FIGURE 5. The prompt increase in the output of small recirculating lymphocytes into efferent lymph of a sheep popliteal lymph node associated with deep hyperthermia only in the study-lymph-node lymphatic drainage area. The increase in efferent lymph flow was slower and to a lesser degree than the increase in lymphocyte output. The increase in efferent lymph flow increased after the termination of deep hyperthermia in the nodal drainage area, while the lymphocyte output returned toward prehyperthermia levels. (From Moore, T. C., et al., *J. Surg. Oncol.*, 35, 75, 1987. With permission.)

III. PAIN

A. PAIN

A 1987 study by Fugiwara and Orita[28] of the University of Okayama (Japan), published in the *Journal of Immunology,* addressed the role of pain in the immune response. These authors found a clear enhancement of the immune response by pain stimulation in mice. Their study involved SRBC immunization and determinization of levels of antibody production by the plaque-forming cell assay. Peripheral pain was induced by subcutaneous injection of either hypotonic distilled water or a hypertonic solution of 4% sodium chloride, or by pulling out body hairs or by tail pinching with a forceps until the animal squealed, bit the forceps, or both. The pain stimuli were given once a day for 4 d after SRBC immunization. Plaque-forming cells were found to be increased following both osmotic and physical-pain stimulations. Serum obtained from the mice from 30 to 60 min following a physical-pain stimulus was found to increase the level of plaque-forming cells of normal mouse spleen cells *in vitro*. An increase in plaque-forming cells also was found in nonimmunized mice when spleen cells from pain-stimulated mice were cultured *in vitro* with SRBC.

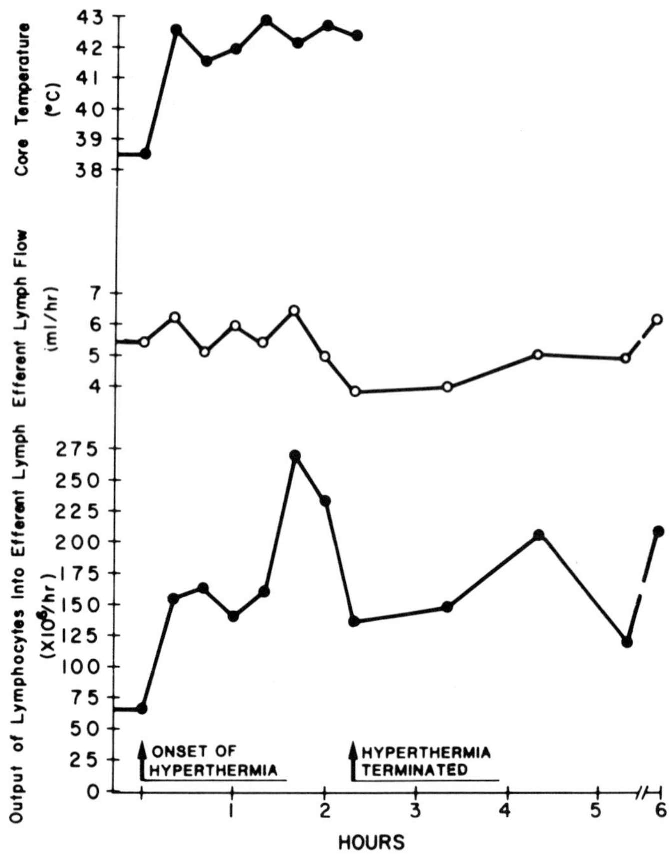

FIGURE 6. An increase in the output of small recirculating lymphocytes into the efferent lymph of a sheep popliteal lymph node with deep hyperthermia to the study-lymph-node drainage area only. Here the increase in efferent lymph flow was modest in degree and decreased after the termination of deep hyperthermia. (From Moore, T. C., et al., *J. Surg. Oncol.*, 35, 75, 1987. With permission.)

B. ANTIPAIN

Antipain drugs and anesthesia have a long history of modulating and inhibiting important components of the cellular immune response. Cantacuzene,[29] working in the laboratory of Elie Metchnikoff at the Pasteur Institute in Paris, reported in 1898 that opium depressed macrophage phagocytosis and speculated that these cells were ''drugged''.

At the turn of the century, Snel,[30] in 1903, and Rubin,[31] in 1904, described increased susceptibility of guinea pigs and rabbits to infection after the administration of general anesthetic agents such as ether, chloroform, and chloral hydrate. In 1911 a publication in the *Journal of Infectious Diseases,* Graham,[32] destined to become America's preeminent academic surgeon of the 20th century, presented experimental evidence of ether-anesthesia-induced immunosuppression. He described a severe depression of phagocytosis produced both by ether *in vitro* and by ether anesthesia in human subjects and in rabbits. The depression in rabbits was the same whether anesthesia with ether was induced alone or with surgery, thus differentiating the effect of ether anesthesia alone from a possible contribution of surgical trauma and tissue injury.

In 1916 Gaylord and Simpson[33] found that daily anesthesia of mice for 10 d with cloroform and ether, briefly to the point of unconsciousness, increased the incidence of ''takes'' of transplanted mammary carcinoma.

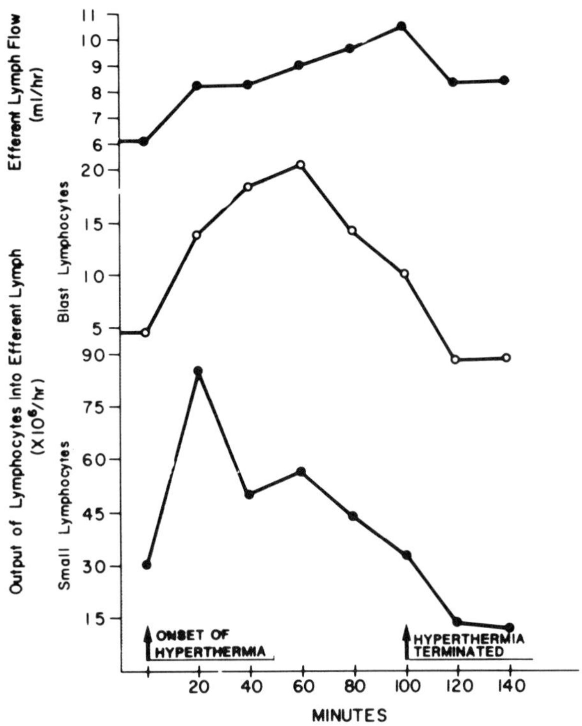

FIGURE 7. Findings in a drainage-area deep hyperthermia study in which blast lymphocytes were present prehyperthermia, as well as small recirculating lymphocytes. The increased output into the efferent lymph of the sheep popliteal lymph node occurred with the small recirculating lymphocytes before the marked increase in the output of blast lymphocytes peaked. There was a slow but steady increase in efferent lymph flow, which remained elevated above prehyperthermia levels after the termination of deep hyperthermia. (From Moore, T. C., et al., *J. Surg. Oncol.*, 35, 75, 1987. With permission.)

A particularly important observation was made in a 1919 publication in the *Journal of Physiology* by Dale and Laidlaw,[34] who established that anaphylaxis could not be produced during general anesthesia. This was the first demonstration of an important connection and interaction of the nervous system and the immune response system, and at the molecular level involving a major transmitter and mediator molecule of both systems (histamine).

Anesthesia *alone* has been found in a number of important studies to have an immunosuppressant/modulatory effect on a number of major actions of the immune response system. Espanol, Todd, and Soothill,[35] of the Great Ormond Street Hospital for Sick Children in London, made a particularly important observation in 1974 that, in tonsilectomy children before the cutting started, anesthesia alone depressed human lymphocyte transformation after mitogen stimulation. This observation is significant because most anesthesia and surgery studies involve both anesthesia and surgical trauma, and the findings most often are attributed (perhaps and likely erroneously) to the surgical trauma.

Important data also has been obtained, in comparatively recent years, in studies of anesthesia relating to the development, "take", and spread/metastasis of malignant cancerous tumors in man and experimental animals. In 1961 Griffiths, Hoppe, and Warren Cole[36] of the Department of Surgery of the University of Illinois–Chicago reported that anesthesia *alone* increased the incidence of "takes" and metastases of transplanted cancer cells. They found that pentobarbitol-induced anesthesia *alone* increased the "takes" of Walker 256 carcinosarcoma cells from 22 to 65% after their inoculation into the peripheral lymph node lymph drainage areas of rats. In this regard it is of interest that Professor Warren Cole,

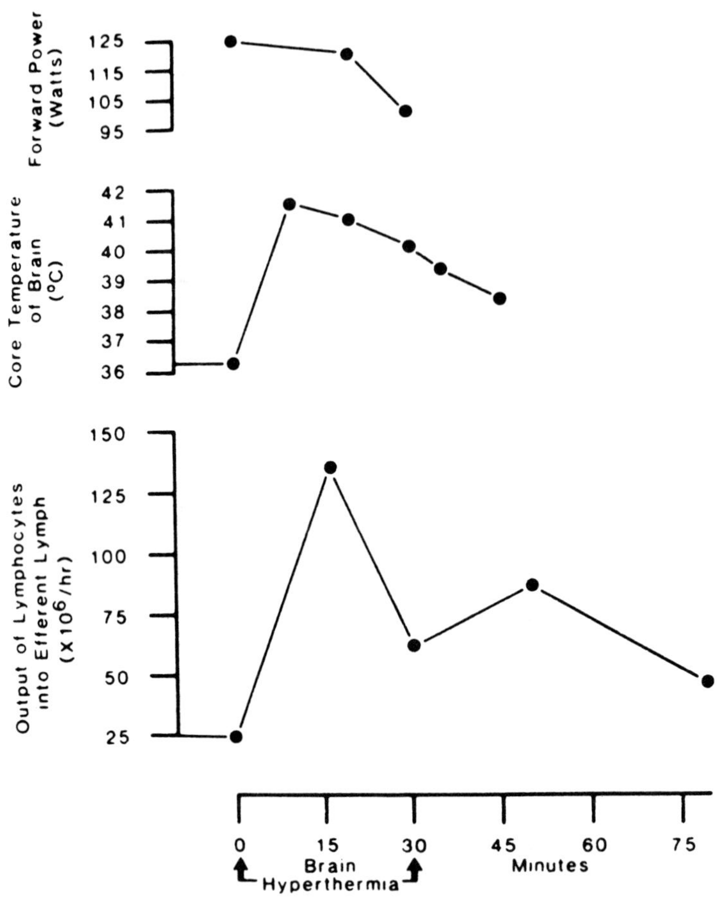

FIGURE 8. The prompt and sharp increase in the output of small recirculating lymphocytes into a study-sheep peripheral hind-limb lymph node (pre-femoral) efferent lymph following the onset of localized deep hyperthermia (magnetic-loop induced) to the sheep brain, with a burr-hole inserted thermal probe into the sheep's brain (''fever'' localized to the brain). (From Moore, T. C., et al., *Med. Sci.*, 13, 1194, 1985. With permission.)

Chairman of the Department of Surgery at the University of Illinois School of Medicine in Chicago, was a major and highly productive protege of Professor Graham, referred to earlier in this section. In 1964 Agostino and Cliffton,[37] cancer surgeons at the Memorial Hospital in New York City, described the effects of anesthesia *alone* in increasing pulmonary metastases of cancer in rats. The greatest and most significant responses occurred with chloroform and ether anesthesia. Lundy et al.,[38] in 1977, reported in *Surgery* an anesthesia-induced increase in pulmonary metastases of cancer in experimental animals (mice). The tumor cells used were 20-methylcholanthrene (MCA)-induced syngeneic fibroblasts and were given intravenously. In these studies, they also showed that anesthesia *alone* also depressed sensitization to DNCB when utilized at the time of sensitization to DNCB, but not when anesthesia was given 5 d after sensitization. These studies suggested an anesthesia-associated effect on the efferent (sensitization) rather than the afferent (effector) arm of the immune response. An additional study by this group demonstrated an anesthesia-induced reduction in MLC (mixed lymphocyte culture) reactivity. The anesthesia was thiopental, and the authors warned of a potentially dangerous consequence of anesthetic-induced immunosuppression.

There is evidence for apprehension in the human experience also. In 1948 Sir Gordon Gordon-Taylor[39] included a section on ''cancer resistance or immunity'' in his Royal College

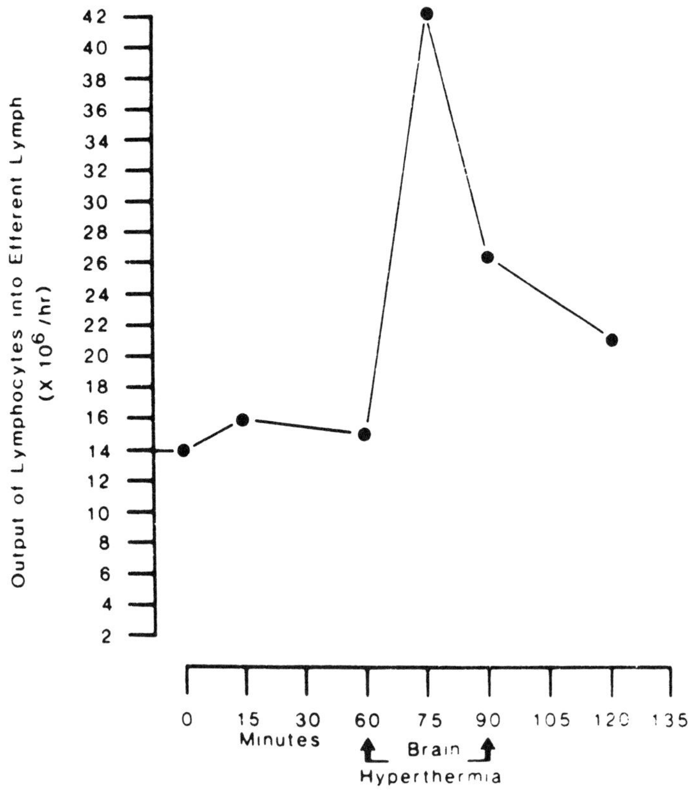

FIGURE 9. A prompt and marked increase in lymphocyte traffic and in the output of small recirculating lymphocytes into lymph efferent from a peripheral lymph node of a sheep, accompanying deep hyperthermia localized to the brain (magnetic-loop induced). (From Moore, T. C., et al., *IRCS Med. Sci.*, 13, 1194, 1985. With permission.)

of Surgeons lecture on breast cancer. In this section he described his personal experience with five cases of breast cancer in which the breast cancer recurred shortly after anesthesia and surgery which were carried out for reasons totally unconnected with the original breast cancer. Also somewhat disturbing with respect to humans and the effects of anesthesia *alone* and in immunosuppression is the 1968 report in *Anesthesiology* by Bruce et al.[40] involving a 20-year survey of the causes of death among anesthesiologists. They found an increase in lymphoid and reticuloendothelial-tissue cancer deaths among anesthesiologists. These are immunosuppression-induced types of human cancer and have been found quite increased in incidence in immunosuppressed human-organ-transplant patients.

In their 1962 ground-breaking publication that opened the new and extremely important field of lymphocyte traffic studies in large animals (sheep), Morris and Hall[41] made the important observation that general anesthesia (cyclopropane in their studies) in sheep produced a profound and prolonged depression in lymphocyte traffic through primary peripheral lymph nodes, as mirrored in the output of lymphocytes into nodal efferent lymph, chronically cannulated. In their studies they found the profound depression to last from 24 to 48 h and cautioned against any studies during this period because of the complicating and confounding effects of this anesthesia (and surgery)-induced depression in lymphocyte outputs.

The author[42-45] and the author and associates[46-49] have carried out a number of studies of anesthesia-induced depression of lymphocyte traffic in the sheep and related events from 1980 to the present. The traffic depressions with barbiturate-halothane general anesthesia

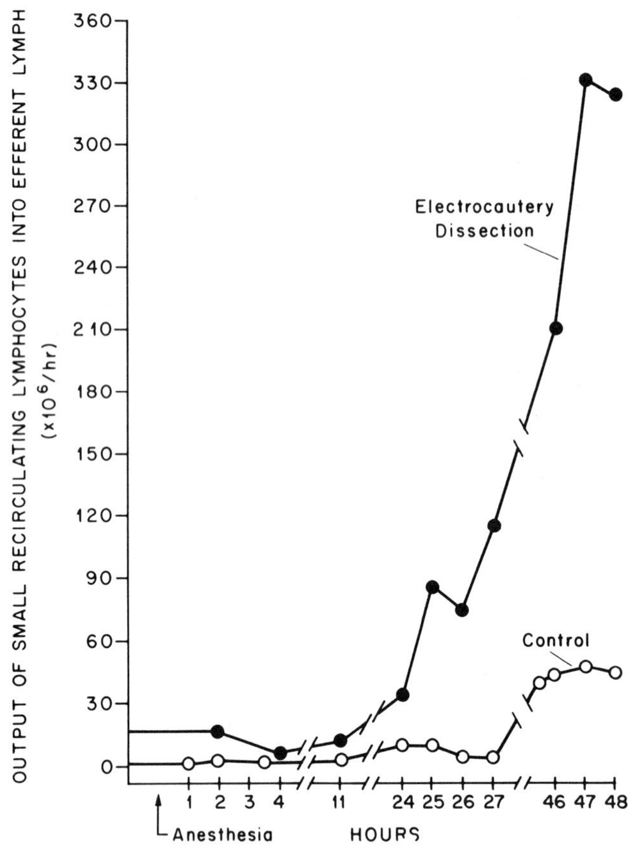

FIGURE 10. A marked lymphocyte-traffic- and output-stimulating effect of popliteal lymph-node drainage-area electrocautery dissection in comparison with control (cold-knife) dissection in this area. This effect sharply modifies the early (11 to 27 h) lymphocyte-traffic-depressing effect of general anesthesia. In the period after 48 h there is a marked increased in electrocautery-dissection-induced increases in small-recirculating-lymphocyte outputs into nodal efferent lymph above control and normal (30 to 60 $\times$ 10^6 cells per hour) levels.

and ketamine-xylazine general anesthesia were marked, prolonged, and comparable. The degree and duration of depression were affected by the length of general anesthesia in the barbiturate-halothane studies (Figure 12). It was uniformly marked and prolonged in the ketamine-xylazine studies (Figure 13). This depression during operative lymphatic cannulation could be overcome briefly by digital manipulation of inflammatory tissue in the popliteal lymph node lymph drainage area (Figure 14). It also could be overcome by an acute infusion of bradykinin into a cannulated afferent lymphatic (see Figure 9 in Chapter 9). The depression in lymphocyte traffic and outputs into efferent lymph was less with epidural anesthesia than with general anesthesia (Figure 15). The administration of the local anesthetic agent lidocaine (lignocaine) into the drainage area of a study lymph node also produced prompt and sharp (and at times prolonged) depressions in lymphocyte outputs (Figures 16 to 18). Central venous infusion of lidocaine into a cannulated jugular vein also produced a sharp, prompt, and prolonged depression in lymphocyte traffic and outputs from popliteal lymph nodes (Figure 19).

In all of ten sheep studied, the depressions in lymphocyte traffic were found to be associated by prompt and marked increases in prostaglandin E_2 (PGE_2) levels in study-lymph-node (popliteal) efferent lymph (Table 1). The depressions in lymphocyte traffic were

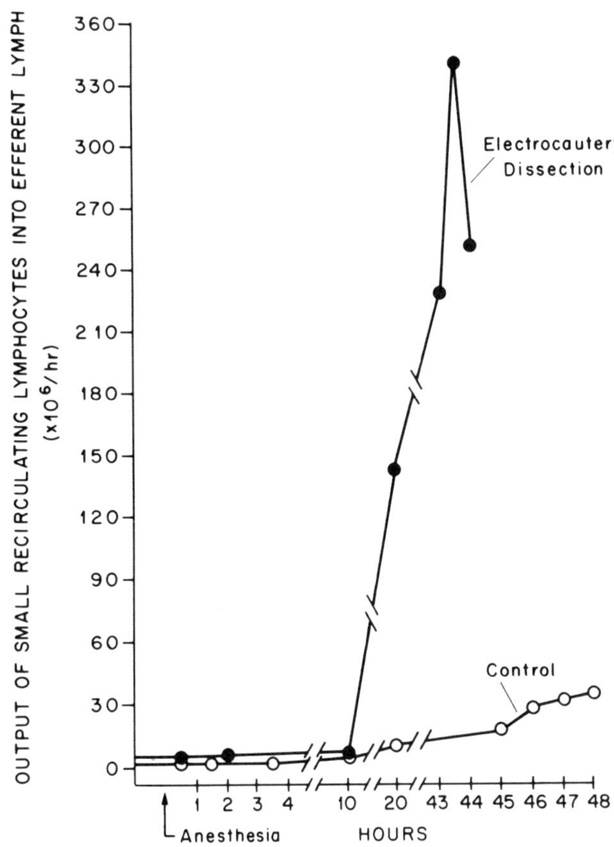

FIGURE 11. Again, the marked lymphocyte-traffic-enhancing effect of lymph-node (popliteal) drainage-area electrocautery dissection.

correlated time-wise with altered (increased) levels of efferent lymph PGE_2 levels (Figures 20 and 21). These traffic depressions and associated marked elevations in efferent lymph levels of PGE_2 also were accompanied by sharp depressions in antibody levels to *Salmonella muenchen* bacteria in animals that, prior to anesthesia, had received study-node lymph-drainage-area immunization with killed *S. muenchen* (Figure 22).

IV. MEMORY

> "It would be interesting to know how multitransmitter neurons respond to tetanic stimulation. Where monoamines and peptides coexist, for example, one would predict that if the neuron was activated for a prolonged period, it might reach a point where the peptide reserves in nerve terminals were exhausted, while the monoamine could still be replenished by local biosynthesis. With prolonged use, there could be a different ratio of release of the monoamine and the peptide, and this might confer some potentially interesting adaptive properties. Such a system would have a form of memory for the previous history of that pathway, and this might be particularly relevant to pathways that exhibit rapid adaption."
>
> L. L. Iversen[50]

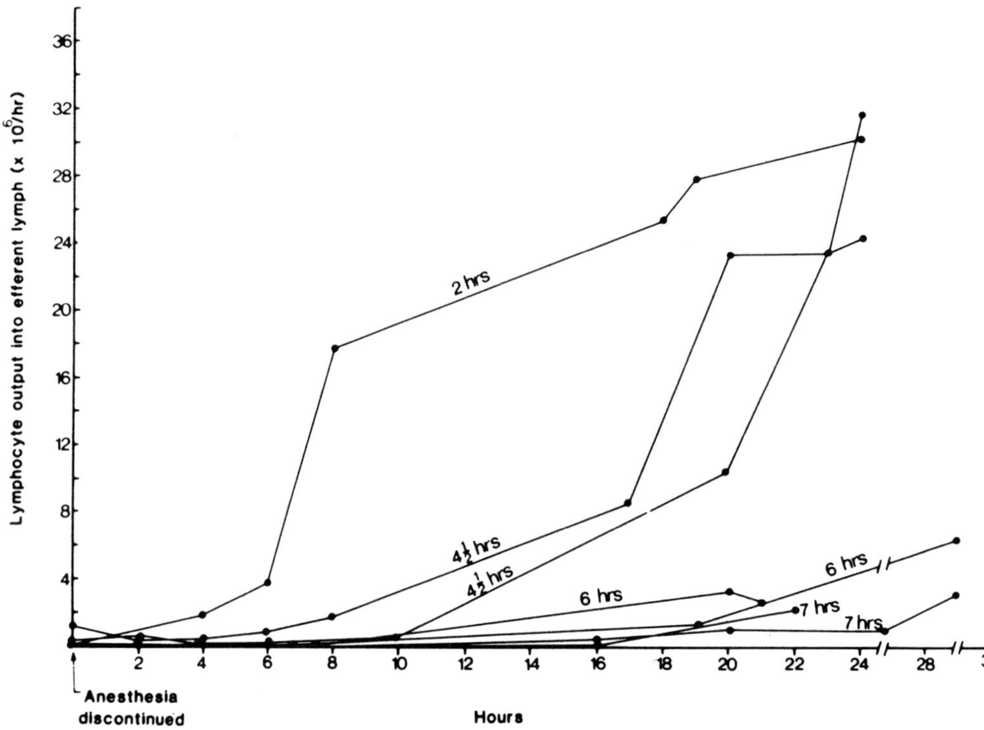

FIGURE 12. The profound and prolonged depression in the output of small recirculating lymphocytes into peripheral nodal efferent lymph of sheep following barbiturate-halothane anesthesia. Longer anesthetics are associated with more prolonged depressions in lymphocyte traffic. (From Moore, T. C., *Am. J. Surg.*, 147, 807, 1984. With permission.)

A. INTRODUCTION

Molecular plasticity, with receptor and signaling-cascade-ancillary amplification, likely is central to the learning and memory phenomena of multicellular living organisms (plant as well as animal). The base and building blocks and survival, growth, and replication/reproduction molecules of these living organisms are amino acids. It is unlikely accidental that the principal communications, transport, and defense molecules are quite small/"primitive" vasoactive neurotransmitter molecules such as those discussed above — amino acids, decarboxylated amino acids (amines), and short sequences of amino acids (short-chain peptides) — as well as short-chain lipids (eicosanoids). It is unrealistic to consider that the basic memory molecules of central nervous system (CNS) and immunological memory differ or that they have different CNS areas and mechanisms of storage and recall. Different types of "memory" (CNS and immunological) at the molecular level likely have much more in common than in discordance.

The last paragraph of the summary of the author's 1982 Ph.D. thesis[42] includes the following:

The relative vascular and neuronal contributions to the alterations in lymphocyte traffic observed in these studies are not identified. Both likely are of importance not only in contributing to on-site traffic regulation, but also in local (? "axon reflex") and more central (spinal and intracranial) coordination and integration and possibly even in memory. Immunologic memory may have something in common with other types of memory. A functioning immune response nervous system with considerable local action and integration may exist.

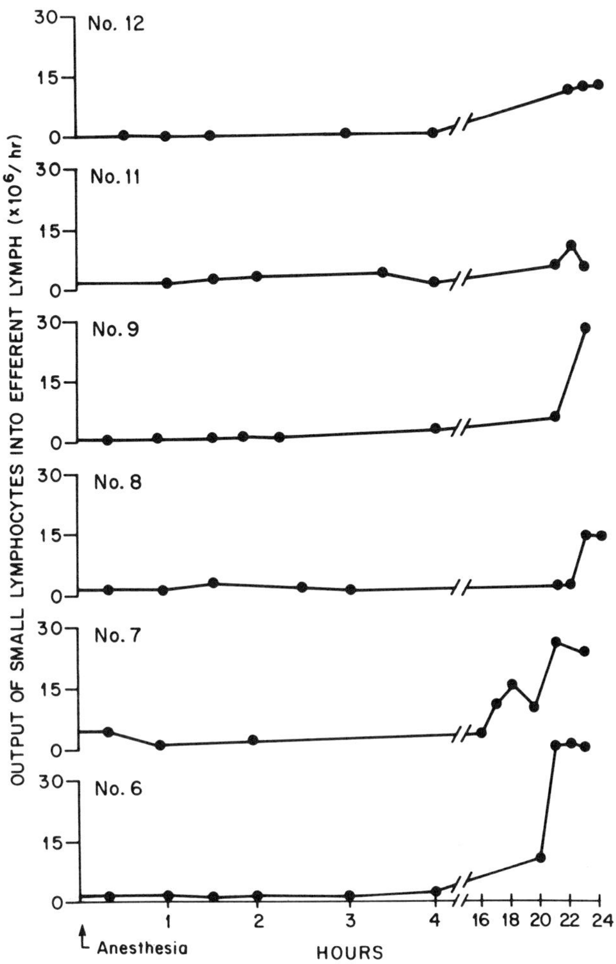

FIGURE 13. Marked and prolonged depression in lymphocyte traffic and in the output of small recirculating lymphocytes into the efferent lymphatics of popliteal lymph nodes of sheep in six studies with ketamine-xylazine general anesthesia. (From Moore, T. C., et al., *Immunology,* 63, 139, 1988. With permission.)

The contributions of specific ''memory molecules'' to the molecular component of memory, immunological as well as CNS memory, are dealt with in the chapters of this book specifically relating to these molecules, especially substance P, bradykinin, serotonin, met-enkephalin, CCK-8, bombesin, and CGRP. Other molecules and specialized receptors, such as histamine at H_1 receptors, PGE_2, GABA at $GABA_B$ receptors, glycine at strychnine-insensitive receptors, and others such as EDRF/nitric oxide/NO, warrant high priority on the to-be-studied agenda.

B. MAMMALIAN MEMORY AND LTP

Long-term potentiation (LTP) is a fascinating and suggestive (memory-wise) use-dependent increase in synaptic efficacy (plasticity), which was first demonstrated in the hippocampus by Lomo[51] in 1966 and explored in depth in the early 1970s by Bliss, Lomo, Gardner-Medwin, and associates.[52-54] These investigators established that a brief tetanic stimulus to any one of several monosynaptic excitatory pathways in the hippocampus could enhance the amplitude of evoked responses in the tetanized pathway for hours or days

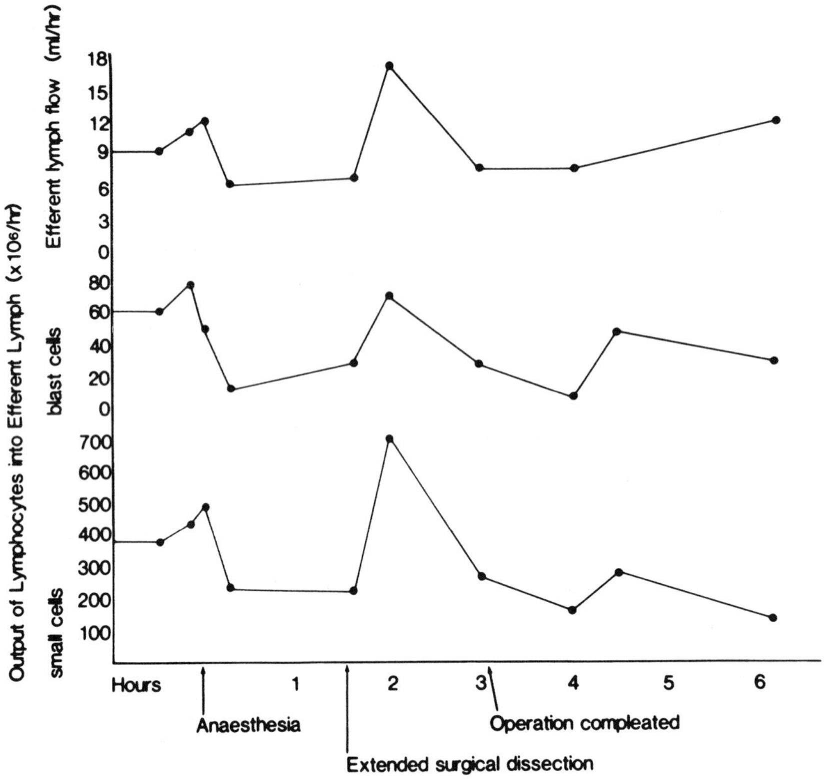

FIGURE 14. Demonstration of the capability of vigorous manipulation of drainage-area inflammatory tissue in a sheep to increase lymphocyte traffic and lymphocyte outputs during general anesthesia. (From Moore, T. C., *Am. J. Surg.*, 147, 807, 1984. With permission.)

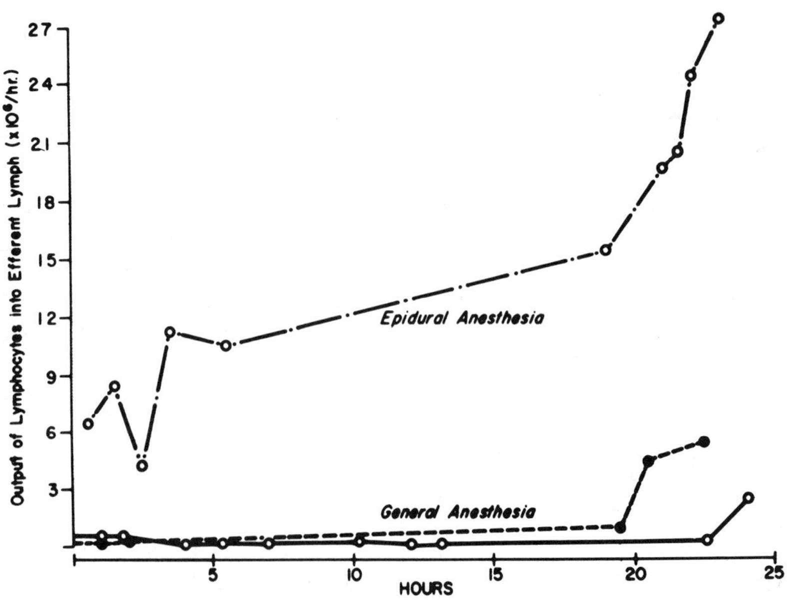

FIGURE 15. There is less depression in lymphocyte traffic and lymphocyte outputs in sheep with epidural, rather than with general, anesthesia. (From Moore, T. C., *Am. J. Surg.*, 148, 71, 1984. With permission.)

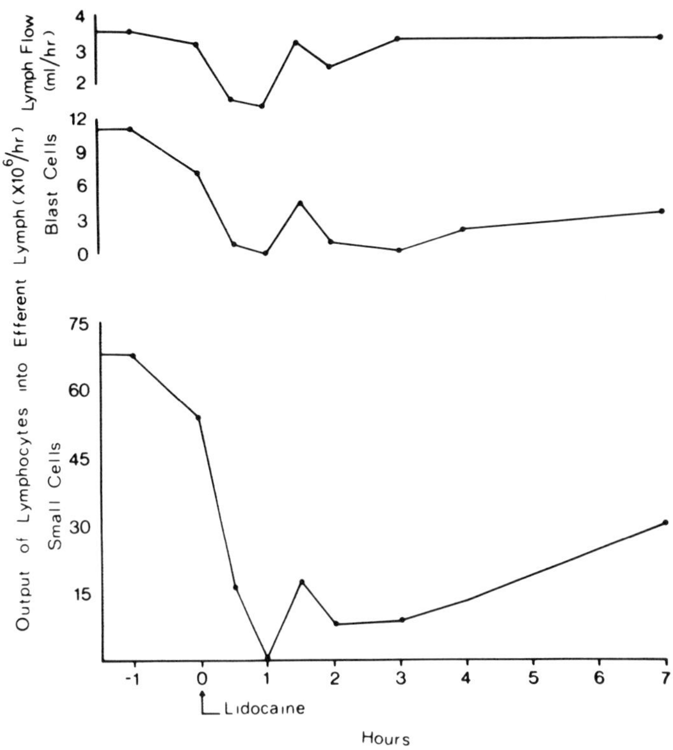

FIGURE 16. A prompt and prolonged depression in lymphocyte traffic and the output of both small recirculating and blast lymphocytes into a sheep popliteal lymph node efferent lymph following injection of lidocaine into the nodal lymphatic drainage area. (From Moore, T. C., *Am. J. Surg.*, 147, 807, 1984. With permission.)

thereafter — a sharp contrast to the seconds-to-minutes range encountered with "classical" forms of facilitation and potentiation. The occurrence of this remarkable response in the hippocampus has generated a large amount of interest in LTP as a synaptic model of memory. This interest derived from the long and well-established involvement of the hippocampus in memory, generated by the 1957 report of Scoville and Milner[55] (and subsequent studies of Milner[56,57]) which described the profound memory deficits in man resulting from bilateral surgical resections of the hippocampus in efforts to control severe intractable and life-threatening epilepsy.

The initial studies by the early investigators demonstrated that brief tetanic stimulation of the perforant path from the entorhinal cortex to the hippocampus resulted in enduring increases in the amplitude of population responses evoked in granule cells of the dentate gyrus by single shocks delivered to the tetanized fibers, with effect persisting for hours in anesthetized rabbits and for days and weeks in nonanesthetized rabbits with chronically implanted electrodes.[52,53] A unique component of LTP is the brevity of the tetanic stimulation required to induce it. Induction of LTP appears to require a combination of postsynaptic depolarization and presynaptic activity. The relative contributions of pre- and postsynaptic components to the initiation and maintenance of the LTP phenomenon continue to be disputed. A clear calcium dependence is beyond question.

In 1986 Malenka, Madison, and Nicoll[58] of the University of California–San Francisco made a major contribution to an increased understanding of LTP by demonstrating that phorbol esters profoundly potentiate excitatory synaptic transmission in rats hippocampus (hippocampal slice preparation) in a manner that appeared to be indistinguishable from LTP.

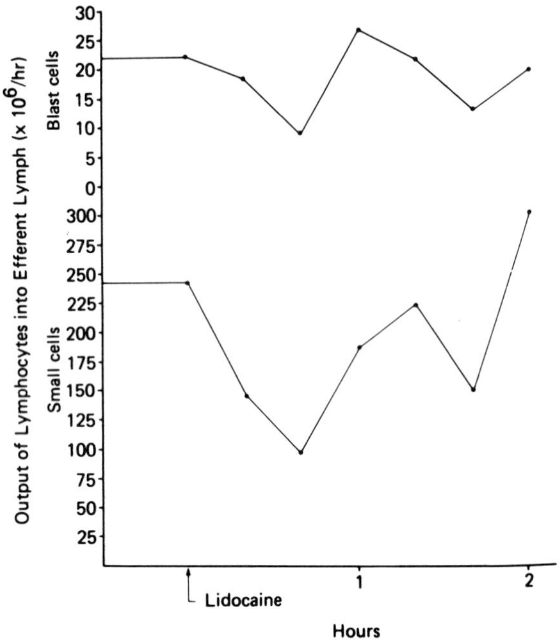

FIGURE 17. When lidocaine is injected into a sheep popliteal lymph-node drainage area where there is considerable inflammation, as evidenced by high preinjection blast- and small-recirculating-lymphocyte outputs into nodal efferent lymph, the traffic-depressing effects are prompt, but less profound than shown in Figure 16. (From Moore, T. C. and Khan, F., *J. Surg. Res.*, 41, 279, 1986. With permission.)

They also showed that after maximal synaptic enhancement with phorbol esters, LTP could no longer be elicited. From the involvement of phorbol esters in protein kinase C (PKC) activation, the authors speculated that, in mammals, PKC may play a role in transmitter release in the hippocampus and in the generation of LTP. A 1987 publication by this group[59] described enhanced phorbol-ester-stimulated transmitter release by two measurements: occurrence of postsynaptic potentials and potassium-stimulated release of endogenous glutamate. These findings are of particular interest with respect to the actions of "memory molecules", such as substance P, bradykinin, serotonin, bombesin, and others, in PKC activation via the PI transmembrane signaling mechanism.

In 1988 Buzsaki of Pec, Hungary,[60] observed that a series of brief trains of high-frequency tetanic stimulation to the perforant path (PP) to the rat hippocampus was capable of producing long-term enhancement of neuronal excitability within the hippocampus beyond the PP-granule cell synapse of the dentate gyrus, including both the CA3 and CA1 neurons.

In 1989 Malinow, Schulman, and Tsien,[61] of Stanford, published evidence suggesting that a postsynaptic requirement of both PKC and Ca^{++}-calmodulin-dependent protein kinase (CaMKII) and a presynaptic requirement for protein kinase are necessary for the expression of LTP.

Also in 1989, Bliss and associates[62] at Mill Hill, London, reported some interesting and important findings in *Nature*. Accepting that LTP is triggered by the postsynaptic entry of calcium through the channel associated with the NMDA glutamate receptor and that LTP maintenance is mediated, at least in part, by presynaptic mechanisms, they tried to explain how postsynaptic events can lead to an increase in neurotransmitter release. They postulated the existence of a retrograde messenger to carry information from the postsynaptic side of the synapse to the recently activated presynaptic terminals and considered arachidonic acid or a metabolite (from the work of Linden et al.[63]) to be a strong candidate for retrograde

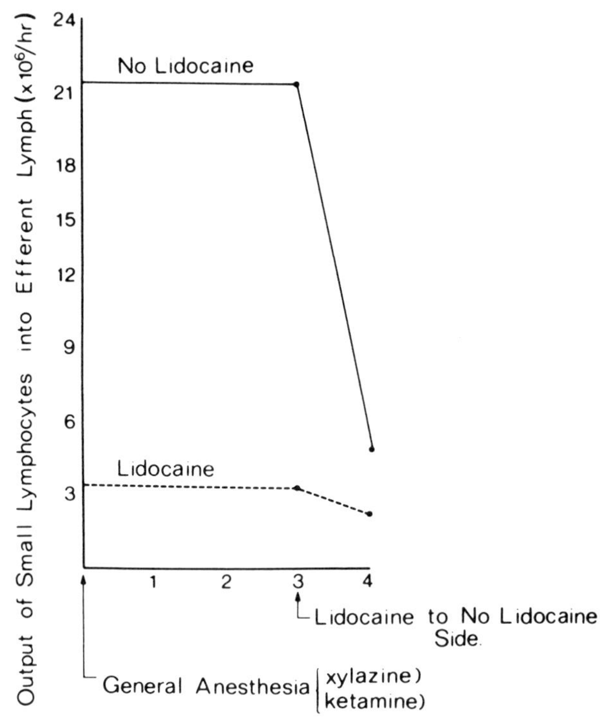

FIGURE 18. Simultaneous outputs of small recirculating lymphocytes into bilaterally cannulated popliteal lymph-node efferent lymphatics. Outputs from both sides were depressed from normal levels, in association with a light general anesthesia, with ketamine and xylazine, but markedly more so on one side where the general anesthesia was supplemented by local infiltration anesthesia with lidocaine for an operative procedure in the drainage area. When this interesting bilateral output data became available, the same amount of lidocaine was given arbitrarily into the nonoperative side without prior lidocaine, with a resulting sharp fall in the output of small recirculating lymphocytes on this side as well. (From Moore, T. C. and Khan, F., *J. Surg. Res.*, 41, 279, 1986. With permission.)

messenger activity. From their studies, they reported that weak activation of the hippocampal perforant path in the presence of arachidonic acid leads to a slow-onset persistent increase in synaptic efficacy (both *in vivo* and *in vitro*) and that the activity-dependent potentiation thus produced is accompanied by an increase in the release of glutamate and is nonadditive with tetanus-induced LTP. The weak high-frequency trains given in this study were considered subthreshold for the induction of LTP. The authors interpreted their findings as indicating a role for arachidonic acid or a metabolite as a retrograde messenger in the later, but not initial, stages of LTP. They also concluded that in the CA1 area, in the dentate gyrus, or more probably in both, potentiation induced by arachidonic acid is associated with an increase in glutamate release. It is the author's hypothesis that arachidonic acid is released from a postsynaptic site and acts on presynaptic terminals to increase transmitter release. They also suggested EDRF/nitric oxide as a putative early messenger from the 1988 findings of Garthwaite et al.[64] The EDRF/nitric oxide ''connection'' with LTP is of especial interest, as bradykinin, a ''memory molecule'', was used by Moncada and associates[65] to identify nitric oxide (NO) as EDRF.

Reports in 1990 by Bekkers and Stevens,[66] of LaJolla (Salk Institute), and by Malinow and Tsein,[67] of Stanford, suggested a presynaptic mechanism for expression of LTP involving an increase of transmitter release. These authors and Bliss[68] considered these findings to support the concept that LTP maintenance is presynaptic. Another 1990 report from Toronto, by Goh and Pennefather,[69] described pertussis toxin prevention of the induction of hippocampal LTP in the stratum radiatum and stratum oriens inputs to CA1 neurons.

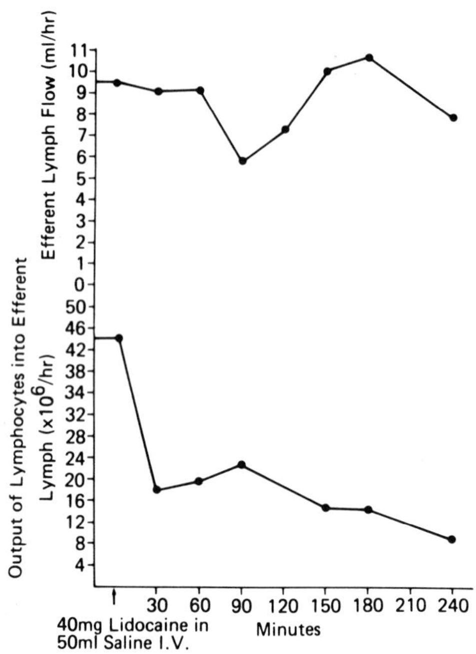

FIGURE 19. A bolus infusion of 40 mg of lidocaine in 50 ml of saline infused into a chronically cannulated jugular vein in an awake and unsedated sheep produced a prompt and sustained depression in the output of small recirculating lymphocytes into the popliteal lymph node efferent lymphatic, chronically cannulated. There was a delayed and short-term depression in the volume of efferent lymph. There was no alteration in the state of alertness of the animal. (From Moore, T. C. and Khan, F., *J. Surg. Res.*, 41, 279, 1986. With permission.)

TABLE 1
Prostaglandin E_2 Levels of Efferent Lymph in Consecutive Samples Obtained Following Efferent Lymphatic Cannulation under Anaesthesia and with No Prior Drainage-Area Stimulation

Study	PGE_2 (pg/50 μl)[a] sample no.[b]							
no.[c]	1	2	3	4	5	6	7	8
1	44	140	143	52	7	ON[d]	1	1
2	100							
3	>500	>500		>500		270		ON
6		185		>500		>500		
8	12 · 4	11 · 4		34 · 5		58	ON	11 · 5
11	22 · 7	58		121	116		ON	31
12	69	73	96	ON				<1
13	68	100		490	>500			
14	>555	>455		>555		150	ON	27
15	75	183		415	265			

[a] Normal PGE_2 levels of efferent popliteal lymph from nonanaesthetized and nonimmunized sheep = <1 pg/50 μl.

[b] Consecutive samples from time of efferent lymphatic cannulation under anesthesia (range 30–60 min).

[c] Studies as numbered in Table 2 of Reference 47.

[d] ON = overnight sample (range 15–18 h).

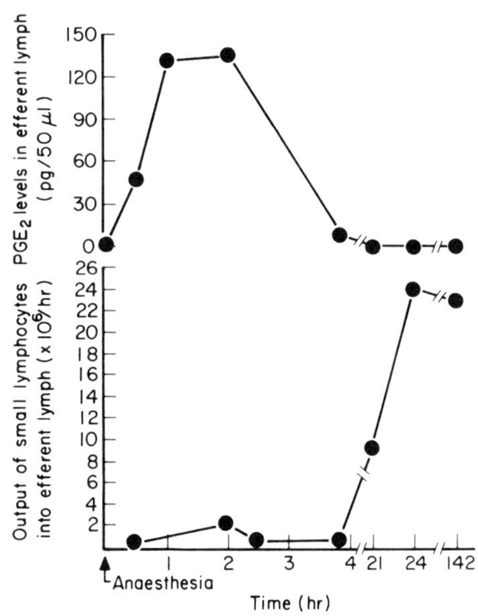

FIGURE 20. Anesthesia-induced depressions in sheep popliteal lymph node lymphocyte traffic and outputs were associated with marked and equally time-limited elevations in study-node efferent lymph levels of prostaglandin E_2 (PGE_2). (From Moore, T. C., et al., *Immunology,* 63, 139, 1988. With permission.)

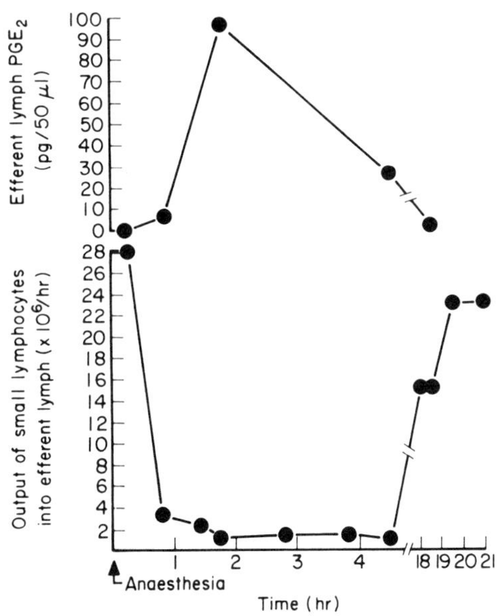

FIGURE 21. Anesthesia-induced depressions in sheep popliteal lymph node lymphocyte traffic and outputs of small recirculating lymphocytes are associated time-wise with marked elevations in nodal lymph PGE_2 levels. (From Moore, T. C., et al., *Immunology,* 63, 139, 1988. With permission.)

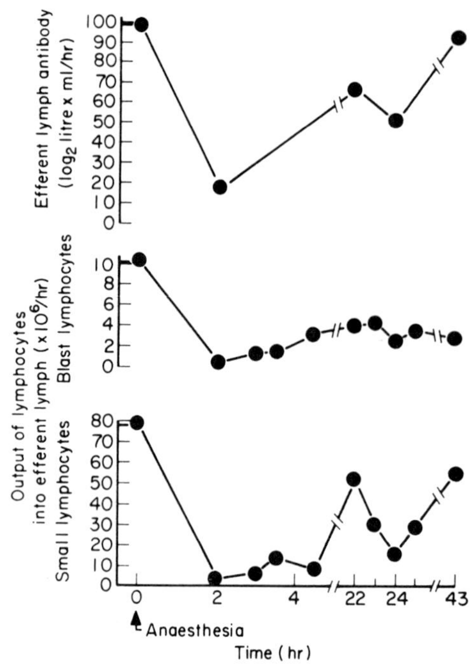

FIGURE 22. General anesthesia-induced depressions of small- and blast-lymphocyte outputs into a sheep popliteal lymph-node efferent lymphatic in the presence of drainage-area stimulation from bacterial immunization *(S. muenchen)*. These anesthesia-induced depressions are associated time-wise with a depression in antibody output into this nodal efferent lymph and lymphatic. (From Moore, T. C., et al., *Immunology,* 63, 139, 1988. With permission.)

Also in 1990, Meshul and Hopkins[70] presented interesting evidence from presynaptic ultrastructural studies of calcium distribution to suggest that LTP may be associated with, and accompanied by, long-term changes in intraterminal calcium homeostasis and the number of synaptic vesicles containing calcium deposits. Another 1990 publication, from Kyoto, Japan, by Ishikara, Satoh, and associates,[71] described different drug susceptibilities of LTP in three input systems to the CA3 region of the guinea pig hippocampus in an *in vitro* study. Drugs blocking NMDA glutamate receptors influence (block) LTP in the commissural/associational and the fimbrial pathways, while non-NMDA-receptor blockade suppressed LTP in the mossy-fiber pathway to CA3. Also relating to these observations, Zalutsky and Nicoll,[72] of San Francisco, reported that while induction of associational-commissural LTP required, in the postsynaptic cell, NMDA receptor activation, membrane depolarization, and a rise in calcium, mossy-fiber LTP does not. They also report that paired-pulse facilitation (an index of increased transmitter release) is unaltered during associational-commissural LTP, but reduced during mossy-fiber LTP. Mossy fibers derive from dentate gyrus granule cells. They conclude that both the induction and persistent change may be presynaptic in mossy-fiber LTP, but not in associational-commissural LTP.

In 1991 Alzheimer et al.[73] from Munich described adenosine depression of the induction of LTP at the mossy-fiber-CA3 synapse *in vitro,* and Aniksztejn and Ben-Ari,[74] of Paris, described a novel form of LTP which they called LTP_k because of associated delayed rectifier block (I_k). This novel form of LTP is induced by enhanced release of glutamate, which generates a depolarization by way of the non-NMDA receptors and consequent activation of voltage-dependent Ca^{++} channels.

A 1991 publication in *Nature* (from Oxford University by Larkman et al.[75]) questions the use of quantal analysis in the evaluation of mechanisms of LTP cited above (Bekkers and Stevens[66] and Malinow and Tsein[67]). A discussant from University College, London,

Edwards[76] agrees with this assessment and suggests, as does Larkman et al., "a more complete quantal analysis will be required to quantify the relative importance of pre- and postsynaptic changes." The resolution of the debate as to whether LTP of synaptic transmission in the brain results from a change in the release of transmitter(s) from presynaptic terminals or from a change in postsynaptic response(s) may be resolved like similar heated debates in the past — such as whether the immune response was humoral or cellular (it was both) or whether neurotransmission was electrical or chemical (it was both) — and so with LTP.

The most recent LTP blockbuster, a publication by Kandel and associates,[77] of Columbia University in New York, appeared in the October 10, 1991 issue of *Nature*. The authors reported that LTP in the hippocampus was blocked by inhibitors of tyrosine kinase, the principal activation molecule of growth factors and oncogenes, including DNA and protein synthesis. Their findings suggested that tyrosine kinase could be required postsynaptically for long-term synaptic plasticity in the hippocampus. As Ca^{++} calmodulin kinase II and protein kinase C also seem to be required, they speculated that tyrosine kinase could participate postsynaptically in a kinase "network", together with the serine and threonine kinases. If this were not enough, the authors concluded with the observation from their laboratory that injection of a specific NO/EDRF synthase inhibitor into the hippocampal postsynaptic cell blocks the induction of LTP and that the application of NO enhances spontaneous transmitter release from hippocampal neurons in culture. Of further interest, they cited the interesting observation that NO synthase, a highly regulated enzyme, is a major substrate for all three major serine-threonine kinases (Cam KII, PKC, and PKA) and has, in addition, putative phosphorylation sites for tyrosine kinase. Again, we have the neurovascular and immune response "connections" with CNS memory (LTP) and the "memory molecules" bradykinin and substance P and others that produce EDRF/NO, act as growth factors, etc.

C. MAMMALIAN MEMORY AND SEIZURE KINDLING

Like LTP, epileptiform seizure "kindling" involves excitatory stimulation of neuronal pathways, repeated stimulations, and kindling by electrical and/or chemical stimulation. The process has been shown to involve protein synthesis, and at a point its memory becomes permanent. In man its anatomical site of activation has included the hippocampus, and to come full circle, surgical attempts to control the most severe forms of epilepsy led to bilateral resection of the hippocampus and the resulting severe memory deficits that, in turn, were a stimulation to the discovery of LTP and the profound fascination neuroscientists have found with the remarkable phenomenon of LTP and the secrets of memory it tantalizingly holds — *and let us add immunological "memory" to the exploration of the molecular basis and underpinnings of these likely related memory phenomena.*

D. INVERTEBRATE MEMORY

Marine molluscs have been used extensively, and quite effectively and productively in the study of memory-like responses and the involvement of vasoactive neurotransmitter "memory molecules", such as serotonin, in these responses. The marine gastropod mollusc *Aplysia californica* has been studied in considerable depth by the group of Kandel of Columbia University. The group of Alkon at the N.I.H. (Bethesda, MD) has carried out equally intense and important explorations of memory-type responses in the nudibranch gastropod mollusc *Hermissenda crassicornis*. These highly significant and revealing studies and findings are considered in detail in Chapter 5 ("Serotonin"). It would be of considerable interest to know how other vasoactive neurotransmitter molecules of a stimulatory nature would function in these established molluscan gastropod responses as well as in other invertebrate and vertebrate memory study models.

E. MACROPHAGES AND VASCULAR ENDOTHELIAL CELLS GENERATE MEMORY-TYPE MOLECULES

A 1990 report in *Science,* by Sokolet et al.[78] from Harvard, described a protein — secreted by a mouse macrophage cell line — that changed the developmental fate of *Xenopus* embryonic cells. Exposure to this macrophage-generated protein caused presumed ectodermal explants to form anterior neural and mesodermal tissues, including the brain and eye, instead of ciliated epidermis. These results indicated that a mouse-macrophage-generated protein can cause the formation of central nervous system tissues, including the brain and eye in the frog, and that it may also cause organization of the *Xenopus* African clawed-frog body plan: truly a most remarkable finding and demonstration of a cross-species immune response cell and nervous system ''connection''.

In the November 29, 1991 issue of *Science,*[79] there was a new report of recent and in-press studies suggesting a role for EDRF/nitric oxide/NO and its forming enzyme NO synthase in the establishment of LTP. As different NO synthase molecules from neuronal cells, endothelial cells, and macrophages may all make NO, the ''plot'' further thickens.

As cited in the ''EDRF/Nitric Oxide/NO'' section of Chapter 17, both the macrophage-derived NO synthase and the neuronal cell NO synthase have been cloned and sequenced and have both homologies and differences. In addition, antibodies to the two molecules do not cross-react. As the November 1991 *Science* article indicated, the neuronal cell NO synthase cloned by the Snyder group does not appear in LTP-associated hippocampal pyramidal cells, it would be of particular interest if the macrophage-generated NO synthase occurs in these cells and plays a role in LTP. As substance P is a strong activator of NO synthase in vascular endothelial cells and likely also in macrophages and as both enzymes may appear in some endothelial cells, these data may provide a mechanism for the action of substance P in both CNS memory and in immunological memory and may lend support, at the molecular level, for a molecular commonality of substance P-induced memory in both the immune response and CNS memory mechanisms.

REFERENCES

1. **Kavaliers, M., Hirst, M., and Teskey, G. C.,** A functional role for an opiate system in snail thermal behavior, *Science,* 220, 99, 1983.
2. **Kluger, M. J., Ringler, D. H., and Anver, M. R.,** Fever and survival, *Science,* 188, 166, 1975.
3. **Covert, J. B. and Reynolds, W. W.,** Survival value of fever in fish, *Nature,* 267, 43, 1977.
4. **Nahas, G. G., Tannieres, M. L., and Lennon, J. F.,** Direct measurement of leukocyte motility: effects of pH and temperature (35894), *Proc. Soc. Exp. Biol. Med.,* 138, 350, 1971.
5. **Duff, G. W. and Durum, S. K.,** Fever and immunoregulation: hyperthermia, interleukins 1 and 2, and T-cell proliferation, *Yale J. Biol. Med.,* 55, 437, 1982.
6. **Yerushalmi, A., Tovey, M. G., and Gresser, I.,** Antitumor effect of combined interferon and hyperthermia in mice (41367), *Proc. Soc. Exp. Biol. Med.,* 169, 413, 1982.
7. **Jampel, H. D., Duff, G. W., Gershon, R. K., Atkins, E., and Durum, S. K.,** Fever and immunoregulation. II. Hyperthermia augments the primary *in vitro* humoral immune response, *J. Exp. Med.,* 157, 1229, 1983.
8. **Gupta, S. and Agarwal, S. S.,** Effect of hyperthermia on *in vitro* proliferative response of lymphocyte subpopulations, *Fed. Proc.,* 39, 916, 1980.
9. **Dinarello, C. A., Bernheim, H. A., Duff, G. W., Le, H. V., Nagabhushan, T. L., Hamilton, N. C., and Coceani, F.,** Mechanisms of fever induced by recombinant human interferon, *J. Clin. Invest.,* 74, 906, 1984.
10. **Hayaishi, O.,** Sleep-wake regulation by prostaglandins D_2 and E_2, *J. Biol. Chem.,* 263, 14593, 1988.
11. **Heron, I. and Berg, K.,** The actions of interferon are potentiated at elevated temperature, *Nature,* 274, 508, 1978.

12. **Ritossa, F.**, A new puffing pattern induced by temperature shock and DNP in *Drosophila, Experientia,* 18, 571, 1962.
13. **Polla, B. S.**, A role for heat shock proteins in inflammation? *Immunol. Today,* 9, 134, 1988.
14. **Koller, M. and Konig, W.**, Arachidonic acid metabolism in heat-shock treated human leucocytes, *Immunology,* 70, 458, 1990.
15. **Stanley, S. K., Bressler, P. B., Poli, G., and Fauci, A. S.**, Heat shock induction of HIV production from chronically infected promonocytic and T cell lines, *J. Immunol.,* 145, 1120, 1990.
16. **Ferris, D. K., Harel-Bellan, A., Morimoto, R. I., Welch, W. J., and Farrar, W. L.**, Mitogen and lymphokine stimulation of heat shock proteins in T lymphocytes, *Proc. Natl. Acad. Sci. U.S.A.,* 85, 3850, 1988.
17. **Spector, N. L., Freedman, A. S., Freeman, G., Segil, J., Whitman, J. F., Welch, W. J., and Nadler, L. M.**, Activation primes human B lymphocytes to respond to heat shock, *J. Exp. Med.,* 170, 1763, 1989.
18. **Perl, E. R., Kumazawa, T., Lynn, B., and Kenins, P.**, Sensitization of high threshold receptors with unmyelinated (C) afferent fibers, *Prog. Brain Res.,* 43, 263, 1976.
19. **Adrian, E. D. and Zotterman, Y.**, The impulses produced by sensory nerve-endings. II. The response of a single end-organ, *J. Physiol. (Lond.),* 61, 151, 1926.
20. **Adrian, E. D. and Zotterman, Y.**, The impulses produced by sensory nerve endings. III. Impulses set up by touch and pressure, *J. Physiol. (Lond.),* 61, 465, 1926.
21. **Moore, T. C., Nur, K., Yunes, K., and Storm, F. K.**, Peripheral node lymphocyte traffic effects of local magnetic-loop induction hyperthermia of the sheep brain, *IRCS Med. Sci.,* 13, 1194, 1985.
22. **Moore, T. C., Nur, K., and Storm, F. K.**, Localized deep hyperthermia increases the traffic of lymphocytes through peripheral lymph nodes of sheep *in vivo, J. Surg. Oncol.,* 35, 75, 1987.
23. **Moore, T. C., Suede, M., and Lami, J. L.**, Electrocautery dissection modulates anesthesia induced depression of lymphocyte traffic in sheep: a saga of heat, pain and memory, *Immunology,* submitted for publication.
24. **Storm, F. K., Harrison, W. H., Elliott, R. S., Silberman, A. W., and Morton, D. L.**, Thermal distribution of magnetic-loop induction hyperthermia in phantoms and animals: effect of the living state and velocity of heating, *Int. J. Radiat. Oncol. Biol. Phys.,* 8, 865, 1982.
25. **Elliott, R. S., Harrison, W. H., and Storm, F. K.**, Hyperthermia: electromagnetic heating of deep-seated tumors, *IEEE Trans. EME,* 29, 61, 1982.
26. **Storm, F. K., Harrison, W. H., Elliott, R. S., and Morton, D. L.**, Clinical RF hyperthermia by magnetic-loop induction: a new approach to cancer therapy, *IEEE Trans. MTT,* 30, 1149, 1982.
27. **Moore, T. C.**, Cannulated afferent lymphatic infusion studies indicate output of lymphocytes into efferent lymph independent of fluid volume input, *IRCS Med. Sci.,* 9, 585, 1981.
28. **Fugiwara, R. and Orita, K.**, Enhancement of the immune response by pain stimulation in mice, *J. Immunol.,* 138, 3699, 1987.
29. **Cantacuzene, J.**, Novelles recherches sur le mode de destruction des vibrions dans l'organisme, *Ann. Inst. Pasteur (Paris),* 12, 273, 1898.
30. **Snel, J. J.**, Immunitat und narkose, *Klin. Wochenschr.,* 40, 212, 1903.
31. **Rubin, G.**, The influence of alcohol, ether and chloroform on natural immunity in its relation to leukocytosis and phagocytosis, *J. Infect. Dis.,* 1, 424, 1904.
32. **Graham, E. A.**, Influence of ether and ether anesthesia on bacteriolysis, agglutination and phagocytosis, *J. Infect. Dis.,* 8, 147, 1911.
33. **Gaylord, H. R. and Simpson, B. T.**, The effect of certain anesthetics and loss of blood upon the growth of transplanted mouse cancer, *Cancer Res.,* 1, 379, 1916.
34. **Dale, H. H. and Laidlaw, P. P.**, Histamine shock, *J. Physiol. (Lond.),* 52, 355, 1919.
35. **Espanol, T., Todd, G. B., and Soothill, J. E.**, The effect of anesthesia on the lymphocyte response to phytohemagglutinin, *Clin. Exp. Immunol.,* 18, 73, 1974.
36. **Griffiths, J. D., Hoppe, E., and Cole, W. H.**, The influence of thermal stress and changes in body temperature on the development of carcinoma 246 Walker in rats, *Cancer,* 14, 111, 1961.
37. **Agostino, D. and Cliffton, E. E.**, Anesthetic effect on pulmonary metastases in rats, *Arch. Surg.,* 88, 735, 1964.
38. **Lundy, J., Lovett, E. J., and Conran, P.**, Pulmonary metastases: a potential biological consequence of anesthetic-induced immunosuppression by thiopental, *Surgery,* 82, 254, 1977.
39. **Gordon-Taylor, G.** On cancer of the breast, *Ann. R. Coll. Surg.,* 2, 60, 1948.
40. **Bruce, D. L., Eide, K. A., Linde, H. W., and Eckenhoff, J. E.**, Causes of death among anesthesiologists: a 20-year survey, *Anesthesiology,* 29, 565, 1968.
41. **Hall, J. G. and Morris, B.**, The output of cells in lymph from the popliteal nodes of sheep, *Q. J. Exp. Physiol.,* 47, 360, 1962.
42. **Moore, T. C.**, The effects of inflammatory mediators and anaesthetic agents upon lymphocyte traffic of sheep *in vivo,* Ph.D. thesis, University of Cambridge (Trinity College), England, 1982.

43. **Moore, T. C. and Pearson, J. D.,** Effects of general anaesthesia on efferent lymph PGE_2 levels and modulation by drainage area surgery, *IRCS Med. Sci.,* 12, 269, 1984.
44. **Moore, T. C.,** Anesthesia associated depression in lymphocyte traffic and its modulation, *Am. J. Surg.,* 147, 807, 1984.
45. **Moore, T. C.,** Anesthesia associated depression in lymphocyte traffic less with regional than with general anesthesia in sheep, *Am. J. Surg.,* 148, 71, 1984.
46. **Moore, T. C. and Khan, F.,** Lymphocyte traffic effects of systemic and local lidocaine in sheep, *J. Surg. Res.,* 41, 279, 1986.
47. **Moore, T. C., Spruck, C. H., and LeDuc, L. E.,** Depression of lymphocyte traffic in sheep by anesthesia and associated changes in efferent lymph PGE_2 and antibody levels, *Immunology,* 63, 139, 1988.
48. **Spruck. C. H. and Moore, T. C.,** Anesthesia associate depression of peripheral node lymphocyte traffic and antibody production in sheep accompanied by elevations in arachidonic acid metabolites in efferent lymph, *Transplant. Proc.,* 20, 1169, 1988.
49. **Moore, T. C., Suede, M., and Lami, J. L.,** Electrocautery dissection modulates anesthesia induced depression of lymphocyte traffic in sheep: a saga of heat, pain and memory, *Immunology,* submitted for publication.
50. **Iversen, L. L.,** *General Discussion, Coexistence of Transmitter, Substance P in the Nervous System,* Ciba Foundation Symposium No. 91, Pitman, London, 1982, 124.
51. **Lomo, T.,** Frequency potentiation of excitatory synaptic activity in the dentate area of the hippocampal formation, *Acta Physiol. Scand.,* 68(Suppl. 277), 128, 1966.
52. **Bliss, T. V. P. and Lomo, T.,** Long-lasting potentiation of synaptic transmission in the dentate area of the anaesthetized rabbit following stimulation of the perforant path, *J. Physiol. (Lond.),* 232, 331, 1973.
53. **Bliss, T. V. P., Gardner-Medwin, A. R., and Lomo, T.,** Synaptic plasticity in the hippocampal formation, in *Macromolecules and Behaviour,* Ansell, G. B. and Bradley, P. B., Eds., MacMillan, London, 1973, 193.
54. **Bliss, T. V. P. and Lynch, M. A.,** Long-term potentiation of synaptic transmission in the hippocampus: properties and mechanisms, in *Long-Term Potentiation: From Biophysics to Behavior,* Landfield, P. W. and Deadwyler, S. A., Eds., Alan R. Liss, New York, 1988, 3.
55. **Scoville, W. B. and Milner, B.,** Loss of recent memory after bilateral hippocampal lesions, *J. Neurol. Neurosurg. Psychiatry,* 20, 11, 1957.
56. **Milner, B.,** Amnesia following operation on the temporal lobes, in *Amnesia,* Whitty, C. W. M. and Zanglwill, O. L., Eds., Butterworths, London, 1966, 109.
57. **Milner, B.,** Disorders of learning and memory after temporal lobe lesions in man, *Clin. Neurosurg.,* 19, 421, 1972.
58. **Malenka, R. C., Madison, D. V., and Nicoll, R. A.,** Potentiation of synaptic transmission in the hippocampus by phorbol esters, *Nature,* 321, 175, 1986.
59. **Malenka, R. C., Ayoub, G. S., and Nicoll, R. A.,** Phorbol esters enhance transmitter release in rat hippocampal slices, *Brain Res.,* 403, 198, 1987.
60. **Buzsaki, G.,** Polysynaptic long-term potentiation: a physiological role of the perforant path-CA3/CA1 pyramidal cell synapse, *Brain Res.,* 455, 192, 1988.
61. **Malinow, R., Schulman, H., and Tsien, R. W.,** Inhibition of postsynaptic PKC or CaMKII blocks induction but not expression of LTP, *Science,* 245, 862, 1989.
62. **Williams, J. H., Errington, M. L., Lynch, M. A., and Bliss, T. V. P.,** Arachidonic acid induces a long-term activity-dependent enhancement of synaptic transmission in the hippocampus, *Nature,* 341, 739, 1989.
63. **Linden, D. J., Sheu, F.-S., Murakami, K., and Routenberg, A.,** Enhancement of long-term potentiation by cis-unsaturated fatty acid: relation to protein kinase C and phospholipase A_2, *J. Neurosci.,* 7, 3783, 1987.
64. **Garthwaite, J., Charles, S. L., and Chess-Williams, R.,** Endothelium-derived relaxing factor release of activation of NMDA receptors suggests role as intercellular messenger in brain, *Nature,* 336, 385, 1988.
65. **Palmer, R. M. J., Ferrigo, A. G., and Moncada, S.,** Nitric oxide release accounts for the biological activity of endothelium-derived relaxing factor, *Nature,* 327, 524, 1987.
66. **Bekkers, J. M. and Stevens, C. F.,** Presynaptic mechanism for long-term potentiation in the hippocampus, *Nature,* 346, 724, 1990.
67. **Malinow, R. and Tsien, R. W.,** Presynaptic enhancement shown by whole-cell recordings of long-term potentiation in hippocampal slices, *Nature,* 346, 177, 1990.
68. **Bliss, T. V. P.,** Maintenance is presynaptic, *Nature,* 346, 698, 1990.
69. **Goh, J. W. and Pennefather, P. S.,** Pertussis toxin prevents induction of long-term potentiation in the stratum radiatum and stratum oriens inputs to CA1 neurons, *Brain Res.,* 511, 345, 1990.
70. **Meshul, C. K. and Hopkins, W. F.,** Presynaptic ultrastructural correlates of long-term potentiation in the CA1 subfield of the hippocampus, *Brain Res.,* 514, 310, 1990.

71. **Ishihara, K., Katsuki, H., Sugimura, M., Kaneko, S., and Satoh, M.,** Different drug-susceptibilities of long-term potentiation in three input systems to the CA3 region of the guinea pig hippocampus *in vitro, Neuropharmacology,* 29, 487, 1990.
72. **Zalutsky, R. A. and Nicoll, R. A.,** Comparison of two forms of long-term potentiation in single hippocampal neurons, *Science,* 248, 1619, 1990.
73. **Alzheimer, C., Rohrenbeck, J., and ten Bruggencate, G.,** Adenosine depresses induction of LTP at the mossy fiber-CA3 synapse *in vitro, Brain Res.,* 543, 163, 1991.
74. **Aniksztejn, L. and Ben-Ari, Y.,** Novel form of long-term potentiation produced by a K^+ channel blocker in the hippocampus, *Nature,* 349, 67, 1991.
75. **Larkman, A., Stratford, K., and Jack, J.,** Quantal analysis of synaptic action and depression in hippocampal slices, *Nature,* 350, 344, 1991.
76. **Edwards, F.,** LTP is a long-term problem, *Nature,* 350, 271, 1991.
77. **O'Dell, T. J., Kandel, E. R., and Grant, S. G. N.,** Long-term potentiation in the hippocampus blocked by tyrosine kinase inhibitors, *Nature,* 353, 558, 1991.
78. **Sokol, S., Wong, G. G., and Melton, D. A.,** A mouse macrophage factor induces head structures and organizes a body axis in *Xenopus, Science,* 249, 562, 1990.
79. **Barinaga, M.,** Is nitric oxide the retrograde messenger? *Science,* 254, 1296, 1991.

Chapter 20

FUTURE DIRECTIONS

I. INTRODUCTION

Primary objectives for future studies should include the establishment, with greater clarity and precision, of the interrelationships of the nervous, vascular, and immune response systems in the *in vivo* responses to various types of immunological challenges, including peripheral, gut, and tracheo-bronchial immunizations and bone marrow, skin, neuronal, and organ allografting. While *in vitro* studies may provide data of some interest, the absence of both nervous system and vascular system connections generate a certain level of ''unrealness'' in responses and present a troublesome component of artifactuality. In addition, responses to vasoactive neurotransmitter molecules involving *in vitro* stimulation have produced difficulty in the evaluation of responses and in confusing and inconclusive reactions.

Underinvestigated molecules, especially those related to learning and memory, should receive greater attention. These include the excitatory amino acids (glutamate and aspartate) and their receptor agonists and antagonists, agonists and antagonists for stimulatory ''inhibitory'' amino acid $GABA_B$ and strychnine-insensitive glycine receptors, substance P, bradykinin, serotonin, met-enkephalin, CCK_8, bombesin, and CGRP, as well as many others yet to be identified and characterized. Molecular and receptor plasticity and involvement with transmembrane signaling, with cascade and with ancillary molecules, should prove to be fruitful areas of investigation.

Identification of anatomic levels, gross and microscopic (including ultrastructural), of chemical, molecular and structural changes in the peripheral and central nervous systems in response to various immune system challenges and the anatomical nerve tracts and intercommunications and projections involved and stimulated should prove to be most useful and revealing.

Alterations in the substantia gelatinosa of lamina I of the dorsal horn of the spinal cord and in the CA3 and CA1 levels of the hippocampus following peripheral tissue immune and inflammatory challenges and allografting would appear to be appropriate early investigational starting points.

Priority *in vitro* studies include continuations of important studies of receptor types, subtypes, and molecular characterization as well as their anatomical distribution and interconnections.

During the past 20 years, the author has published theoretical articles relating to the topic of this book.[1-6] These articles also included suggestions for future studies (as of the time of publication) and some of these have already been carried out.

II. TARGETED RESEARCH FOR CLINICAL OBJECTIVES

A. CANCER CONTROL/CURE

The remarkable effect of repeated intralesional injections of bradykinin (observed by the author, Koppelman, and associates) in slowing and halting the growth of malignant tumors and in elevating depressed tumor and tumor-host lymphoid-tissue depleted levels of histidine decarboxylase (the histamine-forming enzyme) as well as enhancing the antitumor cytolytic activities of tumor-host spleen cells rather urgently needs revisiting. This is particularly true with respect to comparatively recent confirmation of these findings by Japanese investigators (cited in Chapter 9) and the important observations of Professor Robert Fauve of Paris (at the Pasteur Institute) regarding the production of an anti-bradykinin substance by malignant tumors.

In a 1979 theoretical article,[5] the author suggested the use of vasoactive neurotransmitter molecules (such as bradykinin) which also are mediators of inflammation to stimulate the afferent and efferent arcs of the immune response. In tumor-bearing large animals, such as man, cannulation of vascular access sites such as lymphatics and arteries would be useful in delivering stimulated immune response cells and substances to local and discrete targets whose vascularity, permeability, and local inflammation had been enhanced by intralesional injections of bradykinin.

In the 13 years since these suggestions, major advances in arterial catheter placement and CT-scan-directed needle placement have increased the potentials and safety of this approach. As spontaneous malignant tumors do not occur in other large animals with sufficient frequency or in anatomical locations appropriate for experimental use, the limited exploration of this approach in man is worthy of careful consideration and study — and perhaps use.

The demonstrated value of substance P infusions into cannulated afferent lymphatics in targeted increasing of the output of CD4-T lymphocytes and the central role of this subset of lymphocytes in the lymphoid immune response suggests the potential usefulness of this molecule as an immunostimulant when immunosuppression and immunodepression are a part of the malignant tumor process.

B. AIDS CONTROL/CURE

The assault of the HIV/AIDS virus on the T-4 (CD4) lymphocyte, with its selective depression, is a major pathologic effect of the HIV virus and its gp120 envelope protein. The opposite effect of substance P, with a specific T-4 traffic-enhancing and altered immune response to immunization, and a putative memory effect in both the immune response system and the nervous system suggest an exploration of a potential value of substance P in clinical AIDS and in appropriate experimental models *in vivo* and *in vitro*.

VIP (vasoactive intestinal peptide) is another of these vasoactive neurotransmitter molecules that has been suggested to be involved in HIV T-4 cell entry by way of a gp120 attachment to and the use of a VIP receptor on T-4 lymphocytes. Core octapeptides and pentapeptides from VIP and the HIV-gp120 have been found to have interesting responses which need further study and evaluation.

C. TISSUE- AND ORGAN-SPECIFIC LOCAL MOLECULAR IMMUNOSUPPRESSION

Molecules with immunomodulatory and immunosuppressive capabilities such as VIP, prostaglandins (E2 and D2), neurotensin, and histamine at the H2 receptor, alone or in combination with other molecules by intra-arterial and by CT-guided local application, might offer promise in achieving local rather than global immunosuppression and thereby diminish the hazards and life-compromising risk of total-body overimmunosuppression. PGD_2 would appear to be a prime molecule to explore in this area.

D. CONTROL OF ANESTHESIA-INDUCED IMMUNOSUPPRESSION

Anesthesia-induced immunosuppression was considered in some detail in the ‘‘Heat, Pain, and Memory’’ chapter (Chapter 19). This control would be especially important in cancer surgery and surgery in circumstances where systemic or local sepsis were major problems. The demonstration of both immunosuppression and marked elevations in study-lymph-node efferent lymph levels of prostaglandin E_2 (PGE_2) suggests attempts to control the elevated PGE_2 (with persistent pain control), but with no immunosuppression in lymphocyte traffic depression or in antibody production in the supported-sheep lymphocyte traffic model is worthy of further exploration. To dissociate these two responses (hopefully) would be of considerable importance. A search for other possible anesthesia-induced immunosuppression molecules in the sheep and other systems should also be given high priority from an urgent clinical standpoint.

III. BASIC RESEARCH

A. EXCITATORY AND "INHIBITORY" AMINO ACIDS

An exploration of potential stimulatory and excitatory short- and long-term effects on the immune response system and cells, *in vitro* and *in vivo,* of excitatory amino acids and "inhibitory" amino acids which, via their $GABA_B$-receptor agonists and their strychnine-insensitive glycine-receptor agonists have been found to produce long-term potentiation (LTP) in the hippocampal area of the central nervous system, would be of particular interest. Electrophysiological and other studies would be of particular value here, as would antibody probe studies of dorsal-horn substantia gelatinosa, hippocampus, and other sensory transmission areas accompanying peripheral immunization and immunostimulation with allografted cells, tissues and, organs. Responses of immune response cells to these molecules also would be worthy of study, *in vivo* and *in vitro.*

B. AMINES

Dopamine, a mystery molecule of sorts, with immense importance in the central nervous system, is somewhat underinvestigated with respect to the immune response system. Its effects on individual cells and tissues of the lymphoid immune response system, including lymphocyte types and subtypes, macrophages, mast cells, and platelets, are worthy of exploration. The responses to its central venous administrations in the supported-sheep lymphocyte traffic model have been discussed in the chapter on dopamine (Chapter 6). A good starting point in the sheep model, at this point, would be to administer it by acute infusion into a cannulated afferent lymphatic of a popliteal lymph node in order to evaluate its effects on traffic and related events including antibody production.

Further study of both serotonin and histamine in the supported-sheep lymphocyte traffic model, including investigation of agonists and antagonists of the various identified receptors, would be of considerable interest as would studies of the effects of peripheral immunostimulation by immunization and allografting on central levels of serotoninergic, histaminergic, and dopaminergic nervous systems of the spinal cord, brain stem, basal ganglia, hippocampus, and cerebral cortex.

C. IMMUNO- AND NEUROPEPTIDES

Studies similar to those suggested for the amino acids and amines would be of value for the peptides, especially the "memory molecule" peptides such as met-enkephalin, CCK-8, bradykinin, substance P, bombesin, and CGRP. Immunohistochemical studies, including electron microscopic studies, would be of value concerning the effects of peripheral lymph-drainage-area immunizations on immunoreactivity for substance P and other "memory molecules" in regional draining lymph nodes and in sensory axons leading from these lymph nodes, as well as in dorsal roots and dorsal horns and in major ascending sensory tracts and their ultimate distribution sites including the basal ganglia, hippocampus, and cerebral cortex.

REFERENCES

1. **Moore, T. C.,** A theory of the role of histamine metabolism in transplant rejection, *Surg. Gynecol. Obstet.*, 132, 489, 1971.
2. **Moore, T. C.,** Vascular component of transplant rejection, *Bull. Soc. Int. de Chir.*, 30, 322, 1971.
3. **Moore, T. C.,** Role of vasoactive substances in afferent and efferent arcs of the immune response, *IRCS Med. Sci.*, 2, 46, 1974.
4. **Moore, T. C.,** Role of vasoactive substances in the rejection of tumor and transplant tissue, *Bull. Soc. Int. de Chir.*, 24, 467, 1975.
5. **Moore, T. C.,** A role for mediators of inflammation in the cellular immune response to tumor and transplant tissue, *IRCS Med. Sci.*, 7, 273, 1979.
6. **Moore, T. C.,** Neurovascular immunology: a 25 year odyssey, *Ann. N.Y. Acad. Sci.*, 650, 30, 1992.

Chapter 21

CONCLUSION

Neurovascular immunology involves the study and understanding of the defense-survival-propagation, learning, and memory components of the vascular, nervous, and immune response systems of plants, invertebrates, and lower and higher vertebrates, including man, and of their basic and shared molecules (amino acids, amines, and short-chain peptides and lipids).

The vascular system was the first to appear in the evolutionary development of multicellular living organisms (vascular plants). The assumption of geographic motility by multicellular animals necessitated the addition of a nervous system with appropriate mediators and modulators. The lymphoid immune response system of vertebrates engrafted itself upon the more "primitive" existing vascular and nervous systems, yet retained the shared effector and modulator molecules of antiquity.

The siren song of the "heat, pain, and memory" scenario is most compelling and most intriguing. The author proposes, as the Prometheus Thesis of Moore, that in the highways and byways and in the hinterlands and molecular underbush and underpinnings of this scenario, the molecular basis for immunological memory and nervous system memory ultimately will be found to be homologous, if not identical.

> "The continual interaction of lymphocytes, mononuclear phagocytes and endothelial cells is a theme that runs without interruption through every phase of cellular immunology, and it is in the hinterland between the biology of these cell types and the physiology of the microcirculation that many answers will be found."
>
> J. G. Hall, 1974

> "Immune reactions are the result of interacting systems, not just interacting cells."
>
> Cahill, Hay, Frost, and Trnka, 1974

> "The universities do not teach all things, so a doctor must seek out old wives, gypsies, sorcerers, wandering tribes, old robbers and such outlaws, and take lessons from them. A doctor must be a traveller, — knowledge is experience — he who is born in imagination discovers the latent forces of Nature."
>
> Paracelsus (1493–1541)
> Renaissance surgeon and chemist
> First to recommend ether for surgical anesthesia
> Established the role of chemistry in medicine

Appendix A

SUPPORTED-SHEEP LYMPHOCYTE TRAFFIC MODEL

It is virtually impossible to obtain long-term studies of frequent (every 10 to 15 min) determinations of efferent lymph output from cannulated popliteal efferent lymphatics if the sheep is permitted to roam and sit and stand at will (Figure 1). It was determined earlier that lymphocyte output "shutdown" often correlated with sheep sitdown regardless of any experiment hoped to be underway. It also was found to be virtually impossible to maintain chronic cannulation of small popliteal afferent lymphatics for substance infusion if the sheep did not have some limitations on movements and activities.

The supported-animal approach to delicate *in vivo* physiological studies was pioneered many years ago by Pavlov in his classical studies of gastric secretion and conditioned reflexes. For sheep, metabolic cages are quite important for lymphocyte traffic studies, and their framework may provide useful areas to establish girth sling supports and attachments (Figure 2). These components of the metabolic cages also provide secure points for the attachment of a digital syringe pump (or pumps if bilateral studies are to be undertaken). The girth supports are remarkably well tolerated by the sheep. When they wish a change of pace, they simply relax and use the girth supports as hammock-type rests.

Lacking metabolic cages and commercially available custom-made girth supports, it is still possible to employ this methodology with standard sheep cages and the conversion of cloth laundry bags into girth supports (Figure 3). Still, nothing beats the luxury and convenience of proper equipment (Figures 4 and 5).

While the girth supports are absolutely essential for long-term studies of lymphocyte traffic in the sheep with frequent (10 to 15 min or shorter) sample collections, the anesthesia methodology and microsurgical techniques for lymphatic cannulations (generally popliteal) and the securing of these lymphatic cannulas are also of great importance to the carrying out of successful and meaningful experiments.

The combination of ketamine and xylazine provides the best and safest general anesthesia for lymphatic cannulations. With this combination, endotracheal intubation (and associated deeper anesthesia and respiratory paralysis) is not required, and the sheep may safely be kept on its side for the best and most appropriate operative exposure of the major peripheral lymph nodes to be explored for lymphatic (efferent and afferent) cannulation. The principal peripheral nodes used for study are the popliteal, the prefemoral, and the prescapular. For chronic use and cannulation of both efferent and afferent lymphatics, the popliteal lymph node is by far the best one to use.

Standard microsurgical instrumentation is essential for successful cannulation of both efferent and afferent popliteal lymph node lymphatics and for the others as well. Most of these instruments can be obtained from eye surgery instrument suppliers and some from the makers of instruments for heart coronary artery bypass surgery. The Dietrick coronary artery bulldog atramatic clamp (by Codman and Shurtleff, Boston, MA) is of particular use in popliteal lymph node afferent lymphatic cannulations.

State-of-the-art electrocautery equipment is essential for careful microsurgical operation in a blood-free field. We have found Force One Valleylab electrocautery equipment to be quite satisfactory. This is made by Valley Lab Inc. (Boulder, CO) and is in routine use in our hospital operating rooms for human surgery and very popular with our housestaff and faculty. The Force One electrocautery unit is a microbipolar unit with cautery, dissection, and blend capability.

Also of great importance is adequate and optimal lighting. We have employed optical magnification (2.5×) with prescription fashioned glasses and intense fiberoptic lighting

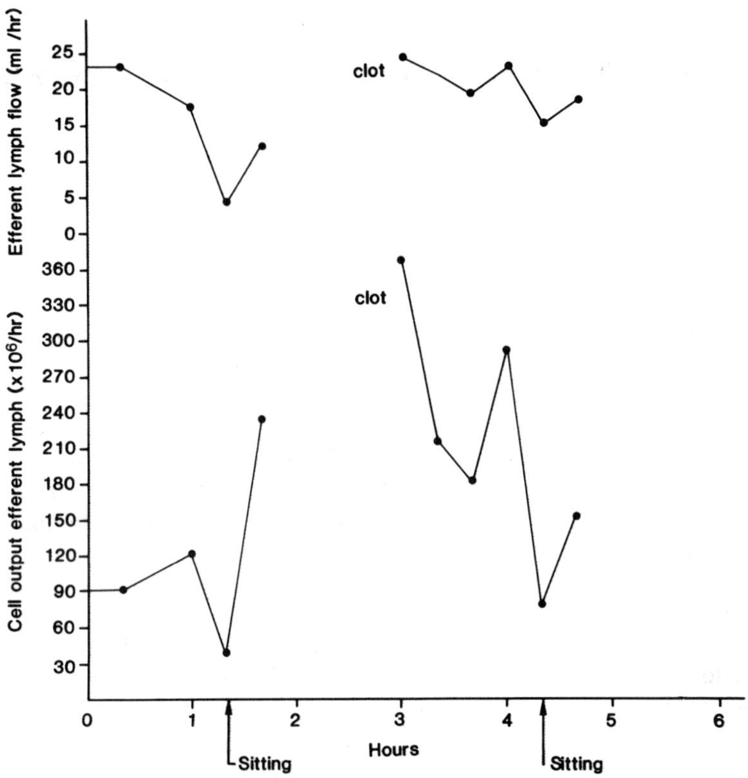

FIGURE 1. Output of small recirculating lymphocytes into the efferent lymphatic of a sheep popliteal lymph node (chronically cannulated) of an unsupported sheep that was permitted to sit or stand at will during sequential efferent lymph collections and study. The output is too erratic and variable in this "control"/no-agent-use situation to be of any short-term study value. Clearly here "sit down" equals "shut down".

(Design for Vision Inc., Ronkonkoma, NY) and are quite pleased with the excellent lighting of both deep and superficial tissues. This lighting and magnification are used clinically by our heart coronary artery bypass faculty surgeons.

Identification of lymphatics, both efferent and afferent, is facilitated greatly by the use of subcutaneous lymphatic-seeking colored-dye injection into the study-lymph-node drainage area. For the most frequently used popliteal lymph node, the drainage-area site for dye injection is into the lower lateral aspect of the leg, into the lateral aspect of the cannon between the fetlock and the hock. For lymphatic identification by blue coloration, we have employed Patent Bleu V (Laboratorie Guerbet, Aulnay-Sous-Bois, France) given subcutaneously immediately after anesthesia induction and just prior to incisions for lymphatic cannulations.

We have employed silastic cannulas (2.8 or 3.6 French, Dow Corning, Midland, MI) for popliteal efferent lymphatic cannulation. The efferent lymphatic is ligated proximal to the intended cannulation site to prevent retrograde lymph leak and to distend the lymphatic at the incision site to facilitate cannulation. In order to permit the insertion of an adequate length of end-tapered silastic cannula, a silver 0000 lacrimal duct probe (Dixie Optical, London, U.K.) is inserted distally into the lymphatic to break (with extreme care) three or four lymphatic valves. The silastic tube in the cannulated efferent lymphatic is secured by two nonoccluding 6-0 silk ligatures to prevent dislodgement and lymp leak as well as lymph-sample loss and inaccuracy of collection volume determination. Popliteal lymph nodes of

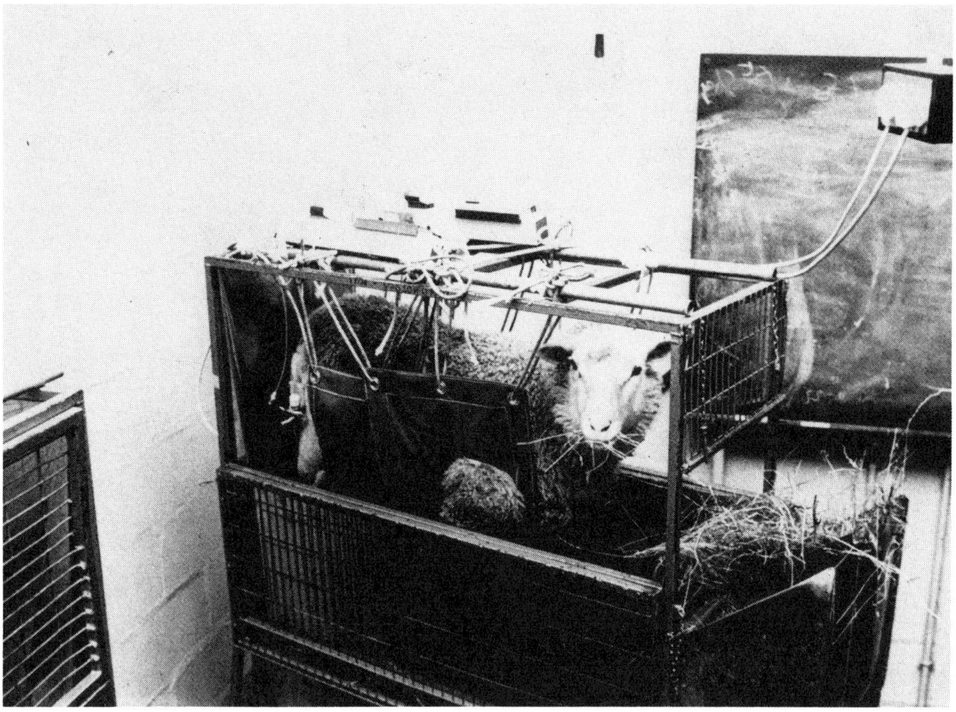

FIGURE 2. Photograph of a supported sheep in a metabolic cage during a bilateral popliteal lymph node afferent lymphatic infusion study with both popliteal lymph-node efferent lymphatics also cannulated at the University of Cambridge in England during the author's Ph.D. thesis research. The girth supports were commercially prepared for the university research laboratories in Cambridge on the author's specification. Clothesline is used to "fine tune" the girth supports. Twin electrically driven digital syringe pumps (Vickers Medical Treonic IP3) can be seen secured to the top of the metabolic cage. The relaxed, meal-munching, and unsedated sheep happily enjoys a bit of hay during an important and nondisturbing afferent lymphatic infusion study. Care must be taken to keep the electrical cords out of munching range, as one sheep in Cambridge was electrocuted while munching on the digital syringe electrical cord of the sheep in the metabolic cage in front.

sheep are discrete, single lymph nodes and are buried quite deeply in the high. They weigh approximately 1 g and generally have a single efferent lymphatic. When more than one efferent lymphatic is encountered on careful dissection, the larger one is cannulated, and the smaller one (or ones) is (are) ligated in continuity. The silastic cannula is brought to the outside through a tiny skin stab wound and secured to the skin with 6-0 silk suture ligatures. The end of the silastic tube is placed into a heparin-containing plastic test tube or a larger plastic bottle (for long or overnight collections). The collection tubes or bottles are secured by umbilical tape attached to a plastic shield that is secured to the animal's rump with multiple (usually four) heavy ligatures at the shield quadrants.

Sheep popliteal lymph node afferent lymphatics are multiple and quite small. Their cannulation is difficult and quite demanding. The first cannulation attempt is directed at the largest of the small blue-containing afferent lymphatics. Afferent lymphatic distention by a delicate Dietrick human coronary artery clamp was found to be of considerable assistance in identifying the largest afferent lymphatic and in its cannulation. If the first attempt at afferent lymphatic cannulation fails, the operating surgeon moves on to the next largest afferent lymphatic in appearance, and so on. Proximal and distal ligatures with 6-0 silk are employed at all failed afferent lymphatic cannulation sites, with distal ligatures only at successful cannulation sites. Successfully cannulated afferent lymphatics are secured with

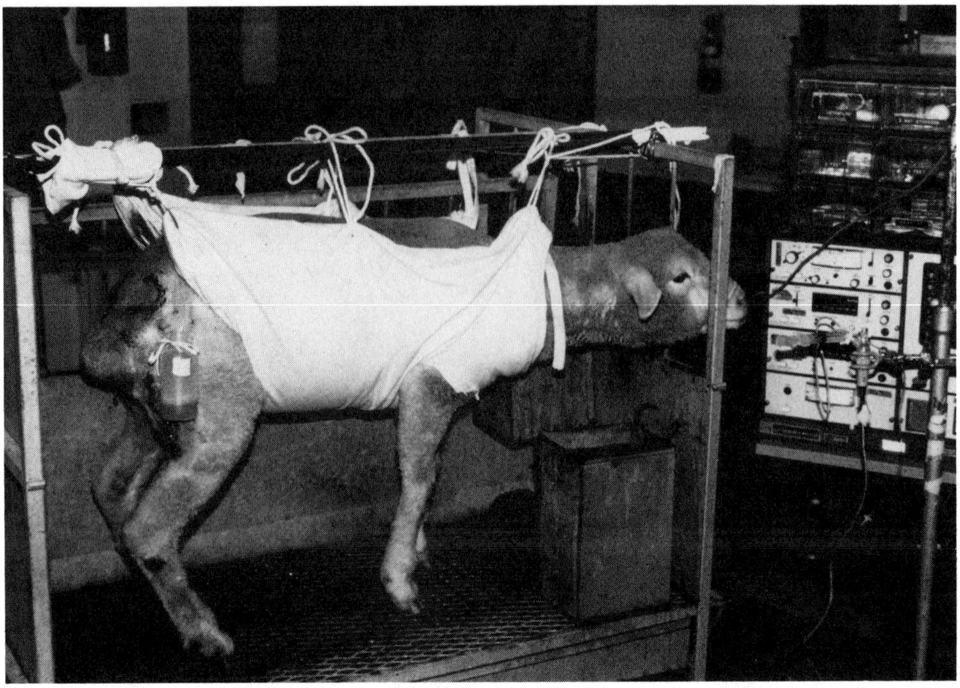

FIGURE 3. Do-it-yourself rig employed by author in early days at his research laboratory at the Harbor–UCLA Medical Center in California. An ordinary, nonmetabolic sheep cage is in use with an altered laundry bag for girth support, with suspension from two fence poles obtained from a local farm-supply outlet. The efferent lymph collection bottle may be seen suspended from the rump. The blood-pressure-monitoring equipment and pressure transducer stimulate the sheep's curiosity during an infusion and efferent lymph collection study.

two nonoccluding 6-0 silk ligatures to prevent cannula dislodgment and lymph leak and loss. Very small end-tapered polythene tubes (size 800/100/100/100, Portex, Hythe, Kent, U.K.) are used for afferent lymphatic cannulation. As with the efferent lymphatic cannulas, the afferent cannulas are brought to the outside through tiny stab wounds and secured to the skin in the area of the stab wound, with multiple 6-0 silk suture ligatures. Vicker Medical IP3 digital perfusion pumps are used for precision afferent lymphatic infusions. All wounds are closed with care with small silk sutures, and all skin incisions are closed with small Michel clips that may be easily and atraumatically removed when wound healing is complete or when limited wound exploration is desired, under light ketamine-xylazine anesthesia, to check cannula placement or for any other reason. This can easily be achieved with the sheep still in metabolic cage girth support.

Small recirculating lymphocyte outputs generally are quite profoundly depressed for the initial 24 and 48 h after anesthesia for lymphatic cannulation — near zero. At least 48 h should be permitted to pass before any nonanesthesia studies are attempted. All studies are carried out on unsedated and awake sheep.

All lymph collections are sequential, precisely timed, and measured with care for volume. Lymphocytes are counted at magnification $\times$ 25 in improved Neubauer hemocytometers and differentiated visually into small and blast lymphocytes. The small lymphocytes are of the recirculating type, and the blast lymphocytes are not recirculating and originate in the study lymph node. Lymphocyte outputs into efferent lymph are expressed as cells $\times 10^6$ per hour. Normal small-recirculating-lymphocyte outputs in awake and nonsedated adult sheep are generally in the 30 to 50 $\times 10^6$ per hour range. Higher outputs suggest drainage-

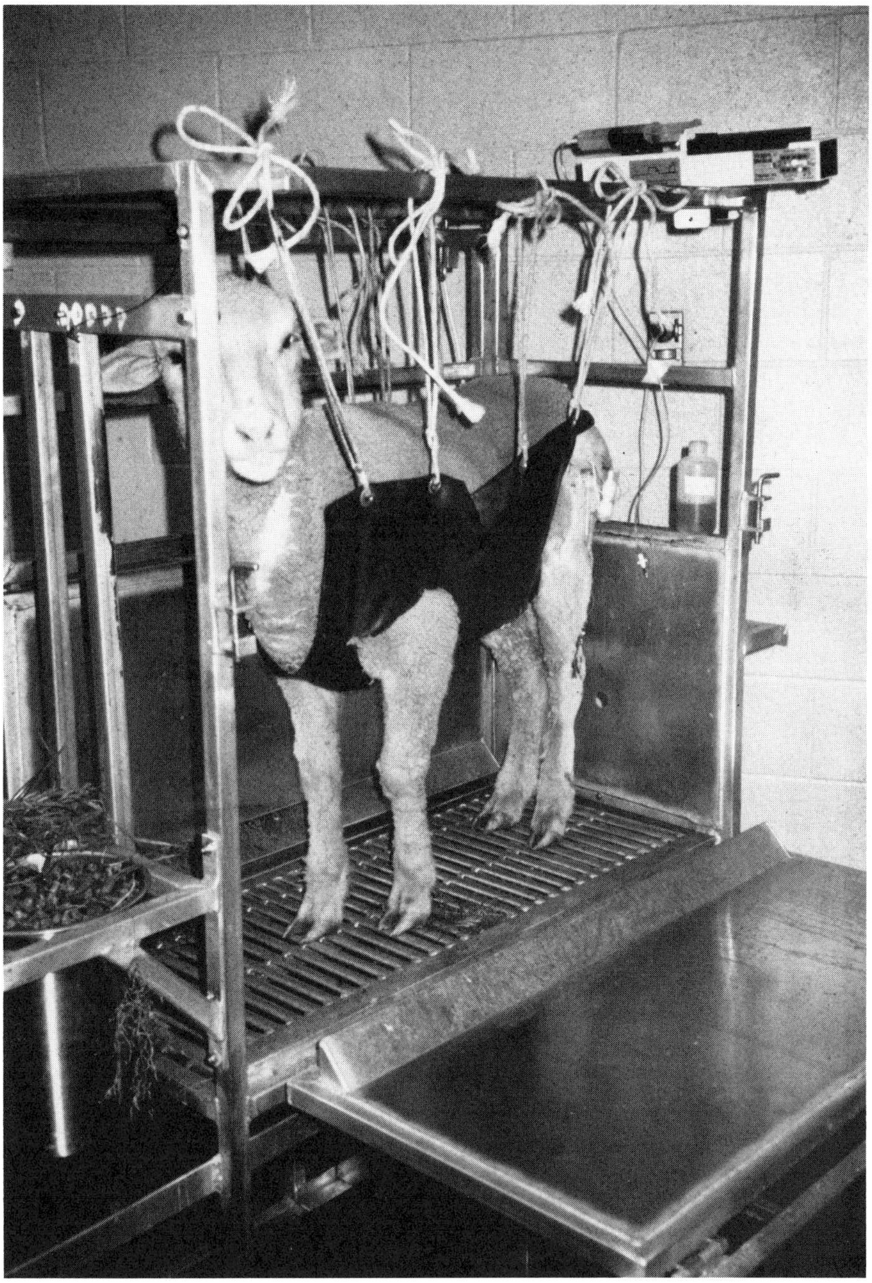

FIGURE 4. Happy days back at the author's Harbor–UCLA Medical Center research laboratory as first-class metabolic cages arrive with pull-down side panels. The Vickers IP3 digital syringe pump, altered for American current, is at the top rear of the cage with an afferent lymphatic infusion study in progress and efferent lymph collections in small (10 ml) plastic test tube. The overnight lymph collection in the overnight collection bottle (250 ml) may be seen at the rear of the cage.

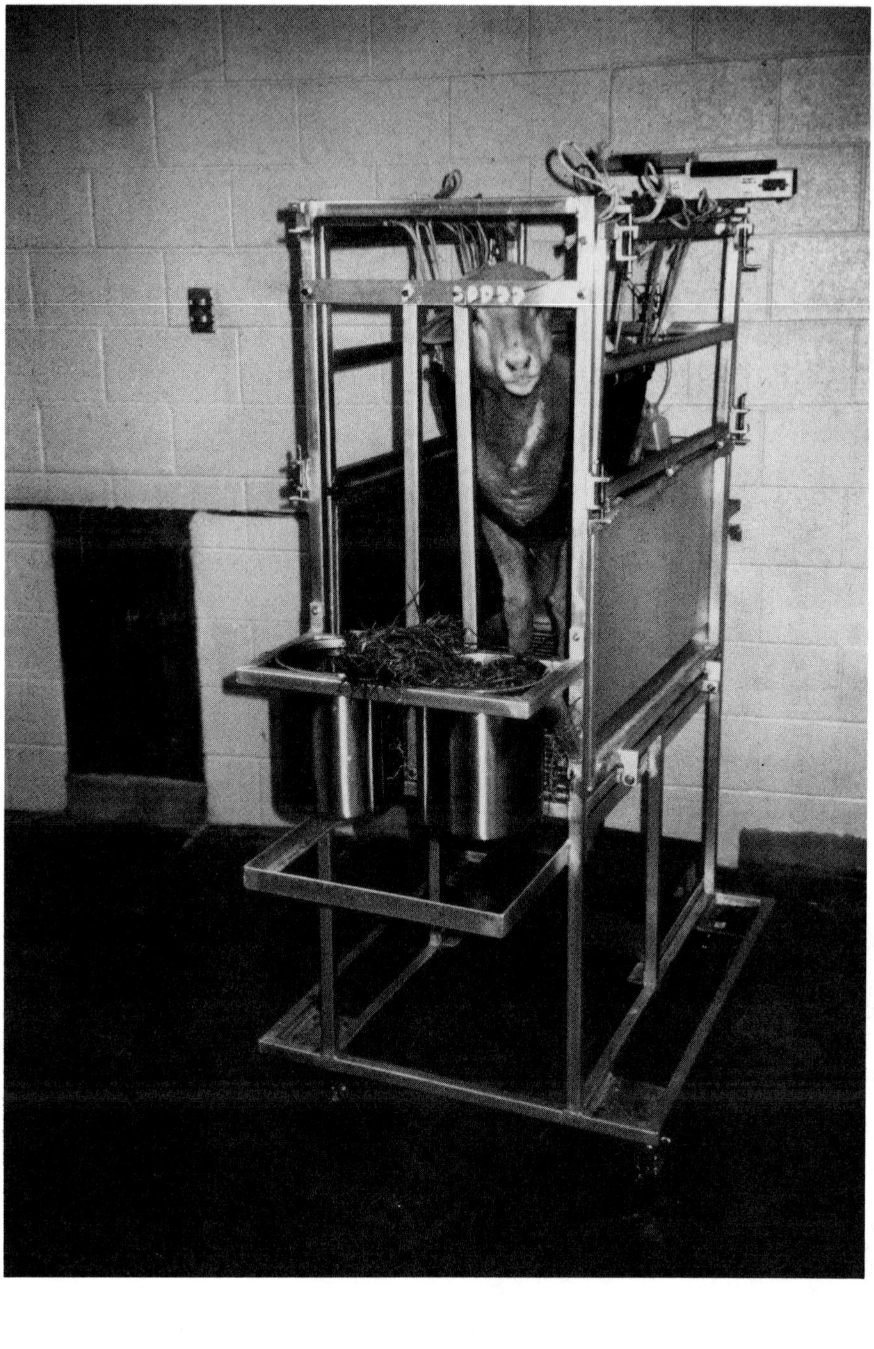

FIGURE 5. Full view of the sheep metabolic cage shown in Figure 4, with the side panel up in standard infusion and collection ''mode''.

area inflammatory change. Blast-lymphocyte outputs generally are quite low, but may be quite elevated in the presence of severe drainage-area inflammation or the drainage-area injection of inflammation-generating agents.

The supported-sheep approach and skin closure with easily removed Michel clips provide a unique and important opportunity to remove popliteal lymph nodes under local anesthesia as a part of ongoing studies to assess lymph node size, shape, and weight and the nature and number of contained lymphocytes and other immune response and vascular system cells and substances (including vasoactive neurotransmitter and neuromodulator and related molecules, as considered in this book). As the popliteal lymph node efferent lymphatic cannulation incision is at the level of the popliteal lymph node, local infiltration anesthesia in this area is not in the nodal lymphatic drainage area and hence will not produce anesthesia-induced alterations in lymph node function or biochemistry in the study lymph node.

The important and extremely valuable sheep peripheral lymphocyte traffic study model was developed in the early 1960s[1,2] in Canberra, Australia, at the John Curtin School of the Australian National University, by Professor J. G. (Joe) Hall and the late Professor Bede Morris in Professor Morris' research laboratory. It was early recognized as important and invaluable for the study of important components of the all-important lymphocyte traffic *in vivo* in the smallest animal and most tractable animal in which these studies could be practically carried out.[3,4] A particularly important early observation was that when a particulate antigen, such as killed *Salmonella muenchen* or other comparable killed *Salmonella* bacteria, was used for popliteal lymph node lymphatic drainage-area immunization and all efferent lymph was diverted by chronic efferent lymphatic cannulation, the immune response could be confined to the study hind limb, and that control contralateral cannulation studies could be carried out.[5]

REFERENCES

1. **Hall, J. G. and Morris, B.,** The output of cells in lymph from the popliteal nodes of sheep, *Q. J. Exp. Physiol.*, 47, 360, 1962.
2. **Hall, J. G. and Morris, B.,** The lymph-born cells of the immune response, *Q. J. Exp. Physiol.*, 48, 235, 1963.
3. **Hall, J. G. and Morris, B.,** The origin of cells in the efferent lymph from a single lymph node, *J. Exp. Med.*, 121, 901, 1965.
4. **Hall, J. G. and Morris, B.,** The immediate effect of antigens on the cell output of a lymph node, *Br. J. Exp. Pathol.*, 46, 450, 1965.
5. **Hall, J. G., Morris, B., Moreno, G. D., and Bessis, M. C.,** The ultrastructure and function of cells in lymph following antigenic stimulation, *J. Exp. Med.*, 125, 91, 1967.

Appendix B

PUBLICATIONS OF AUTHOR RELATING TO THE VASCULAR SYSTEM, THE IMMUNE RESPONSE SYSTEM, AND VASOACTIVE NEUROTRANSMITTER AND NEUROMODULATOR MOLECULES

1. **Shumacker, H. B., Jr. and Moore, T. C.,** Surgical management of traumatic chylothorax, *Surg. Gynecol. Obstet.,* 93, 46, 1951.
2. **Shumacker, H. B., Jr. and Moore, T. C.,** Leg and thigh amputations in obliterative arterial disease, *A.M.A. Arch. Surg.,* 63, 458, 1951.
3. **Moore, T. C.,** Congenital cysts of the mesentery, *Q. Bull. Ind. Univ. Med. Center,* 14, 29, 1952.
4. **Moore, T. C. and Shumacker, H. B., Jr.,** Unsuitability of transventricular autogenous slings for diminishing valvular insufficiency, *Surgery,* 33, 173, 1953.
5. **Moore, T. C. and Shumacker, H. B., Jr.,** Congenital and experimentally produced pericardial defects, *Angiology,* 4, 1, 1953.
6. **Shumacker, H. B., Jr. and Moore, T. C.,** Surgical closure of atrial septal defects; a preliminary report, *Q. Bull. Ind. Univ. Med. Center,* 15, 3, 1953.
7. **Moore, T. C. and Shumacker, H. B., Jr.,** Experimental creation of atrial septal defects with some notes on the production of a right to left atrial shunt, *Angiology,* 41, 244, 1953.
8. **Shumacker, H. B., Jr., Moore, T. C., and King, H.,** The experimental closure of atrial septal defects, *J. Thoracic Surg.,* 26, 551, 1953.
9. **Shumacker, H. B., Jr., Moore, T. C., and Campbell, J. A.,** Functional venography of the lower extremities, *Surg. Gynecol. Obstet.,* 98, 257, 1954.
10. **Moore, T. C. and Harris, E. J.,** Saddle embolus of the aorta as an immediate complication of mitral commissurotomy, *Ann. Surg.,* 140, 223, 1954.
11. **Moore, T. C.,** Functional venography with injection into superficial veins, *Surg. Gynecol. Obstet.,* 100, 721, 1955.
12. **Harris, E. J., Shumacker, H. B., Jr., Siderys, H., III, Moore, T. C., and Grice, P.,** Experimental comparison of certain pliable plastic materials as aortic grafts, *A.M.A. Arch. Surg.,* 71, 449, 1955.
13. **Moore, T. C. and Kirshman, F. E.,** Serum protein and cephalin flocculation changes accompanying thrombophlebitis, *Q. Bull. Ind. Univ. Med. Center,* 17, 51, 1955.
14. **Moore, T. C.,** The study of peripheal venous physiology and pathology by functional venography, *Bull. Hoosier State M.A.,* 6, 5, 1955.
15. **Moore, T. C. and Shumacker, H. B., Jr.,** Adrenalin producing tumors in childhood, *Ann. Surg.,* 143, 256, 1956.
16. **Moore, T. C.,** Functional venography as an aid in the study of peripheral venous disorders, *A.M.A. Arch. Surg.,* 72, 122, 1956.
17. **Moore, T. C., Riberi, A., and Kajijuri, H.,** Freeze-dried and alcohol preserved homografts for replacement of small arteries, *Surg. Gynecol. Obstet.,* 103, 155, 1956.
18. **Moore, T. C.,** Congenital cysts of the mesentery; report of four cases, *Ann. Surg.,* 145, 428, 1957.
19. **Moore, T. C., Martinez, H., and Enerson, D. H.,** Intrathoracic mixed neuroblastoma and ganglioneuroma in childhood, *J. Indiana State M. A.,* 50, 564, 1957.
20. **Shumacker, H. G., Jr., Kajijuri, H., Grice, P., Rodriguez, R., Riberi, A., Moore, T. C., and Siderys, H., III,** Comparative experimental study of aortic grafts, *Surgery,* 41, 945, 1957.
21. **Riberi, A. and Moore, T. C.,** Superior vena caval replacement with a pedicle tube of atrium, a preliminary report, *Q. Bull. Ind. Univ. Med. Center,* 19, 35, 1957.
22. **Moore, T. C. and Riberi, A.,** Use of autogenous aorta for superior vena caval replacement, a preliminary report, *Q. Bull. Ind. Univ. Med. Center,* 19, 59, 1957.
23. **Moore, T. C., Riberi, A., Kajikuri, H., and Galante, N.,** Myocardial revascularization by means of direct systemic to coronary artery anastomosis, *Q. Bull. Ind. Univ. Med. Center,* 19, 66, 1957.
24. **Moore, T. C.,** Recent advances in cardiovascular surgery, *Bull. Hoosier State M.A.,* 8, 5, 1958.
25. **Riberi, A. and Moore, T. C.,** Superior vena caval replacement. I. Unsuitability of free tubes of autogenous pericardium, *A.M.A. Arch. Surg.,* 76, 384, 1958.
26. **Moore, T. C. and Riberi, A.,** Maintenance of coronary circulation during systemic to coronary artery anastomosis, *Surgery,* 43, 245, 1958.
27. **Riberi, A. and Moore, T. C.,** Superior vena caval replacement. II. Successful use of a pedicle tube created from right atrium and appendage, *Surgery,* 43, 355, 1958.
28. **Moore, T. C., Riberi, A., and Hardin, R. A.,** Small artery grafts and shunts, *Bull. Soc. Int. Chir.,* 17, 92, 1958.

29. **Moore, T. C.,** What's new in cardiovascular surgery, *Q. Bull. Ind. Univ. Med. Center,* 20, 10, 1958.
30. **Moore, T. C.,** Acute arterial obstruction due to traumatic circumferential intimal fracture, *Ann. Surg.,* 148, 111, 1958.
31. **Moore, T. C.,** Para-cardiac thymoma simulating pericardial cyst, *J. Indiana State M. A.,* 51, 23, 1958.
32. **Moore, T. C., Riberi, A., Kajikuri, H., and Shumacker, H. B., Jr.,** Use of impedance plethsymography in evaluation of peripheral vascular response to arteriography, *Surgery,* 44, 345, 1958.
33. **Moore, T. C. and Riberi, A.,** Superior vena caval replacement. III. Successful use of fresh, autogenous aorta, *Surgery,* 44, 898, 1958.
34. **Moore, T. C. and Riberi, A.,** Replacement of thoracic aorta with alcohol preserved homografts, *J. Thoracic Surg.,* 37, 510, 1959.
35. **Moore, T. C.,** Esophageal obstruction due to anomalous right subclavian artery, *J. Indiana State M.A.,* 52, 1117, 1959.
36. **Riberi, A. and Moore, T. C.,** Superior vena caval replacement. IV. Experimental use of alcohol preserved heterografts of bovine aorta, *J. Thoracic Surg.,* 33, 172, 1959.
37. **Moore, T. C.,** Traumatic pericardial diaphragmatic hernia, *A.M.A. Arch. Surg.,* 79, 827, 1959.
38. **Moore, T. C., Heimberger, I., and Teramoto, S.,** Experimental replacement of superior vena cava, *Bull. Soc. Int. Chir.,* 19, 340, 1960.
39. **Moore, T. C., Teramoto, S., and Heimberger, I.,** Pulmonary artery replacement. I. Successful use of telfon and systematic venous grafts, *Surgery,* 47, 804, 1960.
40. **Moore, T. C., Heimberger, I., and Teramoto, S.,** Pulmonary artery replacement. II. Use of fresh autogenous systemic artery, *A.M.A. Arch. Surg.,* 80, 1033, 1960.
41. **Moore, T. C., Teramoto, S., and Heimberger, I.,** Successful use of teflon grafts for superior vena caval replacement, *Surg. Gynecol. Obstet.,* 3, 475, 1960.
42. **Moore, T. C., Teramoto, S., Heimberger, I., and Harshman, J. A.** Superior vena caval replacement. V. Use of fresh homografts of superior vena cava, *J. Thoracic Cardiovasc. Surg.,* 42, 475, 1961.
43. **Siderys, H., Moore, T. C., and Shumacker, H. B., Jr.,** Left hepatic lobectomy for hemangioma of the liver in the newborn, *Surgery,* 52, 502, 1962.
44. **Moore, T. C. and Stout, F. E.,** Postpartum adnexal phlegmasia cerulea dolens with adexal infarction, *Obstet. Gynecol.,* 21, 584, 1963.
45. **Moore, T. C., Normell, L., and Eiseman, B.,** Effect of serotonin loading on histamine release and blood flow of isolated perfused liver and lung, *A.M.A. Arch. Surg.,* 87, 42, 1963.
46. **Moore, T. C. and Mandelbaum, I.,** Superior vena caval replacement. VII. Experimental use of toluene diisocyanate treated dacron grafts, *Surgery,* 54, 340, 1963.
47. **Moore, T. C.,** Experimental grafting of the venous system, in *The Fundamentals of Vascular Grafting,* Weselowski, S. A. and Denis, C., Eds., McGraw Hill, New York, 1963, 278–287.
48. **Moore, T. C., Normell, L., and Eiseman, B.,** Histamine metabolism in the isolated perfused canine liver, *Surg. Forum,* 14, 78, 1963.
49. **Eiseman, B., Moore, T. C., and Normell, L.,** Histamine metabolism in the isolated perfused pig liver, *Surg. Gynecol. Obstet.,* 118, 69, 1964.
50. **Yong, N. K., Klein, N., and Moore, T. C.,** Superior vena caval replacement; experimental use of fresh autologous vein grafts, *Br. J. Surg.,* 51, 374, 1964.
51. **Moore, T. C. and Yong, N. K.,** Experimental replacement and bypass of large veins, *Bull. Soc. Int. Chir.,* 23, 274, 1964.
52. **Moore, T. C., Normell, L., and Eiseman, B.,** Effect of serotonin and histamine on blood flow and tracheal resistance in isolated perfused lung, *Surgery,* 57, 730, 1965.
53. **Yong, N. K., Klein, N., and Moore, T. C.,** Superior vena caval replacement. IX. Experimental use of gelatin impregnated teflon grafts, *Am. Surg.,* 31, 318, 1965.
54. **Moore, T. C. and Eiseman, B.,** Serotonin metabolism in the isolated perfused canine liver, *Surgery,* 59, 531, 1966.
55. **Moore, T. C.,** Histidine decarboxylase inhibitors and the survival of skin homografts, *Nature,* 215, 871, 1967.
56. **Moore, T. C. and Chang, J.,** Urinary histamine excretion in the rat following skin homografting and autografting, *Ann. Surg.,* 167, 232, 1968.
57. **Moore, T. C.,** Immunosuppressive action of histidine decarboxylase inhibitors (Eighth Congress of International Cardiovascular Society). *J. Cardiovasc. Surg.,* Special Issue, 63–69, 1967.
58. **Moore, T. C., Chang, J., and Hume, D. M.** Urinary excretion of histamine in renal transplant patients, *Surg. Gynecol. Obstet.,* 127, 1023, 1968.
59. **Millington, G. M. A. and Moore, T. C.,** A rapid method for mass pattern skin grafting in the rat, *J. Surg. Res.,* 8, 379, 1968.
60. **Yong, N. K. and Moore, T. C.,** Endothelial survival in autologous vein grafts, *Aust. N.Z. J. Surg.,* 37, 316, 1968.

61. **Smellie, W. A. B. and Moore, T. C.,** Effect of semicarbazide and pyridoxine deficiency on survival of canine renal homografts, *Surg. Gynecol. Obstet.*, 128, 81, 1969.
62. **Moore, T. C. and Schayer, R. W.,** Histidine decarboxylase activity of autografted and allografted rat skin, *Transplantation*, 7, 99, 1969.
63. **Moore, T. C. and Chang, J.,** Effect of portal venous hepatic bypass on urine histamine excretion in the rat, *J. Cardiovasc. Surg.*, July/August, 299, 1969.
64. **Moore, T. C. and Lawrence, W., Jr.,** Suppression of antibody formation by histidine decarboxylase inhibitors, *Transplantation*, 8, 224, 1969.
65. **Moore, T. C.,** Histidine decarboxylase inhibitors and second-set allograft survival, *Arch. Surg.*, 99, 470, 1969.
66. **Moore, T. C. and Hume, D. M.,** The period and nature of hazard in clinical renal transplantation. I. The hazard to patient survival, *Ann. Surg.*, 170, 1, 1969.
67. **Moore, T. C. and Hume, D. M.,** The period and nature of hazard in clinical renal transplantation. II. The hazard to transplant kidney fucntion, *Ann. Surg.*, 170, 12, 1969.
68. **Moore, T. C. and Hume, D. M.,** The period and nature of hazard in clinical renal transplantation. III. The hazard to transplant kidney survival, *Ann. Surg.*, 170, 25, 1969.
69. **Moore, T. C.,** Role of histidine decarboxylase inhibitors in the suppression of transplant rejection, in *Pharmacological Treatment in Organ and Tissue Transplantation, Proc. Int. Symposium*, International Congress Series No. 197, Bertelli, A. and Monaco, A. P., Eds., Excerpta Medica Foundation, Amsterdam, 1970, 60–71.
70. **Moore, T. C. and Schayer, R. W.,** Histidine decarboxylase activity of the rat spleen following skin allografting, *Ann. Surg.*, 171, 609, 1970.
71. **Moore, T. C. and Thompson, D. P.,** Prolongation of rat skin allograft survival with isonicotinic acid hydrazide and methapyrilene hydrochloride, *Surg. Gynecol. Obstet.*, 132, 275, 1971.
72. **Moore, T. C., Thompson, D. P., and Hayes, M.,** Correlation of renal blood flow indices and histamine metabolism data after canine renal allografting, *Arch. Surg.*, 101, 45, 1970.
73. **Thompson, D. P., Moore, T. C., and Fingerhut, A. G.,** A method for high resolution renal microangiography, *Surg. Gynecol. Obstet.*, 132, 101, 1971.
74. **Moore, T. C., Cleveland, R. J., Thompson, D. P., and Nelson, R. J.,** Vasoactive amine metabolism following canine and human heart transplantation, *Ann. Surg.*, 174, 51, 1971.
75. **Moore, T. C., Thompson, D. P., and Glassock, R. J.,** Elevation in urinary and blood histamine following clinical renal transplantation, *Ann. Surg.*, 173, 381, 1971.
76. **Moore, T. C.,** A theory of the role of histamine metabolism in transplant rejection, *Surg. Gynecol. Obstet.*, 132, 489, 1971.
77. **Moore, T. C., Lemmi, C. A. E., Orlando, J. C., and Pinkerton, W.,** Correlation of splenic histidine decarboxylase activity and antibody formation following SRBC immunization, *Transplantation*, 11, 347, 1971.
78. **Moore, T. C.,** Effective use of isoniazid and an antihistamine in clinical renal transplantation, *Surg. Gynecol. Obstet.*, 133, 75, 1971.
79. **Hayes, M. and Moore, T. C.,** Early detection and classification of renal transplant rejection by B/K scan ratio and blood isotope clearance data, *Transplantation*, 12, 139, 1971.
80. **Hayes, M. and Moore, T. C.,** Early detection of canine renal allograft rejection by reduction in the scan bladder/kidney ratio, *Surgery*, 71, 60, 1972.
81. **Orlando, J. C. and Moore, T. C.,** Splenectomy for trauma in childhood; report of 36 cases, *Surg. Gynecol. Obstet.*, 134, 94, 1972.
82. **Moore, T. C.,** Role of histidine decarboxylase enzyme activity in transplant rejection, in *Proc. XIIth Int. Cong. Pediatrics*, Vol. IV-107, 1971, 437.
83. **Moore, T. C.,** Vascular component of transplant rejection, *Bull. Soc. Int. Chir.*, 30, 322, 1971.
84. **Moore, T. C., Berne, T. V., Martin, D. C., and English, T. S.,** Multi-institutional use of a mobile kidney preservation unit, *Transplantation*, 12, 527, 1971.
85. **Hayes, M., Moore, T. C., and Thompson, D. P.,** Correlation of scans with laboratory data following canine renal allografting, *J. Nucl. Med.*, 11, 390, 1970.
86. **Moore, T. C., Berne, T. V., and English, T. S.,** Use of mobile transplant kidney preservation unit in a large metropolitan area, *Am. J. Surg.*, 124, 229, 1972.
87. **Hayes, M., Moore, T. C., and Taplin, G. V.,** Radionuclide procedures in predicting early renal transplant rejection, *Radiology*, 103, 627, 1972.
88. **Moore, T. C., English, T. S., Berne, T. V., Beck, T. A., Martin, D. C., Sinclair, M. C., and Lemmi, C. A. E.,** Experience with perfusion preservation of 189 cadaveric human kidneys for transplantation, in *Transplantation Abstracts*, Grune and Stratton, New York, 1972, 204.
89. **Lemmi, C. A., Cooper, E. L., and Moore, T. C.,** Immune challenge: effect of histamine and kinin metabolism in earthworms, *IRCS Med. Sci.*, 73(3), 15-11-5, 1973.

90. **Lemmi, C. A., Cooper, E. L., and Moore, T. C.,** Histamine and kinin metabolism in earthworms: techniques for their study, *IRCS Med. Sci.*, 73(3), 15-11-4, 1973.
91. **Kauffman, H. M., Rodgers, R. E., Gross, W. S., Shrettler, B. M., Ivanovich, R., Merkel, F. K., Bergan, J. J., and Moore, T. C.,** Human kidney preservation and shipping, *Transplantation*, 15, 465, 1973.
92. **Moore, T. C. and Sinclair, M. C.,** Is conservative care of minor splenic injury feasible? *Forum Mod. Med.*, Feb. 19, 1973, p. 99.
93. **Moore, T. C., Sinclair, M. C., and Lemmi, C. A. E.,** Effect of renal allografting on rat appendiceal histidine decarboxylase activity, *IRCS Med. Sci.*, 73(5), 13-1-4, 1973.
94. **Moore, T. C., Sinclair, M. C., and Lemmi, C. A. E.,** Stimulation of murine gastric histidine decarboxylase activity by SRBC immunization, *Surgery*, 74, 444, 1973.
95. **Thompson, D. P. and Moore, T. C.,** Blood histamine levels in anephric and in bilaterally nephrectomized and renal allografted dogs, *IRCS Med. Sci.*, 73(6), 20-11-3, 1973.
96. **Weber, G. A., Finklestein, J. Z., Lemmi, C. A. E., and Moore, T. C.,** Effect of repeated SRBC immunization on mouse splenic histidine decarboxylase response to Gross leukemia virus tumor transplantation, *IRCS Med. Sci.*, 73(6), 42-2-10, 1973.
97. **Ban, J. L., Hirose, F. H., Lachman, R. S., and Moore, T. C.,** Pedunculated cystic lymphagioma of the splenocolic ligament in childhood, *Am. J. Surg.*, 24, 410, 1972.
98. **Leemmi, C. A. E., Moore, T. C., and Davison, W. D.,** Use of radiorespirametry to evaluate in rates the effects of feeding and fasting on the decarboxylation of ^{14}C histidine, *Am. J. Surg.*, 126, 627, 1973.
99. **Sinclair, M. C., Moore, T. C., Lemmi, C. A. E., and Orlando, J. C.,** Increase in the urinary excretion of histamine following SRBC immunization of rats, *Am. J. Surg.*, 126, 624, 1973.
100. **Moore, T. C., Johnson, D. E., and Cleveland, R. J.,** Radioisotope entrapment in the rejecting canine cardiac allograft, *IRCS Med. Sci.*, 3(12), 11-27-3, 1973.
101. **Hayes, M., Taplin, C. V., and Moore, T. C.** Quantitated scintiphotos in predicting early renal transplant rejection, in *Medical Radioisotope Scintography 1972*, Vol. II, IAEA-SM-164/139, International Atomic Energy Agency, Vienna, 1973, 211–223.
102. **McAlpin, C. D., Moore, T. C., Sinclair, M. C., and Lemmi, C. A. E.,** Effect of feeding and of fasting with and without water on rat gastric histidine decarboxylase activity, *IRCS Med. Sci.*, 2, 1115, 1974.
103. **Moore, T. C., Lemmi, C. A. E., and Weber, G. A.,** Effect of age on rat thymus histidine decarboxylase activity, *IRCS Med. Sci.*, 2, 1115, 1974.
104. **Lemmi, C. A. E., Cooper, E. L., and Moore, T. C.,** An approach to studying evolution of cellular immunity, in *Contemporary Topics in Immunobiology*, Vol. 4, Cooper, E. L., Ed., Plenum Press, New York, 1974, 109.
105. **Moore, T. C., Sinclair, M. C., and Lemmi, C. A. E.,** Elevation in host splenic and allograft parenchymal and capsular histidine decarboxylase activity and weights following renal allografting in rats, *Surgery*, 75, 203, 1974.
106. **Moore, T. C., Berne, T. V., and English, T. S.,** Influence of duration of hemodialysis on early post-transplant function of machine preserved human cadaveric kidneys, *Ann. Surg.*, 179, 499, 1974.
107. **Moore, T. C., English, T. S., and Berne, T. C.,** Machine preservation of 302 human cadaveric kidneys for transplantation, *Surg. Gynecol. Obstet.*, 138, 239, 1974.
108. **Moore, T. C., Berne, T. V., and English, T. S.,** Accelerated rejection of 36 immediate functioning machine-preserved human cadaveric kidneys, *Surgery*, 75, 357, 1974.
109. **Moore, T. C., English, T. S., and Berne, T. V.,** Early posttransplant renal nonfunction, *Arch. Surg.*, 108, 685, 1974.
110. **Moore, T. C., Berne, T. V., and English, T. S.,** Early post-transplant slow function of 26 machine preserved human cadaveric kidneys, *Am. J. Surg.*, 127, 725, 1974.
111. **Moore, T. C., English, T. S., Conovaloff, A. J., Berne, T. S., and Lemmi, C. A. E.,** Factors involved in discarding cadaveric kidneys preserved by machine perfusion, *Surg. Gynecol. Obstet.*, 138, 545, 1974.
112. **Moore, T. C., Sinclair, M. C., and Lemmi, C. A. E.,** Elevation in rat plasma kallikrein activity following renal allografting, *Am. J. Surg.*, 127, 287, 1974.
113. **Johnson, D. and Moore, T. C.** Large intra-thoracic superior sulcus cystic mass arising from brachial plexus in childhood, *J. Pediatr. Surg.*, 9, 403, 1974.
114. **Koppelman, L., Moore, T. C., Lemmi, C. A. E., and Porter, D. D.,** Effects of bradykinin injection into syngenic SV-40 hamster tumors on rat of tumor growth and splenic histidine decarboxylase activity, *IRCS Med. Sci.*, 2, 1272, 1974.
115. **Jaffe, B. M. and Moore, T. C.,** Elevated prostaglandin E activity of rat heart allograft coronary sinus venous blood, *IRCS Med. Sci.*, 2, 1417, 1974.
116. **Moore, T. C., Sinclair, M. D., Lemmi, C. A. E., and McAlpin, C. D.,** Elevation in gastric histidine decarboxylase activity following renal allografting in the rat, *IRCS Med. Sci.*, 2, 1304, 1974.

117. **Moore, T. C., Granger, G. A., and Lemmi, C. A. E.,** In vivo elevation in rat splenic histidine decarboxylase activity by intra-splenic injection of Con A and Con A activated lymphokines, *IRCS Med. Sci.,*, 2, 1458, 1974.
118. **Sinclair, M. C., Lemmi, C. A. E., and Moore, T. C.,** Elevation in urinary excretion of histamine following renal allografting in rats, *J. Surg. Res.*, 17, 43, 1974.
119. **Cooper, E. L., Lemmi, C. A. E., and Moore, T. C.,** Agglutinins and cellular immunity in earthworms, *Ann. N.Y. Acad. Sci.*, 234, 34, 1974.
120. **Moore, T. C., Fine, R. N., Asch, M. J., Berne, T. V., and English, T. S.,** Early post transplant function in childhood of 33 machine preserved human cadaveric kidneys, *Pediatrics,* 54, 773, 1974.
121. **Moore, T. C., Berne, T. V., English, T. S., and Lemmi, C. A. E.,** Early post-transplant function of machine preserved kidneys in sensitized recipients, *Surg. Gynecol. Obstet.*, 139, 45, 1974.
122. **Moore, T. C. and Lemmi, C. A. E.,** Effects of heparin and ALG with anti-platelet activity on blood and urinary histamine levels of human renal transplant recipients, *Am. Surg.*, 40, 453, 1974.
123. **McAlpin, C. D., Moore, T. C., and Lemmi, C. A. E.,** Gastric histamine metabolism following acute ligation of the pylorus in the Shay rat, *Arch. Surg.*, 109, 65, 1974.
124. **Moore, T. C., Berne, T. V., English, T. S., and Martin, D. C.,** Effect of early transplant function as an index of recipient presensitization on long term function of machine preserved cadaveric kidneys, *Am. J. Surg.*, 128, 293, 1974.
125. **Granger, G. A. and Moore, T. C.,** Accumulation of lymphotoxin secreting lymphoid cells in rat heart allografts, *Surgery,* 76, 542, 1974.
126. **Moore, T. C., Sinclair, M. C., McAlpin, C. D., and Lemmi, C. A. E.,** Effect of aging, of organ allografting and of surgical stress on rat thymus histidine decarboxylase activity, *Surgery,* 76, 733, 1974.
127. **Moore, T. C. and Hayes, M.,** Combined use of ^{131}I hippurate blood clearance and B/K ratio data in the early detection, classification and management of accelerated rejection of human renal transplants, *Surgery,* 76, 587, 1974.
128. **Asch, M. J., Cohen, A. H., and Moore, T. C.,** Hepatic and splenic lymphangiomatosis with skeletal involvement: report of a case and review of the literature, *Surgery,* 76, 334, 1974.
129. **Moore, T. C., Johnson, D. E., and Cleveland, R. J.,** Myocardial ECG voltage and 99mTc albumin clearance times of heterotopic canine heart allografts, *J. Thorac. Cardiovasc. Surg.*, 68, 419, 1974.
130. **Moore, T. C. and Jaffe, B. M.,** Prostaglandin E levels of heterotopic rat heart allografts and of host lymphoid tissues at intervals post-grafting, *Transplantation,* 18, 383, 1974.
131. **Moore, T. C.,** Role of vasoactive substances in the afferent and efferent arcs of the immune response, *IRCS Med. Sci.*, 2, 46, 1974.
132. **Moore, T. C., Vigran, T. S., and Lemmi, C. A. E.,** ^{131}I hippurate isotope accumulation in rejecting heterotopic rat heart allografts, *J. Thorac. Cardiovasc. Surg.*, 69, 402, 1975.
133. **Kwong, M., Moore, T. C., Lemmi, C. A. E., Oh, W., and Thibeault, D. W.,** Histidine decarboxylase in fetal intrauterine growth-retarded rats, *Pediatr. Res.*, 9, 278, 1975.
134. **Koppelmann, L. E., Moore, T. C., Lemmi, C. A. E., and Porter, D. D.,** Accumulation of mononuclear cells in tumors with growth slowing and elevation in host splenic histidine decarboxylase activity following repeated tumor injections with bradykinin, *Surgery,* 78, 181, 1975.
135. **Jaffe, B. M., Moore, T. C., and Vigran, T. S.,** Tissue levels of prostaglandin E following heterotopic rat heart allografting, *Surgery,* 78, 481, 1975.
136. **Lemmi, C. A. E., Vigran, T. S., Koppelmann, L. E., Yu, B. M., and Moore, T. C.,** Alterations in kinin metabolism of coronary sinus venous blood of heterotopic rat heart allografts and isografts, *Arch. Surg.*, 100, 1070, 1975.
137. **Moore, T. C.,** Role of vasoactive substances in the rejection of tumor and transplant tissue, *Bull. Soc. Int. Chir.*, 24, 467, 1975.
138. **Fiala, M., Payne, J. E., Berne, T. V., Moore, T. C., Henle, W., Montgomerie, J. Z., Chatterjee, S. N., and Guze, L. B.,** Epidemiology of cytomegalovirus infection after transplantation and immunosuppression, *J. Infect. Dis.*, 132, 421, 1975.
139. **Kwong, M. S., Moore, T. C., Lemmi, C. A. E., Oh, W., and Thibeault, D. S.,** Histidine decarboxylase activity in fetal intrauterine growth-retarded rats, *Pediatr. Res.*, 10, 737, 1976.
140. **Koppelmann, L. E. and Moore, T. C.,** Bradykinin-induced enhancement *in vitro* of sensitized hamster spleen cell anti-tumor cell cytotoxicity, *IRCS Med. Sci.*, 4, 524, 1976.
141. **Moore, T. C. and Leake, R. D.,** Infra-umbilical entry into the peritoneal cavity through a small incision for the identification and cannulation on an umbilical artery, *J. Pediatr. Surg.*, 12, 147, 1977.
142. **Moore, T. C., Koppelmann, L. E., and Lemmi, C. A. E.,** Effect of tumor cell rejection and progressive tumor growth on hamster thymic histidine decarboxylase, *IRCS Med. Sci.*, 5, 493, 1977.
143. **Moore, T. C., Koppelmann, L. E., and Lemmi, C. A. E.,** An experimental study of tumor growth and splenic histamine enzymes, *Surg. Gynecol. Obstet.*, 146, 559, 1978.

144. **Moore, T. C., Koppelmann, L. E., and Lemmi, C. A. E.,** Decrease in histamine forming enzyme activity of non-metastasizing fibrosarcomas in hamsters with progressive tumor growth, *Ann. Surg.*, 188, 175, 1978.
145. **Moore, T. C.,** Atresia of the colon at the splenic flexure with absence of the distal colon and ischemic destruction of the proximal colon, *J. Pediatr. Surg.*, 8, 89, 1978.
146. **Koppelmann, L. E., Moore, T. C., and Porter, D. D.,** Increased plasma kallikrein activity and tumor growth suppression associated with intralesional bradykinin injections in hamsters, *J. Pathol.*, 126, 1, 1978.
147. **Moore, T. C., Koppelmann, L. E., and Lemmi, C. A. E.,** Splenic and thymic histamine forming enzyme activity and weights following organ allografting and tumor cell transplantation, *Neoplasma,* 26, 73, 1979.
148. **Ohtawa, T. and Moore, T. C.,** Use of reserpine and cyproheptadine in rat heart allografting, *IRCS Med. Sci.*, 7, 35, 1979.
149. **Moore, T. C. and Ohtawa, T.,** Use of histamine receptors blockade by cimetidine and mepyramine in rat heart allografting, *IRCS Med. Sci.*, 7, 147, 1979.
150. **Moore, T. C. and Peter, M.,** Thru-and-thru gunshot penetration of distal abdominal aorta in a 4-year-old child managed by aortic transection, debridement and reanastomosis with survival, *J. Trauma,* 19, 537, 1979.
151. **Moore, T. C. and Lemmi, C. A. E.,** Exposure of histamine receptors on lymphocytes at intervals following heart allografting in the rat, *IRCS Med. Sci.*, 7, 148, 1979.
152. **Ohtawa, T. and Moore, T. C.,** Effects of cyclosporin A dose reduction and antibiotics on hemorrhage and rejection in rat heart allografting, *IRCS Med. Sci.*, 7, 256, 1979.
153. **Moore, T. C.,** A role for mediators of inflammation in the cellular immune response to tumor and transplant tissue, *IRCS Med. Sci.*, 7, 273, 1979.
154. **Moore, T. C., Hopkins, J., McConnell, I., and Lachmann, P. J.,** Efferent lymph output of lymphocytes following infusions of prostaglandin E_2 and other vasoactive mediators into cannulated afferent popliteal lymphatics in sheep, *IRCS Med. Sci.*, 8, 272, 1980.
155. **Moore, T. C.,** Cannulated afferent lymphatic infusion studies indicate output of lymphocytes into efferent lymph independent of fluid volume input, *IRCS Med. Sci.*, 9, 585, 1981.
156. **Moore, T. C.,** Reductions in efferent lymph output of lymphocytes following infusions of histamine into cannulated afferent lymphatics in sheep, *IRCS Med. Sci.*, 9, 588, 1981.
157. **Moore, T. C.,** The effects of inflammatory mediators and anaesthetic agents upon lymphocyte traffic of sheep *in vivo,* Ph.D. thesis, University of Cambridge (Trinity College), England, 1982.
158. **Moore, T. C., Pearson, J. D., Gyure, L. A., and Hall, J. G.,** Increased concentrations of prostaglandin E_2 in lymph efferent from tumours, *IRCS Med. Sci.*, 10, 345, 1982.
159. **Moore, T. C. and Lachmann, P. J.,** Cyclic AMP reduces and cyclic GMP increases the traffic of lymphocytes through peripheral lymph nodes of sheep in vivo, *Immunology,* 47, 423, 1982.
160. **Moore, T. C. and Lachmann, P. J.,** Chronic depression of lymphocyte traffic by longterm administration of prostaglandins to peripheral lymph nodes of sheep, *Surgery,* 95, 45, 1984.
161. **Moore, T. C.,** The modulation by prostaglandins of increases in lymphocyte traffic induced by bradykinin, *Immunology,* 51, 455, 1984.
162. **Moore, T. C.,** Chronic administration of prostaglandins to peripheral lymph nodes of sheep alters lymphocyte traffic response to viral immunization, *IRCS Med. Sci.*, 12, 271, 1984.
163. **Moore, T. C. and Pearson, J. D.,** Effects of general anaesthesia on efferent lymph PGE_2 levels and modulation by drainage area surgery, *IRCS Med. Sci.*, 12, 269, 1984.
164. **Moore, T. C.,** Anesthesia associated depression in lymphocyte traffic and its modulation, *Am. J. Surg.*, 147, 807, 1984.
165. **Moore, T. C.,** Modification of lymphocyte traffic by vasoactive neurotransmitter substances, *Immunology,* 52, 511, 1984.
166. **Moore, T. C.,** Anesthesia associated depression in lymphocyte traffic less with regional than with general anesthesia in sheep, *Am. J. Surg.*, 148, 71, 1984.
167. **Moore, T. C., Lippmann, M., Khan, F., and Locke, R. L.,** Depression of lymphocyte traffic in sheep by induced systemic arterial hypotension and by acute arterial occlusion, *Immunology,* 53, 677, 1984.
168. **Moore, T. C.,** Drainage area complement activation by injections of inulin and cobra venom factor alter regional lymph node lymphocyte traffic, *IRCS Med. Sci.*, 13, 681, 1984.
169. **Moore, T. C. and Cobo, J. C.,** Massive symptomatic cystic hygroma confined to the thorax in early childhood, *J. Thorac. Cardiovasc. Surg.*, 89, 459, 1985.
170. **Moore, T. C.,** Mediator and physiological influences on lymphocyte traffic in sheep, in *Immunology of the Sheep,* Morris, B. and Miyasaka, M., Eds., Editiones Roche, Basel, 1985, 327.
171. **Moore, T. C., Nur, K., Yunes, S., and Storm, F. K.,** Peripheral node lymphocyte traffic effects of localized magnetic-loop induction hyperthermia of the sheep brain, *IRCS Med. Sci.*, 13, 1194, 1985.
172. **Moore, T. C. and Khan, F.,** Lymphocyte traffic effects of systemic and local lidocaine in sheep, *J. Surg. Res.*, 41, 279, 1986.

173. **Moore, T. C.,** Jejunal atresia with midgut deletion, *J. Pediatr. Surg.*, 21, 951, 1986.
174. **Moore, T. C., Nur, K., and Storm, F. K.,** Localized deep hyperthermia increases the traffic of lymphocytes through peripheral lymph nodes of sheep *in vivo, J. Surg. Oncol.*, 35, 75, 1987.
175. **Moore, T. C., Lippmann, M., Spruck, C. H., and Gamal, R.,** Increased outputs of lymphocytes in lymph efferent from the nodes of sheep during systemic arterial hypertension induced by phenylephrine or dopamine, *Immunopharmacology,* 14, 151, 1987.
176. **Moore, T. C.,** Immunology surgery, in *History of the Pacific Coast Surgical Association,* Vol. 11, Brewer, L. A., III, Ed., Pacific Coast Surgical Association, 1988.
177. **Moore, T. C., Spruck, C. H., and LeDuc, L. E.,** Depression of lymphocyte traffic in sheep by anesthesia and associated changes in efferent lymph PGE_2 and antibody levels, *Immunology,* 63, 139, 1988.
178. **Moore, T. C., Spruck, C. H., and Said, S. I.,** Depression of lymphocyte traffic in sheep by vasoactive intestinal peptide (VIP), *Immunology,* 64, 475, 1988.
179. **Moore, T. C., Spruck, C. H., Lami, J. L., and Nomicos, N. E.,** Central venous whole blood transfusion in sheep depresses lymphocyte traffic through primary peripheral lymph nodes, *Med. Sci. Res.*, 16, 567, 1988.
180. **Gamal, R. and Moore, T. C.,** Massive acquired omental cyst as a complication of ventriculo-peritoneal shunting, *J. Pediatr. Surg.*, 23, 1041, 1988.
181. **Spruck, C. H. and Moore, T. C.,** Anesthesia associated depression of peripheral node lymphocyte traffic and antibody production in sheep accompanied by elevations in arachidonic acid metabolites in efferent lymph, *Transplant. Proc.*, 20, 1169, 1988.
182. **Moore, T. C., Spruck, C. H., and Said, S. I.,** *In vivo* depression of lymphocyte traffic in sheep by VIP and HIV(AIDS) related peptides, *Immunopharmacology,* 16, 181, 1988.
183. **Moore, T. C., Lami, J. L., and Spruck, C. H.,** Substance P increases lymphocyte traffic and lymph flow through peripheral lymph nodes of sheep, *Immunology,* 67, 109, 1989.
184. **Dubrow, T. J., Gorrin, N. R., Wackym, P. A., Abdul-Rasool, I. H., and Moore, T. C.,** Malignant hyperthermia: experience in the prospective management of eight children, *J. Pediatr. Surg.*, 24, 163, 1989.
185. **Moore, T. C. and Gamal, R.,** Early and late (15–17 year) transplant functions of 33 machine preserved cadaveric kidneys from pediatric age donors 13 and under, *Transplantation,* 47, 720, 1989.
186. **Moore, T. C., Spruck, C. H., Lami, J. L., and Said, S. I.,** Prompt elevations of PGE_2 and thromboxane B_2 in peripheral node efferent lymph of sheep following drainage area immunization, *Immunopharmacology,* 17, 73, 1989.
187. **Moore, T. C., Spruck, C. H., Lami, J. L., and Said, S. I.,** Depression of primary peripheral node lymphocyte traffic in sheep *in vivo* following central venous whole blood transfusion, *Transplantation,* 49, 838, 1990.
188. **Baumgartner, F. and Moore, T. C.,** Angiographic embolization of post-traumatic hematobilia in a six year old boy, *Digest. Dis. Sci.*, 35, 261, 1990.
189. **Moore, T. C., Whitley, G. A., Lami, J. L., and Said, S. I.,** Neuro-vascular immunology: vasoactive neurotransmitters and cellular immunity, *Ann. N.Y. Acad. Sci.*, 594, 403, 1990.
190. **Moore, T. C.,** Neurovascular immunology: a 25 year odyssey, in *Proc. 4th Int. Workshop on Neuroimmunomodulation,* Italian National Research Council and Fidia Research Biomedical Education, 1990, 424.
191. **Moore, T. C., Whitley, G. A., Lami, J. L., and Said, S. I.,** Substance P increases and prolongs increased output of T4 (CD4) lymphocytes from lymph nodes of sheep *in vivo:* is it a mediator of immunological memory? *Immunopharmacology,* 20, 207, 1990.
192. **Baumgartner, F., Moore, T. C., and Mitchner, J.,** Recurrent ventriculoperitoneal shunt pseudocyst in a nine year old girl, *Klin. Wochenschr.*, 68, 485, 1990.
193. **Moore, T. C., Lami, J. L., and Said, S. I.,** *In vivo* influences of phorbol ester and calcium ionophore on lymphocyte traffic, lymph flow and efferent lymph levels of thromboxane B2 in sheep, *Immunopharmacology,* 21, 1, 1991.
194. **Moore, T. C.,** Neurovascular immunology: a 25 year odyssey, *Ann. N.Y. Acad. Sci.*, 650, 30, 1992.
195. **Moore, T. C., Suede, M., and Lami, J. L.,** Electrocautery dissection modulates anesthesia induced depression of lymphocyte traffic in sheep: a saga of heat, pain and memory, *Immunology,* submitted for publication.

INDEX

I

K

L

M

N

O

P

Q

R